D1327818

Ultrasonics

Fundamentals, Technologies, and Applications

THIRD EDITION

MECHANICAL ENGINEERING
A Series of Textbooks and Reference Books

Founding Editor

L. L. Faulkner
*Columbus Division, Battelle Memorial Institute
and Department of Mechanical Engineering
The Ohio State University
Columbus, Ohio*

RECENTLY PUBLISHED TITLES

Ultrasonics

Fundamentals, Technologies, and Applications

THIRD EDITION

Dale Ensminger and Leonard J. Bond

CRC Press
Taylor & Francis Group
Boca Raton London New York

CRC Press is an imprint of the
Taylor & Francis Group, an **informa** business

CRC Press
Taylor & Francis Group
6000 Broken Sound Parkway NW, Suite 300
Boca Raton, FL 33487-2742

© 2012 by Taylor & Francis Group, LLC
CRC Press is an imprint of Taylor & Francis Group, an Informa business

No claim to original U.S. Government works

International Standard Book Number: 978-0-8247-5889-9 (Hardback)

Library of Congress Cataloging-in-Publication Data

Ensminger, Dale.
    Ultrasonics : fundamentals, technologies, and applications / Dale Ensminger and Leonard J. Bond. -- 3rd ed.
       p. cm. -- (Mechanical engineering)
    Summary: "The book provides a unique and comprehensive treatment of the science, technology, and applications for industrial and medical ultrasonics, including low and high power implementations. The discussion of applications is combined with the fundamental physics, the reporting of the sensors/ transducers and systems for the full spectrum of industrial, nondestructive testing and medical/ bio-medical uses. It includes citations of numerous references and covers both main stream and the more unusual and obscure applications of ultrasound"—Provided by publisher.
    Summary: "There have been significant advances in the science, the technology employed in commercial systems, and in both the scope and the range of applications of ultrasound since the second edition of this book was published in January 1988. The present edition includes discussion of many of these advances and improvements; additional figures and numerous new references are provided. In 1988 it became a requirement to write technical papers using SI (Standard International) units rather than British units. The parameters reported are now completely described by SI units: Meters now replace feet. Velocity is now in kilometers/sec. Frequency is in Hertz rather than cycles per second and so on. These up-dates and new material, combined with the core content developed for the 2nd edition make it, in effect, a new book"-- Provided by publisher.
    Includes bibliographical references and indexes.
    ISBN 978-0-8247-5889-9 (hardback)
    1. Ultrasonic waves--Industrial applications. 2. Ultrasonics. I. Bond, Leonard J. II. Title.

TA367.E568 2011
620.2'8--dc23
                                                 2011018392

Visit the Taylor & Francis Web site at
http://www.taylorandfrancis.com

and the CRC Press Web site at
http://www.crcpress.com

Contents

Preface to the Third Edition

There have been significant advances in the science and technology employed in commercial systems and in both the scope and the range of applications of ultrasound since the second edition of this book was published in January 1988. The present edition includes discussion of many of these advances and improvements; additional figures and numerous new references are provided. In 1988, it became a requirement to write technical papers using the international system of units (abbreviated as "SI" from the French Systeme International d'unites) rather than British units. All the parameters reported are now completely described using SI units: Meters have replaced feet, velocity is now expressed in kilometers per second, frequency is expressed in hertz rather than in cycles per second, and so on. These updates and inclusion of new material combined with the core content developed in the second edition make this edition, in effect, a new book.

Chapter 1 provides a summary of the history of the development of ultrasonics, which has been expanded when compared with that provided in the second edition. It looks at where ultrasonics comes from and what has developed in the field, incorporating content that discusses developments made since the second edition was published.

The scope of modern ultrasonics is set in the context of what is now an interdisciplinary field. Developments in ultrasonic systems have benefitted greatly from advances in electronics: The devices used in power electronics and made from electronic chips and software have become more powerful, compact, and versatile, and these advances have resulted in new and powerful computer-based devices for real-time data capture, storage, analysis, and display.

Chapters 2, 3, and 4 review the basic equations of acoustics, starting from basic wave equations and their applications and designs of vibratory systems and horns used in high-power applications of ultrasonics. Chapter 2 has additional content, compared to the previous edition, on scattering and numerical modeling of wave propagation, as well as a new subsection titled "Measurement System Models" and a number of new figures.

Chapter 3 contains added material on Rayleigh waves, and there are discussions of the wave fields that transducers deliver and the subsequent interactions of these fields with materials and discrete "targets." Chapter 4 now includes additional material on advanced horn and system designs. Improvements in transducer design coincide with advances in and new applications for ultrasonics in various fields. A new "giant magnetostriction" material, terfenol-D, has enabled considerable progress in the use of magnetostriction in transducer development for underwater sound and in certain commercial applications.

Chapter 5 discusses basic designs of ultrasonic transducers including piezoelectric, magnetostrictive, aerodynamic, mechanical, electromechanical, and surface acoustic wave devices. New and improved ceramic piezoelectric materials are becoming available on the market. Some polymeric materials also have piezoelectric characteristics. The best known of these is probably poly (vinylidene fluoride). More figures are added with a discussion of transducer wave fields and equivalent circuits. In the discussion of high-frequency transducers, content is added on electromagnetic acoustic transducers, laser ultrasonics, arrays, and wave fields.

Chapter 6 discusses properties of materials, which are important in the design of transducers and which influence the transmission and characteristics of ultrasonic waves. New discussion of methods to determine "approximate properties of materials," together with the consideration of water as a reference material, is included in Chapter 6. Also, Chapter 6 includes new material on ultrasonic resonance spectroscopy and new reports on the use of ultrasound to provide process parameters such as density, viscosity, and particle size for emulsions, colloids and slurries. A large portion of the second edition Chapter 6, which includes four tables covering the acoustical properties of solids, liquids, and gases, has been expanded and moved to a new appendix, Appendix A. Liquids are divided into two separate tables: tables on (1) common liquids and molten materials and (2) organic liquids.

Chapter 7 is an introduction to nondestructive testing (NDT) dealing with basic methods and general considerations of ultrasonic techniques used in nondestructive testing. Additional nonconventional ultrasonic methods, including ultrasonic spectroscopy, are discussed. New material on flaw size determination and standards for transducer evaluation and characterization are included.

Chapter 8 discusses the NDT of metallic structures. It is a general update of the previous Chapter 8 and now includes solid-state bond inspection and acoustic microscopy together with new material on in-service inspection of nuclear reactors. Many additional references and new illustrations are added.

Chapter 9 covers the inspection of nonmetallic products and the special considerations required for using ultrasonics in this area. Chapter 9 is updated in all sections with some reordering. New material on the use of ultrasonics to characterize polymer membranes, energetic materials (including solid rocket motors), and low-density foams (areo-gel) is also included.

Chapter 10 describes some early methods of visualizing ultrasonic fields. Since the second edition was written, acoustic holography, acoustic microscopy, phased arrays, as well as digital B and C scan systems have all been developed to a much higher level, particularly through the use of computer hardware and software capabilities. However, the earlier measurement methodologies remain interesting and it is worth taking a look at them. Acoustic imaging is now much more widely used in the medical diagnostics area utilizing pulse-echo techniques and visual display systems. Many new application examples have been added. The second part of the chapter is restructured and discusses process monitoring measurement and control. The scope of applications for process use is significantly expanded in this edition. Discussion of high-pressure gas, to provide transducer–sample coupling, has been added to supplement the second edition discussion of ultrasonics in air, which has been expanded considerably.

Chapters 11 through 13 discuss many of the high-intensity applications of ultrasonic energy. Chapter 11 provides an introduction to high-power mechanical effects, chemical effects including what is now known as "sonochemistry," and metallurgical effects. Advances in theory and experiments for single- and multibubble phenomena are included. A new section on diffusion through membranes has been added.

Chapter 12 has received a general update and now includes discussion of flip chip bonding used in electronics. Also added are topics such as particle manipulation, separation, and sorting, which are receiving much attention at the present time in terms of the application of ultrasound for nanotechnology. Chapter 13 has a new title "Application of Ultrasonics Based on Chemical Effects–Sonochemistry." It has been reordered and includes a new discussion of the development of sonochemistry and the field's major advances and trends.

Chapter 14 has been completely rewritten and reordered. It now includes the application of ultrasound in tissue, power, and beam measurements; and discussion of topics grouped as diagnosis, therapy, and surgery. New topics under therapy include sonicated gene and drug delivery and activation. Under surgery, high-intensity focused ultrasound (HIFU), shock wave lithotripsy, and a range of methods for tissue dissection and ablation, together with dentistry, are discussed. The final section of the chapter briefly reviews ancillary applications of biomedical ultrasound including ultrasonic cleaners, emulsification, lysis (cell and spore disruption), bioreactor and enhanced growth effects, biofilm imaging and characterization, and cell sorting.

The new Appendix A includes the updated set of tables on materials properties, which is from Chapter 6 of the second edition, and Appendix B lists some Web-based and other resources that the authors have found to be particularly useful.

Preface to the Second Edition

A person is indeed fortunate if he or she is able to devote an entire professional life to a field that is always interesting, challenging, and enjoyable. I have had this privilege. During the past 40 years, I have participated in more than 925 research studies in ultrasonics. The study of ultrasonics is interesting because it is applied in so many different fields of science and industry. In ultrasonic research, there is a need to learn something about each of the fields in which it is applied. Research helps to satisfy the innate curiosity of humankind. My curiosity certainly has been stimulated by ultrasonic research in diverse fields such as materials forming, metallurgy, medicine, agriculture, chemistry, electronics and communications, underwater sound, intrusion detection, flowmetering, dewatering, and many others. These studies have helped me maintain an enthusiasm toward my field that has led me to write this book. Perhaps by reading this book the reader will conceive new ideas for applications in fields familiar to him or her.

The first edition of *Ultrasonics: Fundamentals, Technologies, and Applications* was published several years ago. Although basic principles do not change over the years, technology and applications do. Many of the advances that took place in the field of ultrasonics since the publishing of the first edition are directly related to developments in electronics, data processing techniques, and market demands.

Nearly every chapter of this second edition has been expanded and revised from the first edition to include recent developments in the field of ultrasonics. Equations for wide horns and large area horns are included in Chapter 4. Large horns have been developed particularly to meet needs in the ultrasonic bonding of plastic materials. Design equations that include the damping term have been revised in order to use the same definition of R_m that was used in equations for damping in a simple oscillator. Compared with the first edition, new material is included in the tables of acoustical properties in Chapter 6. The material is presented in such a manner that the reader can expand it if necessary.

Presented in another chapter are a few basic principles of transducer design. No proprietary information has been included, however. The material should be useful to any beginner in the field of ultrasonics, but readers should consult other reference works, including the literature provided by manufacturers of piezoelectric materials, for more thorough discussions on transducer design. To avoid the publishing of proprietary information and because oscillator and amplifier circuits are common knowledge among electronics engineers, no circuit diagrams are offered. The matching circuitry for ultrasonic devices varies with the transducer, and the number of circuit arrangements for the various applications can be as varied as the number of applications.

A few block diagrams that describe the functions of elements of the circuit at various points are presented in the chapters devoted to low-intensity applications of ultrasonics. This topic too is the basis of several books. For example, books devoted to the use of ultrasonics in nondestructive testing, of physical acoustics and the properties of matter, or acoustical holography are available. The applications of microwave acoustics are bypassed and surface-wave applications are treated with brevity in this book, although both topics are receiving considerable attention at the present time. It was believed that these subjects could not be covered adequately in the time and space available. Tables of acoustical properties are included.

Originally, the plan was to include a chapter on underwater sound. However, several good books on this subject are available, and certain restrictions make whatever information that could be included with the necessary brevity of little value. Those who desire this type of information may find it in books such as *Fundamentals of Sonar* by J. W. Horton, a Naval Institute publication, Annapolis, Maryland; *Principles of Underwater Sound for Engineers* by Robert J. Ulrick, published by McGraw- Hill Book Company; and *Underwater Acoustics Handbook-II* by Vernon M. Albers, published by The Pennsylvania State University Press. Anyone who has the need to design equipment for underwater sound will have available the necessary material for reference.

There are other chapters in this book devoted to high-intensity applications of ultrasonics. A table of solvents used in ultrasonic cleaning is also provided. The final chapter is devoted to the medical applications of ultrasonics, which include both low- and high-intensity applications. The reader should find much useful information in this book.

The author expresses his appreciation to his employer and many of his colleagues for their encouragement and assistance in preparing the manuscript and to the industrial organizations that contributed photographs for this book, as acknowledged throughout the book. He is especially grateful to H. R. Ball for the many hours he spent drawing the illustrations for this book.

Dale Ensminger
1988

Preface to the First Edition

The fact that ultrasonic energy has applications in a large number of areas makes it a most fascinating subject for study. This book is intended to highlight the scope of these applications and to provide readers with information that the author has found very helpful in his research. With a subject so broad, it has been necessary to omit many topics and much information, but what has been included should help the individual reader, whether a novice in the field or an experienced ultrasonics engineer or physicist. Perhaps this book will even stimulate the development of new applications.

It is impossible to present the complete fundamental physics of ultrasonics, describe in detail the equipment and procedures involved, and discuss thoroughly all the applications of ultrasound in one small book. On the other hand, it is not sufficient to name the potential and practical applications of ultrasound without giving at least some theoretical background to show that these applications are indeed possible. Hopefully, the reader will find in this book the necessary information for developing his or her own specific area of interest.

For those who may not be familiar with the terms used in ultrasonics, a glossary is included.

To use ultrasonic energy to the best advantage, whether in designing transducers or applying the energy to some process, it is necessary to understand the basic principles of waves and wave propagation. Three chapters are devoted to the presentation and derivation of many useful equations to this end. The derivations are carried out in such a manner that they can be followed by the average engineering student or graduate. These chapters alone can form the basis for a separate book, but with the limitation imposed by a book devoted to applications it was necessary to abbreviate the material. However, the material is presented in such a manner that the reader can expand it if necessary.

Presented in another chapter are a few basic principles of transducer design. No proprietary information has been included, however. The material should be useful to any beginner in the field of ultrasonics, but readers should consult other reference works, including the literature provided by manufacturers of piezoelectric materials, for more thorough discussions on transducer design. To avoid the publishing of proprietary information and because oscillator and amplifier circuits are common knowledge among electronics engineers no circuit diagrams are offered. The matching circuitry for ultrasonic devices varies with the transducer, and the number of circuit arrangements for the various applications can be as varied as the number of applications.

A few block diagrams that describe the functions of elements of the circuit at various points are presented in the chapters devoted to low-intensity applications of ultrasonics. This topic too is the basis of several books. For example, books devoted to the use of ultrasonics in nondestructive testing, of physical acoustics and the properties of matter, or acoustical holography are available. The applications of microwave acoustics are bypassed, and surface-wave applications are treated with brevity in this book, although both topics are receiving considerable attention at the present time. It was believed that these subjects could not be covered adequately in the time and space available. Tables of acoustical properties are included.

Originally, the plan was to include a chapter on underwater sound. However, several good books on this subject are available, and certain restrictions make whatever information that could be included with the necessary brevity of little value. Those who desire this type of information may find it in books such as *Fundamentals of Sonar* by J. W. Horton a Naval Institute publication, Annapolis, Maryland; *Principles of Underwater Sound for Engineers* by Robert J. Ulrick, published by McGraw-Hill Book Company; and *Underwater Acoustics Handbook-II* by Vernon M. Albers, published by The Pennsylvania State University Press. Anyone who has the need to design equipment for underwater sound will have available the necessary material for reference.

There are other chapters in this book devoted to high-intensity applications of ultrasonics. A table of solvents used in ultrasonic cleaning is also provided.

The final chapter is devoted to the medical applications of ultrasonics, which include both low- and high-intensity applications.

The reader should find much useful information in this book.

The author expresses his appreciation to his employer and many of his colleagues for their encouragement and assistance in preparing the manuscript and to the industrial organizations that contributed photographs for this book, as acknowledged throughout the book. He is especially grateful to H. R. Ball for the many hours he spent drawing the illustrations, to Dolores Landreman for her help in the organization and to Dorothy Wallace for her careful and detailed editing of the draft of the manuscript. Finally, the author expresses his heartfelt thanks to his patient wife who was forced to assume many trying responsibilities during the time the manuscript was being written.

Dale Ensminger
1973

Acknowledgments

I (LJB) wish to acknowledge the contributions made to the development of my understanding of ultrasonics by my numerous students and colleagues in the ultrasound community, including Dale Ensminger. Dale Ensminger provided me with the first and second editions of this book, which I have used during my time as an academic and a researcher in London (United Kingdom), Colorado, and now Washington.

I particularly acknowledge the numerous contributions made to the research reporting, information gathering, and chapter reviewing by my colleagues in the Applied Physics Group at the Pacific Northwest National Laboratory. The projects performed by the members of the group over many years have contributed significantly to advancing the field of ultrasonics. The work provides examples and illustrations in the book, and the specific sources from which they are taken are recognized through citations given in the text.

I specifically acknowledge the contributions made to the development of my understanding of ultrasonics and ultrasonic transducers by Jerry (Gerald) Posakony, a pioneer in ultrasonics and my colleague, mentor, and friend who, in addition to our professional interactions and collaborations on many projects, reviewed and provided helpful suggestions for several chapters of this book.

I thank and acknowledge the many students (too numerous to list individually) who, over a period of more than 30 years, have taken the various courses on aspects of ultrasonics that I have taught and the more than 20 PhD and numerous MS students who have been part of the diverse range of projects that I have performed. The students made me think in many new and different ways and from my experience I can truly say that it is only in teaching that you really come to understand a topic. In many cases in this book, the results from such projects are cited as references.

I acknowledge the various institutions for their support and the opportunities that I was provided with at the City University, London; University College London; the National Institute of Standards and Technology (Boulder, Colorado); the University of Colorado, Boulder, Colorado; the University of Denver, Colorado; Denver Research Institute, Colorado; and those who have supported the grants and contracts that funded the research. Specific and detailed acknowledgments to programs and activities, including details of sponsors, are provided in the references cited throughout the text.

Finally I acknowledge the help, support, and encouragement of my managers at the Pacific Northwest National Laboratory, operated by Battelle, for the U.S. Department of Energy, who supported me in pursuing this project and permitted the use of numerous illustrations in this book.

We wish to include our respects and thanks for our wonderful, devoted wives who have mostly patiently awaited the completion of this book: Dr. Bond's wife, Janet Bond, my first wife Lois Ensminger, who died shortly after we started this work, and my present wife, Patricia Ensminger, who had been my WWII sweetheart.

1 Ultrasonics
A Broad Field

1.1 INTRODUCTION

The study of ultrasonics is the investigation of the effects of propagation, interaction with matter, and the application of a particular form of energy—sound waves—at frequencies above the limits of human perception.

Ultrasonics includes the basic science of the energy–matter interaction, the associated technologies for generation and detection, and an increasingly diverse range of applications, which are now encountered in almost every field of engineering, many of the sciences and in medicine. It encompasses a citation literature with more than 200,000 references. Students in basic physics and engineering courses may receive some introductions to sound, vibrations, and waves [1], but very few undergraduate programs provide classes specifically on ultrasonics. There are some specialist courses and text books available in many of the specialty application areas, such as underwater acoustics, medical imaging, nondestructive testing/evaluation, and surface acoustic wave (SAW) devices for electronics. Such courses, however, seldom include a comprehensive treatment of the field. The result is that ultrasonics and its applications, though a science or technology that many graduate students, scientists, and engineers encounter and need to utilize during their career, is also a topic in which they have received no formal academic education.

This book seeks to meet the needs of undergraduate and graduate students, professional scientists, and engineers who seek to understand and employ the science and technology of ultrasound, its breadth of applications, and one or more of its many specific fields of use. It deals primarily with industrial and medical applications of both low-intensity and high-intensity ultrasonic energy and fundamental principles pertaining to these applications.

Ultrasonics, which is a specific branch of acoustics, deals with vibratory waves in solids, liquids, and gases at frequencies above those within the hearing range of the average person, that is, at frequencies above 16 kHz (16,000 cycles per second). Often, the hearing range of a young person extends above 20 kHz, so the setting of the lower limit for the ultrasonic range is somewhat arbitrary. The acoustic frequency scale covers the range from below 15 Hz to above 1 THz (10^3 GHz) [2], and this is illustrated with examples of ultrasonic system operating frequency ranges shown in Figure 1.1.

Ultrasonics, originally called supersonics, has a long history but can be said to have only become a subject of scientific research since World War I (1918) when Langevin invented the quartz ultrasonic transducer. It is now based on well-established fundamentals in physical acoustics and is used in scientific studies of

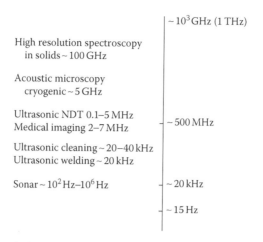

High resolution spectroscopy
in solids ~ 100 GHz

Acoustic microscopy
cryogenic ~ 5 GHz

Ultrasonic NDT 0.1–5 MHz
Medical imaging 2–7 MHz

Ultrasonic cleaning ~ 20–40 kHz
Ultrasonic welding ~ 20 kHz

Sonar ~ 10^2 Hz–10^6 Hz

~ 10^3 GHz (1 THz)

~ 500 MHz

~ 20 kHz

~ 15 Hz

FIGURE 1.1   Acoustic frequency scale and selected applications.

material properties, a diverse range of industrial applications, a wide range of sensing modalities, and in medical applications for both diagnostics and therapy. It is now employed in almost all branches of engineering for sensing and at higher powers for processing. It has become the basic science and technology for specialist areas that include underwater acoustics (SONAR), medical ultrasonics, physical acoustics, sonochemistry, nondestructive testing/evaluation, and material characterization. All of these areas of study are dependent on some form of ultrasonic transducer, which is the "heart" of ultrasonic work, and many types can function as both sender and receiver. Ultrasonic transduction can be achieved at frequencies that are still within the higher audio range (~16 kHz) to above 100 GHz and can employ a range of methods and phenomena including the piezoelectric effect, lasers, electromagnetic coupling, and mechanical coupling. For many industrial ultrasonic applications, an operating frequency near 20 kHz is employed. Ultrasonics has been demonstrated to operate from cryogenic temperatures to above 1500°C.

It was the introduction of the first piezoelectric ceramic, barium titanate, in 1947, which revolutionized industrial applications of ultrasonics by permitting the use of higher powers at reduced costs [2]. The centrality of the ultrasonic transducer—the electromechanical conversion element—is illustrated in Figure 1.2, which shows the range of scientific and engineering disciplines where ultrasound technologies are being employed. Over the past 60 years, a wide range of new and improved piezoelectric transducer materials have been developed. Much of the materials research in this area has been performed to meet the needs of the navy SONAR community.

Since the 1970s, advances in ultrasound technologies have been facilitated by major advances in theoretical analysis, including computer modeling of wave propagation and scattering, where the research was supported so as to meet the needs of the aerospace, nuclear, defense, oil, and gas communities. Fundamental here was the goal to make nondestructive testing (NDT) a more quantitative science-based inspection capability, which has been renamed as nondestructive evaluation (NDE) and quantitative nondestructive evaluation (QNDE).

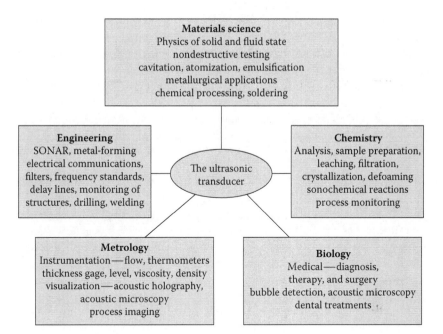

FIGURE 1.2 Scientific and engineering disciplines overlapping with ultrasonics. (After Stephens, R. W. B., An historical review of ultrasonics, *Proceedings, Ultrasonics International 1975*, 9–19, ICP Press, Guildford, UK, 1975.)

Recent decades have seen a second revolution in ultrasonics as modern electronics used in transmitters and receivers has been combined with increasing digitization and computer-based data processing and imaging capability to deliver unprecedented near-real-time processing and advanced display/visualization capabilities. These advances have combined to facilitate new applications at both low and high powers using modern instrumentation and analysis capabilities, and the result has been major growth and diversification of applications in both industrial and medical ultrasonics.

In reviewing the early literature, it is sometimes found that the term *ultrasonics* is used loosely and erroneously to refer to applications of sound waves at lower frequencies (significantly below ~20 kHz) where sound energy is used to produce results similar to those that ultrasonic energy is capable of producing. In the literature, the term *macrosonics* has also historically been applied to high-intensity applications of sonic energy regardless of the frequency being employed.

Ultrasonic waves are stress waves, and for this reason, they can exist only within mass media: sound (including ultrasound) does not transmit through a vacuum. The energy in ultrasonic waves is transmitted from one "mass" or element of material by direct and intimate contact between the masses. In this respect, they differ from light and other forms of electromagnetic radiation, which travel freely through a vacuum. In other respects, these two forms of energy, ultrasonic and electromagnetic, obey many similar laws of propagation [3].

Ultrasonic waves are also termed both *acoustic* and *elastic* waves. Acoustic waves are compression waves that propagate in gases and many liquids. This gives a shared theoretical base for many medical imaging, SONAR, and industrial applications involving fluids. The term elastic waves is, in many cases, considered to be applied just to waves in solids and with propagation properties that depend upon the elastic properties of the medium as they move in response to vibrations. Such waves can occur in many forms including compression, shear, surface, and interface vibrations. These waves again have a shared theoretical base, which goes well outside ultrasonics and is encountered in fields of study as diverse as NDT/NDE, seismology, and electronics. In interactions with structure, scattering, and the underlying theory, depends upon the ratio between feature size and wavelength of the radiation. Distinct scattering regimes are identified: for small features, where dimensions are a small fraction of a wavelength, "Rayleigh scattering" applies; for larger features, where dimensions are several wavelengths, "long wavelength scattering and ray theory" apply; and in the interim regime, "mid-frequency scattering" features are of the order of a wavelength and numerical methods have been used to solve many otherwise analytically intractable problems. Recent years have seen a cross fertilization in both forward and inverse scattering theories in several fields of study and across many wavelength scales, and the diversity of these interactions is illustrated in Figure 1.3 (after [4]). Fundamentals of ultrasonic wave phenomena are discussed in Chapter 2.

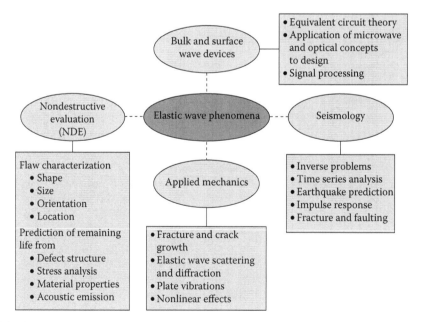

FIGURE 1.3   Fields of study involving elastic wave phenomena. (After Kraut, E. A., Applications of elastic waves to electronic devices, nondestructive evaluation and seismology. *Proceedings IEEE Ultrasonics Symposium 1976*, 1–7, 1976.)

The division of ultrasonics into categories can be achieved in several ways: one that considers applications in terms of low intensity and high intensity is natural, and it is this which is used in this book. Low-intensity applications are those wherein the primary purpose is transmitting energy through a medium. The objective may be to learn something about the medium or to pass information through the media, but this is to be achieved without causing a change in the state of the medium in which there is propagation. Typical low-intensity applications include nondestructive characterization of materials, electronic devices using surface waves, sensing in industrial processes, measurements of fundamental properties of materials, medical diagnosis, and livestock and meat grading. Although many underwater sound generators require significant power, marine applications such as depth sounding, echo-ranging, communications, and both fish and submarine detection are also most appropriately included in this low-intensity category.

High-intensity applications are those wherein the purpose is to produce an effect on a medium, or its contents, through which the wave propagates. In many cases, the interactions are nonlinear and can involve thermal, mechanical, shear force, and cavitation phenomena. Typical high-intensity applications of ultrasonics include a wide range of medical therapy and surgery applications, sonochemistry, sonoluminescence, sorting and particle motion in fluids, atomization of liquids, machining of brittle materials, cleaning, disruption of biological cells (lysis), formation and processing of nanomaterials, welding of plastics and metals, homogenization, emulsification, and mixing of materials.

1.2   BRIEF EARLY HISTORY

The early history of ultrasonics is a part of the history of acoustics. If the field of musical acoustics is excepted, then the historical development of acoustics and in particular ultrasonics, until relatively recently, can be said to have been comparatively unchronicled. A survey of pre–World War II texts (~1939) on the history of science reveals little specific mention, if any, of science and technology of sound [2]. The foundations for ultrasonics as a field of science and application were in large part laid between about 1930 and 1960. Recent decades have seen major developments in theory, in the range of ultrasonic applications, and in data processing and display, which have been facilitated by modern electronics and computers. Two brief histories of ultrasonics were provided in the 1970s by Stephens [2] and Graff [5]. A set of reviews covering the aspects of acoustics, including ultrasonics, were written in 1980, marking the fiftieth anniversary of the Acoustical Society of America; these address research in the physical sciences [6], applications of acoustic (ultrasonic) phenomena [7], and ultrasonic properties of gases, liquids, and solids [8]. A more detailed history was provided in a 1982 book chapter by Graff [9].

The earliest record of a scientific examination of music is probably that of Pythagoras, who, in the sixth century BC, discovered that the shorter of two similar stretched strings of unequal length emits a higher note than the longer one and that if one string is twice the length of the other, their pitch differs by one octave. However, records of the use of stringed instruments predate Pythagoras by several centuries. Aristotle (fourth century BC) assumed that a sound wave resonates in air through motion of the air. Vitruvius (first century BC) determined the correct mechanism for the movement of sound waves. Boethius (sixth century AD) documented several ideas relating science to music, including that perception of pitch is related to the physical property of frequency.

With the Renaissance, there came the real start of the development of a science to understand music and musical instruments. The first explicit mention of sound velocity appears to be by Sir Francis Bacon in 1627 [10]. A notable contribution to the science of acoustics is a published observation by Galileo in 1638 that reports pitch is associated with vibration. A Franciscan friar named Mersenne, who was a contemporary of Galileo, was the first to actually measure the frequency of a long vibrating string and to calculate the frequency of shorter ones from his observations. Mersenne is also reported to have been the first to measure the speed of sound in air. This was one of its earliest properties of air to be measured, and this is reported in works published in 1635 and 1644 [11]. The need for a material medium to propagate sound was demonstrated with a classic experiment by Sir Robert Boyle [12] who placed a ticking clock in an air-filled glass vessel, which he evacuated.

A number of physicists and mathematicians contributed to laying the ground-work for developing the science of acoustics during the seventeenth and eighteenth centuries. Among the familiar names are Robert Hooke (1635–1703), the author of Hooke's law; Joseph Saveur (1653–1716), who first suggested the name acoustics for the science of sound; Brook Taylor (1685–1731), the author of Taylor's theorem of infinite series; Sir Isaac Newton (1642–1727), who derived an equation for the velocity of sound in air and his law of motion is basic to the wave equations used in ultrasonics; Jean d'Alembert (1717–1783); Joseph Lagrange (1736–1813); and Jean Fourier (1768–1830). The elementary wave equation of d'Alembert and the more elaborate wave equation of Lagrange are basic to an understanding of wave motion. The mathematical series proposed by Fourier, which is a method of expressing an arbitrary function as a series of sine and cosine terms, is ideally suited to the analysis of ultrasonic waves.

Sound in nature was also being investigated. Lazzaro Spallanzani, an Italian biol-ogist, discovered in 1794 that the ability of bats to navigate in the dark was through the use of high-frequency echoes achieved with inaudible sound [13]. Starting from this early observation, the insights gained from the study of the occurrence of ultrasound in the animal world, particularly with marine mammals, have become increasingly important to give insights used in SONAR and other applications of ultrasonics.

In 1822, Daniel Colladen used an underwater bell and successfully estimated the speed of sound in the waters of Lake Geneva. In the later part of the 1800s, physicists were working on transmission, propagation, and refraction of waves. In 1877, Lord Rayleigh published his famous work *The Theory of Sound* [14]. This was a milestone in the development of the science of acoustics. Rayleigh had worked in his home with apparatus that was crude according to modern standards, but much of his treatise that resulted from his research remains valid today.

Three nineteenth century discoveries were significant in the development of methods for generating and detecting ultrasonic energy: first, very high-frequency sound generated by Francis Galton (Galton whistle); second, magnetostriction revealed by Joule in a paper published in 1847 [15], which involves a change in the dimensions of a magnetic material under the influence of a magnetic field; and third, piezoelectricity, discovered by the Curie brothers in 1880 [16]. Piezoelectricity is related to the electric charges developed on the surfaces of certain types of crystals when the crystals are subjected to pressure or tension. The inverse effect in which a voltage impressed across two surfaces of a piezoelectric crystal induces stresses in the material is presently the most commonly used method for generating ultrasonic energy in commercially available systems. It was the introduction of the first polarized ceramic, barium titanate, in 1947, which revolutionized the industrial applications for ultrasonics by permitting higher power generation and significantly reduced device costs. High-power horns employing both magnetostrictive units and piezoelectric stacks for process applications are discussed in Chapter 4. Basic design of ultrasonic transducers, including the wide range of piezoelectric materials now employed, are discussed in Chapter 5.

1.3   UNDERWATER SOUND (SONAR)

Ultrasonics as a specific branch of the science of acoustics had its birth in the study of underwater sound. Underwater detection systems were developed for the purpose of underwater navigation by submarines in World War 1. Within a week of the sinking of the Titanic in 1912, Lewis Richardson filed a patent application with the British Patent Office for echo-location/ranging in air and the following month followed with a patent for the underwater equivalent. The first functioning echo-range finder was patented in the United States of America in 1914 by a Canadian Reginald Fessenden. *So*und *na*vigation and *r*anging (SONAR), however, has its origins deep in the past. One of the earliest references to the fact that sound exists beneath the sea, as well as in the air, occurs in a notebook of Leonardo da Vinci. In 1490, two years before Columbus discovered America, he wrote: "If you cause your ship to stop, and place the head of a long tube in the water and place the outer extremity to your ear, you will hear ships at a great distance from you" [17].

Since World War II, there have been remarkable advances in the exploitation of underwater sound in both active and passive systems for both military and nonmilitary applications. Theoretical and experimental studies have investigated ultrasound propagation in the oceans, which is complicated by variation in salinity and temperature, interactions with sea-bed and shore, including the effects of sea depth, and interactions with "targets" as diverse as fish, ships, torpedoes, and mines. The development of submarine "stealth" capabilities using a combination of application of sound matching materials applied to submarines and need for better sensitivity from SONAR have caused advances in theory, new materials, and sensors to be developed.

There is significant ongoing research in the various aspects of the science and technology used in underwater acoustics, and this field has its own literature, user, and R&D community. An introduction to underwater acoustics is provided by Urick [18]. Some examples of nonmilitary uses of SONAR are shown in Figure 1.4. The specific topic of underwater telemetry was reviewed by Kilfoyle and Baggeroer in 2000 [19]. Both professional and more serious leisure seafarers now almost universally employ SONAR as a navigation aid for both depth and fish finding.

Function	Description
Depth sounding:	
Conventional depth sounders	Short pulses down—time echo return
Sub bottom profilers	Use lower frequencies—bottom penetration
Side-scan sonar	Side looking for mapping sea bed at right-angles to ships track
Acoustic speedometers	Use doppler shift to get speed over sea bed
Fish finding	Location of schools of fish and other targets: e.g. jumbo shrimp (schools float above bottom)
Divers' aids	Hand-held sonar and underwater cameras
Position marking:	
Beacons	Transmit sound continuously
Transponders	Respond when interrogated
Communication and telemetry	Use for wireless communication and data transmission
Under water equipment control	Sound-activated underwater equipment: e.g. well head control

FIGURE 1.4   Examples of some nonmilitary uses of SONAR.

1.4   MEDICAL AND BIOLOGICAL ULTRASONICS

Ultrasound has become pervasive in medicine: it is employed at low powers extensively in diagnosis; at intermediate powers, it is used in therapy; and at higher powers, in several forms, it is used in surgery. It is also employed at a range of powers in biomedical research, animal husbandry and related food industry, veterinary medicine, biotechnology industry, and defense applications, such as spore lysis to facilitate enhanced pathogen identification.

The destructive ability of ultrasound was noted by Langevin in the 1920s, who reported the destruction of schools of fishes in the sea and pain induced in the hand when placed in a water tank and sonicated with high intensities. Research into ultrasonics for medical applications can be said to have really started with the work of R. W. Wood and A. L. Loomis [20] who made a comprehensive study of the physical and biological effects of *supersonic* radiation. They demonstrated that vegetable and animal tissues were all able to be disrupted by high-power ultrasonic energy. The heating and disruptive effects of ultrasound in tissue were soon employed for therapy. High-intensity ultrasound became an experimental tool in neurosurgery.

In the early 1940s, attention within the biomedical ultrasound community turned to consider the use of lower powers for diagnosis. Developments in RADAR during World War II were the direct precursors of two-dimensional SONAR and medical imaging systems. The development of compound B-scan imaging was pioneered by Douglass Howry at the University of Colorado, Denver, and an immersion tank ultrasound system was produced in 1951. This apparatus was featured in *LIFE* magazine in 1954 [21]. Although there were others also working in the field, this activity can be said to have provided the basis for modern diagnostic ultrasound.

Numerous researchers rapidly recognized the potential for ultrasound in medical diagnosis, and a wide array of devices followed. By 1956, Wild and Reid were reporting applications to breast examination [22]. Application to investigate abdominal masses was reported in 1958, and clear echoes of a fetal head were obtained in 1959. Since these early beginnings, ultrasound has now become an essential and a routine diagnostic tool in obstetrics and gynecology and is employed in a growing number of other diagnostic applications.

Diagnostic applications of ultrasound share much in common with many of the principles of NDE. There is transmission and reflection of ultrasonic energy from interfaces, scattering from smaller structures, and selective absorption. Soft-tissue imaging has required the development of numerous specialized transducers to meet the requirements for access to image particular parts and organs within the human body. Phased arrays, which share some common history with SONAR, are now used in many applications both externally and for insertion into body cavities, including those for transvaginal and transrectal scanning, and in catheters for access into blood vessels.

In addition to mechanical and electronic scanning to give images, technologies such as ultrasonic holography have been adapted to meet diagnostic challenges, such as breast imaging. Modern ultrasonic imaging incorporates data display using advanced computer systems, false color imaging, and multidimensional near-real-time images of body processes. Blood flow is also measured using the Doppler and displayed to indicate the blood flow and motions of internal organs such as the heart.

With the growing use of ultrasonic imaging in gynecology and obstetrics, there was also significant interest in ensuring the safety of ultrasound, the development of exposure criteria and standards for power and beam measurements. The interaction of ultrasound with living tissue is more complex than that with engineering materials. Living tissue heat transfer properties are related to blood flow, and living systems also have an ability to respond to and recover from a modest ultrasound dose. As a consequence, the in vitro and in vivo responses are different. Ultrasonic effects on tissue are a function of both irradiation time and intensity. The intensity, time, biological effect parameter space, including regimes where bioeffects are believed to be possible, that where no effect is expected, and the power regime used in diagnostic ultrasound are all shown in Figure 1.5. Thermal mechanisms dominate in the ultrasound diagnostic power ranges, and there are well-established safety criteria, dose models, and standards for equipment evaluation [23]. The physical principles for ultrasonic diagnosis were presented by Peter Wells [24] in 1969, in what has become a classic text.

The potential therapeutic and disruptive properties achievable using high-power ultrasound were also investigated. As power is increased, ultrasound–tissue interactions become increasingly complex and nonlinear. It can involve thermal, cavitational, mechanical, and sonochemical interactions. Such interactions with tissue become complex functions of energy and its spatial distribution, which are optimized to give desired effects, and minimize collateral damage. These phenomena are all achieved through attention to the transducer designs employed and the electronics and wave forms used to energize them. At higher powers, biomedical/biochemical effects are induced that enable applications that include transdermal drug delivery, hyperthermia for cancer treatment, and controlled and targeted drug release. Ultrasound can be employed with cell suspensions to induce concentration, sorting, gene transfection, and both cell and spore lysis.

Therapeutic ultrasound is employed at various power levels. Ultrasonic wands are employed in physical therapy to treat soreness and injuries including relieving cramps

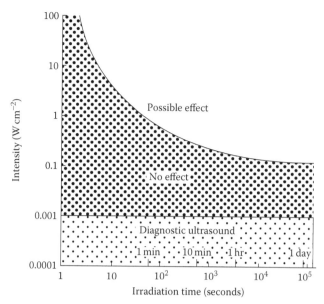

FIGURE 1.5 Bioeffect intensity/time domain plot. (After NCRP. *Exposure Criteria for medical diagnostic ultrasound: I. Criteria based on thermal mechanisms*, National Council Radiation Protection Report # 113, Bethesda, MD, 1992.)

and for a related family of procedures that involve application near injection sites to aid diffusion of subcutaneous injections. Phonophoresis is used to give transdermal drug delivery, through enhanced membrane transport, and with focused applications onto implanted or selectively delivered drugs to achieve controlled release.

Ultrasonic systems using horns for high power delivery were developed into a wide variety of medical systems. Horns are used for emulsification of biological samples. Ultrasonic descalers and drills have been used in dentistry, and horns combined into ultrasonic scalpels have been used in surgery. Tissue fragmentation, using a hollow horn, with aspiration is employed in a wide range of soft-tissue surgeries: in phacoemulsification for cataract removal and in some forms of ultrasound-assisted lipoplasticy or liposuction. Horns in the form of catheters have been developed for experimental studies of blood clot disruption and for arterial plaque removal. They have also been adapted for specialized procedures such as treatment of Menier disease. High-power pulsed, focused systems are used in lithotripsy for fragmentation of kidney and other stones.

Ultrasound is now used in home and clinical humidifiers. Atomization of water and inhalants by ultrasonic means has clinical value. At high frequencies, the nebulized particles may be small enough to be transported into the alveoli, where they are absorbed.

Ultrasonic systems are now commonly encountered in many nonclinical ancillary biomedical and research applications. The imaging technologies, including acoustic microscopy, are now used to provide images, based on acoustic contrast mechanisms for structure at various scales including in developmental and systems biology. The higher power applications include cleaners and emulsifiers for sample preparation.

There are now an emerging range of applications for both high- and low-power ultrasound in biomedical research, animal husbandry and related food industry, veterinary medicine, biotechnology industry, including enhanced growth in bioreactors and imaging processes in biofilms, and defense applications, such as spore lysis to facilitate enhanced pathogen identification.

The fundamental aspects of ultrasonics for medical applications together with some examples of emerging biomedical and biotechnology applications are considered in Chapter 14.

1.5 INDUSTRIAL ULTRASONICS

Industrial applications of ultrasound go back to the early years in the development of technologies. Hubbard in 1932 commented on the apparent limitless possibilities of applications of ultrasonics including for measurements and processing [2]. The scope of industrial ultrasonics divides into two main areas: *processing*, which needs now to be further divided into sonochemistry and other high-power applications, and *measurement and control* within which traditional NDT has now evolved into a major family of activities, including the more science-based NDE and QNDE. The scope of these activities is most simply illustrated in Figure 1.6. Recent advances in digitization and real-time data processing have now enabled new families of systems with near-real-time capability to be deployed. Ultrasonics as a measurement modality has major advantages, when compared with alternate measurement modalities, for many applications, and these are stated in Figure 1.7.

The first chemical process to be monitored with ultrasound in 1946 was the extent of polymerization in a condensation process [25]. Ultrasonic measurements for process control are discussed by Lynnworth [26], and some advances

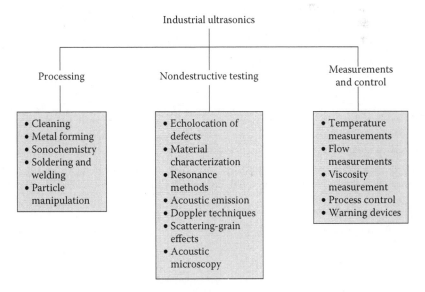

FIGURE 1.6 Illustration of the scope for industrial ultrasonics.

Flexible
Only one side access needed
Volumetric scanning
Superior penetrating power
High sensitivity
Greater accuracy for location (than other methods)
Potential for characterization
Portable (in many cases)
Non-hazardous (except at very high powers)
Electronic and electrical system
Digitized data and images—analysis can be automated

FIGURE 1.7   Some of the advantages of ultrasonic measurements.

Partial List: Industrial Measurements

- Flow
- Density
- Viscosity
- Porosity
- Pressure
- Thermometry
- Dynamic force, vibration
- Level
- Position
- Nondestructive testing

- Composition
- Particle size—(colloids, slurry, suspensions)
- Bubble detection (and size)
- Grain size and strain
- Acoustic emission
- Imaging: holography
- Acoustic microscopy
- Elastic properties
- Gas leaks
- Surface acoustic wave sensors

FIGURE 1.8   Partial list of test, measurement, and process control applications of ultrasonics. (After Lynnworth, L. C. 1989. *Ultrasonic Measurements for Process Control*. Boston: Academic Press.)

and developments in sensors and systems are discussed by Lynnworth and Magori [27]. The topic of ultrasonic sensors for chemical and process plant is discussed by Asher [28]. A comprehensive review of ultrasonic analysis as an element within process analytical chemistry is given by Workman et al. [29], and the narrower topic of process monitoring is covered by Hauptmann et al. [30]. Theses types of measurements are also seeing some application in the emerging field of chemometrics.

The scope of measurements that have been made using ultrasonics is truly extensive. A partial list of measurements that can be made is given in Figure 1.8 (after [26]). Applications as diverse as monitoring solidification in metals, characterization of multiphase fluids and flow, particle sizing, velocity profiles in food products and rheology, and process visualization have all been developed. Measurements have

---

Acoustic field visualization

- Particle visualization (in fluids – liquid + gas)
- Dye visualization
- Moire fringe
- Interferometer techniques
- Shadow photography
- Schlieren
- Thermal techniques
- Sonogram technique
- Ultrasonic scanning
- Acoustic microscopy
- Holography
- Computer models (field simulation)

FIGURE 1.9 Visualization of acoustic/ultrasonic fields.

now been made over the temperature range from temperatures encountered with cryogenic liquid gases (few degrees Celsius) to above 1500°C. Passive monitoring has also been developed to monitor acoustic emissions in the ultrasonic frequency range.

Advances in medical ultrasonics and high-speed electronics have facilitated advances in a diverse range of imaging employed in industrial applications. The numerous forms of ultrasonic imaging are illustrated in Figure 1.9 (after [2]). Acoustic microscopy has become its own specialist area with the ability to achieve submicron resolution with a 15 GHz unit operating at cryogenic temperatures [31]. Tomography has been adapted to process applications [32] and arrays, including using synthetic aperture focusing (SAFT), adapted from both NDT and SONAR, to meet process imaging needs. These areas are discussed further in Chapter 10, which considers imaging, process control, and other low-intensity applications.

High-power industrial applications have employed different forms of transducers from those employed in measurements. High-power transducers have traditionally used magnetostriction as the energy conversion mechanism [33]. Advances in transducer materials and power electronics have also had significant impact at higher powers, with the adoption of piezoelectric stacks to drive some horn configurations.

Traditional areas of high-power ultrasonics have been cleaning, emulsification, atomization, and welding. At intermediate powers, standing wave systems have been developed for particle manipulation, concentration, and sorting [34]. From the early days of ultrasonics, the ability of ultrasound to induce chemical phenomena in some systems has been known. Exploiting combinations of thermal, mechanical, and cavitational energy to enhance or facilitate reactions that otherwise do no occur has now developed into the field of sonochemistry [35, 36].

Cavitation, the generation and then rapid collapse of a bubble in a fluid, is known for its ability to cause erosion in systems such as pumps and ship propellers. The acoustic bubble, driven at ultrasonic frequencies, is a phenomenon where energy has been estimated to be concentrated by as much as 13 orders of magnitude. The collapsing bubble generates short duration high pressures and temperatures, and the emission of light, in what is called sonoluminescence [37]. With the temperatures and pressures that are generated in cavitational collapse, there has also been an ongoing interest in the possibility for fusion to occur [38].

The basic mechanisms occurring in high-intensity ultrasonics are discussed in Chapter 11. Applications based on mechanical effects are covered in Chapter 12, and applications based on chemical effects, now evolved into the field of sonochemistry, are considered in Chapter 13.

1.6   NONDESTRUCTIVE TESTING/EVALUATION

The use of sound for testing pottery, glass, and bells has a long history, going back hundreds if not thousands of years. The birth of ultrasonic NDT can be considered to have occurred in 1928/29 when the concept of ultrasonic metal flaw detection was first suggested by Soviet scientist Sergei Sokolov when working at the Electrotechnical Institute, Leningrad. He subsequently demonstrated that a transmission technique, using continuous waves, could detect internal flaws in metal and later suggested a reflection method. The first application of ultrasonic waves for fault detection appears in a German patent in 1931, but it was the Russian, Sokoloff, who over the next four years worked at the technique in detail [39].

Around 1939, the "Hair-line Crack Committee of British Iron and Steel Institute" faced the problem of how to detect fine cracks. With their encouragement, Sproule used the principle of the echo-sounding technique of Wood for testing [40]. This was at the same time that the equipment for generating and detecting short pulses became available. Although not known in the United Kingdom at that time, Firestone, in the United States of America, was working on similar problems and used a single quartz transducer in what became know as pulse-echo testing for crack detection [41].

The technology was rapidly employed with an early application of pulse-echo inspection being used for air force bomber landing gear. Contact testing was soon seen as extremely efficient in detecting defects in welds and large sections of metals, and an ability to penetrate to depths of more than 13 m was demonstrated. Thickness gages soon followed, as did the use of water tanks for immersion testing and the use of multiple reflections to investigate the resonant and damping characteristics of castings [42]. Ultrasonic NDT in both manual and automated forms was rapidly developed. It was adopted and deployed in many then emerging high-technology industries, including for aircraft testing and, in the form of special rail cars, for automated track inspection.

In the 1970s, more advanced inspection systems were needed to support developments in high-risk technologies, in particular aerospace, nuclear power, and offshore oil and gas. It was recognized that a new science base was needed for NDT to become the quantitative science required to give the defect and material data needed by emerging testing and life philosophies such as damage tolerance. A major U.S. program was sponsored by the Defense Advanced Research Projects Agency (DARPA)

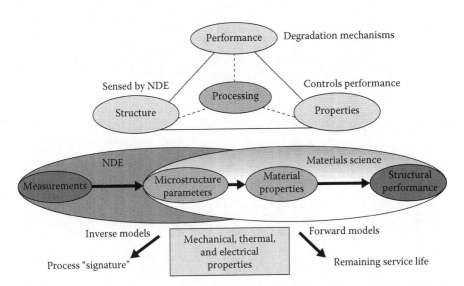

FIGURE 1.10 **(See color insert.)** The intimate relationship between NDE and materials science. NDE = nondestructive evaluation. (Courtesy of Pacific Northwest National Laboratory.)

and the U.S. Air Force, and there were parallel activities in most major industrialized nations [43, 44]. The intimate relationship between NDE and materials science is illustrated in Figure 1.10. Ultrasonics, in many forms, has now become a key technology in advanced diagnostics and on-line monitoring, where it is providing data for use in prognostics and remaining useful life estimation. The basic methods used for ultrasonic NDT/NDE are discussed in Chapter 7. The applications to metals are reviewed in Chapter 8, and those for nonmetals are discussed in Chapter 9.

1.7 ULTRASONICS IN ELECTRONICS

Ultrasonics is intimately linked into electronics, through the circuits and systems employed and with the transducers and sensors. Many of the most recent advances in ultrasonics have been enabled by modern computers, high-speed analogue to digital capabilities, and advanced data processing and imaging software. Ultrasonics is also a part of electronics with its use in surface acoustic and bulk acoustic wave devices, and in acousto–optic devices.

Delay lines: surface and bulk acoustic wave delay lines were once common in televisions.

SAW devices include high-frequency chirp and bandpass filters, convolvers, and acousto–optic devices [45] and have developed into a distinct field of study. A history of SAW electronics is provided by D. P. Morgan [46], and some examples of several classes of acoustic sensor configurations are shown in Figure 1.11. Devices can include the quartz crystal microbalance sensor, the SAW device, used as a delay line and sensor, the flexural plate wave sensor, and the acoustic plate mode

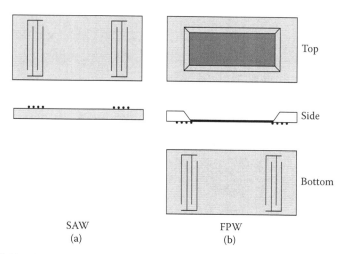

Top

Side

Bottom

SAW
(a)

FPW
(b)

FIGURE 1.11 Schematic illustrating structures for SAW and FPW sensors. SAW = surface acoustic wave; FPW = flexural plate wave.

sensor. All of these types of devices now come in a multitude of forms and can include coatings to provide selective sensing and improve functionality. An increasing diverse range of SAW sensors are being developed using various forms of thin film elements, and these sensors are now becoming a significant element within industrial ultrasonics.

Ultrasonics has entered electronics with applications of acoustic microscopy for semiconductor module and dia-attach inspection. At higher powers, ultrasonic bonding is being used for flip-chip attachment. Extensive coverage of ultrasonics with electronics is to be found in the Ultrasonics, Ferroelectrics, and Frequency Control Society (UFFC) of the Institute of Electrical and Electronics Engineers (IEEE) conference series *Ultrasonics Symposium* [47] and papers published in the related IEEE Transactions, ultrasonics, ferroelectrics, and frequency control. The ultrasonic transducer is discussed in Chapter 5, sensors are considered in Chapter 10, and high-power applications for bonding in electronics are included in Chapter 12.

1.8 PHYSICAL ACOUSTICS

The fundamental foundations of ultrasonics are in the field of physical acoustics and firmly based in physics. Mechanical vibrations—ultrasonic waves—propagate through and interact with solids, liquids, and gases. Ultrasonic propagation and interaction characteristics are intrinsically linked with fundamental mechanical and thermal properties of the propagation medium. The attenuation of energy as it propagates is potentially due to a diverse family of mechanisms: *absorption*, where energy is converted into heat because of elastic motion, heat flow in grains in polycrystalline media, domain movements, and inelastic motion; *scattering* that occurs because material is not truly homogeneous, and effects are due to crystal discontinuities, grain boundaries, inclusions, particles, and voids. Scattering in general increases with frequency and is a strong function of the wavelength—feature size ratio. In measurement systems, there are also effects due to inherent phenomena of the wave field. Diffraction, reflection, and refraction at interfaces of the propagating energy need consideration, particularly when fundamental property measurements are being made.

Some of the many early major discoveries and insights in physical ultrasonics are summarized in Figure 1.12. A brief overview of some of the fundamental aspects of physical acoustics, waves in solids, liquids, and gases, was provided by Greenspan at the fiftieth Meeting of the Acoustical Society of America [8]. A review, covering the status of understanding for elastic waves and scattering in solids, was provided by Pao [48]. In these relatively early years, ultrasonics was being seen mostly as an element within solid state physics and as a measurement approach for the determination of thermal–mechanical properties [49].

With the major advances in scattering theory (forward scattering) [48] and advances in NDE needing defect sizing, attention has also focused on the related problem of inverse scattering, the determination of scatter characteristics based on measurements of the scattered waves [50]. Such characterizations of targets continue to present significant fundamental challenges because of restrictions imposed by sparse data, limited viewing and measurement angles, and the inherent ill-posed nature of the problems being addressed.

In the interactions between ultrasonic waves and media, the phenomena can be both linear and nonlinear. For example, the higher order elastic constants affect wave velocity, and in more complex problems, grain characteristics, stress, strain, material aging and damage, and directionality all become functions of propagating frequency, temperatures, and applied loads. A major body of work seeks to address the fundamental physics for nonlinear acoustics, and a good overview is provided by Beyer [51].

For ultrasonic measurement in fluids, pure water provides a good reference standard where the relationships between measured properties and fundamental theory are well understood, both in terms of velocity of propagation and attenuation. In investigating media, the ultrasonic waves, particularly in fluids, can excite translational, vibrational, and rotational modes in molecules. In the 1970s, many of these relationships were the foundations for what was known as molecular acoustics, which became a form of spectroscopy, linked to fundamental physical acoustics [52]. There are useful data to be obtained in systems such as those with supercritical fluids and in the singularities that occur at phase transition [6].

Ultrasonics can make major contributions to the determination of fundamental physical properties with rich data sets for solids, liquids, and gases. Many good review papers on physical acoustics at ultrasonic frequencies are included in the more than 20 volume series started by Mason in 1964 [53]. The richness of the ultrasonic data set is illustrated with just two examples of frequency-dependant attenuation data: for polycrystalline aluminum, when attenuation is considered as a function of frequency (Figure 1.13) [54], data cover six orders of magnitude as you traverse the ultrasonic frequency range and the data presented include both experimental and theoretical work by Merkulov [55]; and for air attenuation as a function of frequency, data illustrate the occurrence of fundamental absorption peaks (Figure 1.14) [56].

Many of the fundamental equations for ultrasonics are given in Chapter 2, and methods for determining properties of materials are covered in Chapter 6.

Date	Pioneer (s)	Comment
	Gases	
1911	N. Neklepajew	Found absorption in air to be twice classical
1925	G. W. Pierce	First acoustic interferometer (CO_2 dispersion)
1933	H. O. Knesser	Measured sound absorption in multi-atomic gases
1934	A. Eucken and R. Becker	Showed dispersion and absorption depend on mean free path
1936	P. Biquard	Directed attention to importance of non-linearity
1950	V. Timbrell	Absolute measure of sound pressures at high frequency
1952	E. F. Smiley, E. H. Winkler and Z. I. Slawaky	Measured vibrational relaxation in CO_2
1961	T. L. Cottrell, J. C. McCubrey	Molecular energy transfer in gases
1965	K. Takayanagi, also K. P. Lawley and J. Ross	Calculations on rotational transitions
	Liquids	
1927	C. C. Hubbard and A. L. Lomas	A sonic interferomreter for liquids
1931	P. Biquard	Notable improvements in methods for measuring absorption for liquids
1942	C. S. Venkateswaran	Interferometric studies—light scattering in liquids
1943	P. Biquard and G. Ahler	First pulse methods in ultrasonics
1946	F.E. Fox and G. D. Rock	Found excess attenuation in water 2–40 C
1948	I. N. Liebermann	Explained increased absorption in sea water (as compared with distilled water)
1958	K. F. Herzfeld	Bulk viscosity and shear viscosity damping in fluids
1963	J. Jarzynski	Used quartz buffer rod between molten metals and transducer to measure velocity and attenuation
	Solid State	
1937	I. Landau and G. Rumer	Sound absorption in insulating crystals at high frequencies: multi-phonon process
1954	H. E. Bommel	Ultrasonic attenuation in superconducting lead
1956	H. E. Bommel, W. P. Mason and A. W. Warner	Dislocations, relaxations and anelasticity of quartz
1958	R. W. Morse, H. V. Bohn and J. D. Gavenda	Electron resonances with ultrasonic waves: copper
1959	I. G. Merkulow	Ultrasonic absorption in crystals to 2 GHz
1960	R. W. Morse, A. Myers and C. T. Walker	Fermi surfaces Au and Ag determined by ultrasonic attenuation measurements
1964	E. H. Jacobsen	Microwave ultrasonics

FIGURE 1.12 Some key milestones in physical acoustics applicable to ultrasonics: gases, liquids, and solid state. (After Stephens, R. W. B., An historical review of ultrasonics. *Proceedings, Ultrasonics International 1975*, 9–19, ICP Press, Guildford, UK, 1975.)

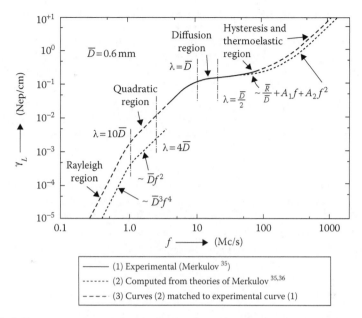

FIGURE 1.13 Attenuation of longitudinal waves in polycrystalline aluminum. (From Smith, R. T., and R. W. B. Stephens, Effects of anisotropy on ultrasonic propagation in solids, *Progress in Applied Materials Research*, ed. E. G. Stanford, J. H. Fearon, and W. J. McGonnagle, Vol 5., Gordon and Breach, London, 1964. With permission.)

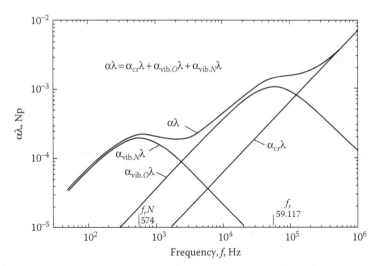

FIGURE 1.14 Absorption per unit wavelength for acoustic waves in air at 293.15 K (20°C), 101.325 kPa, and 70% relative humidity. Here, $f_{r,N}$ is the vibrational relaxation frequency for nitrogen; $f_{r,O}$ is the vibrational relaxation frequency for oxygen. (From American National Standards Institute, (ANSI) Standard 1.26-1978A, *Standard Method for the Calculation of the Absorption of Sound by the Atmosphere*, ANSI, New York, 1978.)

1.9 ULTRASONIC SYSTEMS: TRANSMITTERS AND RECEIVERS

The senses of the ultrasonic system are the transmitters and receivers that couple energy into material or receive energy that has been propagating in a medium. Ultrasonic systems are best understood as simply a specialized application of the approaches employed in all measurement systems [57]. A schematic for such an ultrasonic measurement system is shown in Figure 1.15. In many low-power applications, a single piezoelectric element is used in a pulse-echo mode, analogous to active SONAR, and the properties of the generated ultrasonic wave field are well described by antenna theory. Two and multi-transducer systems can operate in transmission, a line of sight configuration between transducers, or pitch–catch, where scattered or reflected energy is detected. Many arrays operate in pulse-echo, particularly those used in medical imaging systems, and use multi-element transducers, with either electrical or mechanical scanning (or a combination). There are also systems that employ transducers as passive receivers, that is, in NDT acoustic emission, with listening in a fashion that is analogous to passive SONAR.

Ultrasonic systems have benefited from major advances in system analysis, power electronics, improved transducer materials, analogue electronics, digitization, and both computer hardware and software for signal processing. In system development, there has been an interesting interplay between equipment used for industrial and medical applications, which have also benefited from advances in SONAR, particularly with respect to transducer materials.

The two most common technologies employed to generate ultrasonic waves are those based on piezoelectric and magnetostrictive devices. The design of a transmitter depends upon the application, temperature, material to be sonified and required

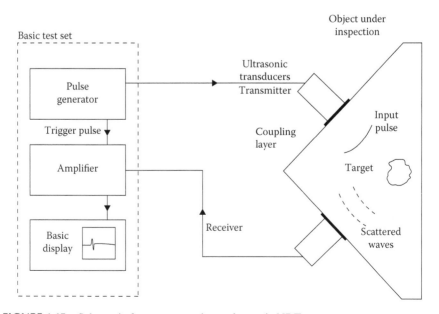

FIGURE 1.15 Schematic for a two transducer ultrasonic NDT system.

ultrasonic frequency, radiation field, and power level. Piezoelectric materials have been developed to give operation at elevated temperatures and in harsh environments, such as those near nuclear plants where a high radiation flux is encountered. To improve energy transmission, matching layers, with intermediate acoustic impedance properties, can be used particularly to increase the efficiency of energy transmission and reception, particularly when working in fluids and gases.

For high powers, energy is delivered with magnetostrictive or piezoelectric stacks driving a horn or piston, a diaphragm driven electromechanically, a siren or whistle, or various forms of mechanical devices. Many high-power applications are at 20 kHz, with a few specialized applications going to frequencies as high as 120 kHz.

In pulse-echo operation with piezoelectric materials, the same element is used to both transmit and receive and reciprocity applies. Receivers are the "ears" of a system, and a passive device responds to sounds and converts the information into signals for recording and processing in the system "brain." All forms of higher animal life have either ears or the ability to detect vibrations. Devices that detect ultrasound are called receivers or sensors. They serve the same function as the ear, receiving the signal and transforming into another form of energy (usually electrical), where it can be processed and analyzed as desired. In the lower ultrasonic frequency range in air, microphones are typical receivers, and magnetostrictive units have been used occasionally. Advances in piezoelectric materials, for particularly SONAR application, combined with new matching layers using micromachined silicon have now developed to give new classes of air-coupled transducers. In water and many fluids, piezoelectric polymer films and other piezoelectric materials are used as hydrophones to make field measurements.

All forms of ultrasonic transducers have band-pass characteristics, which limit the frequency spectrum generated and filter received energy. The effectiveness of ultrasonic systems is best estimated using system models, which enable understanding of conversion efficiencies and the losses that occur between transmission and reception. The simplest models use an approach adopted from what are called the SONAR equations [58]. More advanced models have now been developed that also include the generation, propagation, and interaction characteristics of ultrasonic waves [59].

The basic design of ultrasonic transmitters and receivers and the various forms technologies including lasers and electromagnetic acoustic transducers are discussed in Chapter 5. System models and related topics are discussed in several places in this text, including Chapter 2, Section 2.10.5.

1.10   LOW-INTENSITY APPLICATIONS

There are numerous low-intensity applications of ultrasonic energy, some of which have already been mentioned, and these include process control, NDT, intrusion detection, measurement of elastic properties, medical diagnosis, delay lines, and signal processing.

Ultrasonic devices that use low-intensity principles also include some door openers, height detectors, liquid-level gages, detectors for paper web breaks in mills, and sensors characterization of two-phase material systems. Also they are used in sensing, where acoustic emissions from crack growth in metals, from cavitation induced in process pumps, and emissions from bubble nucleation in super-heated liquids induced by particle interactions. These applications are all analogous to passive SONAR. Low-intensity applications are discussed in more detail beginning with Chapter 6.

1.11   HIGH-INTENSITY APPLICATIONS

High-intensity applications are also numerous. These are discussed beginning with Chapter 10. The leading commercial high-intensity applications are cleaning, welding plastics, machining, and soldering, and there is growing interest in deploying ultrasound for sonochemistry. These effects are based on thermal, mechanical, cavitation, and fluid shear forces.

Ultrasonic cleaning provides one of the largest markets for high-intensity ultrasound equipment. There is also extensive use of ultrasound for emulsification and sample preparation. Standing wave systems are being developed for particle and cell sorting and concentration in fluid streams.

Ultrasonic soldering irons were available before 1950, and the semiconductor industry is now using ultrasound in flip-chip bonding. Phenomena such as acousto-plasticity, softening and flow of metals, including gold, under the influence of high-power ultrasound, remain of fundamental scientific interest.

Several of the more interdisciplinary effects of ultrasonic energy, as in acousto–optic interactions, are continuing to receive significant attention in electronics. There also still remain fundamental aspects of cavitational collapse that generate light with sonoluminescence, and there is speculative interest in the possibility of ultrasonically induced fusion, following from the high pressures and temperatures seen with cavitational collapse.

In sensing systems, there are acoustic emissions from crack growth in metals, from cavitation induced in process pumps, and from bubble nucleation in super-heated liquids induced by particle interactions. The full scope and depth of high-power ultrasonics is considered starting with Chapter 11.

1.12   MODERN ULTRASONICS: AN INTERDISCIPLINARY FIELD

Since the 1970s, there have been rapid and extensive developments in the various subspecialties that combine to form "Ultrasonics." Many fields within ultrasonics are now treated as distinct fields of study or subspecialties within other fields: underwater acoustics; ultrasonics for NDT/NDE; applications for process monitoring, measurement, and control; imaging in its many forms; sonochemistry; medical ultrasonics for therapy, diagnosis, and surgery; the emerging areas within biotechnology; aspects of particle formation and manipulation in nanotechnology; animal medicine; and defense applications are all areas of activity and growth.

The core of ultrasonics remains within physical acoustics. The technological applications are leveraging advances in computers, digitization, signal capture, and processing and enabling near-real-time data display, imaging, and analysis. Many of the more fundamental insights gained 30 or more years ago are now transitioning from laboratory to industry. New materials are giving better transducers. The theory of elastic wave and wider classes of ultrasonic wave propagation, scattering, and inversion are all contributing to new measurement approaches.

Recent years have seen unprecedented growth within the multidisciplinary field that encompasses ultrasonics in terms of the numbers and scopes for both publications and patents. In 2005, more than 10,000 new papers were published on aspects of ultrasonics, including the various specialized fields, and recorded in the "ISI Web of Science," and in addition, more than a further 4,000 papers were recorded in the "ISI Proceedings" data base. Growth in the literature continues apace.

Ultrasonics remains a combination of "art and science": the understanding of the relevant physical acoustics, transducer selection, and good experimental design are critical. Many applications fall firmly within the scope of measurement science. The analysis of complete ultrasonic systems must include the transducers and wave-material interactions. Such analysis is critical for the design and optimization of effective measurement systems. The adoption of computer modeling of system performance has aided in giving insights, particularly where no closed analytical solution is available or there is an ill-posed inverse problem.

More than 90 years have passed since ultrasonics began its real development from scientific curiosity to critical enabling technology in basic science and in both industrial applications and biomedical sciences. As the reader ventures into ultrasound and its application, he should be sure to completely search the literature, looking across the numerous interrelated fields. A diverse array of problems has been analyzed over the years, and many applications tested in laboratory studies, that could not at their time be implemented, due to instrumentation, theoretical, or data processing limitations.

This third edition of *Ultrasonics* provides a comprehensive introduction to the full width and diversity of this rich and fascinating field. It discusses the basic science of waves, wave fields, and transducers. It then moves on to consider both low- and high- power ultrasonics focusing on industrial and then medical fields. This includes discussions of some of the less well-known, yet highly useful, diversity of applications that have been reported over the decades. It also provides numerous references to guide the reader as to where further details are available.

REFERENCES

1. Feynman, R., R. B. Leighton, and M. Sands. 1977. *Feynman Lectures on Physics.* Vol. 1, Chap. 47–51. Redwood City, CA: Addison-Wesley.
2. Stephens, R. W. B. 1975. An historical review of ultrasonics. *Proceedings, Ultrasonics International, 1975.* 9–19. Guildford, UK: ICP Press.
3. Pain, H. J. 1999. *The Physics of Vibrations and Waves.* 5th ed. Chichester: Wiley.
4. Kraut, E. A. 1976. Applications of elastic waves to electronic devices, nondestructive evaluation and seismology. *Proceedings IEEE Ultrasonics Symposium, 1976,* eds. J. deKlerk and B. McAvoy, Annapolis, Maryland, Sept. 29–Oct. 1, 1–7. New York: IEEE, Sonics and Ultrasonics Group.
5. Graff, K. F. 1977. Ultrasonics: Historical aspects. *Proceedings IEEE Ultrasonic Symposium,* eds. J. deKlerk and B. R. McAvoy, Phoenix, AZ, Oct. 26–28, 1–10. New York: IEEE, Sonics and Ultrasonics Group.

6. Rudnick, I. 1980. *J Acoust Soc Am* 68(1):36–45.
7. Mason, W. P. 1980. *J Acoust Soc Am* 68(1):29–35.
8. Greenspan, M. 1980. *J Acoust Soc Am* 68(1):29–35.
9. Graff, K. F. 1981. History of ultrasonics. In *Physical Acoustics*, ed. W. P. Mason, Vol. 15, Chap. 1, 2–90. New York: Academic Press.
10. Bacon, F. 1627. *Sylva Sylvarum*.
11. Lenihan, J. M. A. 1951. *Acoustics* 2:96–9.
12. Boyle, R. 1662. *New Experiments, Physico-Mechanical, Touching the Spring of the Air*. 2nd ed. Oxford: Oxford University. Experiment 27, modern interpretation: R. B. Lindsay. 1948. Transmission of sound through air at low pressures. *Am J Phys* 16:371–7.
13. Sales, G., and D. Pye. 1974. *Ultrasonic Communication by Animals*. London: Chapman & Hall.
14. Rayleigh, J. W. S. 1945. *The Theory of Sound*. Vols. 1 and 2. New York: Dover Publications.
15. Joule, J. P. 1847. *Philsoc Mag (III)* 30:76.
16. Curie, J. P. 1880. *C R Acad Sci Paris* 91:294.
17. Urick, R. J. 1983. *Principles of Underwater Sound*. 3rd ed., 2. New York: McGraw-Hill.
18. Urick, R. J. 1983. *Principles of Underwater Sound*. 3rd ed. New York: McGrawHill.
19. Kilfoyle, D. B., and A. B. Baggeroer. 2000. *IEEE J Oceanic Eng* 25(1):4–27.
20. Wood, R. W., and A. L. Loomis. 1927. *Philos Mag* 4:417.
21. Life Magazine. 1954. 37(12): 71–2.
22. Wild, J. J., and J. M. Reid. 1957. *IRE Trans Ultrason Eng* 5:44–56.
23. NCRP. 1992. Exposure criteria for medical diagnostic ultrasound: I. Criteria based on thermal mechanisms. National Council Radiation Protection Report # 113.
24. Wells, P. N. T. 1969. *Physical Principles of Ultrasonic Diagnosis*. London: Academic Press.
25. Sokolov, S. I. 1946. *Zh Tekh Fiz* 16:283.
26. Lynnworth, L. C. 1989. *Ultrasonic Measurements for Process Control*. Boston: Academic Press.
27. Lynnworth, L. C., and V. Magori. 1999. Industrial process control sensors and systems. In *Physical Acoustics*, ed. R. N. Thurston, and A. D. Pierce, Vol. 23, Chap. 4, 276–470. New York: Academic Press.
28. Asher, R. C. 1997. *Ultrasonic Sensors*. Bristol: Institute of Physics Publishing.
29. Workman, J., D. J. Veltkamp, S. Doherty, B. B. Anderson, K. E. Creasy, M. Koch, J. F. Tatera et al. 1999. *Anal Chem* 71(12):121R–80R.
30. Hauptmann, P., N. Hoppe, and A. Puttmer. 2002. *Meas Sci Technol* 13(8):R73–83.
31. Heiserman, J., D. Rugar, and C. F. Quate. 1980. *J Acoust Soc Am* 67(5):1629–37.
32. Plaskowski, A., M. S. Beck, R. Thorn, and T. Dyakowski. 1995. *Imaging Industrial Flows*. Bristol: Institute of Physics.
33. Crawford, A. E. 1955. *Ultrasonic Engineering*. New York: Academic Press.
34. Groschl, M. 1998. *Acustica* 84(3):432–47.
35. Suslick, K. S. 1988. *Ultrasound: Its Chemical, Physical and Biological Effects*. New York: VCH Publishers.
36. Mason, T. J., and J. P. Lorimer. 1988. *Sonochemistry*. Chichester, England: Ellis Horwood.
37. Leighton, T. G. 1995. *The Acoustic Bubble*. San Diego: Academic Press.
38. Camara, C. G., S. D. Hopkins, K. S. Suslick, and S. J. Putterman. 2007. *Phys Rev Lett* 98:064301.
39. Mullins, L. 1964. The evolution of non-destructive testing. In *Progress in Applied Materials Research*, eds. E. G. Stanford, J. H. Fearon, and W. J. McGonnagle, Vol. 5, 207–37. London: Gordon and Breach.
40. Desch, C. H., D. O. Sproule, and W. J. Dawson. 1946. *J Iron Steel Inst* 153:316.
41. Firestone, F. A. 1945. *Metall Prog* 48:505.

42. Murray, A., and L. Frommer. 1954. *J Iron Steel Inst* 1:45.
43. Thompson, D. O., and D. E. Chimenti, eds. 2008. *Review of Progress in Quantitative Nondestructive Evaluation*. Vols. 1–35. New York: Plenum. now *Am Inst Phys Conf Ser* (1982–2008): 35th Annual Review of Progress in QNDE, AIP Conf. Proceedings # 1096.
44. Sharpe, R. S., ed. 1970–1985. *Research Techniques in NDT*, Vols. 1–8. London: Academic Press.
45. Kino, G. S. 1987. *Acoustic Waves*. Englewood Cliffs, NJ: Prentice-Hall.
46. Morgan, D. P. 1998. History of SAW devices. In *Frequency Control Symposium*, Proceedings of the 1998 IEEE International, Volume, Issue, 27–29, 439–60; Morgan, D. P. 2000. A history of SAW devices. In *Advances in SAW Technology, Systems and Applications* ed. C. C. W. Ruppel, and T. A. Fjeldy, Vol. 1, 1–50. Singapore: World Scientific.
47. IEEE. Proceedings Ultrasonics Symposium (1962–2009) http://ewh.ieee.org/conf/ius_2010/z_web_partial/misc_00_past_ius.htm (accessed 2008–2010).
48. Pao, Y.-H. 1983. *Trans ASME* 50(12):1152–64.
49. Turrell, R., C. Elbaum, and B. B. Chick. 1969. *Ultrasonic Methods in Solid State Physics*. New York: Academic Press.
50. Rose, J. H. 1989. Elastic wave inverse scattering in nondestructive evaluation. *PAGEOPH* 131(4):716–39.
51. Beyer, R. T. 1997. *Non-Linear Acoustics*. New York: Acoustical Soc. Am.
52. Matheson, A. J. 1971. *Molecular Acoustics*. London: Wiley.
53. Mason, W. P., and R. N. Thurston. 1964–1999. *Physical Acoustics*. Vols. 1–25. New York: Academic Press.
54. Smith, R. T., and R. W. B. Stephens. 1964. Effects of anisotropy on ultrasonic propagation in solids. In *Progress in Applied Materials Research*, ed. E. G. Stanford, J. H. Fearon, and W. J. McGonnagle, Vol 5. London: Gordon and Breach.
55. Merkulov, L. G. 1957. *Sov Phys Tech Phys* 1:59–69.
56. ANSI: 1 .26-1978A SA 23-78. 1978. *Method for the Calculation of the Absorption of Sound by the Atmosphere*. 1989, 40. New York: American National Standards Institute.
57. Doeblin, E. O. 1990. *Measurement Systems*. 4th ed. New York: McGraw-Hill.
58. Bond, L. J., and N. Saffari. 1984. Mode conversion ultrasonic testing. In *Research Techniques in Non-Destructive Testing*, ed. R. S. Sharpe, Vol. 7, Chap. 5, 145–89. London: Academic Press.
59. Thompson, R. B., and H. N. G. Wadley. 1989. *Critical Reviews in Solid State and Materials Science* 16(1):37–89.

2 Elastic Wave Propagation and Associated Phenomena

2.1 INTRODUCTION

Designing for the effective use of ultrasonic energy requires an understanding of the basic principles of wave propagation and associated phenomena. Because of their importance in the applications of ultrasonics, these principles are presented in some detail. This chapter is devoted to describing these phenomena. The succeeding chapter is devoted to derivations and solutions of those equations, which describe propagation characteristics of various wave types and which are used in ultrasonic system designs and analyses. Chapter 4 includes equations and procedures for designing certain components, particularly solid horns, or mechanical amplifiers, for high-intensity applications. The reader should find the derivations of the equations as well as the equations themselves of considerable value in ultrasonic designs and in clarifying various statements and equations encountered in this book or elsewhere.

Ultrasonics touches on nearly every aspect of human endeavor. The phenomena that make this wide range of interest and application possible have a firm basis in physical acoustics. The factors that affect the propagation of ultrasonic energy are basic to designing systems that will produce the phenomena desired and will do this efficiently. A study of the propagation characteristics themselves yields much information about the media through which the wave travels. This chapter is intended to acquaint the reader with some of the fundamental principles involved in the various applications of ultrasonic energy discussed in later chapters.

In many respects, ultrasound is similar to light and other forms of electromagnetic radiation in that it is a wave motion and obeys a general wave equation. Each wave type in a given homogeneous medium travels at a velocity that is dependent upon the properties of the medium. Similarly to light, ultrasound is reflected from surfaces, refracted when going from one medium into another which effects change in the velocity of sound, and diffracted at the edges of surfaces or around obstacles. Energy is scattered from particles or rough surfaces just as light is scattered from similar surfaces or dust particles.

Ultrasound is a form of mechanical energy, a vibration or wave-field, that propagates through solids, liquids, and gases. In many respects, it can be considered to be analogous to light or an electromagnetic wave [1]. Forces acting across an area at a given point in the wave are analogous to electrical voltage, and the velocity potential at that point is analogous to current. Just as the ratio of voltage to current is an electrical impedance, the ratio of force to velocity is an acoustical impedance. The impedances in the two cases are used in similar ways, for example, to calculate reflections and transmissions at an impedance discontinuity or to match components for effective energy transfer from one element to another. When a stationary impedance in a transmission line produces a reflection that is everywhere in phase with an incident, steady-state wave, the two waves reinforce each other to produce a standing wave. This is the condition at resonance, which is important in certain interferometer types of measurements and in some types of high-intensity applications.

At a given frequency, the particle motion at any point in an ultrasonic wave is sinusoidal if the stresses developed in the waves remain in the linear, elastic range of the medium. If two sinusoidal waves of slightly differing frequencies are superimposed, their amplitudes are alternately added and subtracted so that the overall effect is a wave with an amplitude equal to the sum of the amplitudes of the individual waves and a frequency equal to the difference in the frequencies of the same two waves. The superposition can cause beats, and the new frequency is called a beat frequency. The beat frequency is used effectively with the Doppler effect in many applications such as measuring the velocity of blood flow, measuring the velocity of heart actions, or detecting the movement of intruders. The Doppler effect is a change in frequency between an initial wave and a received wave occurring as a result of a changing path length between a source of sound and an observer and can be caused by a moving reflector even when the source and the receiver are stationary. The beats are produced between the initial outgoing wave and the received wave in which the frequency has been shifted by the Doppler effect.

As a wave propagates through a medium, its amplitude decreases, or attenuates. There are several causes of this attenuation, such as spreading of the wave front, conversion of the acoustical energy to heat (absorption), and scattering from irregular surfaces. One factor that affects attenuation is relaxation, a term that describes the lag between an initiating disturbance and a readjustment of energy distribution induced by the disturbance—for example, heat flow from a region under compression to a lower pressure region. A finite time is involved, and the exchange of energy approaches equilibrium value exponentially. A study of relaxation phenomena yields considerable information regarding the nature of the solid, liquid, and gaseous states of matter (e.g., [2]).

At high intensities of ultrasound, become increasingly nonlinear [3,4] the absorbed energy can produce considerable heat; in fact, at extremely high intensities, it can produce enough heat to rapidly melt glass or steel. Another phenomenon associated with high intensities in liquids is cavitation, to which many of the effects produced are related. Cavitation is associated with the stresses generated in the ultrasonic waves and their rate of change. Free chemical radicals are produced with cavitation, which promotes certain chemical reactions. The stresses developed are sufficiently strong to erode hard and tough materials. Cavitation is especially helpful in ultrasonic cleaning processing and sonochemistry [5–7].

These principles are discussed more fully in later sections of this chapter.

2.2 POWER DELIVERED TO AN OSCILLATING SYSTEM

An ultrasonic wave may be considered to consist of an infinite number of oscillating masses, or elements, connected by means of elastic springs so that each element is influenced by the motion of its nearest neighbor. To induce oscillation in each mass requires an initial power input, that is, introduction of energy at a given rate.

Consider a simple oscillating system, driven by an external source of power. The instantaneous power delivered to the system by a driving force is given by

$$P_i = F_i v_i \tag{2.1}$$

where F_i is the instantaneous driving force and v_i is the resulting velocity. The instantaneous power, P, dissipated is the product of the real parts of F_i and v_i or, for the simple oscillator of Equation 2.1,

$$P = \left(\frac{F^2}{Z_m}\right)\cos \omega t \cos(\omega t - \phi) \tag{2.2}$$

where ω is $2\pi f$, ϕ is the phase angle between the driving force F and the velocity v, and f is the frequency of oscillations.

The average power P_{ave} dissipated in the system is equal to the total work done per cycle divided by the period T, or

$$
\begin{aligned}
P_{ave} &= \frac{\int_0^T P\, dt}{T} \\
&= \frac{F^2}{TZ_m}\int_0^T \cos \omega t \cos(\omega t - \phi)\, dt \\
&= \left(\frac{F^2}{2Z_m}\right)\cos \phi
\end{aligned}
\tag{2.3}
$$

where Z_m is the complex mechanical impedance of the system.

When a vibratory system is driven by a constant oscillatory force, the initial energy is partitioned between a component that is used to overcome the losses of the system and a second component that is used to change the vibratory state of the system. Energy from the latter component is stored in the system until the system reaches an equilibrium where the energy supplied exactly equals the energy dissipated.

In Equation 2.3, the term $\cos\phi$ $(= R_m/Z_m)$ is the mechanical power factor.

2.3 VELOCITY OF SOUND

In keeping with the analogy of a series of an infinite number of mass elements in an ultrasonic wave, when a disturbance is induced in a mass, the first element affected transfers energy to the next one in line, the second influences the third in a similar manner, and so on as the energy introduced passes from element to element (propagates) through the medium until it is dissipated. The elements of the entire mass do not move in unison. The reasons for this are (1) that a mass has inertia and accelerates at a rate corresponding to the applied force according to Newton's laws of motion and (2) that it is elastic and deforms under stress. Thus the disturbance propagates through the medium at the velocity of sound.

The term *velocity of sound* requires some qualification. The nature of the disturbance is a factor in this velocity; that is, the type of oscillatory motion that the disturbance produces is a factor in the speed at which the disturbance travels through the medium. The rate of propagation depends upon the type of wave, the elastic properties of the medium, the density of the medium, and, in some cases, the frequency and amplitude. Since sound, or ultrasound, represents energy transmitted as stress waves, the velocity of sound also depends upon the mode of vibration with which it is associated. Gases are capable of transmitting only compressional (or longitudinal) waves. Most liquids transmit only longitudinal waves and surface waves. Solids may transmit a large variety of wave modes, and systems comprised of several media or phases can also propagate interface and leaky waves. The particle trajectories for the main types of waves used in ultrasonic research are shown in Table 2.1. The particle displacements for three important types of ultrasonic waves, compressional (longitudinal), shear, and Rayleigh waves, are shown in Figures 2.1 through 2.3, respectively. For the Rayleigh wave, the particle displacement is shown for a Ricker-type pulse and as a function of depth (Z), which decays rapidly below the surface, and most energy is concentrated within about 1.5 wavelengths of the surface.

2.3.1 Velocity of Sound in Solids

The velocity of sound in a long, slender bar in which the particle motion is parallel to the axis of the bar (longitudinal thin-bar velocity) is given by

$$c_0 = \left(\frac{E}{\rho}\right)^{1/2} \tag{2.4}$$

where E is Young's modulus of elasticity (N/m²) and ρ is the density of the bar (kg/m³). The dimensions of c_0 are in meter per second.

TABLE 2.1

Particle Trajectories for the Majority of Types of Waves Used in Ultrasonic Research, Measurement, Devices, Nondestructive Testing, and Process Control

Wave Type	Particle Motion
Longitudinal and extensional	To and fro—parallel to direction of propagation
Transverse shear	Orthogonal to direction of propagation
Rayleigh	Ellipse in direction perpendicular to surface—parallel to propagation (for sinusoidal)
Torsion	Circular path in plane perpendicular to axis
Lamb (symmertical)	Edge of plate view shows pattern regular bulges
Lamb (asymetrical)	Edge of plate flexes—sinusoidal
Love	Parallel to surface—perpendicular to direction of propagation in layered media—coating velocity less than substrate

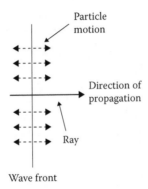

FIGURE 2.1   Longitudinal (compressional)-wave particle motion.

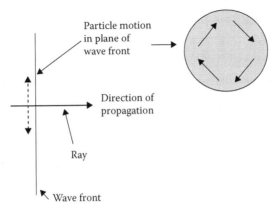

FIGURE 2.2   Shear-wave particle motion.

When an element in a slender bar is compressed axially, it is free to expand radially, and when it is in tension, it contracts radially. If a small element in a large mass of large cross-sectional area and length is compressed in one direction, its expansion in all other directions is limited by the surrounding mass. The effect is to increase the bulk modulus over the Young's, or bar, modulus by a function of Poisson's ratio. There is a similar effect on the velocity of sound, which becomes

$$c_B = c_0 \left[\frac{1-\sigma}{(1+\sigma)(1-2\sigma)} \right]^{1/2} \tag{2.5}$$

where c_B is longitudinal bulk velocity and σ is Poisson's ratio. The corresponding longitudinal velocity in thin plates is

$$c_P = c_0 (1-\sigma^2)^{-1/2} \tag{2.5a}$$

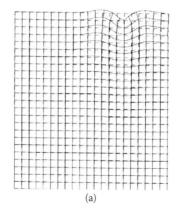

(a)

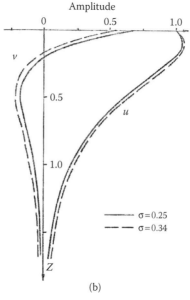

(b)

FIGURE 2.3 Rayleigh wave. (a) Particle displacement model for a Ricker-type pulse on a free surface, showing pulse lobes, and near-exponential decay with depth below surface. (b) Rayleigh-wave displacements with depth (Z) given by analytic equations, at the point of maximum surface displacement for two media (σ is Poisson's ratio for material).

When the particle motion in the wave is normal to the direction of propagation, the energy passes from element to element by shear stresses and the wave is called a shear wave. In this case, the shear modulus, μ, is a factor in the velocity. Shear modulus may be related to Young's or bulk modulus as functions of Poisson's ratio. The relationships result in the following equation for shear-wave velocity, c_s:

$$c_s = c_0[2(1+\sigma)]^{-1/2} = \left(\frac{\mu}{\rho}\right)^{1/2} \qquad (2.6)$$

Equation 2.6 also applies to torsional waves since these are shear waves.

Some ultrasonic waves follow the contour of the surface of a medium and penetrate the medium only to shallow depths. These waves are known as *surface waves*. Two types are of interest in the application of ultrasonics. These are Love waves and Rayleigh waves. The particle motion in Love waves is normal to the direction of propagation and is polarized in the plane of the surface. Rayleigh wave motion combines a longitudinal component with a shear component, which is normal to the surface, and the particle motion is elliptical. Since the restraints differ somewhat from those within the medium, the velocity of surface waves also differs from that of either shear or longitudinal waves. The velocity of Rayleigh waves, which are of the type most commonly used in applications of ultrasonic energy, is slightly lower than that of shear waves in the same medium. The velocity of Rayleigh waves is given by

$$c_R = K_R c_s \tag{2.7}$$

where K_R is a constant determined by the relationship between the longitudinal velocity and the shear velocity, and, therefore, Poisson's ratio. The relationship between c_B, c_s, and c_R is given by

$$\frac{c_R^6}{c_s^6} - \frac{8c_R^4}{c_s^4} + c_R^2 \left(\frac{24}{c_s^2} - \frac{16}{c_B^2} \right) - 16 \left(1 - \frac{c_s^2}{c_B^2} \right) = 0 \tag{2.8}$$

where c_R is the particular solution for which its value is slightly less than c_s. The effective depth of penetration of surface waves is approximately one wavelength.

Also used in ultrasonic applications are Lamb waves, which propagate in thin plates. There is nearly infinite number of velocities for Lamb waves since the velocity of propagation of these waves depends upon the relationship between the wavelength and the thickness of the material [8]. These waves are discussed in greater detail in Chapter 3, and various advanced texts (e.g., [9]).

2.3.2   Velocity of Sound in Liquids

Generally, shear waves are rapidly attenuated in liquids. Surface waves at ultrasonic frequencies in liquids are seldom of interest to the ultrasonics engineer. Therefore, the velocity of sound in a liquid usually refers to a longitudinal type of wave mode. The velocity of sound, c (m/s), in liquids is

$$c = \left(\frac{\gamma}{\rho \beta_i} \right)^{1/2} = (\rho \beta_s)^{-1/2} \tag{2.9}$$

where $\gamma = c_p/c_v$ is the ratio of specific heats, c_p is specific heat at constant pressure, c_v is specific heat at constant volume, and β_s is the adiabatic compressibility.

Equation 2.9 applies at low pressure amplitudes (acoustic intensities). At high pressure levels (finite amplitude wave intensities), propagation characteristics become more complex and nonlinear. Finite amplitude wave theory is applied to shock waves, and the velocity of these waves is always higher than the acoustic velocity. In general, only acoustic velocities will be considered in this book.

Measurement of the velocity of sound in a liquid provides a means of determining the adiabatic compressibility of a liquid and the ratio of specific heats of the liquid.

Transverse shear waves may be generated in a liquid with phase velocity given by

$$c_{vis} = \left(\frac{2\omega\pi}{\rho_0} \right)^{1/2} = (2\omega v)^{1/2} \tag{2.10}$$

and with an attenuation in nepers per meter (Np/m) given by

$$\alpha_{vis} \left(\frac{\omega\rho_0}{2\eta} \right)^{1/2} = \left(\frac{\omega}{2v} \right)^{1/2} \tag{2.10a}$$

where η is the shear viscosity, ρ_0 is the density of the liquid at static conditions, and v is the kinematic viscosity of the liquid.

In practice, the attenuation factor for the amplitude is measured by determining the amplitude ratio of the wave at two different positions x_1 and x_2.

Hence,

$$A = \exp\alpha(x_2 - x_1) \tag{2.10b}$$

The attenuation in nepers $\equiv \ln(A) = \alpha(x_2 - x_1)$ so that α is measured in Np/m.

It is more common to use the decibel (dB) scale to compare acoustic intensity level; the attenuation in dB is defined as

$$\begin{aligned} \text{attenuation (dB)} &= 10\log_{10}(r_{12})^2 \\ &= 20(\log_{10} e)\alpha(x_2 - x_1)\, \text{dB} \end{aligned} \tag{2.10c}$$

where α is in dB/m.

Hence, the relation between the two units is

$$\alpha(\text{dB/m}) = 20(\log_{10} e)\alpha(\text{Np/m}) = 8.686\alpha(\text{Np/m}) \tag{2.10d}$$

Further discussion of viscoelastic properties of liquids and solids, their measurement, and their importance appears in Chapter 6, and various advanced texts (e.g., [10]).

2.3.3 VELOCITY OF SOUND IN GASES

The equation for the velocity of sound in gases is derived from the equation of state. All fluids (gases and liquids) and solids are characterized by an equation of state of the form

$$f(p,V,T) = 0 \tag{2.11}$$

where p is the pressure exerted by the system (gas, liquid, or solid), V is its volume, and T is its temperature on some suitable scale. It can be shown [11] from Equation 2.11 that

$$\left(\frac{\partial p}{\partial V}\right)_T \left(\frac{\partial V}{\partial T}\right)_p \left(\frac{\partial T}{\partial p}\right)_v = -1 \tag{2.12}$$

Generally, the propagation of ultrasonic energy in gases (or in liquids) is adiabatic because the variations in pressure and corresponding volume changes are so rapid that heat does not have time to flow from high-pressure regions to rarefaction regions, and, therefore, no heat is lost in the process.

For an ideal gas, the equation of state is

$$pV = \frac{N}{N_0} RT \tag{2.13}$$

where p is the pressure (Pa), V is the volume of the gas (m^3), N is the number of molecules in V, N_0 is the number of molecules in a gram-molecular weight of the gas, R is the gas constant, and T is the absolute temperature (K). The ratio R/N_0 is called Boltzmann's constant.

The velocity of sound, c, in a gas is given by the relation

$$c^2 = \gamma \left(\frac{\partial p}{\partial \rho}\right)_T \tag{2.14}$$

where c is velocity in meters per second (m/s).

Therefore, from Equation 2.13, the equation for the velocity of sound in an ideal (perfect) gas is

$$c^2 = \gamma \frac{RT}{M} \tag{2.15}$$

where M is the molecular weight.

Roberts and Miller [12] give the following relationship for the velocity of sound, c, in a real gas

$$c^2 = \gamma \frac{RT}{M} \left[1 + \frac{9}{64} \frac{p}{p_c} \left(1 - 6 \frac{T_c^2}{T^2} \right) \frac{T_c}{T} \right] \tag{2.16}$$

where T_c is the critical temperature of the gas (K) and p_c is the critical pressure of the gas (kg/m^2 [Pascal]). Measurement of the velocity of sound in a gas is one of the most convenient and accurate methods of measuring the ratio of specific heats.

2.4 IMPINGMENT OF AN ULTRASONIC WAVE ON A BOUNDARY BETWEEN TWO MEDIA

When an ultrasonic wave encounters an interface between two media, the energy of the wave is partitioned in a manner that depends upon the type of incident wave, upon how the wave approaches the interface, and upon the acoustic properties of the two media. The condition is generally more complicated than the optical analogue of a light beam incident on the surface of a transparent object. As with light, Snell's law is used to determine the angle of reflection and refraction, but the acoustic problem is complicated by the greater number of wave modes and by the longer wavelengths usually associated with ultrasonic energy, which are important factors in the applications discussed in this book.

The characteristic that leads to Snell's law is that, along a nonslip boundary, the phase velocity along the interface must be the same for each wave in order to fulfill the boundary conditions on displacements and stresses (see Figure 2.4). The phase velocity is the rate at which a point of constant phase travels along the boundary and is

$$c_\mathrm{p} = \frac{c_\mathrm{a}}{\sin \alpha_\mathrm{a}} \qquad (2.17)$$

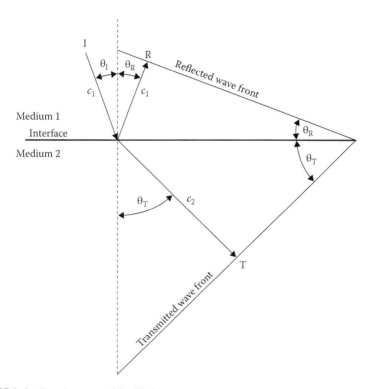

FIGURE 2.4 Development of Snell's law.

where c_a is the velocity of propagation of the particular wave under consideration and α_a is the angle between the wave front and the boundary surface. According to Equation 2.17,

$$c_p \geq c_a \tag{2.18}$$

Mode conversion occurs at boundaries in solids because solids can sustain both shear and dilatational stresses. The types of waves that can be produced by a wave incident on an interface may be determined by resolving the stress and displacement components of the incident wave along the coordinates of the boundary. These stress and displacement requirements fix the boundary conditions for the particular solutions of the general wave equation by which it is possible to determine the parameters of the waves reflected from the interface and of the waves transmitted across the interface. Four separate boundary conditions exist for the general case. On both sides of the interface, the following quantities must be equal:

- Normal displacements
- Tangential displacements
- Normal stresses
- Tangential stresses

2.4.1 Simple Reflection and Transmission at Normal Incidence

When a plane wave impinges at normal incidence on an interface between two semi-infinite media, as in Figure 2.5, the energy of the incident wave I is partitioned between a reflected wave R and a transmitted wave T. At the boundary, the total displacement is the same in either medium. Then

$$\xi_T = \xi_i + \xi_R \tag{2.19}$$

where ξ_T, ξ_i, and ξ_R are potential functions defined as

$\xi_T = \xi_{T0} \sin\omega(t - x/c_2)$, the displacement of the transmitted wave in medium 2
$\xi_i = \xi_0 \sin\omega(t - x/c_1)$, the displacement of the incident wave
$\xi_R = \xi_{R0} \sin\omega(t + x/c_1)$, the displacement of the reflected wave

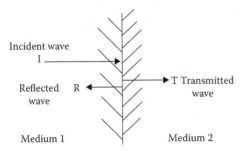

FIGURE 2.5   Reflection and transmission at an interface between two media with different acoustic impedances.

Similarly, the stresses on one side of the interface are equal to those on the other side, that is,

$$P_T = P_i + P_R \tag{2.20}$$

where P_T, P_i, and P_R also are potential functions of stress and are related to the displacements according to the following equation:

$$P = -j \frac{\rho c^2}{\omega} \frac{\partial \dot{\xi}}{\partial x} \tag{2.21}$$

where $\dot{\xi} = \partial \xi / \partial t$.

From Equations 2.20 and 2.21, we obtain

$$\left. \begin{array}{l} P_T = -j\omega\rho_2 c_2 \xi_T \\ P_i = -j\omega\rho_2 c_1 \xi_i \\ P_R = j\omega\rho_1 c_1 \xi_R \end{array} \right\} \tag{2.22}$$

Substituting Equation 2.22 in Equation 2.20 gives

$$\rho_2 c_2 \xi_T = \rho_1 c_1 (\xi_i - \xi_R) \tag{2.23}$$

where $\rho_1 c_1$ and $\rho_2 c_2$ are the specific acoustic impedances of medium 1 and medium 2, respectively, and c_1 and c_2 are the velocities of wave propagation in medium 1 and medium 2, respectively.

At the interface, $x = 0$, and

$$\xi_{T0} = \xi_0 + \xi_{R0} \tag{2.24}$$

Then

$$\frac{\xi_T}{\xi_i} = \left(1 - \frac{\xi_R}{\xi_i}\right) \frac{\rho_1 c_1}{\rho_2 c_2} \tag{2.25}$$

From Equation 2.19

$$\frac{\xi_R}{\xi_i} = \frac{\xi_T}{\xi_i} - 1 \tag{2.19a}$$

Therefore,

$$\frac{\xi_T}{\xi_i} = \frac{2\rho_1 c_1}{\rho_1 c_1 + \rho_2 c_2} \tag{2.26}$$

Again, from Equation 2.19a

$$\frac{\xi_R}{\xi_i} = \frac{\xi_T}{\xi_i} - 1 = \frac{2\rho_1 c_1}{\rho_1 c_1 + \rho_2 c_2} - 1 = \frac{\rho_1 c_1 - \rho_2 c_2}{\rho_1 c_1 + \rho_2 c_2} \qquad (2.27)$$

The relationships between the stresses in the waves are found by equating the stresses to their corresponding displacements and impedances and substituting in Equations 2.23 and 2.24. Then

$$\frac{P_T}{P_i} = \frac{\xi_T \rho_2 c_2}{\xi_i \rho_1 c_1} = \frac{2\rho_2 c_2}{\rho_1 c_1 + \rho_2 c_2} \qquad (2.28)$$

and

$$\frac{P_R}{P_i} = -\frac{\xi_R}{\xi_i} = \frac{\rho_2 c_2 - \rho_1 c_1}{\rho_1 c_1 + \rho_2 c_2} \qquad (2.29)$$

According to Equations 2.26 through 2.29, no phase shift occurs between the incident wave and the transmitted wave either in the displacement or in the stress, regardless of which medium has the higher acoustic impedance. However, when the acoustic impedance of medium 2 is greater than that of medium 1, the displacement of the reflected wave is 180° out of phase with that of the incident wave, but the stresses are in phase. The reverse is true if the acoustic impedance of medium 2 is less than that of medium 1.

More general equations that apply in the case of finite media are

$$\frac{\xi_R}{\xi_i} = \frac{z_1 - z_2}{z_1 + z_2} \qquad (2.27a)$$

$$\frac{P_t}{P_i} = \frac{2z_2}{z_1 + z_2} \qquad (2.28a)$$

$$\frac{P_R}{P_i} = \frac{z_2 - z_1}{z_1 + z_2} \qquad (2.29a)$$

where z_1 and z_2 are the complex acoustic impedances of the respective media.

It is the acoustic energy and not the pressure that is partitioned at an interface. The acoustic energy transmitted at an interface between two semi-infinite solids, as a function of acoustic impedance ratio, is shown in Figure 2.6 [13]. The intensity is related to the amplitude of the particle vibrations. The acoustic pressure is used to denote the amplitude of the alternating stress, and the intensity is proportional to the square of the acoustic pressure. A piezoelectric transducer element senses the acoustic pressure.

In practice, the actual situations are often considerably more complicated than the ideal conditions described in this section. Reflections and refraction at interfaces between media enter into every aspect of the applications of ultrasonic energy. For this reason, this topic is given extra consideration here.

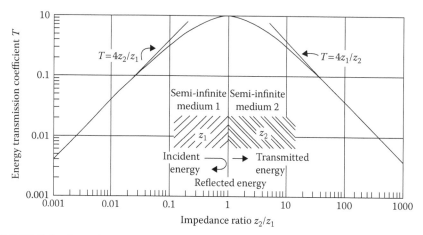

FIGURE 2.6 Acoustic energy transmission across an interface between two semi-infinite media, calculated at normal incidence. (From Lynnworth, L. C., *IEEE Trans Sonics Ultrason*, SU-12(2):37–48. © 1965 IEEE. With permission.)

2.4.2 SOME BASIC MECHANICS

As a preface to the analysis of stress waves incident at any angle on an interface between two media, consider the mechanics of stresses in a block. An elemental block of width ℓ, of height dy, and of length dx (see Figure 2.7) is subjected to a compressive stress p on faces AB and CD. The resultant stresses on the plane AC are p' and s'. The total applied force is $p\ell dx$. To maintain equilibrium, the force pdx is resisted by a force $(p'\ell ds \cos\theta + s'\ell ds \sin\theta)$, neglecting Poisson's ratio, or

$$p'\ell ds \cos\theta + s'\ell ds \sin\theta = p\ell dx \tag{2.30}$$

where p' is the compressive stress normal to the plane AC and s' is the shear stress parallel to the plane. By substituting $(dx/ds) = \cos\theta$, Equation 2.30 can be rewritten

$$p' = p - s'\left(\frac{\sin\theta}{\cos\theta}\right) \tag{2.30a}$$

In the horizontal direction, normal to the direction of p

$$p'\ell ds \sin\theta = s'\ell ds \cos\theta \tag{2.31}$$

from which

$$p' = s'\left(\frac{\cos\theta}{\sin\theta}\right) \tag{2.31a}$$

Substituting Equation 2.31a into Equation 2.30a and solving for s' in terms of p gives

$$s' = \frac{p\sin 2\theta}{2} \tag{2.32}$$

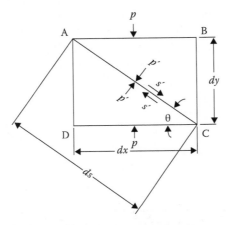

FIGURE 2.7 Stresses on a plane through a block subjected to compression between two parallel faces.

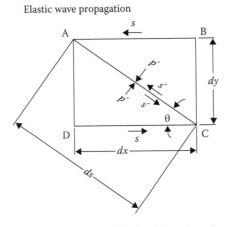

FIGURE 2.8 Stresses on a plane through a block subjected to shear between two parallel faces.

Substituting Equation 2.32 into Equation 2.31a and solving for p' in terms of p, we obtain

$$p' = p\cos^2\theta = p\left(\frac{1+\cos 2\theta}{2}\right) \tag{2.33}$$

Now assume that the same two faces AB and CD are subjected to a pure shear stress, s, instead of p, as shown in Figure 2.8. The horizontal forces acting on the diagonal AC at equilibrium are

$$(p'\sin\theta + s'\cos\theta)\ell ds = s\ell dx \tag{2.34}$$

and the vertical forces are

$$p'\ell ds \cos\theta = s'\ell ds \sin\theta \tag{2.35}$$

From Equations 2.34 and 2.35

$$s' = s\cos^2\theta = s\left(\frac{1+\cos 2\theta}{2}\right)$$

and   (2.36)

$$p' = \frac{s}{2}\sin 2\theta$$

The same procedure can be used to determine the stresses produced on any plane by stresses acting on any of the other faces of the block.

If both p and s were acting on faces AB and CD simultaneously, the resultant stresses acting on the plane AC would be the sum of those caused by p and by s, as though each were applied independently of the other. However, it is essential that the directions of the stresses be identified by proper signs. For instance, combining p and s of Figures 2.7 and 2.8 and defining the positive directions above the diagonal AC as being downward and toward the left, we see that the normal stress on AC is the sum of the stresses p' given by Equations 2.33 and 2.36, or

$$p'_{(sum)} = p\left(\frac{1+\cos 2\theta}{2}\right) + s\frac{\sin 2\theta}{2} \tag{2.37}$$

The total shear stress on diagonal AC is s' calculated by Equation 2.36 minus s' calculated by Equation 2.32 since these are in opposite directions. Then

$$s'_{(sum)} = s\left(\frac{1+\cos 2\theta}{2}\right) - p\frac{\sin 2\theta}{2} \tag{2.38}$$

If $s'_{(sum)}$ is positive, its direction is toward the left, as previously defined, and if it is negative, its direction is toward the right. When $\theta = 0$, $s' = s$ and $p' = p$. Numerous relationships are possible between ultrasonic waves incident on an interface between two media and the reflected and refracted waves that result. These are dependent upon the directions of the stresses and strains associated with the various waves and the acoustic properties of the materials. Only a brief discussion of these possibilities is presented in this chapter. For a detailed analysis, the reader should consult [9,14,15].

2.4.3 GENERAL CONSIDERATIONS OF INCIDENT WAVES

Figures 2.9 and 2.10 show the possible directions of strain in waves incident on a plane boundary between two media and the resulting reflected and refracted waves. Figure 2.9 illustrates a longitudinal wave impinging on the interface, and Figure 2.10 illustrates an incident shear wave. The two conditions are shown separately for clarity.

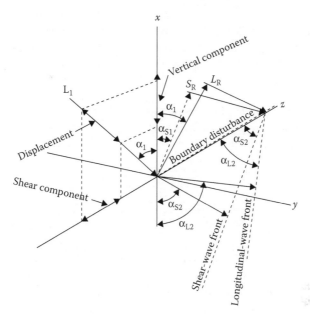

FIGURE 2.9　Longitudinal-wave incident on a boundary between two media.

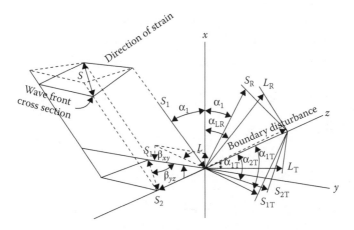

FIGURE 2.10　Shear-wave incident on a boundary between two media.

A single equation may be derived by which the amplitude of each wave reflected or refracted may be related to that of the incident wave by applying the four boundary conditions of continuity of stress and strain across the interface to appropriate solutions to the wave equations (Section 2.4.1). This equation may be used to obtain a specific relationship by applying further conditions related to the media and to the particular incident wave mode. For instance, a vacuum can propagate no ultrasonic energy. Therefore, waves incident on an interface between a vacuum and any other medium are reflected completely. In this case, both S_t and L_t are zero.

Generally, liquids can propagate no shear waves. Viscous waves generated in liquids according to Equation 2.10 are considered to be very highly attenuated and weak and are, therefore, negligible for purposes of the present discussion. Therefore, all shear-wave components are zero in such a liquid medium.

Both shear and longitudinal waves can propagate through isotropic solids. The elastic constants in an isotropic solid are equal in all directions. Therefore, only one shear-wave velocity and one longitudinal-wave velocity are possible in an unbounded isotropic medium, assuming that the elastic properties are indeed constant. The ratio of the velocity of propagation of the longitudinal wave to that of the shear wave is

$$\frac{c_L}{c_S} = \sqrt{\frac{2(1-\sigma)}{1-2\sigma}} \tag{2.39}$$

where σ is Poisson's ratio.

Both shear and longitudinal waves can propagate through anisotropic solids. The velocities of propagation through such media depend upon the orientations of the strains with respect to the crystallographic orientations of the media. Macroscopic properties of the media depend upon the grain orientation (random or oriented). Propagation characteristics, attenuation, and velocity are affected by the relationship between the wavelength and the grain size [16,17].

2.4.4 DEVELOPMENT OF GENERAL EQUATIONS FOR REFLECTION AND REFRACTION WHERE MODE CONVERSION IS POSSIBLE

A wave equation may be written for each of the types of waves, which may be propagated in the media. Redwood [15] uses the scalar potential ϕ corresponding to longitudinal waves and the vector potential ψ corresponding to shear waves. The corresponding wave equations

$$\nabla^2\phi = \frac{1}{c_L^2}\frac{\partial^2\phi}{\partial t^2} \tag{2.40}$$

$$\nabla^2\phi = \frac{1}{c_S^2}\frac{\partial^2\psi}{\partial t^2} \tag{2.41}$$

where $c_L^2 = (\lambda' + 2\mu)/\rho$, $c_S^2 = \mu/\rho$, and λ' and μ are Lamé's elastic constants, μ being the elastic shear modulus mentioned in Section 2.3.1 of this chapter.

To obtain the general solutions to the wave equations, using rectangular coordinates, we recall that all the waves are related to the phase velocity c_p. Referring to Equation 2.17 and Figures 2.5 and 2.6, the phase velocity is

$$c_p = \frac{c_{1L}}{\sin\alpha_1} = \frac{c_{1s}}{\sin\alpha_{s1}} = \frac{c_{2L}}{\sin\alpha_{L2}} = \frac{c_{2s}}{\sin\alpha_{s2}} \tag{2.42}$$

The most general solutions for the potential functions are

$$\phi_1 = \left[\xi_{Li}\, e^{-j\left(\frac{\omega}{c_{1L}}\cos\alpha_1\right)x} + \xi_{Lr}\, e^{j\left(\frac{\omega}{c_{1L}}\cos\alpha_1\right)x} \right] \times e^{-j\left(\frac{\omega}{c_{1L}}\sin\alpha_1\right)x}\, e^{j\omega t} \tag{2.43}$$

$$\psi_1 = \left[\xi_{si}\, e^{-j\left(\frac{\omega}{c_{1s}}\cos\alpha_{s1}\right)x} + \xi_{sr}\, e^{j\left(\frac{\omega}{c_{1s}}\cos\alpha_{s1}\right)x} \right] \times e^{-j\left(\frac{\omega}{c_{1s}}\sin\alpha_{s1}\right)z}\, e^{j\omega t} \tag{2.44}$$

$$\phi_2 = \xi_{LT}\, e^{-j\left(\frac{\omega}{c_{2L}}\cos\alpha_{L2}\right)x}\, e^{-j\left(\frac{\omega}{c_{2L}}\sin\alpha_{L2}\right)z}\, e^{j\omega t} \tag{2.45}$$

$$\psi_2 = \xi_{ST}\, e^{-j\left(\frac{\omega}{c_{2S}}\cos\alpha_{S2}\right)x}\, e^{-j\left(\frac{\omega}{c_{2S}}\sin\alpha_{S2}\right)z}\, e^{j\omega t} \tag{2.46}$$

in which the normals to the wave fronts are in the xz plane, the interface lies in the yz plane, and

ξ_{Li} is the displacement amplitude of the incident longitudinal wave in the direction of wave propagation.

ξ_{Lr} is the displacement amplitude of the reflected longitudinal wave in the direction of wave propagation.

ξ_{si} is the displacement amplitude of the incident shear wave in a direction normal to the direction of wave propagation.

ξ_{sr} is the displacement amplitude of the reflected shear wave in a direction normal to the direction of wave propagation.

ξ_{LT} is the displacement amplitude of the transmitted longitudinal wave in the direction of wave propagation.

ξ_{ST} is the displacement amplitude of the transmitted shear wave normal to the direction of wave propagation.

Referring to Figure 2.7, the displacements are assumed to be parallel to the xz plane and the interface is in the yz plane. According to the boundary conditions (at $x = 0$)

$$(\xi_{Li} + \xi_{Lr})\sin\alpha_1 + (\xi_{si} - \xi_{sr})\cos\alpha_{s1} = \xi_{sT}\cos\alpha_{s2} + \xi_{LT}\sin\alpha_{L2} \tag{2.47}$$

(displacements parallel to interface)

$$(\xi_{Li} + \xi_{Lr})\cos\alpha_1 - (\xi_{si} - \xi_{sr})\sin\alpha_{s1} = \xi_{LT}\cos\alpha_{L2} - \xi_{sT}\sin\alpha_{s2} \tag{2.48}$$

(displacements normal to interface)

The shear stresses and the tensile stresses due to both shear and longitudinal waves are determined (1) by algebraically summing the stresses parallel to the interface and acting through the interface, that is, those components of stress causing

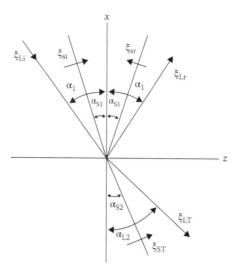

FIGURE 2.11 Displacements in incident, reflected, and refracted waves at an interface between two solid media.

shear on the interface, and (2) by summing the stresses acting on a normal to the interface.

First, consider the conditions for which all displacements and the direction of propagation are parallel to the xz plane and for which the interface lies in the yz plane (Figure 2.11).

The components of stress acting parallel to the interface are given by

$$S_s = -j \frac{\rho c_s^2}{\omega} \left(\frac{\partial^2 \psi}{\partial x \partial t} + \frac{\partial^2 \phi}{\partial x \partial t} \right) z \qquad (2.49)$$

Using the directions for displacements of Figure 2.7, the shear displacements parallel to z caused by the longitudinal waves are $\xi_L \sin \alpha_L$ and those caused by the shear waves are $\xi_s \cos \alpha_s$. In the equations that follow, the positive direction is considered to be to the right and downward in both displacements and propagation.

From Equation 2.43 and Figure 2.7

$$\frac{\partial^2 \phi_1}{\partial x \partial t} \bigg|_z = \left(\frac{\omega^2}{c_{1L}} \cos \alpha_1 \right) \xi_{Li} \, e^{-j\left(\frac{\omega}{c_{1L}} \cos \alpha_1\right)x} - \xi_{Lr} \, e^{j\left(\frac{\omega}{c_{1L}} \cos \alpha_2\right)x} \times e^{-j\left(\frac{\omega}{c_{1L}} \sin \alpha_1\right)z} \, e^{j\omega t} \sin \alpha_1 \qquad (2.50a)$$

and at $x = 0$

$$\frac{\partial^2 \phi_1}{\partial x \partial t} \bigg|_z = \frac{\omega^2}{2 c_{1L}} \left[(\xi_{Li} - \xi_{Lr}) \sin 2\alpha_1 \right] e^{-j\left(\frac{\omega}{c_p}\right)z} \, e^{j\omega t} \qquad (2.50b)$$

Similarly, from Equations 2.44 through 2.46 and Figure 2.7, and at $x = 0$

$$\frac{\partial^2 \psi_1}{\partial x \partial t}\bigg|_z = \frac{\omega^2}{2c_{1s}}[(\xi_{si} + \xi_{sr})(1 + \cos 2\alpha_1)]e^{-j\left(\frac{\omega}{c_p}\right)z}e^{j\omega t} \tag{2.51}$$

$$\frac{\partial^2 \phi_2}{\partial x \partial t}\bigg|_z = \frac{\omega^2}{2c_{2L}}[\xi_{LT} \sin 2\alpha_{L2}]e^{-j\left(\frac{\omega}{c_p}\right)z}e^{j\omega t} \tag{2.52}$$

and

$$\frac{\partial^2 \psi_2}{\partial x \partial t}\bigg|_z = \frac{\omega^2}{2c_{2s}}[\xi_{ST}(1 + \cos 2\alpha_{s2})]e^{-j\left(\frac{\omega}{c_p}\right)z}e^{j\omega t} \tag{2.53}$$

where c_p is phase velocity as defined previously, subscripts 1 and 2 refer to waves in medium 1 and medium 2, respectively, and subscripts L and s refer to longitudinal and shear waves, respectively. The displacements ξ are the actual amplitudes of the displacements.

Applying Equation 2.49 to Equations 2.50 through 2.53 and setting the sum of the stresses due to the waves in medium 1 equal to the sum of those in medium 2, we obtain

$$\rho_1 c_{1s}^2 \left[(\xi_{Li} - \xi_{Lr}) \sin 2\alpha_1 + \left(\frac{c_{1L}}{c_{1s}}\right)(\xi_{si} + \xi_{sr})(1 + \cos 2\alpha_{s1}) \right]$$
$$= \rho_2 c_{2s}^2 \left[\left(\frac{c_{1L}}{c_{2L}}\right)\xi_{LT} \sin 2\alpha_{L2} + \left(\frac{c_{1L}}{c_{2s}}\right)\xi_{ST}(1 + \cos 2\alpha_{s2}) \right] \tag{2.54}$$

for stresses parallel to the interface.

When $\alpha_1 = \alpha_{s1} = \alpha_{L2} = \alpha_{s2} = 0$

$$\rho_1 c_{1s}(\xi_{si} + \xi_{sr}) = \rho_2 c_{2s}\xi_{sT} \tag{2.54a}$$

The components of stress acting in a direction normal to the interface are

$$P_L = -j\frac{\rho}{\omega}\left[c_L^2 \left(\frac{\partial^2 \phi}{\partial x \partial t} + \frac{\partial^2 \psi}{\partial x \partial t}\right)_x + (c_L^2 - 2c_s^2)\left(\frac{\partial^2 \phi}{\partial z \partial t} + \frac{\partial^2 \psi}{\partial z \partial t}\right) \right] \tag{2.55}$$

Following the procedures used in obtaining Equation 2.54 and using Figure 2.7, at $x = 0$

$$\frac{\partial^2 \phi_1}{\partial x \partial t}\bigg|_x = \left[\left(\frac{\omega^2}{c_{1L}}\right)(\xi_{Li} + \xi_{Lr})e^{-j\left(\frac{\omega}{c_p}\right)z}e^{j\omega t} \right]\cos^2 \alpha_1 \tag{2.56}$$

$$\left.\frac{\partial^2 \psi_1}{\partial x \partial t}\right|_x = \left[-\left(\frac{\omega^2}{c_{1s}}\right)(\xi_{si} + \xi_{sr})e^{-j\left(\frac{\omega}{c_p}\right)z}e^{j\omega t}\right]\sin\alpha_{s1}\cos\alpha_{s1} \tag{2.57}$$

$$\left.\frac{\partial^2 \phi_2}{\partial x \partial t}\right|_x = \left[\left(\frac{\omega^2}{c_{2L}}\right)\xi_{LT}e^{-j\left(\frac{\omega}{c_p}\right)z}e^{j\omega t}\right]\cos^2\alpha_{L2} \tag{2.58}$$

$$\left.\frac{\partial^2 \psi_2}{\partial x \partial t}\right|_x = -\left(\frac{\omega^2}{c_{2s}}\right)\xi_{st}e^{-j\left(\frac{\omega}{c_p}\right)z}e^{j\omega t}\sin\alpha_{s2}\cos\alpha_{s2} \tag{2.59}$$

$$\left.\frac{\partial^2 \phi_1}{\partial z \partial t}\right|_z = \left(\frac{\omega^2}{c_{1L}}\right)(\xi_{Li} + \xi_{Lr})e^{-j\left(\frac{\omega}{c_p}\right)z}e^{j\omega t}\sin^2\alpha_1 \tag{2.60}$$

$$\left.\frac{\partial^2 \psi_1}{\partial z \partial t}\right|_z = \left(\frac{\omega^2}{c_{1s}}\right)(\xi_{si} - \xi_{sr})e^{-j\left(\frac{\omega}{c_p}\right)z}e^{j\omega t}\cos\alpha_{s1}\sin\alpha_{s1} \tag{2.61}$$

$$\left.\frac{\partial^2 \phi_2}{\partial z \partial t}\right|_z = \left(\frac{\omega^2}{c_{2L}}\right)\xi_{LT}e^{-j\left(\frac{\omega}{c_p}\right)z}e^{j\omega t}\sin^2\alpha_{L2} \tag{2.62}$$

and

$$\left.\frac{\partial^2 \psi_2}{\partial z \partial t}\right|_z = \left(\frac{\omega^2}{c_{2s}}\right)e^{-j\left(\frac{\omega}{c_p}\right)z}e^{j\omega t}\sin\alpha_{s2}\cos\alpha_{s2} \tag{2.63}$$

Equating the stresses acting in a direction normal to the interface in medium 1 to those in medium 2 determined according to Equation 2.55 gives

$$\rho_1 c_{1L}(\xi_{Li} + \xi_{Lr}) - 2\rho_1 c_{1s}(\xi_{Li} + \xi_{Lr})\sin\alpha_{s1}\sin\alpha_1 + \rho_1 c_{1s}\xi_{sr}\sin 2\alpha_{s1} \tag{2.64a}$$
$$= \rho_2 c_{2L}\xi_{Lt} - 2\rho_2 c_{2s}\xi_{Lt}\sin\alpha_{s2}\sin\alpha_{L2} - \rho_2 c_{2s}\xi_{sT}\sin 2\alpha_{s2}$$

for incident longitudinal wave, reflected shear and longitudinal waves, and transmitted shear and longitudinal waves, and

$$\rho_1 c_{1L}\xi_{Lr} - 2\rho_1 c_{1s}\xi_{Lr}\sin\alpha_{s1}\sin\alpha_1 - \rho_1 c_{1s}(\xi_{si} - \xi_{sr})\sin 2\alpha_{s1}$$
$$= \rho_2 c_{2L}\xi_{LT} - 2\rho_2 c_{2s}\xi_{LT}\sin\alpha_{s2}\sin\alpha_{L2} - \rho_2 c_{2s}\xi_{sT}\sin 2\alpha_{s2} \tag{2.64b}$$

for incident shear wave, reflected shear and longitudinal waves, and transmitted shear and longitudinal waves.

Equations 2.42, 2.47, 2.48, 2.54, and 2.64 may be used to determine the directions and relative amplitudes of waves reflected or refracted at an interface between two isotropic media if the incidence angle and the relative velocities of sound are known. In using these equations, it is necessary to consider:

- Which waves can be propagated in the media
- Whether the critical angle of any wave mode has been exceeded, in which case that particular mode is completely reflected
- In which direction the incident wave is polarized if it is a shear wave

Other phenomena also occur under various conditions. For instance, boundary disturbances or surface waves may be generated, which travel along a boundary or surface of a medium. The existence of these waves has been observed by schlieren techniques in steel and aluminum immersed in water even at normal incidence. There are several types of these boundary and surface waves. The two best known types are Love waves, in which the particle motion is parallel to the surface and normal to the direction of propagation, and Rayleigh waves, in which the particle motion is elliptical and parallel to a plane normal to the surface. Both of these types of waves were mentioned in Section 2.3.1.

The effective depth of penetration of a Rayleigh surface wave is approximately one wavelength, the intensity dropping exponentially with depth.

Rayleigh waves are often used, as are longitudinal and shear waves, for nondestructive testing (NDT).

When the direction of polarization of a shear wave is parallel to the plane of the interface, the components of stress in a direction normal to the interface are zero and no longitudinal waves are generated.

When the direction of polarization of an incident shear wave is at an angle both to the plane of the interface and to a plane normal to the interface, which contains the normal to the wave front, the solutions describing reflection and refraction involve displacements corresponding to all three directions in a rectangular coordinate system. The solutions may be obtained in the manner in which Equations 2.47, 2.48, 2.54, and 2.64 were derived. In isotropic media, these formulas may be used by applying an angular correction to compensate for the third dimension. Anisotropic media, however, present a much more difficult problem in that the elastic constants vary with the direction. This gives rise to birefringence, that is, the formation of two waves having the same mode and frequency but traveling at different velocities.

A shear wave traveling through an oriented anisotropic material and polarized at an angle to the orientation undergoes a change in polarization because of birefringence. This phenomenon has been investigated as a means of measuring residual stresses in materials since stress induces anisotropy.

The response and energy partition for various combinations of boundary conditions and media are reported in numerous papers and texts. For example, the variation of amplitude for reflected compressional wave at a solid–air interface for media with varying Poisson's ratio is shown in Figure 2.12 [18].

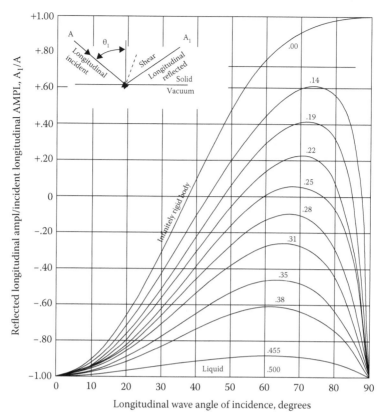

FIGURE 2.12 Variations of amplitude of reflected longitudinal waves at a solid–air interface, for varying Poisson's ratios. (Reprinted from Arenberg, D. L., *J Acoust Soc Am*, 20(1):1–16, 1948. With permission.)

2.4.5 WAVE INCIDENT ON A LIQUID–SOLID PLANE INTERFACE, SEMI-INFINITE MEDIA

To demonstrate the use of Equations 2.47, 2.48, 2.54, and 2.64 to calculate the relative amplitudes of reflected and transmitted waves, we use the example of an ultrasonic wave traveling through a liquid medium and encountering a plane solid surface at an angle α_1. The solid is isotropic. Under these conditions, the waves transmitted into the solid may be one shear wave, a longitudinal wave, and a boundary disturbance under certain conditions. Only longitudinal waves exist in the liquid. Therefore, for the case of the longitudinal wave in liquid incident on a liquid–solid interface, Equations 2.47, 2.48, and 2.64 become

$$(\xi_{Li} + \xi_{Lr}) \sin \alpha_1 = \xi_{sT} \cos \alpha_{s2} + \xi_{LT} \sin \alpha_{L2} \tag{2.47a}$$

$$(\xi_{Li} + \xi_{Lr}) \cos \alpha_1 = \xi_{sT} \cos \alpha_{L2} + \xi_{sT} \sin \alpha_{s2} \tag{2.48a}$$

$$\rho_1 c_{1L}(\xi_{Li} + \xi_{Lr}) = (\rho_2 c_{2L} - 2\rho_2 c_{2s} \sin \alpha_{s2} \sin \alpha_{L2}) \xi_{LT}$$
$$- \rho_2 c_{2s} \xi_{sT} \sin 2\alpha_{s2} \tag{2.64c}$$

The ratios of the displacements in terms of the characteristic impedances (ρc), velocities (c), and angles of incidence, reflection, and refraction are obtained by solving Equations 2.47a, 2.48a, and 2.64c simultaneously.

When $\alpha_{s2} < \alpha_{L2} < 90°$ assuming negligible boundary disturbances, solving Equations 2.47a, 2.48a, and 2.64c for the displacement amplitudes of the transmitted and reflected waves relative to the displacement amplitude of the incident wave gives

$$\frac{\xi_{Lr}}{\xi_{Li}} = \frac{\begin{array}{l}\rho_2 c_{2L}[\cos(\alpha_{s2} + \alpha_1) + \sin \alpha_{s2}] + \rho_2 c_{2s} \sin 2\alpha_1 \sin \alpha_{s2} \\ \times [\cos(\alpha_1 + \alpha_{s2}) - \sin \alpha_{s2} \sin \alpha_{L2} - \cos(\alpha_{s2} - \alpha_{L2})] \\ \quad - \rho_1 c_{1L}[\cos(\alpha_{s2} - \alpha_{L2}) + 2\sin \alpha_{s2} \sin \alpha_{L2}]\end{array}}{\begin{array}{l}[\rho_1 c_{1L} \cos(\alpha_{s2} + \alpha_{L2}) + \rho_2 c_{2L} \cos(\alpha_{s2} + \alpha_1) + 2\rho_2 c_{2s} \\ \times \sin \alpha_1 \sin \alpha_{s2} \cos(\alpha_1 + \alpha_{s2})]\cos \alpha_1\end{array}} \tag{2.65}$$

$$\frac{\xi_{LT}}{\xi_{Li}} = \frac{2[\rho_1 c_{1L} \cos \alpha_{s2} + \rho_2 c_{2s} \sin \alpha_1 \sin 2\alpha_{s2}]\cos \alpha_1}{\begin{array}{l}\rho_1 c_{1L} \cos(\alpha_{s2} + \alpha_{L2}) + \rho_2 c_{2L} \cos(\alpha_{s2} + \alpha_1) + 2\rho_2 c_{2s} \\ \times \sin \alpha_1 \sin \alpha_{s2} \cos(\alpha_1 + \alpha_{s2})\end{array}} \tag{2.66}$$

$$\frac{\xi_{sT}}{\xi_{Li}} = \frac{\begin{array}{l}\rho_2 c_{2L} - 2\rho_2 c_{2s} \sin \alpha_{s2} \sin \alpha_{L2} \sin 2\alpha_1 \\ \quad - 2\rho_1 c_{1L} \sin \alpha_{L2} \cos \alpha_1\end{array}}{\begin{array}{l}\rho_1 c_{1L} \cos(\alpha_{s2} + \alpha_{L2}) + \rho_2 c_{2L} \cos(\alpha_{s2} + \alpha_1) \\ \quad + 2\rho_2 c_{2s} \sin \alpha_1 \sin \alpha_{s2} \cos(\alpha_1 + \alpha_{s2})\end{array}} \tag{2.67}$$

At the critical angle for longitudinal waves in the solid, $\alpha_{L2} = 90°$ and

$$\sin \alpha_1 = \frac{c_{1L}}{c_{2L}}$$

where c_{1L} is the velocity of sound in the liquid and c_{2L} is the velocity of longitudinal-wave propagation in the solid. Neglecting the boundary disturbance, the energy is partitioned between a reflected longitudinal wave and a transmitted shear wave, and the relative amplitudes are given by

$$\frac{\xi_{Lr}}{\xi_{Li}} = \frac{\begin{array}{l}\rho_2 c_{2L}[\cos(\alpha_{s2} + \alpha_1) + \sin \alpha_{s2}] + \rho_2 c_2 \sin 2\alpha_1 \sin \alpha_{s2} \\ \times [\cos(\alpha_1 + \alpha_{s2}) - 2\sin \alpha_{s2}] - 3\rho_1 c_{1L} \sin \alpha_{s2}\end{array}}{\begin{array}{l}[-\rho_1 c_{1L} \sin \alpha_{s2} + \rho_2 c_{2L} \cos(\alpha_{s2} + \alpha_1) + 2\rho_2 c_{2s} \sin \alpha_1 \\ \times \sin \alpha_{s2} \cos(\alpha_1 + \alpha_{s2})]\cos \alpha_1\end{array}} \tag{2.65a}$$

$$\frac{\xi_{sT}}{\xi_{Li}} = \frac{\rho_2 c_{2L} - 2\rho_2 c_{2s} \sin \alpha_{s2} \sin 2\alpha_1 - 2\rho_1 c_{1L} \cos \alpha_1}{\begin{array}{l}-\rho_1 c_{1L} \sin \alpha_{s2} - \rho_2 c_{2L} \sin \alpha_{s2} + 2\rho_2 c_{2s} \sin \alpha_1 \\ \times \sin \alpha_{s2} \cos(\alpha_1 + \alpha_{s2})\end{array}} \tag{2.67a}$$

At normal incidence, the shear-wave components vanish from Equations 2.47, 2.48, and 2.64, and all other angles are zero. Three waves remain: an incident longitudinal wave, a reflected longitudinal wave, and a transmitted longitudinal wave. Equations 2.48 and 2.64 become

$$\xi_{Li} - \xi_{Lr} = \xi_{LT} \tag{2.48b}$$

$$\rho_1 c_{1L} (\xi_{Li} + \xi_{Lr}) = \rho_2 c_{2L} \xi_{LT} \tag{2.64c}$$

from which

$$\frac{\xi_{LT}}{\xi_{Li}} = \frac{2\rho_1 c_{1L}}{\rho_2 c_{2L} + \rho_1 c_{1L}} \tag{2.66a}$$

$$\frac{\xi_{Lr}}{\xi_{Li}} = \frac{\rho_2 c_{2L} - \rho_1 c_{1L}}{\rho_2 c_{2L} + \rho_1 c_{1L}} \tag{2.65b}$$

Equations 2.66a and 2.65b differ from Equations 2.26 and 2.27 in that the latter give the ratios of the displacements in terms of the potential functions while Equations 2.66a and 2.65b give the relationships between the displacement amplitudes.

At the critical angle for shear waves in the solid $\alpha_{s2} = 90°$ and $\sin \alpha_1 = c_{1L}/c_{2s}$, where c_{2s} is the velocity of shear-wave propagation in the solid. At this point, the energy is partitioned between a reflected longitudinal wave and a surface or boundary wave. Actually, these boundary waves may occur at smaller angles of incidence and often cause annoying interferences in certain types of measurements.

The relative power for the transmission and reflection coefficient as a function of incident angle for a compressional wave incident on a solid–water interface and the corresponding data for shear wave incidence are shown in Figure 2.13 [19]. The corresponding data for the shear wave at the solid–water interface are shown in Figure 2.14 [19]. The case commonly encountered in ultrasonic testing is a plate in liquid. The waves and the energy partition for a water–steel combination are shown in Figure 2.15 [20].

2.4.6 SHEAR WAVE AT A SOLID–SOLID INTERFACE POLARIZED PARALLEL TO THE PLANE OF THE INTERFACE

When a shear wave in which the displacement is parallel to the interface is incident on a plane interface between two solids, only shear waves are reflected and refracted. If the two media are isotropic, only one shear wave is transmitted across the boundary and one is reflected. In this case, all terms pertaining to longitudinal components in Equations 2.47, 2.48, 2.54, and 2.64 vanish, leaving

$$(\xi_{si} - \xi_{sr}) \cos \alpha_{s1} = \xi_{sT} \cos \alpha_{s2} \tag{2.47b}$$

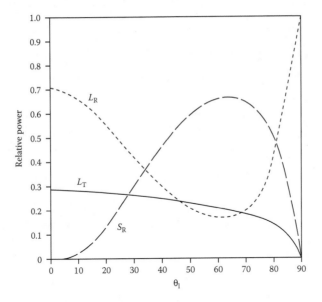

FIGURE 2.13 Transmission and reflection coefficient as a function of incident angle θ_1 for a longitudinal-wave incident on a solid–water interface. L_R = reflected longitudinal wave; L_T = transmitted longitudinal in water; S_R = reflected shear. (From Kino, G. S., *Acoustic Waves*, Prentice-Hall, Englewood Cliffs, NJ, 1987. With permission.)

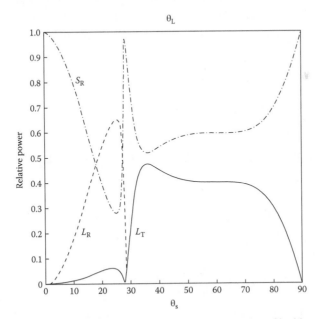

FIGURE 2.14 Transmission and reflection coefficient as a function of incident angle θ_s for a shear-wave incident on a solid–water interface. S_R = reflected shear; L_T = transmitted in water; L_R = reflected longitudinal. (From Kino, G. S., *Acoustic Waves*, Prentice-Hall, Englewood Cliffs, NJ, 1987. With permission.)

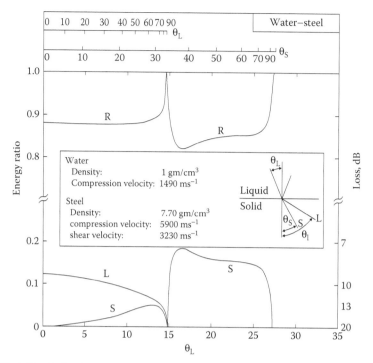

FIGURE 2.15   Reflection and transmission of ultrasonic waves at a liquid–solid (water–steel) interface showing energy transmitted as a function of angle. R = reflected; L = transmitted longitudinal; S = transmitted shear. (From Lynnworth, L. C., and J. N. C. Chen, Energy transmission coefficient at liquid/solid interfaces, *Ultrason Symp Proc IEEE*, 574, © 1975 IEEE. With permission.)

$$\rho_1 c_{1s}(\xi_{si} + \xi_{sr})\cos^2 \alpha_{s1} = \rho_2 c_2 \xi_{sT} \cos^2 \alpha_{s2} \tag{2.64d}$$

Since the displacements are parallel to the interface, the cosine terms, relating the displacements to the interface, are unity, so that

$$\frac{\xi_{sT}}{\xi_{si}} = \frac{2\rho_1 c_{1s}}{\rho_2 c_{2s} + \rho_2 c_{1s}} \tag{2.68}$$

for the transmitted wave, and

$$\frac{\xi_{sr}}{\xi_{si}} = \frac{\rho_2 c_{2s} - \rho_1 c_{1s}}{\rho_2 c_{2s} + \rho_1 c_{1s}} \tag{2.69}$$

for the reflected wave. Equations 2.68 and 2.69 apply at any incidence angle below the critical angle.

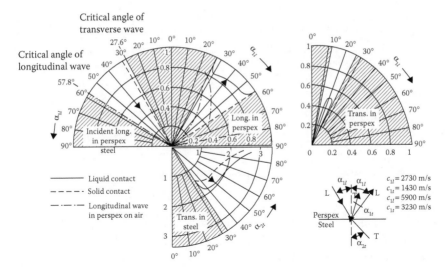

FIGURE 2.16 Pressure ratios for refracted shear (transverse) wave for Plexiglas and steel interface. (Reproduced with kind permission from Springer Science+Business Media: *Ultrasonic Testing of Materials*, 4th ed., 1990, page 29, J. Krautkramer and H. Krautkramer, Figure 2.12, copyright Springer-Verlag, Berlin, 1969, 1977, 1983, and 1990.)

In ultrasonic testing, waves are mode-converted at solid–solid interfaces. One combination of materials commonly encountered is steel and Plexiglas, where there is a thin layer of couplant between the two media. The angular response and the pressure ratios, including identification of critical angles for the case of a Plexiglas–steel system for this case, are shown in Figure 2.16 [21].

2.4.7 REFLECTION, REFRACTION, AND MODE CONVERSION IN GENERAL APPLICATIONS OF ULTRASONIC ENERGY

The many equations in this chapter represent an attempt to present to the reader the basic mechanics of wave propagation and their influence on various conditions faced in the actual applications of ultrasonic energy. A wave propagating through common media may encounter interfaces of every conceivable geometrical configuration. A multiplicity of incidence angles exist for the common geometries in which the interface is neither plane nor infinite. In such cases, energy may be scattered, refracted, and converted to other wave forms (mode conversion), as a review of the equations presented in this chapter would reveal.

Wave motion in solids is reviewed in the text by Graff [9], and many practical aspects of NDT are considered by Krautkramer and Krautkramer [21]. The science base for ultrasonic measurements in solid state physics is provided by Truell et al. [22] and in various chapters compiled in the text by Edmonds [23].

Equations pertaining to anisotropic conditions have not been presented. These can be derived following the techniques used in deriving the equations for isotropic conditions and considering the directional characteristics of the elastic constants. Acoustic waves in anisotropic solids are considered in detail by Auld [16], Kolsky [24], and Redwood [15]. The reader may consult those works for more detailed information. There are 21 independent elastic constants relating stress to strain in anisotropic crystals. All 21 are necessary to define the stress–strain relations in the most complex crystals. Crystals with planes of symmetry have fewer independent elastic constants; for instance, a cubic crystal has three and an isotropic material has only two.

A multicrystalline anisotropic solid may be treated as an isotropic solid if the wavelength is large compared to the size of the crystals and if the crystals are randomly oriented. The anisotropy will, however, affect the attenuation of waves passing through the material. The general topic of propagation and scattering in polycrystalline media is considered in detail by Papadakis [17].

If the crystals in the material are oriented in the material so that the elastic constants in the direction of orientation differ from those in a direction normal to the orientation, the velocity of wave propagation also will differ in these two directions.

2.5 TRANSMISSION THROUGH THIN PLATES

When an acoustic plane wave of finite cross section traveling in a medium encounters a plate that is very thin relative to the wavelength, the plate flexes with the medium. A plate can flex in either extensional or a flexural mode, as shown in schematic form in Figure 2.17. The pressures on the two faces of the plate are approximately equal in magnitude. Only that energy associated with flexing the plate is lost and the intensity of the transmitted wave approximately equals that of the incident beam if the medium on either side of the plate is the same.

When the plate thickness is of the same order of magnitude as the wavelength in the plate, internal reflections from the interfaces affect the amplitude of the transmitted and reflected waves. For instance, reflections of a longitudinal wave from the back surface of a plate one-quarter wavelength thick are 180° out of phase with the incident wave at normal incidence. No energy is transmitted through the plate under these conditions. If the plate is one-half wavelength thick, transmission is maximum and only those losses associated with attenuation within the material of the plate affect the amplitude of the transmitted wave. No reflections occur at boundaries between two media with equal acoustic impedances.

The extent of the interference of pulses of ultrasonic energy within plates in which the wavelength and the thickness are of the same order of magnitude depends upon the ratio of thickness to wavelength (d/λ), the coupling to the surrounding medium, the pulse length and shape, the spectrum of the pulse, and the internal absorption losses. When d/λ equals 1/4, 3/4, and so on (to a limit depending upon pulse length), maximum destructive interference occurs within the plate. When d/λ is 1/2, 1, and so on, maximum constructive interference (resonance) occurs and the energy is transferred through the plate with minimum loss.

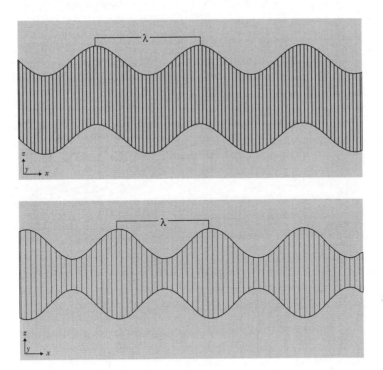

FIGURE 2.17 Schematic showing plate and particle displacement for flexural and extensional Lamb waves.

The relative transmission is given by

$$\frac{I_T}{I_I} = \frac{e^{-2\alpha d}}{\left[1 + \frac{1}{4}\left(\frac{Z_2}{Z_1} - \frac{Z_1}{Z_2}\right)^2 \sin^2 \frac{\omega d}{c_2}\right]} \qquad (2.70)$$

If d is small and α is negligible

$$\frac{I_T}{I_I} = \frac{1}{\left[1 + \frac{1}{4}\left(\frac{Z_2}{Z_1} - \frac{Z_2}{Z_2}\right)^2 \sin^2 \frac{\omega d}{c_2}\right]} \qquad (2.70a)$$

where Z_1 is the acoustic impedance of the surrounding medium and Z_2 is the acoustic impedance of the plate.

In recent years, there has been a growing interest in the use of plate and guided waves in NDT [8] and also in surface acoustic wave (SAW) devices [16,19]. As an example, the wave speed data for Lamb waves in a steel plate with normalized plate thicknesses are shown in Figure 2.18 [25].

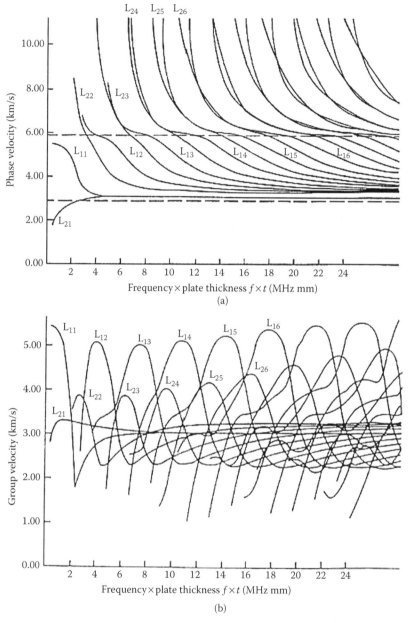

FIGURE 2.18 Lamb wave velocities. (a) Phase speed for Lamb waves in steel with excitation frequency f (MHz) and plate thickness t (mm). (b) Group speeds for Lamb waves in steel with excitation frequency f (MHz) and plate thickness t (mm). (From Egle, D. M., and D. E. Bray, *Nondestructive Measurement of Longitudinal Rail Stresses*, Report FRA-ORD-76-270, Pb-272061, Federal Railroad Administration, Springfield, VA, 1975; Bray, D. E., and R. K. Stanley, *Nondestructive Evaluation*, McGraw-Hill, New York, 1989. With permission.)

2.6 DIFFRACTION

Diffraction refers to the modification of wave fronts in passing by edges of opaque bodies, passing through narrow slits, or being reflected or transmitted from surfaces. In the diffraction of light at the edge of an opaque object, rays appear to be deflected, producing fringes of light and dark or of colored bands. A similar modification of other waves (as sound and electromagnetic waves) occurs and causes the curvature of waves around objects in their paths. Diffraction prevents full utilization of a wave front or bringing the wave front to a focus.

Diffraction effects are of concern in many ultrasonic applications such as in the measurement of attenuation of sound in materials, nondestructive inspection of materials, underwater sound, and imaging. Diffraction accounts for the distribution of the wave amplitude in the sound field radiating from a source and in the spreading of the beam from the source.

2.6.1 HUYGENS' PRINCIPLE

Equations for describing diffraction are derived by the use of Huygens' principle, according to which every vibrating point on a wave front is regarded as the center of a new disturbance. The new disturbances act as point sources, each emitting a spherical wave, which is assumed to produce an effect only along its wave front. Propagation from those sources is always assumed to be forward. In the use of Huygens' principle, it is assumed that the wavelength is much smaller than the dimensions of the surfaces or obstacles encountered.

Huygens' principle is illustrated in Figure 2.19. Spherical waves vibrating sinusoidally are shown emanating from points A and B on a wave front. The two waves are of equal amplitude and phase in the direction of the wave front. The arrival of the two new wave fronts at points P_1 and P_4 is such that they are in phase and thus reinforce each other (neglecting the fact that, at these points, the waves meet obliquely). At points P_2 and P_5, they are 180° out of phase, and, therefore, destructive interference occurs at these points. At point P_3, the wave from B leads that from A by 90° and the amplitude of the disturbance at that point is equal to the amplitude of either wave taken separately.

Huygens' principle is used to calculate radiation patterns from transmitters of various configurations. Consider the radiation from a circular piston source of radius a, Figure 2.20. The pressure dp at the point $P(0, x)$ due to the radiation from the incremental area ds is

$$dp = -j \frac{\rho f v_0 (r dr\, d\phi)}{d} e^{j\frac{\omega}{c}(d-ct)} \tag{2.71}$$

where d is the distance from ds to $P(0, x)$, r is the distance from the origin to ds, and v_0 is the particle velocity at ds.

The total contribution to the radiation pressure at $P(0, x)$ of all increments ds is, therefore,

$$p = -j(\rho f v_0) \int_0^{2\pi} \int_0^a \frac{e^{j\frac{\omega}{c}(d-ct)}}{d} r dr\, d\phi \tag{2.72}$$

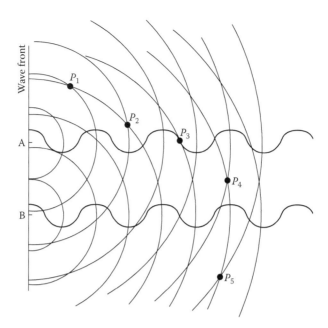

FIGURE 2.19   Demonstration of Huygens' principle.

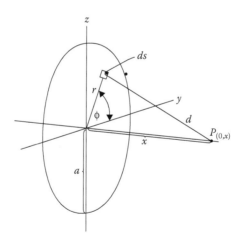

FIGURE 2.20   Radiation from a circular piston.

From Figure 2.9, for small values of x/a

$$d = \sqrt{r^2 + x^2}$$

or

$$p = -j(\rho f v_0)\int_0^{2\pi}\int_0^a \frac{e^{j\left(\frac{\omega}{c}\right)\sqrt{r^2+x^2}-ct}}{\sqrt{r^2 + x^2}}\, r\,dr\,d\phi \tag{2.72a}$$

Therefore,

$$p = -j2\rho c v_0 \left[\cos\frac{\omega}{2c}(\sqrt{a^2 + x^2} + x) + j\sin\frac{\omega}{2c}(\sqrt{a^2 + x^2} + x) \right]$$

$$\times e^{-j\omega t} \sin\left(\frac{\omega}{2c}\right)(\sqrt{a^2 + x^2} - x) \qquad (2.73)$$

From Equation 2.73, it can be seen that p is a *maximum* when

$$\left(\frac{\omega}{2c}\right)(\sqrt{a^2 + x^2} - x) = \frac{\pi}{2}, \frac{3\pi}{2}, \frac{5\pi}{2}, \cdots, \frac{2m+1}{2}\pi$$

or solving for x

$$x = \frac{4a^2 - (2m+1)^2\lambda^2}{4(2m+1)\lambda} \quad (m = 0,1,2,3,\ldots) \qquad (2.74)$$

The maximum closest to the radiator occurs when $2a = (2m + 1)\lambda$. Likewise, p is a *minimum* when

$$x = \frac{a^2 - n^2\lambda^2}{2n\lambda} \quad (n = 1,2,3,4,\ldots) \qquad (2.75)$$

The minimum closest to the radiator occurs when $a = n$. Equation 2.73 thus describes the pressure distribution *along the axis* of the radiated beam in the *near field*. The on-axis pressure, shown against normalized distance, has been reported by numerous authors including [21,26] and is shown in Figure 2.21. A similar approach is used in determining the pressure distribution at any point, on or off the axis.

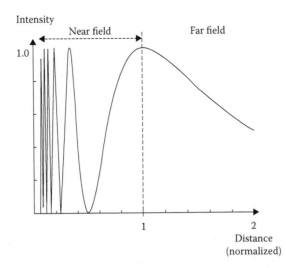

FIGURE 2.21 On-axis acoustic intensity distribution for a disc transducer.

The value of x at $m = 0$ is the extent of the near field. When $\lambda << a$, the near field extends to a distance that is approximately equal to a^2/λ. The far field begins at this point. Pressure distribution at greater distances may be calculated on or off the axis by assuming that $x \approx d$.

Calculations of sound-field patterns can become increasingly complex. However, a number of cases have now been described in analytical form, and the corresponding equations are summarized by Kino [19]. Transducer wave fields can now be calculated using numerical methods.

Consider, now, diffraction around a disk-shaped obstacle at the center of a circular beam (Figure 2.22).

According to Huygens' principle, the plane in which the obstruction (radius a_2) is located may be considered a new source of sound. Following the same procedure as before, the pressure distribution along the axis is

$$p = -\rho V_0 f e^{-j\omega t}\left(\frac{c}{\omega}\right)\int_0^{2\pi}\int_{a_2}^{a_1}\frac{e^{j\frac{\omega}{c}\sqrt{r^2+x^2}}}{\sqrt{r^2+x^2}}j\frac{\omega}{c}r\,dr\,d\phi$$

$$= -\rho c V_0\left(e^{j\frac{\omega}{c}\sqrt{a_1^2+x^2}} - e^{j\frac{\omega}{c}\sqrt{a_2^2+x^2}}\right)e^{-j\omega t}$$

(2.76)

The *pressure maxima* along the axis in the near field occur when

$$\frac{\omega}{2c}\left(\sqrt{a_1^2 + x^2} - \sqrt{a_2^2 + x^2}\right) = \frac{\pi}{2}, \frac{3\pi}{2}, \frac{5\pi}{2}, \cdots, \frac{2m+1}{2}\pi$$

that is, at

$$x = \frac{\sqrt{\left(a_1^2 - a_2^2\right)^2 - \left(a_1^2 + a_2^2\right)(\lambda^2/2)(2m+1)^2 + (\lambda^4/16)(2m+1)^4}}{\lambda(2m+1)}$$

(2.77)

$$(m = 0, \pi, 2\pi, \ldots, n\pi)$$

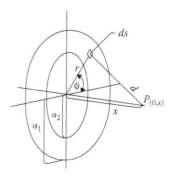

FIGURE 2.22 Pressure distribution on the axis of a circular sound beam obstructed by a disk-shaped obstacle.

The *pressure minima* along the axis in the near field occur when

$$\frac{\omega}{2c}\left(\sqrt{a_1^2 + x^2} - \sqrt{a_2^2 + x^2}\right) = 0, \pi, 2\pi, \ldots, n\pi$$

from which

$$x = \frac{\sqrt{\left(a_1^2 - a_2^2\right)^2 - 2\left(a_1^2 + a_2^2\right)\lambda^2 n^2 + \lambda^4 n^4}}{2\lambda n} \quad (n = 1, 2, 3, \ldots) \tag{2.78}$$

The start of the far field occurs at the final maximum, that is, when $m = 0$.
    The maximum closest to the obstruction occurs when

$$\left(a_1^2 - a_2^2\right) = \left(a_1^2 + a_2^2\right)\frac{\lambda^2}{2}(2m + 1)^2 + \frac{\lambda^4}{16}(2m + 1)^4 \tag{2.79}$$

The minimum closest to the obstruction occurs when

$$\left(a_1^2 - a_2^2\right) = 2\left(a_1^2 + a_2^2\right)\lambda^2 n^2 + \lambda^4 n^4 \tag{2.80}$$

Equations 2.79 and 2.80 determine the distance of the "dark-shadow" zone due to the disk-shaped obstruction. The distance corresponding to the axial fluctuations between maxima and minima might be called a "gray-shadow" zone. These two zones would correspond to the umbra and penumbra observed in shadows in the path of a light beam. These conditions are of special interest for any application based on shadow techniques, such as through-transmission testing for delaminations in thin plates or imaging of a sound field.

2.6.2 DIFFRACTION IN THREE-DIMENSIONAL SPACE

From the previous analysis and the premises upon which it was based, one may quickly discern that along any plane in the near field normal to the axis of the propagating beam, the pressure distribution is not uniform. In the case of a piston-type radiator, which is several wavelengths in diameter, the pattern appears as a series of concentric rings, corresponding to pressure maxima and minima. The result is the production of a main beam in the direction of the axis and of side lobes in which energy propagates from the source at angles to the axis. The spread of the main beam from a disk-shaped radiator is determined by the relation

$$\sin\frac{\theta}{2} = 0.6\lambda/a \tag{2.81}$$

where θ is the angle describing the spreading of the effective part of the beam, a is the radius of the emitter, and λ is the wavelength of sound in the propagating medium.

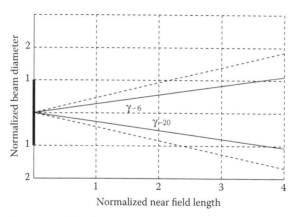

FIGURE 2.23 Beam spread—ultrasonic transducer beam spread—showing −6 dB and −20 dB half angles.

The beam divergence for the −6 dB and −20 dB cases can be approximated as Equation 2.81b and are shown in Figure 2.23. The pressure field divergence angle is constant because the sound pressure amplitude normal to the acoustic axis for a disc-shaped source follows the relationship given as Equation 2.81c,

$$\gamma_{20} = \arc\ \sin 0.87\lambda/D \qquad (2.81b)$$

$$\frac{p}{p_0} = \frac{2J_1\left(\dfrac{\pi}{\lambda}D\sin\gamma\right)}{\left(\dfrac{\pi}{\lambda}D\sin\gamma\right)} \qquad (2.81c)$$

where $J_1(x)$ is the first order Bessel function.

2.6.3 DIRECTIVITY PATTERN

Directivity pattern (or beam pattern) refers to the relative response of a receiver (or pressure or intensity of the radiated wave), plotted as a function of position with respect to the transmitter. The pattern in a specified plane is usually given and may be referred to either polar or rectangular coordinates. The directivity patterns are functions of the wavelength in the medium into which the wave is radiated and of the dimensions of the radiation surface.

2.6.4 FOCUSING

As is true of light, ultrasonic energy may be focused. A significant difference results from the longer lengths of sound waves. Because of the longer wavelengths, sound focuses in a region rather than at a point, and the smallest practical focal region is a sphere one wavelength in diameter. In any type of focusing device, the sharpness of the focus is proportional to the ratio of the aperture of the device to the wavelength.

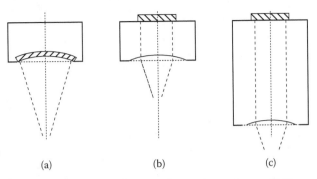

(a)   (b)   (c)

FIGURE 2.24 Examples of acoustic lenses and curved transducers. (a) Curved element. (b) Flat transducer, with lens. (c) Flat transducer on buffer-rob, with lens.

The best focus is obtained when the focal length is the distance to the final pressure peak in the near field.

Focused transducers are often used in NDT and the growing field of acoustic microscopy to improve resolution of small defects. Focusing has also been used to obtain very high intensities for processing applications and in medical applications such as neurosurgery. Focusing of ultrasound at low powers commonly employs acoustic lens and curved radiation sources, and examples are shown in Figure 2.24. Developments in electronics have resulted in increasing use of phased array systems in both medical and NDT applications, and they are employed operating at both low and higher powers. At higher powers, parabolic reflectors are also used to achieve focusing, such as in lithotripters, which are used for kidney stone disruption.

2.7   STANDING WAVES

Standing waves are periodic waves having a fixed distribution in space, which is the result of interference of progressive waves of the same frequency and kind. Such waves are characterized by the existence of nodes or partial nodes and antinodes that are fixed in space. Their use is common in ultrasonic applications.

The equation for the acoustic pressure of an infinite and progressive wave is

$$p = P_0\, e^{-ax+j(\omega t - kx)} \tag{2.82}$$

where P_0 is the pressure amplitude of the wave as it leaves the source, α is the attenuation of the sound in the medium, and k is ω/c. Upon encountering an infinite plane surface that is parallel to the front, part of the energy is reflected, depending upon the relationship between acoustic impedances of the two media. During the reflection process, the wave also may undergo a shift in phase. If the distance between the source and the reflecting surface is L, the pressure amplitude of the incident wave at L is

$$P_1 = P_0\, e^{-\alpha L} \tag{2.83}$$

The pressure amplitude of the reflected wave at L is

$$P_2 = P_0\, e^{-\alpha L - 2\alpha_0 - 2j\beta} \tag{2.84}$$

and the pressure of the reflected wave is

$$P_2 = P_0\, e^{-2\alpha L - 2\alpha_0 + ax + j(\omega t - 2\beta - kL + kx)} \tag{2.85}$$

where a_0 accounts for the loss of energy on reflection and β is the phase shift due to reflection. The pressure p at a given distance x from the source is the algebraic sum of all contributing waves. Considering only a single reflection, the pressure p is

$$p = P_0\left[e^{-\alpha x + j(\omega t - kx)} - e^{-2\alpha_0 - 2\alpha L + \alpha x + j(\omega t - kL - 2\beta + kx)}\right] \quad 0 < x \le L \tag{2.86}$$

Since $e^{j\theta} + e^{-j\theta} = 2\cos\theta$ and $e^{j\theta} - e^{-j\theta} = 2j\sin\theta$, Equation 2.86 may be written

$$\begin{aligned}
p = P_0\Big\{&e^{-\alpha x}\left[\cos(\omega t - kx) + j\sin(\omega t - kx)\right] - e^{-2\alpha_0 - 2\alpha L + \alpha x}\\
&\times \cos(\omega t - kL - 2\beta + kx) + j\sin(\omega t - kL - 2\beta + kx)\Big\}
\end{aligned} \tag{2.86a}$$

The pressure amplitude fluctuates along x owing to the reflected waves. If the reflecting surfaces were located an infinite distance from the source or if the impedances at the interface between the two media were perfectly matched, the pressure amplitude would decrease only as a function of α. The standing-wave ratio (SWR) is the sum of the two amplitudes of Equation 2.86a divided by the difference between these two amplitudes. Then

$$\mathrm{SWR} = \frac{e^{-\alpha x} + e^{-2\alpha_0 - 2\alpha L + \alpha x}}{e^{-\alpha x} - e^{-2\alpha_0 - 2\alpha L + ax}} \tag{2.87}$$

If α is negligible, Equation 2.87 reduces to

$$\mathrm{SWR} = \frac{1 + e^{-2\alpha_0}}{1 - e^{-2\alpha_0}} \tag{2.87a}$$

(see Figure 2.25).

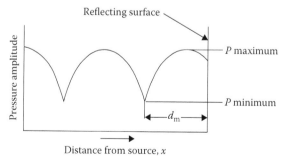

FIGURE 2.25   Dimensions for measuring standing wave ratio.

The fundamentals for ultrasonic standing waves applied to the case of separation of suspended particles are presented in a series of three 1998 articles [27]. Devices using standing waves are being increasingly developed for biotechnology and nano-technology applications and these topics are considered further in Chapter 12.

2.8 DOPPLER EFFECT

The Doppler effect is a change in frequency from that of a transmitted wave detected by a receiver because of various conditions of relative motion between the receiver, the transmitter, the medium, and reflecting surfaces. Several ultrasonic applications involve the Doppler effect. The change in frequency is due to the effect of the motion on the rate at which a complete wave (successive pressure peaks) passes the receiver; that is, the period of the wave relative to the receiver may be compressed or expanded, depending upon the nature of the motion. The various possibilities are (1) stationary transmitter and medium and moving receiver, (2) stationary receiver and medium and moving source, (3) stationary source and receiver and moving medium, (4) moving source and receiver in stationary medium, (5) moving source and medium and stationary receiver, (6) moving medium and receiver and stationary source, (7) moving source, medium, and receiver, and (8) any of the conditions (1) through (7) plus a moving reflector (target).

When the *receiver approaches a transmitter* at a vector velocity $\bar{v}_r$ in a stationary medium, the time between pressure peaks at the receiver is

$$T_D = \frac{\lambda}{(c_0 + |\bar{v}|)} \tag{2.88}$$

where λ is the wavelength in the medium, T_D is the Doppler period of the sound wave, and c_0 is the velocity of sound in the medium. Therefore, the frequency at the receiver is

$$f_r = \frac{1}{T_D} = \frac{(c_0 + |\bar{v}_r|)}{\lambda} = \frac{f_0}{c_0}(c_0 + |\bar{v}_r|) \tag{2.89}$$

where f_0 is the transmitted frequency. If the velocity of the receiver, $\bar{v}_r$, is in the opposite direction

$$f_r = \frac{1}{T_D} = \frac{(c_0 - |\bar{v}_r|)}{\lambda} = \frac{f_0}{c_0}(c_0 - |\bar{v}_r|) \tag{2.89a}$$

If the velocity of the receiver, $\bar{v}_r$, is at an angle θ_r to a line through the receiver and transmitter, the receiver frequency is

$$f_r = \frac{f_0}{c_0}(c_0 + |\bar{v}_r| \cos \theta_r) \tag{2.89b}$$

where, if $\theta_r < \pi/2$, the sign of the cosine is positive, indicating that the motion brings the transmitter and receiver closer together, and, if $\theta_r < \pi/2$, the sign of the cosine is negative and the source and receiver are moving farther apart.

If the medium and the receiver are stationary and the source is moving at a velocity $\bar{v}_s$, the distance between the source and the receiver changes by $(|\bar{v}_s|\cos\theta_s)T_0$ during one period of the source frequency, T_0. Therefore, the distance between successive peaks in the Doppler wave is

$$\lambda_D = (c_0 - |\bar{v}_s|\cos\theta_s)T_0 \tag{2.90}$$

and the received frequency is

$$\begin{aligned} f_r &= \frac{c_0}{\lambda_D} = \frac{c_0}{(c_0 - |\bar{v}_s|\cos\theta_s)T_0} \\ &= \frac{f_0 c_0}{(c_0 - |\bar{v}_s|\cos\theta_s)} \end{aligned} \tag{2.91}$$

where the sign of the cosine is opposite to the direction of the source relative to the receiver.

When both the receiver and the transmitter are moving

$$f_r = \frac{(c_0 + |\bar{v}_r|\cos\theta_r)f_0}{(c_0 - |\bar{v}_s|\cos\theta_s)} \tag{2.92}$$

In Equation 2.92, if both θ_s and θ_r are less than 90°, relative motion between source and receiver is generally toward each other and the signs of the cosine terms are positive. For example, if both are moving on a line toward each other, $\cos\theta_r = \cos\theta_s = 1$ and

$$f_r = \frac{(c_0 + |\bar{v}_r|)f_0}{(c_0 - |\bar{v}_s|)} \tag{2.92a}$$

If they are moving in opposite directions, the signs of the velocity terms in the numerator and denominator of Equation 2.92a are reversed

If the transmitter and the receiver are stationary and the medium is moving at a constant velocity $\bar{v}_m$, the effect of $\bar{v}_m$ is to change the apparent velocity of sound in the medium. The period of the wave passing the receiver is the same as the period of the transmitter frequency, and no Doppler shift occurs. If the velocity of the medium varies, the wavelength expands and contracts with the variation and $(d|\bar{v}_m|/d\lambda)\cos\theta_m = f_0$. However, the time that it takes a wave front to travel from the transmitter to the receiver depends upon the sum of the velocities $(c_0 + |\bar{v}_m|\cos\theta_m)$, where θ_m is the angle between the direction of the velocity vector and a straight line connecting the transmitter and the receiver. The travel time t is

$$t = \frac{d}{(c_0 + |v_m|\cos\theta_m)} \tag{2.93}$$

When $\theta_m < 90°$, $\cos\theta_m$ is positive and $|v_m|\cos\theta_m$ is directed toward the receiver. When $\theta_m > 90°$, $\cos\theta_m$ is negative and directed toward the transmitter.

Although there is no Doppler effect due to a moving medium between stationary transmitter and receivers, a moving medium does affect the Doppler frequency when either transmitter or receiver or both are moved relative to the other. The equation is

$$f_r = f_0 \frac{(c_0 + |\bar{v}_m|\cos\theta_m + |\bar{v}_r|\cos\theta_r)}{(c_0 + |\bar{v}_m|\cos\theta_m - |\bar{v}_s|\cos\theta_s)} \tag{2.94}$$

When a Doppler technique is used to detect a moving object, the frequency of the wave received at the transmitter after the wave has reflected from the object is approximately

$$f_r = f_0 \left[1 + \frac{2(|\bar{v}_s|\cos\theta_s - |\bar{v}_t|\cos\theta_t)}{c_0} \right] \tag{2.95}$$

where $\bar{v}_t$ is the velocity of the target and θ_t and θ_s are the instantaneous angles between the directions, which the corresponding velocity vectors $\bar{v}_t$ and $\bar{v}_s$ of the target and the source, respectively, make with a straight line through the source and target. Or if the fluid is moving at a constant velocity $\bar{v}_m$

$$f_r = f_0 \left[1 + \frac{2(|\bar{v}_s|\cos\theta_s - |\bar{v}_t|\cos\theta_t)}{c_0 + |\bar{v}_m|\cos\theta_m} \right] \tag{2.95a}$$

where θ_m is the instantaneous angle between the velocity vector of the fluid and the radius vector from the target to the source.

When a stationary transmitter and stationary receiver are used in a bistatic arrangement, that is, the transmitter and receiver are separated by a distance, the Doppler frequency caused by a moving target in a stationary medium is

$$f_D = \frac{f_0 |\bar{v}_t|}{c_0} (\cos\theta_{s'} + \cos\theta_{r'}) \tag{2.95b}$$

where $\theta_{s'}$ is the angle between the radius vector from the source s to the target t and the velocity vector, and $\theta_{r'}$ is the angle between the radius vector from T to the receiver R and the velocity vector. As a result, the frequency of the wave as detected by the receiver is

$$f_r = f_0 \left[1 + \frac{|\bar{v}|(\cos\theta_{s'} + \cos\theta_{r'})}{c_0} \right] \tag{2.95c}$$

A medium moving with varying velocities and directions may cause variations in the total path length of an ultrasonic beam incident on a reflecting surface and cause an effect similar to the effects caused by a moving surface.

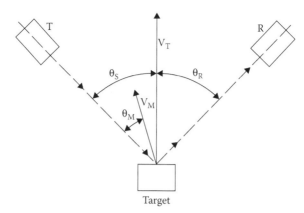

FIGURE 2.26 Conditions for producing the Doppler effect in a bistatic arrangement of transducers in a flowing medium.

The effect of a constant fluid velocity on the Doppler frequency of a bistatic arrangement of transmitter and receiver is

$$f_D = f_0 \mid \bar{v}_t \mid \left[\frac{\cos\theta_{s'}}{c_0 - \mid \bar{v}_m \mid \cos\theta_{m'}} + \frac{\cos\theta_{r'}}{c_0 + \mid \bar{v}_m \mid \cos(\theta_{r'} + \theta_{s'} - \theta_{m'})} \right] \quad (2.95d)$$

where $\theta_{m'}$ is the angle between the fluid velocity vector and the radius vector from the source to the target. The receiver frequency, f_r,

$$f_r = f_0 \left\{ 1 + \mid \bar{v}_t \mid \left[\frac{\cos\theta_{s'}}{c_0 - \mid \bar{v}_m \mid \cos\theta_{m'}} - \frac{\cos\theta_{r'}}{c_0 + \mid \bar{v}_m \mid \cos(\theta_{r'} + \theta_{s'} - \theta_{m'})} \right] \right\} \quad (2.95e)$$

(see Figure 2.26).

2.9 SUPERPOSITION OF WAVES

When two simple-harmonic sound waves of nearly the same intensity but of slightly different frequencies are combined linearly, the superposition of the amplitudes of vibration shows maxima and minima at periodic intervals. The result is a third wave, which is approximately simple harmonic at a frequency equal to the difference between the frequencies of the initial waves. This interference or superposition phenomenon, called *beating*, is sometimes used in ultrasonic applications.

If the instantaneous amplitudes of two superimposed sinusoidal waves are y_1 and y_2 and the corresponding angular frequencies are ω_1 and ω_2, the net amplitude is the sum of the two waves or

$$y = y_1 + y_2 \quad (2.96)$$

Letting $y_1 = A \cos(\omega_1 t + \phi_1)$ and $y_2 = A_2 \cos\omega_2 t$, where t is so chosen that $\phi_2 = 0$, and ϕ_2 being the phase angles, gives

$$\begin{aligned} y &= A_1 \cos(\omega_1 + \phi_1) + A_2 \cos\omega_2 t \\ &= A_1 \cos(\omega_1 t + \phi_1) + A_2 \cos(\Delta\omega + \omega_1)t \\ &= (A_1 \cos\phi_1 + A_2 \cos\Delta\omega t)\cos\omega_1 t - (A_1 \sin\phi_1 + A_2 \sin\Delta\omega t) \times \sin\omega_1 t \end{aligned} \tag{2.97}$$

where $\Delta\omega = \omega_2 - \omega_1$.

Equation 2.97 can be written in the form

$$\begin{aligned} y &= A \cos(\omega_1 t + \phi) \\ &= A(\cos\omega_1 t \cos\phi - \sin\omega_1 t \sin\phi) \end{aligned} \tag{2.97a}$$

The constants A and ϕ can be determined by first substracting Equation 2.97a from Equation 2.97 to obtain

$$\begin{aligned} (A_1 \cos\phi_1 + A_2 \cos\Delta\omega t - A\cos\phi)\cos\omega_1 t \\ - (A_1 \sin\phi_1 + A_2 \sin\Delta\omega t - A\sin\phi)\sin\omega_1 t = 0 \end{aligned} \tag{2.98}$$

For Equation 2.98 to hold true for all values of t, the coefficients of $\cos\omega_1 t$ and of $\sin\omega_1 t$ must equal 0, that is,

$$A_1 \cos\phi_1 + A_2 \cos\Delta\omega t = A\cos\phi \tag{2.99a}$$

and

$$A_1 \sin\phi_1 + A_2 \sin\Delta\omega t = A\sin\phi \tag{2.99b}$$

Equations 2.99a and 2.99b show that the values of A and ϕ of Equation 2.97a are given by

$$A^2 = A_1^2 + A_2^2 + 2A_1 A_2 \cos(\phi_1 - \Delta\omega t) \tag{2.100}$$

and

$$\tan\phi = \frac{A_1 \sin\phi_1 + A_2 \sin\Delta\omega t}{A_1 \cos\phi_1 + A_2 \cos\Delta\omega t} \tag{2.101}$$

Equation 2.97a describes all the possible effects of superimposing two sinusoidal waves.

The amplitude of the beats fluctuates between $A_1 + A_2$ and $A_1 - A_2$ at a frequency of $\Delta\omega/2\pi$.

The phase presents a more complicated effect. It causes a non-constant vibration rate with an average lying somewhere between ω_1 and ω_2, depending upon the relative magnitudes of A_1 and A_2.

2.10 ATTENUATION OF AN ULTRASONIC WAVE

Attenuation refers to the diminishing of intensity of a wave front as it progresses through a medium. In practice, several factors contribute to attenuation. These factors include the spreading of the beam, scattering, absorption due to various mechanisms, and mode conversion resulting in partitioning of the energy among two or more wave modes each traveling at its own velocity.

The effect of attenuation on a periodic wave is demonstrated by the equation

$$\xi = \xi_0 \, e^{-\alpha x} \, e^{j\omega(t-x/c)} \tag{2.102}$$

where α is the loss coefficient in nepers per meter (N/m).

The intensity I is proportional to the square of the displacement. Therefore, if I_0 is the initial intensity, in watts per square meter, of a progressive ultrasonic wave, the intensity at distance x is

$$I = I_0 \, e^{-2\alpha x} \tag{2.103}$$

2.10.1 ATTENUATION DUE TO BEAM SPREADING

If a spherical wave front of uniform intensity (Figure 2.27) suffers no loss in traveling from its source, the total energy (P_T) at each spherical surface of total area S at distance r from the center of the source is always

$$P_T = I_r S_r$$

where I_r is the intensity at distance r and S_r is the total area of a sphere or segment of the sphere through which the total energy P_T passes.

The area of a sphere is $4\pi r^2$. Let r_1 and r_2 correspond to the radii of two concentric spheres containing a point source at their centers radiating a total energy P_T. If there are no losses, the total energy passing through S_1 corresponding to r_1 is equal to that through S_2 corresponding to r_2. From Equation 2.104

$$P_T = I_1 S_1 = I_2 S_2 = 4\pi r_1^2 I_1 = 4\pi r_2^2 I_2 \tag{2.104a}$$

Therefore,

$$I_1 r_1^2 = I_2 r_2^2 \tag{2.105}$$

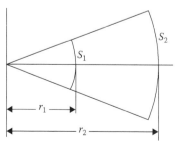

FIGURE 2.27   Spreading of a spherical wave front.

Equation 2.105 is true for all radiation patterns of spherical segments if the conditions of uniform intensity, no losses, and constant direction are met.

Corresponding relationships exist for other geometric spreading functions, and these are summarized in Table 2.2. For the case of an ultrasonic transducer, the diffraction loss usually follows a more complex relationship, which depends on the operating frequency and the transducer geometry. The diffraction loss for a circular piston-source transducer is shown in Figure 2.28 [28]. This loss can significantly impact quantitative measurements for both defect size and attenuation estimation. Transducers are discussed further in Chapter 5.

2.10.2   ATTENUATION DUE TO SCATTERING

Scattering is a common cause of attenuation of ultrasonic energy. The basic principles that cause this scattering of ultrasonic energy were discussed previously in the sections on refraction and reflection. A precise formula for scattering including all conditions of wave modes, obstacle configurations, and material properties is impractical. However, analyses based upon scattering from basic geometrical forms such as spheres, cylinders, and cubes are usually sufficient.

TABLE 2.2
Spreading Laws

Type	Intensity Relationship	Transmission Loss, B	Configuration
No spreading	r^0	0	Tube
Cylindrical	r^{-1}	$10 \log r$	Between parallel planes
Spherical	r^{-2}	$20 \log r$	Free field

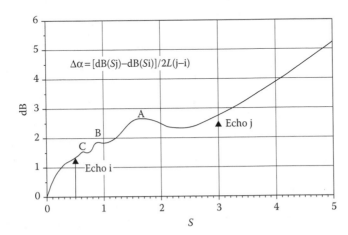

FIGURE 2.28 Loss for a circular piston radiating longitudinal waves into isotropic media according to Seki, Granto, and Truell. (Reprinted from Papadakis, E. P., *J Acoust Soc Am*, 40(4):863–67, 1966. With permission.)

For a particular "scatterer" or target geometry, the scattering responses depend on two fundamental relationships. These are the relative "scale" for the size of the target and the wavelength for the radiation in the incident wave field. This relationship is usually defined in terms of either d/λ, where d is the target characteristic dimension and λ is the wavelength or in terms of "ka," where k is the wave number and "a" is the characteristic dimension. There is also a basic relationship between k and wavelength: $k = 2\pi/\lambda$. When considering the scatterer–wavelength relationships, the three most commonly defined scattering regimes are: (1) long wavelength, (2) mid-frequency, and (3) short wavelength. When used to discuss grain scattering, the three regimes are commonly known as Rayleigh (long wavelength), stochastic (mid-frequency), and diffuse (short wavelength), and the scale relationships are presented in Table 2.3.

In addition to the relative size scale, the acoustic impedance contrast across the boundary defines scattering class. For normal incidence, the energy transmission across the interface between two semi-infinite media is shown in Figure 2.6 [29]. For a low impedance contrast system, such as that found in tissue, the so called "Weak Scattering" response applies. Where there is a high acoustic impedance contrast, such as with a metal ball bearing in water, the scattering is described as "Strong Scattering." In strong scattering, there is commonly reverberation within a target, and a significant fraction of energy is scattered. Further classes of scattering are for the case of (1) in the two scattering cases single scattering or (2) multiple scattering. In single scattering, energy is deflected once by an obstacle in the path of the beam and is thus lost to the main beamy. In multiple scattering, a ray of energy may be scattered many times by successive scatterers with the following results: (1) all the energy may be lost to the main beam, (2) part of the energy may be lost and part returned, and (3) all the energy may be returned to the main beam.

When energy in a beam has undergone multiple scattering, but remains in the beam, phase relationships between the components of the wave are affected. This is due to the differences in path length traveled by components of the energy and the fact that some of the energy may be converted to slower wave modes.

TABLE 2.3

Scattering of Ultrasonic Waves

Rayleigh	$\lambda \gg D$	$\alpha = A_1 D^3 f^4$
Stochastic (mid-frequency)	$\lambda \approx D$	$\alpha = A_2 D f^2$
Diffusive (long wavelength)	$\lambda \ll D$	$\alpha = A_3 D$

where

λ is the Wavelength of ultrasonic wave

D is the Average grain diameter

α is the Ultrasonic attenuation

A_1, A_2, A_3 are the Coefficients, dependent on elastic moduli

f is the Utrasonic frequency

Scattering of waves from an obstacle can also be considered as a function of the *scattering cross section* S_{cs} of the obstacle. Scattering cross section is defined as the ratio of the *total energy scattered per unit time* to the *energy per unit area carried per unit time by the incident wave*, the unit area being normal to the direction of propagation of the plane incident wave. The ultrasonic attenuation is directly proportional to the scattering cross section. From the definition of S_{cs}, it follows that IS_{cs} is the total energy scattered per single scatterer per unit time at any point x where the intensity is I.

In recent years, there have been numerous advances in the theory of scattering in solids applied to ultrasonic problems, and this has been in large part driven by moves to provide a science base for NDT and for it to evolve into quantitative nondestructive evaluation (QNDE). There has also been a growing recognition of the commonality of the fundamental science for elastic wave theory in electronic devices, NDT, and seismology [30,31]. The research contributions to the theory of elastic waves in solids are reviewed in a comprehensive article by Pao [32], which cites 160 references that cover the theory and analysis of elastodynamics over a 50-year period. The use of elastic-wave material structure interaction theories in NDE modeling are presented in a review by Thompson and Wadley [33] that cites some 335 references. This latter review provides an overview of the fundamental theories that are now supporting QNDE.

The theory for scattering by targets in fluids, particularly for underwater acoustics, is summarized by Urick [34]. A brief summary of the theory for both small and large objects is given by both Urick [34] and Kino [19]. A good summary of the mechanisms for attenuation of ultrasound in solids, liquids, and gases is given as an appendix in the book by Asher [35]. Some selected fundamental scattering relations are now presented.

2.10.2.1 Scattering from a Cylindrical Obstruction in a Homogeneous Medium

The total ultrasonic power scattered by a hollow cylinder in a homogeneous medium in which the wave front is parallel to the axis of the cylinder is [36,37]

$$P_s = \begin{cases} (6\pi^5 a^4/\lambda^3)I_0 & (\lambda \gg 2\pi a) \\ (2aI_0) & (\lambda_1 \leq 2\pi a \leq \lambda_2) \\ (4aI_0) & (\lambda \ll 2\pi a) \end{cases} \tag{2.106}$$

where I_0 is the intensity of the incident wave within the material (W/m²); P_s is the total power scattered per unit length (W); and a is the radius of the cylinder (m). The limiting value of total scattered power for very short wavelengths is the power contained in a beam twice as wide as the cylinder.

2.10.2.2 Scattering by a Sphere in a Homogeneous Medium

The scattering response of a sphere depends upon material property contrast between matrix and the inclusion and the relationship between its size and the wavelength of the incident ultrasound, which is typically given in terms of d/λ or ka. Analytical relationships exist for both small target—long wavelength (Rayleigh scattering) and large target—short wavelength (Geometrical Theory of Diffraction [GTD]) combinations. For the case of spheres with radius of the order of a wavelength, the response is more complex (mid-frequency scattering).

The normalized pulse-echo reflection coefficient for a plane wave incident on a sphere is shown in Figure 2.29. Three responses are shown: the specular reflection, which is the response for an infinite flat reflector; the Rayleigh response, which is that for a small target and which is commonly calculated using perturbation theory; and the oscillatory mid-frequency response. The Rayleigh scattering response is in close agreement with the full analytical solution for ka up to about 0.5, and the specular reflection is an asymptote approached when the radius of the scatterer is much larger than the wavelength (i.e., the response of an infinite plane surface is approximated and where the magnitude of reflection and transmission is determined by the acoustic impedance ratio).

The theory for scattering by a small, fixed rigid sphere was first derived by Rayleigh. By small is meant a sphere whose ratio of circumference to wavelength is much less than unity ($ka = 2\pi a/\lambda \ll 1$); by fixed, a sphere that does not partake in the acoustic motion of particles of the fluid in which the sphere is embedded; by rigid, a sphere that is nondeformable by the incident acoustic waves and into which the sound field does not penetrate. Under these conditions, Rayleigh showed that the ratio of the scattered intensity P_s at a large distance r to the intensity I_0 of the incident phase wave is

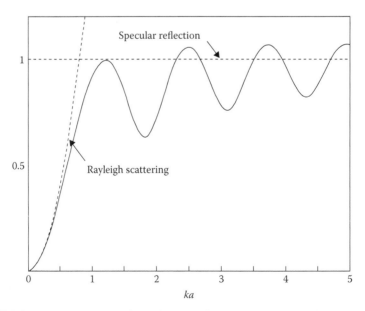

FIGURE 2.29 Normalized amplitude for the reflection from a sphere in a homogenous fluid. (From Kino, G. S., *Acoustic Waves*, Prentice-Hall, Englewood Cliffs, NJ, 1987. With permission.)

$$\frac{P_s}{I_0} = \frac{\pi^2 T^2}{r^2 \lambda^4} \left(1 + \frac{3}{2}\mu\right)^2 \qquad (2.107)$$

where

T is the volume of sphere $(1/3\ \pi a^3)$
λ is the wavelength
μ is the cosine of angle between scattering direction and reverse direction of incident wave

2.10.2.3   Scattering from a Disk-Shaped Cavity in the Path of an Ultrasonic Beam

If a disk-shaped cavity of radius a_1 obstructs a collimated beam of ultrasonic energy of radius a_2 and the axis of the disk coincides with the axis of the collimated beam, the approximate total energy remaining within the beam past the obstruction is

$$P_t = \frac{(a_2^2 - a_1^2)P_i}{a_2^2} \lambda < a_1 \qquad (2.108)$$

where P_i is the total energy of the incident beam.

If the disk lies in a plane that is other than normal to the direction of propagation, the radius a_1 in Equation 2.108 is the radius of the normal projection on the axis of the beam, except as the angle between the two axes approaches $\pi/2$.

2.10.2.4   Scattering from an Elastic Isotropic Sphere in a Homogeneous Medium

From the previous discussions regarding reflections and refraction at an interface between media of different elastic properties, one may realize that scattering from elastic obstructions within the path of an ultrasonic beam may become quite complicated. Waves are generated inside the obstacle, and these may include both longitudinal waves and shear waves.

The scattering cross section S_{cs} of an elastic sphere is [39]

$$S_{cs\ elastic sphere} = \frac{4\pi}{9} g_e k_1^4 a^6 \qquad (2.109)$$

where

$$g_e = \left\{ \frac{3(\kappa_1/k_1)^2}{3(\kappa_2/k_2)^2 - 4(\mu_2/\mu_1) + 4} - 1 \right\}^2 + \frac{1}{3}\left[1 + 2(\kappa_1/k_1)^3\right]$$
$$\times \left[(\kappa_2/\kappa_1)(\mu_2/\mu_1) - 1\right]^2 + 40\left[2 + 3(\kappa_1/k_1)^5\right]$$
$$\times \left\{ \frac{(\mu_2/\mu_1) - 1}{2[3(\kappa_1/k_1)^2 + 2](\mu_2/\mu_1) + 9(\kappa_1/k_1)^2 - 4} \right\}^2$$

and where $k_m = \omega_{1m}$, c_{1m} is the bulk longitudinal velocity, $\kappa_m = \omega/c_{sm}$, c_{sm} is the shear-wave velocity, μ_m is the shear modulus of elasticity, subscript 1 refers to the medium in which the scatterer is embedded, and subscript 2 refers to the material of the obstacle. The expression for g_e also may be given in terms of acoustic impedances, densities, and moduli or velocities and densities.

For very small spherical cavities, $\dfrac{\kappa_1}{k_1} a \ll 1$, the scattering cross section is

$$S_{cs \text{ spherical cavity}} = (4\pi/9)g_c(k_1 a)^6/k_1^2 \tag{2.110}$$

where

$$g_c = (4/3) + 40\left\{\frac{2 + 3(\kappa_1/k_1)^5}{\left[4 - 9(\kappa_1/k_1)^2\right]^2}\right\} - \frac{3}{2}(\kappa_1/k_2)^2 + \frac{2}{3}(\kappa_1/k_1)^3 + \frac{9}{16}(\kappa_1/k_1)^4$$

Similar equations may be derived for cylinders and other configurations.

Various authors have tabulated scatter responses, and a selection of simple target strength relationships are summarized in Table 2.4 [34].

2.10.2.5 Numerical Techniques to Study Wave Propagation and Scattering

Modern computers and graphics capabilities have made the use of numerical methods for investigating fundamental ultrasound–material interactions, and in particular scattering an attractive option. They can be used to investigate a wide range of classes of individual interactions and also model complete inspection systems.

The range of problems for which analytical solutions are available, although extensive, remains limited, in terms of the degree of complexity that can be modeled [9,32,37,40], and it is quickly seen from the recent literature that many important ultrasonic problems are analytically intractable [32]. To overcome these limitations with theory, a variety of approximate solutions are employed [32,40–42]; for example, Born approximation for weak scattering and Kirchhoff approximation for strong scattering in the high-frequency (long wavelength) regime. It has however been found that the various approximate models also fail to provide adequate descriptions of interactions, particularly in the mid-frequency scattering regime. To provide the necessary analytical models/theory, a range of numerical models has been developed [41,43,44].

In a review of the literature, at least seven families of modeling techniques can be identified, and the range of theories and classes of problems for which they have been used is given in Table 2.5 [40]. All methods calculate components of displacement in a wave field. In the high ka regime, there are some analytical solutions, and both the Kirchhoff and GTD methods apply to give far-field solutions. In the low ka regime, other analytical techniques are available together with the Born approximation that can again give far-field solutions. There are three remaining families of techniques, and these are T-matrix, Boundary Integral Equation (BIE) or Boundary Element Methods (BEM), and the finite difference (FD) and

TABLE 2.4
Target Strength Simple Formula

Form	Target Strength = 10 log t	Symbols	Direction of Incidence	Conditions	References
Any convex surface	$\dfrac{a_1 a_2}{4}$	$a_1 a_2$ = principal radii of curvaturer; r = range, $k = 2\pi/$wavelength	Normal to surface	$ka_1, ka_2 \gg 1 \quad r > a$	[67]
Sphere					
Large	$\dfrac{a^2}{4}$	a = radius of sphere	Any	$ka \gg 1 \quad r > a$	[67]
Small	$61.7 \dfrac{V^2}{\lambda^4}$	V = volume of sphere; λ = wavelength	Any	$ka \ll 1 \quad kr \gg 1$	[38. p. 277]
Cylinder Infinitely Long					
Thick	$\dfrac{ar}{2}$	a = radius of cylinder	Normal to axis of cylinder	$ka \gg 1 \quad r > a$	[67]
Thin	$\dfrac{9\pi^4 a^4}{\lambda^2} r$	a = radius of cyliner	Normal to axis of cylinder	$ka \ll 1$	[38. p. 311]
Finite	$aL^2/2\lambda$	L = length of cylinder, a = radius of cylinder	Normal to axis of clyinder	$ka \gg 1 \quad r > \dfrac{L^2}{\lambda}$	[68]
Plate infinite (plane surface)	$aL^2/2\lambda(\sin\beta/\beta)^2 \cos^2\theta$ $\dfrac{r^2}{4}$	a = radius of cylinder, $\beta = kL \sin\theta$	At angle θ with normal Normal to plane		
Finite any shape	$\left(\dfrac{A}{\lambda}\right)^2$	A = area of plate, L = greatest linear dimension of plate, l = smallest linear dimension of plate	Normal to plane	$r > \dfrac{L^2}{\lambda} \quad kl \gg 1$	[69]
Rectangular	$\left(\dfrac{ab}{\lambda}\right)^2 \left(\dfrac{\sin\beta}{\beta}\right)^2 \cos^2\theta$	a, b = side of rectangle, $\beta = ka \sin\theta$	At angle θ to normal in plane containing side a	$r > \dfrac{a^2}{\lambda} \quad kb \gg 1 \quad a > b$	[68]

(Continued)

TABLE 2.4

Target Strength Simple Formula (Continued)

Form	t Target Strength = 10 log t	Symbols	Direction of Incidence	Conditions	References
Circular	$\left(\dfrac{\pi a^2}{\lambda}\right)^2 \left(\dfrac{2J_1(\beta)}{\beta}\right)^2 \cos^2\theta$	a = radius of plate, $\beta = 2ka\sin\theta$	At angle θ to normal	$r > \dfrac{a^2}{\lambda}$ $ka \gg 1$	[68]
Ellipsoid	$\left(\dfrac{bc}{2a}\right)^2$	a, b, c = semimajor axes of ellipsoid	Parallel to axis of a	$ka, kb, kc \gg 1$ $r \gg a, b, c$	[70]
Circular disk (average over all aspects)	$\dfrac{a^2}{8}$	a = radius of disk	Average over all directions	$ka \gg 1$ $r > \dfrac{(2a)^2}{\lambda}$	[69]
Conical tip	$\left(\dfrac{\lambda}{8\pi}\right)^2 \tan^4\psi \left(1 - \dfrac{\sin^2\theta}{\cos^2\psi}\right)$	ψ = half angle of cone	At angle θ with axis of cone	$\theta < \psi$	[71]
Any smooth convex object	$\dfrac{S}{16\pi}$	S = total surface area of object	Average over all directions	All dimensions and radii of curvature large compared with λ	[68,71]
Triangular corner reflector	$\dfrac{L^4}{3\lambda^2}(1 - 0.00076\theta^2)^2$	L = length of edge of reflector	At angle θ to axis of symmetry	Dimensions large compared with λ	[69]
Any elongated body of revolution	$\dfrac{16\pi^2 V^2}{\lambda^4}$	V = body volume	Along axis of revolution	All dimensions small compared with λ	[72]
Circular plate	$\left(\dfrac{4}{3\pi}\right)^2 k^4 d^5$	a = radius, $k = 2\pi/\lambda$	Perpendicular to plate	$ka \ll 1$	[72]
Infinite plane strip	$\dfrac{1}{4\pi k}\left[\dfrac{\cos\theta\sin(2ka\sin\theta)}{\sin\theta}\right]^2$	$2a$ = width of strip, θ = angle to normal	At angle θ	$ka \gg 1$	[72]
	$\dfrac{ka^2}{\pi}$		Perpendicular to strip	$ka \gg 1$ $\theta = 0$	

Source: Reproduced from Urick, R. J., *Principles of Underwater Sound*, McGraw-Hill, New York, 1983. With permission.

TABLE 2.5

Summary of the Range of Theories Used to Describe Elastic Wave Scattering Problems

Method	Frequency Range	Type of Object	Reference
Wave function expansion	Low frequency	Scatters with X-sections of a circle, ellipse, or parabola	[73]
Perturbation techniques	Low frequency	Spheroids, symmetric flaws that lie on a curvilinear coordinate system	[73]
Flilpczynski approximation	Low frequency	Spheroids	[74]
Born approximation	Low frequency up to $ka \sim 2$	Arbitrary-shaped flaw	[75]
Quasistatic approximation	Low frequency	Generally shaped flaws of finite volume	[76]
Power series expansion	Low frequency	Spheroidal voids	[77]
Boundary integral equation (BIE methods)			
Hilbert–Schmidt method	Low to intermediate	Flaw must have surface on a curvilinear coordinate	[73]
Finite-difference method	Low to intermediate	General shapes	[73]
Finite-element method	Low to intermediate	General shapes	[78]
Finite-sum approximation	Low to intermediate	Generally shaped surface	[73]
Kirchoff and modified kirchoff approximation	High frequency	Flat cracks	[76,79]
T-Matrix method	$0.1 < ka \lesssim 10$	General shapes	[81]
Method of optimal truncation (MOOT)	$0.1 < ka \lesssim 10$	General shapes	[80]
Geometrical theory of diffraction (GTD)	High frequency down to $ka \approx 2$	Flat cracks and slits	[82,83]
Numerical modeling: finite–difference technique	$0.1 < ka < 10$	Regular shapes	For example, [84]

Source: Reprinted from Bond, L. J., and N. Saffari, Mode conversion ultrasonic testing. In *Research Techniques in Non-Destructive Testing*, Vol. 7, Chap. 5, 160–61, Academic Press, London, 1984.

finite element (FE) methods. This last group FE and FD are generally considered separate, but they have, in some cases, been combined into systems that use FE to model geometry in spatial dimensions and FD for the time-dependent evolution of pulsed wave fields.

A comparison of various modeling techniques applied to ultrasonic wave propagation and scattering is given in Table 2.6 [43]. The table reviews strengths and weaknesses of seven families of techniques. A comparison of finite difference and related modeling techniques is provided in Table 2.7 [43].

TABLE 2.6
Comparison of Various Modeling Techniques Applied to Ultrasonic Wave Propagation and Scattering and Rating System

Technique	Analytical Methods	Finite Differences	Elastic Kirchhoff	Born Approx	Geometrical Theory Diffraction	T-Matrix	BIE or BEM
Property							
ka Restriction	$ka \ll 1$	$0.1 < ka < 20$	$ka \gg 1$	$ka < 1$	$ka \gg 1$	$0 < ka < 15$	$0 < ka < 6$
Field region	Far	Near or Far	Far	Far	Far	Either	Near or Far
Dimension	3-D	2-D (3-D)	3-D	3-D	2.5-D	3-D	Most 3-D
Shape of scatterer	Circle, cylinder sphere	Limited (square)*	Small surface slope	Good range**	Crack-like	Ellipsoidal cavity	Arbitrary
Included material	Restricted	Most***	**	Weak	×	***	Strong or weak
Mode conversion	**	***	***	**	*	***	***
Incident wave	Plane or spherical	Arbitrary 2-D	Plane or spherical	**	**	**	Arbitrary
Short pulse	**	***	*	+	*	×	FFT used
Multiple scattering (2 body)	+	***	+	+	*	**	**

Source: Reprinted from *Proc. IUTAM Symposium on Elastic Wave Propagation and Nondestructive Evaluation* (invited paper), L. J. Bond, ed. S. K. Datta, J. D. Achenbach, and Y. D. S. Rajapakse, Numerical techniques and their use to study wave propagation and scattering, 17–28. Copyright 1990, with permission from Elsevier.

Rating system: *** Very good; ** Good; * Copes; + Poor; × Very poor.

TABLE 2.7

Comparison of Various Finite Difference and Related Modeling Techniques Applied to Ultrasonic Wave Propagation and Scattering and Rating System

Technique	Finite Difference	FE/FD	Hybrid FD/ GTD	Lumped Mass	FD + Perturbation
Property					
ka restriction	$0.1 < ka < 20$	$0.1 < ka < 20$	$0.1 < ka < 20$	$0.1 < ka < 20$	$0.01 < ka < 20$ **
Field region	Near or far	Near or far	Near or far**	Near or far	Near or far.
Dimension	2-D and 3-D	2-D	2-D	2-D	2-D
Shape of scatterer	Limited (square)*	***	Limited	**	**
Included material	Most	Most	Most	Most	Most
Mode conversion	***	***	***	***	***
Incident wave	Arbitrary 2-D	Arbitrary 2-D	Arbitrary 2-D	Arbitrary 2-D	Arbitrary 2-D
Short pulse	***	***	***	***	***
Multiple scattering (2 body)	***	***	***	***	***

Source: Reprinted from *Proc. IUTAM Symposium on Elastic Wave Propagation and Nondestructive Evaluation* (invited paper), L. J. Bond, ed. S. K. Datta, J. D. Achenbach, and Y. D. S. Rajapakse, Numerical techniques and their use to study wave propagation and scattering, 17–28, Copyright 1990, with permission from Elsevier.

Rating system: *** Very good; ** Good; * Copes; + Poor; × Very poor.

These various modeling methods have now provided full wave solutions, including mode conversion, for a wide range of propagation and scattering problems. Commercial ray-tracing models are available, and they are being increasingly used for NDT technique development in complex geometries, such as nuclear plant inspection [45]. The modeling literature now includes solutions to a range of canonical problems, involving compression, shear, Rayleigh, and Lamb waves for geometries such as wedge interactions, and the insights gained have significantly expanded the understanding of forward-scattering systems encountered and used in NDE.

2.10.2.6 Scattering in Practice

In practice, an ultrasonic wave propagating in a scattering medium encounters many scatterers. Typical examples are grain boundaries in metals, porosity in any media, and bubbles in liquids. The total effect on the attenuation is a summation of individual effects. The mechanisms that occur in the interaction between an ultrasonic wave and a polycrystalline material are a function of the scale, which again can be defined in terms of ka or D/λ, where D is a grain diameter.

In measurements, as the ultrasonic frequency is scanned, the attenuation can vary over many orders of magnitude. This is illustrated with the case of a polycrystalline aluminum, mean grain diameter 0.6 mm for frequencies between 100 kHz and 1 GHz. Data are shown in Figure 1.13 [46], and this includes data from Merkulov [47,48]. It is seen that as the D/λ ratio changes from long wavelength, with Rayleigh scattering, and where attenuation is 10^{-5} Np/cm to 1 GHz, where hysteresis and thermoelastic effects dominate and attenuation is up to 10 Np/cm.

With the need to understand damage/crack precursors, early crack growth, and cumulative damage, as well as to provide more quantitative material structure characterization, there has been growth in the investigation of ultrasound-grain response. Such signals were previously considered to be "just" noise in traditional NDT [49], are giving material signatures, and are being investigated to give what is now called back-scatter and diffuse field ultrasound response, and these signatures are found as different parts of the pulse-excitation response time window that is investigated [50].

According to Mason and McSkimin [51], the attenuation in a metal is

$$\alpha = \beta_1 f + \beta_2 f^4 \quad \lambda \geq 3D \tag{2.111}$$

where D is the average diameter of the grains, β_1 is a constant corresponding to hysteresis, and β_2 is a constant corresponding to scattering.

Scattering loss from grain boundaries in metals results from the random orientation of the grains with respect to the direction of wave propagation. Since, in an anisotropic material, the elastic constant varies with orientation, scattering occurs at intersections between grain boundaries. Materials at the grain boundaries, which differ in density and elasticity from the body of the grains, also contribute to scattering.

According to Rayleigh [38], scattering of energy from a single particle is given by the formula

$$\frac{SA}{IA} = \frac{\pi V}{R\lambda^2} \left(\frac{\Delta\kappa}{\kappa} + \frac{\Delta\rho}{\rho} \right) \cos\theta \tag{2.112}$$

where SA/IA is the ratio of the amplitude of the scattered wave to that of the incident wave, V is the volume of the scatterer, R is the radius of the scatterer, κ is the elasticity of the medium, $\Delta\kappa$ is the difference in elasticity between the particle and the medium, ρ is the density of the medium, and $\Delta\rho$ is the difference in density between the particle and the medium.

According to Mason and McSkimin [51], the Rayleigh scattering law leads to the following scattering term for longitudinal waves in metals:

$$\alpha = \frac{\pi^4 D^3 f^4}{3c_L^4} \left\langle \left(\frac{c_{11}' - <c_{11}'>_{av}}{<c_{11}'>_{av}} \right)^2 \right\rangle_{av} \tag{2.113}$$

and for shear waves:

$$\alpha = \frac{4\pi^3 D^3 f^4}{9c_s^4} \left\langle \left(\frac{c_{44}' - <c_{44}'>_{av}}{<c_{44}'>_{av}} \right)^2 \right\rangle_{av} \tag{2.114}$$

where c_L is the longitudinal-wave velocity in the metal, c_S is the shear-wave velocity in the metal, and a is attenuation (Np/m).

$$c'_{11} = c_{11}(\ell_1^2 + m_1^2)^2 + c_{33}m_1^4 + (2c_{13} + 4c_{44})n_1^2(\ell_1^2 + m_1^2)$$

(for hexagonal crystals, any orientation)

$$c'_{44} = c_{11}\left[(\ell_1\ell_2 + m_1m_2)^2 \left(\frac{\ell_1m_2 - m_1\ell_2}{2}\right)^2\right]$$

$$- \frac{c_{12}}{2}(\ell_1m_2 - m_1\ell_2) + 2c_{13}n_1n_2(\ell_1\ell_2 + m_1m_2)$$

$$+ c_{33}n_1^2n_2^2 + c_{44}(m_1n_2 + n_1m_2)^2 + (n_1\ell_2 + \ell_1n_2)^2$$

(for hexagonal crystals, any orientation)

$$c'_{11} = c_{11}(\ell_1^4 + m_1^4 + n_1^4) + (2c_{12} + 4c_{44})$$
$$\times (\ell_1^2m_1^2 + \ell_1^2n_1^2 + m_1^2n_1^2)$$

(for cubic crystals, any orientation)

$$c'_{44} = c_{11}(\ell_1^2\ell_2^2 + m_1^2m_2^2 + n_1^2n_2^2) + 2c_{12}[\ell_1\ell_2m_1m_2 + n_1n_2$$
$$\times (\ell_1\ell_2 + m_1m_2)] + c_{44}[(\ell_1\ell_2 + m_1m_2)^2 + (\ell_1\ell_2 + n_1n_2)^2$$
$$+ (m_1m_2 + n_1n_2)^2]$$

(for cubic crystals, any orientation)

where $\ell_1, \ell_2, \ell_3, m_1, ..., n_3$ are the direction cosines between the new set of axes and the crystallographic axes x, y, and z according to the relation

	x	y	z
x'	ℓ_1	m_1	n_1
y'	ℓ_2	m_2	n_2
z'	ℓ_3	m_3	n_3

and $c_{11}, c_{12}, ..., c_{44}$ are the elastic constants corresponding to the various crystallographic orientations. Mason and McSkimin [51] give the scattering factors for *cubic metals* as

$$\left\langle \left(\frac{c'_{11} - <c'_{11}>_{av}}{<c'_{11}>_{av}}\right)^2 \right\rangle_{av} = \frac{4}{21}\left[\frac{2(c_{12} - c_{11}) + 4c_{44}}{5c_{11} + 2(c_{12} - c_{11}) + 4c_{44}}\right]^2 \tag{2.115}$$

$$\left\langle \left(\frac{c'_{44} - <c'_{44}>_{av}}{<c'_{44}>_{av}}\right)^2 \right\rangle_{av} = \frac{17}{5760}\left[\frac{2(c_{11} - c_{12}) + 4c_{44}}{<c'_{44}>_{av}}\right]^2 \tag{2.116}$$

and for *hexagonal metals*

$$\left\langle\left(\frac{c_{11}' - <c_{11}'>_{av}}{<c_{11}'>_{av}}\right)^2\right\rangle_{av} = \frac{4}{1575}\left[48c_{11}^2 - 64c_{11}c_{33} + 28c_{33}^2\right]$$

$$-16c_{11}(2c_{13} + 4c_{44}) + 4c_{33}(2c_{13} + 4c_{44}) \qquad (2.117)$$

$$+3(2c_{13} + 4c_{44})^2]/<c_{11}'>_{av}^2$$

and

$$\left\langle\left(\frac{c_{44}' - <c_{44}'>_{av}}{<c_{44}'>_{av}}\right)^2\right\rangle_{av} = \frac{1}{45}\left(\frac{c_{11} - c_{12} - 2c_{44}}{<c_{44}'>_{av}}\right)^2 \qquad (2.118)$$

For *cubic metals*

$$<c_{11}'>_{av} = c_{11} + \frac{2(c_{12} - c_{11}) + 4c_{44}}{5} \qquad (2.119)$$

and

$$<c_{44}'>_{av} = c_{44} + \frac{c_{11} - c_{12} + 2c_{44}}{12} \qquad (2.120)$$

and for *hexagonal metals*

$$<c_{44}'>_{av} = \frac{8}{15}c_{11} + \frac{3}{15}c_{33} + \frac{2}{15}(2c_{13} + 4c_{44}) \qquad (2.121)$$

and

$$<c_{44}'>_{av} = \frac{1}{3}\left(\frac{c_{11} - c_{12}}{2}\right) + \frac{2}{3}c_{44} \qquad (2.122)$$

Later studies have shown Equation 2.111 to be applicable to a limited range of ratios D/λ. Discrepancies between scattering factors α given by Equations 2.133 through 2.122 and later studies are common. Papadakis (1981 and 1968) [52,53] gives a more thorough discussion of such scattering. These discrepancies show the extreme difficulty of accounting for all practical conditions in a simple theory. However, the ultrasonic specialist should be aware of the factors that influence attenuation, and Equations 2.111 through 2.122 serve this purpose for hysteresis and scattering from grain boundaries in metals. Scattering is also caused by inclusions, microcracks, and other forms of discontinuities.

The functional form for the attenuation due the grain scattering in a polycrystalline media, in the different scattering regimes, is shown in Table 2.8. From a practical perspective in determining response, a simple approach is to measure the attenuation as a function of frequency. However, before using attenuation data, it is necessary to determine if there is a need to correct for transducer diffraction or beam spread effects. When the attenuation is at a low level, when compared with diffraction loss, correction is essential. For media with very high attenuation, the

TABLE 2.8

Functional Dependence of Attenuation

Range	Dependence[a]
$\lambda > 2\pi D$	$B_1 f + A_4 D^3 f^4$
$\lambda > 2\pi D$	$A_2 D f^4$
$\lambda \ll D_{min}$	$B_1 f + B_2 f^2 + A_s / D$

After [51,52]

[a] $B_1 f$ is elastic hysteresis loss and $B_2 f^2$ is thermoelastic loss
D is mean grain diameter.

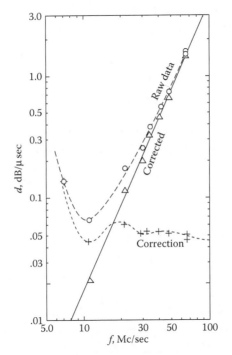

FIGURE 2.30 Example of the correction of attenuation data for diffraction loss. The diffraction corrections change the apparent frequency dependence of the attenuation. (Reprinted from Papadakis, E. P., *J Acoust Soc Am*, 40(4):863–67, 1966. With permission.)

diffraction loss effects can, in many cases, be negligible. It is however advisable to evaluate the relative magnitudes for both attenuation and beam effects when data are collected. An example of the effect of correction for beam spread on a measured attenuation is illustrated with the data given in Figure 2.30 [28]. In this case, the raw data and the magnitude of the apparent attenuation are significantly impacted when the correction for beam spread is applied.

2.10.3 ATTENUATION DUE TO HYSTERESIS

Hysteresis in an ultrasonic wave refers to a lag between the imposed stress and the resulting strain. Consequently, a plot of strain with increasing stress does not coincide with that for decreasing stress. As the stress is varied over a complete cycle, the stress–strain curve forms a characteristic hysteresis loop. The area inside this loop corresponds to energy lost by hysteresis during the cycle. Therefore, this type of absorption is proportional to the frequency (corresponding to the term $\beta_1 f$ in Equation 2.111).

The significance of hysteresis effects is shown in the data for aluminum given in Figure 1.13. In metals, this effect becomes significant at higher frequencies. In some materials, a variety of viscoelastic phenomena are encountered. In polymers, this typically becomes significant at low megahertz frequencies [2].

2.10.4 ATTENUATION DUE TO OTHER MECHANISMS

Attenuation is caused by many factors other than scattering. Any mechanism that results in the removal of energy or conversion of energy from the original state to a new form is a cause of attenuation. Some of these causes are (1) frictional losses due to relative motion between adjacent surfaces, for instance, adjacent surfaces in laminated structures and powder metal compactions of less than 100% theoretical density; (2) conduction of heat from high-stress regions to low-stress regions and to regions adjacent to the ultrasonic beam; (3) micro eddy currents; (4) motions of atoms in a lattice caused by stresses, which effect a divergence from a preferential distribution of the atoms; (5) viscosity (gases and liquids); and (6) dislocations in solids.

Mason [54] and others (e.g., [2,35,55]) discuss many of these mechanisms in some detail.

2.10.5 MEASUREMENT SYSTEM MODELS

Recent advances in computer hardware and digitization capabilities are enabling new generations of measurement tools to be developed. Much of the fundamental physics and many measurement methodologies are long established. They have been employed to meet needs in solid state physics [22] and to solve a diverse range of process measurement problems (e.g., [56]). The full range of ultrasonic methods used in process monitoring and chemical analysis has been extensively reviewed by Asher [35] and Hauptmann [57].

To facilitate the evaluation of feasibility and design of systems for many of the new applications of ultrasonics, both a metrology and tools are needed. The needed framework can best be understood by starting by considering the example of a general pulse-echo and two-transducer system, as shown in Figure 1.15. This general two-transducer model can be considered a linear system, and its response has been modeled in terms of a series of convolutions [38]. Such a one-dimensional model was presented by Newhouse and Furgason [58] as an extension of work by McGillen and Cooper [59] and Seydel [60]. In this work, the various elements in the system, namely the basic impulse generator, transducers, and the flaw response, are all represented as linear filters with effects that are combined by a series of convolutions.

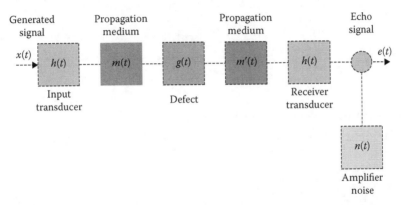

FIGURE 2.31 Schematic for a linear system model for basic ultrasonic NDT. (After Rayleigh, L., *The Theory of Sound, 1st Am. ed.*, Vol. 2., Dover, New York, 1945, and Newhouse, V. L., and E. S. Furgason, Ultrasonic correlation techniques, In *Research Techniques in Non-Destructive Testing*, ed. R. S. Sharpe, Vol. 3, 101–34, Academic Press, London, 1984.)

The elements in such a model are shown in Figure 2.31 and can be described with an equation using notation similar to that of Newhouse and Furgason [58] that can be given as

$$e(t) = x(t) * h(t) * m(t) * g(t) * m'(t) * h'(t) + n(t) \qquad (2.123)$$

Where * denotes convolution and the various terms are identified with the components in Figure 1.15. For the case of a single transducer in pulse-echo operation, reciprocal action for the transmitter and receiver can be assumed. In such a system, at least in principle, the desired, and usually unknown, defect or flaw response can be extracted from the complex echo signal by a series of deconvolutions.

The application of a Fourier transformation and deconvolution procedure to Equation 2.123 can be shown for a system with sufficient bandwidth to give improved resolution [58].

The overall response of a two-dimensional pulse-echo system has been considered by Staphanishen [61], who combined the transducer and acoustic media response in a "transfer function" model. The main aim of this model is to combine both the transducer's dynamic response and the acoustic diffraction, which cause the near- and far-field phenomena. A Fourier transform formulation is applied, and the effects of the system elements are combined as convolutions so as to calculate the receiver voltage that is given for a particular form of input voltage.

A new model [40] has recently been developed by Berkhout et al. [62] for the description of acoustic wave phenomena in inhomogeneous media and demonstrated for the case of scatters in a water tank. The model considers a forward-traveling pressure or acoustic wave problem in a layered media, and it is formulated in terms of a matrix equation representation. The system is described by a matrix equation of the form

$$P(z_0) = D(z_0) \left[\sum_m W(z_0, z_m) R(z_m) W(z_m, z_0) \right] S(z_0) \qquad (2.124)$$

where

$S(z_0)$ is the source vector representing the source configuration at the surface (z_0)

$W(z_m, z_0)$ is the downward propagation matrix representing the propagation properties from the surface (z_0) to depth level z_m

$R(z_m)$ is the scattering matrix representing the scattering properties of the inhomogeneities at depth level z_m

$W(z_0, z_m)$ is the upward propagation matrix representing the propagation properties from depth level z_m to the surface (z_0). It can be shown that for a time-invariant medium $W(z_m, z_0) = W^T(z_0, z_m)$, where T denotes a matrix transposition.

$D(z_0)$ is the detector matrix representing the detector configuration at the surface (z_0)

$P(z_0)$ is the data vector representing the single-scattered echo data at the surface (z_0) due to source $S(z_0)$

In Equation 2.124, W is defined by the wave equation for a pressure wave in inhomogenous absorptive fluids, R is defined by the elastic boundary conditions, and S and D are defined by the geometrical and the acoustical properties of the transducer in emission and in reception, respectively.

This model has been used by Berkhout et al. [62] to form inversion techniques based on focusing and deconvolution [62].

This general approach is analogous to the so called SONAR equations [34] and models that have been used in NDT [40]. Such linear models use data in dB and are the easiest ways to estimate signal-to-noise, range and detection sensitivity.

2.10.5.1 Resolution

The basic equations and phenomena of the pulse-echo NDT systems are well known, and the concept of resolution has been treated by various workers (e.g., Newhouse and Furgason, [58]). In outline, for a pulse-echo system in normal operation, the time delay (τ) between the transmitted and returned pulse from a target at a given range (R) is

$$\tau = \frac{2R}{V} \qquad (2.125)$$

where V is the wave velocity in the test piece. The pulse-repetition frequency for the system must be such that there is no overlap in wave trains, therefore,

$$\tau_{max} > \frac{2R_{max}}{V} \qquad (2.126)$$

Assuming no frequency dependence in the reflection coefficients of the waves, for two features with separation (ΔR), the time spacing in the wave train for resolution must be the length of the pulse used $(\Delta \tau)$, therefore,

$$\Delta\tau = \frac{2\Delta R}{V} \qquad (2.127)$$

It can also shown that the half-power bandwidth and the length of the pulse are related as

$$B = 1/\Delta\tau \qquad (2.128)$$

and the resolution and bandwidth are related by

$$\Delta R = V/2B \qquad (2.129)$$

Therefore, mode-conversion to a wave-type with a lower velocity (e.g., shear wave) can give better resolution. The advantages of using the compression wave as the input pulse are to give higher power inputs for the same initial electrical impulse, or in those cases where media encountered in the system have complex properties.

2.10.5.2 Signal-to-Noise and Measurement Window

For all ultrasonic NDT systems, there are fundamental limits for the "detectable" signal. The relationships and phenomena that determine this limit are a complex function of many systems parameters [40].

To try to quantify systems performance, the concept of a "measurement window," defined in terms of signal-to-noise, was developed for systems involving scattering by volume defects by Thompson [63]. Such a concept can usefully be extended to the case of mode-conversion testing, and show some of its advantages.

The factors that control the signal-to-noise ratio in effect determine the window in the frequency spectrum in which useful information can be obtained.

The signal-to-noise ratio was expressed by Thompson, for the case of a transducer used in pulse-echo measurements, as

$$S/N = \frac{A(f)\exp[-2B(f)l]}{N_b + N_g + N_e} \qquad (2.130)$$

where $A(f)$ is flaw scattering amplitude; $B(f)$ is material attenuation in nepers per unit length; l is distance from transducer to the flaw; N_b is noise associated with scattering from nearby part surfaces, which depends on details of transducer temporal response and is related to k, a transducer parameter; N_e is electronic receiver noise; and N_g is grain boundary scattering.

For the case of a volume flaw, such as a spherical void, the frequency dependence of the functions, which combine to form Equation 2.130, is shown in Figure 2.32a. A schematic measurement window is shown in Figure 2.32b, where it is assumed that the terms in Equation 2.130 have the following forms:

- At low frequencies, $A(f) \propto f^2 a^3$, where f is the frequency and a is the flaw radius.
- At higher frequencies, $A(f)$ approaches a constant.

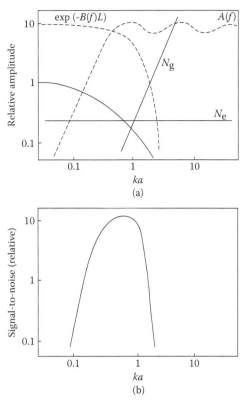

FIGURE 2.32 Schematic view of measurement windows for Equation 2.130. (a) Frequency dependence of individual factors affecting signal-to-noise ratio. (b) Signal-to-noise ratio for hypothetical case. (From Thompson, R. B., Evaluation of inversion algorithms [Project 1 – unit 1 – Task IVe], Ames Laboratory, Iowa State University, Iowa, DARPA Report # 4106. With permission.)

- N_e is also a constant for all frequencies when measurements are made with a fixed bandwidth.
- In most metals, both $B(f)$ and N_g are dominated by grain scattering.
- When ultrasonic frequencies are sufficiently low that the wavelength is large with respect to grain size, $B(f) \propto f^4$ and $N_g \propto f^2$.
- N_b depends upon details of transducer temporal response. A first estimate suggests that it will vary as the function $e^{-kf/l}$, where k is a transducer parameter.

2.11   RELAXATION

For many ultrasonic transducers, the electrical equivalent circuits and their characteristics can be modeled in terms of simple electrical equivalent circuits and their response to an applied voltage [1]. Relaxation phenomena are also encountered in some material responses [2,3].

When a constant dc voltage is applied across an RC series circuit or an RL series circuit (Figure 2.33), the final values of charge and current are not obtained instantaneously, but they are arrived at after a finite period of time. For instance, for the circuit of Figure 2.33a, the charge buildup on the condenser after closing the switch is

$$q = CE(1 - e^{-t/RC}) \tag{2.131}$$

where q is the quantity of charge on the capacitor at time t after closing the switch, C is the capacitance of the capacitor, E is the voltage across the circuit, and R is the resistance. Similarly, the current buildup in the RL circuit of Figure 2.33b is

$$i = \frac{E}{R}(1 - e^{-Rt/L}) \tag{2.132}$$

where i is the current through the circuit at time t and L is the circuit inductance.

From Equations 2.133 and 2.134, it is obvious that the quantities RC and L/R govern the time required for the charge or current, respectively, to reach a certain percentage of its final value. Because of their influence on the time required to reach a percentage of the final value, these quantities, RC and L/R, are called time constants. The time constants also are equal to the time it would take the charge or current to reach its final value if these quantities continued to change at their initial rates.

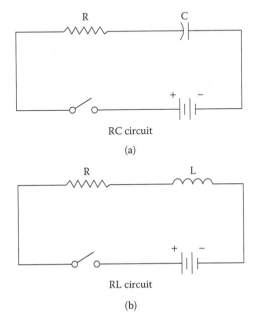

RC circuit

(a)

RL circuit

(b)

FIGURE 2.33   (a) An RC circuit. (b) An RL circuit.

When the time t exactly equals the time constant, the quantities in the parentheses of Equations 2.131 and 2.132 are

$$1 - e^{-1} = 0.632 \tag{2.133}$$

In acoustics, various phenomena cause a lag between an applied strain and the resulting stress, or vice versa. These phenomena are called relaxation phenomena. In comparison with the electrical analogues of Equations 2.133 and 2.134, we define a relaxation time as the time required for the dependent variable to increase from zero to $(1 - e^{-1})$ or 0.632 of its final value. The acoustical equivalent of Equations 2.133 and 2.134 is

$$S = \frac{p}{\rho c^2}(1 - e^{-t/\tau}) \tag{2.134}$$

where S is the applied strain, τ is the relaxation time, and p is the steady-state value of pressure.

Relaxation causes absorption, and this type of absorption is the one most commonly observed at high ultrasonic frequencies (megahertz range). At low frequencies, absorption due to relaxation is negligible. But, as the frequency increases, the absorption also increases, rising to a peak at a relaxation frequency f, and then decreases. Absorption due to relaxation goes to zero at a higher frequency corresponding to a condition where energy transfer from the sound wave to the particular effect to which the relaxation is related (for instance, vibrational motion of molecules) has no time to occur. At this same frequency, the elastic modulus of the medium increases, causing a corresponding increase in the velocity of sound. This increase in velocity of sound is called velocity dispersion.

Relaxational types of absorption include (1) those due to conduction of heat, (2) those due to viscosity, or (3) those due to any other process that transfers acoustical energy to other modes, such as rotation or vibration of atoms within a molecule, or that transfers energy from the acoustical wave to another energy form where it is temporarily stored and returned. Some materials exhibit more than one relaxation frequency.

2.12 HIGH-POWER PHENOMENA

For the wave propagation and scattering phenomena considered thus far in this chapter, the wave amplitudes have been implicitly assumed to be of small amplitude. This implies that for an ultrasound wave that is introduced as a sine wave, it will, with the exception of some viscoelastic absorption phenomena, remain in that form, even when there are scattering phenomena, which are encountered.

When the energy in the applied wave is increased, there is a diverse range of ultrasound phenomena that can be involved. These high-power interactions can in general be classed as "nonlinear" interactions. Such interactions may involve shock waves, harmonic generation, and a wide range of other phenomena that are thermal, mechanical, and chemical in nature.

2.12.1 CAVITATION

Many of the useful effects of ultrasonic energy are associated with cavitation, a term used to describe the formation of cavities, or bubbles, in a liquid medium. During the rarefaction portion of the cycle, when the pressure in the wave is below ambient, gas pockets may form and expand with the impressed field. These gases may be of two types: (1) those that have been dissolved or trapped in minute bubbles in the liquid or on surfaces in contact with the liquid and (2) vapors of the liquid itself. The first of these produces *gaseous cavitation*, and this form is of relatively low intensity. The second type, called *vaporous cavitation*, is of fairly high intensity.

Not all phenomena associated with cavitation appear to be explained completely by either vaporous or gaseous cavitation. If the pressure within the cavity is lower than the vapor pressure of the liquid during the expansion phase, the bubble is a result of fragmentation due to the tensile stress imposed by the ultrasonic wave being equal to the tensile strength of the liquid. This type of cavitation is very intense. The tensile strength of the liquid imposes an upper limit on the amplitude of the stress of the ultrasonic wave used to produce cavitation.

It is unlikely that the true tensile strength of the liquid is ever reached in practice because most liquids contain nuclei about which cavitation bubbles originate. These nuclei may consist of dispersed dust particles, prominences on immersed surfaces, or minute gas bubbles.

High localized stresses are developed during the formation and subsequent collapse of cavitation bubbles. Free chemical ions are produced within the vicinity of the bubble walls. High localized temperatures may also be present. Sometimes, weak flashes of light are produced—an effect called sonoluminescence. The peak values of either the temperature, sometimes estimated to be as high as 7200°C (1300°F), or the pressures (possibly as high as 520 MPa (75,00 psi) can be determined only theoretically because if these temperatures and pressures are obtained, their duration is extremely brief and they are minutely localized.

Some effects produced in the presence of cavitation include increased chemical activity (including reactions that would not occur in the absence of cavitation), erosion of surfaces, rupture or fragmentation of suspended particles, emulsification of liquid mixtures, and dispersion of small particles in the liquid. Applications based on these phenomena will be discussed in later chapters.

The importance of cavitation in ultrasonic processing has prompted a considerable amount of research and many publications with respect to the physics and associated effects of this phenomenon [7,64,65].

The onset of cavitation occurs at intensities, or cavitation thresholds, that depend upon such factors as the sizes of nuclei, ambient pressure, amount of dissolved gases, vapor pressure, viscosity, surface tension, and the frequency and duration of the ultrasonic energy.

When cavitation occurs, it not only dissipates the ultrasonic energy but also impedes transmission of the sound past it. Each bubble is a scattering site, and the scattering cross section includes both the bubble and the surrounding medium that is most directly affected by the bubble oscillations.

To be effective, that is, to cause the effects associated with the expansion and violent collapse of cavitation bubbles, the bubble must be capable of expanding with the rarefaction part of the cycle of the impressed field and of collapsing before the total pressure reaches its minimum value. That is, the bubble must reach the size where it will collapse catastrophically in less than one-quarter cycle of the impressed wave. Therefore, generation of intense cavitation depends upon the relationship between the dimensions of the nuclei, the wavelength of the sound field, and the intensity of the sound field. Bubbles larger than a critical radius, R_c, will not expand to an unstable size for catastrophic collapse before the pressure in the wave starts to increase. Frederick [64] gives the following relationships for R_c:

$$R_c = \frac{1}{\omega}\sqrt{\frac{3\gamma P_0}{\rho^-}} = \frac{326}{f}P_0^{1/2} \quad P_0 \gg \frac{2\sigma}{R_c} \tag{2.135}$$

$$R_c = \frac{1}{\omega}\sqrt{\frac{6\gamma c}{\rho R_0}} \approx \left(\frac{3.9}{f}\right)^{2/3} \quad P_0 \ll \frac{2\sigma}{R_0} \tag{2.136}$$

where R_c is the critical bubble radius (cm), R_0 is the initial bubble radius, γ is the ratio of specific heats of the gas in the bubble, σ is the surface tension of the liquid (dyn/cm), P_0 is the hydrostatic pressure (atm), and ρ is the density of liquid (g/cm³).

The intensity, or violence, of the cavitation bubble is a function of the ratio of the bubble radius at the instant before collapse to the minimum radius before expansion.

The effect of surface tension on bubble size is indicated by the equation

$$P_i = P_0 + \frac{2\sigma}{R_0} \tag{2.137}$$

where P_i is the vapor pressure inside the bubble and R_0 is the radius of a bubble, which is stable under these conditions.

Viscosity affects the rate of growth and collapse of cavitation bubbles. For this reason, very high viscosities may preclude the generation of cavitation.

In some applications, high ambient pressures have been used to suppress cavitation in coupling fluids used to transmit focused energy into specimens located in the focal region. This technique eliminates losses due to scattering from cavities.

Figure 2.34 is a photograph of ultrasonic horns designed to produce high intensities for processing in liquids (a) after very little service, (b) after exposure to ultrasonic cavitation for approximately four hours in detergent and water, and (c) after exposure to ultrasonic cavitation for only a few minutes in an acidic environment. Figure 2.34b is an example of cavitation-induced erosion. Figure 2.34c is an example of cavitation-induced corrosion. The frequency was 20 kHz.

In addition to cavitational phenomena in fluids, high-power ultrasound can also cause phenomena such as acoustoplasticity, or softening, in metals [66]. The topics relating to high-power ultrasound interactions in both industrial and medical applications are discussed in subsequent chapters of this text.

(a)   (b)   (c)

FIGURE 2.34 Effect of cavitation on the radiating surfaces of Monel horns vibrating at 20 kHz. (a) After little use. (b) After approximately 4 hours exposure to intense cavitation in detergent and water. (c) After a few seconds in dilute acid.

REFERENCES

1. Pain, H. J. 1999. *The Physics of Vibrations and Waves*. 5th ed. Chichester, New York: Wiley.
2. Matherson, A. J. 1971. *Molecular Acoustics*. London, New York: Wiley.
3. Beyer, R. T. 1997. *Non-Linear Acoustics*. New York: Acoustical Soc. Am.
4. Hamilton, M. F., and D. T. Blackstock, eds. 1998. *Nonlinear Acoustics*. San Diego, CA: Academic Press.
5. Leighton, T. G. 1994. *The Acoustic Bubble*. San Diego, CA: Academic Press.
6. Mason, T. J., and J. P. Lorimer. 1988. *Sonochemistry*. Chichester, England: Ellis Horwood.
7. Suslick, K. S. 1988. *Ultrasound: Its Chemical, Physical and Biological Effects*. New York: VCH Publishers.
8. Rose, J. L. 1999. *Ultrasonic Waves in Solid Media*. Cambridge: Cambridge University Press.
9. Graff, K. F. 1975. *Wave Motion in Elastic Solids*. Oxford: Clarendon Press.
10. Povey, M. J. W. 1997. *Ultrasonic Techniques for Fluids Characterization*. San Diego, CA: Academic Press.
11. Pierce, A. D. 1994. *Acoustics, American Acoustical Soc.* New York: Through the Am. Institute of Physics.
12. Roberts, J. K., and A. R. Miller. 1960. *Heat and Thermodynamics*. New York: Interscience.
13. Lynnworth, L. C. 1965. *IEEE Trans Sonics Ultrason* SU-12(2):37–48.
14. Ewing, W. M., W. S. Jardetzky, and F. Press. 1957. *Elastic Waves in Layered Media*. New York: McGraw-Hill.
15. Redwood, M. 1960. *Mechanical Waveguides*. New York: Pergamon Press.
16. Auld, B. A. 1973. *Acoustic Fields and Waves in Solids*. Vol I & II. New York: John Wiley.
17. Papadakis, E. P. 1981. Scattering in polycrystalline media. In *Ultrasonics, Methods of Experimental Physics*, ed. P. D. Edmonds, Vol. 19, Chap. 5, 237–298. New York: Academic Press.
18. Arenberg, D. L. 1948. *J Acoust Soc Am* 20(1):1–16.
19. Kino, G. S. 1987. *Acoustic Waves*. Englewood Cliffs, NJ: Prentice-Hall.
20. Lynnworth, L. C., and J. N. C. Chen. 1975. Energy transmission coefficient at liquid/solid interfaces. *Ultrason Symp Proc IEEE* 575–8.
21. Krautkramer, J., and H. Krautkramer. *Ultrasonic Testing of Materials*. 4th ed. Berlin: Springer-Verlag.

22. Turrell, R., C. Elbaum, and B. B. Chick. 1969. *Ultrasonic Methods in Solid State Physics*. New York: Academic Press.
23. Edmonds, P. D., ed. 1981. *Ultrasonics, Methods of Experimental Physics*. Vol. 19. New York: Academic Press.
24. Kolsky, H. 1953. *Stress Waves in Solids*. London: Oxford at the Claredon Press.
25. Egle, D. M., and D. E. Bray. 1975. Nondestructive measurement of longitudinal rail stresses. Report FRA-ORD-76-270, Pb-272061. Springfield, VA: Federal Railroad Administration, & Bray, D. E., and R. K. Stanley. 1989. *Nondestructive Evaluation*. New York: McGraw-Hill.
26. *The Krautkramer Booklet* Sonics and Ultrasonics Group. 1983. Cologne, W. Germany: Krautkramer GmbH.
27. Groschl, M. 1998. *Acustica* 84(3):432–47.
28. Papadakis, E. P. 1966. *J Acoust Soc Am* 40(4):863–67.
29. Kino, G. S. 1987. *Acoustic Waves*. 103. Englewood Cliffs, NJ: Prentice-Hall.
30. Achenbach, J. D., Y. H. Pao, and H. F. Tiersten, eds. 1976. Application of elastic waves in electrical devices, non-destructive testing and seismology. Report on a workshop: Northwestern University, May 24–26, 1976, National Science Foundation, Evanston, Il.
31. Datta, S. K., J. D. Achenbach, and Y. D. S. Rajapakse, eds. 1989. *Proc. IUTAM Symposium on Elastic Wave Propagation and Nondestructive Evaluation*. Boulder, CO: Elsevier (Amsterdam).
32. Pao, Y.-H. 1983. Elastic waves in solids. *Trans ASME* 50(12):1152–64.
33. Thompson, R. B., and H. N. G. Wadley. 1989. Critical reviews in solid state and materials science. 16(1):37–89.
34. Urick, R. J. 1983. *Principles of Underwater Sound*. 3rd ed. New York: McGraw-Hill.
35. Asher, R. C. 1997. *Ultrasonic Sensors*. Bristol: Institute of Physics Publishing.
36. Morse, P. M. 1948. *Vibration and Sound*. New York: McGraw-Hill.
37. Morse, P. M., and K. U. Ingard. 1968. *Theoretical Acoustics*. New York: McGraw-Hill.
38. Rayleigh, L. 1945. *The Theory of Sound, 1st Am. ed.*, Vol. 2. New York: Dover.
39. Ying, C. F., and R. Truell. 1956. *J Appl Phys* 27:1086–97.
40. Bond, L. J., and N. Saffari. 1984. Mode conversion ultrasonic testing. In *Research Techniques in Non-Destructive Testing*, ed. R. S. Sharpe, Vol. 7, Chap. 5, 145–89. London: Academic Press.
41. Harker, A. H. 1988. *Elastic Waves in Solids, with Applications to NDT of Pipelines*. Bristol: Adam Hilger, in association with British Gas.
42. Temple, J. A. G. 1987. *Int J Press Ves Pip* 28(1–5):277–97.
43. Bond, L. J. 1990. Numerical techniques and their use to study wave propagation and scattering. Proc. IUTAM Symposium on Elastic Wave Propagation and Nondestructive Evaluation (invited paper). ed. S. K. Datta, J. D. Achenbach, and Y. D. S. Rajapakse, July 30-Aug. 3, 1989, 17–28. Boulder, CO: Elsevier (Amsterdam).
44. Schmerr, L. W. 1998. *Fundamentals of Ultrasonic Nondestructive Evaluation*. New York: Plenum.
45. Calmon, P., A. Lhemery, I. Lecoeur-Taibi, and R. Raillion. 2008. Integrated models of ultrasonic examination for NDT Expertise. In *Review of Progress in QNDE*, ed. D. O. Thomposon, and D. E. Chimrenti, Vol. 16, 1861–8. New York: Plenum Press. (1997): Modeling software—current version CIVAnde version 9.
46. Smith, R. T., and R. W. B. Stephens. 1964. Effects of anisotropy on ultrasonic propagation in solids. In *Progress in Applied Materials Research*, ed. E. G. Stanford, J. H. Fearon, and W. J. McGonnagle, Vol. 6. New York: Gordon and Breach.
47. Merkulov, L. G. 1956. *Sov Phys Tech Phys* 1:59.
48. Merkulov, L. G. 1957. *Sov Phys Tech Phys* 2:953.

49. Goebbels, K. 1981. Structural analysis by scattered radiation. In *Research Techniques in NDT*, ed. R. S. Sharpe, Vol. 4, 109–57. London: Academic Press.
50. Weaver, R. L., and O. I. Lobkis. 2006. *Geophysics* 71(4):S15–9.
51. Mason and McSkimin. 1947. W. P. Mason and H. J. McSkimin. 1947. *J Acoust Soc Am* 19(3): 464–73.
52. Papadakis, E. P. 1968. In *Physical Acoustics, Principles and Methods*, ed. W. P. Mason, Vol. 4B. New York: Academic.
53. Papadakis, E. P. 1981. Scattering in polycrystalline media. In *Ultrasonics, Methods of Experimental Physics*, ed. P. D. Edmonds, Vol. 19, Chap. 5, 237–98. New York: Academic Press.
54. Mason, W. P. 1958. *Physical Acoustics and the Properties of Solids*. Princeton, NJ: Van Nostrand.
55. Bhatia, A. B. 1967. *Ultrasonic Absorption: An Introduction to the Theory of Sound Absorption and Dispersion in Gases, Liquids and Solids*. Oxford: Clarendon Press.
56. Lynnworth, L. C. 1989. *Ultrasonic Measurements for Process Control*. Boston: Academic Press.
57. Hauptmann, P., N. Hoppe, and A. Puttmer. 2002. *Meas Sci Technol* 13(8):R73–83.
58. Newhouse, V. L., and E. S. Furgason. 1984. Ultrasonic correlation techniques. In *Research Techniques in Non-Destructive Testing*, ed. R. S. Sharpe, Vol. 3, 101–34. London: Academic Press.
59. McGillen, C. D., and G. R. Cooper. 1971. *Continuous and Discrete Signal Systems Analysis*. New York: Holt, Rinehardt and Winston.
60. Seydel, J. 1973. *Computerized Enhancement of Ultrasonic Nondestructive Testing Data*, Ph.D. Thesis. Ann Arbor, MI: University of Michigan.
61. Staphanishen, P. R. 1981. *J Acoust Soc Am* 69(6):1815–27.
62. Berkhout, A. J., J. Riddler, and M. P. de Graaff. 1982. In *Acoustical Imaging*, ed. E. A. Ash, and C. R. Hill, Vol. 12, 269–80. New York: Plenum Press.
63. Thompson, R. B. Evaluation of inversion algorithms, (Project 1—unit 1—Task IVe) Ames Laboratory, Iowa State University, Iowa, USA, DARPA Report # 4106.
64. Frederick, J. R. 1965. *Ultrasonic Engineering*. New York: Wiley.
65. Thompson, L. H., and L. K. Doraiswamy. 1999. *Ind Eng Chem Res* 38(4):1215–49.
66. Langenecker, B., and O. Vodep. 1975. Metal plasticity in macroscopic fields. In *Proceedings, Ultrasonics International 1975*, 202–5. Guildford, UK: IPC Press.
67. Physics of Sound in the Sea. 1946. pt. III, *Nat Def Res Comm Div 6 Sum Tech Rep* 8:358–62. (Note: Equations. 49, 50, 53, and 56 in this reference are in error.)
68. Kerr, D. E., ed. 1951. *Propagation of Short Radio Waves*. M.I.T. Radiation Laboratory Series, vol. 13, 445–69. New York: McGraw-Hill.
69. Propagation of Radio Waves. 1946. Committee on propagation. *Nat Def Res Commun Sum Rep* 3:182.
70. Willis, H. F. 1941. Unpublished (British) report.
71. Spencer, R. C. 1950. *Backscattering from Conducting Surfaces, RDB Committee on Electronics, Radar Reflection Studies*. Washington, DC: U.S. Department of Defense.
72. Ruck, G. T., and others. 1970. *Radar Cross-Section Handbook*, Vols. I and II. New York: Plenum Press.
73. Pao, Y.-H., and C. C. Mow. 1973. *Diffraction of Elastic Waves and Dynamic Stress Concentrations*. New York: Crane and Russak.
74. Filipczynski, L. 1961. In *Proc. Vibration Problems*, Vol 2, # 1(6):41–56, Warsaw, Poland.
75. Gebernatis, J. E., E. Domany, and J. A. Krumhansl. 1977. *J Appl Phys* 48(7):2804–11.
76. Domany, E., J. A. Krumhansl, and S. Teitel. 1978. *J Appl Phys* 49(5):2599–604.

77. Richardson, J. M. 1979. *Proc. ARPA/AFML Review of Progress in Quantitative NDE, AFML-TR-78-205*, 332–40. Dayton, OH: Air Force Materials Laboratory.
78. Hess, J. L. 1973. *Comp Meth Appl Mech Eng* 2:1–15.
79. Kraut, E. A. 1976. *IEEE Trans Sonics Ultrason* SU-23(3):162–7.
80. Vissher, W. M. 1980. *J Appl Phys* 51(2):825–34.
81. Waterman, P. C. 1976. *J Acoust Soc Am* 60:567–80.
82. Keller, J. B. 1957. *J Appl Phys* 28:426–44.
83. Achenbach, J. D., and A. K. Gautesen. 1977. *J Acoust Soc Am* 61(2):413–21.
84. Bond, L. J. 1982. In *Research Techniques in Nondestructive Testing*, ed. R. S. Sharpe, Vol. 6, 107–50. London: Academic Press.

3 Fundamental Equations Employed in Ultrasonic Design and Applications

3.1 INTRODUCTION

To cover the wide range of uses for ultrasonic energy, the design of ultrasonic systems and of their applications involves the use of many equations. The most significant of these in ultrasonics is the general wave equation, to which many of the others are related. The derivation of the general wave equation and specific solutions are presented in this chapter. These derivations and some solutions are intended to help the reader in understanding the principles involved in the propagation of ultrasonic waves. They also serve as a guide in deriving other equations that may not be readily available to the reader. In addition to its tutorial aspects, this chapter provides a large number of the types of equations for, which the designer of ultrasonic equipment and applications may have a need. The material in this chapter is complementary to equations given in both Chapter 2 and Chapter 4 and supports material provided in the remainder of the text. It is hoped that these collections of equations will prove to be time-savers to those who have such needs.

The oscillating mass-loaded spring is commonly considered in engineering. In some systems, the combination is used to isolate vibrations of one mass from another, as in quieting machinery or household appliances, and resonance in these cases is to be avoided. In other systems, the combination is used to produce high-intensity vibrations, as in electrodynamic vibrators, and in these cases, the system is driven at resonance. The mass-loaded spring, or simple spring–mass oscillator, has engineering significance. It also serves as a fitting introduction to the derivation of the wave equation.

As mentioned in Chapter 2, an ultrasonic wave may be considered as consisting of a series of infinitesimally small mass elements elastically connected to its nearest neighbors. The spring constants are the elastic moduli. The wave equations are developed according to this concept.

By applying appropriate boundary conditions (restrictions) to the wave equation, solutions are obtained for designing various components of ultrasonic systems, for analyzing waves or causes of changes in wave structure during propagation, or for making measurements of engineering quantities.

Acoustic impedance must be considered in the design of every ultrasonic system. It is a primary factor controlling the efficiency and operation of a system. Therefore, the concepts of mechanical and acoustical impedances are introduced early in this chapter and are defined in terms that are related to the solutions of the wave equations.

Acoustic impedances were used in Chapter 2 with respect to reflection and transmission of ultrasonic energy at a boundary between two media. They are also used extensively in Chapter 5, which is devoted to the design of ultrasonic transducers.

This chapter includes the derivation of the equation for the simple spring–mass oscillator and applies the same principles to the derivation of the general wave equation. The general solution to the plane-wave equation is given and applied to longitudinally vibrating uniform bars.

The transverse-wave equation is introduced in order to present the important functions of vibrations of bars vibrating in transverse modes (flexure). There is considerable interest in the measurement of flexural vibrations in bars as a means of obtaining useful engineering data, particularly elastic moduli. To obtain these data requires an understanding of the role of these quantities in the vibrations of the bars. Also, flexural modes are often used in ultrasonic transducers or systems.

Consideration of the general wave equation and of the particular form applied to transverse waves provides a background for investigating plate waves (or so-called Lamb waves). These waves are used widely in the evaluation of materials (nondestructive testing). Lamb waves are of two types: symmetrical and asymmetrical. The symmetrical waves correspond to longitudinal waves, and asymmetrical waves are flexural waves. Hence, discussion of these waves is reserved until after the discussion related to the general wave equation and its derivative, the transverse-wave equation. The chapter concludes with equations describing flexural vibrations in plates.

The design formulas presented in Chapter 4 for horn or wave-guide systems for energy delivery are specific solutions to the plane-wave equation.

3.2 SIMPLE SPRING–MASS OSCILLATOR

3.2.1 IDEAL CONDITION—SIMPLE HARMONIC MOTION

An acoustic wave can exist in a medium only because the medium has mass and elastic properties. Sound cannot travel through a vacuum. The incremental mass affected by an ultrasonic wave and its elastic couplings are analogous to an oscillating mass attached to springs. Therefore, as a preliminary to developing the wave equations, consider the action of such a mass attached to two springs, as illustrated in Figure 3.1. For simplicity, the system of Figure 3.1 is assumed to be ideal; that is, no friction or other losses and hence no energy is dissipated when the mass is set in motion. In actual practice, the ideal conditions can be approximated, and the equations derived in this section can be applied to engineering designs in which energy losses are very small. Where these losses are significantly large, the equations developed in Section 3.2.2 apply.

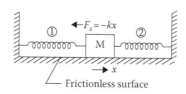

FIGURE 3.1   Simple spring–mass oscillator.

To fulfill the hypothetical conditions, it must be assumed that the mass rests on a frictionless surface. Since the surface is horizontal, the gravitational pull on the mass is constant, regardless of the position of the mass. In addition, assume that the system is located in a vacuum so that when the mass moves there is no loss of energy due to a resistance. Under such conditions, when the mass is displaced to the right by a distance x, the spring to the left is subjected to tension and the spring on the right to compression, the force tending to restore the mass to its original position. If the total force exerted by the springs on mass M (in kg) is F_x newtons and the combined *spring constant* (defined as the force required to displace the springs over unit distance) is k, the magnitude of the total force F_s exerted on the mass M is

$$F = -kx \qquad (3.1)$$

where x is the distance in meter that the mass is displaced from rest position. The negative sign is used because the force is in a direction opposite to the displacement x. If the mass M is released, the force F_s moves the mass back toward the rest position. Since mass has inertia, the restoring force causes an acceleration given by

$$F = Ma = M\frac{dv}{dt} = M\frac{d^2x}{dt^2} \qquad (3.2)$$

where a is the acceleration of the mass (m/s^2), M is the mass (kg), v is the instantaneous velocity of the mass (m/s), w is the weight of the mass (kg), x is the displacement of the mass (meter), and t is time (sec).

The force F_M is directly opposed to the force F_s and equal to F_s in magnitude, that is,

$$F_M - F_s = 0 \qquad (3.3)$$

or

$$M\frac{d^2x}{dt^2} + kx = 0 \qquad (3.3a)$$

The solution for Equation 3.3a is

$$x = A\cos\sqrt{\frac{k}{M}}t + B\sin\sqrt{\frac{k}{M}}t \qquad (3.4)$$

Therefore, under ideal conditions (vacuum, frictionless horizontal surface), the simple oscillator of Figure 3.1 will vibrate sinusoidally with an angular frequency $\omega = 2\pi f = \sqrt{k/M}$, where f is the frequency of oscillation. This is a condition known as simple harmonic motion, which can be approximated with both mechanical and electrical systems. Since the system is lossless, the total potential energy stored in the spring at the maximum displacement is converted to kinetic energy when the mass returns to rest position. This stored kinetic energy of the mass causes a displacement in the opposite direction, which is equal in magnitude to the initial displacement.

Without energy losses, the operation would be repeated indefinitely. Often the mass of the spring is not negligible. In this case, one-third of the mass of the spring is added to the M of Equation 3.4. This equation also applies to an oscillating mass suspended by a spring.

In the real situation in which energy is lost from the system, for instance, through friction, the oscillations become damped. The amplitude of vibrations decreases exponentially if the damping factor remains constant.

3.2.2 REAL CONDITION—DAMPED SIMPLE HARMONIC MOTION

To consider the real condition with losses, assume that the mass is suspended on a spring. In this case, the resistance to motion imposed by the air tends to remove energy from the system that is otherwise stored to promote oscillation. Slight losses in the spring, which tend to heat the spring, also contribute to the removal of such energy. The obvious result is that, unless energy lost is replaced by a driving force, the vibratory system will eventually stop oscillating.

The losses thus described are dependent upon the velocity of the body in vibration. Usually, the resistive force F_r is directly proportional to the velocity. Therefore, it can be expressed mathematically as

$$F_r = -R_m \frac{d\xi}{dt} \tag{3.5}$$

where R_m is called the mechanical resistance of the system and ξ is the displacement of the system. The dimensions of R_m are kilogram per second.

Introducing the resistance term in Equation 3.3a produces the following relationship describing the motion of a simple oscillator:

$$M \frac{d^2\xi}{dt^2} + R_m \frac{d\xi}{dt} + k\xi = 0 \tag{3.6}$$

The general solution to Equation 3.6 is

$$\begin{aligned}
\xi &= e^{-R_m t/2M} \left(k_1 \cos \sqrt{\frac{k}{M'}} t + k_2 \sin \sqrt{\frac{k}{M'}} t \right) \\
&= e^{-\alpha t} \left(k_1 \cos \sqrt{\frac{k}{M'}} t + k_2 \sin \sqrt{\frac{k}{M'}} t \right)
\end{aligned} \tag{3.7}$$

where

$$M' = \frac{4M^2 k}{4kM - R_m^2}$$

$$\alpha = \frac{R_m}{2M} \text{(the damping factor)}$$

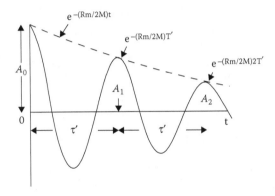

FIGURE 3.2 Logarithmic ratio of any two amplitudes; one period apart is the logarithmic decrement, defined as $\delta = \log_e(A_n/A_{n+1}) = R_m\tau'/2M$, and τ' is the period for the damped frequency. (From Pain, H. J., *The Physics of Vibrations and Waves*, 5th Ed., John Wiley, Chichester, UK, 1999. With permission.)

Therefore, the damped angular frequency, ω_d, is

$$\omega_d = \sqrt{\frac{k}{M'}} = \sqrt{\omega_0^2 - \alpha^2} = 2\pi f_d \tag{3.8}$$

where ω_0 is $\sqrt{k/M}$ (the undamped angular frequency).

When $t = 0$, $k_1 = \xi$, the displacement from the equilibrium position at time zero. The maximum amplitude depends upon the initial conditions and the manner of initiating the oscillations.

The solution of Equation 3.6 may also be written

$$\xi = Ke^{-\alpha t}\cos(\omega_d t + \phi) \tag{3.9}$$

where K and ϕ are amplitude and phase constants, respectively, and depend upon the initial conditions. Therefore, by measuring the rate of decay logarithmic decrement of such a vibratory system, it is possible to determine α and, hence, R_m. The logarithmic decrement and related parameters are shown for damped simple harmonic motion in Figure 3.2.

Equation 3.8 shows that the actual resonant frequency of a mass on a spring is less than the resonant frequency of a theoretically undamped system by a factor which depends upon the damping resistance. When this damping resistance is very small, $\omega_0 \approx \omega_d$.

3.2.3 EFFECT OF DAMPING ON PHASE RELATIONSHIPS—THE FORCED OSCILLATOR

The simple oscillator (mass-loaded spring) may serve to illustrate effects to be expected in more complicated systems. For instance, assume that the simple damped oscillator, whose motion is described by Equation 3.6, is driven by a force $Fe^{j\omega t}$. The equation describing this condition is

$$M\frac{d^2\xi^*}{dt^2} + R_m\frac{d\xi^*}{dt} + k\xi^* = Fe^{j\omega t} \tag{3.10}$$

where ξ^* is a complex displacement. The solution of Equation 3.10 is the sum of two parts: a steady-state term, which depends upon F, and a transient term containing two arbitrary constants. The transient term is obtained by setting $F = 0$, which gives Equation 3.6, for which the general solution is Equation 3.7. After a period of time, the damped portion of the solution becomes negligible, leaving only the steady-state solution.

If the steady-state solution is assumed to be of the form

$$\xi^* = K^* e^{j\omega t}$$

where K^* is a complex constant, substituting in Equation 3.10 gives

$$K^*(-\omega^2 M + j\omega R_m + k) = F \tag{3.11}$$

For Equation 3.11 to be true for all times, t,

$$K^* = \frac{F}{(k - \omega^2 M) + j\omega R_m}$$

and, therefore, the complex displacement ξ^* may be written

$$\xi^* = \frac{-jFe^{j\omega t}}{\omega R_m - j(k - \omega^2 M)} = \frac{-jFe^{j\omega t}}{\omega[R_m + j(\omega M - k/\omega)]}$$
$$= \frac{-jFe^{j(\omega t - \phi)}}{w\sqrt{R_m^2 + (\omega M - k/\omega)^2}} \tag{3.12}$$

where ϕ is the phase angle between the driving force and the velocity and is given by

$$\phi = \tan^{-1}\left(\frac{\omega M - k/\omega}{R_m}\right)$$

By definition, the complex velocity $v^* = d\xi/dt$, or

$$v^* = \frac{d\xi^*}{dt} = \frac{Fe^{j(\omega t)}}{R_m + j(\omega M - k/\omega)} \tag{3.13}$$

Also, the mechanical impedance Z_m is the ratio between the driving force and the velocity. From Equation 3.13, the impedance is

$$Z_m = R_m + j\left(\omega M - \frac{k}{\omega}\right) \tag{3.14}$$

and Equation 3.13 may be rewritten

$$v^* = \frac{Fe^{j(\omega t)}}{Z_m} \tag{3.15}$$

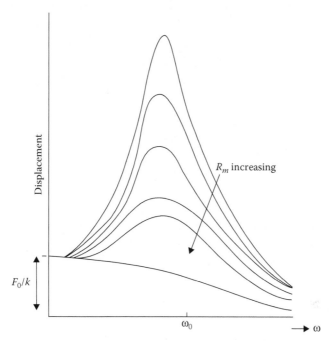

FIGURE 3.3 Variation of the displacement of a forced oscillator versus driving force frequency $\dot{\omega}$ for various values of R_m. (From Pain, H. J., *The Physics of Vibrations and Waves*, 5th Ed., John Wiley, Chichester, UK, 1999. With permission.)

The real term R_m of Equation 3.14 corresponds to a resistance, and the imaginary term corresponds to reactive impedance.

If the driving frequency is varied from zero to above the resonant frequency, the form of the displacement will depend upon the initial driving force and damping and have the relationship as shown in Figure 3.3.

The electrical analogs for simple harmonic motion, damped harmonic motion, and the forced oscillator are found in the simple "LCR" circuits. Electrical resistance "R" is the analog of R_m, inductance L is analogous to mass M, and the inverse of the spring constant "k" is analogous to capacitance "C." Further treatment of both the electrical and the mechanical systems responding to give harmonic motion is found in most standard physics texts on vibrations and waves (e.g., [1]).

3.3   WAVE EQUATIONS

The term *wave equation* is applied to a wide range of phenomena encountered in the realm of physics. It is used in one form or another in the study of electromagnetic radiation, to describe phenomena in solid-state physics, seismology, and geophysics, and, of course, in the study of acoustics. The following section is devoted to the derivation of the plane-wave equation in terms applicable to ultrasonics. The form of the general equation is then inferred from the derivation of the plane-wave equation.

3.3.1 Plane-Wave Equation

An elastic medium may be considered as consisting of a series of homogeneous, incremental elements or masses, each of density ρ, thickness dx, and cross-sectional area S, shown in Figure 3.4. When a net longitudinal force is applied to the element, the element accelerates according to Equation 3.2. Therefore, a finite period of time is required for the element to acquire a velocity other than its initial velocity.

As soon as the element moves from its initial position, it applies a force to succeeding elements, which, in turn, undergo acceleration. At point x, the force on the face of the element located between x and $x + dx$ is

$$F_x = -YS\frac{d\xi}{dx} \tag{3.16}$$

where Y is the modulus of elasticity of the medium in kg/m² and ξ is the displacement in meter. At distance $x + dx$, the force on the opposite face of the element is

$$-\left(F_x + \frac{\partial F_x}{\partial x}dx\right)$$

Therefore, the net force across the element is

$$F_x - \left(F_x + \frac{\partial F_x}{\partial x}dx\right) = -\frac{\partial F_x}{\partial x}dx = YS\frac{\partial^2\xi}{\partial x^2}dx \tag{3.17}$$

By replacing M with $\rho S\,dx$ and a with $\partial^2\xi/\partial t^2$ in Equation 3.2, the inertial force of the element is

$$F_m = \rho S\frac{\partial^2\xi}{\partial t^2}dx \tag{3.18}$$

where ρ is the density of the medium.

For equilibrium to exist, the elastic forces on the element must equal the inertial forces imposed by the element. Therefore, from Equations (3.17) and (3.18)

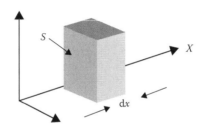

FIGURE 3.4　Coordinates and volume element.

$$\rho S \frac{\partial^2 \xi}{\partial t^2} dx = YS \frac{\partial^2 \xi}{\partial x^2} dx \qquad (3.19)$$

or

$$\frac{\partial^2 \xi}{\partial t^2} = \frac{Y}{\rho} \frac{\partial^2 \xi}{\partial x^2} = c^2 \frac{\partial^2 \xi}{\partial x^2} \qquad (3.20)$$

which is the longitudinal- or plane-wave equation with no losses. Because c is the velocity at which the stress moves from element to element in the system, it is the speed of sound in the medium. Here, Y is Young's modulus of elasticity if the medium is a solid slender bar, small in cross section compared to a wavelength.

Equation 3.20 is the one-dimensional plane-wave equation. Transverse waves on a string obey the same wave equation; however, the constants are different.

A method similar to that used in deriving Equation 3.20 can be used to derive equations governing acoustic behavior in configurations in which S, the cross section, is variable with x. For example, in Chapter 4, a horn equation, a form of the wave equation,

$$\frac{1}{c^2} \frac{\partial^2 \xi}{\partial t^2} - \frac{1}{S} \frac{\partial S}{\partial x} \frac{\partial \xi}{\partial x} - \frac{\partial^2 \xi}{\partial x^2} = 0$$

or

$$\frac{\partial^2 v}{\partial x^2} + \frac{1}{S} \frac{\partial S}{\partial x} \frac{\partial v}{\partial x} + \frac{\omega^2}{c^2} v = 0 \qquad (3.21)$$

is used to derive formulas for designing various solid horn or wave-guide configurations when the cross-sectional dimensions at any point x are small compared to a wavelength.

3.3.2 GENERAL WAVE EQUATION

Both Equation 3.20 and Equation 3.21 describe wave motion in only one spatial dimension (here defined as the x direction). The more general case considers all possible directions of propagation. The general wave equation, then, is

$$\frac{\partial^2 \xi}{\partial t^2} = c^2 \left(\frac{\partial^2}{\partial x^2} + \frac{\partial^2}{\partial y^2} + \frac{\partial^2}{\partial z^2} \right) \xi = c^2 \nabla^2 \xi \qquad (3.22)$$

The term $\nabla^2 \xi$ may be expressed in any system of coordinates that best suit the configuration of the boundaries, that is, Cartesian coordinates for a rectangular configuration or spherical coordinates for a spherical system.

In a nonhomogeneous medium, velocity c is a function of spatial coordinates. When velocities are considered in structured inhomogeneous media, there can be strong dependence on direction [2] and material loading or damage. Many different solutions to the wave equation for nonhomogeneous media are possible. The velocity c as used in this book, in general, will be an integrated average within certain coordinates. However, it is often necessary for various applications to consider non-homogeneous systems.

3.4   SOLUTION OF THE PLANE-WAVE EQUATION, LINEAR SYSTEM

3.4.1   GENERAL SOLUTION

The general solution for the plane-wave equation, given as Equation 3.20, is

$$\xi = f_1(ct - x) + f_2(ct + x) \tag{3.23}$$

which mathematically may or may not be a periodic function. The solution $f_1(ct - x)$ represents a wave moving in the positive x direction, and $f_2(ct + x)$ represents a wave moving in the negative x direction. The functions f_1 and f_2 are determined by the boundary conditions of a specific problem.

As a practical example, let Equations 3.20 and 3.23 describe longitudinal-wave motion in a thin bar. Assuming that the solution of Equation 3.20 can be rewritten as a product of two functions that individually depend on one variable only,

$$\xi = X(x)T(t) \tag{3.24}$$

Differentiating Equation 3.24 gives

$$\frac{\partial^2 \xi}{\partial t^2} = X \frac{\partial^2 T}{\partial t^2} \quad \text{and} \quad \frac{\partial^2 \xi}{\partial x^2} = T \frac{\partial^2 X}{\partial x^2} \tag{3.25}$$

Substituting Equation 3.25 in Equation 3.20 gives

$$X \frac{\partial^2 T}{\partial t^2} = c^2 T \frac{\partial^2 X}{\partial x^2} \tag{3.26}$$

The variables may be separated by dividing both sides of Equation 3.26 by the product XT; therefore

$$\frac{1}{T} \frac{\partial^2 T}{\partial t^2} = \frac{c^2}{X} \frac{\partial^2 X}{\partial x^2} \tag{3.27}$$

To be valid for all possible values of x and t, it follows that both sides of Equation 3.27 must be equal to a common constant. For this example, let

$$\frac{1}{T} \frac{\partial^2 T}{\partial t^2} = \frac{c^2}{X} \frac{\partial^2 X}{\partial x^2} = -p^2 \tag{3.27a}$$

Then the following ordinary differential equations can be written:

$$\frac{d^2T}{dt^2} + p^2T = 0 \quad \text{and} \quad \frac{d^2X}{dx^2} + \frac{p^2}{c^2}X = 0 \tag{3.28}$$

The solutions of Equation 3.28 are

$$T = A \cos pt + B \sin pt$$
$$X = C \cos\frac{p}{c}x + D \sin\frac{p}{c}x \tag{3.29}$$

where p has the dimensions of radians per second and is identical to the angular frequency $\omega = 2\pi f$ defined previously.

Therefore, the solution of Equation 3.20 may be written as follows:

$$\xi = X(x)T(t)$$
$$= \left(C\cos\frac{\omega x}{c} + D\sin\frac{\omega x}{c}\right)(A\cos\omega t + B\sin\omega t)$$
$$= \frac{1}{2}\left[(AC+BD)\cos\left(\frac{\omega x}{c} - \omega t\right) + (AC-BD)\cos\left(\frac{\omega x}{c} + \omega t\right)\right. \tag{3.30}$$
$$\left. +(AD+BC)\sin\left(\frac{\omega x}{c} + \omega t\right) + (AD-BC)\sin\left(\frac{\omega x}{c} - \omega t\right)\right]$$

3.4.2 FREE–FREE LONGITUDINALLY VIBRATING UNIFORM BAR

The simplest example of the use of Equation 3.30 is in describing the vibratory characteristics of a slender bar, free at each end, and in longitudinal half-wave resonance (Figure 3.5). In this example, the boundary conditions are

- At $x = 0$ and at $x = \ell$, $\partial\xi/\partial x = 0$.
- At $x = 0$, $\xi = \xi_0$ and $v = v_0$.

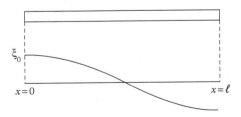

FIGURE 3.5   Slender uniform bar in longitudinal half-wave resonance (free–free bar).

Differentiating Equation 3.30 with respect to x gives

$$\frac{\partial \xi}{\partial x} = \frac{\omega}{2c}\left[-(AC + BD)\sin\left(\frac{\omega x}{c} - \omega t\right) - (AC - BD)\sin\left(\frac{\omega x}{c} + \omega t\right) \right.$$
$$\left. + (AD + BC)\cos\left(\frac{\omega x}{c} + \omega t\right) + (AD - BC)\cos\left(\frac{\omega x}{c} - \omega t\right) \right] \tag{3.31}$$

Applying the boundary condition that at $x = 0$, $\partial \xi / \partial x = 0$ reduces Equation 3.31 to

$$\frac{\partial \xi}{\partial x} = \frac{\omega}{c}(BC \cos \omega t + BD \sin \omega t) = 0 \tag{3.32}$$

Since Equation 3.32 must be true for all values of t, $BD = 0$ and $BC = 0$. Substituting these values into Equation 3.30 gives

$$\xi = (AC)\cos\left(\frac{\omega x}{c}\right)\cos(\omega t) + (AD)\sin\left(\frac{\omega x}{c}\right)\cos(\omega t) \tag{3.33}$$

When $x = 0$

$$\xi = \xi_0 = (AC)\cos(\omega t) = \xi_m \cos(\omega t) \tag{3.34}$$

where ξ_m is the maximum displacement of the end of the bar where $x = 0$. Therefore

$$\xi = \left[\xi_m \cos\left(\frac{\omega x}{c}\right) + (AD)\sin\left(\frac{\omega x}{c}\right) \right]\cos(\omega t) \tag{3.35}$$

$$v = \frac{\partial \xi}{\partial t} = -\omega\left[\xi_m \cos\left(\frac{\omega x}{c}\right) + (AD)\sin\left(\frac{\omega x}{c}\right) \right]\sin(\omega t) \tag{3.36}$$

According to the boundary conditions, at $x = 0$

$$v = v_0 = -\omega \xi_m \sin(\omega t) \tag{3.37}$$

or

$$v_m = \omega \xi_m$$

The remaining unknown quantity is (AD). This quantity is determined by applying the boundary condition that at $x = \ell$

$$\frac{\partial \xi}{\partial x} = 0$$

which gives

$$\frac{\partial \xi}{\partial x} = \left[-\frac{\omega}{c} \xi_m \sin\left(\frac{\omega x}{c}\right) + (AD)\frac{\omega}{c}\cos\left(\frac{\omega x}{c}\right) \right]\cos(\omega t)$$

$$= (AD)\frac{\omega}{c}(-1)\cos(\omega t) = 0 \tag{3.38}$$

Therefore, $(AD) = 0$ and

$$\xi = \xi_m \cos\left(\frac{\omega x}{c}\right)\cos(\omega t) \tag{3.39}$$

describes the displacement along a slender bar in longitudinal-half-wave resonance. The particle velocity at any point x along the bar is

$$v = -\omega\xi_m \cos\left(\frac{\omega x}{c}\right)\sin(\omega t) = j\omega\xi \tag{3.40}$$

or writing the displacement in exponent form

$$\xi = \xi_m e^{j\omega t}\cos\frac{\omega x}{c}$$

leads to the velocity

$$v = \frac{d\xi}{dt} = j\omega\xi_m e^{j\omega t}\cos\left(\frac{\omega x}{c}\right) = j\omega\xi \tag{3.40a}$$

The particle acceleration at any point x along the bar is

$$a = \frac{\partial v}{\partial t} = -\omega^2\xi_m \cos\left(\frac{\omega x}{c}\right)\cos(\omega t) = -\omega^2\xi \tag{3.41}$$

3.4.3 STRESS IN A VIBRATING UNIFORM BAR

Formulas were derived in the previous sections on the assumption that stresses were developed in proportion to the strain coincident with the acoustic wave. Equation 3.16 defined the assumed condition mathematically and may be written

$$\frac{F_x}{S} = -Y\frac{\partial \xi}{\partial x} = s = stress$$

$$= j\frac{Y}{\omega}\frac{\partial v}{\partial x} \tag{3.42}$$

Equations 3.16 and 3.42 assume that Y is constant regardless of whether the material is in compression or in tension. It must also be assumed that the material is isotropic. If Y is defined as Young's modulus, in the dynamic case the bar must be slender compared to the wavelength (which, for the half-wave resonant bar, is twice the length of the bar). Other moduli will be considered in later sections as the need arises.

From Equations 3.39, 3.40, and 3.42, it follows that

$$s = -Y\frac{\partial\xi}{\partial x} = Y\frac{\omega\xi_m}{c}\sin\left(\frac{\omega x}{c}\right)\cos(\omega t) \tag{3.42a}$$

or

$$s = j\frac{Y}{\omega}\frac{\partial v}{\partial x} = jY\frac{\omega\xi_m}{c}\sin\left(\frac{\omega x}{c}\right)\sin(\omega t) \tag{3.42b}$$

3.4.4 MECHANICAL IMPEDANCE

Impedance is the primary factor controlling the efficiency and operation of a vibrating system, whether the system is electrical mechanical or acoustical.

Impedances in an electrical circuit limit the flow of electrical current in accordance with the magnitude and type of impedances. The analog for an acoustical system relates force (or stress) to velocity.

A force F impressed on an increment of an acoustical system imparts a velocity of motion, v, to the increment in proportion to F. Expressed mathematically,

$$F = Zv \tag{3.43}$$

Equation 3.43 is analogous to the electrical case

$$E = Z_e I \tag{3.44}$$

For the electrical analog, the force F corresponds to the voltage E, velocity potential v to current I, and proportionality constant Z to electrical impedance $Z_e = R_e + jX_e$. Therefore, $Z = R + jX$ may be defined as the mechanical impedance.

A long, slender bar is the acoustical equivalent of an electrical transmission line with distributed constants. The impedance at any point x is determined by dividing the force at x by the velocity potential at x. For the free–free bar without losses

$$Z = \frac{F}{v} = \frac{sS}{v} = -j\frac{SY}{c}\tan\left(\frac{\omega x}{c}\right) = -jS\rho c\tan\left(\frac{\omega x}{c}\right) \tag{3.45}$$

where $c = \sqrt{Y/\rho}$. The product, ρc, of Equation 3.45 is called the characteristic impedance of the medium. Standing alone, it is a pure resistance.

Equation 3.45 shows a purely reactive impedance. The reason for this is that losses have been ignored in deriving the wave equation and the equations of vibration leading to Equation 3.45. Equation 3.45 is a close approximation to the actual impedance if the loss factors are very small compared to the reactive components of the system. However, if the losses are significant, the wave equation must take them into account. The plane-wave equation can be modified to account for losses as follows:

$$\frac{\partial^2 \xi}{\partial t^2} + \frac{R_m}{\rho S}\frac{\partial^2 \xi}{\partial x \partial t} = c^2 \frac{\partial^2 \xi}{\partial x^2} \tag{3.46}$$

or

$$\frac{\partial^2 v}{\partial x^2} - \frac{j\omega}{c^2}\frac{R_m}{\rho S}\frac{\partial v}{\partial x} + \frac{\omega^2}{c^2}v = 0 \tag{3.46a}$$

where ω is the angular frequency that would occur if there were no losses in the system and ρ is the density in kg/m^3.

Equation 3.46a is a form of the equation for damped vibration:

$$\frac{d^2 y}{dx^2} + 2b\frac{dy}{dx} + a^2 y = 0 \tag{3.46b}$$

where

$$b = \frac{-j\omega R_m}{2c^2 \rho S} \quad \text{and} \quad a = \frac{\omega}{c}$$

The solution to Equation 3.46b depends upon the relative magnitude of the constants a and b.

Case 1. If $a^2 - b^2 > 0$, let $m = \sqrt{(a^2 - b^2)}$.

$$y = C_1 e^{-bx} \sin(mx + C_2)$$

Case 2. If $a^2 - b^2 = 0$,

$$y = e^{-bx}(C_1 + C_2 x)$$

Case 3. If $a^2 - b^2 < 0$, let $n = \sqrt{(b^2 - a^2)}$ and

$$y = C_3 e^{-(b+n)x} + C_4 e^{-(b-n)x}$$

Since $b^2 = -\omega^2 R_m^2/4c^4\rho^2 S^2$, obviously $a^2 - b^2 > 0$. Therefore, the solution to Equation 3.46b is

$$y = e^{-bx}\, C_1 \sin(mx + C_2) \tag{3.47}$$

or

$$y = e^{-bx}[C_3 \sin(ms) + C_4 \cos(mx)]$$

Therefore, the solution to Equation 3.46a is

$$v = C_1\, e^{j\omega R_m x/2c^2\rho S}\, \sin\left[\left(\frac{\omega}{c}\sqrt{1+\frac{R_m^2}{4c^2\rho^2 S^2}}\right)x + C_2\right] \tag{3.47a}$$

or

$$v = e^{(j\omega R_m/2c^2\rho S)x}\left[C_3 \sin\left(\frac{\omega}{c}\sqrt{1+\frac{R_m^2}{4c^2\rho^2 S^2}}\right)x + C_4 \cos\left(\frac{\omega}{c}\sqrt{1+\frac{R_m^2}{4c^2\rho^2 S^2}}\right)x\right] \tag{3.47b}$$

The values of the constants C_3 and C_4 are determined by the boundary conditions. The constant C_1 obviously is velocity V_m defined in Equation 3.37, and the constant C_2 is a phase angle.

Equations 3.47a and 3.47b can be written in the form

$$v = C_1\, e^{j(\omega R_m/2c^2\rho S)x}\, \sin\left[\left(\frac{\omega}{c'}\right)x + C_2\right]c$$

or

$$v = e^{j(\omega R_m/2c^2\rho S)x}\left[C_3 \sin\left(\frac{\omega}{c'}\right)x + C_4 \cos\left(\frac{\omega}{c'}\right)x\right]$$

where

$$c' = \frac{c}{\sqrt{1 + R_m^2/4c^2\rho^2 S^2}}$$

3.4.5   QUALITY FACTOR (Q)

If it were possible to have a vibratory system in which no loss of energy occurs, the system would vibrate indefinitely. Energy introduced into such a system would be stored and would increase the intensity of the oscillations. As soon as energy is extracted from the system, whether to do work or by absorption, the vibratory characteristics are affected. Energy losses are always present in a real system. These losses are functions of velocities in the vibratory system so that, when energy is supplied to the system, the amplitude rises to an equilibrium level where the energy supplied exactly equals the energy dissipated. Thus, it can be shown that when the driving force is suddenly removed from an oscillatory system, the amplitude of vibration decreases exponentially since the energy losses decrease with decreasing amplitude of vibration, as shown in Figure 3.2. The decrease in amplitude under these conditions is given by

$$A_n = A_0\, e^{-n\delta} \tag{3.48}$$

where

A_n is the amplitude of the nth cycle
A_0 is the amplitude of a cycle near the beginning of decay
n is the number of cycles following the reference cycle at which A_n is measured
δ is the logarithmic decrement

As has been shown, the ratio of energy stored to energy dissipated affects the amplitude of vibration and the decay rate of the oscillations. In addition, it controls the bandwidth of the driven system, defined as the total width of the steady-state response curve measured in hertz between the frequency below resonance, f_1, and that above resonance, f_2, at which the power dissipated is one-half the power at the resonance peak. The ratio of energy stored to energy dissipated is represented by the ratio of the reactive component of the impedance of the system to the resistive component. This ratio is called the Q of the system. In mechanical terms,

$$Q = \frac{\omega_r^M}{R_m} = \frac{f_r}{(f_2 - f_1)} = \frac{\pi}{\delta} \tag{3.49}$$

where

ω_r is the angular frequency at resonance ($=2\pi f_r$)
M is the equivalent mass of the system
R_m is the mechanical loss factor, or equivalent resistance

If the resistive component is very small compared with the reactance, the bandwidth is approximately the product of the angular frequency at resonance and the inverse of the Q.

The effect of Q on both the time and the frequency domain characteristics is illustrated in Figure 3.6. For high-Q the bandwidth is narrow, whereas for low-Q bandwidth is wide. The corresponding forms of the time-domain signals are shown in Figure 3.6c and d. The low-Q gives a short pulse, whereas the high-Q gives a long pulse. Examples of normalized displacements as a function of Q are shown in Figure 3.7. The significance of Q for transducer characteristics is discussed further in Chapter 5.

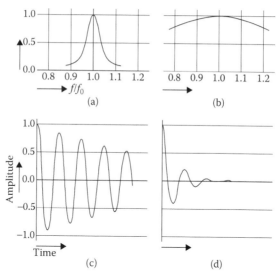

FIGURE 3.6 Effect of Q on oscillator characteristics (a) Narrow spectrum with corresponding long lightly damped (high Q) tone–burst, shown as time domain signal in (c). (b) Wide spectrum with corresponding highly damped (low Q) short pulse time domain signal shown in (d). (a) High Q and (b) low Q: frequency response curves (c) High Q and (d) Low Q: decrement curves. $Q = 20$ for (a) and (c); $Q = 2$ for (b) and (d). (From Wells, P. T. N. *Physical Principles of Ultrasonic Diagnosis*, Academic Press, New York, 1969. With permission.)

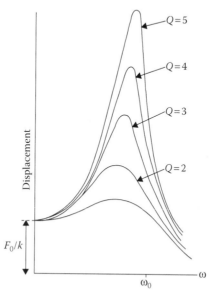

FIGURE 3.7 Response in terms of quality factor Q of the system, where Q is amplification at resonance of low-frequency response $x = F_0/s$. (From Pain, H. J., *The Physics of Vibrations and Waves*, 5th Ed., John Wiley, Chichester, UK, 1999. With permission.)

3.5   TRANSVERSE-WAVE EQUATION

The form of the wave equation that applies to bars vibrating in flexure (see Figure 3.8), that is, the transverse-wave equation, is

$$\frac{\partial^2 \xi}{\partial t^2} = -\kappa^2 c^2 \frac{\partial^4 \xi}{\partial x^4} \tag{3.50}$$

where ξ is the transverse displacement of the bar from its normal configuration (m), κ is the radius of gyration of the cross section of the bar (m), and c is $\sqrt{Y/\rho}$ (m/s). Equation 3.50 is derived in much the same manner as was Equation 3.20 except that bending moments (M) and shear forces (F_y) are considered rather than the simple longitudinal forces considered previously. Even by limiting the displacement and slope of the bar to small values so that variations in angular momentum are negligible, Equation 3.50 is still an approximation in which second-order terms involving $(dx)^2$ are ignored. The procedure is as follows (referring to Figure 3.8). At static equilibrium, moments about the left end of the segment of Figure 3.8 are

$$M_x - M_{(x+dx)} - (F_y)_{(x+dx)}\, dx = 0 \tag{3.51}$$

For small dx

$$M_{(x+dx)} = M_x + \frac{\partial M}{\partial x} dx$$

and

$$(F_y)_{(x+dx)} = (F_y)_x + \frac{\partial F_y}{\partial x} dx$$

These expressions introduced into Equation 3.51 give

$$M_x - M_x - \frac{\partial M}{\partial x} dx - (F_y)_x\, dx - \left(\frac{\partial F_y}{\partial x} dx\right) dx = 0$$

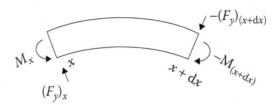

FIGURE 3.8   Bending moments and shear forces in a bar in flexure.

from which

$$F_y = \frac{\partial M}{\partial x} = YS\kappa^2 \frac{\partial^3 \xi}{\partial x} \tag{3.52}$$

if the second-order term involving $(dx)^2$ is neglected. Equation 3.52 approximates the true relationship between F_y and y only if the displacement and slope of the bar are small enough that variations in angular momentum may be neglected.

The net upward force dF_y acting on the segment dx is

$$dF_y = (F_y)_x - (F_y)_{(x+dx)} = -\left(\frac{\partial F_y}{\partial x}\right)dx = -YS\kappa^2 \frac{\partial^4 \xi}{\partial x^4}dx \tag{3.53}$$

which will give the segment an upward acceleration, or

$$dF_y = (\rho S\,dx)\frac{\partial^2 \xi}{\partial t^2} = -YSk^2 \frac{\partial^4 \xi}{\partial x^4}dx \tag{3.54}$$

or

$$\frac{\partial^2 \xi}{\partial t^2} = -\kappa^2 c^2 \frac{\partial^4 \xi}{\partial x^4} \tag{3.50}$$

3.6　SOLUTION OF THE TRANSVERSE-WAVE EQUATION

The general solution of the transverse-wave equation (Equation 3.50) is

$$y = e^{j\omega t}\left(\bar{A}e^{\omega x/v} + \bar{B}e^{-\omega x/v} + \bar{C}e^{j(\omega x/v)} + \bar{D}e^{-j(\omega x/v)}\right) \tag{3.55}$$

where

$\bar{A}, \bar{B}, \bar{C},$ and $\bar{D}$ are complex amplitude constants
v is $\sqrt{\omega c \kappa}$ (phase velocity)
c is $\sqrt{Y/\rho}$ (wave velocity)
κ is radius of gyration of the cross section [3].

It can be seen that phase velocity v is a function of frequency as well as geometry. Since this is true, when a wave of transverse vibrations containing many frequency components is impressed on a bar, the wave shape changes as the wave progresses along the bar because the higher frequency components travel faster than the lower frequency components.

Only the real part of Equation 3.55 is of any consequence in practical applications. Equation 3.55 may, therefore, be rewritten in the form

$$y = (\cos \omega t + j \sin \omega t)\left[\bar{A}\left(\cosh \frac{\omega x}{v} + \sinh \frac{\omega x}{v}\right)\right.$$

$$+\bar{B}\left(\cosh \frac{\omega x}{v} - \sinh \frac{\omega x}{v}\right) + \bar{C}\left(\cos \frac{\omega x}{v} + j \sin \frac{\omega x}{v}\right)$$

$$\left.+\bar{D}\left(\cos \frac{\omega x}{v} - j \sin \frac{\omega x}{v}\right)\right] \tag{3.55a}$$

$$= \cos(\omega t + \phi)\left[(\bar{A} + \bar{B})\cosh \frac{\omega x}{v} + (\bar{A} - \bar{B})\sinh \frac{\omega x}{v}\right.$$

$$\left.+(\bar{C} + \bar{D})\cos \frac{\omega x}{v} + j(\bar{C} - \bar{D})\sin \frac{\omega x}{v}\right]$$

Letting

A equal the real component of $(\bar{A} + \bar{B})$
B equal the real component of $(\bar{A} - \bar{B})$
C equal the real component of $(\bar{C} + \bar{D})$
D equal the real component of $j(\bar{C} - \bar{D})$

we can write

$$\xi = \cos(\omega t + \phi)\left[A\cosh \frac{\omega x}{v} + B\sinh \frac{\omega x}{v} - C\cos \frac{\omega x}{v} + D\sin \frac{\omega x}{v}\right] \tag{3.56}$$

which contains only real amplitude terms A, B, C, and D. This simplification is justified by the harmonic form of ξ. In Equation 3.55a, ϕ is the initial phase angle of the motion. An example of the solution of these equations is for the case of a clamped–free uniform bar.

3.6.1 CLAMPED–FREE UNIFORM BAR

A clamped–free bar, such as a cantilever, is rigidly mounted at one end so that it is perfectly restrained from moving at that end but unrestrained and free to vibrate at the other end. Under these conditions, at the clamped end ($x = 0$), the displacement ξ is zero and the slope $\partial \xi / \partial x$ is zero for all values of t. At the free end ($x = \ell$), no external force is applied and both the bending moment and the shearing force are zero in a plane through the end face of the bar. Therefore, at the free end ($x = \ell$), $\partial^2 \xi / \partial x^2 = 0$ and $\partial^3 \xi / \partial x^3 = 0$.

Applying the boundary conditions at $x = 0$ to Equation 3.56, we obtain

$$\xi = 0 = A + C \quad \text{or} \quad A = -C$$

and

$$\frac{\partial \xi}{\partial x} = 0 = B + D \quad \text{or} \quad B = -D$$

from which

$$\xi = \cos(\omega t + \phi)\left[A\left(\cosh\frac{\omega x}{v} - \cos\frac{\omega x}{v} \right) + B\left(\sinh\frac{\omega x}{v} - \sin\frac{\omega x}{v} \right) \right] \quad (3.57)$$

Applying the boundary conditions at $x = \ell$ to Equation 3.57 gives

$$\frac{\partial^2 \xi}{\partial x^2} = 0 = \cos(\omega t + \phi)\left(\frac{\omega}{v} \right)^2 \left[A\left(\cosh\frac{\omega \ell}{v} + \cos\frac{\omega \ell}{v} \right) \right.$$
$$\left. + B\left(\sinh\frac{\omega \ell}{v} + \sin\frac{\omega \ell}{v} \right) \right] \quad (3.58)$$

and

$$\frac{\partial^3 \xi}{\partial x^3} = 0 = \cos(\omega t + \phi)\left(\frac{\omega}{v} \right)^3 \left[A\left(\sinh\frac{\omega \ell}{v} - \sin\frac{\omega \ell}{v} \right) \right.$$
$$\left. + B\left(\cosh\frac{\omega \ell}{v} + \cos\frac{\omega \ell}{v} \right) \right] \quad (3.59)$$

Therefore

$$A\left(\cosh\frac{\omega \ell}{v} + \cos\frac{\omega \ell}{v} \right) = -B\left(\sinh\frac{\omega \ell}{v} + \sin\frac{\omega \ell}{v} \right) \quad (3.58a)$$

and

$$A\left(\sinh\frac{\omega \ell}{v} - \sin\frac{\omega \ell}{v} \right) = -B\left(\cosh\frac{\omega \ell}{v} + \cos\frac{\omega \ell}{v} \right) \quad (3.59a)$$

which can be true only at discrete frequencies. These discrete, or allowed, frequencies are determined by dividing Equation 3.58a by Equation 3.59a, as follows:

$$\frac{\cosh(\omega \ell/v) + \cos(\omega \ell/v)}{\sinh(\omega \ell/v) - \sin(\omega \ell/v)} = \frac{\sinh(\omega \ell/v) + \sin(\omega \ell/v)}{\cosh(\omega \ell/v) + \cos(\omega \ell/v)}$$

from which

$$\cosh^2\frac{\omega \ell}{v} + 2\cosh\frac{\omega \ell}{v}\cos\frac{\omega \ell}{v} + \cos^2\frac{\omega \ell}{v} = \sinh^2\frac{\omega \ell}{v} - \sin^2\frac{\omega \ell}{v} \quad (3.60)$$

or

$$\cosh\frac{\omega \ell}{v}\cos\frac{\omega \ell}{v} = -1$$

Thus the resonance frequencies of a clamped–free bar resonating in flexure are determined by Equation 3.60a.

From the identities

$$\cosh 2\theta = \cosh^2 \theta + \sinh^2 \theta$$

$$\cos 2\theta = \cos^2 \theta - \sin^2 \theta$$

Equation 3.60a may be rewritten

$$\left(\cosh^2 \frac{\omega\ell}{2v} + \sinh^2 \frac{\omega\ell}{2v}\right)\left(\cos^2 \frac{\omega\ell}{2v} - \sin^2 \frac{\omega\ell}{2v}\right) = -1 \tag{3.61}$$

Dividing both sides of Equation 3.61 by $\cosh^2(\omega\ell/2v)\cos^2(\omega\ell/2v)$ gives

$$\left(1 + \tanh^2 \frac{\omega\ell}{2v}\right)\left(1 - \tan^2 \frac{\omega\ell}{2v}\right) = -\frac{1}{\cosh^2(\omega\ell/2v)\cos^2(\omega\ell/2v)}$$

$$= -\left(1 - \tanh^2 \frac{\omega\ell}{2v}\right)\left(1 + \tan^2 \frac{\omega\ell}{2v}\right) \tag{3.62}$$

or

$$\tan^2 \frac{\omega\ell}{2v} \tanh^2 \frac{\omega\ell}{2v} = 1$$

or

$$\cot \frac{\omega\ell}{2v} = \pm \tanh \frac{\omega\ell}{2v}$$

The intersections of the curves corresponding to $\cot \omega\ell/2v$ and $\pm\tanh \omega\ell/2v$ plotted against $(\omega\ell/2v)$ give the frequencies corresponding to the allowed modes of vibration of the clamped–free uniform bar. These curves are plotted in Figure 3.9.

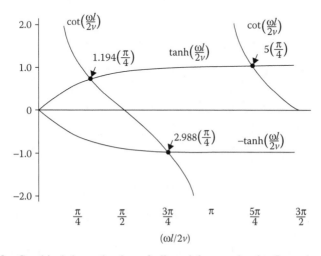

FIGURE 3.9 Graphical determination of allowed frequencies for flexural resonances in fixed–free uniform bars.

The intersections occur at

$$\frac{\omega \ell}{2v} = \frac{\pi}{4}(1.194, 2.988, 5, 7, \ldots) \tag{3.63}$$

The numerals 5, 7, ... follow from the fact that the hyperbolic tangent is nearly unity for all angles greater than π.

Since

$$v = \sqrt{\omega c \kappa} \quad \text{or} \quad v^2 = \omega c \kappa = 2\pi f c \kappa$$

it follows from Equation 3.63 that

$$f = \frac{\pi c \kappa}{8\ell^2}(1.194^2, 2.988^2, 5^2, 7^2, \ldots) \tag{3.64}$$

are allowed resonant frequencies of a clamped–free bar vibrating in flexure.

The nodes of the clamped–free bar (corresponding to $\xi = 0$) are not evenly distributed along the bar. Since the hypothesis states that the bar is clamped at one end ($x = 0$), the clamped end is a nodal position. Here, both $\xi = 0$ and $\partial\xi/\partial x = 0$. This is the only node occurring at the fundamental frequency. One additional node occurs near the free end at the first overtone. This node is characterized by $\xi = 0$ but $\partial\xi/\partial x \neq 0$ and $\partial^2\xi/\partial x^2 \neq 0$. Therefore, two different types of nodes exist for this mode. The second overtone includes nodes of these two types, one at the clamped end and the other near the free end, plus a third node characterized by $\xi = 0$ and $\partial^2\xi/\partial x^2 = 0$.

Each successive overtone adds an additional node characterized by the conditions $\xi = 0$ and $\partial^2\xi/\partial x^2 = 0$, while also retaining the two types of nodes near the extremities of the lower modes.

These intermediate nodes associated with the second overtone and higher modes $(f_3, f_4, f_5, \ldots, f_n)$ can be determined by the equations

$$\xi_n = \cos(\omega t + \phi_n)\left[A\left(\cosh\frac{\omega x}{v} - \cos\frac{\omega x}{v}\right) + B\left(\sinh\frac{\omega x}{v} - \sin\frac{\omega x}{v}\right)\right] \tag{3.57a}$$

$$= 0 \ (n = 3, 4, 5, \ldots)$$

and

$$\frac{\partial^2\xi_n}{\partial x^2} = \cos(\omega t + \phi_n)\left(\frac{\omega}{v}\right)^2\left[A\left(\cosh\frac{\omega x}{v} + \cos\frac{\omega x}{v}\right)\right.$$

$$\left. + B\left(\sinh\frac{\omega x}{v} + \sin\frac{\omega x}{v}\right)\right] = 0 \tag{3.58a}$$

Therefore,

$$A\left(\cosh\frac{\omega x}{v} - \cos\frac{\omega x}{v}\right) = -B\left(\sinh\frac{\omega x}{v} - \sin\frac{\omega x}{v}\right) \tag{3.57b}$$

$$A\left(\cosh\frac{\omega x}{v}+\cos\frac{\omega x}{v}\right)=-B\left(\sinh\frac{\omega x}{v}+\sin\frac{\omega x}{v}\right) \qquad (3.58b)$$

Dividing Equation 3.57b by Equation 3.58b and simplifying gives

$$2\cosh\frac{\omega v}{v}\sin\frac{\omega x}{v}=2\sinh\frac{\omega x}{v}\cos\frac{\omega x}{v} \qquad (3.65)$$

or

$$\tan\frac{\omega x}{v}=\tanh\frac{\omega x}{v}$$

Equation 3.65 is solved graphically as was Equation 3.62 for discrete values of x corresponding to $y=0$. The quantity $\omega x/v$ is greater than π and lies only in the quadrants where $\tan \omega x/v$ is positive, that is, where

$$\frac{\omega x}{v}=\frac{5\pi}{4},\frac{9\pi}{4},\frac{13\pi}{4},\dots \qquad (3.66)$$

and

$$\frac{\omega v}{v}=\frac{\pi}{2\ell}(5,7,9,\dots) \qquad (3.63a)$$

corresponding to the second, third, fourth, and so on, overtones. For example, the nodal positions for the second overtone are as follows:

First nodal position: $x=0$
Second nodal position:

$$\frac{\omega x}{v}=\frac{5\pi}{4}$$

and

$$\frac{\omega}{v}=\frac{5\pi}{2\ell} \quad \text{(from Equation 3.63)}$$

and

$$\frac{5\pi}{2\ell}x=\frac{5\pi}{4}$$

$x=0.5\,\ell$ (corresponding to f_3).

The third nodal position of the second overtone is not accurately determined by Equation 3.65 because the condition that $\partial^2\xi/\partial x^2 = 0$, on which Equation 3.65 is based, does not exist at the node nearest the free end. Rather $\partial^2\xi/\partial x^2 = 0$ occurs at the free end. Neither is $\partial\xi/\partial x = 0$. However, ξ must be zero for a node to exist. Equations 3.63a and 3.66 may be used to obtain a first approximation to the location of the third node of the second overtone, and this approximate value is $x = 0.9\,\ell$, which would be the true value if $\partial^2\xi/\partial x^2 = 0$ at that position. The actual location of the third node is at $x = 0.868\,\ell$.

TABLE 3.1

Locations of Nodes and Corresponding Phase Velocities of a Clamped–Free Uniform Bar in Transverse Vibration

Frequency	Phase Velocity	Nodal Positions
f_1	v_1	$x = 0$
$f_2(6.267f_1)$	$2.50v_1$	$x = 0, 0.774\,\ell$
$f_3(17.55f_1)$	$4.18v_1$	$x = 0, 0.58\,\ell, 0.868\,\ell$
$f_4(34.39f_1)$	$5.87v_1$	$x = 0, 0.356\,\ell, 0.644\,\ell, 0.905\,\ell$

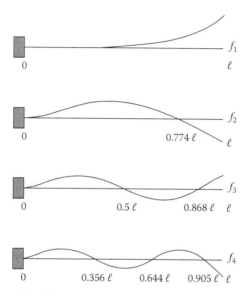

FIGURE 3.10   The modes of transverse vibration of a clamped–free uniform bar.

Actual nodal positions, corresponding frequencies, and phase velocities are given in Table 3.1. The modes of transverse vibrations of a clamped–free bar are shown in Figure 3.10.

3.6.2 FREE–FREE BAR (BAR FREE AT BOTH ENDS)

Again referring to the general solution of the wave equation for a bar in transverse vibration (Equation 3.56), the boundary conditions for a free–free bar are $\partial^2\xi/\partial x^2 = 0$ and $\partial^3\xi/\partial x^3 = 0$ at both $x = 0$ and $x = \ell$.

Applying these conditions at $x = 0$,

$$\frac{\partial^2\xi}{\partial x^2} = \cos(\omega t + \phi)\frac{\omega^2}{v^2}(A\cosh 0 + B\sinh 0 - C\cos 0 - D\sin 0) = 0$$

and

$$\frac{\partial^3 \xi}{\partial x^3} = \cos(\omega t + \phi)\frac{\omega^3}{v^3}(A \sinh 0 + B \cosh 0 + C \sin 0 - D \cos 0) = 0 \qquad (3.67)$$

from which $A - C = 0$ and $B - D = 0$. Therefore,

$$\xi = \cos(\omega t + \phi)\left[A\left(\cosh\frac{\omega x}{v} + \cos\frac{\omega x}{v}\right) + B\left(\sin\frac{\omega x}{v} + \sin\frac{\omega x}{v}\right)\right] \qquad (3.68)$$

is the equation of motion of a free–free bar in flexure.

At $x = \ell$,

$$\frac{\partial^2 \xi}{\partial x^2} = \cos(\omega t + \phi)\frac{\omega^2}{v^2}\left(A\cosh\frac{\omega\ell}{v} + B\sinh\frac{\omega\ell}{v} - A\cos\frac{\omega\ell}{v} - B\sin\frac{\omega\ell}{v}\right) = 0$$

and

$$\frac{\partial^3 \xi}{\partial x^3} = \cos(\omega t + \phi)\frac{\omega^3}{v^3}\left(A\sinh\frac{\omega\ell}{v} + B\cosh\frac{\omega\ell}{v} - A\sin\frac{\omega\ell}{v} - B\cos\frac{\omega\ell}{v}\right) = 0 \quad (3.69)$$

Using the procedure in Section 3.6.1 for the clamped–free bar, we obtain

$$\cosh\left(\frac{\omega\ell}{v}\right)\cos\left(\frac{\omega\ell}{v}\right) = 1$$

$$\tan\left(\frac{\omega\ell}{2v}\right) = \pm\tanh\left(\frac{\omega\ell}{2v}\right) \qquad (3.70)$$

from which the allowed frequencies are obtained. The allowed frequencies corresponding to the intersection of $\tan(\omega\ell/2v)$ with $\pm\tanh(\omega\ell/2v)$ plotted against $\omega\ell/2v$ are given by

$$\frac{\omega\ell}{2v} = \frac{\pi}{4}(3.0112, 5, 7, 9, \ldots) \qquad (3.71)$$

and since

$$v = \sqrt{\omega c \kappa}$$

$$f = \frac{\pi c \kappa}{8\ell^2}(3.0112^2, 5^2, 7^2, 9^2, \ldots) \qquad (3.72)$$

As with the clamped–free bar, the overtones are not harmonics. The nodes are of two types. Those nearest the ends are characterized by $\xi = 0$, but $\partial\xi/\partial x \neq 0$ and $\partial^2\xi/\partial x^2 \neq 0$. The nodes between those nearest the ends are characterized by $\xi = 0$ and $\partial^2\xi/\partial x^2 = 0$, and the same procedure as used to determine the locations of the nodes of a clamped–free bar is used here. In comparison with the clamped–free bar, two nodes instead of one are associated with the fundamental frequency. The first overtone produces three, and so on. Table 3.2 gives the relative frequencies, phase velocities, and positions of nodes of the fundamental and the first three overtones.

The corresponding modes are plotted in Figure 3.11. As one would expect, the positions of the nodes are symmetrical about the center of the bar for all modes. The shapes of the displacement curves for all modes corresponding to odd-numbered frequencies, including the fundamental f_1, are symmetrical about the center. However, the modes corresponding to even-numbered frequencies are asymmetrical, that is, not symmetrical about the center. The latter modes exhibit true nodes at the center.

TABLE 3.2

Vibration Characteristics of a Free–Free Uniform Bar in Flexure

Frequency	Phase Velocity	Nodal Positions
f_1	v_1	$x = 0.224\ell, 0.776\ell$
$f_2 = (2.756f_1)$	$1.66v_1$	$x = 0.132\ell, 0.5\ell, 0.868\ell$
$f_3 = (5.404f_1)$	$2.32v_1$	$x = 0.0942\ell, 0.356\ell, 0.644\ell, 0.906\ell$
$f_4 = (8.933f_1)$	$2.99v_1$	$x = 0.073\ell, 0.277\ell, 0.5\ell, 0.723\ell, 0.927\ell$

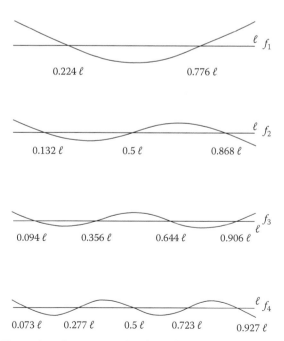

FIGURE 3.11   The modes of transverse vibrations of a free–free uniform bar.

3.6.3 CLAMPED–CLAMPED BAR (BAR CLAMPED AT BOTH ENDS)

A clamped–clamped bar, that is, a bar which is rigidly clamped at each end, will vibrate at the same frequencies as the free–free bar of equal dimensions. Here, again, two types of nodes occur. The nodes at the ends are characterized by $\xi = 0$ and $\partial\xi/\partial x = 0$. Those in between are characterized by $\xi = 0$ and $\partial^2\xi/\partial x^2 = 0$. The nodal positions do not coincide with those of the free–free bar but are as shown in Table 3.3.

3.6.4 EFFECT OF GEOMETRY ON TRANSVERSE VIBRATIONS OF BARS

The effect of geometry on the vibratory characteristics of a bar depends upon the corresponding distribution of the mass of the bar.

The fundamental resonant frequencies of various tapered bars, which are rigidly mounted at their large end, are listed in Table 3.4. In Table 3.4, a is the radius of the cone at the base, in meter, Y is Young's modulus, ρ is the density (kg/m³), and b is the thickness of the bar at the base in the direction of vibration, in meter.

Further discussion of wave equations, vibrations and waves in rods and bars are given in references [1,4].

TABLE 3.3
Locations of Nodes and Corresponding Phase Velocities of a Fixed–Fixed Uniform Bar Resonant in Transverse Vibration

Frequency	Phase Velocity	Nodal Positions
f_1	v_1	$0, \ell$
$f_2 = (2.756f_1)$	$1.66v_1$	$0, 0.5\ell, \ell$
$f_3 = (5.404f_1)$	$2.32v_1$	$0, 0.357\ell, 0.643\ell, \ell$
$f_4 = (8.933f_1)$	$2.99v_1$	$0, 0.277\ell, 0.5\ell, 0.723\ell, \ell$
$f_5 = (13.3446f_1)$	$3.653v_1$	$0, 0.227\ell, 0.409\ell, 0.591\ell, 0.773\ell, \ell$

TABLE 3.4
Resonant Frequencies of Tapered Bars Rigidly Mounted at the Large End

Geometry	Direction of Vibration	Frequency (Hz)
Wedge-shaped bar	Normal to parallel sides	$\dfrac{1.14}{\ell^2}\sqrt{\dfrac{Yb^2}{12\rho}}$
Wedge-shaped bar	Parallel to parallel sides	$\dfrac{0.85}{\ell^2}\sqrt{\dfrac{Yb^2}{12\rho}}$
Conical bar		$\dfrac{1.39}{\ell^2}\sqrt{\dfrac{Ya^2}{4\rho}}$

3.7 PLATE WAVES

3.7.1 GENERAL

The form of the wave equation for plates is

$$\nabla^4\xi \frac{3\rho(1-\sigma^2)}{Yh^2}\frac{\partial^2\xi}{\partial t^2} = 0 \qquad (3.73)$$

where

ξ is the displacement amplitude
ρ is the density of the material of the plate
Y is the modulus of elasticity (Young's modulus)
σ is the Poisson's ratio
h is the half-thickness of the plate
t is time

The equation for the bar vibrating in flexure, Equation 3.50 (Section 3.5), may be derived from Equation 3.73 by applying the boundary conditions of a thin bar in flexure. The thin bar in flexure is not constrained laterally by adjacent material. The stresses are assumed to be directed in planes that are parallel to the direction of bending. The directions of stress are (1) parallel to the length of the bar and (2) parallel to the direction of bending.

The stresses developed in plates in flexure are more complicated than those for the flexural bar because each incremental element in a plate under stress may be subjected to constraints in all directions. Hence, the term involving Poisson's ratio, which is incorporated in Equation 3.73.

Solutions to the plate equation allow an infinite number of vibrational conditions. For instance, the simple harmonic solution given by Morse [5] is

$$\xi(r,\phi) =^{\cos}_{\sin} (m\Phi)[AJ_m(\gamma r) + BI_m(\gamma r)] \qquad (3.74)$$

where

J_m is a cylindrical Bessel function
I_m is a hyperbolic Bessel function
γ is the wavelength constant for a plate

For the circular plate of uniform thickness and radius a clamped at the edge, the boundary conditions are that both ξ and $\partial\xi/\partial r$ are zero at $r = a$. To satisfy these conditions, from Equation 3.74

$$\xi(a,\phi) =^{\cos}_{\sin} (m\Phi)[AJ_m(\gamma a) + BI_m(\gamma a)] = 0$$

$$B = -A\left[\frac{J_m(\gamma a)}{I_m(\gamma a)}\right] \qquad (3.75)$$

and

$$I_m(\gamma a)\frac{d}{dr}J_m(\gamma r) = J_m(\gamma a)\frac{d}{dr}I_m(\gamma r) \tag{3.76}$$

As with the flexural bar, these conditions can be satisfied only at discrete frequencies. Again referring to Morse [5], these frequencies are fixed by the values

$$\beta_{01} = 1.015 \quad \beta_{02} = 2.007 \quad \beta_{03} = 3.00$$
$$\beta_{11} = 1.468 \quad \beta_{12} = 2.483 \quad \beta_{13} = 3.490$$
$$\beta_{21} = 1.879 \quad \beta_{22} = 2.992 \quad \beta_{23} = 4.00 \tag{3.77}$$
$$\beta_{mn} \xrightarrow[n\to\infty]{} n + \frac{m}{2}$$

where $\gamma_{mn} = (\pi/a)\beta_{mn}$. Therefore, the allowed frequencies for the normal plate clamped at its circumference are

$$f_{mn} = \frac{\pi h}{2a^2}\sqrt{\frac{Y}{3\rho(1-\sigma^2)}}(\beta_{mn})^2 \tag{3.78}$$

where the number of nodal circles is n, including the clamped edge as a nodal circle, and the number of nodal diameters is m.

From the values for β_{mn} given in Equation 3.77, the allowed frequencies are

$$f_{01} = 0.9342\left(\frac{h}{a^2}\right)\sqrt{\frac{Y}{\rho(1-\sigma^2)}} \tag{3.79}$$

and

$$f_{mn} = \left(\frac{\beta_{mn}}{\beta_{01}}\right)^2 f_{01} = 0.9707(\beta_{mn})^2 f_{01}$$

From the β_{mn} values in Equation 3.77 and from Equation 3.79, one may easily see that the overtones of the clamped plate are not harmonics. It can be shown similarly that this is generally true for all other boundary conditions for vibrating plates. Therefore, the velocity of propagation of plate waves is dependent not only upon the elastic properties and the density but also upon the frequency of the oscillations.

3.7.2 LAMB WAVES

Lamb waves are a form of plate wave and are discussed in detail by several authors [6]. As may be deduced from the previous discussions, there is an infinite number of Lamb waves, each propagating at a velocity dictated by its frequency, thickness, density, and elasticity. In fact, two types are recognized—symmetrical and asymmetrical—and there also exists an infinite number of each of these types. Figure 3.12 illustrates the particle motion in each of the two types.

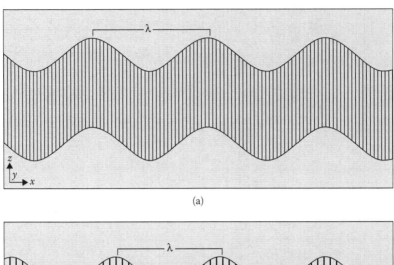

(a)

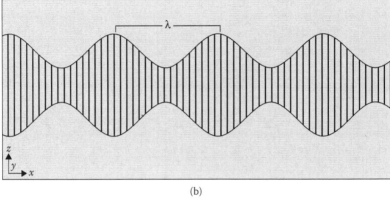

(b)

FIGURE 3.12 Displacement characteristics of Lamb waves. (a) A flexural, antisymmetric or asymmetrical mode in a plate. (b) An extensional or compressional (symmetric), mode in a plate.

Lamb-wave equations may be derived either from Equation 3.73 or from the general wave equation (Equation 3.22 in Section 3.3). Equations for shear waves and for longitudinal (or compressional) waves again may be written in the form of Equations 3.80 and 3.81, that is,

$$\nabla^2 \Phi = \frac{1}{c_L^2} \frac{\partial^2 \phi}{\partial t^2} \quad (compressional\ waves) \tag{3.80}$$

$$\nabla^2 \psi = \frac{1}{c_s^2} \frac{\partial^2 \psi}{\partial t^2} \quad (shear\ waves) \tag{3.81}$$

Solutions to Equations 3.80 and 3.81 for infinite plates bounded by two planes parallel to and equidistant from the yz axis are [7]

$$\phi_1 = \left(A \cosh \frac{2\pi}{\lambda} \sqrt{1 - \frac{c_p^2}{c_L^2}}\, x \right) e^{j(\omega t - 2\pi z/\lambda)}$$

(symmetrical waves)

$$\psi_1 = \left(B \sinh \frac{2\pi}{\lambda} \sqrt{1 - \frac{c_p^2}{c_s^2}} x \right) e^{j(\omega t - 2\pi z/\lambda)} \tag{3.82}$$

$$\phi_2 = \left(A \sinh \frac{2\pi}{\lambda} \sqrt{1 - \frac{c_p^2}{c_L^2}} x \right) e^{j(\omega t - 2\pi z/\lambda)}$$

(asymmetrical waves)

$$\psi_2 = \left(B \cosh \frac{2\pi}{\lambda} \sqrt{1 - \frac{c_p^2}{c_s^2}} x \right) e^{j(\omega t - 2\pi z/\lambda)} \tag{3.83}$$

where c_L is the velocity of a longitudinal wave in the material of the plate, c_s is the velocity of shear waves, and c_p is the phase velocity.

The boundary conditions are that the stresses at the surfaces of the plate ($x = \pm h$) are zero. As used in Chapter 2 in the derivations of the relative amplitudes of reflected and refracted waves from a boundary, Hooke's law for isotropic media may be written

$$P_{xx} = (\lambda' + 2\mu)\frac{\partial v}{\partial x} + \lambda'\frac{\partial u}{\partial z}$$

$$= \rho\left[c_L^2 \frac{\partial v}{\partial x} + (c_L^2 - 2c_s^2)\frac{\partial u}{\partial z} \right] \tag{3.84}$$

$$P_{zx} = \rho c_s^2 \left(\frac{\partial u}{\partial x} + \frac{\partial v}{\partial z} \right)$$

where

λ' is Lamé's constant
μ is the shear modulus
u is $\partial\phi/\partial z + \partial\psi/\partial x$
v is $\partial\phi/\partial x - \partial\psi/\partial z$

Substituting Equation 3.82 in Equation 3.84 and applying the boundary condition that the stress at $x = \pm h$ is zero gives the frequency equation for symmetrical Lamb waves, which is

$$\frac{\tanh \omega h\sqrt{(c_s^2 - c_p^2)/c_L^2 c_p^2}}{\tanh \omega h\sqrt{(c_L^2 - c_p^2)/c_s^2 c_p^2}} = 4\sqrt{\frac{(c_L^2 - c_p^2)(c_s^2 - c_p^2)c_s^2}{(2c_s^2 - c_p^2)^4 c_L^2}} \tag{3.85}$$

$$c_p < c_s < c_L$$

When c_p is greater than c_s or both c_s and c_L, Equation 3.85 becomes imaginary on both sides. The solutions then may be written

$$\frac{\tanh \omega h \sqrt{(c_p^2 - c_s^2)/c_s^2 c_p^2}}{\tanh \omega h \sqrt{(c_L^2 - c_p^2)/c_L^2 c_p^2}} = 4 \sqrt{\frac{(c_L^2 - c_p^2)(c_L^2 - c_s^2)c_s^2}{(2c_s^2 - c_p^2)^4 c_L^2}}$$

$$c_s < c_p < c_L$$ (3.85a)

and

$$\frac{\tanh \omega h \sqrt{(c_p^2 - c_s^2)/c_s^2 c_p^2}}{\tanh \omega h \sqrt{(c_p^2 - c_L^2)/c_L^2 c_p^2}} = -4 \sqrt{\frac{(c_p^2 - c_L^2)(c_p^2 - c_s^2)c_s^2}{(2c_s^2 - c_p^2)^4 c_L^2}}$$

$$c_s < c_L < c_p$$ (3.85b)

Frequency equations for asymmetrical Lamb waves are obtained by substituting Equation 3.83 in Equation 3.84 and applying the boundary conditions of zero stress at $x = \pm h$. For the *asymmetrical Lamb waves*, the frequency equations are

$$\frac{\tanh \omega h \sqrt{(c_s^2 - c_p^2)/c_s^2 c_p^2}}{\tanh \omega h \sqrt{(c_L^2 - c_p^2)/c_s^2 c_p^2}} = \frac{1}{4} \sqrt{\frac{(2c_s^2 - c_p^2)^4 c_L^2}{(c_s^2 - c_p^2)(c_L^2 - c_p^2)c_s^2}}$$

$$c_p < c_s < c_L$$ (3.86a)

$$\frac{\tanh \omega h \sqrt{(c_p^2 - c_s^2)/c_s^2 c_p^2}}{\tanh \omega h \sqrt{(c_L^2 - c_p^2)/c_L^2 c_p^2}} = -\frac{1}{4} \sqrt{\frac{(2c_s^2 - c_p^2)c_L^2}{(c_p^2 - c_s^2)(c_L^2 - c_p^2)c_s^2}}$$

$$c_s < c_p < c_L$$ (3.86b)

and

$$\frac{\tanh \omega h \sqrt{(c_p^2 - c_s^2)/c_s^2 c_p^2}}{\tanh \omega h \sqrt{(c_p^2 - c_L^2)/c_L^2 c_p^2}} = -\frac{1}{4} \sqrt{\frac{(2c_s^2 - c_p^2)^4 c_L^2}{(c_p^2 - c_s^2)(c_p^2 - c_L^2)c_s^2}}$$

$$c_s < c_L < c_p$$ (3.86c)

In Equation 3.84, given the plate thickness $2h$ and the material of the plate, all quantities become constant except c_p and the frequency f, and these are interrelated. The allowed frequencies (f) and velocities (c_p), which also are the velocities of the Lamb waves, are those combinations of f and c_p, which satisfy Equations 3.85 and 3.86.

When a symmetrical or asymmetrical Lamb wave propagates in a plate with a phase velocity c_p equal to either the shear velocity c_s or the longitudinal velocity c_L, the displacement normal to the plate surfaces is zero and the surface vibrates in the horizontal direction only.

When a symmetrical or asymmetrical Lamb wave propagates through a plate with a phase velocity equal to $\sqrt{2}$ times the shear velocity, the displacement of the surfaces is in the vertical direction only, that is, normal to the planes of the surfaces.

The phase velocity c_p of both the first symmetrical mode and the asymmetrical mode approaches the velocity of Rayleigh waves as a lower limit.

Lamb waves are constructed through the superposition of longitudinal and shear waves as shown in Figure 3.13. And an example of a dispersion curve for an isotropic plate is given in Figure 3.14 [8]. Plate waves have received extensive treatment in several texts including those by Redwood [9], and the more mathematical treatment by Miklowitz [10].

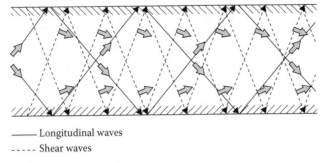

———— Longitudinal waves

- - - - - Shear waves

FIGURE 3.13   Lamb waves: (Rayleigh plate vibrations of the second kind, 1889) constructed by superposing longitudinal and shear partial waves.

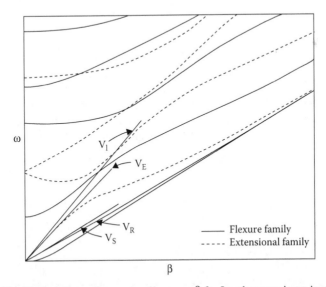

FIGURE 3.14   Typical dispersion curves (ω versus β for Lamb waves in an isotropic plate). (With kind permission from Springer Science+Business Media: Springer Series on Wave Phenomena, eds. E. A. Ash and E. G. S. Paige, Rayleigh wave theory and application, Vol. 2, 1985, B. A. Auld, Figure 5, © Springer-Verlag, Berlin, 1985.)

3.7.3 RAYLEIGH WAVES

It is also possible for ultrasonic waves to propagate on the free-surface of an elastic solid. This form of wave is known as the Rayleigh wave, named after Lord Rayleigh who published his theoretical proof for the existence of such waves in 1885 [11]. When stress-free boundary conditions are applied, a characteristic equation is obtained, which after transformation, reduces to the form given as Equation 3.87: form;

$$\eta^6 - 8\eta^4 + 8(3 - 2v^2)\eta^2 - 16(1 - v^2) = 0$$

where

$$\eta = V_r/V_s; \ v = V_s/V_c. \tag{3.87}$$

This equation is cubic in η^2 and is known as the Rayleigh wave equation. The Rayleigh wave velocity is given from the real roots.

For all real media, Poisson's ratio is subject to the restriction;

$$0 < \sigma < 0.5$$

and this condition ensures that only one root will satisfy the restriction on the values for the V_r/V_s ratio.

A useful approximation for the value of the Rayleigh wave root has been given by Bergmann [12], which provides a method for rapid calculation of the Rayleigh wave velocity.

$$\frac{V_r}{V_s} = \left[\frac{0.87 + 1.12\sigma}{1 + \sigma} \right]$$

Particle displacement at the surface is elliptical, and the energy in a Rayleigh wave decays rapidly with depth. The Rayleigh wave displacement/depth profiles for two different media are shown in Figure 3.15 [13]. Rayleigh waves are important in ultrasonic nondestructive testing/evaluation (NDT/NDE), in geophysics, and in surface acoustic wave devices, which are used in electronics. Extensive treatments of the theory and application of surface/Rayleigh waves are given in several texts (e.g., [14,15]).

3.7.4 FLEXURAL PLATES

Flexural plates used in industrial applications may be any of several designs. The derivation of equations for some of these designs is relatively straightforward, starting with the wave equation for plates, Equation 3.73. Other solutions become extremely cumbersome and difficult without simplifying approximations.

We will not enter into derivations beyond those already covered in this chapter but will present the solutions to some of the more important types that might be used in designing ultrasonic devices. Equations 3.77 through 3.79 are sufficient for designing circular plates clamped at the circumference. In those equations and those to follow,

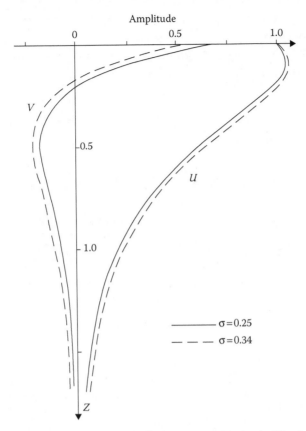

FIGURE 3.15 Rayleigh wave decay of displacements with depth (Z), given by analytic equations, at the points of maximum surface displacement for two media (σ is Poisson's ratio for material).

the plate is assumed to consist of a perfectly elastic, homogeneous, isotropic material of uniform thickness. The thickness is considered to be small in comparison with all other dimensions, that is, width and length or radius.

3.7.4.1 Rectangular Plate with Simply Supported Edges

The equation for the frequencies of all flexural modes of vibration of the rectangular plate simply supported is

$$\omega = \pi^2 h \sqrt{\frac{Y}{3\rho(1-\sigma^2)}} \left(\frac{m^2}{a^2} + \frac{n^2}{b^2} \right) \tag{3.89}$$

where, as before,

h is the half-thickness of the plate
Y is Young's modulus of the material of the plate
ρ is the density of the material of the plate

σ is Poisson's ratio of the material of the plate

m and n are integers

a and b are the lengths of the sides of the rectangle

For example, for the lowest mode of vibration of a square plate simply supported at its edges,

$$\omega_1 = \frac{2\pi^2 h}{a^2} \sqrt{\frac{Y}{3\rho(1-\sigma^2)}} \qquad (3.89a)$$

Other conditions, such as (1) the case of forced vibrations of a rectangular plate with simply supported edges and (2) the case of a rectangular plate of which two opposite edges are supported while the other two edges are free or clamped, are fairly easy to solve.

Vibration of rectangular plates with all edges free or clamped is very complicated. Approximate solutions have been derived using a method attributed to Ritz [16]. These solutions will not be presented here. The reader may obtain help from references such as Timoshenko's *Vibration Problems in Engineering* [17].

Flexural bars designed according to equations provided in Section 3.6 of this chapter will fulfill most needs requiring flexural applications of ultrasonic energy.

3.7.4.2   Free Circular Plate

The frequencies of any vibrating circular plate can be calculated using Equations 3.78 and 3.79, provided the applicable constants, β_{mn}, for each mode and each set of boundary conditions are known.

For a free circular plate with m nodal diameters and n nodal circles, the values of β_{mn} to use with Equation 3.78 are

		$\beta_{20} = 0.7294$	$\beta_{30} = 1.1132$
$\beta_{01} = 0.9590$	$\beta_{11} = 1.4419$	$\beta_{21} = 1.8896$	$\beta_{31} = 2.3154$
$\beta_{02} = 1.9756$	$\beta_{12} = 2.4627$	—	—

assuming Poisson's ratio is 0.333.

3.7.4.3   Circular Plate with Its Center Fixed

The values of β_{mn} for a circular plate with its center fixed and with n nodal circles are

$$\beta_{00} = 0.6164$$
$$\beta_{01} = 1.4556$$
$$\beta_{02} = 2.4796$$
$$\beta_{03} = 3.4825$$

The frequencies of vibration for circular plates with center fixed and having nodal diameters are the same as those for vibrations in a free plate.

Flexural free plates are useful for continuous processes such as continuous seam welding. Derivations of these quantities have not taken into account the effects of rotational inertia of the moving point contact at the periphery of the disks.

3.7.4.4   Finite Exciting Sources (Transducers)

Recent years have seen a significant development in the theoretical analysis for the vibration of, and wave propagation resulting from, a range of finite sources. Such sources are fundamental to the analysis of transducers and wave fields used in both NDT and medical ultrasound. Extensive treatments of vibrating sources are given by Kino [18], and this topic is considered further in Chapter 5, which considers transducers and the wave fields that are generated.

REFERENCES

1. Pain, H. J. 1999. *The Physics of Vibrations and Waves*. 5th ed. Chichester, UK: Wiley.
2. Auld, B. A. 1973. *Acoustic Fields and Waves in Solids*. Vol. I and II. New York: John Wiley.
3. Wells, P. N. T. 1969. *Physical Principles of Ultrasonic Diagnosis*. London and New York: Academic Press.
4. Kinsler, L. E., A. R. Frey, A. B. Coppens, and J. V. Sanders. 2000. *Fundamentals of Acoustics*. 4th ed. New York: Wiley.
5. Morse, P. M. 1948. *Vibration and Sound*. New York: McGraw-Hill.
6. Rose, J. L. 1999. *Ultrasonic Waves in Solid Media*. Cambridge: Cambridge University Press.
7. Worlton, D. C. 1969. Lamb waves at ultrasonic frequencies. Hanford Atomic Products Operation, Report HW-60662, Richland, WA (U.S. Atomic Energy Commission) June 6, 1969.
8. Auld, B. A. 1985. Rayleigh wave propagation. In *Rayleigh Wave Theory and Application*, ed. E. A. Ash and E. G. S. Paige, Vol. 2. Berlin: Springer Series on Wave Phenomena, Springer-Verlag.
9. Redwood, M. 1960. *Mechanical Waveguides*. New York: Pergamon Press.
10. Miklowitz, J. 1978. The theory of elastic waves and waveguides. In *North-Holland Applied Mathematics and Mechanics Series*, ed. H. A. Lauwerier and W. T. Koiter. Amsterdam: North-Holland.
11. Lord Rayleigh, S. 1885. *Proc Lond Math Soc* 17:4.
12. Bergmann, L. 1949. Ultraschall und seine anwendung in wissenschaft und technik. Edwards Ann. Arbor Mich. (In German) 1949.
13. Bond, L. J. 1978. *Surface Cracks in Metals and their Characterization Using Rayleigh Waves*. Ph.D. Thesis. London: The City University.
14. Viktorov, I. A. 1967. *Rayleigh and Lamb Waves*. New York: Plenum Press.
15. Ash, E. A., and E. G. S. Paige, eds. 1985. *Rayleigh Wave Theory and Application*, Vol. 2. Berlin: Springer Series on Wave Phenomena, Springer-Verlag.
16. Ritz, W. 1909. *Ann Phys* 28:737.
17. Timoshenko, S. 1955. *Vibration Problems in Engineering*. 3rd ed. New York: Van Nostrand.
18. Kino, G. S. 1987. *Acoustic Waves*. Englewood Cliffs, NJ: Prentice Hall.

4 Design of Ultrasonic Horns for High Power Applications

4.1 INTRODUCTION

In machining, materials forming, sonochemistry, some forms of ultrasonic surgery, and a diverse range of other high-power applications, ultrasonic energy is transferred from the transducer into the medium subjected to treatment through "horns" or transmission lines of various configurations. The amplitude of vibration at the radiating surface depends upon the geometry of the transmission line, its energy losses, and the amplitude of vibration at the driven end. A tapered transmission line produces a displacement amplification; that is, the amplitude of vibration at resonance at the small end is greater than that at the larger end. These tapered elements are called mechanical amplifiers or horns.

The design of transducers for high-intensity use often involves combining segments of transmission lines that are composed of different materials; therefore, the same equations used in the design of ultrasonic horns, or mechanical amplifiers, are useful in designing transducers and other types of transmission lines. The so-called Langevin sandwich is an example of such a combination [1]. The Langevin-sandwich type of transducer consists of piezoelectric elements sandwiched between two masses so that the fundamental resonance frequency is lower than the resonance frequency of any of the elements taken separately. These transducers are often designed by assuming that the masses are lumped and that the piezoelectric plates are stiff springs. If the piezoelectric element is thin and the masses are comparatively long, it is more accurate to assume that the mass is distributed, and this means that the equations for transmission lines must be used.

Because of the importance of the horns in the application of high-intensity ultrasonic energy, this chapter is devoted to the development of equations for the design of these elements.

4.2 HORN EQUATIONS

The wave equation was derived in Chapter 3, and the specific form, the horn equation (3.21), is reproduced here in the following forms:

$$\frac{1}{c^2}\frac{\partial^2 \xi}{\partial t^2} - \frac{1}{S}\frac{\partial S}{\partial x}\frac{\partial \xi}{\partial x} - \frac{\partial^2 \xi}{\partial x^2} = 0 \tag{3.21}$$

and

$$\frac{\partial^2 v}{\partial x^2} + \frac{1}{S}\frac{\partial S}{\partial x}\frac{\partial v}{\partial x} + \frac{\omega^2}{c^2}v = 0$$

Equation 3.21, a special form of the wave equation, reduces to the plane-wave equation, Equation 3.20, if the taper is zero, that is, $\partial S/\partial x$ is zero. Thus, the uniform bar may be considered to be a horn with zero taper and with an amplification factor of unity.

4.3 TYPES OF HORNS

4.3.1 CYLINDER OR UNIFORM BAR AS AN ULTRASONIC HORN

The simplest horn is a cylinder, or uniform bar. Its characteristics are described by Equations 3.39 through 3.42 and 3.45 in Chapter 3. The length of the horn for resonance is determined by Equation 4.1:

$$c = \lambda f \qquad (4.1)$$

where λ is the wavelength of the ultrasound. The length ℓ of a half-wave cylinder is

$$\ell = \frac{\lambda}{2} = \frac{c}{2f} \qquad (4.2)$$

Since the cylinder is a uniform bar, the displacements at the two ends are equal in magnitude when the bar is in half-wave resonance, that is, $\xi_0 = \xi_\ell$.

4.3.2 STEPPED HORN (DOUBLE CYLINDER)

The stepped horn is a modification of the uniform bar (Figure 4.1). The diameter at one end of the bar differs from that at the other end. Referring to Figure 4.1, $D_1 > D_2$. Conservation of momentum would indicate that when the horn is resonant, v_2 must be greater than v_1 for equilibrium to be maintained. Since the displacement is proportional to the velocity potential v, $\xi_1 < \xi_2$. These statements may be proved by referring to the formulas for uniform bars (free–free bars), which are also applicable for the stepped horn when the different areas are taken into account.

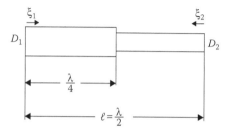

FIGURE 4.1 Stepped horn.

Referring to Figure 4.1 and assuming that the horn is homogeneous, the velocity potential v_1 at $x = \lambda/4 - dx$ is $v_1 = F_x/Z_1$. At $x = \lambda/4 + dx$, $v_2 = F_x/Z_2$. Therefore,

$$\frac{v_1}{v_2} \approx \frac{Z_2}{Z_1} = \frac{S_2 Y_2 c_1 \tan(\omega \ell/2c_2)}{S_1 Y_1 c_2 \tan(\omega \ell/2c_1)} = \frac{S_2}{S_1} \qquad (4.3)$$

since for this case $Y_2 = Y_1$ and $c_2 = c_1$.

Equation 4.3 shows that the velocities, and hence the displacements at the ends of the stepped horn of Figure 4.1, are inversely proportional to the areas of the ends. Therefore, for cylindrical stepped horns, if $D_1/D_2 = 2$, $v_2/v_1 = 4$. In other words, the amplification factor is the ratio of the areas of the ends of the horn, with the higher amplitude being at the smaller end.

The plane of maximum stress in a half-wave uniform bar and of a half-wave stepped horn where the step occurs at the center (quarter-wave position) is located at the node. A horn fabricated without a fillet at the junction of the two cylinders, as in Figure 4.1, would experience such high concentrations of stress at the node where the dimensions change abruptly that, at high intensities, it might soon fail in fatigue.

Common practice in designing horns for developing high sound intensities calls for sufficient filleting at the junction to prevent fatigue failure. An arbitrary rule is to make the radius of the fillet equal to the difference between the radii of the two cylinders. This may be impractical if the ratio of diameters is much greater than two. Scratches or machine marks on the fillet may also cause failure for the same reason.

4.3.3 EXPONENTIALLY TAPERED HORN

The exponential horn (Figure 4.2) is described mathematically by

$$S = S_0 e^{-\gamma x} \qquad (4.4)$$

where γ is a taper factor [2].

Differentiating S with respect to x and substituting in Equation 3.21 gives the differential equation for the exponential horn:

$$\frac{1}{c^2} \frac{\partial^2 \xi}{\partial t^2} + \gamma \frac{\partial \xi}{\partial x} - \frac{\partial^2 \xi}{\partial x^2} = 0 \qquad (4.5)$$

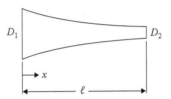

FIGURE 4.2 Exponential horn.

If the motion is simple harmonic

$$\frac{\partial \xi}{\partial t} = v = j\xi\omega$$

or

$$\xi = -j\frac{v}{\omega} \tag{4.6}$$

Substituting Equation 4.6 in Equation 4.5 gives

$$\frac{\partial^2 v}{\partial x^2} - \gamma\frac{\partial v}{\partial x} + \frac{\omega^2}{c^2}v = 0 \tag{4.7}$$

Equation 4.7 may be solved by assuming a solution of the form

$$v = K_0\, e^{\gamma x/2}\, U(x) \tag{4.8}$$

where K_0 is a constant. Then

$$\frac{\partial v}{\partial x} = \frac{K_0}{2}\gamma\, e^{\gamma x/2}\, U + K_0\, e^{\gamma x/2}\frac{\partial U}{\partial x}$$

and

$$\frac{\partial^2 v}{\partial x^2} = \frac{K_0}{4}\gamma^2\, e^{\gamma x/2}\, U + K_0\gamma\, e^{\gamma x/2}\frac{\partial U}{\partial x} + K_0\, e^{\gamma x/2}\frac{\partial^2 U}{\partial x^2}$$

Substituting these quantities for $\partial v/\partial x$ and $\partial^2 v/\partial x^2$ in Equation 4.7 gives the ordinary differential equation

$$\frac{\partial^2 U}{\partial x^2} + \left(\frac{\omega^2}{c^2} - \frac{\gamma^2}{4}\right)U = 0 \tag{4.9}$$

for which the solution is (assuming that $\gamma < 2\omega/c$)

$$U = k_1\cos\left(\frac{\omega^2}{c^2} - \frac{\gamma^2}{4}\right)^{1/2}x + k_2\sin\left(\frac{\omega^2}{c^2} - \frac{\gamma^2}{4}\right)^{1/2}x$$
$$= k_1\cos\frac{\omega x}{c'} + k_2\sin\frac{\omega x}{c'} \tag{4.10}$$

where

$$c' = \frac{c}{\sqrt{1 - \gamma^2 c^2/4\omega^2}} = c\sqrt{1 + \left(\frac{\ln(s_o/s\ell)^2}{2\pi}\right)}$$

Substituting Equation 4.10 in Equation 4.8 gives Equation 4.11:

$$v = e^{\gamma x/2} \left(K_{01} \cos \frac{\omega x}{c'} + K_{02} \sin \frac{\omega x}{c'} \right) \tag{4.11}$$

where $K_{01} = K_0 k_1$ and $K_{02} = K_0 k_2$.

The coefficients K_{01} and K_{02} may be determined by applying the appropriate boundary conditions:

$$\text{At } x = 0: \ v = V_0 = K_{01}$$

and

$$\frac{\partial v}{\partial x} = 0 = e^{\gamma x/2} \left[\left(-V_0 \frac{\omega}{c'} \sin \frac{\omega x}{c'} + K_{02} \frac{\omega}{c'} \cos \frac{\omega x}{c'} \right) + \frac{\gamma}{2} \left(V_0 \cos \frac{\omega x}{c'} + K_{02} \sin \frac{\omega x}{c'} \right) \right] \tag{4.12}$$

$$= K_{02} \frac{\omega}{c'} + \frac{\gamma}{2} V_0$$

Therefore,

$$K_{02} = -V_0 \frac{\gamma c'}{2\omega} \tag{4.13}$$

giving

$$v = e^{\gamma x/2} V_0 \left(\cos \frac{\omega x}{c'} - \gamma \frac{c'}{2\omega} \sin \frac{\omega x}{c'} \right) \tag{4.14}$$

as the equation for the velocity potential at distance x from the large end of an exponentially tapered horn.

The particle velocity of the large end of the horn V_0 is described in simple harmonic motion, as previously stated: at $x = \ell$, the velocity potential $V = -e^{\gamma \ell/2} V_0$. Since according to Equation 4.4 $S/S_0 = e^{-\gamma x}$, then

$$V_\ell = -V_0 \left(\frac{S_0}{S_\ell} \right)^{1/2} = -V_0 \left(\frac{D_0}{D_\ell} \right)$$

and

$$\xi_\ell = -\xi_0 \left(\frac{D_0}{D_\ell} \right) \tag{4.15}$$

where D_0 and D_ℓ are diameters at $x = 0$ and $x = \ell$, respectively.

Equation 4.15 indicates that the displacements at the ends of an exponentially tapered horn are inversely proportional to the diameters at the ends.

The velocity node ($v = 0$) occurs where

$$\tan \frac{\omega x}{c'} = \frac{2\omega}{\gamma c'} \tag{4.16}$$

The stress of a horn is

$$s = j\frac{Y}{\omega}\frac{\partial v}{\partial x} \tag{4.17}$$

which, for the exponential horn, is

$$s = -j\frac{Y}{\omega}V_0\left(\frac{\omega}{c'} + \frac{\gamma^2 c'}{4\omega}\right)e^{\gamma x/2}\sin\frac{\omega x}{c'} \tag{4.18}$$

Maximum stress occurs when $\partial s/\partial x = 0$, that is, when

$$\tan\frac{\omega x}{c'} = -\frac{2\omega}{\gamma c'} \tag{4.19}$$

Comparison of Equations 4.16 and 4.19 shows that the velocity node occurs near the large-diameter end of the horn, while the point of maximum stress lies closer to the small-diameter end of the horn.

The length ℓ of the half-wave exponential horn is given by the formula

$$\ell = \pi\frac{c'}{\omega} = \frac{c'}{2f} \tag{4.20}$$

In this case, $c' > c$ and, therefore, ℓ is longer than a uniform bar vibrating in half-wave resonance at the same frequency as the exponential horn.

The mechanical impedance at x of the exponential horn is

$$\begin{aligned}Z = \frac{sS}{v} &= \frac{-jSY(\omega/c' + \gamma^2 c'/4\omega)}{\omega[\cot(\omega x/c') - \gamma c'/2\omega]}\\ &= \frac{-jSY(\omega/c' + \gamma^2 c'/4\omega)\tan(\omega x/c')}{\omega[1 - (\gamma c'/2\omega)\tan(\omega x/c')]}\end{aligned} \tag{4.21}$$

where $S = S_0 e^{-\gamma x}$ and S_0 is the area at the large end of the horn.

4.3.4 WEDGE-SHAPED HORNS

The equation describing the taper of a wedge is (see Figure 4.3)

$$S = S_1 + k_w x \tag{4.22}$$

where $k_w = (S_2 - S_1)/\ell$.

Assuming again that at resonance the displacement is in simple harmonic motion, performing the essential differentiations, that is,

$$\frac{\partial S}{\partial x} = k_w \quad \text{and} \quad \frac{1}{S}\frac{\partial S}{\partial x} = \frac{k_w}{S_1 + k_w x}$$

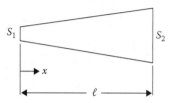

FIGURE 4.3 Wedge-shaped horn.

and substituting these quantities in Equation 3.21 gives the wave equation for the wedge in the form

$$\frac{\partial^2 v}{\partial x^2} + \frac{k_w}{S_1 + k_w x}\frac{\partial v}{\partial x} + \frac{\omega^2}{c^2}v = 0 \tag{4.23}$$

Equation 4.23 may be simplified by letting $u = S_1 + k_w x$.
Then

$$\frac{\partial v}{\partial u} = \frac{1}{k_w}\frac{\partial v}{\partial x} \quad \text{and} \quad \frac{\partial^2 v}{\partial u^2} = \frac{1}{k_w^2}\frac{\partial^2 v}{\partial x^2} \tag{4.24}$$

Substituting Equation 4.24 in Equation 4.23 and letting $p^2 = \omega^2/k_w^2 c^2$ reduces the wedge-shaped horn equation to

$$\frac{\partial^2 v}{\partial u^2} + \frac{1}{u}\frac{\partial v}{\partial u} + p^2 v = 0 \tag{4.25}$$

The solution for Equation 4.25 is

$$v = KJ_0(pu)$$

$$= KJ_0\left[\frac{\omega}{ck_w}(S_1 + k_w x)\right] = KJ_0\left[\frac{\omega}{c}\left(\frac{S_1 \ell}{S_2 - S_1} + x\right)\right] \tag{4.26}$$

where K is a constant of integration and $J_0(pu)$ is a Bessel function of the first kind of order zero.

At $x = 0$, $v = V_1$, the velocity at the small end. Therefore,

$$K = \frac{V_1}{J_0\left[\dfrac{\omega}{c}\left(\dfrac{\ell S_1}{S_2 - S_1}\right)\right]}$$

and

$$v = \frac{V_1 J_0 \left[\dfrac{\omega}{c} \left(\dfrac{\ell S_1}{S_2 - S_1} + x \right) \right]}{J_0 \left[\dfrac{\omega}{c} \left(\dfrac{\ell S_1}{S_2 - S_1} \right) \right]} = j\omega\xi \tag{4.27}$$

where V_1 represents simple harmonic motion at $x = 0$.

The stress is

$$s = j\frac{Y}{\omega}\frac{\partial v}{\partial x}$$

$$= j\rho c V_1 \frac{J_1 \left[\dfrac{\omega}{c} \left(\dfrac{\ell S_1}{S_2 - S_1} + x \right) \right]}{J_0 \left[\dfrac{\omega}{c} \left(\dfrac{\ell S_1}{S_2 - S_1} \right) \right]} \tag{4.28}$$

where J_1 is a Bessel function of the first kind of order one.

The impedance Z at x of the wedge is s

$$Z_x = \frac{sS_x}{v}$$

$$= -j\rho c \left[S_1 + \left(\frac{S_2 - S_1}{\ell} \right) x \right] \left\{ \frac{J_1 \left[\dfrac{\omega}{c} \right] \left(\dfrac{\ell S_1}{S_2 - S_1} + x \right)}{J_0 \left[\dfrac{\omega}{c} \left(\dfrac{\ell S_1}{S_2 - S_1} \right) \right]} \right\} \tag{4.29}$$

The velocity node is at a value of x where

$$J_0 \left[\frac{\omega}{c} \left(\frac{\ell S_1}{S_2 - S_1} + x \right) \right] = 0 \tag{4.30}$$

and the maximum stress occurs when

$$J_1 \left[\frac{\omega}{c} \left(\frac{\ell S_1}{S_2 - S_1} + x \right) \right] = \text{a maximum or a minimum} \tag{4.31}$$

When $S_1 = 0$, Equations 4.27 through 4.31 are

$$v = V_1 J_0 \left(\frac{\omega x}{c} \right) \tag{4.27a}$$

$$s = -j\rho c V_1 \left[J_1\left(\frac{\omega x}{c}\right) \right] \tag{4.28a}$$

$$Z = -j\rho c \left(\frac{S_2 x}{\ell}\right) \left[\frac{J_1(\omega x/c)}{J_0(\omega x/c)} \right] \tag{4.29a}$$

$$J_0\left(\frac{\omega x}{c}\right) = 0 \text{ (velocity node)} \tag{4.30a}$$

$$J_1\left(\frac{\omega x}{c}\right) = \text{a maximum or a minimum (maximum stress)} \tag{4.31a}$$

The length ℓ is determined by the boundary conditions. If the wedge is a resonant, half-wave vibrator, at $x = 0$, ℓ, the stress is zero and $\partial v/\partial x$ is zero. Therefore, the quantity

$$\frac{\omega}{c}\left(\frac{\ell S_1}{S_2 - S_1} + x\right)$$

is a zero of J_1 at $x = 0$ and at $x = \ell$; that is,

$$\frac{\omega}{c}\left(\frac{\ell S_1}{S_2 - S_1}\right) = 0, 3.8317, 7.0156, 10.1735, 13.3237, \dots \text{ when } x = 0 \quad (4.32a)$$

and correspondingly

$$\frac{\omega}{c}\left(\frac{\ell S_2}{S_2 - S_1}\right) = 3.8317, 7.0156, 10.1735, 13.3237, 16.4706, \dots \text{ when } x = \ell \quad (4.32b)$$

To determine the length ℓ for given ratios of area, divide Equation 4.32b by Equation 4.32a to obtain the ratio S_2/S_1; then substitute for S_2 in terms of S_1 in Equation 4.32b and calculate ℓ. The one exception occurs when $S_1 = 0$. In this case, $\omega \ell/c = 3.8317$ determines ℓ directly. Area ratios corresponding to consecutive zeros of Equations (4.32a) and (4.32b) and calculated lengths for half-wave resonances are given in Table 4.1.

The following example illustrates the method of calculating the values given in Table 4.1.

TABLE 4.1

Calculated Lengths for Half-Wave Resonant Wedge-Shaped Horns

S_2/S_1	=	∞	1.832	1.515	1.310	1.236	1.000
ℓ	=	$3.83c/\omega$	$3.20c/\omega$	$3.16c/\omega$	$3.15c/\omega$	$3.14c/\omega$	$\pi c/\omega$

When

$$\frac{\omega}{c}\left(\frac{\ell S_2}{S_2 - S_1}\right) = 7.0156 \tag{4.32b}$$

$$\frac{\omega}{c}\left(\frac{\ell S_1}{S_2 - S_1}\right) = 3.8317 \tag{4.32a}$$

Dividing Equation 4.32b by Equation 4.32a gives

$$\frac{S_2}{S_1} = 7.0156/3.8317 = 1.832$$

Substituting this value in Equation 4.32b gives

$$\ell = 3.2 \frac{c}{\omega}$$

Figure 4.4 is a curve for which $\omega \ell / c$ is plotted as a function of area ratio S_1/S_2. Reasonably accurate lengths of wedge-shaped horns may be calculated by using this curve.

The ratio of velocities at the ends of the horn, and therefore the ratio of the corresponding displacement amplitudes, is determined by dividing the velocity at the small end ($x = 0$) by the velocity V_2 at the large end ($x = \ell$) of the horn. Then

$$\frac{V_1}{V_2} = \frac{J_0\left[\frac{\omega}{c}\left(\frac{\ell S_1}{S_2 - S_1}\right)\right]}{J_0\left[\frac{\omega}{c}\left(\frac{\ell S_2}{S_2 - S_1}\right)\right]} \tag{4.33}$$

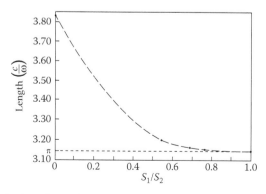

FIGURE 4.4   Wedge length as a function of taper.

When $S_1 = 0$

$$\frac{V_1}{V_2} = \frac{1}{J_0(\omega\ell/c)} = 2.48 \quad \text{(for half-wave)} \tag{4.34}$$

where $J_0(\omega\ell/c)$ is the minimum of $J_0(X_n)$ and is equal to -0.4028.

4.3.5 Conical Horns

The taper of a conical horn, Figure 4.5, is given by [3]

$$\frac{d_1}{d_x} = \frac{\ell d_1}{\ell d_1 + x(d_1 - d_1)} \tag{4.35}$$

where d_1 is the diameter at the small end and d_x is the diameter at distance x from the small end. Then the area S at x is

$$S = \frac{\pi d_x^2}{4} = \frac{\pi}{4}\left[d_1 + \frac{x}{\ell}(d_2 - d_1)\right]^2 = S_1\left[1 + \frac{(N-1)x}{\ell}\right]^2 \tag{4.36}$$

where S_1 is the area at the small end of the horn and $N = d_2/d_1$, the ratio of the diameters at the two ends.

Assuming simple harmonic motion for displacement at resonance and substituting for S, according to Equation 4.36, in the horn equation (Equation 3.21) results in the following equation for the conical horn:

$$\frac{d^2v}{dx^2} + \frac{2(d_2 - d_1)}{d_1\ell + (d_2 - d_1)x}\frac{dv}{dx} + \frac{\omega^2}{c^2}v = 0 \tag{4.37a}$$

or

$$\frac{d^2v}{dx^2} + \frac{2(N-1)}{\ell + (N-1)x}\frac{dv}{dx} + \frac{\omega^2}{c^2}v = 0 \tag{4.37b}$$

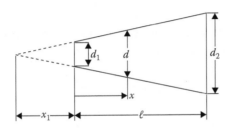

FIGURE 4.5 Conical section.

The solution for Equation 4.37a is

$$
\begin{aligned}
v = \frac{d_2 - d_1}{\ell d_1 + (d_2 - d_1)x} &\left\{ A\cos\frac{\omega}{c}\left[\frac{\ell d_1 + (d_2 - d_1)x}{d_2 - d_1}\right] \right. \\
&\left. + B\sin\frac{\omega}{c}\left[\frac{\ell d_1 + (d_2 - d_1)x}{d_2 - d_1}\right] \right\}
\end{aligned}
\tag{4.38}
$$

The constants A and B can be determined by applying appropriate boundary conditions. At $x = 0$, $v = V_1$, and at $x = \ell$, $v = V_2$. Both V_1 and V_2 are maxima so that at $x = 0$, ℓ, $\partial v / \partial x = 0$. Dividing both sides of Equation 4.38 by $(d_2 - d_1)/[\ell d_1 + (d_2 - d_1)x]$ and differentiating yields

$$
\begin{aligned}
\left[\frac{\ell d_1 + (d_2 - d_1)x}{d_2 - d_1}\right]\frac{dv}{dx} + v &= -A\frac{\omega}{c}\sin\left[\frac{\omega}{c}\frac{\ell d_1 + (d_2 - d_1)x}{d_2 - d_1}\right] \\
&+ B\frac{\omega}{c}\cos\left[\frac{\omega}{c}\frac{\ell d_1 + (d_2 - d_1)x}{d_2 - d_1}\right]
\end{aligned}
\tag{4.39}
$$

By applying the boundary conditions to Equations 4.38 and 4.39, it can be shown that

$$
A = V_2\left(\frac{\ell d_2}{d_2 - d_1}\cos\frac{\omega\ell}{c}\frac{d_2}{d_2 - d_1} - \frac{c}{\omega}\sin\frac{\omega\ell}{c}\frac{d_2}{d_2 - d_1}\right)
\tag{4.40}
$$

and

$$
B = V_2\left(\frac{c}{\omega}\cos\frac{\omega\ell}{c}\frac{d_2}{d_2 - d_1} + \frac{\ell d_2}{d_2 - d_1}\sin\frac{\omega\ell}{c}\frac{d_2}{d_2 - d_1}\right)
\tag{4.41}
$$

Substituting these values for A and B in Equation 4.38 yields

$$
\begin{aligned}
v &= \left[\frac{(d_2 - d_1)V_2}{\ell d_1 + (d_2 - d_1)x}\right]\left[\frac{\ell d_2}{d_2 - d_1}\cos\frac{\omega}{c}(x - \ell) + \frac{c}{\omega}\sin\frac{\omega}{c}(x - \ell)\right] \\
&= \left[\frac{(N-1)V_2}{\ell + (N-1)x}\right]\left[\frac{\ell N}{N-1}\cos\frac{\omega}{c}(x - \ell) + \frac{c}{\omega}\sin\frac{\omega}{c}(x - \ell)\right]
\end{aligned}
\tag{4.42}
$$

Since at $x = 0$, $v = V_1$, the ratio of velocities and, therefore, of the displacement amplitudes at the ends of the horn is

$$
\begin{aligned}
\frac{V_1}{V_2} = \frac{\xi_1}{\xi_2} &= \frac{d_2}{d_1}\cos\frac{\omega\ell}{c} - \frac{c(d_2 - d_1)}{\omega\ell d_1}\sin\frac{\omega\ell}{c} \\
&= N\cos\frac{\omega\ell}{c} - \frac{c}{\omega}\frac{(N-1)}{\ell}\sin\frac{\omega\ell}{c}
\end{aligned}
\tag{4.43}
$$

Also, since at the velocity node, $v = 0$, it follows from Equation 4.42 that the velocity node occurs when

$$\tan\frac{\omega}{c}(\ell - x) = \frac{\omega\ell}{c}\frac{d_2}{d_2 - d_1} = \frac{\omega\ell}{c}\frac{N}{N-1} \tag{4.44}$$

or

$$\tan\frac{\omega}{c}(\ell - x) = \frac{\omega\ell}{c} \quad (\text{when } d_1 = 0) \tag{4.44a}$$

The half-wave resonant length of a conical horn is determined by applying the boundary condition that

$$\left(\frac{\partial v}{\partial x}\right)_{x=0} = 0$$

Differentiating Equation 4.42 and applying the boundary conditions leads to

$$\tan\frac{\omega\ell}{c} = \frac{\omega\ell c(d_2 - d_1)^2}{\omega^2\ell^2 d_1 d_2 + c^2(d_2 - d_1)^2} = \frac{\ell(N-1)^2}{\frac{\omega}{c}\left[\ell^2 N + \frac{c^2}{\omega^2}(N-1)^2\right]} \tag{4.45}$$

or

$$\tan\frac{\omega\ell}{c} = \frac{\omega\ell}{c} \quad (\text{when } d_1 = 0) \tag{4.45a}$$

For the half-wave horns, $\omega\ell/c$ lies in the third quadrant, that is, $\pi \le \omega\ell/c < 3\pi/2$. When $d_1 = d_2$ (infinite taper), $\tan \omega\ell/c = 0$ so that $\omega\ell/c = 0, \pi, \ldots$, which is the equation for a uniform thin bar.

The stress s at x is given by

$$\begin{aligned}
s &= \frac{Y}{\omega}\frac{\partial v}{\partial x} \\
&= j\frac{(d_2 - d_1)V_2 Y}{[\ell d_1 + (d_2 - d_1)x]^2\omega}\left\{(d_2 - d_1)(x - \ell)\cos\frac{\omega}{c}(x - l)\right. \\
&\quad \left. -\frac{\omega}{c}\left[\frac{\ell^2 d_1 d_2}{d_2 - d_1} + \frac{c^2}{\omega^2}(d_2 - d_1) + d_2\ell x\right]\sin\frac{\omega}{c}(x - \ell)\right\} \\
&= j\frac{V_2 Y}{\omega}\frac{(N-1)}{[\ell + (N-1)x]^2}\left\{(N-1)(x - \ell)\cos\frac{\omega}{c}(x - 2)\right. \\
&\quad \left. -\frac{\omega}{c}\left[\frac{\ell^2 N}{N-1} + \frac{c^2}{\omega^2}(N-1) + N\ell x\right]\sin\frac{\omega}{c}(x - \ell)\right\}
\end{aligned} \tag{4.46}$$

or when $d_1 = 0$

$$s = j\frac{YV_2}{\omega x^2}\left[(x-\ell)\cos\frac{\omega}{c}(x-\ell) - \left(\ell x + \frac{c^2}{\omega^2}\right)\frac{\omega}{c}\sin\frac{\omega}{c}(x-\ell)\right] \quad (4.47)$$

Differentiating Equation 4.46 and setting $\partial s/\partial x = 0$ gives

$$\tan\frac{\omega}{c}(\ell-x) = \frac{\omega}{c}\frac{\omega^2\ell D_2[\ell D_2 + (D_2 - D_1)x]^2 + c^2(D_2 - D_1)^3(\ell-x)}{(D_2 - D_1)(\omega^2[\ell D_1 + (D_2 - D_1)x][(D_1 + D_2)\ell + (D_2 - D_1)x] + c^2(D_2 - D_1)^2)}$$

(4.48)

$$\tan\frac{\omega}{c}(\ell-x) = \left(\frac{\omega}{c}\right)\left[\frac{\omega^2\ell x^2 + c^2(\ell-x)}{\omega^2 x(\ell+x) + c^2}\right] \quad (4.48a)$$

at the position of maximum stress.

Therefore, since $Z = F/v = sS/v$, it follows from Equations 4.42 and 4.46 and the equation describing the area S at x (Equation 4.36) that

$$Z = F/v$$

$$= j\frac{\pi[\ell d_1 + (d_2 - d_1)x]y}{4\omega\ell^2\left[\dfrac{\ell d_2}{(d_2 - d_1)} + \dfrac{c}{\omega}\tan\dfrac{\omega}{c}(x-\ell)\right]}\{(d_2 - d_1)(x-\ell)$$

$$-\frac{\omega}{c}\left[\frac{\ell^2 d_1 d_2}{d_2 - d_1} + \frac{c^2}{\omega^2}(d_2 - d_1) + d_2\ell x\right]\tan\frac{\omega}{c}(x-\ell)\}$$

$$= j\frac{\pi(\ell + (N-1)x]Yd_1^2}{4\omega\ell^2\left[\dfrac{\ell N}{N-1} + \dfrac{c}{\omega}\tan\dfrac{\omega}{c}(x-\ell)\right]}\{(N-1)(x-\ell) \quad (4.49)$$

$$-\frac{\omega}{c}\left[\frac{\ell^2 N}{(N-1)} + \frac{c^2}{\omega^2}(N-1) + N\ell x\right]\tan\frac{\omega}{c}(x-\ell)\}$$

or when $d_1 = 0$

$$Z = j\frac{\pi d_2^2 Yx}{4\ell^2}\left[\frac{(x-\ell) - \left[\ell x + \dfrac{c^2}{\omega^2}\right]\dfrac{\omega}{c}\left[\tan\dfrac{\omega}{c}(x-\ell)\right]}{\omega\ell + c\left[\tan\dfrac{\omega}{c}(x-\ell)\right]}\right] \quad (4.49a)$$

As Equations 4.44, 4.45, and 4.48 show, the design of a conical horn involves solving an algebraic function of x and a trigonometric function of x simultaneously to determine the location of the node, the half-wave length, and the position of maximum stress.

When $d_1 = 0$, the quantity $\omega\ell/c$ is constant regardless of frequency or velocity of sound and is approximately 4.5, assuming that d_2 is small compared with a half-wavelength. Then $\ell = 4.5c/\omega$. As Equation 4.45 shows for all values of $d_1 > 0$, $\pi \leq \tan \omega\ell/c$ (π being the limit for an infinite taper, that is, $d_1 = d_2$).

4.3.6 CATENOIDAL HORNS

The equation of the catenary is

$$y = \frac{a}{2}(e^{x/a} + e^{-x/a}) = a \cosh \frac{x}{a} \tag{4.50}$$

(see Figure 4.6). If y corresponds to the radius of the horn at x, and a is the radius at $x = 0$, introducing a taper factor, $k < 1$, the area S_x is

$$
\begin{aligned}
S_x = \pi y^2 &= \left(\frac{\pi a^2}{4}\right) + (e^{2kx/a} + 2 + e^{-2kx/a}) \\
&= \frac{S_1}{4}(e^{2kx/a} + 2 + e^{-2x/a}) \\
&= \frac{\pi d^2}{4}\pi \\
&= S_1 \cosh^2 \frac{kx}{a}
\end{aligned}
\tag{4.51}
$$

$$\frac{\partial S}{\partial x} = \frac{kS_1}{2a}(e^{2kx/a} - e^{-2kx/a}) = \frac{2kS_1}{a}\cos\frac{kx}{a}\sinh\frac{kx}{a} \tag{4.52}$$

$$\frac{d^2v}{dx^2} + \frac{2k}{a}\left(\frac{e^{kx/a} - e^{-kx/a}}{e^{kx/a} + e^{-kx/a}}\right)\frac{dv}{dx} + \frac{\omega^2}{c^2}v = 0$$

or

$$\frac{d^2v}{dx^2} + \left(\frac{2k}{a}\tanh\frac{kx}{a}\right)\frac{dv}{dx} + \frac{\omega^2}{c^2}v = 0 \tag{4.53}$$

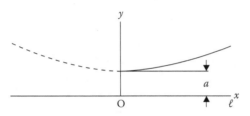

FIGURE 4.6 The catenary.

for the differential equation of the catenoidal horn [4]. The solution to Equation 4.53 is

$$v = \frac{1}{a\cosh(kx/a)}(A\cos k'x + B\sin k'x) \tag{4.54}$$

where

$$k' = \sqrt{\frac{\omega^2}{c^2} - \frac{k^2}{a^2}}$$

$$k = \frac{a}{\ell}\cosh^{-1}(r_2/r_1) = \frac{a}{\ell}\cosh^{-1}(D_2/D_1)$$

At $x = 0$,

$$\frac{\partial v}{\partial x} = 0 \quad \text{and} \quad v = V_1$$

Therefore,

$$V_1 = \frac{A}{a} \quad \text{or} \quad A - aV_1$$

$$\left.\frac{\partial v}{\partial x}\right|_0 = 0$$

and

$$\frac{Bk'}{a} = 0$$

There, $B = 0$, and Equation 4.54 becomes

$$v = \frac{V_1\cos k'x}{\cosh(kx/a)} = j\omega\zeta \tag{4.55}$$

By definition, $D_2/D_1 = \cosh k\,\ell/a$, that is, $D_1 = 2a$ and $D_2 = 2a\cosh k\ell/a$ according to Equation 4.51. Therefore,

$$V_2 = \frac{V_1\cos k'\ell}{D_2/D_1} \tag{4.56}$$

The length ℓ of a half-wave catenoidal horn is

$$\ell = \frac{\pi}{k'} = \frac{c}{\omega}\left[\pi^2 + \left(\cosh^{-1}\left(\frac{kD_2}{D_1}\right)\right)^2\right]^{1/2} \tag{4.57}$$

The velocity and displacement node occurs where $\cos k'x = 0$, that is, where $k'x = \pi/2$. Therefore, the node occurs when

$$x\big|_{v=0} = \frac{\pi}{2k'} = \frac{\pi}{2\sqrt{\omega^2/c^2 - 4k^2/D_1^2}} \tag{4.58}$$

The stress along a catenoidal horn is given by Equation 4.17 and is

$$s = j\frac{Y}{\omega}\frac{\partial v}{\partial xv} = j\frac{YV_1\cos k'x}{\omega a \cosh(kx/a)}\left(\tanh\frac{kx}{a} + ak'\tan k'x\right) \tag{4.59}$$

At the point of maximum stress s_m, $\partial s/\partial x = 0$.

$$\frac{\partial s}{\partial x} = j\frac{YV_1\cos k'x}{\omega a^2 \cosh(kx/a)}\left\{\left(\tanh\frac{kx}{a} + ak'\tan k'x\right)^2 - \left(\mathrm{sech}^2\frac{kx}{a} + a^2k'^2\sec^2 k'^2x\right)\right\} = 0 \tag{4.60}$$

from which

$$\tanh\frac{kx}{a}\left(\tanh\frac{kx}{a} + ak'\tan k' x\right) = 1 + \frac{a^2k'^2}{2} \tag{4.61}$$

where $a = D_1/2$ and $k'^2 = \omega^2/c^2 - 4k^2/D_1^2$ where s is maximum.
   The impedance at x of a catenoidal horn is $Z_x = s_x S_x/v_x$

$$Z_x = -j\frac{YS_1}{\omega a}\left(\tanh\frac{kx}{a} + ak'\tan k'x\right)\cosh^2\frac{kx}{a} \tag{4.62}$$

4.4 COMBINING SECTIONS OF DIFFERENT CONFIGURATIONS FOR PRACTICAL APPLICATIONS

The previous sections contain design formulas for the solid horns, which are most commonly used in the production of high-intensity ultrasonic energy used for materials forming, *sonochemistry*, certain processing, and *medical applications*. In materials-forming applications, such as ultrasonic drilling or welding, usually a tool is silver-soldered to a horn. This horn is then attached by means of a stud to a transducer or to a second horn, which is attached permanently to the transducer

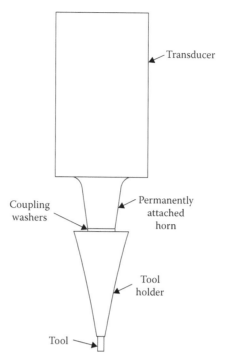

FIGURE 4.7   Vibratory components of conventional ultrasonic processing system.

(see Figure 4.7). Maximum displacement is obtained at the tip of the tool when the resonance frequencies of the transducer, the permanent horn, and the tool/horn combination are the same.

Tool wear during ultrasonic processing causes an upward shift in frequency. Temperature changes in any of the components cause inverse changes in resonance frequency of the corresponding component. For these reasons, it is difficult to maintain a perfect match between elements of the system of Figure 4.7.

The problem of frequency shift is minimized by using transducers with low mechanical Q, which produces a fairly broad tuning curve. The permanently attached horn is tuned to the center frequency of the transducer. The attached toolholder with a new tool attached resonates at a slightly lower frequency. As the tool wears, the frequency increases and may rise to a value slightly above the resonant frequency of the transducer.

Modern commercial and medical applications of high-intensity ultrasonic energy fall generally into two categories: (1) those exhibiting low impedance loading and little wear and (2) those working into high impedance loads requiring short time exposures. Low impedance loads include applications such as homogenization in surgical replacement of lenses of the eyes (phacoemulsification), cleaning, and atomization of liquids. Ultrasonic welding of plastics and metals is an example of high impedance loads. In many cases, energy is applied to the tool prior to applying it to the load. The Q of the system is usually high, and the amplitude of vibration is high. In many cases, as the tool is pressed against the load, the energy to the tool is cut off and the stored energy is expended in the work. In each case, the electronic driver works within a narrow bandwidth, which is of considerable economic benefit.

A practical limit on the distance the resonant frequency of the toolholder may shift from the center frequency of the transducer is imposed by the geometry of the system. When the three sections (transducer, permanent horn, and toolholder plus tool) of Figure 4.7 are resonant at the same frequency, the impedance at each junction would be zero if there were no energy losses in the system. In the practical case, the stress level at the junctions is minimum, being sufficient to compensate for losses in the system and thus to maintain a constant level of vibration. Under this ideal condition, maximum energy is transferable from the transducer to the free end of the tool. Each section terminates at a velocity antinode.

When the various sections are not matched, if a velocity antinode appears at the free end of the tool, the opposite end of the toolholder is not at an antinode. Energy is reflected back to the transducer out of phase with the transmitted wave. The resulting cancellation effect reduces the energy available to the toolholders. In addition, the junction is no longer at a minimum stress level. Driving at high intensities can result in fatigue failure in the coupling stud. This condition is especially critical if the areas of the two surfaces at the junction are not equal and the larger area is on the side of the junction away from the transducer.

When two members of different materials, or different geometries, or both, are combined to form a single half-wave resonant member (as the toolholder combination, Figure 4.7), matching is accomplished by equating the magnitudes of the mechanical impedances of the two sections at the junction and at the desired frequency, that is, $|Z_h| = |Z_\ell|$. The length of the horn ℓ should always be greater than $\lambda/4$ and that of the tool (a uniform bar) always less than $\lambda/4$ for maximum displacement amplification. The junction between the tool and the toolholder should never occur at the point of maximum stress if the tool is to be driven at high intensity. The junction would fail in fatigue within a very short time.

4.5 EFFECT OF DAMPING ON THE OPERATION OF HORNS

The horn equation may be corrected for damping by inserting the term $R_m/\rho c^2 S$ as follows:

$$\frac{1}{c^2}\frac{\partial^2 \xi}{\partial t^2} + \frac{R_m}{\rho c^2}\frac{\partial^2 \xi}{\partial x \partial t} - \frac{1}{S}\frac{\partial S}{\partial x}\frac{\partial \xi}{\partial x} - \frac{\partial^2 \xi}{\partial x^2} = 0 \tag{4.63}$$

or

$$\frac{\partial^2 v}{\partial x^2} + \frac{1}{S}\frac{\partial S}{\partial x}\frac{\partial v}{\partial x} - j\frac{\omega R_m}{\rho c^2}\frac{\partial v}{\partial x} + \frac{\omega^2}{c^2}v = 0 \tag{4.64}$$

where ω is the undamped angular frequency, c is the longitudinal velocity of sound in the medium, and R_m is the loss term with dimensions as defined in Chapter 3, Equation 3.5. When $R_m = 0$, Equations 4.63 and 4.64 reduce to Equation 3.21.

The solution of Equation 4.64 depends upon S. The first step in its solution is to identify the type of horn and thereby determine the quantity $(1/S)\,\partial S/\partial x$.

Equations 4.63 and 4.64 are thus converted to second-order differential equations that can be solved by standard methods. For instance, for the exponentially tapered horn, where $S = S_0 e^{-\gamma x}$, Equation 4.64 reduces to

$$\frac{d^2 v}{dx^2} - \left(\gamma + j\frac{\omega R_m}{\rho c^2}\right)\frac{dv}{dx} + \frac{\omega^2}{c^2} v = 0 \tag{4.65}$$

The solution to Equation 4.65 is

$$v = e^{\left(\gamma + j\frac{\omega R_m}{\rho c^2}\right) x/2} V_o \left\{ \cos\frac{\omega}{c}\left[1 - \frac{\rho c^2 \gamma^2 - j\omega R_m \gamma}{4\omega^2 \rho} \right]^{1/2} x \right.$$

$$\left. - \left[\frac{\rho c^2 \gamma^2 - j\omega R_m \gamma}{2\rho c^2\left(1 - \dfrac{\rho c^2 \gamma^2 - j\omega R_m \gamma}{4\omega^2 \rho}\right)^{1/2} \omega} \right] \times \sin\frac{\omega}{c}\left(1 - \frac{\rho c^2 \gamma^2 - j\omega R_m \gamma}{4\omega^2 \rho}\right)^{1/2} x \right\} \tag{4.66}$$

The apparent phase velocity, c'_d, of an exponentially tapered horn with losses is

$$c'_d = \frac{c}{\sqrt{1 - \left[\dfrac{\rho c^2 \gamma + j\omega R_m}{2\omega\rho c}\right]^2}}$$

which converts to

$$c'_d = \frac{c}{\sqrt{1 - \dfrac{\gamma^2 c^2}{4\omega^2}}}$$

when $R_m = 0$.

4.6 WIDE HORNS AND HORNS OF LARGE CROSS SECTION

In deriving equations for designing the horns described in the previous sections of this chapter, the cross-sectional dimensions were assumed to be considerably smaller than a wavelength. There are many ultrasonic applications for wide horns and horns of large cross section. Two shapes of wide horns that are commonly used for seam welding thermoplastics are stepped wide-blade types (Figure 4.8) and wedge types. Large-area horns may have the shape of a rectangular block (Figure 4.9), or they may have any other convenient shape such as a section of a cylinder.

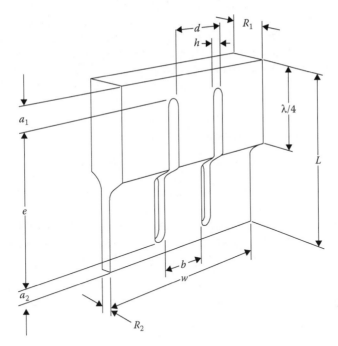

FIGURE 4.8   Wide blade-shaped horn.

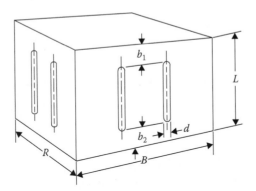

FIGURE 4.9   Large rectangular (block type) horn.

Large commercial horns contain slots running parallel to the direction of longitudinal motion, which produce the effect of arrays of narrow horns. The main reasons for slotting are to improve heat dissipation and thus prevent the creation of thermal "hot" spots in operation and to control modal characteristics. Equations for designing large horns are derived with slender horn principles as a basis. Each individual element is considered as a separate horn carrying a mass load at each end. However, since these elements are tied together through a solid and continuous element of the same material as the horn, additional factors related particularly to various potential spurious modes of vibration must be considered in their design. Poisson's ratio now is a factor in their design and performance. There is always the danger that a large horn will be driven into resonance in a spurious mode rather than the desired resonance when these two resonances occur close together. A spurious resonance is deleterious to the performance of the objective process and may cause product damage. Usually, the spurious modes can be separated from the primary mode by a judicious choice or manipulation of dimensions.

4.6.1 WIDE-BLADE TYPE HORNS

Wide-blade type horns may be of any of several geometrical designs such as arrays of uniform bars (Section 4.3.1), stepped horns (Section 4.3.2), exponentially tapered horns (Section 4.3.3), wedge-shaped horns (Section 4.3.4), and catenoidal horns (Section 4.3.6).

The procedure for designing the wide-blade type of horn will be demonstrated using the uniform bar as an example. In the wide-blade construction, each element is separated by slots from its neighboring element and is joined to its neighbor by a bridging section (Figure 4.8). The bridging section is an additional mass that presents a mechanical impedance to the longitudinal elements of the horn. The magnitudes of the impedances of the bar section and of the mass of the bridging element are equal at resonance. From Equation 3.45, we see that the impedance at any point x for the free–free bar without losses is

$$Z_b = -jS\rho c \ \tan\left(\frac{\omega x}{c}\right) \tag{3.45}$$

and the impedance of the added mass is

$$Z_m = j\omega M \tag{4.67}$$

if the mass is considered as lumped mass, or

$$Z_m = -jS_m\rho c \ \tan\left(\frac{\omega a}{c}\right) \tag{4.68}$$

if the mass is considered to be distributed. The mass M is the product of the density and volume of the end element, S is the area of the cross section of width b, and S_m is the cross-sectional area of the section of width d and length a.

The blade is assumed to be of uniform thickness, R, at all positions. The end elements are equal in length, a, and the longitudinal elements are equal in width, b, and length, e. As shown in Figure 4.8, the dimensions are related as follows:

The total width of the horn is W.
The total length of the horn is L.
The width of the slots is h.
The width of the portion of the face of the horn for each element is d.

The cross-sectional area, S, is equal to Rb.

The density of the horn material is p and the bar velocity of sound is c.

The density of the blade is constant throughout the horn.

The end elements corresponding to the dimension a present an impedance equivalent to a segment of the elements of width b at the junction between the two sections. Since it is assumed that the end elements are of equal length, a, the segments of uniform cross section between slots are less than a half-wavelength of a uniform bar by twice the equivalent length, x, of a. Then the total length, L, is

$$L = \frac{\lambda}{2} - 2x + 2a = \frac{c}{2f} - 2x + 2a \qquad (4.69)$$

Equating Equations 3.45 and 4.68 gives

$$\tan\left(\frac{\omega x}{c}\right) = \left(\frac{S_m}{S}\right)\tan\left(\frac{\omega a}{c}\right) = \left(\frac{d}{b}\right)\tan\left(\frac{\omega a}{c}\right) \qquad (4.70)$$

The equivalent length, x, of a is, therefore,

$$x = \left(\frac{c}{\omega}\right)\tan^{-1}\left[\left(\frac{d}{b}\right)\tan\left(\frac{\omega a}{c}\right)\right] \qquad (4.71)$$

From Figure 4.8, W

$$d = \frac{W}{n+1} = b + h$$

$$b = \frac{W - (n+1)h}{n+1}$$

so that

$$x = \frac{c}{\omega}\tan^{-1}\left[\frac{W}{W - (n+1)h}\tan\left(\frac{\omega a}{c}\right)\right] \qquad (4.71a)$$

Therefore, the total length, L, from Equation 4.69 is

$$L = \frac{c}{2f} + 2a - 2\left(\frac{c}{\omega}\right)\tan^{-1}\left[\frac{W}{W - (n+1)h}\tan\left(\frac{\omega a}{c}\right)\right] \qquad (4.69a)$$

Lateral dimensions in the neighborhood of

$$W = \frac{c}{2f}$$

should be avoided, if possible, to prevent resonances in a lateral direction by virtue of Poisson's ratio. Equations can be derived in a similar manner for other horn shapes.

4.6.2 Horns of Large Cross Section

Horns of large cross section, such as the rectangular (block-type) horn of Figure 4.9, are slotted for the reasons given previously. Therefore, design equations are derived in a manner similar to that followed in deriving the equations for ultrasonic blade-type horns. The equations take into account the mass loading at the ends of the individual segments and correction for lateral inertia caused by Poisson effects. Referring to Figure 4.9, if the lateral dimensions of the individual elements between slots are small compared with a wavelength of sound in the horn material, the corrected value of the velocity of sound, c', resulting from the lateral inertia is given by Derks [5] as

$$\frac{c'}{c} = 1 - \frac{1}{6}\left(\frac{\sigma\pi f}{c}\right)^2\left[\left(\frac{B-n_1d}{n_1+1}\right)^2 + \left(\frac{R-n_2d}{n_2+1}\right)^2\right]$$ (4.72)

The length, L, of the horn is then given by

$$L = \frac{c'}{2f} + b_1 + b_2 - \frac{1}{k'}\left\{\tan^{-1}\left[\left(1+\frac{n_1d}{B-n_1d}\right)\right.\right.$$
$$\left.\times\left(1+\frac{n_2d}{R-n_2d}\right)\tan(k'b_1)\right] + \tan^{-1}\left[\left(1+\frac{n_1d}{B-n_1d}\right)\right.$$ (4.73)
$$\left.\left.\times\left(1+\frac{n_2d}{R-n_1d}\right)\tan(k'b_2)\right]\right\}$$

where

  c' is the corrected value of velocity of sound
  L is the overall horn length, as shown in Figure 4.9
  b_1 and b_2 are the longitudinal lengths of the end masses $k' = \omega/c'$
  n_1 and n_2 are the number of slots through sides B and R, respectively
  σ is Poisson's ratio in the material of the horn
  B is the long side of the face of the horn
  R is the shorter side of the face of the horn
  d is the width of the slots

4.6.3 Rotating Hollow Horn

Figure 4.10 is the design of another large horn that can rotate while performing tasks such as continuous bonding of plastic sheet or strips. This is an example of combining sections of different geometrical cross sections.

  "The horn is a cylinder containing a concentric conically tapered cavity. It is loaded at the large open end by an enlarged section to be used for contacting the workpiece. The conical taper is chosen to make the piece easy to machine."

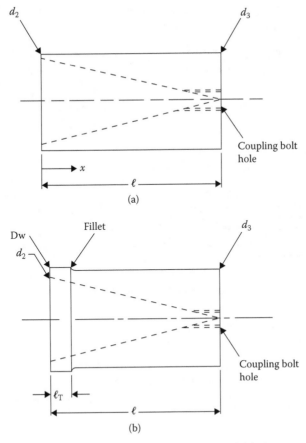

FIGURE 4.10 Horn with a cylindrical shell and cavity in which the cross-sectional area function is equivalent to that of a conical horn. (a) Basic horn. (b) With a wheel-loaded rolling type configuration integrated.

"The cross-sectional area as a function of x (measured from the small diameter end of the cone) is

$$S_x = \frac{\pi}{4}\left[d_3^2 - d_x^2\right] = \frac{\pi}{4}\left[d_3^2 - \left(\frac{d_2^2}{\ell}\right)x^2\right] \tag{4.74}$$

Inserting Equation 4.74 into the wave equation

$$\frac{\partial^2 v}{\partial x^2} + \frac{1}{S}\frac{\partial S}{\partial x}\frac{\partial v}{\partial x} + \frac{\omega^2}{c^2}v = 0$$

gives

$$\frac{d^2 v}{dx^2} - \left[\frac{2d_2^2 x}{d_3^2\ell - d_2^2 x^2}\right]\frac{dv}{dx} + \frac{\omega^2}{c^2}v = 0 \tag{4.75}$$

assuming that d_3 is small enough that the equation applies and that $d_1 = 0$. Assuming $d_3 < \lambda/4$ and $d_1 = 0$, the solution to Equation 4.75 is

$$
v = \left[\frac{(d_3 - \sqrt{d_3^2 - d_2^2})V_1}{\ell\sqrt{d_3^2 - d_2^2} + \left(d_3 - \sqrt{d_3^2 - d_2^2}\right)(\ell - x)} \right] \times \left[\frac{\ell d_3}{\left(d_3 - \sqrt{d_3^2 - d_2^2}\right)} \cos\frac{\omega x}{c} - \frac{c}{\omega}\sin\left(\frac{\omega x}{c}\right) \right]
$$

$$(4.76)$$

where

> ℓ is the half-wave length of the horn
> d_3 is the diameter of the cylinder
> d_2 is the inside diameter of the open end of the horn
> d_1 is the diameter of the small end of the cone
> d_x is the inside diameter of the horn at x
> V_1 is the velocity of the closed end of the horn
> V_2 is the velocity of the open end of the horn

The *velocity node* occurs where

$$
\tan\frac{\omega x}{c} = \frac{\omega\ell}{c}\frac{d_3}{d_3 - \sqrt{d_3^2 - d_2^2}}
$$

$$(4.77)$$

The half-wave resonant length is determined by

$$
\tan\frac{\omega\ell}{c} = \frac{\omega\ell c\left(d_3 - \sqrt{d_3^2 - d_2^2}\right)^2}{\omega^2\ell^2 d_3\sqrt{d_3^2 - d_2^2} + c^2\left(d_3 - \sqrt{d_3^2 - d_2^2}\right)^2}
$$

$$(4.78)$$

The stress at x is given by

$$
S = j\frac{\left(d_3 - \sqrt{d_3^2 - d_2^2}\right)YV_2}{\left[\ell\sqrt{d_3^2 - d_2^2} + \left(d_3 - \sqrt{d_3^2 - d_2^2}\right)(\ell - x)\right]^2 \omega} x
$$

$$
\left\{ \left(d_3 - \sqrt{d_3^2 - d_2^2}\right)(x - \ell)\cos\frac{\omega}{c}(x) - \frac{\omega}{c}\left[\frac{\ell^2 d_3\sqrt{d_3^2 - d_2^2}}{d_3 - \sqrt{d_3^2 - d_2^2}}\right] \right.
$$

$$(4.79)$$

$$
\left. + \frac{c^2}{\omega^2}\left(d_3 - \sqrt{d_3^2 - d_2^2}\right) + d_3\ell x\right]\sin\frac{\omega x}{c}\right\}
$$

The impedance of the horn at x is $Z = F/v = sS/v$

$$Z = j \frac{\pi\left[\ell\sqrt{d_3^2 - d_2^2} + \left(d_3 - \sqrt{d_3^2 - d_2^2}\right)x\right]Y}{4\omega\ell^2\left[\dfrac{\ell d_3}{d_3 - \sqrt{d_3^2 - d_2^2}} + \dfrac{c}{\omega}\tan\dfrac{\omega}{c}(x-\ell)\right]}$$

$$\times\left\{\left(d_3 - \sqrt{d_3^2 - d_2^2}\right)(x-\ell) - \frac{\omega}{c}\left[\frac{\ell^2 d_3 \sqrt{d_3^2 - d_2^2}}{d_3 - \sqrt{d_3^2 - d_2^2}}\right.\right.$$

$$\left.\left. + \frac{c^2}{\omega^2}\left(d_3 - \sqrt{d_3^2 - d_2^2}\right) + d_3\ell x\right]\tan\frac{\omega}{c}(x-\ell)\right\}$$

(4.80)

The impedance equation above is used to match the wheel segment of the horn to the horn where x is the value of x from the end of a full half-wave horn to the junction between the horn and the wheel mass.

"The horn should be well finished all over to remove all stress concentrations."
"The enlarged (wheel) portion should be hard-faced to minimize erosion due to wear."

4.7 ADVANCED HORN AND SYSTEM DESIGN

The analytical equations presented in this chapter remain useful for providing initial designs. Also, with the move to use computer-based tools for design development and evaluation, the analytical equations remain important for model validation.

With improved design and more efficient drive electronics, there has been a significant increase in the range and numbers of high-power horns being used in the research and industrial communities. Some of the diversity in available horns are illustrated with the examples in Figure 4.11: (a) is a standard cylindrical horn used in many sample preparation and small-scale sonochemistry experiments; (b) is a shorter version of the same design; (c) is a microtip—used with small sample containers such as endorf tubes; and (d) are exponential horns with surgical tools.

The horn is only part of a high-power ultrasonic system. It must be integrated with a drive or electro–mechanical converter unit and a power supply to provide excitation power. The converter is typically a piezoelectric stack or magnetostrictive unit of the types that are discussed in Chapter 5. An example of the elements in a high-power sonifier system, designed for cell disruption, is shown in Figure 4.12.

A series of high-power ultrasonic system analyses and the models used for the design of ultrasonic units that employ horns are now reported in the literature. There is increasing use of computer models to simulate both the mechanical and the electrical responses. The capability of finite element analysis to give horn vibration modes is illustrated with the data calculated by Morris et al. [6] for the case of an ultrasonic bonding tool. Examples of the mode shapes that are obtained are shown in Figure 4.13.

There are also models for the electrical drive used with horns and the analysis of impedance loading in ultrasonic transducer systems, such as those used in bonding systems [7], and these include the effect of loading and the interdependence of the

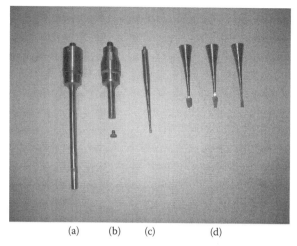

(a)   (b)   (c)   (d)

FIGURE 4.11   Examples of horns. (a) Standard cylindrical horn. (b) Short horn, shown with replaceable titanium tip removed. (c) Wedge-shaped horn (microtip). (d) Exponential solid horns with examples of tools used in surgery.

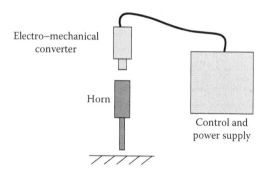

FIGURE 4.12   Elements in an ultrasonic sonifier cell disruption system.

electrical drive, the drive–horn unit, and load, which, in the case of wire bonding, is the wire being joined and the anvil.

The range of models and applications is further illustrated with a less conventional example: a model and computer simulation for a NASA ultrasonic rock drill/corer, which is reported in an article by Bao et al. [8]. This unit employs a PZT ($Pb(Zr_{1-x}Ti_x)O_3$, lead zirconate titanate, which exhibits a piezoelectric effect) stack to drive a horn. The article reports the equivalent electrical circuit, an analysis of the mass–horn interaction, and an integrated computer model used for design development and evaluation.

The effects produced by high-intensity ultrasound and a range of applications are discussed in Chapters 11 through 13. Some high-power medical applications of ultrasound are discussed in Chapter 14.

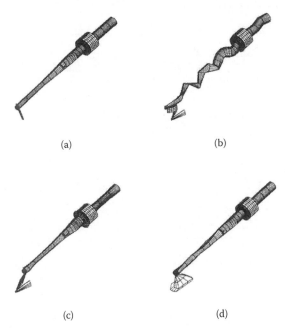

(a)   (b)

(c)   (d)

FIGURE 4.13   Mode shapes computed using finite element analysis. (a) Undeformed. (b) Bending mode (59.7 kHz). (c) Axial mode (61.5 kHz). (d) Tip expansion, torsion (64.6 kHz). (Reprinted with permission from Morris, R. B., P. Carnevali, and W. T. Bandy, *J Acoust Soc Amm* 90(6):2919–23, Acoustical Society of America, 1991.)

REFERENCES

1. Hulst, A. P. 1973. On a family of high-power transducers. In *Proc. Ultrasonics International 1973*, 27–9. Guildford: IPC.
2. Mason, W. P., and R. F. Wick. 1951. *J Acoust Soc Am* 23(2):209–14.
3. Ensminger, D. 1960. *J Acoust Soc Am* 32(2):194–6.
4. Merkulov, L. G. 1957. *Sov Phys Acoust* 3:246–55.
5. Derks, P. L. L. M. 1984. *The Design of Ultrasonic Resonators with Wide Output Cross-Sections*. Eindhoven, The Netherlands: Philips.
6. Morris, R. B., P. Carnevali, and W. T. Bandy. 1991. *J Acoust Soc Am* 90(6):2919–23.
7. Ensminger, D. 2009. Ultrasonic horns, couplers, and tools. In *Ultrasonics: Data, Equations, and Their Practical Uses*, 2009, eds. Ensminger D., and F. B. Stulen, pp. 29–128. New York: CRC Press.
8. McBrearty, M. M., L. H. Kim, and N. M. Bilgutay. 1988. Analysis of impedance loading in ultrasonic transducer systems. In *IEEE Ultrasonics Symposium, 1988*, ed. B. R. McAvoy, Chicago, IL, Oct. 2–5, 1:497–502. New York: Ultrasonics Ferroelectrics and Frequency Control Society, Institute of Electrical and Electronic Engineers.
9. Bao, X., Y. Bar-Cohen, Z. Chang, B. P. Dolgin, S. Sherrit, D. S. Pal, S. Du, and T. Peterson. 2003. *IEEE Trans UFFC* 50(9):1147–60.

5 Basic Design of Ultrasonic Transducers

5.1 INTRODUCTION

Ultrasonic energy is generated and detected by devices called *transducers*. By definition, a transducer is a "device that is actuated by power from one system to supply power in any other form to a second system"; that is, a transducer converts energy from one form to another. In ultrasonics, the most typical conversions are electrical to ultrasonic energy (transmitters) or ultrasonic to electrical energy (receivers). Transducers most often used for generating ultrasonics are piezoelectric, magnetostrictive, electromagnetic, pneumatic (whistles), and mechanical devices. Transducers most often used for detecting or receiving ultrasonic energy are piezoelectric, capacitive or electrostatic, and magnetostrictive devices. For some remote and noncontact applications, particularly in hostile environments, laser-based ultrasound generation and detection systems are also now available. A few basic principles of transducer design are presented in this chapter. Manufacturers of piezoelectric or magnetostrictive materials usually make available specific information on the characteristics of their products, which is helpful in designing and fabricating transducers and ultrasonic systems.

Piezoelectric, magnetostrictive, capacitive or electrostatic, and electromagnetic transducers are electrical devices; that is, they convert electrical energy to ultrasonic energy or ultrasonic energy to electrical energy. They are also mechanical devices in that mechanical oscillations are involved in the generation and detection of ultrasonic energy. Transducer designers must be concerned with the electrical and acoustic characteristics of the transducer in relation to associated electrical equipment such as electronic power sources or receiver amplifiers. They must be concerned with the mechanical properties of the transducer in relation to its vibrational performance, coupling energy, and its ultrasonic wave-field radiation characteristics. The electrical design of these "electromechanical" transducers is emphasized in this chapter. Mechanically, the transducers incorporate the characteristics of ultrasonic transmission lines with various terminations, including reflections and transmissions of energy across boundaries, for which adequate equations are provided in Chapters 2 through 4. A little investigation will reveal that a large number of possible configurations include bars vibrating in longitudinal modes, plates vibrating in thickness modes, bars vibrating in flexure, and so on.

In the transfer of energy from the electrical system to the mechanical system and vice versa, there must be an interaction between the two systems; that is, the impedance measured at the electrical terminals is influenced by the motional characteristics of the transducer and by the type of mechanical loading that it may experience. In other words, the transducer is equivalent to an electrical transformer in which the electrical components, that is, the true electrical resistances and reactances, appear in the primary circuit and the mechanical components appear in the secondary circuit. With respect to their influence on the electrical properties of the transducer, the mechanical components are real impedances, which can be computed by using certain transformation constants, to be discussed later. These properties are convenient for use in the *initial* electrical design of transducers, and it must be the initial design in the case of transducers for high-intensity applications because these devices are nonlinear and their loads may also exhibit nonlinear characteristics as functions of intensity. They also provide means for predicting and measuring efficiencies of transducers on the basis of the electrical characteristics of the transducers. To facilitate accomplishing these tasks, the transducers are represented by "equivalent" electrical circuits, which contain components representing all of the factors known to influence the electrical characteristics of the transducer. These circuits are similar to equivalent circuits of electrical transformers. Representative equivalent circuits for piezoelectric and magnetostrictive transducers are described in this chapter.

In the analysis of ultrasonic systems, various mathematical descriptions and integration models have been developed. These include the SONAR equations and equivalents used in industrial-scale sensing and nondestructive testing (NDT). These equations and some wave-field characteristics were introduced in Chapter 2, and they are discussed further at the end of this chapter.

A brief discussion of the design of each of the following types of transducers is presented in this chapter: piezoelectric, capacitive, magnetostrictive, electromagnetic, pneumatic, sirens, mechanical, and lasers.

Piezoelectric transducers are used at all ultrasonic frequencies for generating and detecting ultrasonic energy at all levels of intensity. These transducers utilize piezoelectric components, such as plates or other suitable configurations, which generate a charge on preferred surfaces under the influence of stresses or which change dimensions when they are subjected to an electric field. The diversity found in the available piezoelectric elements is illustrated with the examples shown in the photograph given as Figure 5.1 [1]. Applications of piezoelectric transducers range from ultrasonic velocity measurements, medical diagnosis, fish detection, and NDT at low intensities to humidifiers, cleaning, and welding at high intensities.

Magnetostrictive transducers are used at the lower ultrasonic frequencies, primarily for generating high-intensity ultrasonic energy. Magnetostrictive transducers utilize materials that exhibit the magnetostrictive effect, that is, materials that change dimensions under the influence of a magnetic field. Typical uses of this type of transducer are driving ultrasonic cleaners and ultrasonic processing units and machine tools.

Electromagnetic acoustic transducers (EMATs) make use of the attractive forces of electromagnets to generate vibrations. EMATs are used to obtain high-amplitude vibrations at frequencies usually below the ultrasonic range and for a special class of ultrasonic transducers employed in NDT, including those used to produce shear horizontal waves.

Capacitive transducers are used primarily for measuring displacements of active elements such as those that occur at the high-amplitude end of ultrasonic horns.

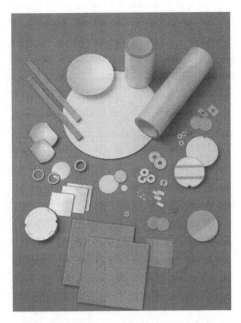

FIGURE 5.1 (**See color insert.**) Example of the range of commercially available piezoelectric element transducers. (Courtesy of Piezotech LLC.)

Reluctance types of transducers have also proved to be effective for such measurements to frequencies close to 100 kHz.

Pneumatic transducers, such as whistles, are useful at frequencies extending into the lower ultrasonic range to produce small-particle sprays of liquids for use in burners, coating materials, humidity control, and certain types of bulk cleaners. They have been used to dry materials and to break up foams resulting from various chemical processes.

Hydraulic transducers have been used for emulsifying materials rapidly. A high-pressure jet of one liquid is forced through a narrow slot-type orifice against the edge of a metal reed. The reed is mounted at two flexural nodal positions and vibrates violently under the influence of the liquid jet. The second liquid is introduced into the active zone of the whistle, where emulsification takes place. The active zone is usually located in a cavity designed to resonate at the frequency of the resonant reed. A second type of hydraulic transducer operates on a water hammer principle. It can be designed in a number of configurations to perform different functions.

Mechanical transducers are devices that are actuated mechanically. Their applications are more correctly classed as macrosonic. They are used to obtain high-amplitude, often high-intensity, vibrations at low sonic frequencies. A typical example is the sonic pile driver in which counterbalanced eccentrics are used to generate vibrations in a direction parallel to the axis of the pile.

Sirens have been made to operate at sonic and low ultrasonic frequencies. These transducers employ rotating perforated disks or cylinders, which effect intermittent opening and closing of ports through which air (or other suitable gas) passes to

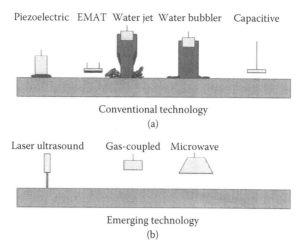

FIGURE 5.2 Conventional and emerging ultrasonic transduction technologies. (After Green, R. E., Emerging technologies for NDE of aging aircraft structures, In Nondestructive Characterization of Materials in Aging Systems, *Mat Res Soc Symp Proc*, Vol. 503, 3–14, 1998.)

generate the sound. Mechanical tolerances must be very close for efficient operation at ultrasonic frequencies. Liquid sirens also have been studied but practical considerations of design, reliability, and application have hindered development of these sirens for industrial use.

Transducers using combustion and detonation principles have been used for specific applications. Combustion can be effective in generating continuous waves by synchronizing the fuel injection and ignition with the reverberating pressure wave in a resonating gas column. A combustion type of transducer is also a low-frequency device, but extremely high intensities can be generated in air at frequencies to 1000 Hz. Detonation-type transducers are useful in generating high-pressure pulses. The blast can be "shaped" into a pulse of a few cycles by using reflectors spaced at regular intervals from the detonation source. The pulse repetition rate is necessarily low, if detonating mixtures of gas are used, because time is required to refill the cavity with the proper mixture to accomplish detonation. A 50–50 mixture of acetylene and oxygen is a typical detonation mixture.

As ultrasonic measurements move into more diverse types of sensing applications, new transduction technologies are being added. For example, laser-based techniques are now available for both generation and detection of ultrasonic waves [2]. The full range of both conventional and emerging transduction/sensing technologies is discussed by Green [3] and is illustrated in Figure 5.2.

5.2 EQUIVALENT CIRCUITS

As mentioned previously, the electrical characteristics of transducers are influenced by their mechanical properties and by the types of mechanical loading to which they are subjected. Since these influences have the characteristics of electrical impedances, they may be represented in a circuit diagram, which also includes all the electrical parameters of the transducer. A circuit diagram of this type is called an equivalent circuit. It is useful for the electrical design of the transducer, for predicting its efficiency and power-handling capabilities, and for analyzing the performance of a transducer after it is constructed.

An equivalent circuit is usually based on the assumption that the transducer operates in a linear manner and, therefore, all circuit values are constant. This assumption is true only to a limited extent. Nonlinearities occur in the stress–strain curve of materials used in the construction of the transducer, in the electrical–mechanical interaction at various driving potentials, and in variations in the acoustical load impedance at various levels of intensity. Although nonlinearities cause alterations in the constants of the circuit, they do not diminish its value for transducer design or analysis. Equivalent circuits that reliably model transducer characteristics can be drawn for specific conditions of operation. If these conditions are changed, the parameters of the circuit must be changed accordingly.

5.3 PIEZOELECTRIC TRANSDUCERS

Many books and articles have been written on the subject of piezoelectricity and piezoelectric properties in various materials and transducers [4]. Therefore, only basic principles of piezoelectricity are discussed in these few paragraphs and only for the purpose of clarifying the design of transducers based on the piezoelectric effect.

As defined by Mason [5], piezoelectricity is "pressure electricity"; a pressure applied along certain crystallographic axes produces electrical charges on preferred crystallographic surfaces. The inverse is also true. A voltage applied between two preferred surfaces produces a stress or strain along axes of the crystal. Piezoelectricity appears only in insulating or dielectric materials.

The *American Institute of Physics Handbook* [6] defines piezoelectric effects as "the phenomena of separation of charge in a crystal by mechanical stresses and the converse." The basic element that causes this effect is the *electrical dipole*, which is defined as a pair of equal and opposite electric charges having their effective centers separated by a finite distance. A stress applied along the dipole changes the distance between charges, thus causing the charge density at each position to increase or decrease depending upon the type of stress, that is, compression or tension, respectively. Thus, if a crystal consists of a group of individual unit cells each of which contains a dipole and if the unit cells are aligned so that the dipoles are effectively oriented parallel to one another and perpendicular to two parallel surfaces of the crystal, a pressure between the parallel surfaces will produce net charges of equal and opposite polarity on these surfaces. The converse is also true: if a voltage is applied between the two surfaces, the crystal will either expand or contract depending upon the polarity of the applied voltage. In certain crystals, such as quartz, dipoles appear naturally in an alignment that produces a piezoelectric effect.

A more general term for the condition found in all dielectric materials in which the material changes dimensions when placed in an electric field is *electrostriction*. In most dielectric materials, this effect is very small. The change in dimension is a function of the field voltage squared, or to some power, and it is always in the same direction regardless of the polarity of the electric field. A random orientation of the dipoles in such materials prevents electrostrictive materials from being piezoelectric until the generally random dipole orientation is transformed into an arrangement favoring a piezoelectric effect, that is, the generation of charges on the faces of a plate of the material when the plate is subjected to compression. Most piezoelectric materials used in ultrasonics, especially for power applications, are of this latter type.

These induced piezoelectric materials are of a class of electrostrictive materials called *ferroelectrics*, which gets its name from similarities in characteristics with those of magnetic materials. The *American Institute of Physics Handbook* [6] defines *ferroelectric materials* as "those in which the electric polarization is produced by cooperative action between groups or domains of collectively oriented molecules." A ferroelectric crystal also may be defined as "a crystal which belongs to the pyroelectric family and of which the direction of spontaneous polarization can be reversed by an electric field" [7]. The usual test for the ferroelectric property of a material is to trace the relationship between the voltage across the faces of a crystal of the material and the charges appearing on these faces. The faces of the crystal are normal to one of the three mutually orthogonal directions, and the applied voltage is usually at a frequency of 50–60 Hz. If the trace shows a hysteresis-type loop, the material is ferroelectric. This is one of the similarities between ferroelectric and ferromagnetic materials. Quartz does not show a hysteresis loop.

The "three mutually orthogonal axes" are those of the coordinate system established by crystallographers for the purpose of unifying crystal terminology. These axes are designated *X*, *Y*, and *Z*. The electrical, mechanical, and electromechanical properties of piezoelectric crystals differ along these different axes; therefore, the constants relating stress and voltage, strain and voltage, electromechanical coupling coefficients, capacitance, and so on differ. Subscripts 1, 2, and 3, corresponding respectively to the axes *X*, *Y*, and *Z*, are used in tabulating or discussing these quantities. The direction in which a piezoelectric ceramic plate is poled is designated the 3 direction. The 1 and 2 directions are normal to each other and arbitrarily located in a plane that is normal to the 3 direction. When a crystal is used in a design in which the motion is parallel to the direction of the applied electrical potential, the crystal is said to operate in the 33 thickness mode. If the voltage is applied along the 3 axis and the motion is along the 1 axis, the crystal is said to operate in the 31 mode.

Ferroelectric ceramic materials, such as lead zirconate titanates, and polymeric materials, such as poly(vinylidene fluoride; PVF_2), are made piezoelectric by a biasing or polarization procedure. The procedure includes elevating the temperature of the material to above the Curie point and allowing it to cool slowly in a high dc electric field, which is oriented along an axis of planned piezoelectric excitation. The Curie point is the temperature above which a material becomes inactive. A crystal may also exhibit a lower Curie temperature below which the material is inactive. If a naturally piezoelectric crystal, such as quartz, is subjected to temperatures above the Curie point, its piezoelectric activity is restored on cooling. If a ferroelectric or piezoelectric material, such as barium titanate or lead zirconate titanate, is subjected to such high temperatures, it loses its piezoelectric property permanently.

Piezoelectric materials, for use in ultrasonic transducers, are now available in a vast range of shapes, sizes, and thicknesses, as is illustrated by those shown in Figure 5.1. The materials are, in many cases, available with electrode material

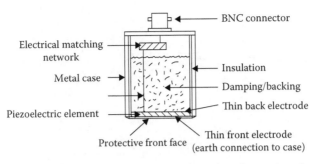

FIGURE 5.3   Basic structure of an ultrasonic transducer, based on a piezoelectric disc.

already applied in the form of thin metallic layers. Materials have been developed to operate at both high (500°C and above) and low (cryogenic) temperatures, and in harsh environments, such as nuclear power plants where there are high doses of radiation [8]. The basic structure of an ultrasonic transducer, using a piezoelectric element, for NDT and many other sensing applications is shown in Figure 5.3. The element is damped with a backing (commonly a particulate-loaded epoxy), the properties of which determine the Q and bandwidth (and hence pulse length) for the resulting acoustic waves. The fabrication of such transducers, although apparently simple, remains a mix of art and science, and the fabrication of "identical" units, even from the same batch, is hard to achieve. The causes of non-ideal behavior was investigated and reported by Bond et al. [9]. Several examples of ultrasonic transducers are shown in Figure 5.4. The internal structure of a 2.25 MHz, 2-cm diameter, focused compression wave immersion transducer is shown in Figure 5.4a. A 5 MHz, 12-mm diameter, 10-cm focal length compression wave immersion transducer is shown in Figure 5.4b, and a 2.25 MHz, 12-mm diameter contact compression wave transducer is shown in Figure 5.4c. A pinducer, a small transducer with an active element less than 2 mm in diameter used for field probing, is shown in Figure 5.4d. A compression wave transducer set on a wedge for shear wave generation is shown in Figure 5.4e.

Ultrasonic transducers, employing some form of piezoelectric material, are now available to operate at frequencies from below 20 kHz to above 2 GHz. Transducers have been designed to operate and give wave fields in solids, liquids (including tissue), and gases. Transducers, such as those shown in Figure 5.4, generate compression waves, and the elements are excited in the crystal thickness mode. Such transducers can be coupled to a wedge (Figure 5.4e) and used to generate shear, Rayleigh, or plate waves, depending on wedge geometry. These forms of wedge transducers are commonly employed in NDT and are discussed in later chapters of this book. The transducers operating in the highest parts of the frequency range commonly employ an internal buffer rod and are used for acoustic microscopy and spectroscopy. Custom transducer designs, with special material selection, have also been developed to meet the needs for operation in a wide range of harsh environments, including with aggressive chemicals, with radiation, and at both low (cryogenic temperatures) and high temperatures (up to and above 1000°C). A recent review of medical and industrial ultrasonic transducers, including arrays, found more than

FIGURE 5.4   (See color insert.) Examples of compression wave transducers. (a) Sectioned 2.25 MHz, 2 cm diameter, focused immersion transducer. (b) 5 MHz, 12 mm diameter, 10 cm focused immersion transducer. (c) 2.25 MHz, 12 mm diameter contact transducer. (d) Pinducer—a small transducer-active element less than 2 mm in diameter. (e) A transducer on a wedge for shear wave generation, using mode conversion, with 1 cent coin shown for scale.

3000 commercially available designs. Many hundreds of these are in the form of standard transducer designs, which are also commercially available and used for NDT and other industrial measurement applications.

5.3.1   Equivalent Circuit of a Simple Piezoelectric Transducer

A simple piezoelectric transducer, consisting of a plate vibrating in the thickness mode, may radiate from both faces (symmetrical load) or from only one face (air backed). An air-backed transducer has only one-half the radiating surface of a symmetrically loaded transducer. Thus, if a transformation factor α is used to convert mechanical impedances to electrical impedances in a symmetrically loaded transducer, a factor 2α must be used for the air-backed transducer.

To develop the equivalent circuit, it is necessary to determine the quantities that influence the overall input impedance to the transducer. In a piezoelectric material, the imposed stress is in phase with the impressed voltage. The piezoelectric element is an electrical capacitor of capacitance C_0 by virtue of the dielectric nature of the transducer material and the electroded surfaces. The electrical resistance of the capacitor is negligible. The capacitance C_0 appears in parallel with a series branch that includes the converted mechanical impedances. These mechanical impedances consist of (1) a resistance R_ℓ corresponding to the losses in the transducer (negligible in a high-Q piezoelectric transducer); (2) a load resistance or radiation impedance $Z_r = \rho_0 c_0 S$, where $\rho_0 c_0$ is the characteristic acoustic impedance of the load and S is the area of the radiating surface; (3) an inductance M due to the mass of the transducer; and (4) a capacitance $1/K$ due to the compliance of the transducer. The equivalent electrical components, assuming $R\ell$ negligible, are

For a symmetrically loaded transducer:

$$R_R = \frac{Z_R}{\alpha^2} \quad L = \frac{M}{\alpha^2} \quad C = \frac{\alpha^2}{K} \tag{5.1a}$$

For an air-backed transducer:

$$R_R = \frac{Z_R}{4\alpha^2} \quad L = \frac{M}{4\alpha^2} \quad C = \frac{4\alpha^2}{K} \tag{5.1b}$$

Since the transducer is a simple plate, a simple equivalent circuit can be drawn, as shown in Figure 5.5.

The remaining discussion applies only to air-backed transducers. Figure 5.5 transforms into Figure 5.6 by virtue of the transformer ratio $N = 1/2\alpha$.

It should be noted that the circuits of Figures 5.5 and 5.6 apply only to a specific type of transducer, a simple piezoelectric plate. Additional components, such as mechanical horns, necessitate additions to the circuit, and the values of these as well as those of the driving element alone depend upon the geometry of the additional element or elements, the mass, the elasticity, and the mode of operation of the overall system.

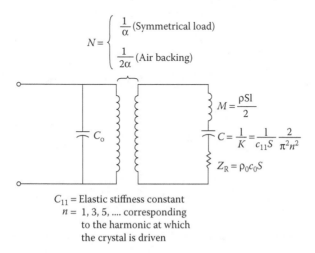

FIGURE 5.5   Equivalent circuit of a simple piezoelectric transducer.

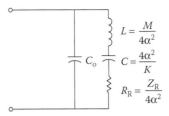

FIGURE 5.6 Equivalent circuit of a simple air-backed piezoelectric transducer.

5.3.2   EFFICIENCY OF A SIMPLE PIEZOELECTRIC TRANSDUCER

The efficiency of any system is the ratio of the power out to the total power into the system. Referring to Figure 5.6, the total power dissipation due to the transducer is that which is radiated into the load through R; that is, the efficiency of such a transducer would be 100%. In actual practice, transducer losses are present, and in the piezoelectric elements, they include dielectric losses and internal losses due to the strain.

At resonance, $\omega L = 1/\omega C$, and inserting the resistances R_D corresponding to the dielectric losses and R_f corresponding to the internal strain losses, Figure 5.6 may be modified according to Figure 5.7.

Referring to Figure 5.7, the power output of the transducer is

$$P_{\text{out}} = I_3^2 R_R \tag{5.2}$$

The total power input of the transducer is

$$P_{\text{in}} = I_1^2 R_D + I_3^2 (R_f + R_R) \tag{5.3}$$

The efficiency of the transducer is

$$\eta = \frac{P_{\text{out}}}{P_{\text{in}}} = \frac{R_D R_R}{(R_f + R_R)(R_f + R_R + R_D)} \tag{5.4}$$

or, in terms of acoustical radiation impedance,

$$\eta = \frac{R_D (\rho_0 c_0 S/4\alpha^2)}{(R_f + \rho_0 c_0 S/4\alpha^2)(R_f + R_D + \rho_0 c_0 S/4\alpha^2)} \tag{5.5}$$

where S is the cross-sectional area of the transducer and also the area of the radiating surface.

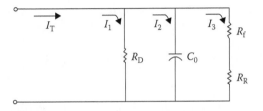

FIGURE 5.7 Equivalent circuit of a simple resonant piezoelectric transducer with losses.

5.3.3 MAXIMUM POWER TRANSFER BETWEEN ELECTRONIC POWER SOURCE AND SIMPLE PIEZOELECTRIC TRANSDUCERS

According to elementary electric-circuit theory, maximum power transfer from a generator to a load occurs as follows:

1. When the internal impedance of the generator is resistive, the load is resistive and equal in magnitude to the internal generator resistance.
2. When the internal generator impedance is complex, the load impedance is the conjugate of the generator impedance.
3. When neither 1 nor 2 is possible, the load impedance is equal in magnitude to the generator impedance.

In the equivalent circuit of Figure 5.7, C_0 causes a low power factor, which usually may be eliminated by "tuning out" C_0 with an inductance in parallel with the transducer. Therefore, for maximum power transfer to the transducer

$$R_i = \frac{R_D(R_f + R_R)}{R_D + R_f + R_R} = \frac{R_D(R_f + \rho_0 c_0 S/4\alpha^2)}{R_D + R_f + \rho_0 c_0 S/4\alpha^2} \tag{5.6}$$

where R_i is the internal resistance of the generator.

Resistances R_D and R_f are determined by the material of the transducer and its dimensions. For the condition of Equation 5.6 to exist

$$R_R = \frac{R_i R_D}{R_D - R_i} - R_f \tag{5.7}$$

The maximum transfer of energy to the acoustic load occurs when $R_R = R_f$, at which time

$$R_i = \frac{2R_D R_R}{R_D + 2R_R}$$

$$R_R = \frac{R_i R_D}{2(R_D - R_i)} = \frac{\rho_0 c_0 S}{4\alpha^2} \quad \text{(air-backed transducer)} \tag{5.7a}$$

for maximum power transfer from the generator to the load. Substituting Equation 5.7a in Equation 5.5 gives an efficiency

$$\eta = \frac{R_D}{2(R_D + 2R_R)} \tag{5.8}$$

when the transducer is matched to the generator and the radiation impedance is equal in magnitude to the internal loss resistance of the transducer. If, in this special case, R_D is negligible, that is, $R_D = \infty$, the transducer efficiency is 50% and the overall efficiency of the ultrasonic system is 25%.

Other equivalent circuits representative of more complex systems can be developed by using the procedures followed in these sections. The previously derived equations for efficiency and maximum power transfer are easily adapted to the special case of a resonant piezoelectric transducer radiating through a half-wave mechanical transformer (solid double cylinders, exponential tapers, etc.) without losses. In each formula, the value of R_R representing the actual radiation impedance is multiplied by the appropriate transformation ratio. Typical ratios are

For the exponential taper, $N_1/N_2 = d_1/d_2$.
For the double cylinder, $N_1/N_2 = d_1^2/d_2^2$.
For the cone, N_1/N_2 depends upon the dimensions; goes to a limit of 4.6.
For the wedge, N_1/N_2 depends upon the dimensions; solution is a Bessel function.

For example, the efficiency of a half-wave resonant piezoelectric transducer radiating through a half-wave lossless exponentially tapered horn is

$$\eta_{exp} = \frac{R_D (d_1/d_2)^2 R_R}{[R_f + (d_1/d_2)^2 R_R][R_f + (d_1/d_2)^2 R_R + R_D]} \tag{5.5a}$$

and maximum power transfer from the generator to the transducer occurs when

$$\left(\frac{d_1}{d_2}\right)^2 R_R = \frac{R_i R_D}{R_D - R_i} - R_f \tag{5.7b}$$

where S_2 is the radiating surface $\pi d_2^2/4$ and $R_R = \rho_0 c_0 S_2/4\alpha^2$. If d_1 of the exponentially tapered horn is equal to the diameter of the piezoelectric transducer

$$\left(\frac{d_1}{d_2}\right)^2 R_R = \frac{\rho_0 c_0 S_1}{4\alpha^2}$$

which is identical to the radiation impedance of the transducer working into the same load without the horn. The exponentially tapered horn is the only one, other than the straight cylinder, that produces this condition.

5.3.4 DETERMINING TRANSFORMATION FACTOR (α) FOR THE PIEZOELECTRIC TRANSDUCER MATERIAL

The transformation factor α relates mechanical impedances to electrical impedances for use in the design of ultrasonic systems. The constant α is a function of the electromechanical coupling factor k_C by the relationship

$$\alpha^2 = \frac{\text{clamped capacitance of the piezoelectric element}}{\text{mechanical compliance of the piezoelectric element}} \times k_C^2$$

$$= \frac{C_0}{1/K} k_C^2$$

(5.9)

where k_C^2 is defined as the energy transformed divided by the total input energy, regardless of whether electrical energy is transformed to mechanical energy or vice versa. The approximate value of k_C^2 may be obtained by measuring the frequency, f_r, at resonance and the frequency, f_a, at antiresonance. Then

$$k_C^2 = \frac{\pi^2}{4} \frac{f_a - f_r}{r_r}$$

(5.10)

5.3.5 QUALITY FACTOR (Q) OF PIEZOELECTRIC TRANSDUCERS

In electrical systems, the quality factor (Q) of an inductance or a capacitance is the ratio of the energy stored in the component to the energy dissipated in an associated resistance. Similarly, the Q of a transducer is the ratio of the energy stored in its reactive components to the energy dissipated in its resistive components, which include the radiation resistance. The Q of a transducer is a measure of the sharpness of resonance; therefore, the operating frequency bandwidth of the transducer is a function of the Q.

Neglecting all electrical dissipative factors, from Figure 5.6 we see that the Q of a simple piezoelectric transducer with air backing is

$$Q = \frac{\omega L}{R_R} = \frac{\omega M/4\alpha^2}{\rho_0 c_0 S/4\alpha^2} = \frac{\omega M}{\rho_0 c_0 S}$$

(5.11)

where M is the equivalent mass of the transducer. ($M = \rho_1 S\ell/2$, where ℓ is the thickness of the transducer.) Then

$$Q = \frac{\pi \rho_1 f\ell}{\rho_0 c_0} = \frac{\pi \rho_1 (\lambda f/2)}{\rho_0 c_0} = \frac{\pi \rho_1 c_1}{2\rho_0 c_0}$$

(5.11a)

or, for the nth. harmonic,

$$Q_n = \frac{n\pi}{2} \frac{\rho_1 c_1}{\rho_0 c_0} \quad \text{(air backed)} \tag{5.11b}$$

Equations 5.11, 5.11a, and 5.11b apply to ideal lossless transducers in which R_D and R_f are negligible.

The effect of "Q" on the frequency and time domain responses of transducers is a key metric used in transducer design and in the assessment of the suitability of a transducer's performance for a particular application. The effect of Q on frequency domain response is illustrated in Figure 3.6 [10]. Figure 3.6a and b show sketches illustrating spectra for high- and low-Q responses, respectively. The corresponding time domain responses and the pulse lengths for the conditions shown in Figures 3.6a and 3.6b ($Q = 20$ and 2 respectively) are shown in Figure 3.6c and d. For high spatial resolution NDT these require wide band-width and low Q, but these conditions reduce both transmitted signal amplitude and receiver sensitivity. Such transducers also typically have a smaller dynamic range. High-Q transducers deliver more energy, which have narrow bandwidth but long pulses and can give less spatial resolution. These transducers do typically deliver more energy and have higher dynamic range [10].

5.3.6 KLM AND EXAMPLES OF DESIGNS USING TRANSDUCER MODEL

The prediction of performance for broadband transducers operating into a fluid or solid has remained difficult and requires the use of models. One such model that does provide insights into performance for such conditions was developed by Krimholtz, Leedom, and Mattaei, and this is commonly called the "KLM Model" [11]. In the KLM model, the transducer is treated as a three-port electrical network where an electrical source is applied to the center of a "transmission line," as shown in Figure 5.8. A summary of the detailed mathematics for this equivalent circuit,

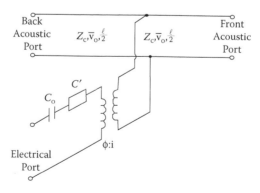

FIGURE 5.8 KLM model of a piezoelectric transducer. (From Krimholtz, R., D. A. Leedom, and G. L. Matthaei, *Electron Lett*, 6(13):398–9, IET, 1970. With permission.)

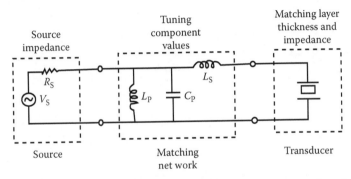

FIGURE 5.9   Equivalent circuit showing driving source, matching network, and simplified transducer. (From Selfridge, A. R., R. Baer, B. T. Khuri-Yakub, and G. S. Kino, Computer-optimized design of quarter-wave acoustic matching and electrical matching networks for acoustic transducers, *Ultrason Symp Proc (IEEE)* 81-CH1689-9, Vol. 2, 644–48, © 1981 IEEE. With permission.)

together with examples of both model and measured data, is given by Kino [4] and for various examples in a number of journal articles.

An ultrasonic transducer's performance is not independent from the characteristics of the electrical source, any tuning components, acoustic matching layers, and the "load" or material into which ultrasonic waves are transmitted. In the design of transducers, it is, therefore, necessary to consider the source, backing, and loading, including any matching network and the transducer elements in a complete simulation. An example of an equivalent circuit, with source and tuning elements, is given in Figure 5.9 [11].

To illustrate the effectiveness of transducer models, one data set, of what are now numerous examples of the comparison of theoretical and experimental transducer performance, that is given in the literature is shown in Figure 5.10 [4]. The figures present data for the case of a 3.4 MHz, 19-mm diameter, PZT transducer element with matching layers of 2.25×10^6 kg/m²-s and 8.91×10^6 kg/m²-s and no backing. The data in the figure are (a) the electrical impedance, (b) the two way insertion loss, and (c) theoretical and experimental impulse response. For a full characterization of a transducer, these are the minimum electrical parameters that should be measured.

Several standards have now been developed for transducer characterization. In addition to the electrical properties it is necessary to characterize the transducer ultrasonics wave field, and this topic is considered further in Section 5.11, of this Chapter.

5.3.7   PIEZOELECTRIC TRANSDUCERS FOR HIGH-INTENSITY APPLICATIONS

Piezoelectric transducers for high-power applications sometimes consist of plates or blocks of piezoelectric materials resonant at the desired frequency, bonded to the surface of a treatment chamber—for example, the bottom of a cleaning tank. A more desirable technique is to clamp the element between two blocks, which become

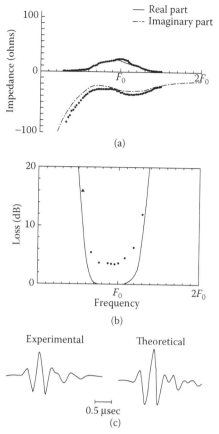

FIGURE 5.10 Comparison of experiment with theory for a 19 mm diameter, 3.4 MHz, PZT transducer with matching layers of impedance, 2.45×10^6 kg/m²-s and 8.91×10^6 kg/m²-s, and no backing (solid curve, theory; dots, experimental results). (a) Electrical impedance. (b) Two-way insertion loss. (c) Theoretical and experimental impulse response. (Courtesy of Kino, G. S., *Acoustic Waves*, Prentice-Hall, Englewood Cliffs, NJ, 1987.)

a part of the resonant system. A typical half-wave transducer is shown schematically in Figure 5.11 and is reminiscent of the Langevin sandwich structure.

Here, two piezoelectric elements C_1 and C_2 are located between two identical metal blocks A and B. The elements C_1 and C_2 are separated by an electrode E connected to the high-voltage lead. The electrode is, therefore, located at a node, and the elements C_1 and C_2 must be polarized in opposite directions for optimum activity.

If the mating surfaces are perfectly flat and highly polished, coupling without bonding between the piezoelectric elements and the metal end pieces may be accomplished by applying high pressure (bias pressure) across the elements. Since these elements are ceramic, usually with tensile strengths within the range of 34–69 MPa, a bias pressure provides a safety factor against fatigue failure. Bias pressure within the range 48–90 MPa also increases the efficiency of the transducer. In Figure 5.11, the elements are located on either side of the plane of maximum stress, which can

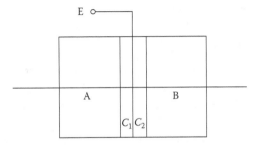

FIGURE 5.11 Typical half-wavelength piezoelectric transducer.

easily exceed the tensile strength of the elements or of the adhesive, which bonds them together, if one is used.

Bonding the elements together on clamping improves coupling and permits operation at lower and more effective bias pressures (48–76 MPa compared with 89–96.5 MPa without bonding in lead zirconate titanate elements).

Clamping may be done in one of two ways: (1) a bolt may be run through the center of the elements, but insulated from the piezoelectric elements; or (2) a flange may be attached to the metal blocks and clamped externally by a series of bolts.

The proper torque to apply to the bolts may be determined by (1) calculating the total force required to produce the desired stress, (2) determining the force F that each bolt must hold, and (3) applying this value of F in the following equation:

$$T = \frac{Fp}{2\pi e} \tag{5.12}$$

where

T is the torque (kg/m)
F is the force per bolt (kg)
p is the pitch of the bolt thread (m)
e is the efficiency of the threaded combinations

The assembly just described will resonate at a frequency below that of either piezolectric element (see Figure 5.11) when it is unloaded and the frequency shift is by an amount controlled by the dimensions and acoustical characteristics of the metal blocks and of the piezoelectric elements.

The dimensions of the transducer elements can be calculated for a given operating frequency, using equations given in Chapter 4 and the known elastic properties of the materials to be employed.

Piezoelectric transducers are high-impedance units, and the voltage used in driving power units may exceed 20 kV. These voltages do not pose a difficult problem but must be taken into consideration in the design of the assembly.

It is good practice to acoustically isolate the transducer from all structures not intended to be part of the vibratory system in order to conserve as much energy as possible for processing. Non-useful energy dissipated in the supporting structure lowers the efficiency of the ultrasonic system. Isolation is usually done (1) by mounting the transducer at nodes through O-ring assemblies, (2) by using the radiating surface as the mounting plate (for example, the bottom of ultrasonic cleaning tanks), or (3) by floating or pinning the transducer in its housing for applications, which do not require an applied external force of any significance.

Several applications of piezoelectric ceramic types of transducers call for larger motions and smaller force than can be obtained from transducers operating in a thickness mode, as discussed previously. Such applications include buzzers, record player pickups, microphones, earphones, optical scanners, accelerometers, stress and strain gauges, transducers for coupling ultrasound to air (for example, intrusion detectors), liquid-level sensors, low-frequency filters, and many others. These larger amplitudes and lower forces are obtained by structures operating in a flexural mode. The structure may be that of a cantilever bar, a flexural bar simply supported at two nodal positions, or a plate supported at its periphery. Flexural transducers make use of the 31 elastic characteristics of piezoelectric elements; a voltage across the thickness of the elements causes expansion or contraction in a direction normal to the direction of the applied field [12]. Whether expansion or contraction occurs depends upon the polarity of the driving voltage relative to the poling direction. Conversely, a mechanical displacement of the element will cause a voltage to appear across the electrodes of the element.

A typical flexural element may consist of (1) a single piezoelectric plate bonded to a matched metallic element, which serves as one electrode (unimorph); (2) two piezoelectric elements bonded to opposite sides of a metal electrode (bimorph); or (3) a single plate containing a series of parallel silver-coated holes through the center plane and extending in the direction parallel to the plane of bending.

The unimorph flexes by the metal plate attempting to constrain the expansion or contraction of the piezoelectric element as it is energized electrically. Or, conversely, flexing the element mechanically produces a voltage across the crystal.

The bimorph flexes by applying a voltage simultaneously across both crystals in such a manner that it is always with the direction of poling in one element and against poling with the other, the situation alternating during a cycle of vibration. Bimorphs may be poled for either series or parallel operation. For series operation, the electrical input (or output) connections are made between the outer electrodes and the crystals are poled in opposite directions. For parallel operation, the directions of polarization are the same; the center electrode is used as one terminal while the two outer electrodes are connected in parallel and form the second terminal.

The electroded center holes of the multimorph are used only for poling the element. These are the simplest of the flexural piezoelectric devices to produce because they do not involve bonding between plates. They are used only in series operation with terminals located at the outer electrodes. Since they are poled between the center and the outer electrodes, the strain relationships are always such as to promote bending or to produce a voltage between outer electrodes on flexure.

The unimorph design is used in many applications including inexpensive types of ultrasonic cleaners.

There are various methods of mounting flexural transducers. For the cantilever, the bar is rigidly mounted at one end and free to vibrate at the other end. Both ends of the flexural bar may be constrained but not clamped in a resilient material to allow free bending. Mounting the flexural bar or plate at nodal positions permits a very light and compact structure in which losses in the mounting are a minimum. Foam rubber and even neoprene O-rings can be used for nodal support.

5.3.8   PULSE-TYPE TRANSDUCERS FOR LOW-INTENSITY APPLICATIONS SENSING

Transducers generally used in NDT are of the piezoelectric type. The piezoelectric elements have high-Q values and, as a result, a short electrical impulse will cause them to vibrate or "ring" for a long period of time. With few exceptions, in pulse-type NDT, such ringing is definitely undesirable. The length of the pulse, which is the length of time required for the transducer to stop ringing, corresponds to a distance of pulse travel in the medium under investigation. In pulse-echo testing, resolution of defects is a function of the pulse length. If an echo is received before the initial pulse dies down, its electrical indication becomes lost in the electrical indication of the initial pulse. Therefore, the ability to detect defects near the surface of a material or to resolve two defects of only slightly different depths is a function of the pulse length.

Shorter pulse lengths also give better defect sensitivity. Shorter pulses contain a wider band of frequencies. The increased sensitivity to small defects is due to the shorter wavelengths of the higher frequencies associated with the wide bandwidth of the shorter pulse.

The bandwidth of a transducer is a function of its Q as follows:

$$Q = \frac{f_C}{f_2 - f_1} \qquad (5.13)$$

where f_C is the center frequency of the pulse, and f_1 and f_2 are the half-power points on each side of f_C, where the amplitude is 0.707 times the amplitude at f_C.

The high Q generally obtained in the piezoelectric materials used to produce high-frequency, short-pulse transducers is offset by bonding a material with high damping characteristics to one face, or back side, of the piezoelectric plate. Ideally, these two materials should have identical characteristic acoustic impedances to eliminate reflections from the interface between them. The ultrasonic energy entering the backing material must be completely absorbed within a convenient distance. The opposite, or radiating, face of the piezoelectric plate may be bonded to a low-loss window, wedge (for generating modes other than longitudinal), lens, buffer rod, wave guide, or other useful structure.

Various materials have high absorption characteristics and are suitable for damping purposes. Suitable mixtures can be formulated with graphite, powdered metals, and metaloxides of random grain sizes either fuzed or adhesively bonded together. Suitable adhesives and bonding materials include Araldite and epoxy cements.

Transducers used for NDT are identified by frequency, the frequency usually being the fundamental resonance frequency of the unloaded crystal and supposedly designating the frequency at the center of the bandwidth, f_C. Often, after the backing material has been added, the center frequency drops significantly below that desired. The proper center frequency can be maintained by properly controlling the dimensions of the backup material. Or, the frequency rating may be determined empirically. Transducers that produce extremely short pulses are not designated by a frequency.

Piezoelectric materials have high impedances and are driven at high voltages. The two major (and opposite) faces are electroded and connected to the electrical leads. The radiating face usually is at ground potential, and the high-potential lead is connected to the opposite face. Proper spacing and insulation between high-potential leads and grounded structures is essential to avoid electrical arcing and damage to the piezoelectric plate.

5.3.9 PIEZOELECTRIC POLYMERS FOR TRANSDUCERS

Several polymeric materials exhibit ferroelectric properties; that is, they contain polar crystals whose direction of polarization can be reversed by the action of an electric field. Poly-vinylidene flouride (PVF_2), is best known and still exhibits among the highest piezoelectric coefficients of the piezoelectric polymers [13]. PVF_2 is marketed under the trade name Kynar by the Pennwalt Company. They recommend many interesting uses for Kynar films, including (1) microphones, stereo pickups, speakers, and headphones; (2) switch elements in finger-pressure touch action keyboards and keypads for typewriters, telephones, calculators, and computer terminals; (3) sensors for instruments used to monitor vibration, deformation, acceleration, pressure, fluid flow, and seismic pulses; and (4) devices for monitoring heartbeat, breathing, and blood flow. The acoustic impedance of PVF_2 is close to that of water and soft body tissue.

Polymeric piezoelectric films in the form of diaphragms are easily coupled to air. They couple well to soft tissue through a coupling medium such as oil or water, and because they are very soft and flexible, they conform to the contour of the body.

In keeping with the convention established for piezoelectric ceramic materials, the 3 axis of a piezoelectric film is the direction of poling or thickness. The 1 axis is parallel to the direction of stretch if the film is stretched in only one direction (uniaxial stretch) while being formed. The d_{31} of a uniaxially stretched film is greater than the d_{32}, the 2 axis being normal to the 1 and 3 axes. If the film was stretched equally along both the 1 and the 2 axes while being formed, $d_{31} = d_{32}$.

Many applications for PVF_2 have been reported in the literature including medical transducers and hydrophones used for transducer evaluation and beam plotting. Marcus [14,15] offers expanded lists of applications, many of which are included in the categories listed previously. He summarizes the effects of orientation and polymer film thickness on suitability for various applications and divides these applications into three categories: pyroelectric, electromechanical, and mechanoelectric applications.

Regarding the effects of orientation and film thickness on suitability for various applications, he provides the following:

Application	Film Type
Flexure-mode transducer (electrical to mechanical)	Thin, uniaxially stretched
Displacement transducer	Thick, uniaxially stretched
Pyroelectric detectors	Thin, biaxially stretched
Hydrophones	Thick, biaxially stretched
Capacitors	Thin, high dielectric constant, high breakdown strength

The pyroelectric applications listed are infrared detector, video camera sensor (vidicon targets), laser-beam profiling, radiometer, heat scanner, intrusion alarm, spectral reference detector, reflectometer, charge separator (filter), and photocopying.

Electromechanical applications of ferroelectric polymers include speakers, headphones, flexure-mode transducers, optical scanners, light deflectors, vibrational fans, position sensors, variable-aperture diaphragms, shutters, deformable mirrors, and ultrasonic light modulators.

Mechanoelectric applications of ferroelectric polymers are microphones, record player cartridge pickups, hydrophones, medical ultrasonic imaging, touch switches, stress gauges, strain gauges, nip pressure transducers, accelerometers, stress wave monitors for seismic studies, impact detectors, fuses, coin sensors, personal verification devices, pulse monitors, blood-flow monitors, pacemakers, physiological implants, and surface acoustic wave (SAW) devices.

5.3.10   PIEZOELECTRIC MATERIALS AND THEIR PROPERTIES

Many publications deal with the theories, structure, and properties of piezoelectric, electrostrictive, and ferroelectric materials. The reader is referred to these publications for more detailed information [5,7,14–18]. Manufacturers of piezoelectric elements used in ultrasonic transducers can provide useful design data applicable to their own products.

Several parameters describe the properties of piezoelectric elements. Those mentioned previously include transformation factor (α) for converting mechanical impedance to electrical impedance and vice versa, capacitance of a piezoelectric plate (C_0), compliance (K), resistance corresponding to dielectric losses (R_D), internal strain losses (R_f), efficiency (η), electromechanical coupling factor (k_C), density (ρ), quality factor (Q), and characteristic impedance (ρc). Additional quantities of importance to the design of piezoelectric transducers include dielectric constant (ϵ), piezoelectric modulus (d), stress constant (g), and Curie temperature.

The *Piezoelectric modulus, d,* is defined as the ratio of the strain developed along or around a specified axis to the field applied parallel to a specified axis when all external stresses are constant. It also expresses the ratio of the short-circuit charge per unit area of electrode flowing between connected electrodes, which are perpendicular to a specified axis, to the stress applied along or around a specified axis when all other external stresses are constant.

The *piezoelectric stress constant*, g, is defined as the ratio of the field developed along a specified axis to the stress applied along or around a specified axis when all other external stresses are constant. The g constants are also the ratio of the strain developed along or around a specified axis to the electric charge per unit area of electrode applied to electrodes, which are perpendicular to a specified axis.

For all the piezoelectric constants, k, d, and g, the directions of the field and the stress or strain are indicated by two subscripts. The first subscript indicates the direction of the electric field, and the second subscript indicates the direction of the stress or strain. The subscripts 1, 2, and 3 refer to the x, y, and z axes, respectively, and subscripts 4, 5, and 6 denote stress or strain around the 1, 2, and 3 axes, respectively. For example, g_{33} denotes the ratio of field developed in the 3 direction to stress applied in the 3 direction when all other external stresses are zero.

The relationship between the g and d constants is

$$g = \frac{d}{8.85 \times 10^{-12} \varepsilon} \frac{\text{volts/meter}}{\text{newtons/meter}^2} \quad \text{or} \quad \frac{\text{meters/meter}}{\text{coulombs/meter}^2}$$

where ε is the free relative dielectric constant measured between electrodes and subscripts for g and d are the same; that is, g_{33} corresponds to d_{33} and g_{31} corresponds to d_{31}.

Typical piezoelectric materials and their properties are summarized in Table 5.1.

5.4 MAGNETOSTRICTIVE TRANSDUCERS

Magnetostriction is a term applied to the change in dimensions of a magnetic material when the impressed magnetic field is varied in magnitude. This phenomenon was discovered by Joule and was described by him in an article published in 1847 [19].

Magnetostrictive transducers are discussed in detail in several texts [20–22] and in trade literature, which can now, in most cases, be found for several manufacturers on the World Wide Web. A good review of the early literature and the fundamentals for the design of magnetostriction transducers is given by Wise [23]. A schematic showing a typical magnetostrictive stack is given in Figure 5.12.

The earliest high-power ultrasonic transducers operating in the low-kilohertz frequency range (below 100 kHz) were made of magnetostrictive materials, high-frequency generation being left to piezoelectric transducers. The trend today is toward piezoelectric ceramic material for both low- and high-frequency applications, but certain applications for ultrasonic energy may continue to favor certain

TABLE 5.1
Selected Piezoelectric Materials and Their Properties

Materials	Free Dielectric Constant $\varepsilon_{33}/\varepsilon_0$	$\varepsilon_{11}/\varepsilon_0$	Coupling Factor k_{31}	k_{33}	Piezoelectric Strain Constant (10^{-12} meters per volt) d_{31}	d_{33}	Piezoelectric Stress Constant (10^{-3}, voltmeters per Newton) g_{31}	g_{33}	Mechanical Q Thin Disc	Frequency Constant N_3 (cycle-meters per second)	Density $\rho\,10^3$ (kg/m^3)	Curie Point (°C)	Source
Barium Titanate (Vernitron Ceramic B)	1200	1300	−0.194	0.48	−58	149	−5.5	14.1	400	2740	5.55	115	a
Lead Zirconate titanates													
PZT-4, Vernitron	1300	1475	−0.334	0.70	−123	289	−11.1	26.1	500	2000	7.5	328	a
PZT-5A, Vernitron	1700	1730	−0.344	0.705	−171	374	−11.4	24.8	75	1890	7.75	365	a
PZT-5H, Vernitron	3400	3130	−0.388	0.752	−274	593	−9.11	19.7	65	2000	7.5	193	a
PZT-8	1000	1290	−0.30	0.64	−97	225	−10.9	25.4	1000	–	7.6	300	a
PXE-5, Philips	2000	1800	−0.37	0.69	−190	390	−10.9	22.0	–	1850	7.7	285	b
PXE-52, Philips	3500	3000	−0.39	0.74	−270	580	−8.7	19.0	–	1900	7.8	165	b
PXE-21, Philips	1750	–	−0.37	0.72	−180	385	−11.6	25.0	–	1900	7.75	270	b
PXE-41, Philips	1200	1400	−0.34	0.68	−119	268	−11.6	25.2	–	2000	7.9	315	b
PXE-42, Philips	1300	–	−0.34	0.68	−130	285	−11.0	25.0	–	2015	7.7	325	b
PXE-43, Philips	1000	–	−0.30	0.63	−95	210	−10.7	25.0	–	2050	7.7	300	b
PXE-71, Philips	1300	1700	−0.35	–	−147	–	−12.8	–	–	–	7.75	270	b
Quartz, X-cut	4.5	–	−0.10	0.10	−2	2.3	−50	58.0	10^6	2870	2.65	575	c,d
Rochelle Salt, 45° X	350	–	−0.73	–	−275	–	−90	–	–	–	1.77	−18, 24	e

(Continued)

TABLE 5.1 (*Continued*)
Selected Piezoelectric Materials and Their Properties

Materials	Free Dielectric Constant		Coupling Factor		Piezoelectric Strain Constant (10^{-12} meters per volt)		Piezoelectric Stress Constant (10^{-3}, voltmeters per Newton)		Mechanical Q Thin Disc	Frequency Constant N_3 (cyclemeters per second)	Density ρ 10^3 (kg/m^3)	Curie Point (°C)	Source
	$\varepsilon_{33}/\varepsilon_0$	$\varepsilon_{11}/\varepsilon_0$	k_{31}	k_{33}	d_{31}	d_{33}	g_{31}	g_{33}					
Lithium Sulfate 0° Y-cut	10.3	–	–	0.33	–	16	–	175.0	–	2730	2.06	–	d
Lithium Niobate 36° Y-cut	39	–	–	0.49	–	8.3	–4	23	1000–5000	3300	4.64	1150	g
Lead Metaniobate	225	–	–	0.42	–	85	–	42.5	11	1400	5.8	550	d
Poly(vinylidene fluoride) (PVF$_2$)	12	–	–0.14	–	20–25	20–22	–230	210.0	–	–	1.76	M.P. 165–180	f

a   Vernitron corporation, 232 Forbes Road, Bedford, OH 44146.

b   Ferroxcube, 5086 Kings Highway, Saugerties, NY 12477.

c   P. E. Bloomfield, "Piezoelectric Polymer Transducers for Detection of Structural Defects in Aircraft," Final Report, Contract No. N62269-77-M-3186, for Naval Air Development Center, Warminster, Pennsylvania, p. 8, July 22m 1977.

d   J. R. Frederick, Ultrasonic Engineering, Wiley New York, 1965, p. 6.

e   W. P. Mason, Piezoelectric Crystals and Their Application to Ultrasonics, Van Nostrand, New York, 1950, pp. 117–118.

d   J. R. Frederick, Ultrasonic Engineering, Wiley New York, 1965, p. 66.

f   Kynar Piezo Group, Pennwalt Corporation, 900 First Avenue, King of Prussia, PA 19406–0018.

g   G. Kino, Acoustic Waves, Prentice Hall, 987 P557.

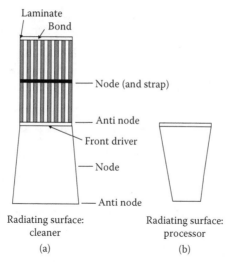

Laminate
Bond
Node (and strap)
Anti node
Front driver
Node
Anti node

Radiating surface:
cleaner
(a)

Radiating surface:
processor
(b)

FIGURE 5.12 Schematic for magnetostrictive stack transducer, with horn: (a) for cleaning application and (b) for processing.

characteristics peculiar to magnetostrictive devices. One such characteristic is the fact that, if the temperature of a magnetostrictive transducer should accidentally exceed the Curie point (temperature above which it becomes inactive), the material loses its magnetostrictive properties while at temperature, but the core is again usable upon cooling. However, the thermal history of the core does influence the magnetostrictive properties of the material. The properties of magnetostrictive materials that have been heated are in contrast to those of a piezoelectric where when the temperature of a ceramic piezoelectric material exceeds the Curie temperature, the material must be polarized anew before it can be used again for ultrasonic generation or reception.

The electrical components of a magnetostrictive transducer consist of a coil wrapped about a core of magnetic material. The coil has an inductance corresponding to the number of turns N, the permeability of the core material μ_0, and the length of the magnetic path I, as follows:

$$L_C = \frac{\mu_i \mu_0 S N^2}{\ell} \tag{5.14}$$

when the magnetic path is a closed loop. If an air gap is present, the reluctance of the air gap must be taken into account.

The coil exhibits the electrical resistance common to any inductor with equivalent length and size of conductor, and also the magnetic circuit is subject to hysteresis and eddy-current losses.

Consider the mechanical components of the magnetostrictive transducer. These are the elastic properties of the material, the mass of the material, the load characteristics (that is, radiation impedance or the mechanical characteristics of a transition

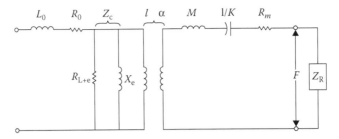

FIGURE 5.13 Equivalent circuit of magnetostrictive transducer.

member such as a horn and its radiation impedance), and the mechanical losses of the material.

The electrical and the mechanical quantities may be represented in an equivalent electrical circuit, as shown in Figure 5.13. In Figure 5.13,

$$\alpha = -\frac{\Lambda}{\omega N} e^{-j\phi}$$

where

Λ is the magnetostrictive constant
$e^{-j\phi}$ is a phase factor accounting for hysteresis
N is the number of turns on the core
ω is $2\pi f$
f is the frequency

The magnetostrictive constant Λ and the transformation coefficients are related to the electromechanical coupling coefficient k_C as follows:

$$k_C = \sqrt{\frac{\mu_i \mu_0 \Lambda^2}{Y_0}} = \frac{\Lambda}{N} = -\omega\alpha\sqrt{\frac{L_C^\ell}{SY_0}} e^{j\phi} \tag{5.15}$$

or

$$\alpha^2 = \frac{jSk_C^2 Y_0 e^{-j2\phi}}{\omega\ell Z_0} \tag{5.16}$$

where $Z = j\omega L$.

The circuit of Figure 5.13 may be drawn in its equivalent form, adding the internal impedance of the generator, Z, as in Figure 5.14. Certain facts must be kept in mind in using such an idealized circuit as that shown in Figure 5.14. First, the use of the circuit implies linearity in the operation of the transducer and implies that each component in the circuit has a constant value. Actually, this assumption approaches validity only when a very small driving magnetomotive force (mmf) is applied. Second, the circuit applies only to closed loops as the elements are defined; however, the circuit can be modified to accommodate other magnetic configurations. Third, due to the skin effect or limitation on flux

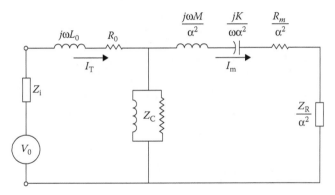

FIGURE 5.14 Equivalent electrical circuit of magnetostrictive transducer.

penetration at high frequencies, the operating frequency f is lower than a critical value given by

$$f_C = \frac{\rho_C}{2\pi^2 \mu_1 \mu_0 d^2} \tag{5.17}$$

where ρ_C is the resistivity of the core (Ω-m) and d is the lamination thickness (m). Therefore, since the transducer element provides a nonlinear impedance/power function, equivalent circuits must be used with care.

Keeping these statements in mind, one may proceed to derive equations for maximum power transfer and efficiency for a magnetostrictive transducer operating at low amplitude. The equations derived for efficiency will be the maximum that can be expected from a magnetostrictive device. At high amplitudes, where the characteristics become nonlinear, the efficiency is always lower.

5.4.1 MAXIMUM POWER TRANSFER TO THE MAGNETOSTRICTIVE TRANSDUCER

Referring to the equivalent electrical circuit shown in Figure 5.14, at resonance, the reactive components of the mechanical branch of the circuit are "tuned out." By adding a tuning capacitor at the output of the generator, $j\omega L_0$ also may be tuned out.

Letting the internal impedance of the generator now be R_i, the condition for maximum power transfer from the generator to the load is

$$R_i = R_0 + \left[\frac{(R_m + Z_R)}{\alpha^2} Z_c \right] \Big/ \left[\frac{R_m + Z_R}{\alpha^2} + Z_c \right] \tag{5.18}$$

or

$$Z_R = \frac{R_0 R_m + \alpha^2 Z_c R_0 + R_m Z_c - R_i R_m - \alpha^2 R_i R_c}{R_i - R_0 - Z_c} \tag{5.19}$$

for a transducer radiating from one end, where $Z_R = (\rho c)_\ell S$ and $(\rho c)_\ell$ is the acoustic impedance of the load medium.

5.4.2 EFFICIENCY OF THE MAGNETOSTRICTIVE TRANSDUCER

The efficiency of the magnetostrictive transducer is

$$\eta = \frac{\text{power output}}{\text{power input}} = \frac{W_O}{W_{in}} \tag{5.20}$$

The power output, W_O, is the power delivered to the acoustic load, or

$$W_O = I_m^2 \frac{(\rho c)_\ell R_R}{\alpha^2} = \frac{I_m^2 R_R}{\alpha^2} \tag{5.21}$$

where I_m is the current through the output branch of Figure 5.14 and is given by

$$I_m = \frac{Z_C I_T}{Z_C + (R_m + Z_R)/\alpha^2}$$

The power input, W_{in}, is

$$W_{in} = I_T^2 \left[R_0 + \frac{Z_C(R_m + Z_R)}{\left(Z_C + \dfrac{R_m + Z_R}{\alpha^2} \right)\alpha^2} \right] \tag{5.22}$$

where I_T is the total current delivered to the transducer. Then

$$\eta = \frac{\alpha^2 (\rho c)_\ell S Z_C^2}{[\alpha^2 Z_C + R_m + (\rho c)_\ell S][\alpha^2 R_0 Z_C + R_0 R_M + R_0 (\rho c)_\ell S + R_m Z_C + (\rho c)_\ell S Z_C]} \tag{5.23}$$

5.4.3 MAGNETOSTRICTIVE TRANSDUCERS FOR HIGH-INTENSITY APPLICATIONS

Magnetostrictive transducers are low-impedance, high-current devices. The impedance of a 20 kHz, 1 kW transducer of A-nickel is approximately 10 Ω if the coil resistance is negligible.

Several materials exhibit magnetostrictive properties. Nickel, Nickel–iron, and iron–cobalt (Permendur) alloys are most commonly used for transducers.

Transducers of A-nickel (commercial, high-purity) are rugged. They are usually made in laminated stacks that are one-half wavelength long cut so that the rolling direction is in the direction of motion.

The bar velocity of sound in nickel is 4787 m/s. Therefore, a half-wavelength stack of nickel with uniform cross section vibrating at 20 kHz is approximately 12 cm long.

The thickness of the laminations used in a magnetostrictive stack is determined by the flux penetration, which is limited by eddy currents in the materials. The total thickness of each lamination is approximately twice the effective penetration. The penetration depth is determined by

$$\delta = 1.98 \sqrt{\frac{\rho_e}{\mu_\Delta f}} \qquad (5.24)$$

where

ρ_e is the electrical conductivity of the magnetic material (Ω-cm)
μ_Δ is the relative magnetic permeability of the magnetic material
f is the frequency (Hz)

Annealing A-nickel in air at 760°C for 1 hour and allowing it to cool very slowly in the furnace produces good magnetostrictive characteristics for ultrasonic transducers. The oxide layer thus produced is tough and has good electrical resistance so that the laminations can be stacked in direct contact with one another.

Magnetostrictive stacks are commonly formed as rectangular stacks with windows, thus giving a closed magnetic flux path. The energizing coils are wound through the windows. The coils are protected from mechanical damage to the insulation by slabs of insulators. The oxide layer may be removed from one end of the stack and the stack silver brazed to a solid half-wave horn, which is used to couple the energy to the work.

A-nickel saturates at a magnetomotive force slightly above 100 Oe and a flux density of approximately 0.6 T. At these levels of magnetization in a dc electric field, the length of the stack shortens by approximately 33 μcm/cm. However, at resonance, the maximum strain is a function of the mechanical Q of the transducer times the dc strain corresponding to the peak ac magnetization.

The efficiency of a magnetostrictive transducer is affected by the magnetic bias. At zero bias, the transducer vibrates at twice the frequency of the driving current. Operation at zero bias is inefficient. The highest efficiency is obtained with A-nickel when it is biased to approximately 0.4 T.

Magnetostrictive transducers are also low-Q devices. This low Q is due partly to electrical eddy currents and hysteresis losses. These losses also produce heat, which must be dissipated. Nickel transducers usually are water-cooled to maintain a temperature at which the transducer will operate at its best. As the temperature approaches the Curie point, the magnetostrictive activity goes to zero. The Curie point for nickel is 358°C.

Magnetostrictive devices may be used at frequencies as high as 100 kHz; however, for practical power-handling capabilities, they are limited to frequencies below 50 kHz.

"The maximum alternating stress that can be produced in a resonating half-wave rod through magnetostriction of nickel has been calculated to be between 2.5 MPa and 10.0 MPa. This corresponds to a contraction of 1.47×10^{-4} cm. (The magentostriction of nickel is negative, that is, nickel contracts in length on being magnetized) This is the stress that occurs at the motional node. "The pressure that could be exerted upon water in an actual case would depend upon a number of factors; for example, the sharpness of the mechanical resonance of the vibrator and the ratio of the cross section in the region of maximum strain in the nickel. In some experiments on the maximum obtainable output of laminated nickel stacks made at the Harvard Underwater Sound Laboratory, acoustic pressures as great as 0.8 MPa, dyne/cm² (approximately 8 atm), were produced in an oil-filled pressure chamber by a stack of nickel laminations driven by magnetostriction. This is roughly 12,000 times the pressure in an atmospheric sound wave that is at the threshold of pain for the human ear." [24]

5.4.4   GIANT MAGNETOSTRICTIVE MATERIALS

In search of materials for higher power SONAR with greater bandwidth and greater reliability, the Naval Ordnance Labs (now the Naval Surface Warfare Center) developed TERFENOL-D. Etrema Products, Inc. [25] holds patents and licenses to many TERFENOL-D applications, including several key patents on the material and the methods of manufacture of TERFENOL-D. Etrema is the world's leading supplier of TERFENOL-D.

TERFENOL-D is called "Smart Material" because of its ability to adapt to outside influences. Its physical properties include those listed in Table 5.2.

5.4.5   COMPARATIVE PROPERTIES BETWEEN SELECTED MAGNETOSTRICTIVE MATERIALS [25]

The properties of some magnetostrictive materials, other than TERFENOL-D are provided in Table 5.3.

TABLE 5.2
Physical Properties of TERFENOL-D

Nominal Composition	$Tb_{0.3}Dy_{0.7}Fe_{1.92}$
Mechanical Properties	
Young's Modulus	25–35 Gpa
Speed of Sound	1640–1940 m/sec
Tensile Strength	28 Mpa
Compressive Strength	700 Mpa
Thermal Properties	
Coefficient of Thermal Expansion	$12\ ppm^0$
Special Heat	0.35 W/m.-k
Electrical Properties	
Resistivity	58×10^{-8} O-m
Curie Temperature	380°C
Magnetostrictive Properties	
Strain (Estimated Linear)	800–1200 ppm
Energy Density	14–25 kJ/m²
Magnetomechanical Properties	
Relative Permeability	3–10
Coupling Factor	

TABLE 5.3
Properties of Magnetostrictive Materials Other Than TERFENOL-D

Material	Heat Treatment	$\rho_e \times 10^6$	H_c	μ	B_o	$\lambda \times 10^{-4}$	k	Max eff.
A-Nickel	Unannealed	8.0	19.0	13	3,200R*	1.02	0.09	0.61
	600C	8.0	14.3	30	3,640R*	1.10	0.15	0.79
	1000C	8.0	0.86	106	2,128R*	0.44	0.14	0.39
	1000C in H_2	8.0	...	78	4,000	1.25	0.27	0.78
	1000C in H_2	8.0	...	41	5,000	1.99	0.32	0.83
45-Permalloy	Unannealed	45	7.60	53	8,600R*	0.31	0.054	0.64
	1000C in H_2	45	0.26	1,768	5,840R*	0.05	0.008	0.045
	1000C in H_2	45	...	372	12,000	0.32	0.16	0.70
2V-Permendur	1000C	25	1.90	42	9,920R*	0.40	0.06	0.76
	800C	25	...	91	20,000	1.37	0.32	0.90
	650C	25	9.1	126	13,950R*	0.85	0.21	0.92
	600C	25	18.4	59	17,500R*	1.23	0.24	0.92

H_c = Coercive force (oersteds)
k = Coupling coefficient
μ = Reversible permeability
B_o = Polarizing flux density

$$k = \sqrt{\frac{4\pi\lambda^2\mu}{E}}$$

ρ_e = Resistivity, Ohm-cm
λ = Magnetostrictive constant
Maximum efficiencies for an assumed $Q = 4$
All tests made on samples 0.002 in. thick.
* Remanent flux density.

5.5 ELECTROMAGNETIC DEVICES

Electromagnetic devices for generating vibratory energy may be produced in many forms, for example, loudspeakers, foundry-flask vibrators, marking tools, and hair clippers. They are fairly simple devices to design and build.

A simple device may consist of an armature mounted on a spring in a magnetic circuit (Figure 5.15).

The frequency at which the armature will vibrate is determined by the equation for the mass-loaded spring or

$$f = \frac{1}{2\pi}\sqrt{\frac{k}{M + m_s/3}} \tag{5.25}$$

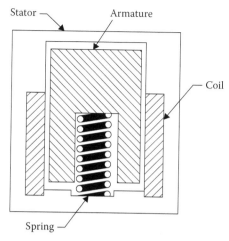

FIGURE 5.15   Simple electromagnetic vibrator.

where

    M is the mass of the armature
    m is the mass of the spring
    k is the spring constant

5.6   PNEUMATIC DEVICES (WHISTLES)

Whistles have been used to generate high-intensity sound both in air and in liquids. Although whistles are capable of generating ultrasonic energy in air to 500 kHz, using helium or hydrogen, a practical upper frequency for power applications of whistles in air appears to be approximately 30 kHz.

Several types of whistles are possible. Most people are familiar with the common whistle based on the organ-pipe effect. A jet of air impinges on a sharp edge and generates a noise, which is shaped into a fairly pure tone through a resonant column. The efficiencies of such whistles are low (below 5%).

A more efficient whistle ($\eta = 5\%$) is based on a design credited to Hartmann [26] (Figure 5.16). A jet of gas of diameter D_1 is directed into a cavity of diameter D_2. The impact of the jet on the bottom of the cavity causes a rise in pressure, which, in turn, causes a counterflow of gas toward the nozzle.

The momentum of this counterflow of gas causes a pressure rarefaction in the cavity, and at the point where the force of the jet overcomes this momentum, the flow direction again reverses toward the bottom of the cavity to complete the pressure cycle. The wavelength of the resulting sound is approximately (see Figure 5.16)

$$\lambda = 4(b + 0.3D_2) \tag{5.26}$$

where b is the depth of the cavity and the quantity $0.3D_2$ is an end correction factor.

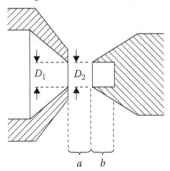

Design of ultrasonic transducers

FIGURE 5.16 Hartmann whistle.

In the Hartmann whistle, the diameter of the cup D_2 is equal to the diameter of the jet D_j. The spacing a between the end of the nozzle and the open end of the cavity is critical, as is the axial alignment between the two sections. According to Hartmann [26], the relationship between wavelength λ, nozzle and resonator diameter $D_1 = D_2 = b$, excess gas pressure p, and nozzle-to-resonator spacing a is

$$\frac{\lambda}{D_1} = \frac{\lambda}{D_2} = 5.8 + 2.5 \left\{ \frac{a}{D_1} - [1 + 0.041(p - 0.9)^2] \right\} \tag{5.27}$$

where p is in kilograms per square centimeter and a, b, D_1, and D_2 are in millimeters. Hartmann shows that a may vary within limits a_0 and a_m given by

$$a_0 = [1 + 0.041(p - 0.9)^2] D_2 \tag{5.28}$$

and

$$\frac{a_m - a_0}{D_2} = 0.13(p - 0.9)^2$$

Within the range a_0 to a_m, the wavelength increases nearly linearly at a rate

$$\frac{d\lambda}{da} = 2.5$$

The permissible range over which the efficiency is maximum is somewhat less than that given by Equation 5.29 and is more nearly

$$\frac{a_2 - a_0}{D_2} = 0.44 \sqrt{p - 1.8} \tag{5.30}$$

According to Hartmann [26], the radiated power P, in watts, is

$$P = 3D_2^2 \sqrt{p - 0.9} \quad (W) \tag{5.31}$$

where p is now given in atmospheres.

The efficiency η is

$$\eta = \frac{P_0}{P_j} = \frac{3D_2^2 \sqrt{p - 0.9}}{P_j} \tag{5.32}$$

where P_j is the theoretical power required to maintain the jet. The 5% efficiency claimed by Hartmann is based on Equation 5.32 and, therefore, does not include the efficiency of the compressor.

Brun and Boucher [24] have developed a modification of the Hartmann whistle for which they claim an efficiency of 14%. In Boucher's whistle, the resonator is mounted on a stem extending axially through the nozzle (Figure 5.17).

Boucher claims to verify Hartmann's equation with his "stem-jet" whistle; however, he finds an improvement in efficiency by making the ratio $D_2/D_1 \geq 1.3$. He also finds that the wavelength-to-distance ratio $d\lambda/da$ over the interval a_0 to a_m is more nearly

$$\frac{d\lambda}{da} = 1.88$$

for the stem-jet whistle.

The depth, b, of the resonator is not critical at low frequencies. Cleanliness is important. A thin film of grease, dirt, or liquid in the cup can lower the efficiency by a considerable amount.

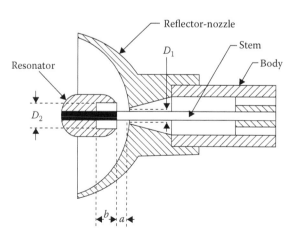

FIGURE 5.17   Modified Hartmann whistle.

FIGURE 5.18   Modified Hartmann whistle for high-intensity applications.

Figure 5.18 is a photograph of two modified Hartmann whistles. These whistles operate at 10 kHz and deliver 100 acoustical watts at an air line pressure of 620 kPa. The smaller reflector is equipped with a series of holes through which liquids are passed to be nebulized.

A second modification of the Hartmann whistle includes a pin extending through the resonator and into the open end of the nozzle. The end of the pin is beveled with an included angle of 20–30°. Alignment of the axes of the pin, the nozzle, and the resonator is very important for efficient operation of the whistle. However, efficiencies claimed for properly designed and assembled "pin-jet" whistles are higher than efficiencies obtained using the "stem-jet" whistles. Equations used in designing the stem-jet are also applicable to the pin-jet whistle.

Whistles for generating sound in liquid have been developed for industrial use. The Hartmann whistle required jet velocities equal to the velocity of sound. The velocity of sound is considerably higher in liquids than in gases. Flow velocities equal to the velocity of sound are impractical in liquids. Therefore, the liquid whistle must operate on a different principle.

The whistles of Janovsky and Pohlmann [27] operate on a jet-edge principle. A high-pressure jet of the liquid or liquids is forced into the edge of a thin plate, which is mounted at the displacement nodes. The plate vibrates in flexure at resonance. The resulting cavitation in a multiliquid mixture causes the liquids to emulsify more effectively than is accomplished by a similar jet impinging on a stationary plate. Again, liquid whistles are low-frequency devices, 5 kHz being typical.

5.6.1   SOME PRACTICAL APPLICATIONS OF PNEUMATIC WHISTLES

5.6.1.1   Coating Fine Particles

A few references to the use of ultrasonic whistles for fuel atomization are described in Chapter 12 of this book. Other significant uses are discussed here.

A project requiring the use of fine coated particles, 5 μm in diameter to be coated with 0.5 μm of epoxy, was satisfied using the whistle shown on the left of Figure 5.18. The particles were sprayed through the outer orifices; the liquid epoxy was sprayed through the inner orifices. The axes of the orifices (inner and outer) intersected in a circle at a position slightly past the mid-plane of the resonator. The whistle produced a fog of the particles and of the epoxy in such a manner that the small particles were individually coated with the epoxy, which were then cured in suspension. No aggregates were produced. They were cured to the point that no two particles stuck together in settling out of suspension. This was the only method known of producing this product.

5.6.1.2 Controlling Foam in Large Industrial Tanks for Liquids

A case where truckers hauling large tanks full of a liquid used in the manufacture of soap were bothered by the huge of volumes of foam being produced as they drove along. Gearing of the effectiveness of ultrasonic energy in controlling foams, they sought help. It was decided to use a copy of the smaller whistle of Figure 5.18 in the large supply tank. It was calculated that five or six whistles would be needed considering the size of the tank. However, the contractor decided to try only one whistle as a test of the process. The whistle operates at 10 kHz and emits 100 watts of energy at 620 kPa of nitrogen.

To the surprise of all involved, the one whistle proved to be adequate for the defoaming operation. After a few months, the performance began to decline. The reasons were probably related to either small amounts of dirt in the resonator or a change in alignment of the stem. The first would require a thorough washing of the resonator (possibly the entire whistle). The second would require a new whistle.

5.7 MECHANICAL DEVICES

High-amplitude vibrations at low frequencies are sometimes generated by rotating counterbalanced weights. Two sets of weights are mounted to balance out forces in all directions but one. The two sets are rotated in opposite directions (Figure 5.19).

In Figure 5.19, the centrifugal force due to the rotation of each weight is given by

$$F = mr_g\omega^2 \tag{5.33}$$

where

F is the force (Newton)
m is the mass of the weight
r_g is the radius of gyration of the weight (meter)
ω is $2\pi n$
n is the number of revolutions the weight makes in 1 second

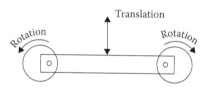

FIGURE 5.19 Counterbalanced mechanical vibration generator.

Devices of this type are limited to low frequencies by mechanical restrictions common to the use of bearings and gears.

From Figure 5.19, it is evident that the sum of forces in the horizontal direction is zero, but that the vertical components are additive and the maximum vertical force equals the sum of the centrifugal forces. Forces about a vertical axis also must cancel out to avoid rotation. This is done by using more than one weight for each direction and balancing these for zero rotation.

5.8 SOME SPECIAL HIGH-FREQUENCY TRANSDUCERS

These "special" transducers include those designed (1) to provide coupling of ultrasonic energy into an object without directly contacting the surface of the object and (2) to generate ultrasonic energy at high frequency in an acoustic transmission line by methods that convert almost all the electrical energy into ultrasonic energy with very little loss due to heat dissipation or energy reflection toward the source. Examples of the former type are noncontacting transducers used in the inspection of certain materials. Examples of the latter type include SAW devices and resistive layer types.

For the noncontacting devices, Szilard [28] gives the basic principles of electromagnetic coupling (EMAT) and electrostatic coupling in ultrasonic NDT. These transducers are used where direct coupling would not be practical using "conventional" transducers. Examples include inspecting through rough surfaces and testing hot materials. Advantages, in addition to being able to inspect at higher surface temperatures, include possible higher testing speeds and elimination of wear on the transducers.

5.8.1 ELECTROMAGNETIC COUPLING

Generation of ultrasonic waves in metals by electromagnetic induction is brought about by holding a coil carrying high frequency in proximity to the surface of the metal. The high-frequency current of the coil induces eddy currents in the surface of the metal to a depth given by the classical skin depth equation

$$\delta = \left(\frac{2}{\mu\mu_0\sigma\omega} \right)^{1/2} \tag{5.34}$$

where

ω is the angular frequency
μ is the magnetic permeability of the material
μ_0 is the permeability of vacuum
σ is the electrical conductivity of the material

When a constant magnetic field is applied to the metal in addition to the high-frequency field, Lorentz forces are produced in the surface of the metal to the effective depth of the alternating field. The result is generation of ultrasonic waves propagating into the metal in a direction normal to the surface. The polarization of the waves (shear or longitudinal) depends upon the direction of the magnetic field. The same process may be used for both transmitting and receiving ultrasonic waves to frequencies as high as 10 MHz. The sensitivity of the electromagnetic transducers is much lower (by 40–100 dB) than sensitivities achieved by conventional contact probes, but the advantage gained by the noncontact feature has led to their fairly wide acceptance in industry. Szilard [28] provides additional equations basic to the design of this type of transducer.

5.8.2 ELECTROSTATIC COUPLING

Szilard [28] also discusses the use of electrostatic transducers for noncontact testing. The sensitivities of these probes are even lower than those of the electromagnetic coupling type.

Electrostatic coupling is accomplished by locating a flat electrode in proximity to the surface of the test object. When a voltage is impressed between the electrode and the surface of the object, it produces an attractive force between the two surfaces given by

$$F = \frac{-V^2 \varepsilon \varepsilon_0 S}{2d} \tag{5.35}$$

where

V is the voltage
ε is the dielectric constant of air or the material between electrodes
ε_0 is the permittivity of vacuum
S is the surface area
d is the distance between electrodes

If a dc voltage and an alternating voltage are applied simultaneously, it is possible to generate ultrasonic waves in the test object at the frequency of the alternating voltage. In order to avoid excessive distortion of the waveform, it is necessary to make the dc voltage very much larger than the peak amplitude of the alternating voltage. Again, the same principles may be used to receive an ultrasonic signal without contacting the surface of the test object.

The capacitive, or electrostatic method, has been used in dry contact applications by using Mylar or other types of plastic films as dielectric materials between the surface of the test object and the surface of the electrode. A major advantage of the electrostatic coupling technique is its broadband capabilities, which include 10 kHz to 10 MHz (within 5 db).

5.8.3 SURFACE ACOUSTIC WAVE DEVICES

SAW transducers are finding wide usage in industry today. Applications include such signal processing functions as delay lines, correlators, and filters. Typical frequencies lie in the range of 50 MHz to gigahertz. Signals are carried as Rayleigh surface waves from a transmitting section to receivers. The design of the transmitter and the receiving elements is determined by the function the device, which is to perform.

A typical SAW device consists of interdigital electrodes deposited on the surface of a piezoelectric crystal. In their simplest form, the interdigital electrodes consist of many parallel fingers (see Figure 5.20), which are alternately connected to opposite

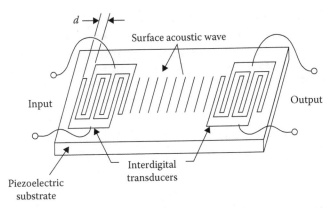

FIGURE 5.20   Constant-period surface acoustic wave device. (Courtesy of BMI.)

electrodes to which the signal is applied in the case of the transmitting section. The reverse effect is used to retrieve the signal for its intended use. In Figure 5.20, the distance between successive electrodes is d. The frequency of the applied field is chosen so that the wavelength is equal to $2d$.

Since SAW technology development started at Bell Telephone Laboratories in 1965, the devices have transformed many research and development projects and applications, particularly for those that are involved in numerous defense projects. The designs and uses of SAW devices are discussed extensively in the technical literature. An introduction to the fundamentals is given by Kino, in a master's course (MS) level text [4], and numerous articles are now found published in the technical literature (e.g., [29,30]), which has become its own specialization among the several that are now found within the wide field of ultrasonics. Typical uses include filters for television, radar, and other communication systems. The range of frequencies within which these devices are used presents new problems in generating the signals with minimum loss of signal as heat and by attenuation and dispersion in transmission and reception. One method of generating these high-frequency waves with minimum loss due to conversion to heat is found in resistive layer transducers.

5.8.4   Resistive Layer Transducers

White [31] gives a good presentation of the basic principles in the design of transducers of the resistive layer types. Their basic construction includes a thin high-resistivity layer of piezoelectric material formed at the surface or inside a larger conducting plate of semiconductor. White classes these transducers as depletion layer transducers, diffusion layer transducers, and epitaxial layer transducers. For a fuller discussion of these types, the reader is referred to White's work and to many more recent works where transducers of the resistive layer types are designed for various applications [32].

The *depletion layer transducer* consists of two dissimilar semiconducting materials such as n-type and p-type materials joined at a common plane. Or, they may include a rectifying metal-to-semiconductor contact. A depletion layer (or space charge layer or exhaustion layer) is formed at the interface between the two layers.

If the semiconductor is piezoelectric, the depletion layer acts as a piezoelectric plate and the maximum piezoelectric stress is produced within the depletion layer. Its thickness can be controlled and varied by applying a variable bias voltage across the assembly. The thickness of the depletion layer increases with voltage to a maximum determined by impurity density. The thickness of the layer also determines the frequency of operation of the device, this thickness being one-half wavelength of the operating frequency.

The diffusion layer transducer consists of a resistive layer formed in the surface of a conductive plate of piezoelectric semiconductor. The semiconductor material is made highly resistive by diffusing impurities into its surface region to a very shallow depth. Since the semiconductor is piezoelectric, the diffused layer can be operated as a piezoelectric transducer by applying a voltage between the outer electrode and the semiconductor or inner, conductive material.

There have been many references in recent years to *epitaxial transducers*. These consist of a thin layer of resistive piezoelectric crystalline material grown on a flat conductive substrate. The crystallographic orientation of the piezoelectric layer is controlled by the substrate, which must be conductive.

5.8.5　Laser Ultrasonics

The use of lasers has now been established, for some applications, as a viable non-contact alternate to piezoelectric transduction including both ultrasound generation and detection. This development resulted from some pioneering work in the 1960s and early 1970s, which was then followed by more systematic studies in the late 1970s and early 1980s [2].

A laser has the ability to deliver concentrated and pulsed light energy onto a surface, which can be absorbed and converted into an ultrasonic wave field. As laser power is increased, there are two interaction mechanisms for a laser material, which generate ultrasonic waves. In a lower power regime there is a thermoelastic ultrasound generation mechanism. This is where a pulsed laser beam shines on a surfaces, is absorbed, and causes rapid local thermal expansion. It is this expansion that in turn generates a displacement "pulse" that moves as spreading ultrasonic waves into the material. Such an ultrasonic wave-field will, in general, consist of a combination of shear, compression, and surface waves. At higher power, there is an ablation regime where either a surface material or a coating is vaporized, causing a net reactive force resulting from the ablated material and plasma–surface interactions. In detection, when ultrasonic waves interact with a free surface, they cause surface displacements, and these small movements can be detected using laser-based interferometers, which now come in several forms [4,33].

Laser systems have now been successfully deployed to meet a number of inspection needs. These tend to be where no direct contact between a surface and a conventional, piezoelectic transducer is possible, such as for parts at high temperature, and/or where they are in rapid motion. A schematic showing an example of the elements that form an ultrasonic laser system used for the study of ultrasonic Rayleigh waves [34] is given in Figure 5.21.

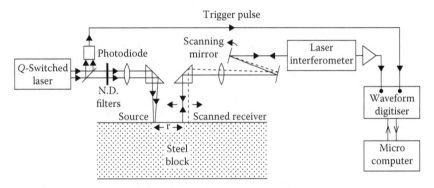

FIGURE 5.21 Apparatus used to study laser generation of ultrasonic Rayleigh waves. (With kind permission from Springer Science+Business Media: *Rayleigh Wave Theory and Applications*, The launching of Rayleigh waves from point sources, 1985, 102–9, C. B. Scruby and B. C. Moss, ed. E. A. Ash and E. G. S. Paige, Springer, Berlin.)

Laser ultrasound does give a remote or stand-off generation and detection capability that can be very useful for selected applications. However, the short pulse from lasers tend to produce high-frequency ultrasonic waves (50 MHz and higher) and ablation mechanisms that are not truly nondestructive, as they do leave small pits on a part's surface. When compared with signals given by piezoelectric transducers, signals tend to be at lower signal levels, and there can be issues with detectability and sensitivity due to poor signal-to-noise. There are also potential safety issues that need to be considered when operating with the high-power lasers needed in some applications.

5.8.6 Ultrasonic Arrays

Ultrasonic arrays are extensively used in both SONAR and medical ultrasonics. They are also now starting to receive more attention and use in NDT and industrial process applications, including deployment in harsh environments.

Ultrasonic arrays can be either mechanically or electronically steered. They can take several forms including being linear (a line of elements), which gives a 2-D wave field that illuminates a thin slice; such as is commonly used in the medical B-scan. It can also be in the form of a 2-D array of elements, analogous to a pattern of small tiles that cover a larger square or rectangular area of floor. Such a 2-D array can be used to form a 3-D beam of ultrasound, which can be steered, and the resulting reflected data can be recorded to form blocks of 4-D data (three spatial dimensions and time) that can be viewed in real time or recorded and then post-processed to extract a wide range of "slices." Some simple conceptual examples of ultrasonic arrays are shown in Figure 5.22 [35].

Specialized industrial applications of ultrasonic arrays have been considered for high-value applications, since the 1970s, and recent advances in computers and electronics have made such applications less expensive to develop. For example, they were extensively studied for imaging in liquid sodium (~260°C) to provide viewing

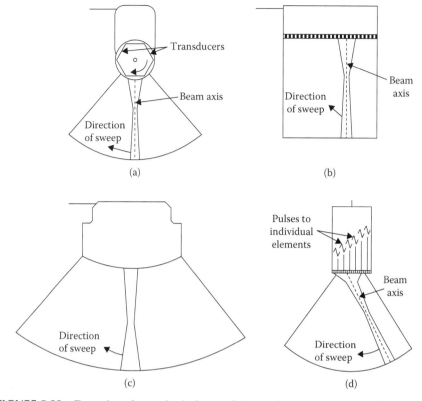

FIGURE 5.22   Examples of some basic forms of ultrasonic scan/array heads. (a) Rotating mechanical device. (b) Linear phased array, scan area the same width as array. (c) Curved linear array, with sector scan. (d) Phased array, sector scan with bean steering. (From Webster, J. G.: *Medical Instrumentation*. 1998. Copyright John Wiley & Sons, Inc. Reproduced with permission.)

for fast reactor applications in the 1970s [36]. With the resurgence of interest in fast spectrum nuclear reactors, the under sodium viewing technology is again being investigated. A Japanese group has developed and tested a 3-D array imaging system (using a 2-D matrix of 36×36 elements) operating at 4.5 MHz for deployment at high temperatures (~200°C) and in a high radiation flux [37]. The full range of industrial applications for imaging systems is discussed in Chapter 10, and the uses of ultrasonic arrays form medical applications is considered in Chapter 14.

5.9   TRANSDUCER-GENERATED WAVE FIELDS

The various forms of ultrasonic transducer deliver mechanical energy that is coupled into a material into which ultrasonic waves will propagate, or they act as a sensor to detect the presence of ultrasonic waves propagating in a material. The transducer is the electro–mechanical or mechanical–electro conversion device and reciprocity applies.

All forms of ultrasonic transducer act as some form of "filter," applied to the spectrum that is generated for the ultrasonic waves seen in the frequency domain, as they convert electrical energy into an acoustic/ultrasonic field or when used to sense ultrasonic energy and provide an electrical signal. Transducers can be designed to give ultrasonic sources and waves that come in a wide range of forms, including point sources (e.g., with a laser), a line source (e.g., with a 1-D array), and when acting as a baffle or disk (e.g., the typical piezoelectric-based compression wave NDT transducer). In addition to the transducer characteristics, it is the loading and the properties of the sonicated medium that also contribute to the determination of the mix and form of waves generated (i.e., fluids only support compression waves, whereas solids can support compression, shear, and surface waves). The energy partition between wave types as well as both the spatial and the temporal compositions, which result from ultrasound–material interaction, can take a near-infinite variety of forms. In many cases, theory is available, which can be used to calculate the ultrasonic wave field.

Some fundamental aspects of wave propagation were addressed in Chapter 2. The general topic of waves in solids is considered in numerous texts, including that by Graff [38]. Waves in fluids are fundamental to underwater acoustics and are discussed by Urick [39]. There are also numerous texts that address what is usually considered "acoustics," covering basic vibrations and waves, including wave propagation in air (e.g., [40]). The specific topic of the theory for transduction and wave fields encountered in ultrasonic NDT is covered by several authors including Schmerr [41], and it is given a more general overview, but still including significant theory, in a text by Kino [4].

The intensity in an ultrasonic beam is related to the amplitude of particle vibrations. The acoustic pressure (A_p) is used to denote the amplitude of the alternating pressures in a fluid or stress in a solid. The proportionality in the pressure relationship also depends on the acoustic impedance of the medium (Z):

$$A_p \propto Zx \quad \{\text{amplitude of motion}\} \tag{5.36}$$

The intensity is then the energy transmitted, which is proportional to the square of the acoustic pressure:

$$I \propto (A_p)^2 \tag{5.37}$$

For piezoelectric transducers, the element senses acoustic pressure and generates a voltage signal. Most instruments used for NDT are designed to give an output that is proportional to the square of the voltage received from the transducer, as shown in Equation 5.38. Therefore, most commercial NDT systems are calibrated to give an output voltage that is proportional to the ultrasonic intensity, as shown with the relationship given as Equation 5.39.

$$\text{Output} \propto \{\text{received voltage}\}^2 \tag{5.38}$$

$$\text{Voltage out} \propto (\text{ultrasonic intensity}) \tag{5.39}$$

When comparing voltage data, care needs to be taken to ensure that calibration differences between different types of instruments do not impact the comparisons: for example, in the case of comparing a signal from a digital oscilloscope received from a transducer connected to a laboratory amplifier, with that received displayed with most commercial NDT instrumentation. Also, it should be remembered that it is the acoustic energy and not acoustic pressure, which is partitioned at many ultrasonic interactions, that is, that for signals reflected and transmitted at an interface.

For low-power industrial applications, many ultrasonic transducers, particularly those that use a baffle or piezoelectric element, can be analyzed using antenna theory.

For a circular transducer, the resulting wave field is analogous to a flash-light beam. The relationship between frequency and disc diameter controls the beam spread and the degree of divergence, and this is typically calculated and measured for the −6 dB and −20 dB beam boundaries. The relationship for beam spread for −6 dB and −20 dB down beam edges is calculated using Equation 5.40. The normalized beam divergence is illustrated in Figure 5.23.

$$\text{Sin } \gamma_{-6} = 0.51 \; \lambda/D \tag{5.40}$$

$$\text{Sin } \gamma_{-20} = 0.87 \; \lambda/D$$

Another important parameter is the near-field length, which is the distance on-axis from the transducer face to where there is a peak in intensity, followed by a reduction with distance. The near-field length for a circular source is given by Equation 5.41.

$$N_o = \frac{D^2}{4\lambda} = \frac{D^2 f}{4c} \tag{5.41}$$

where

D is the diameter of crystal
λ is the wavelength

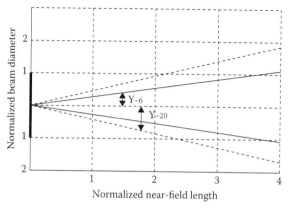

FIGURE 5.23 Ultrasonic field from a disc transducer, showing beam spread. (a) −6 dB. (b) −20 dB, dimensions normalized near-field length and transducer radius.

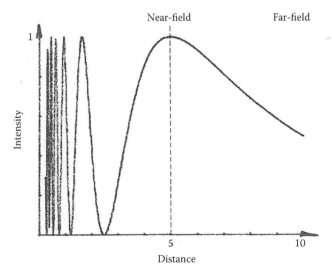

FIGURE 5.24 On-axis pressure for a disc transducer, showing near- and far-field zones.

c is the velocity of sound
f is the frequency

A plot of on-axis pressure, as a function of distance, with near-field length indicated, is shown in Figure 5.24.

The methods for measuring the characteristics of an ultrasonic field are given in several ASTM Standards, and they tend to use reference reflectors such as a small disk (flat end of a rod) or a ball (a ball bearing on a rod). Focused, usually higher frequency, transducers used in imaging, including acoustic microscopy, can also be characterized using descriptors that are analogous to those commonly employed in optics, the point spread function, the line spread function, and the modulation transfer function [42].

The understanding of both beam parameters and their impact on performance is central to many applications of ultrasonics in quantitative material property measurement, which is considered further in Chapter 6, and in NDT for defect sizing and imaging, which are considered in Chapters 7 through 10.

5.10 GENERAL REMARKS

Design of ultrasonic transducers has been treated in an elementary manner in this chapter. A more extensive treatment is not within the scope of this book. Since an ultrasonic transducer is essentially a device for producing ultrasonic vibrations or for detecting ultrasonic energy, any device that is capable of performing these functions is a potentially useful transducer, so that the list of types is not limited to those mentioned in this chapter.

The potential designs of transducers incorporating the principles discussed in this chapter are probably infinite. This might be recognized by considering the transducer elements presented (piezoelectric plates, magnetostrictive stacks, etc.) as motors that energize a piece of machinery and by comparing the possible combinations with the number of ways in which electric motors are used.

A distinction should be made between the terms *effectiveness, efficiency, maximum power transfer,* and *quality factor* (*Q*) of a system, which can affect the designer's approach to the design of transducers.

The *effectiveness* of an ultrasonic transducer or system is determined by the adequacy of the transducer to performing the required task. An effective transducer is not necessarily an efficient transducer. In fact, efficiency may have to be sacrificed to obtain the desired effect, especially in certain high-intensity applications of ultrasonics.

The *efficiency* of a transducer is the ratio of the radiated power to the power delivered to the transducer. The efficiency of the electronic generator is the ratio of the power delivered to the transducer to the power into the generator. These combined efficiencies (product of the two) give the efficiency of the total system.

The *maximum power transfer* occurs when the load impedance matches the internal impedance of a power source. When this match exists, the generator is capable of delivering the maximum amount of power to the load. This condition is often mistakenly considered to be the condition of maximum efficiency.

The *quality factor* (*Q*) of a transducer is the ratio of the total inductive reactance (at resonance) to the total resistance, or the ratio of the energy stored in the system to the energy dissipated. The amplitude of vibration of transducers increases with *Q*. If, under load, the *Q* of a transducer continues to be high, it is an indication that little energy is dissipated in the load and, therefore, the efficiency of the system is correspondingly low. There are applications where high-*Q* operation is required, such as where transducers are used for accurately controlling the frequency of an oscillator. Many applications require low-*Q* operation under load, for example, ultrasonic cleaning, wherein high-efficiency operation is very important.

REFERENCES

1. Piezotech, L. L. C. 2009. Indianapolis, IN: Piezotech, L. L. C.
2. Scruby, C. B., and L. E. Drain. 1990. *Laser Ultrasonics*. Bristol: Adam Hilger.
3. Green, R. E. 1998. Emerging technologies for NDE of aging aircraft structures. In Nondestructive Characterization of Materials in Aging Systems, *Mat Res Soc Symp Proc,* eds. R. L. Crane, J. D. Achenbach, S. P. Shah, T. E. Matikas, P. T. Khuri-Yakub, and R. S. Gilmore, Vol. 503, 3–14.
4. Kino, G. S. 1987. *Acoustic Waves*. Englewood Cliffs, NJ: Prentice-Hall.
5. Mason, W. P. 1950. *Piezoelectric Crystals and Their Application to Ultrasonics*. New York: Van Nostrand.
6. 1957. *American Institute of Physics Handbook*, coord. ed. D. E. Gray. New York: McGraw-Hill.
7. Mitsui, T., I. Tatsuzaki, and E. Nakamura. 1976. In *Ferroelectricity and Related Phenomena*, ed. I. Lefkowitz, and G. W. Taylor, Vol. 1, An Introduction to the Physics of Ferroelectrics. New York: Gordon and Breach.
8. Broomfield, G. H. 1985. The effects of temperature and irradiation on piezoelectric acoustic transducers and materials, UKAEA Report AERE-R 11942. Harwell, UK: United Kingdom Atomic Energy Authority.

9. Jayasundere, N., and L. J. Bond. 1983. Ultrasonic transducer standards. In *Proc. Review of Progress in Quantitative Nondestructive Evaluation (QNDE)*, ed. D. O. Thompson, and D. E. Chimenti, Vol. 2B, Aug 1–6, 1982, 1807–24. San Diego, CA: Plenum Publishing Corp. (New York).

10. Wells, P. N. T. 1969. *Physical Principles of Ultrasonic Diagnosis*. London and New York: Adademic Press.

11. Krimholtz, R., D. A. Leedom, and G. L. Matthaei. 1970. *Electron Lett* 6(13):398–9.

12. Auld, B. A. 1973. *Acoustic Fields and Waves in Solids*. Vol I and II. New York: John Wiley.

13. Yamamura, I., and M. Tamura. 1974. *High Molecular Weight, Thin Film Piezoelectric Transducers*. U.S. Pat. 3,832,580 (August 27, 1974).

14. Marcus, M. A. 1981. Ferroelectric Polymers and Their Applications," presented at the Fifth International Meeting on Ferroelectricity, Pennsylvania State University, August 17—21, 1981.

15. Marcus, M. A. 1984. *Ferroelectrics* 57(1–4):203–20.

16. Mason, W. P., ed. 1964. *Physical Acoustics, Principles and Methods*, Vol. 1, Pt. A, Chap. 3, 169–270. New York: Academic.

17. Kojima, T. 1987. A Review of piezoelectric materials for ultrasonic transducers. In *Proc. Ultrasonics International 2007*, 888–95. Guildford, UK: Butterworth Scientific.

18. Eberle, G., H. Schmidt, and W. Eisenmenger. 1996. *IEEE Trans Dielectr Electr Insul* 3(5):624–46.

19. Joule, J. P. 1847. *Philos Mag III* 30:76.

20. Kuttruff, H. 1991. *Ultrasonics: Fundamentals and Applications*. Amsterdam: Elsevier.

21. Finkelstein, L., and K. T. V. Grattan, eds. 1994. *Concise Encyclopedia of Measurements and Instrumentation*. New York: Pergamon Press.

22. Sachse, W., and N. H. Hau. 1979. Ultrasonic transducers for materials testing and their characterization. In *Physical Acoustics*, ed. W. P. Mason, and R. N. Thurston, Vol. 14, 277–406. New York: Academic Press.

23. Wise, B. A. 1957. *Design of Nickel Magnetostriction Transducers*. New York: Int. Nickel Company.

24. Brun, E., and R. M. G. Boucher. 1957. *J Acoust Soc Am* 29:573–83.

25. Etrema Products Inc. 2009. Product data sheets, http://www.etrema-usa.com/.

26. Hartmann, J. 1939. Construction, performance, and design of the acoustic air-jet generator. *J Sci Instrum* 16:140–9.

27. Janovsky, W., and R. Pohlmann. 1948. *Z Angew Phys* 1:222.

28. Szilard, J., ed. 1982. *Ultrasonic Testing; Non-Conventional Testing Techniques*, 381–99. New York: Wiley.

29. Morgan, D. P. 1998. History of SAW devices. In *Proc. 1998 IEEE International Frequency Control Symposium*, eds. S. C. Schneider, M. Levy, and B. R. McAvoy. Sendai, Miyagi, Japan, Oct. 5–8. 1998, 439–60. New York: Ferroelectrics, and Frequency Control Society, IEEE Institute of Electrical and Electronic Engineers.

30. Morgan, D. P. 2000. A history of SAW devices. In *Advances in SAW Technology, Systems and Applications*, ed. C. C. W. Ruppel, and T. A. Fjeldy, Vol. 1, 1–50. Singapore: World Scientific.

31. White, D. L. 1964. In *The Depletion Layer and Other High-Frequency Transducers Using Fundamental Modes, Physical Acoustics*, ed. W. P. Mason, Vol. 1, Pt. B, Chap. 13, 321–52. New York: Academic.

32. Miragawa, S., T. Okamoto, T. Niitsuma, K. Tsubouchi, and N. Mikoshiba. 1985. Efficient ZnO-SiO-Si Sezawa Wave Convolver. *IEEE Trans Sonics Ultrason* SU-32(5):670–4.

33. Dewhurst, R. J., and Q. S. Shan. 1999. *Meas Sci Technol* 10(11):R139–68.

34. Scruby, C. B., and B. C. Moss. 1985. The launching of Rayleigh waves from point sources. In *Rayleigh Wave Theory and Applications*, ed. E. A. Ash, and E. G. S. Paige, 102–9. Berlin: Springer.

35. Sideband, M. P. 1998. Medical imaging systems. In *Medical Instrumentation*, ed. J. G. Webster, 3rd ed. New York: Wiley.

36. Bond, L. J., S. R. Doctor, K. J. Bunch, M. Good, and A. E. Waltar. 2007. Instrumentation, monitoring and NDE for new fast reactors. In *Proc. GLOBAL 2007, Advanced Nuclear Fuel Cycles and Systems, Boise, ID, September 9–13, 2007, Am. Nuclear Soc*, 1274–9.

37. Karasawa, H., M. Izumi, T. Suzuki, S. Nagai, M. Tamura, and S. Fujimori. 2000. *J Nuclear Sci Technol* 37(9):769–79.

38. Graff, K. F. 1975. *Wave Motion in Elastic Solids*. Oxford: Clarendon Press.

39. Urick, R. 1983. *Principles of Underwater Sound*. 3rd ed. New York: McGraw-Hill.

40. Kinsler, L. E., A. R. Frey, A. B. Coppens, and J. V. Sanders. 2000. *Fundamentals of Acoustics*. 4th ed. New York: Wiley.

41. Schmerr, L. W. 1998. *Fundamentals of Ultrasonic Nondestructive Evaluation*. New York: Plenum Press.

42. Shiloh, K., A. K. Som, and L. J. Bond. 1991. *IEE Proc A* 138(4):205–12.

6 Determining Properties of Materials

6.1 INTRODUCTION

In the previous chapters, it was shown that many factors influence the propagation characteristics of ultrasonic waves. In this and succeeding chapters (through Chapter 10), methods of measuring and evaluating propagation characteristics are presented, demonstrating many varied and practical low-intensity applications of ultrasonic energy. Low-intensity applications are those where the primary purpose is to transmit the energy through a medium and measure a property or parameter, based on analysis of the response. The objective is never to change the state of the medium; that is, the energy level is so low that the medium remains unchanged from its original state after the wave has passed through it. Presented in this chapter are methods of measuring propagation characteristics for obtaining engineering data for materials. For more detailed information on methods, circuitry, accuracies, and applications than is practical in the present discussion, the reader should consult the many excellent treatises cited in this chapter and in Appendix A. The circuits discussed are predominantly those that were originally developed using analogue electronics. Digital circuits and computer-based instruments are increasingly being used, but such circuits can introduce quantization errors that need attention.

Propagation characteristics involve both velocity of sound and attenuation. The elastic properties and density of a medium control the velocity of sound. Mechanisms producing losses of energy control the attenuation. Both velocity and attenuation are measurable quantities, and particularly, attenuation can be a strong function of the frequency of the ultrasound used.

There are many methods of measuring the velocity and attenuation of sound. The choice of method depends upon the nature of the material and its environment and upon the accuracy required. Various methods are presented including an easily implemented approach that can give a good estimate for velocity and attenuation. Included in the precise methods for velocity determination discussed are those which use an interferometer, resonance, "sing around," pulse-superposition, and pulse-echo overlap. These are followed by a brief discussion of attenuation measurement. Methods used at high temperatures and those used at high pressures are presented as special cases of interest.

Measurements of flexural, longitudinal, and torsional resonant frequencies of bars provide means of determining the elastic moduli and Poisson's ratio of materials, which are quantities of engineering importance. These measurements have been valuable in determining the quality, including strength, of ceramic materials and of powder–metal compacts.

219

Ultrasonic methods of determining viscosity and of determining the properties of fluids and slurries and plastic and high polymers are also presented.

Speed of sound and attenuation is a significant source of data that can be related to thermal and mechanical properties, the molecular structure of liquids, and physical and chemical processes that are occurring in reacting systems. Methods have also been implemented in many forms and under a wide range of environmental conditions, including many cases of elevated and low temperatures and also at both high and low pressures (e.g., [1–5]). It can also provide data that relate to composition in mixtures, and for two phase systems, such as slurries, ultrasound can be used to measure both the particle size and the composition.

The acoustical properties for some materials are given in Appendix A. Data for solids (Table A.1), common liquids and molten salts (Table A.2), some organic liquids (Table A.3), and some gases (Table A.4) are tabulated. A selected bibliography for some sources of additional data and Web databases are identified and provided as Appendix B.

6.2 APPROXIMATE METHODS FOR MEASUREMENT OF VELOCITY AND ATTENUATION

A wide range of methods have been developed that can be used to give sound velocity and attenuation in solids, liquids, and gases. These methods can give data to various degrees of both accuracy and precision. In many industrial and research situations where a new material is encountered, a method is needed that can be easily implemented at low cost to give approximate properties.

6.2.1 MEASUREMENT OF VELOCITY AND ATTENUATION IN ISOTROPIC SOLIDS

In many studies, it is necessary to evaluate the acoustic properties of samples of material. A commonly cited approximate method for property determination was described in detail by Selfridge [6], who also provided an extensive listing of properties for isotropic materials in the same paper. The basic method reported has been used both before and since the 1985 paper by numerous other researchers (e.g., [1] to give reasonable estimates for property data.

In this method, a transducer and parallel-sided sample are immersed in a small measurement tank, such as that shown in Figure 6.1. The sample is mounted in a gimble and aligned so that the front face is perpendicular to the wave field from the transducer. The transducer–sample separation should be at least one near-field length, and path in water should also be such that the propagation time is four or five times longer than the reverberation time in the sample.

A tone burst (or a pulse) is then applied to the transducer, and pulse-echo data are used to provide arrival time and amplitudes, which are then used to give both velocity and attenuation. An example of a measured pulse-echo signal is shown in Figure 6.2. The series of responses A_1–A_4 correspond to A_1 reflection from the front face, A_2 reflection from the back of the specimen, and A_3 and A_4 reverberations in the specimen. Depending on the attenuation, a long series of reverberations may be observed. The series of echoes are shown in schematic form in Figure 6.3, in which

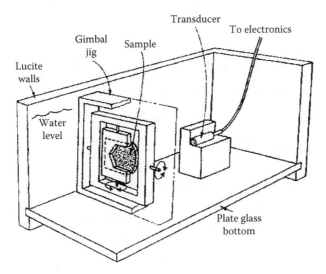

FIGURE 6.1 Measurement tank for approximate velocity and attenuation measurement. (From Selfridge, A. R., *IEEE Trans Sonics and Ultrason*, SU-32(3):381–94, © 1985 IEEE. With permission.)

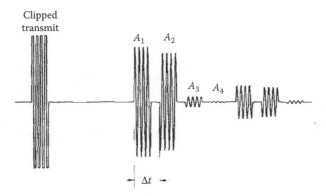

FIGURE 6.2 Measured pulse-echo signal for flat solid sample perpendicular to beam. (From Selfridge, A. R., *IEEE Trans Sonics and Ultrason*, SU-32(3):381–94, © 1985 IEEE. With permission.)

are separated in time. In some cases the interpretation can be more complex and care needs to be taken if super-position of pulses is seen.

The difference in the arrival times for the compression waves reflected at the front and back sample surfaces is Δt, as shown in Figure 6.2. This time interval can be conveniently measured using a digital oscilloscope. It is then used to give the compression wave velocity from the relationship:

$$V = \frac{2d}{\Delta t} \qquad (6.1)$$

where V is the velocity and d is the sample thickness.

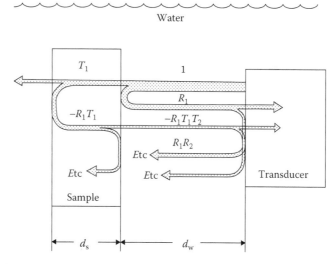

FIGURE 6.3   Schematic showing reverberation path between transducer and sample. (From Selfridge, A. R., *IEEE Trans Sonics and Ultrason*, SU-32(3):381–94, © 1985 IEEE. With permission.)

The sample density, which is needed to calculate acoustic impedance (Z), can be determined either by weighing a known volume of the material or by using Archimedes method [6], and in that case, density ρ is calculated as

$$\rho = \frac{W_a}{W_a - W_w} \tag{6.2}$$

where W_a is the weight in air and W_w the weight of the object, when immersed in water.

The acoustic impedance can then be calculated using density and compression wave velocity:

$$Z_s = \rho V \tag{6.3}$$

If the velocity is calculated in millimeter per microsecond and density is expressed in gram per milliliter, the product gives impedance in units of MRayls (kg/[s × m²]) × 10⁶. When data are taken at or near 4°C, the impedance of water in these units is 1.5 MRayls.

In the analysis of the pulse propagation, the loss at interfaces and in the sample is calculated using reflection and transmission equations discussed in Chapter 2. When the acoustic impedance is calculated, the reference loss at interfaces can then be calculated, and this enables the excess attenuation (or loss) due to attenuation in the sample to be estimated. At a water–solid interface, the reflection coefficient is obtained using

$$R_1 = \frac{Z_s - Z_w}{Z_s + Z_w} \tag{6.4}$$

The corresponding transmission coefficient is calculated using

$$T_1 = \frac{2Z_s}{Z_s + Z_w} = 1 + R_1 \qquad (6.5)$$

The expressions corresponding to reflections at subsequent interfaces can then be calculated.

$$T_2 = \frac{2Z_w}{Z_s + Z_w} = 1 - R_1 \qquad (6.6)$$

Experimentally, the loss in the sample can be determined from comparison of the measured amplitudes A_1 and A_2 (Figure 6.2) and the calculated ratio based on reflection coefficients calculated using acoustic impedance data. The calculated ratio of amplitudes is given from transmission and reflection coefficients as

$$\text{Calculated } \frac{A_2}{A_1} = T_1 * T_2 = 1 - R**2 \qquad (6.7)$$

The actual loss (in decibel per centimeter) is then given by combining the calculated and measured amplitudes using the relationship:

$$\text{Loss in decibel per centimeter} = 20 * \log \left(\frac{\text{Calculated } \dfrac{A_2}{A_1}}{\text{Measured } \dfrac{A_2}{A_1}} \right) \Big/ (2 * d) \qquad (6.8)$$

It is important to ensure that the transducer is of reasonable diameter. Commercially available transducers may have distorted beam characteristics, and also for low-attenuation samples, it may be necessary to add a correction for the effects of beam divergence. To ensure reliable data, a uniform and low-attenuation block should be used as a calibration sample. If needed, an empirical correction for beam spread can be estimated and used to correct measured data.

This basic approach described above can be extended in several ways to give higher accuracy, for example, through the use of a pulse excitation and measurement of the decay of a series of echoes in the sample and/or through the use of spectral analysis to give data at several frequencies.

6.2.2 MEASUREMENT OF VELOCITY AND ATTENUATION IN FLUIDS

The approach, which, in many ways, is similar to that used with solids and discussed in Section 6.2.1, can be adapted and used with fluids. The measurement system is fabricated using a transducer and buffer-rod to replace the water path between sample and transducer, and there is then a reflector set perpendicular to the beam. The dimensions of the cell depend on the frequency at which measurements are to be made. A schematic for a typical fluid characterization cell is shown in Figure 6.4.

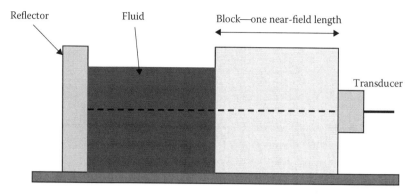

FIGURE 6.4   Simple cell for fluid characterization.

The properties obtained for fluids are significantly more sensitive to experimental conditions, in particular the temperature, than those for solids. A simple implementation of a pulse-echo measurement at either a single or multiple frequencies can be performed to give approximate data. Establishing a system to give an absolute measurement for the attenuation value can be quite complex; however, determination of the value relative to that for a reference fluid, such as water, is more easily made. For low-attenuation fluids, it is commonly necessary to account for a beam spread and apply a diffraction correction.

Using various approaches, the properties of fluids have been measured using ultrasound for many decades. There is extensive data in the literature, and those for some materials are given in Appendix A.

6.3   METHODS OF MEASURING VELOCITY OF SOUND

Measurements conducted over a period of many years by numerous investigators have resulted in an extensive accumulation of data on the velocity of sound. However, there is continued interest in such measurements in new materials and in materials to be used in hostile or abnormal environments—high temperatures and high pressures, for example.

There are basically three methods of measuring the velocity of sound: (1) by determining the wavelength of a continuous wave (CW) at a known frequency, (2) by determining the time required for a progressive wave to travel a known distance, and (3) by determining the angle of refraction of a beam of sound incident on an interface between two media in one of which the velocity of sound is known. These methods have been implemented in numerous ways. A few of these are described in the following sections.

All measurements of velocity of sound and attenuation are related to the basic equation for a progressive wave propagating in the x direction in a medium:

$$A = A_0\, e^{-\alpha x} \cos(kx - \omega t) \tag{6.9}$$

where

      A is the amplitude of the sound wave at x
      A_0 is the amplitude of the sound wave at the origin $(x = 0)$
      α is the attenuation coefficient
      k is the propagation constant $(= 2\pi/\lambda = \omega/c)$
      ω is the angular frequency $(= 2\pi f)$
      c is the phase velocity

6.3.1 INTERFEROMETER METHOD

The interferometer is a continuous-wave (CW) device, which has been used for accurately measuring the velocity and attenuation of sound in liquids and gases in which standing waves can be sustained. It can be in its simplest form, as shown in Figure 6.5, which consists of a fixed, air-backed piezoelectric transducer located at one end of a fluid column and a moveable rigid reflector at the other end. The technique described here is for fixed frequency. The reflector is moved toward or away from the transducer by a micrometer adjustment mechanism. As the reflector moves, the reflected wave becomes periodically in phase and out of phase with the transmitted wave, thus causing a corresponding constructive and destructive interference. The reaction of the interference on the crystal influences the load impedance reflected into the electronic system. The load current of the electronic amplifier is caused to fluctuate accordingly. The wavelength of the sound is determined by the distance the micrometer moves the reflector during one cycle of load current fluctuation and by the distance between two successive maxima being equal to $\lambda/2$. Optical interference methods have also been used to accurately measure the length of standing waves, especially at high frequencies (near or above 1.0 MHz). The accuracy of the measurement depends upon the accuracy of the micrometer readings,

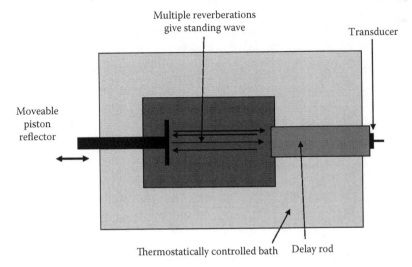

FIGURE 6.5 Schematic showing an ultrasonic interferometer.

the parallelism between reflector and transducer surfaces, and the accuracy of the frequency determination. An accuracy of 0.05% is typical for the interferometer. The velocity of sound is given by multiplying the wavelength times the frequency. The attenuation is determined by the decay of the maxima of the periodic curve of amplitude formed as the distance x between transmitter and reflector increases [4].

6.3.2 RESONANCE METHOD

The resonance method is similar to the interferometer method and is applicable to the measurement of the velocity of sound in gases, liquids, or solids.

The resonance method involves the use of a fixed transducer and a fixed reflector or two transducers spaced a known distance apart. It is implemented with the configuration shown in Figure 6.5 when the transducer separation is fixed. The transducer is driven through a range of frequencies to determine successive resonances. The difference between two successive resonant frequencies in a nondispersive medium is equal to the fundamental resonant frequency of the medium, or

$$c = 2\ell\Delta f \tag{6.10}$$

where ℓ is the distance between the transducer and the reflecting surface and Δf is the difference between successive resonant frequencies.

The accuracy of the resonance method depends upon the accuracy of the frequency determination and of the measurement of the distance between reflecting surfaces. Phase shifts at reflecting interfaces must be taken into account [2,7].

6.3.3 "SING-AROUND" METHOD

The "sing-around" method of measuring the velocity of sound involves the use of two piezoelectric transducers, one at each end of the specimen, as shown in Figure 6.6. One transducer receives an impulse from the electronic source and converts it into an ultrasonic pulse in the specimen. The pulse of ultrasonic energy, after passing through the specimen, is detected by the receiving transducer. The received pulse triggers the electronic generator to initiate a succeeding pulse. The velocity is related to the pulse repetition rate as follows:

$$c = \frac{\ell \times \text{PRR}}{(1 - e \times \text{PRR})} \approx \ell \times \text{PRR} \tag{6.11}$$

where

   PRR is pulse repetition rate per second
   ℓ is the length of the specimen
   e is a correction factor for delays in the transducers, in coupling between the
      transducers and the specimen, and in the electrical components

The pulse repetition rate is measured by a frequency counter. The electrical delays are associated with triggering of the transmitting transducer, rise time of the

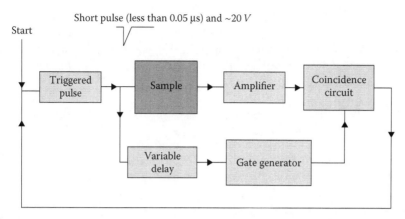

FIGURE 6.6 Simplified block diagram for sing-around technique.

amplified pulse, and generation of the trigger signal. In homogeneous nondispersive specimens, the effects of the acoustical and electrical delays can be minimized by measurements through several different path lengths. In liquids, distances traveled in specimens can be varied by changing the distance between transmitter and receiver, being extremely careful to maintain parallelism and accurately measuring the path lengths. By comparing successive values of ℓ and corresponding pulse repetition rate, it is possible to obtain fairly accurate values of e in Equation 6.11. A similar method can be used to measure the velocity of sound accurately in nondispersive solids by using several samples of different thicknesses if there is reasonable assurance that the material properties are uniform from specimen to specimen and care is taken to have exactly the same coupling conditions for all specimens.

The sing-around method is useful for measuring the velocity of sound to moderately high accuracy. However, it is good for monitoring changes in ultrasonic velocity to very high accuracy. Papadakis [7] credits Forgacs [8] with developing "what would seem to be the ultimate improvement in the sing-around system" with a precision of 1 part of 10^7. Forgacs takes advantage of multiple echoes in materials of low acoustic attenuation to increase the ratio of the acoustic-to-electronic delay, thus minimizing the importance of the electronic delay. Pulses of CW (in Forgacs' system, the frequency was 10 MHz) bounce back and forth between the transmitter surface and the receiver surface. Thus, successive received pulses have traveled odd multiples of the thickness ℓ (ℓ, 3ℓ, 5ℓ, …). Rather than using the leading edge of the first cycle through the specimen to retrigger the transmitter, Forgacs selects a certain cycle of a particular echo for this purpose. In addition to increased path length, the method also provides a fast-rising trigger pulse resulting in minimum jitter and drift and reduces extraneous effects resulting from the presence of echoes from previous sing-around cycles.

Forgacs' method of measuring the sing-around time helps maximize the accuracy of the technique. One counter is used to count a number of sing-around cycles, each cycle corresponding to the predetermined odd number of passes through the specimen. A second counter measures the elapsed time corresponding to the count of the first counter. The start of both counters is synchronized. The first counter is preset to count 10^N cycles. The 10^N cycle is the STOP pulse for the timer. This is accomplished by injecting an extra pulse into the input of the counter a few microseconds after the measurement interval begins so that the counter always reads one more than the number of sync pulses, which have arrived. Thus, on the arrival of the $(10^N - 1)$st pulse, the counter readout is 10^N and the actual 10^Nth pulse is passed to the STOP terminal of the timer so that there exists complete synchronism between START and STOP of the counters.

Hirao et al. [9] used the sing-around method to measure the velocity of Rayleigh and surface-horizontal (SH) waves in 7075-T651 aluminum. They used broadband pulses, which give group velocity rather than phase velocity. The center frequency of the Rayleigh pulses was 5 MHz, and that of the SH pulses was 2 MHz. Elapsed time was measured for transducer separations ranging from 25 to 70 mm in nine steps. The velocity of propagation was determined by the inverse of the slope of elapsed time versus transducer separation.

The group velocity in a nondispersive medium such as aluminum is essentially equal to the phase velocity, that is, the velocity associated with a discrete frequency. However, since velocity is a function of frequency in a dispersive medium, group velocity of a broadband pulse may vary significantly from phase velocity of any frequency included within the bandwidth of the pulse. In the case of dispersive media, CW pulses must be used in order to obtain accurate information regarding the acoustic properties of the test specimen. Thus, we are introduced to the fact that the method of measuring the velocity of sound must be chosen according to need and with knowledge of the capabilities and limitations of candidate methods.

6.3.4 PULSE-SUPERPOSITION METHOD

The pulse-superposition method [14,15] is capable of measuring the velocity of sound to accuracies of 1 part in 5000 or better. A piezoelectric transducer initiates radio-frequency (rf) pulses of ultrasonic energy into a specimen, as shown in Figure 6.7. These pulses echo back and forth within the specimen. Each succeeding echo is constructively added to the previous echo of a given pulse by controlling the pulse repetition rate at the reciprocal of the travel time in the specimen. This superposition

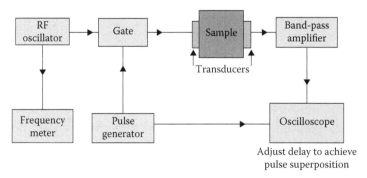

FIGURE 6.7   Block diagram of pulse-superposition method.

of the pulses within the specimen gives the technique its name. The method is capable of measuring accurately from one cycle of one echo to the corresponding cycle of the succeeding echo. A CW oscillator is used to control the pulse repetition rate. When the repetition rate is adjusted so that the initiation of a pulse coincides with the return of the first echo from the preceding pulses, the change in the signal amplitude indicates superposition. The CW oscillator frequency is monitored by a frequency counter. The pulse repetition rate is the reciprocal of the CW frequency and is a measure of the travel time within the specimen (twice the thickness of the specimen). The velocity is twice the specimen thickness times the pulse repetition rate after correcting for phase shift at the bond areas. Excellent descriptions of the pulse-superposition technique and procedures are given in [2,7,10,11].

6.3.5 PULSE-ECHO-OVERLAP METHOD

The pulse-echo-overlap method [7] is similar in many respects to the pulse-superposition method but is much more versatile. The pulse-echo-overlap method utilizes either rf bursts (to measure phase velocity) or broadband pulses (for group velocity), whereas the pulse-superposition method uses only rf bursts. Like the pulse-superposition method, the pulse-echo-overlap method is capable of measuring accurately from any cycle of one echo to the corresponding cycle of the next echo. Another difference between the two methods is in the fact that the pulse-superposition method lends itself to automation by feedback mechanisms to monitor velocity or velocity changes, whereas the pulse-echo-overlap method does not. The overlap is accomplished by visual observation by the technician performing the measurements.

The pulse-echo overlap is accomplished in analogue implementation by driving the x-axis of an oscilloscope with a variable frequency CW oscillator. By adjusting the frequency of the oscillator, one echo of interest is caused to appear on one sweep of the x-axis and the next echo on the succeeding sweep. When the CW oscillator frequency is adjusted so that the two pulses exactly overlap, the period of the oscillations is the travel time between the signals of interest. The repetition rate of the input pulse is generated from the phase of the CW oscillator through a frequency divider. This synchronism between the pulse triggering and the x-axis sweep eliminates jitter. By dividing by a large enough number, echoes from one pulse will be attenuated before the next one is triggered.

An additional advantage of the pulse-echo-overlap method is that the transducer may be coupled directly to the specimen or may be coupled through a delay line such as a buffer rod or a liquid column. In the pulse-superposition method, the transducer is coupled directly to the specimen. The method also may be used to make through-transmission measurements using a transmitting and a receiving transducer.

For a more complete discussion of the pulse-echo-overlap method, its advantages and limitations, and typical electronic circuitry for implementing the method, the reader should consult Papadakis' work [7]. Equipment for this type of measurement is commercially available.

Both pulse-superposition and pulse-echo-overlap methods have been successfully used to give accurate data. For example, the pulse-echo-overlap method has been used to give higher order elastic constants with the maximum error estimated

TABLE 6.1

Advantages and Disadvantages of the Pulse-Echo Overlap and Sing-Around Methods for Velocity Determination

Pulse-Echo Overlap	Sing-Around
Parameter measured	*Parameter measured*
• Phase speed	• Group speed
Advantages	*Advantages*
• High versatility	• Simple automation
• More accurate	• More precise
Limitations	*Limitations*
• Measuring process more complicated and time consuming	• Calibration more complicated and time consuming
• Subjective condition of the overlap	• Variable time delay
• Phase jitters	• Threshold error

Source: After Zorebski, E., M. Zorebski, and S. Ernst, *J De Physique* IV, 129:79–82, 2005.

at four parts in 10^5 [12]. However, each of these two methods has its own specific strengths and weaknesses. And the features of the two approaches are summarized in Table 6.1 [13].

6.3.6 Measurements in Materials of High Attenuation

The pulse-superposition and pulse-echo-overlap techniques are useful for measuring the velocity of sound in materials in which multiple echoes can be established. However, because these methods rely upon multiple echoes, they are not applicable to measuring velocities of high-absorption materials in which these pulses attenuate rapidly. Matsuzawa et al. [14] have developed a technique for measuring both velocity of sound and attenuation in highly absorptive materials. The measurement system consists of (1) an ultrasonic cell in which the specimen liquid is placed, (2) an air-backed quartz crystal located at one end of the cell used as the transmitting transducer, (3) an air-backed quartz crystal at the opposite end of the cell used as a receiving transducer, and (4) appropriate electronic circuitry. Spacing, L, between transmitter and receiver is 65 mm.

In many respects, the method of Matsuzawa et al. is similar to the pulse-echo-overlap technique. They use a two-channel oscilloscope in place of the single-channel oscilloscope of the overlap technique. Their system includes a CW oscillator operating in the range 15–25 kHz, which provides the frequency, f_t, determining the pulse travel time within the specimen material; that is, pulse travel time is $1/f_t$. They use a frequency division factor of 2^7 to prevent confusion by echoes from previous pulses. The rf frequency is 3 MHz.

The voltage pulse, e_T, is applied to the transmitting transducer and simultaneously to one of the input channels of the two-channel oscilloscope. The output voltage from the receiving transducer is fed through an integrator, which is an amplifier that produces a phase shift at 3 MHz of −90° and a voltage gain of 25 dB. The output voltage, e_R, after passing through the integrator, is applied to the second input terminal of the oscilloscope.

Measurements are made by displaying the rf signals of e_T and e_R on the oscilloscope. The position of e_R relative to e_T is made by varying f_t. Then the frequency of e_T is varied to cause the horizontal length of e_T to coincide with that of eight waves of e_R. The part of e_R used is selected so as to contain the maximum eight positive peaks. Coincidence occurs at the positions where the crossover points of $e_T = e_R = 0$. At this point, the frequency, f_t, is read accurately from a counter connected to the CW oscillator. The velocity of sound in the material is given to a fair accuracy by $c = L/f_t$.

The technique previously described can be used to measure attenuation also to a high accuracy. If e_T and e_R are the maximum amplitudes of $e > p$ and $e^\wedge$, respectively, attenuation can be determined from the equation

$$\frac{E_R}{E_T} = \frac{-(2SA_m^2 \rho c \omega_0 G/C_0)e^{-\alpha L}}{(j\omega)^2(Z_m + S\rho c)[Z_m + A_m^2/(j\omega C_0) + S\rho c]} \tag{6.12}$$

where

ρ is the density of the medium under test
α is the absorption coefficient of the specimen
S is the effective area of the quartz transducers
G is the amplification factor of the integrator
ω_0 is the angular frequency at which Im(z_m) = 0
A_m is a force factor
Z_m is the mechanical impedance of the transducer with its electrical terminal closed
ω is the angular frequency
C_0 is the capacitance of the cable between the receiving transducer and the integrator

The attenuation of sound in air is usually considered to be low. However, at frequencies above 1 MHz, it increases rapidly, and at higher frequencies, it must be treated as a material with high attenuation. This phenomenon, and a method used to measure attenuation, is illustrated with a study that investigated properties of air between 10 and 30 MHz [15].

The measurement system to measure both velocity and attenuation is shown in Figure 6.8. The mechanical system used to achieve the needed transducer alignment is shown as Figure 6.9 [15].

The air-backed wide band (nominally 20 MHz) transducers are mounted on buffer rods that are initially placed in contact. As the transducers are separated, the amplitude of a tone-burst signal was measured. An example of data for 20 MHz is shown in Figure 6.10. Data exhibit three characteristic zones: (1) interference—when transducers are close, there is reverberation and superposition of the pulses, (2) regular

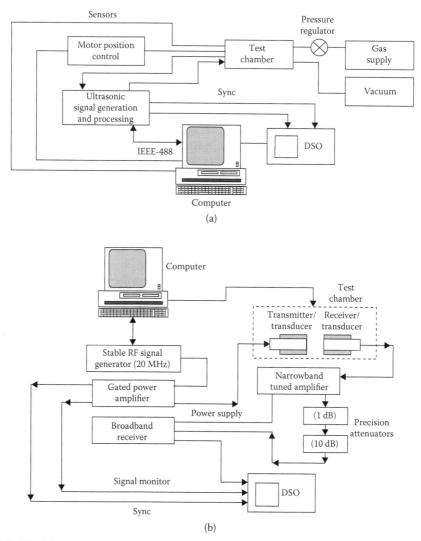

FIGURE 6.8 Schematic for high-attenuation measurement system using transmission. (a) Overall measurement system. (b) System for generating and processing ultrasonic signals. (Reprinted with permission from Bond, L. J., C-H Chiang, and C. M. Fortunko, *J Acoust Soc Am*, 92(4):2006–15, Acoustical Society of America, 1992.)

distance–amplitude relationship—there is a linear region in which the slope gives the attenuation, and (3) a noise limited region—the extent of which depends on the system signal-to-noise capabilities.

The temperature, (19°C), pressure (83.58 kPa—measured in Boulder, CO at an elevation of 1680 m), and humidity (52%) were measured, and both absorption and velocity (346 m/s) were calculated. The measured and calculated values for the acoustic parameters were in good agreement [15].

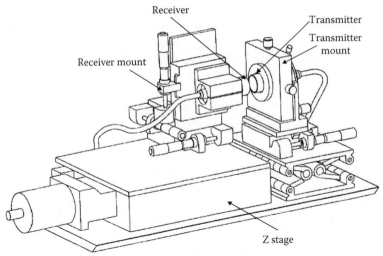

FIGURE 6.9 Transducer alignment system, used with buffer rod transducers. (Reprinted with permission from Bond, L. J., C-H Chiang, and C. M. Fortunko, *J Acoust Soc Am*, 92(4):2006–15, Acoustical Society of America, 1992.)

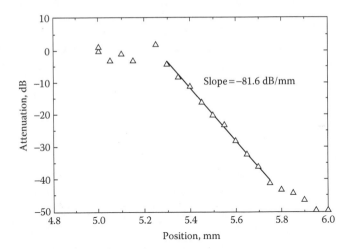

FIGURE 6.10 Amplitude–distance data measured in air at 20 MHz to determine attenuation. (Reprinted with permission from Bond, L. J., C-H Chiang, and C. M. Fortunko, *J Acoust Soc Am*, 92(4):2006–15, Acoustical Society of America, 1992.)

6.3.7 MEASUREMENTS AT HIGH TEMPERATURES

The velocity of sound in materials can be measured at elevated temperatures by means of transmission lines (or buffer rods). Wall effects and phase shifts due to coupling media must be taken into consideration.

When the attenuation in the rod is low and the material is plentiful, the specimen can be used to provide its own transmission line from the transducer located outside the furnace to the constant-temperature zone of the furnace. A short length of the end of the rod may be machined to provide a shoulder from which energy is reflected back into the transmission line to the transducer. Echoes from this shoulder and from the end of the rod return to the transducer at different times corresponding to the length of the specimen section and the velocity of sound in the reduced section. Machined notches in the rod at the high-temperature section may provide additional reverberations for more accurate measurements of both velocity and attenuation [11].

The delay time may be measured by comparator methods in which a pulse of energy is passed through a suitable calibrated delay line and displayed on an oscilloscope to coincide consecutively with each echo from the specimen area. Time interval counters are available, which make possible accurate measurements of the delay times.

Similar measurements are possible in high-loss materials by attaching the specimen to low-loss buffer rods. McSkimin [11] discusses these techniques in greater detail.

The thermal expansion of the specimen and phase shifts at bond areas must be taken into account in these measurements. In addition, wall effects, causing mode conversion and consequently interferences between modes of different velocities, can cause problems and should be prevented. One method often used to reduce wall effects in the buffer rods is to roughen or thread the surfaces of the transmission line segments of the rods.

Similar techniques can be used to measure the velocity of sound in liquids at high temperatures. Figure 6.11 shows a method used by Higgs and Litovitz [16] to measure ultrasonic absorption and velocity in molten salts. The buffer rods are fused

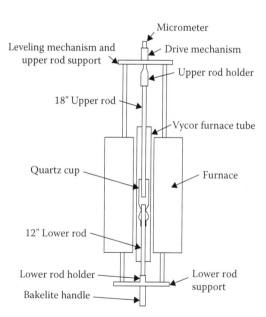

FIGURE 6.11 Equipment for measuring the ultrasonic absorption and velocity in molten salt. (Data from Higgs, R. W., and T. A. Litovitz, 1115, 1960.)

quartz to which piezoelectric transducers are bonded. The upper end of the lower rod has a machined shoulder to provide mechanical support for the sample cup. The sample cup is sealed to the quartz rod at the shoulder by means of Sauereisen cement.

The sample cup consists of a 24/40 standard taper joint (fused quartz) mated with a cup made of 35-mm-ID quartz tubing. Parallelism between the adjacent ends of the rods is important for accuracy.

Measurements of absorption and velocity were made by comparator methods [17]. The amplitude of the echoes for absorption measurements was measured by feeding the calibrated pulsed output of a standard signal generator into the receiver and delaying it in time so that it was adjacent to the echo under consideration.

Velocity measurements were made by comparing the velocity of sound through the specimen with the velocity of sound in water in which the temperature was controlled to 0.1°C. By making the travel time in the water identical with that in the specimen and determining the path length in each, it was possible to determine the velocity of sound as follows:

$$c_x = \left(\frac{\ell_x}{\ell_w}\right) c_w \qquad (6.13)$$

where

c_x is the velocity of sound in the specimen
ℓ_x is the distance the pulse travels in the specimen in time T
ℓ_w is the distance the pulse travels in the water in time T
c_w is the velocity of sound in the water or comparator fluid

The accuracy of the velocity measurement depends upon the accuracy of the micrometer readings on the comparator fluid path and the specimen path and also upon the accuracy to which the velocity of sound in the comparator fluid is known.

Another factor to be considered in measuring velocity and absorption of sound in liquids at high temperatures is the compatibility of the materials used with both the environment and the specimen materials. For example, fused quartz cannot be used at temperatures above which it devitrifies. Also, certain salts rapidly attack fused quartz at elevated temperature.

Figure 6.12 is a modification of Higgs and Litovitz's [16] method of measuring velocities and attenuations of sound in molten salts that react with fused quartz. Dense alumina rods are used instead of quartz. The cup is also alumina and is attached to the lower rod by means of an alumina cement. An alumina collar is fitted near the upper end of the rod and cemented to it to provide additional strength to support the weight of the specimen. A hole is machined through the center of the bottom of the cup, which rests on and is cemented to the collar. The cement is slowly and carefully cured to avoid damage from vapor-pressure buildup when the temperature is raised.

A pulse generator provides an impulse to excite the transmitter transducer and simultaneously trigger one sweep of a dual-beam oscilloscope. (Adjustable delay no. 1 is optional and may be eliminated if adjustable delay no. 2 has sufficient delay capability.) The transmitted pulse travels through the buffer rods and the specimen

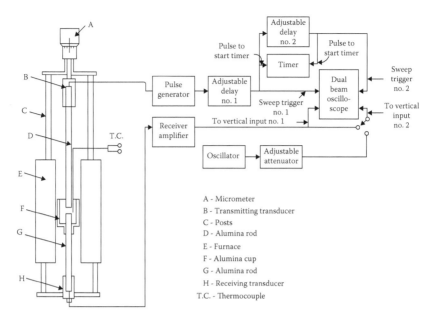

FIGURE 6.12   Schematic of apparatus for ultrasonic measurements in molten salts.

and is detected by the receiver. After amplification, the received pulse is displayed on the oscilloscope screen at a time corresponding to the total delay time.

The same electrical impulse that excites the transmitter also triggers a time interval counter and afterward passes through a variable delay. The output from the variable delay is used to stop the counter and to trigger the sweep of the second trace on the dual-beam oscilloscope. Adjusting the delay time on the second trace to coincide with the delay through the rods and the specimen gives a reference time reading T_1. Changing the path length through the specimen material and repeating the procedure gives a new time T_2. Both T_1 and T_2 are presented on the digital counter. The difference in path length, $\Delta\ell$, is the difference in the two micrometer readings, and the velocity of sound is determined by the ratio

$$c = \frac{\Delta\ell}{\Delta T} \tag{6.14}$$

where

$$\Delta T = T_1 - T_2 \quad (T_1 > T_2)$$
$$\Delta\ell = \ell_1 - \ell_2 \quad (\ell_1 > \ell_2)$$

The attenuation is obtained by comparing the calibrated output signal from a CW oscillator with the received pulse height corresponding to each of the positions at which T_1 and T_2 are obtained. The attenuation is given by

$$\alpha = \frac{\ln A_1 - \ln A_2}{\Delta\ell} \quad (A_1 > A_2) \tag{6.15}$$

where A_1 and A_2 are the amplitudes of the received signal corresponding to positions 1 and 2, respectively. Correction for diffraction is necessary for absolute attenuation measurements.

The thermal coefficient of expansion of alumina must be considered in making measurements with the setup of Figure 6.12. Sufficient spacing can be maintained by running the two rods together at each temperature and adjusting the carriage to the minimum spacing desired.

Rugged micrometers accurate to the nearest 25.4 micron are available. A less expensive micrometer accurate to the nearest 63.5 micron and having total range of 2.54 cm will provide accuracies that are more than adequate compared with the capabilities of the rest of the system.

Several systems have been developed that are used to measure on line-properties at elevated temperature, and it is the application of pressure to a buffer rod which couples sound from the cooled transducer assembly into a product such as red hot steel [18].

6.3.8 MEASUREMENTS AT HIGH PRESSURES

Velocity of sound and absorption can be measured at high pressures by techniques similar to those used at high temperatures. Dimensional changes due to the compressibilities of the materials must be taken into consideration. Also, the design of the equipment must be such that there is no interference between sound bypassing the specimen and waves passing through the specimen.

Peselnik et al. [19] used the phase-comparison method of McSkimin [20] to measure the velocity of sound in fused quartz to pressures of 1000 MPa. Figure 6.13 is the pressure vessel apparatus used.

With the setup of Figure 6.13, separate transducers are used to generate the shear and the longitudinal pulses. A stable CW oscillator, periodically gated, drives the transducer. The resulting ultrasonic pulses are transmitted through the tool–steel cap and into the specimen, where they reverberate and return to the transducer, where they are detected. When the reverberations are in phase, corresponding to a resonant condition within the specimen, the velocity of sound is

$$c = \frac{2\ell f}{N + (\gamma/360)} = \lambda f \tag{6.16}$$

where

  c is the velocity of sound, longitudinal or shear wave
  ℓ is the thickness of the specimen
  f is the frequency at which the in-phase condition occurs
  N is the number of half-wavelengths in ℓ, an integer
  γ is the phase shift due to the couplant between the cap and the specimen
  λ is the equivalent wavelength at frequency f

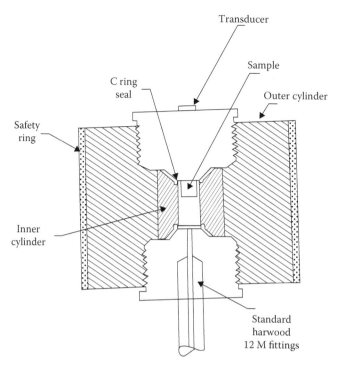

FIGURE 6.13   Pressure vessel apparatus for measuring velocity of sound in fused quartz to pressures of 1,000 MPa. (Data from Peselnik, L., R. Meister, and W. H. Wilson, *J Phys Chem Solids*, 28:635–9, 1967.)

The fundamental frequency is the difference, Δf, between two successive in-phase frequencies. Therefore, $N = f/\Delta f$ at any given frequency.

Measurements in geological specimens are subject to hysteresis effects. As a result, making two series of measurements in order to obtain both the longitudinal-mode and the shear-mode data may be unsatisfactory and result in considerable error. These measurements can be made while subjecting the specimen to only one pressure cycle by a technique described by Gregory and Podio [21]. Figure 6.14 shows the apparatus used by Gregory and Podio. The longitudinal wave is generated by a piezoelectric transducer vibrating in its thickness mode and attached to the vertical side of the aluminum block. The ultrasonic wave impinges on the 45° face of the aluminum block and is reflected downward and toward the specimen.

The shear wave is generated by mode conversion at the interface between the aluminum and an oil bath containing the shear-wave transducer. The angle of incidence is greater than the critical angle for longitudinal waves in the aluminum block. The shear wave is refracted at 45° to the sloping side of the block. This refraction angle also causes the shear wave to be directed toward the specimen.

Figure 6.15 is a schematic of the electronic–acoustic system of Gregory and Podio.

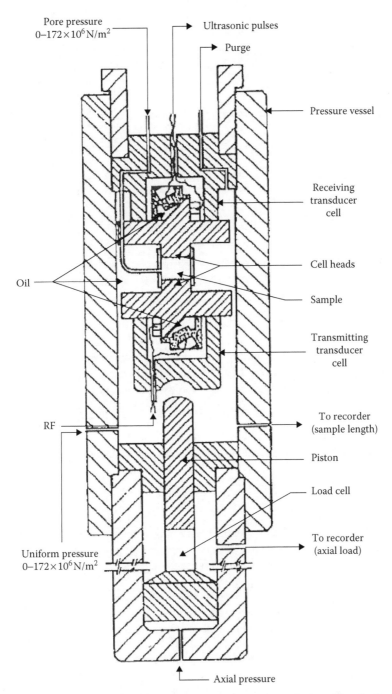

FIGURE 6.14 Dual-mode ultrasonic apparatus for measuring velocity of longitudinal and shear waves in specimens under pressure. (Data from Lubbers, J., and R. Graaff, *Ultrasound Med Biol*, 24(7):1065–8, 1998.)

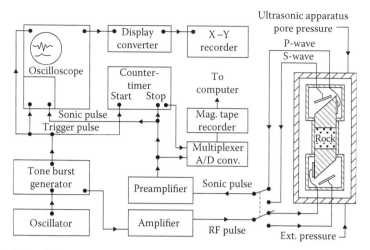

FIGURE 6.15 Schematic of the electronic–acoustic system of Gregory and Podio. (Data from Lubbers, J., and R. Graaff, *Ultrasound Med Biol*, 24(7):1065–8, 1998.)

6.3.9 WATER AND OTHER REFERENCE MATERIALS

Ultrasonic methods have a long and successful history of making measurements and relating them to fundamental material constants including thermophysical and elastic properties. There is however little in the literature that provides an assessment of capabilities and limitations of various ultrasonic techniques or discussion of the most appropriate applications, accuracy, and achievable sensitivity for various techniques.

Measurements have been made on many solids: various different metals have velocities typically in the 5–6000 m/s range, and for some ceramics, velocities are as high as 12,000 m/s. Many metal alloys (and other solids) can be differentiated based on ultrasound velocity as it is the elastic properties that control velocity, and in many cases, alloys of the same metal have significantly different elastic properties. The attenuation is also highly variable and depends on grain size. Precision velocity measurements on samples and transducers with defined geometry do enable higher order elastic constants to be measured. Elastic properties are dependent on material temperature. Therefore, for precision measurements sample temperature control is required. In practice, unless a very high precision temperature control and measurement method is employed it is only possible to achieve an accuracy ±1 m/s for velocity.

There have also been numerous measurements of velocity in gases. Such measurements are more difficult than those for solids. For composition analysis, it is necessary to control temperature and pressure to provide conditions that enable velocity and attenuation to be related to composition. Models are available for predicting properties in a number of two and three component gas mixtures. Where there is significant difference in velocity between materials in a two component mixtures, e.g., as for hydrogen and nitrogen, resonant cells have provided measurements that give composition to about 1:10,000. However, under reasonably controlled temperature and pressure data to achieve 0.1% velocity resolution is usually the practical limit. For frequencies up to 1 MHz, properties for air and its constituent parts are discussed in detail in an ANSI standard [22].

In looking at many measurement systems, it is necessary to have a reference material. One such fluid that is universally available is pure water for which the acoustic properties can be related to fundamental thermophysical properties. In spite of the apparent simplicity in making a measurement of the speed of sound in a liquid using techniques such as the pulse-echo-overlap and the sing-around methods, property values given in the literature are not consistent and there are discrepancies, which commonly exceed the declared accuracy. The variability in ultrasonically derived properties appears to be due the use of different measurement methods and the inherent systematic errors, in each approach, which are not adequately considered. A useful discussion of the issue of errors in precision ultrasound measurements is provided by Zorebski et al. [13]. There are also some basic principles that are commonly forgotten when measuring and interpreting velocity data for fluids:

- There is a need to control temperature. With temperature control of ±5°C, this gives approximately a 20 m/s change in a nominal velocity of 1481 m/s. At 15°C, this range, from 10°C to 20°C, causes a spread of nearly 40 m/s in velocity. With 1°C temperature control, the variation in expected velocity is approximately 4 m/s, which is similar to the experimental error seen in many practical velocity measurements.
- Effects of composition change. Measurements of velocity for determination of composition of fluids work well for compositions in the few percent by weight range. For most materials, detection limits are at about 1/10%, but to achieve this, there needs to be prior knowledge of the materials involved; that is, a model is available for data interpretation.
- For high precision measurements in fluids, temperature stability to better than 1/10°C is needed. Also, the material needs to be given time to achieve thermal equilibrium.
- The sample geometry needs to be well defined.
- Transducers need well characterized wave fields, and the beam spread/ diffraction and other effects require to be understood and considered in system calibration. The differences in response between near-field and far-field as well as diffraction corrections need to be applied, particularly for materials with low attenuation.

The early data for an extensive range of fluid materials are given in a review paper by Markham et al. [23]. Velocity for water has been discussed extensively, and relationships have been developed that can be used to correct for temperature variations [24]. Data are also available for a number of salt water solutions (e.g., [4,25,26]).

For many fluids, a velocity can be measured, but without prior knowledge of composition, you have an ill-posed inverse problem and any measured variability is commonly lost in measurement scatter in the data. Data interpretation—based on velocity—is plagued by issues of "uniqueness." Additional data, potentially attenuation, and data at many frequencies are needed to aid interpretation. But even when both velocity and attenuation are known at a range of frequencies, the inverse problem, the determination of composition, remains hard, and in many cases, it is simply impossible to be performed.

With mixtures and systems with known components that enable a velocity/attenuation data set, as a function of frequency, to be developed compositional analysis to about 1/10% can be achieved, given good temperature control.

6.4   LOW-FREQUENCY MEASUREMENTS OF ELASTIC MODULI AND POISSON'S RATIO

The elastic moduli—Young's modulus, shear modulus, and Poisson's ratio—can be determined by fairly simple devices and equipment. In this section are described one method for obtaining Young's modulus, one for obtaining shear modulus, and the use of these data to obtain Poisson's ratio.

6.4.1   MEASURING FLEXURAL AND LONGITUDINAL RESONANT FREQUENCIES OF BARS

Figure 6.16 is a schematic of a simple method of measuring the flexural resonance frequency of a uniform bar. In Chapter 3, Section 3.6.2, it was shown that the fundamental resonant frequency of a free–free bar is

$$f_1 = \frac{\pi c \kappa}{8\ell^2}(3.0112)^2 \qquad (3.72)$$

where

κ is the radius of gyration of the cross section
c is the bar velocity of sound (thin bar compared to the length)
ℓ is the length of the bar

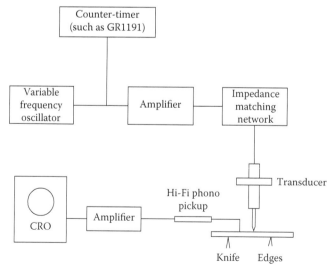

FIGURE 6.16   Block diagram of the circuit for measuring flexural resonance frequency in bars (fundamental frequency only).

At this frequency, the nodal points occur 0.224ℓ from either end. Therefore, if the knife edges are located at 0.224ℓ from either end of the bar (Figure 6.16), a simple electromagnetic, piezoelectric, or magnetostrictive transducer driving a stylus located at the center of the bar can set the bar in flexural resonance. The true resonant frequency can be determined by scanning the length of the bar with the phonograph pickup. At the nodes, the signal amplitude will be minimum, and at the antinodes, the signal amplitude will be maximum. When the minima coincide with the positions of the knife edges, the maxima occur at the center and at each end, and no other maxima or minima can be identified by the pickup, and then the frequency as shown by the counter is the fundamental resonant frequency of the bar. Young's modulus can be calculated from the following equation:

$$c = \sqrt{\frac{Y}{\rho}} = \frac{8\ell^2 f_1}{\pi\kappa(3.0112)^2} \tag{6.17}$$

The radius of gyration, κ, depends upon the geometry of the cross section of the bar. For a square or rectangular section, $\kappa = 0.289a$, where a is the thickness of the bar in the direction of vibration motion. For a circular cross section

$$\kappa = 0.5r$$

where r is the radius of the cross section. For a hollow cylinder

$$\kappa = 0.5\sqrt{r^2 + r_0^2}$$

where r is the radius of the outer circle and r_0 is the radius of the inner circle.

Longitudinal resonant frequencies of bars may be measured by a similar method, except that the exciter is located at one end of the bar and the bar is laid on soft rubber or a similar material. A microphone with wide response is located above the bar. The driving potential and the amplified signal from the microphone are applied, respectively, to the horizontal and vertical sweep terminals of an oscilloscope. The resulting Lissajous patterns can be used to locate the fundamental longitudinal frequency of the bar. The exciter generates a wide band of frequencies in striking the bar so that, for example, if the driver oscillates at one-half the resonance frequency, a double loop is formed, which indicates that the microphone detects a frequency which is double that of the exciter. Increasing the frequency of the exciter by a factor of two should verify the resonance frequency.

Bars used in this manner to determine the elastic modulus should be slender, uniform in dimension, and homogeneous. On the other hand, the method is often used to detect differences in homogeneity and quality of materials because variations in density and strength and the presence of defects affect the vibrational characteristics and the resonance frequencies.

6.4.2 Measuring Torsional Resonant Frequencies of Isotropic Bars

The torsional method has been used for measuring shear modulus of isotropic bars such as glass [27]. The elastic relations for torsional vibrations in crystalline specimens are complicated by the anisotropy of the crystals. In a circular bar of length ℓ, the fundamental frequency at torsional resonance is

$$f_r = \frac{c_s}{2\ell} = \frac{(\mu/\rho)^{1/2}}{2\ell} \tag{6.18}$$

where

> μ is the shear modulus of elasticity
> ρ is the density of the material
> c_s is the velocity of a shear wave in the material of the bar

The device used by Spinner [27] to excite both flexural and torsional vibrations in glass rods consisted simply of an electromagnetic driver, which excited the bar from a point located near one end and offset from the axis of the bar. The bar was mounted at the center. The nodes were identified by scanning the rod with a phonograph pickup.

Several precautions should be observed in making torsional resonance measurements, and most of these apply to flexural measurements as well:

- Identification of the mode of vibration should be done with care to be certain that the vibration is torsional.
- Interference due to structural resonances should be prevented to avoid erroneous interpretations.
- The specimen should be mounted so that it approaches a free state or so that any necessary clamping occurs at the preferred locations of the nodes and these nodal clamps approach line-contact conditions. The objective of the measurement is to identify a quarter-wavelength of the proper mode in the specimen bar at resonance.

It may be desirable to bond the specimen to another rod, which may be more readily vibrated in a torsional mode. If, at resonance, a quarter-wavelength in the specimen can be identified, one may easily calculate c_s and, therefore, μ.

6.4.3 Determining Poisson's Ratio, Young's Modulus, and Shear Modulus from Flexural and Torsional Resonance Data

Having determined the bar velocity c, by Equation 6.17, and the shear-wave velocity c_s, by Equation 6.18, Poisson's ratio may be calculated by

$$\sigma = \frac{(c/c_s)^2 - 2}{2} \tag{6.19}$$

Also from Equations 6.17 and 6.18, Young's modulus is

$$Y = \rho c^2 \tag{6.20}$$

and the shear modulus is

$$\mu = \rho c_s^2 \tag{6.21}$$

6.5 DENSITY, VISCOSITY AND PARTICLE SIZE MEASUREMENTS

Advanced ultrasonic sensor technologies are providing a powerful tool set that can provide physical property monitoring to give data including composition, flow characteristics, viscosity, and density. Ultrasonic sensor technologies are non-invasive, on-line, and low maintenance and can provide real-time data for large volume samples. The sensors are based on one or more of four principal ultrasonic metrics: velocity, attenuation, impedance, and scattering as functions of frequency. Their form factors are inherently robust and have been deployed at low and elevated temperatures and in both highly corrosive (high and low pH) environments and high radiation environments [28,29]. The data provided are now being used, in combination with chemical parameters, in process analytical chemistry and incorporation into emerging chemometric tool sets [30].

6.5.1 ULTRASONIC DEVICE FOR QUANTITATIVE DENSITY MEASUREMENTS OF SLURRIES

The need for a compact, non-invasive, real-time measurement of the density led to the development of an ultrasonic sensing technique based on the reflection of ultrasound at the solid–fluid interface, as shown in Figure 6.17. The transducer is mounted on the outside of the pipeline wall, and an ultrasonic pulse is reflected at the solid–fluid interface and again reflected at the solid–transducer interface and makes over 15 so-called echoes with the pipeline wall. These echoes are recorded by the same transducer, as shown in illustration. Each time the ultrasound strikes

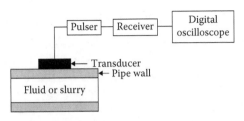

FIGURE 6.17 Concept drawing of ultrasonic density meter. (From Pappas, R. A., L. J. Bond, M. S. Greenwood, and C. J. Hostick, On-line physical property process measurements for nuclear fuel recycling, In *Proc GLOBAL 2007, Advanced Nuclear Fuel Cycles and Systems*, Boise, ID, 1808–16, © 2007 by the American Nuclear Society, La Grange Park, IL. With permission.)

the solid–fluid interface, some fraction is reflected and some fraction is transmitted into the fluid. These fractions are dependent upon the properties of the fluid or slurry—its density and velocity of sound. The many echoes act as an amplifier of this effect, and observation of the echo yields the product of the density and the velocity of sound in the fluid. Observation of the time-of-flight across the pipeline leads to a determination of the velocity. The combination of the data from the two measurements gives the density of the fluid. The uncertainty in the density measurement is about 0.2%–0.5% [31].

The ultrasonic sensor unit consists of transducers with a 6-mm-thick disk of solid material cemented to the front face of each transducer. The disk can be made from stainless steel, aluminum, or quartz; but the lower density material, such as aluminum or quartz, is preferred. By measuring the diminishing amplitudes of successive echoes within the solid disk and using the relationships for reflection coefficients, the acoustic impedance (defined as the product of the density and the velocity of sound in the fluid) is determined. By measuring the time-of-flight through the fluid or slurry, the density of the material is determined, which for a slurry can be converted to a volume percentage or a weight percentage.

The acoustic impedance is determined by plotting the natural logarithm of the amplitude versus the echo number, and a straight line is fitted through these points. The slope is related to the acoustic impedance. The uncertainty of the slope is determined in the fitting procedure and leads to an uncertainty in the density. So, both the density and its measurement uncertainty are determined. In previous applications, with transducers mounted on the outside of a 6-mm wall thickness, stainless steel pipeline, the uncertainty in the density measurement ranged from 0.1% to 0.5% error. Using quartz, which has a low density, should result in consistent measurements of 0.1% uncertainty.

The robustness of the density measurement is due to the patented self-calibrating feature [32]. This means that the slope of the line does not change if pulser voltage (or other electronics) changes. A water calibration can be carried out in the laboratory as a function of temperature and remains valid. Recalibration is not needed during the tests. It does not require any measurements with the slurries beforehand in this mode of operation. The field deployable unit, in the form of a spool piece, is shown in Figure 6.18.

6.5.2 Viscosity Measurements by Ultrasonics

Methods of measuring viscosity by means of ultrasonics have been implemented. One commercial device makes use of the damping effect of a viscous liquid on a pulsed, resonant, magnetostrictive strip [33]. The system consists of the magnetostrictive probe and an electronic computer. The probe is about the size of a fountain pen. Its operating frequency is approximately 28 kHz.

The computer is an analog type and its function is to produce an electrical output that is a function of the viscosity times the density.

The viscometer measures viscosity by comparing the damping effects of the viscous medium with those of an unloaded probe. The transducer is energized with a

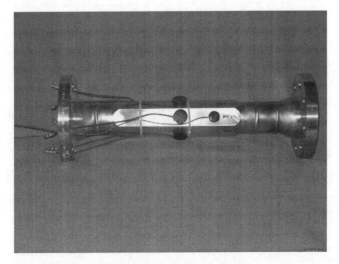

FIGURE 6.18 **(See color insert.)** Field deployment configuration of density meter. (From Pappas, R. A., L. J. Bond, M. S. Greenwood, and C. J. Hostick, On-line physical property process measurements for nuclear fuel recycling, In *Proc GLOBAL 2007, Advanced Nuclear Fuel Cycles and Systems*, Boise, ID, 1808–16, © 2007 by the American Nuclear Society, La Grange Park, IL. With permission.)

pulse. Between pulses, the vibration amplitude decays at a rate proportional to e^{at}, where α is a function of viscosity and is given by

$$\alpha = \left(\frac{n\pi c_m \rho \eta}{8 L \rho_m^2 d^2} \right)^{1/2} \quad n = 1, 3, \dots \tag{6.22}$$

where

c_m is the bar velocity of the probe strip
ρ is the density of the viscoelastic material
η is the shear-viscosity coefficient
L is the length of the probe strip
ρ_m is the density of the material of the probe strip
d is the thickness of the strip

The probe is calibrated for the true viscosity of Newtonian liquids. However, for non-Newtonian liquids, which exhibit a shear modulus, the indicated viscosity is in error by a function of the shear modulus of the material under test.

A method developed by Mason and McSkimin [34] is based on measuring the acoustic shear-load impedance of a film of the test liquid on the surface of a transmission line (Figure 6.19). The crystals are Y-cut quartz oriented to produce a shear wave in which the particle motion is parallel to the surface of the bar on which the fluid is located. The shear impedance of the test liquid is measured by comparing

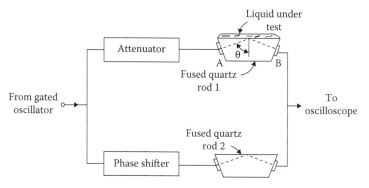

FIGURE 6.19 Shear wave reflection technique of McSkimin. (Data from Mason, W. P., and H. J. McSkimin, *Bell Syst Tech J*, 31:122–71, 1952.)

signals transmitted through two identical rods, one containing the liquid load and the other remaining empty. The load impedance offered by the liquid is

$$Z_L = Z_q \left(\frac{1 - r^2 + 2jr\sin\phi}{1 + r^2 + 2r\cos\phi} \right) \cos\theta \tag{6.23}$$

where

Z_q is the shear-wave impedance of the rod material

r is the ratio of the amplitude of the attenuated signal to the amplitude of the unattenuated signal

ϕ is the phase shift between the two branches

θ is the angle of incidence of the shear wave on the surface containing the liquid

This method is applicable to measurements from 3 to 100 MHz. At lower frequencies, Mason and McSkimin used shear loading on torsional crystals to measure dynamic viscosities ranging from 10 to 10^3 p to an accuracy of approximately 10%.

The complex impedance of the liquid, whether measured by the shear-wave-reflection method of Figure 6.19 or by the torsional-wave method, may be written in the usual form:

$$Z_L = R + jX = (j\omega\rho\bar{\eta})^{1/2} = [j\omega\rho(\eta_1 - j\eta_2)]^{1/2} \tag{6.24}$$

where $\bar{\eta}$ is the complex viscosity of the liquid, which is

$$\bar{\eta} = \eta_1 - j\eta_2 \tag{6.25}$$

For a simple Newtonian liquid, $\eta_2 = 0$, and the real and imaginary parts of Equation 6.24 are equal and

$$Z_L = (1 + j) \left(\frac{\omega\rho\eta_1}{2} \right)^{1/2} \tag{6.26}$$

For a viscoelastic liquid $2RX$

$$\eta_1 = \frac{2RX}{\omega\rho} \tag{6.27}$$

$$\eta_2 = \frac{R^2 - X^2}{\omega\rho} \tag{6.28}$$

An on-line computer-controlled sensor has been developed at the Department of Energy's Pacific Northwest National Laboratory (PNNL) to measure the product of the viscosity and density ($\rho\eta$) of a liquid or slurry for Newtonian fluids and the shear impedance of the liquid for non-Newtonian fluids [35]. Measurements have been carried out for a quartz wedge having a base angle of 70°, as shown in Figure 6.20, and also for one with a base angle of 45°.

The multiple reflections of shear horizontal waves within the quartz wedge are illustrated in Figure 6.20, where the vibrations are perpendicular to the plane of the paper. No mode-converted longitudinal waves are produced upon reflection. At the opposite side, 100% of the ultrasound is reflected from the quartz–air interface. Liquids do not support shear waves easily. However, as the viscosity increases, a larger fraction of the incident ultrasound is transmitted into the liquid. Therefore, the reflection coefficient at the quartz–liquid interface is a measure of the viscosity, and, as the viscosity increases, the reflection coefficient decreases. The measurement of the reflection coefficient is difficult because it is close to 1.0. The smallest reflection coefficient in this research is about 0.97 and, furthermore, one needs to measure small changes in the reflection coefficient in order to measure small changes in the viscosity. By observing multiple reflections, the effect is amplified. By observing 10 reflections, one compares 0.970^{10} (0.737) with 0.975^{10} (0.776), which are easily distinguishable. The viscosity values are shown to be in good agreement with those obtained independently using a laboratory viscometer. The measurement of the density results in a determination of the viscosity for Newtonian fluids or the shear wave velocity for non-Newtonian fluids. The sensor can be deployed for process control in a pipeline, with the base of the wedge as part of the pipeline wall, or immersed in a tank.

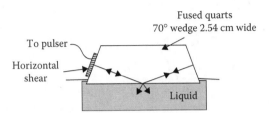

FIGURE 6.20 Diagram of fused quartz wedge and illustration of multiple reflections of a horizontal shear wave. (From Pappas, R. A., L. J. Bond, M. S. Greenwood, and C. J. Hostick, On-line physical property process measurements for nuclear fuel recycling, In *Proc GLOBAL 2007, Advanced Nuclear Fuel Cycles and Systems*, Boise, ID, 1808–16, © 2007 by the American Nuclear Society, La Grange Park, IL. With permission.)

6.5.3 Ultrasonic Diffraction Grating Spectroscopy for Particle Size and Viscosity

There are a variety of techniques for measuring the particle size and viscosity of a slurry in a laboratory setting, but few methods for on-line and real-time measurements. During the last 6 years, ultrasonic diffraction grating spectroscopy (UDGS) has been investigated at the Pacific Northwest National Laboratory, with funding from the department of energy (DOE) Environmental Management Science Program. The results demonstrate a new technique for the measurement of particle size and show effects of the viscosity of the slurry, using a system shown in Figure 6.21 [36,37]. The velocity of sound in the slurry is also measured by this technique.

The ultrasonic diffraction grating was formed by machining triangular grooves with a periodicity of 483 μm on the flat surface of an aluminum unit with send and receive transducers fastened to it. The unit has a height of 3.81 cm. A plastic cup contains the slurry, in contact with the grating surface. The ultrasonic beam from the send transducer A strikes the back of the grating at an incident angle of 30°, and the reflected beam travels to receive transducer B. Beams of various transmitted orders

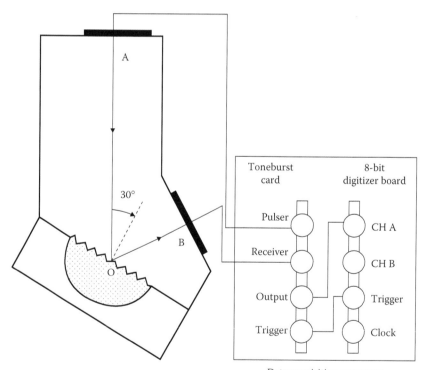

FIGURE 6.21 Diffraction grating for particle size and viscosity. (From Pappas, R. A., L. J. Bond, M. S. Greenwood, and C. J. Hostick, On-line physical property process measurements for nuclear fuel recycling, In *Proc GLOBAL 2007, Advanced Nuclear Fuel Cycles and Systems*, Boise, ID, 1808–16, © 2007 by the American Nuclear Society, La Grange Park, IL. With permission.)

m are also transmitted into the slurry, where $m = 0, \pm1, \pm2$, and so on. The attenuation is focused on the $m = 1$ transmitted beam that increases in angle, relative to a normal to the grating surface, as the frequency of the beam decreases. The frequency where it reaches 90° is called the critical frequency. At this transition point, the transmitted beam becomes evanescent and interacts with the particles in the slurry, reducing the amplitude of the wave. Essentially, an evanescent wave is one that travels in a direction parallel to the grating surface, but its amplitude decreases exponentially with the distance from the grating surface. Slightly below the critical frequency, the existence of an $m = 1$ transmitted beam is no longer possible. The energy is transferred to other beams, such as the $m = 0$ reflected beam, which is observed by the receive transducer. Thus, a peak is expected to be observed by the receive transducer at the critical frequency. The height of this peak depends upon the particle size since the interaction of the evanescent with the particle also depends upon particle size.

The critical frequency can be used to determine the velocity of sound in a liquid or slurry. Experimental data have been obtained for nine particle sizes of polystyrene spheres, ranging from 45 to 467 μm, for concentrations up to a volume fraction of 0.2. The data for each particle size have a different slope and, thus, can be used for particle sizing. Theoretical lines are obtained from the viscous–inertial model, using only one constant of proportionality.

The effects of the viscosity of the liquid were also investigated by using slurries of 275-μm-diameter polystyrene spheres in sugar water solutions of various concentrations and also in mineral oil, using the same grating described above. Polystyrene spheres were added to each solution in small amounts up to a volume fraction of 0.20. The peak heights for $m = 1$ and $m = 2$ were measured at each concentration, and the natural logarithm of the peak height is plotted versus the volume fraction for water and sugar water solutions. The data for mineral oil showed slopes that were very different from those of water. The results show distinctive slopes for water, 11 Wt% sugar water, and 30 Wt% sugar water even though the viscosities are given by 1.0, 1.38, and 3.18 m-Pa-s, respectively. More experimental measurements are planned, as well as comparison with various theories.

The ultrasonic diffraction grating could form part of the wall of a pipeline or be immersed in a tank. In addition, UDGS provides a means to probe the viscous–inertial model, with the simultaneous observation of scattering effects [36–38]. Such studies will provide a means for improving the theoretical interpretation of standard attenuation measurements.

6.5.4 PARTICLE SIZE IN EMULSIONS, COLLOIDS, AND SLURRIES

There are a near infinite number of systems encountered in industrial and manufacturing processes where there is interest in characterization of particles or droplets suspended in a second fluid phase. As ultrasound is increasingly seen as being a chemometric tool, giving data for characterization that can be used for on-line monitoring and for process control; this area has become and remains a topic of significant research interest [28]. In looking at the range of challenges at one extreme, there is the single solid (strong scattering) particle that can vary in size, normalized in terms of wavelength, from the long wave to the short wavelength limits. At the other extreme are multiple particle systems—dense slurries and porous media—which again can cover a wide range of acoustic impedance contrast regime, as well as relative size ranges. The systems and theory involved in strong scattering, by either one or many particles (metal particles in a polymer) are very different from those that apply to weak scattering, with minimal acoustic contrast (polystyrene beads in a polymer matrix or soft tumor in healthy tissue). In this section, the reader is introduced to this topic and referred to some of the reviews, of what is a very extensive, and in many cases mathematical literature.

Modern computer systems that have the ability to perform spectral analysis are enabling the potential use of ultrasonic spectroscopy for analysis to be employed. However, central to data analysis is the availability of a model for the system that is being interrogated with ultrasound.

The topics of the effects of particle size and concentration and sizing were reviewed by Kytomma [39]. Scattering by particles, together with the major developments in scattering theory, is reviewed by Dukhin and Goetz [40]. A major review of ultrasound for the characterization of colloids is provided by Challis et al. [41]. A wider ranging and comprehensive review of many scattering theories and their relationship to the characterization of liquid and multiphase systems is included in the texts by Povey [4] and Asher [42].

6.6 DETERMINING PROPERTIES OF PLASTICS AND HIGH POLYMERS

Sonic techniques are used to determine certain mechanical properties of high polymers and to locate flaws in a variety of manufactured plastic products. By measuring the velocity of sound, or resonance characteristics of beams under controlled conditions, it is possible to determine the physical constants characteristic to the viscoelastic behavior of the material. Lohr et al. [43] claim that the long-time mechanical and thermal stability of polymeric materials can be predicted from experiments conducted over relatively short periods of time. The materials that they tested included Plexiglas, polyvinyl chloride, polystyrene, Delrin, and Lexan.

The velocity of sound in polymers may be used to determine the degree of polymerization or cure of plastics, which is a measure of the molecular weight [44].

Pulse-echo and through-transmission techniques have both been used in nondestructive testing (NDT) of plastics and plastic laminates. Pulse-echo techniques are methods involving the detection of pulses of ultrasonic energy reflected from impedance discontinuities. Through-transmission techniques are methods involving the use of a transmitter at one end or side of a specimen and a receiver at the other end whereby it is possible to evaluate the material on the basis of the nature of the effect produced on the propagating wave. Properties evaluated include delamination, bond condition, material thickness, and bond strength [45].

Zurbrick [46] has used ultrasonics to determine the performance reliability of glass-reinforced laminates designed for structural, primary-load-bearing components. Bobrikov and Medredev [47] discuss the possibility of using ultrasonic energy for determining the binder content and the porosity of monodirectional glass fiber–reinforced plastic materials. They give semiempirical relations between the propagation rate of longitudinal ultrasonic oscillations and the binder content or porosity of such material. Schultz [48] shows correlation between ultrasonic longitudinal bulk-wave velocity and bulk density and correlation between the velocity and the elastic modulus for laminates of approximately 2.54-cm thick, used as heat shields. The material for laminates was carbon fabric impregnated with phenolic resin. The correlations exhibited variabilities (attributed to fabrication processing) of a few percent over the density range 0.28–1.44 g/cm^3 and the elastic modulus range 8,300–12,065 MPA. From a velocity–density correlation, the variability of a velocity–modulus correlation can be determined, and if an approximate value of Poisson's ratio can be estimated, the velocity–modulus correlation itself can be determined. If the velocity–density and velocity–modulus correlations are known, Poisson's ratio can be determined from equations describing the propagation of ultrasonic waves.

6.7   GENERAL COMMENTS ON MEASURING ACOUSTICAL PROPERTIES OF MATERIALS

In practical science and engineering, it is often necessary to develop approaches to measuring acoustic properties of materials that differ from the "conventional." The high accuracies possible with some of the techniques described in the present chapter may not be necessary, and the methods themselves may not be practical. However, by understanding certain factors related to the need, new or unique techniques might be applied to obtain the necessary information. Two such methods are described in the present section for illustration.

In many instances, comparator methods provide the most practical means of measuring both velocity of sound and attenuation in materials. Both velocity of sound and attenuation data for pure (distilled) water are easily available and, therefore, water is a good medium to use as a standard. Other materials, solid or liquid, can be used for this purpose provided sufficient velocity and attenuation data for the materials are available.

One comparator approach to measuring both longitudinal (bulk) velocity and shear velocity in solid specimens and, from these data, determining Poisson's ratio involves establishing separately multiple echoes of each mode in a representative test specimen. To make the measurements, a transducer is tightly coupled to one flat surface of the specimen to excite a predetermined wave mode—shear or longitudinal. In nondispersive materials, broadband pulses may provide sufficient accuracies. However, pulsed CW is necessary in order to obtain reliable accuracies in dispersive media.

Measurements in a single solid specimen in which it is impractical to change spacing between reverberating surfaces can be made by displaying multiple echoes obtained from the specimen and echoes obtained from a reflector located in a water path between a transducer and a reflector on the same horizontal axis of an oscilloscope. The reflector or the transducer in the water is mounted on a base fitted to a micrometer adjustment. Measurements are made by comparing the length of the specimen with the distance the water path changes in moving the water path echo from one stationary specimen echo to the next on the oscilloscope screen. The National Bureau of Standards has published values for the velocity of sound in water over a wide range of temperatures [49].

Dynamic shear modulus, G^*, storage (or elastic) modulus, G', loss modulus, G'', and loss tangent, tan δ, in a fiber-and-water pulp product have been measured using a unique technique developed at Battelle's Columbus Laboratories. Measurements of these viscoelastic properties were obtained as a step in optimizing a process for refining these pulps. The apparatus is illustrated in Figure 6.22. The device consists of a cylinder in which are mounted (1) a platform, or anvil, for holding the pulp specimen, (2) an axial shaft with a "foot" on its internal end through which pressure and shear are applied to the specimens, (3) a magnetostrictive transducer coupled to the foot to generate the shear wave transmitted to the specimen, and (4) a piezoelectric shear plate mounted flush with the base of the anvil that is used to measure the stress transmitted through the specimen. An accelerometer mounted on the foot of the driver is used to measure the motion of the foot.

Compressive stress is applied to the specimen through the axial shaft connected to the driven foot. A calibrated coil spring controls the pressure applied.

Steam may be used to heat the specimen in the apparatus. A thermocouple located in the housing measures the exact temperature attained during the measurements.

The apparatus is capable of providing attenuation and shear modulus under a wide range of temperatures and pressures.

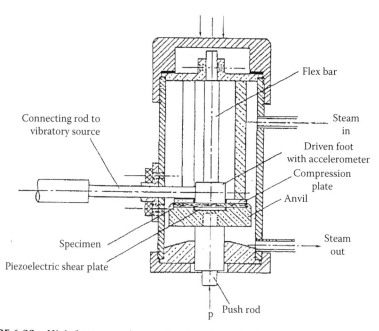

FIGURE 6.22 High-frequency characterization of wood pulp.

The *complex shear modulus*, G^*, is defined as the ratio of *complex stress amplitude* to *complex strain amplitude*. Written in its complex form,

$$G^* = G' + jG''$$ (6.29)

where G' is the storage modulus previously defined and G'' is the corresponding loss modulus. Referring to Equation 6.9 and assuming that the wave is completely attenuated within twice the thickness of the test object, G^* can be written in terms of the propagation constant in the form

$$G^* = \frac{-\rho\omega^2}{\gamma^2}$$ (6.30)

where $\gamma = \alpha + j\omega/c_s = \alpha + j2\pi/\lambda_s$. Thus, if γ can be measured, G', G'', and, therefore, $\tan\delta$ can be calculated. The relationships that apply are

$$G^* = -\frac{\rho\omega^2}{\gamma^2} = \frac{-\rho\omega^2}{(\alpha + j\omega/c)^2}$$ (6.31)

$$G' = \frac{\rho\omega^2[(\omega/c)^2 - \alpha^2]}{[(\omega/c)^2 + \alpha^2]^2}$$ (6.32)

$$G'' = \frac{\rho\omega^2[2\alpha\omega/c]}{[(\omega/c)^2 + \alpha^2]^2}$$ (6.33)

and

$$\tan\delta = \frac{G''}{G'} = \frac{2\alpha\omega/c}{(\omega^2/c^2) - \alpha^2}$$ (6.34)

Attenuation α and $\tan\delta$ are obtained by comparing stress or displacement amplitudes at the surfaces of two identical specimens of different thicknesses. These values are obtained by processing the output signals from the accelerometer on the foot of the driver and the piezoelectric shear sensor seated in the anvil, taking into account that, in a sinusoidally vibrating surface, acceleration is 180° out of phase with the displacement and ω^2 times its amplitude. Attenuation α is obtained using Equation 6.9 and maintaining constant displacement amplitude and density ρ.

Velocity of shear waves in the materials is obtained by any of three different but related methods as follows:

1. Using constant frequency, f, difference in thickness, $\Delta\ell$, and corresponding difference in phase, $\Delta\phi$, where ϕ is the phase relationship between the stress in the specimen at the driver surface as indicated by the accelerometer and the stress at the piezoelectric shear sensor in the anvil, the equation for velocity of sound is

$$c_s = 360° \left(\frac{\Delta\ell}{\Delta\phi}\right) f$$ (6.35)

2. Using constant thickness, ℓ, difference in frequency, Δf, and difference in phase, $\Delta\phi$, the velocity of sound is

$$c_s = 360° \left(\frac{\Delta f}{\Delta\phi} \right)$$ (6.36)

3. Using constant phase, ϕ, and density, ρ, and differences in thickness, $\Delta\ell$, and frequency, Δf, and letting

λ_1 = wavelength at the lower frequency, f_1
λ_2 = wavelength at the upper frequency, f_2
$\Delta\ell = \ell_1 - \ell_2$ = difference in the thickness of the two specimens

$$\lambda_1 = \lambda_2 + 2\Delta\ell$$ (6.37)

Then

$$c_s = (\lambda_2 + 2\Delta\ell)f_1 = \lambda_1 f_1 = \lambda_2 f_2$$ (6.38)

For a given phase, ϕ, the following points are identified:

ℓ_1 corresponding to f_1
ℓ_2 corresponding to f_2

(especially between sections of slopes of ϕ/f curves that are essentially identical). By calculating $\Delta\ell$ and using the corresponding values of f_1 and f_2 for constant ϕ values for the two thicknesses, λ_1 and λ_2 can be calculated using Equations 6.37 and 6.38. The velocity of sound is then calculated using Equation 6.38. These methods take into account shifts in ϕ, and so on, attributable to structural effects.

Certain rules apply to all measurements of velocity and attenuation. These rules are meant to ensure that the measurements will provide reliable data. They apply to (1) parallelism and alignment between transmitter and receiver or reflector to avoid dimensional and associated errors such as amplitude loss due to acoustic path and phase differences, (2) structural and sensor effects to avoid extracting energy from the system and giving false indications of velocity and attenuation, (3) electrical matching and electromechanical factors selected to avoid an implication of an attenuation that is in excess of the actual value, and (4) effects of phase shifts at reflecting interfaces. In general, accurate measurements of attenuation and velocity of sound are made at discrete frequencies using either steady-state or pulsed CW modes. However, for some applications (in nondispersive media), use of broadband pulses may be adequate. Diffraction, beam spread, velocity dispersion, scattering, and absorption must enter design considerations to various degrees depending upon the method and materials under study. For measuring attenuation at high frequencies using a pulsed planar source generating planar waves, it is necessary to use a transducer having a diameter that is equal to many wavelengths at the applied frequency.

REFERENCES

1. Truell, R., C. Elbaum, and B. B. Chick. 1969. *Ultrasonic Methods in Solid State Physics*. New York: Academic Press.
2. Edmonds, P. D., ed. 1981. *Ultrasonics, Methods of Experimental Physics*. Vol. 19. New York: Academic Press.
3. Trusler, J. P. M. 1991. *Physical Acoustics and the Metrology of Fluids*. Bristol: Adam Hilger.
4. Povey, M. J. W. 1997. *Ultrasonic Techniques for Fluids Characterization*. San Diego, CA: Academic Press.
5. Bhatia, A. B. 1967. *Ultrasonic Absorption*. Oxford: Clarendon Press.
6. Selfridge, A. R. 1985. *IEEE Trans Sonics Ultrason* SU-32(3):381–94.
7. Papadakis, E. P. 1976. Ultrasonic velocity and attenuation: Measurement methods with scientific and industrial applications. In *Physical Acoustics, Principles and Methods*, ed. W. P. Mason, and R. N. Thurston, Vol. XII, 277–373. New York: Academic.
8. Forgacs, R. L. 1960. *IRE Trans Instrum* 1–9:359–67.
9. Hirao, M., K. Aoki, and H. Fukuoka. 1987. *J Acoust Soc Am* 81(5):1434–40.
10. McSkimin, H. J. 1961. *J Acoust Soc* 33(1):12–16.
11. McSkimin, H. J. 1961. *J Acoust Soc* 33(5):606–15.
12. Hsu, N. N. 1974. *Exp Mech* 14(5):169–76.
13. Zorebski, E., M. Zorebski, and S. Ernst. 2005. *J De Physique IV* 129:79–82.
14. Matsuzawa, K., N. Inoue, and T. Hasegawa. 1987. *J Acoust Soc Am* 81:947–51.
15. Bond, L. J., C.-H. Chiang, and C. M. Fortunko. 1992. *J Acoust Soc Am* 92(4):2006–15.
16. Higgs, R. W., and T. A. Litovitz. 1960. *J Acoust Soc Am* 32(9):1108–15.
17. Litovitz, T. A., T. Lyon, and L. Peselnick. 1954. *J Acoust Soc Am* 26(4):566–76.
18. Clark, A. V., M. Lozev, B. J. Filla, and L. J. Bond. 1983. Sensor system for intelligent processing of hot-rolled steel. In *Proceedings 6th International Symposium on NDE of Materials*. ed. R. Green et al., 29–36. New York: Plenum Press.
19. Peselnik, L., R. Meister, and W. H. Wilson. 1967. *J Phys Chem Solids* 28:635–9.
20. McSkimin, H. J. 1957. *IRE Trans Ultrasonics Eng* PGUE- 5:25.
21. Gregory, A. R., and A. L. Podio. 1969. Paper D-10 presented at the 1969 IEEE Ultrasonics Symposium, St. Louis, Missouri, September 24–26. New York: Sonics and Ultrasonics Group, IEEE Institute of Electrical and Electronic Engineers.
22. ANSI. 1978. Method for the calculation of the absorption of sound by the atmosphere, ANSI, American National Standards Institute S1.26-1978.
23. Markham, J. J., R. T. Beyer, and R. B. Lindsay. 1951. *Rev Mod Phys* 23(4):353–411.
24. Lubbers, J., and R. Graaff. 1998. *Ultrasound Med Biol* 24(7):1065–8.
25. Urick, J. R. 1983. *Principles of Underwater Sound*. 3rd ed. New York: McGraw-Hill.
26. Kleis, S. J., and L. A. Sanchez. 1991. *Solar Energy* 46(6):371–5.
27. Spinner, S. 1956. *J Am Ceram Soc* 39:113–8.
28. Pappas, R. A., L. J. Bond, M. S. Greenwood, and C. J. Hostick. 2007. On-line physical property measurements for nuclear fuel recycling. In *Proc GLOBAL 2007, Advanced Nuclear Fuel Cycles and Systems*, held September 2007, 1808–16. Boise, ID: Am. Nuclear Soc.
29. Lynnworth, L. C. 1989. *Ultrasonic Measurements for Process Control*. Boston: Academic Press.
30. Workman, J., D. J. Veltkamp, S. Doherty, B. B. Anderson, K. E. Creasy, M. Koch, J. F. Tatera et al. 1999. *Anal Chem* 71(12):121R–80R.
31. Bamberger, J. A., and M. S. Greenwood. 2004. *Ultrasonics* 42(1–9):563–7.
32. Greenwood, M. S. 2004. Self Calibrating System and Technique for Ultrasonic Determination of Fluid Properties, US Patent 6,763,698.
33. Roth, W., and S. R. Rich. 1953. *J Appl Phys* 24(7):940–50.

34. Mason, W. P., and H. J. McSkimin. 1952. *Bell Syst Tech J* 31:122–71.
35. Greenwood, M. S., and J. D. Bamberger. 2005. US Patent 6, 977, 375.
36. Greenwood, M. S., and S. Ahmed. 2006. *Ultrasonics* 44(Suppl. 1):e1385–93.
37. Greenwood, M. S., A. Brodsky, L. Burgess, and L. J. Bond. 2006. Investigating ultrasonic diffraction grating spectroscopy and reflection techniques for characterizing slurry properties. In *Nuclear Waste Management: Accomplishments of the Environmental Management Science Program, Symposium Series No. 943*, ed. P. W. Wang, and T. Zachry, Chap. 6, 100–32. New York: Proceedings, Fall Meeting, September 2003 Am. Chemical Soc.
38. Greenwood, M. S., J. D. Adamson, and L. J. Bond. 2006. *Ultrasonics* 44(Suppl. 1): e1031–6.
39. Kytömaa, H. K. 1995. *Powder Technol* 82(1):115–21.
40. Dukhin, A. S., and P. J. Goetz. 1998. *Colloids Surf A* 144(1–3):49–58.
41. Challis, R. E., M. J. W. Povey, M. L. Mather, and A. K. Holmes. 2005. *Rep Prog Phys* 68(7):1541–637.
42. Asher, R. C. 1997. *Ultrasonic Sensors*. Bristol: Institute of Physics Publishing.
43. Lohr, J. J., D. E. Wilson, F. M. Hamaker, and W. J. Stewart. 1968. *J Spacecr Rockets* 5(1):68–74.
44. Anon. 1951. *Technol Rev* S3:467.
45. Hitt, W. C., and J. B. Ramsay. 1963. *Rubber Plast Age* 44:411–413.
46. Zurbrick, J. R. 1968. *Mater Res Stand* 8(7):25–36.
47. Bobrikov, L. P., and M. Z. Medredev. 1968. *Uekh Polim* 4:547–54.
48. Schultz, A. W. 1967. *Mater Res Stand* 7(8):341–5.
49. Greenspan, M., and C. E. Tschiegg. 1959. *J Acoust Soc Am* 31(1):75–6.

7 Nondestructive Testing
Basic Methods and
General Considerations

7.1 INTRODUCTION

Ultrasonic methods for nondestructive testing (NDT) and material characterization are perhaps the most versatile of the testing techniques available to industry. Ultrasonic NDT, therefore, provides one of the largest non-medical areas for applications of low-intensity ultrasonic energy.

As discussed in Chapter 1, there have been major advances in NDT in the past 30 years. Traditional manual testing, using skilled operators, who perform standardized inspections, has been supplemented by an increasing range of automated and semi-automated inspections. It was recognized in the mid-1970s that for NDT to meet the needs of high-technology industries, such as aerospace, nuclear power, defense, oil, and gas, it needed to become a more quantitative science-based technology. Research was initiated in several countries, and as a result new capabilities were developed that are part of what is now called "quantitative nondestructive evaluation," (QNDE) and is commonly referred to as nondestructive evaluation (NDE) and quantitative nondestructive evaluation (QNDE) [1]. Recent years have seen better understanding of the measurement processes and quantification of performance in terms of a probability of detection (POD), rather than performance being stated as a single number or ultimate detection limit. As a part of this development process, the phenomena and material interactions were better understood and it was increasingly shown how performance limits are, in many cases, set by the part inspected, the material used, its geometry, the defects involved, and the fundamental physics of wave–feature interactions [2].

A handbook for ultrasonic testing is found in the classic text by Krautkramer and Krautkramer [3]. The fundamentals for many ultrasonic tests are also provided in standards that are to be found in the various standards series, including the American Society for Testing and Materials (ASTM) series [4]. Discussion of implementation and training materials are provided by many companies and international organizations (e.g., International Atomic Energy Agency [5]), and further material can be identified on national NDT society web sites and in journals and conference proceedings series. The subject of ultrasonic NDT/NDE is also included in more general texts that cover all types of NDT and quality control/quality assurance (e.g., [6,7]).

At the same time, as the science base has been developed, there has been significant progress made in the performance and manufacture of ultrasonic equipment. The NDT equipment industry started, in many cases, in peoples' home workshops or basements. This industry has matured and the quality of equipment has improved.

The global market in NDT equipment, for the five major technologies (radiography, ultrasonics, eddy current, magnetic particle, and penetrant testing) was in 2001 reported as being about $1 billion [8] of which a significant fraction is ultrasonic. More recent reports have put the 2003 total NDT equipment market size as $723 million [9], with the world ultrasonic NDT equipment market representing 26.6% of total NDT equipment in 2005 [10]. The total cost to users of implementing ultrasonic testing is significantly higher than just equipment purchases since much inspection is still labor intensive and for many in-service inspections significant preparation is required prior to testing. With the moves toward more digital systems and this market growth, there has been significant consolidation, with many well-known brands becoming parts of larger companies.

This chapter discusses ultrasonic inspection/measurement methods, factors that affect the capabilities to detect defects, and some basic aspects of ultrasonic NDT instrumentation.

7.2 BASIC METHODS

Most commonly used ultrasonic NDT methods may be grouped under three headings: (1) resonance methods, (2) pulse methods, and (3) acoustic emission methods. There are also some additional non-conventional approaches that have been reported in the research literature (e.g., [1,11], and some of these are identified when specific applications are discussed in subsequent chapters.

7.2.1 Resonance Methods

Resonance methods have a long history, and they been used to measure the thickness of plates, pipe, and tank/vessel walls and for the detection of delaminations when only one surface is available. For many years, they were considered to be mostly just an historic technique of limited usefulness, which were largely replaced by pulse methods. However, there have been major developments with some forms of ultrasonic spectroscopy [11] and, in particular, resonant ultrasonic spectroscopy (RUS) that is now being used both to give elastic properties and to inspect parts with simple geometry for small defects [12]. The basic theory for RUS, following the approach given by Migliori and Sarrao [12], is summarized in an article by Zadler et al. [13]. However, the principles applied are of sufficient usefulness to justify their inclusion here. The methods involve determining the fundamental frequency calibrated in terms of thickness or other dimensions. The thickness is one half-wavelength at the fundamental frequency, and at the overtones, the thickness is a multiple of half-wavelengths. A piezoelectric transducer used to excite ultrasonic waves in the specimen material is driven by a frequency-modulated electrical signal through a range of frequencies usually covering one or more octaves. Changes in impedance due to the in-phase or out-of-phase conditions within the specimen are reflected through the transducer into the electronic system. The effects are indicated by a suitable display unit (usually an oscilloscope) or computer. When an oscilloscope is used, the sweep triggering and the sweep rate are synchronized with the variations in frequency of the oscillator. Therefore, resonance conditions are indicated on the oscilloscope screen as peaks at positions that correspond to multiples of half-wavelengths in the material. The thickness of the material is given, within the accuracy of the test conditions, by

$$\ell_T = \frac{c}{2f_1} \tag{7.1}$$

where f_1 is the fundamental resonant frequency of the test material. Usually, the display screen is equipped with scales calibrated in units of length corresponding to half-wavelengths within the material being tested. The important factor in this type of measurement is to identify the fundamental resonance frequency. When resonance occurs at a harmonic frequency, more than one peak occurs on the display screen. The difference between two successive resonant frequencies is the fundamental frequency. Since the calibration is in linear dimensions, it is possible to determine the actual dimension from the indicated dimensions on the display screen from the relationship

$$f_1 = f_{n+1} - f_n \tag{7.2}$$

where n is the order of the harmonic. If two successive peaks occur on the display screen at ℓ_n and ℓ_{n+1} corresponding to f_n and f_{n+1}, the thickness ℓ is

$$\ell = \frac{\ell_n \ell_{n+1}}{\ell_n - \ell_{n+1}} \tag{7.3}$$

within the accuracy of the measurement technique. Calibrated oscilloscope screens or computer displays scales corresponding to the various harmonics are sometimes used. By aligning the various peaks with corresponding values on each scale, it is possible to read the thickness directly without resorting to Equation 7.3.

Since the elastic properties and velocity of sound are different for different materials, the instrumentation must be calibrated using calibration blocks made of the same material as that to be tested. Resonance techniques have been used to measure elastic properties in the laboratory since the 1920s [14]. Theoretical work by Visscher et al. [15], using a variational technique, liberated the technique from the limitation of only being able to analyze properties of samples of simple 3-D geometry to being able to consider bodies of arbitrary shape. The developments made in computational analysis and improved experimental systems have formed the basis for modern RUS [12].

7.2.2 PULSE METHODS

Ultrasonic pulse methods of NDT are used much more extensively than resonance methods in industry. Waves of various modes are used, including longitudinal, shear, Rayleigh, and Lamb waves. Ultrasonic pulses can penetrate to considerable depths—several meters in low-carbon steel and aluminum.

Pulse methods may be classed under three categories: (1) pulse-echo, (2) pitch–catch, and (3) through-transmission, or transmission. These basic configurations have been implemented in numerous forms. These approaches are illustrated with the series of Figures 7.1 through 7.3. The principles of pulse-echo ultrasound are shown in Figure 7.1, which gives responses from a plate with compression waves for no defect, a small defect, and a larger defect in the form of simplified A-scan traces. An example of a pitch–catch configuration is shown in Figure 7.2, where reflections from both defect and back wall reach the receiver. A transmission configuration with a pair of compression wave transducers is shown in Figure 7.3. The responses for no defect, a small defect, and a larger defect with simplified A-scan screen displays are shown.

Pulse-echo methods involve the use of transducers, which act first as emitters of ultrasonic pulses and then as receivers to detect echoes from defects or other interfaces. This is illustrated in Figure 7.1 for the case of no defect, together with those for a small defect and a larger defect.

Pitch–catch methods involve the use of two identical piezoelectric elements, often, but not necessarily, mounted in the same holder, with one element serving to emit ultrasonic pulses and the other to receive the reflected pulses. An example of a pitch–catch configuration is shown in Figure 7.2.

Through-transmission methods involve the use of two transducers located relative to each other and to the specimen in such a manner that one transducer receives the energy transmitted from the other after the energy has passed through a region

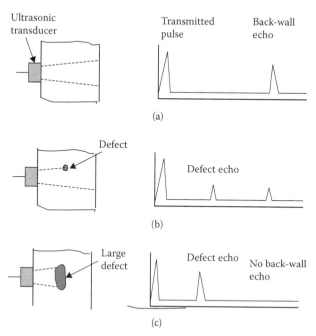

FIGURE 7.1   The principles of the pulse-echo methods of ultrasonic testing. (a) Defect-free specimen. (b) Specimen with small defect. (c) Specimen with larger defect.

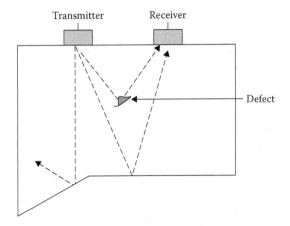

FIGURE 7.2   An example of a pitch–catch configuration—sound reflected to receiver from defect and back wall.

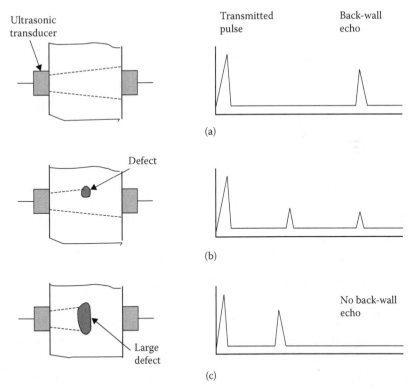

FIGURE 7.3   Transmission configuration and simplified screen displays. (a) Defect-free specimen. (b) Specimen with small defect. (c) Specimen with larger defect.

of interest. Discontinuities cause losses of energy characterized by drops in output voltage from the receiver. When the longitudinal mode is used, the transmitter and receiver are located on opposite sides of the specimen. When shear waves are used, the transmitter and receiver may be located on the same face of the specimen, or on the opposite sides, but at opposite ends of the test area. When surface waves or Lamb waves are used, the transducers are located on the same surface, but at opposite ends of the test area. An example for the case of a longitudinal wave in a simple geometry is shown in Figure 7.3. The presence of a defect is indicated by a reduction in the transmitted signal. For no defects, two pulses are typically displayed, and this is shown in schematic form in Figure 7.3a. Pulse (1) is the excitation pulse, and the second pulse (2) is the direct transmission, transmitter–receiver. The introduction of a small defect (shown in Figure 7.3b) reduces the amplitude of the transmitted pulse and will add intermediate pulses in reverberations. The case of a large defect blocks the transmitted signal, as is illustrated with Figure 7.3c.

7.2.3   ACOUSTIC EMISSION TECHNIQUE

Acoustic emission is the basis of an important class of NDT methods. Acoustic emission is the spontaneous emission of sound pulses from materials subjected to external stress as a result of sudden relaxation of stresses within the material. Assume that a material is subjected to tension and that this material contains a point of stress concentration. As the average stress increases, the stress in the material about the point of stress concentration approaches the elastic limit before that of the remaining material. The material subjected to the concentrated stress is no longer capable of carrying its share of the load, and the adjacent fibers suddenly experience an increase in stress. According to the development of the wave equation, in order for equilibrium to obtain, the stress distributes throughout the remaining material at the velocity of sound.

Stress relaxation producing acoustic emission can occur as a result of the nucleation and propagation of cracks or as a result of such elastic and plastic deformation processes as the slip of existing dislocations in metal, the activation of dislocation sources, twinning, phase change, and the slip of grain boundaries [16].

The emitted energy consists of two components: (1) a low-level, high-frequency component, which appears as an increasing noise as the stress continues to be applied, and (2) a discontinuous, burst-type component or pulse. The high-frequency component is believed to be induced by dislocation pinning during slip deformation. The burst type of pulse is believed to be related to the formation of stacking faults and mechanical twins and to the formation of cracks.

The sound level of acoustic emission is very low and depends upon the stress level at which it is produced, neglecting strain rates and volume effects [17]. The loading mechanism used to stress the material must be carefully designed so that it does not generate sufficient noise to mask the acoustic emission.

The frequency spectrum of acoustic emissions depends upon how the sound is produced. The frequencies range from audible to ultrasonic frequencies in the megahertz range. Sometimes frequency is used to denote rate of emission or the number of bursts or counts per second [16].

The microstructure and, therefore, the processing treatment that a metal has been subjected to affect the acoustic emission. Aging, either at zero stress or under an applied stress, precipitation hardening, and cold work affect the emission [18]. Deformation may cause a random signal, which is dependent on strain rate, with an average frequency being discernible. Along with the random signal may occur sporadic bursts of higher frequency. Brittle fracture produces burst signals of much higher frequency [19].

When a metal is stressed to a value near or above the yield stress, unloaded, and then restressed to this same level again, if there had been emission during the first load cycle, it would be absent or very much reduced during the second application of a load up to the previous stress level. However, when the previous stress level is exceeded, acoustic emission is again evident [16]. This phenomenon has been called the Kaiser effect. In composites, emission usually occurs before the previously applied load level is attained. The ratio of the stress at which emissions occur divided by the previously attained maximum stress is called the Felicity ratio and is indicative of the amount of damage [20].

Basic equipment required for applying acoustic emission (AE) to NDT comprises a highly sensitive microphone and an amplifier. If possible, detection is done at frequencies at which the background noise is minimum. This may be done by filtering or by choosing a transducer with a response limited to a small range of frequencies within the range of interest.

In 1985, the relationship between an acoustic emission source and the wave modes that are excited and propagate from the source began to be developed [21]. These insights coupled with digital instrumentation, as well as an increased capability to analyze plate modes, including Lamb wave propagation, have resulted in new families of monitoring systems. This so-called "modal AE" enables analysis of waves to give insights into sources that is analogous to characterization that can be applied to earthquakes [22].

Although most applications of acoustic emission for NDT have been in metals— pressure vessels, nuclear components, aerospace materials, and pipes—it is gaining favor as a means of testing ceramic materials as well. Romrell and Bunnell [23] were among the first to indicate the feasibility of monitoring the growth of cracks in ceramic materials. They used a lead–zirconate–titanate transducer with a primary frequency response to acoustic emission at 600 kHz. They increased the signal-to-noise ratio by using a preamplifier with a 600-kHz tuned circuit. For additional rejection of low-frequency noise, the signal conditioner included a high-pass filter adjusted to 450 kHz. Materials studied were thorium–yttrium oxide (tubes) and aluminum oxide rods containing magnesium oxide to allow sintering to nearly theoretical density while maintaining fine grain structure. The acoustic emission signals were counted by both emission rate and summation circuitry, which operated in parallel.

Other applications of acoustic emission include in-process inspection of welds (if cracks form during cooling, they can be detected as they form), monitoring of crack formation during hydrotesting of solid rocket chambers, and monitoring of reactor pressure systems. Acoustic emission testing is finding wide usage in industry, and further examples of more recent applications including for more comprehensive structural health monitoring are presented in Chapters 8 and 9.

7.3   FACTORS AFFECTING RESOLUTION AND SENSITIVITY

The *sensitivity* of an ultrasonic transducer refers to the relationship between the amplitude of the electrical voltage generated by the transducer and the magnitude of an ultrasonic signal impinging on its surface. It, therefore, is a determining factor in how small a defect can be detected. The sensitivity of transducers used for NDT is measured by the amplitude of their response to energy reflected from a standard discontinuity (for instance, a hole in a standard reference block or a round rod in a liquid bath). Several factors influence the sensitivity of a transducer and thus not all of a given type and make have similar sensitivities.

Resolution refers to the capability to separate echoes from two discontinuities located at only slightly different depths. A long pulse has poor spatial resolving power. When a second echo arrives at the receiver before the first one dies out sufficiently for the second one to be observed, the second signal is masked by the first. Very short pulses are desirable for high resolution. This requires broad bandwidth, low-Q transducers, where

$$Q = \frac{f_0}{f_2 - f_1} \qquad (7.4)$$

where

f_0 is the center frequency of the transducer

$f_2 - f_1$ is the bandwidth

f_1 and f_2 are the frequencies below and above f_0, respectively, where the amplitude of vibration is only 0.707 times the maximum amplitude.

Resolving power of broadband transducers decreases with distance in materials, and the rate of decrease depends upon the attenuation characteristics of the material. For example, reflections at grain boundaries cause attenuation due to scattering. The amount of scattering is a function of the anisotropy of the grains, of the metallurgical conditions at the grain boundaries, and of the relationship between the wavelength and the grain dimensions. Higher frequencies are attenuated more rapidly than lower frequencies, and this preferred attenuation tends to reduce the bandwidth and lower the center frequency of the pulse, thus causing the loss in resolving power.

7.3.1   NEAR-FIELD EFFECTS

Diffraction in the near-field of a transducer produces variations in the distribution of sound pressure along and parallel to the axis and also in a direction normal to the axis. In NDT, the amplitude of waves reflected from defects located within the near-field is affected by the diffraction pattern and depends upon the location of the defect within the near-field. In the far-field, the amplitude of the reflected wave decreases approximately exponentially as the distance from the source increases. However, in the near-field, the amplitude of the reflected wave may experience at least one maximum and one minimum as the distance from the source increases to a point corresponding to the start of the far-field, where maximum amplitude is usually obtained. For this reason, it is difficult to determine the sizes of defects located in the near-field by the amplitude of reflected waves alone.

(a)   (b)

FIGURE 7.4 **(See color insert.)** Example of measured intensity cross section through a transducer beam. (a) Section parallel to beam axis. (b) Transverse to beam, measured near peak on-axis intensity (arrow in [a])—false color bands indicate intensity zones.

The ultrasound beam produced by a transducer is 3-D, and when probed, the intensity structure is complex, particularly in the near-field. An example of cross-sectional beam plots for a 1 MHz, 24-mm-diameter, unfocused compression wave transducer, measured in pulse-echo using a 3-mm-diameter wire target and displayed at 3 dB intensity increments, is shown in Figure 7.4.

7.3.2 Properties of the Materials

The properties of the materials that influence defect resolution and sensitivity include geometry of the part, metallurgical conditions such as grain size and grain-boundary conditions, anisotropy, surface roughness, and the presence of internal scatterers and their scattering cross section.

Scattering from grain boundaries may completely obscure conditions within a material if the frequency chosen exceeds that for which the wavelength is greater than three times the grain dimensions. Therefore, grain size and the conditions that exist at the grain boundaries are determining factors in the choice of frequency and the size of defect that can be detected ultrasonically. Defects of sizes corresponding to the grain dimensions may be impossible to detect unless these indication defects appear in clusters such that the combination of scattering cross sections appears as a unit, which is larger than the average grain sizes. For example, shrink porosity in wrought-iron castings is easily detected by pulse-echo and through-transmission methods because the clusters of small pores produce echoes and scattering that are similar to those produced by larger defects with continuous boundaries.

The geometry of the part determines the accessibility of the ultrasonic beam to the defects. It also is a determining factor in the choice of modes for the inspection. The surface contour, in particular, affects the resolution and sensitivity of the inspection by its effect on the incidence angle at any given position within the cross section of the beam. For instance, when a plane wave approaches by means of a liquid couplant, a curved surface, which has a radius of the same order of magnitude as the cross section of the beam, the incident angle of approach may range from 0°, where the radius of the surface is parallel to the direction of propagation, to a maximum of 90°, where the radius is normal to the direction of propagation. Angles of reflection and refraction vary accordingly. Consequently, both shear and longitudinal waves are generated within the material. The effect of this mode conversion is to partition and scatter the energy within the part and, since the different modes travel at different velocities, to decrease the sensitivity to defects because of the confusion that results.

Surface roughness affects the resolution and sensitivity of an ultrasonic nonde-structive test in much the same manner as grain conditions. Scattering from rough surfaces can significantly reduce signals and, in some cases, make inspection at a particular frequency impossible if the wavelength is similar to the dimensions of the asperities. Some fundamental aspects of wave–material interaction were discussed in Chapter 2. However, the analysis considered only plane-surface interactions. Scattering from many real surfaces is more complex and depends on the depth and characteristic dimensions of roughness expressed as a function of the incident wavelength: random rough surfaces is considered extensively by Ogilvy [24]. The interaction of ultrasound with internal structures, such as grains, is treated and discussed by Goebbels [25,26] and Papadakis [27,28], and a more comprehensive review of scattering theory, for both canonical problems and a range of classes of targets, is provided in an article by Pao, which cites many references [29].

A summary of the theory for reflection and scattering by small and large objects was provided in a text by Kino [30], and this is similar to treatments that are also found in the SONAR literature [31]. In the theory for interactions encountered in NDT, as it has evolved into NDE and QNDE, the approaches used for defect characterization, material structure, and signature analysis, on the microscale, have in general evolved from simple defect sizing approaches [3,4] to now include digital data capture and the application of inversion algorithms for sizing of isolated targets [32] and imaging applied to larger features. Not only does a rough surface cause scattering and mode conversion on entry, but it also may affect the coupling of energy into the part.

It is possible to define a critical surface roughness, which corresponds to a condition of maximum destructive interference in the wave within the part. This condition is brought about by the difference in the velocity of sound in the liquid couplant and in the solid. The velocity of sound in a solid is greater than in a liquid. Waves entering the part at a roughness peak travel the distance to the bottom of the valley at a higher velocity than the waves that travel through the liquid and enter the part at the bottom of the valley. When the phase difference between the two components after both enter the part is 180°, or an odd multiple thereof, maximum cancellation occurs. To obtain complete cancellation would require a unique situation that is not obtained with broadband, short pulses. However, refraction, diffraction, and the difference in phase change the shape and affect the pattern of the pulse within the specimen.

The theoretical critical roughness, R_c, is given by the equation

$$R_c = \frac{\lambda_1 c_2}{2(c_2 - c_1)} = \frac{\lambda_2 c_1}{2(c_2 - c_1)} \tag{7.5}$$

where
    λ_1 is the wavelength in the liquid
    c_1 is the velocity of sound in the liquid
    λ_2 is the wavelength in the solid
    c_2 is the velocity of sound in the solid

Anisotropy in a multicrystalline material contributes to scattering losses. This is particularly so when the grain direction is distributed randomly with respect to the direction of wave propagation. Since the elastic constants vary with orientation, scattering occurs as a result of reflections at intersections between grain boundaries.

7.4 UNCONVENTIONAL TECHNIQUES USED FOR NONDESTRUCTIVE TESTING

7.4.1 EDDY SONIC INSPECTION METHOD

In 1912, Thomas Fessenden developed the Fessenden oscillator for producing underwater sound signals. The oscillator was based on the principle that the magnetic field developed by eddy currents in a conducting surface reacts with the source magnetic field, which produces the eddy currents in the first place, and as a result, a force is applied to the conductor. Both the eddy currents and the mechanical force are time variant, and thus a sound wave is generated.

Botsco [33] and coworkers made use of the same principle to generate sound in materials to be tested, and the method was applied successfully in testing adhesive-bonded honeycomb composites for the Saturn space vehicle [34]. The method has been named eddy sonic testing. The principal advantage of the method is that it requires no liquid couplant. Some constituent of the material to be inspected must be electrically conductive. High conductivity is preferred; however, low-conductivity materials that are magnetic, such as ferritic stainless steel, can be inspected successfully. Since the structure being examined is essentially the dynamic element of a transducer, any change in its condition affects the character of the sonic energy generated.

Detection of the radiated sound offers no difficulty because inspection is in the audible range—2–4 kHz being typical frequencies.

Commercial eddy sonic units are available with a meter readout. However, C-scan recording also is possible.

Present equipment is practical for large panels such as honeycombs but impractical for small panels. Resonant patterns across small composite samples cause confusion when their signals are similar to those produced by defects.

One of the earliest evaluations of the eddy sonic method for inspecting adhesive-bonded honeycomb structures was made by General American Research Division of General American Transportation Corporation [35]. Their results showed that the technique is capable of resolving near-side unbonded areas approximately 6.0 cm in diameter in test panels with a 0.5-mm-thick aluminum facing sheet over an aluminum core, but far-side unbonds of the same size could not be detected in panels with a core thickness exceeding 6.3 mm. The actual sensitivity depends upon the test conditions—especially the dimensions of the panel, the conductivity of the conductive part of the panel, and the nature of the defect. The method has low sensitivity to tight delaminations.

Types of defects that can be detected are nonbonds, crushed core, fractured core, and voids in the adhesive.

7.4.2 SONIC ANALYSIS

Sonic analysis in the present sense is the analysis of acoustic signals emitted by a device or system to detect the presence of undesirable conditions such as the development of defects that are the first stages of serious failure. Every person who can hear has applied sonic analysis mentally without being conscious of the mechanism. Identification of a person's voice, recognition of a jet flying overhead, and recognition of a discord in one's piano playing are typical of the many ways people use a mental analysis of sounds.

Similar principles may be applied in industry, but here the received signal must be processed in such a manner that identification of an anomaly may be made on the basis of a visual rather than an aural data presentation, if necessary. Any moving system generates a sound. If the system is operating in a normal manner, the signal emitted may be called the "acoustic signature" of that system. The acoustic signature, therefore, identifies the normal condition. Any abnormality will affect the character of the emitted sound. For instance, a mechanic can readily distinguish between the pounding of a broken connecting rod in an internal combustion engine and the sound of a properly functioning engine.

Sonic signature analysis is performed by comparing the known normal signature with the sound emitted by the operating system. The potential methods of instrumentation are numerous, but each has a sound- or vibration-sensing device (microphone or accelerometer). For instance, in the case of leak detection, the only equipment necessary may be a listening device capable of detecting frequencies within a range where the background noise is minimum. The Heco ultrasonic detector [36] is one such device. Gases escaping through a small opening produce a broad band of white noise, that is, noise having a wide range of frequencies. The Heco device detects through a microphone those components of ultrasound falling within the range 35–45 kHz. The detected signal is beat down to the audible range and detected through earphones or through an external amplifier and speaker. Orifice-type gas leaks may be detected at distances greater than 30 m.

Another type of leak test makes use of the noise generated by a leaking liquid, which escapes under conditions of temperature and pressure that cause the liquid to vaporize within the length of the crack. Still another method is based on the detection of bursting microscopic bubbles produced by gas slowly escaping into a fluid coating applied over the area under test. The bursting bubbles release ultrasonic energy that can be detected by an ultrasonic probe. This method can be sensitive to leaks as small as 0.002 cm^3/s.

Ultrasonic signature analysis is used in monitoring the condition of bearings. Defects in the inner or outer races or in a ball bearing produce distinguishable sounds as the balls move in the races. The equations for the rotational frequency, as given by Martin [18], are

$$f_0 = \frac{n}{2}\phi\left(1 - \frac{2r}{D}\cos\beta\right) \quad \text{(flaw on outer race)} \quad (7.6a)$$

where
    n is the number of balls
    ϕ is the relative speed between inner and outer races (rps)
    r is the radius of the ball
    D is the pitch diameter
    β is the contact angle

$$f_i = \frac{n}{2}\phi\left(1 + \frac{2r}{D}\cos\beta\right) \quad \text{(flaw on inner race)} \tag{7.6b}$$

$$f = \frac{D}{2r}\phi\left[1 - \left(\frac{2r}{D}\right)^2 \cos^2\beta\right] \quad \text{(flaw on ball bearing)} \tag{7.6c}$$

Each time a moving ball contacts a crack or pit; for instance, it generates a shock wave that can drive the races and also the balls into resonance. The vibrations that are most easily detected are those that are generated in the races. The races are rings, and resonant frequencies of rings are given by the equation

$$f_r = \frac{w(w-1)}{2(w^2+1)^{1/2}} \frac{1}{a^2}\left(\frac{YI}{m}\right)^{1/2} \tag{7.7}$$

where
   w is the number of waves around the circumference of a ring
   a is the radius of the race (to the neutral axis)
   I is the moment of inertia of the cross section of the race about the neutral axis
   Y is the modulus of elasticity of the material of the race
   m is the mass of the ring per meter

When a defect such as a pit occurs in a ball bearing, the intensity of the radiated sonic, or ultrasonic, energy increases at frequencies corresponding to those determined by Equations 7.6 and 7.7. The amount of the increase in intensity depends upon the relationship between the rotational speed and the resonant frequencies of the races. It can be orders of magnitude higher than the intensity emitted at the same frequencies in the absence of defects.

Signature analysis and sonic detectors have been used to detect electrical noise in transformers, capacitors, or insulators. Other applications include detecting abnormal conditions in hydraulic systems, piston ring blowby and cylinder scoring in internal combustion engines and compressors, incipient boiling in pressurized-water reactors, and broken blades and worn bearings in jet engines, cavitation detection, signals due to impacts by loose parts moving in piping systems as well as monitoring machinery and transmission noises for indications of incipient failures.

7.4.3 ACOUSTIC IMPACT TECHNIQUE

Testing a material by acoustic impact is not uncommon. A banker who drops a coin on a marble countertop to determine its genuineness is performing an acoustic impact test. Thumping a melon to determine whether it is ripe is another acoustic impact test. One may go on with many other illustrations, including the crewmember with the hammer testing the wheels and other critical parts of a locomotive. In each of these examples, the brain processes and analyzes the data and from these arrives at an evaluation.

When a material is shocked by impact, a spectrum of frequencies is contained in the resulting pulse, and the wave constituents of the pulse depend upon the nature of the impact. As the slope of the wave front increases, so does the width of the spectrum. Since the propagation of the waves depends upon the properties of the material, changes in these properties affect the propagation characteristics (velocity and attenuation).

An acoustic impact technique was developed by Schroeer et al. [37] based on the fact that the appearance of a defect changes the mechanical impedance of a structure or part and affects the vibrational response of the structure in the area of the defect. The instrumentation includes an impactor, a sensor, a preamplifier, a tunable band-pass filter, and a computer. The impactor and sensor are included in one assembly called the probe. The impactor is the carbide-tipped core of a solenoid coil. It is capable of providing a single blow or a series of repetitive shocks as desired.

The sensor is a lightweight accelerometer capable of withstanding high acceleration forces. As would be expected, its position on the test specimen affects the accuracy of the test.

Schroeer et al. explained the effectiveness of the acoustic impact technique in terms of two mechanisms of damping of vibrations—hysteresis and shear damping. Hysteresis occurs in materials undergoing cyclic stressing; that is, the stress–strain curve corresponding to decreasing stress does not coincide with the stress–strain curve of increasing stress, and the dissipated energy is proportional to the area between the two curves. In a beam vibrating in flexure, the outer fibers experience the greatest stress and, therefore, the greatest hysteresis loss occurs in these fibers. When a crack forms at the surface in a direction normal to the stress, the stress in the outer fibers is relieved and the bulk of the load is carried by the outermost of the intact fibers. If the amplitude of vibration remains constant, the hysteresis losses drop. In addition, the effective thickness of the beam decreases, which causes a corresponding decrease in stiffness and, therefore, flexural reso-nant frequency. The mass of material between the surface and the depth of the cracks approaches a dead weight as the number of cracks increases. Therefore, the appearance of cracks in a flexing beam will lower its natural resonant frequencies and may decrease its damping by decreasing its hysteresis losses. On the other hand, if the mating surfaces of the cracks are subjected to shear stresses, energy is dissipated because of friction. The shear stress is equal to the product of the coefficient of sliding friction times the compressive stress normal to the mating surfaces. Frictional losses are greatest with the overtones. Therefore, cracks in a beam vibrating in flexure may cause a reduction in damping and the funda-mental frequency but may cause an increase in attenuation of higher frequency components.

Schroeer et al. developed their technique primarily for detecting cracks con-cealed under fastener heads but showed its practicability in the inspection of com-posite materials and fatigue specimens. Many other applications are possible. In fact, similar techniques have been used to separate parts, for example, in the automotive industry, by ringing the test specimen and accepting or rejecting the part on the basis of its ringing frequency.

In order to follow the history of a failure, it is imperative that the probe location be exactly the same during each measurement. Variations in loading can affect the frequency content of the pulse detected by the sensor much as changing the position of the mass on a pendulum changes the frequency of the pendulum.

The acoustic impact echo technique [38] is being increasingly used to inspect concrete in civil infrastructure and composites. The acoustic impact technique detects the presence of a crack by its influence on the vibrational characteristics of the specimen, but it does not delineate the crack or cracks.

7.4.4 ULTRASONIC SPECTROSCOPY

In most conventional applications of ultrasonics in NDT, evaluation is based primarily on amplitude and time information. Transducers are selected according to a nominal test frequency that meets the requirements of the inspection task, but, in general, the relationship between the initial and the received pulse spectra is not recorded. Additional information to be gained by analyzing pulses used in NDT becomes obvious when the frequency dependence of reflection, diffraction, and attenuation (especially in crystalline materials) is concerned. For example, the components of a pulse in which the wavelengths are large compared to the dimensions of a discontinuity will be reflected spherically; those in which the wavelengths are small compared to the dimensions of the discontinuity are reflected according to the contour of the surface of the discontinuity on which the pulse is incident. By diffraction, the longer waves pass the discontinuity without producing a 'shadow" effect while the shorter waves cast a shadow whose nature depends upon the relationship between wavelength, cross section of the beam, and cross section of the discontinuity. These phenomena provide a basis for flaw analysis including size, geometry, and orientation. The frequency dependence of attenuation in crystalline materials can be used to determine certain properties of the grains, particularly the average grain size of certain materials.

Procedures for ultrasonic frequency analysis have been developed and are being used in several areas [39]. Early in 1964, Gericke [40,41] began developing equipment for ultrasonic spectroscopy. Gericke's application was to pulse-echo studies using the direct-contact methods. The essential components for such studies are a broadband transducer, rectangular pulse generators, time gate, broadband amplifiers, a suitable detector, and a spectrum analyzer capable of analyzing typical pulses used in NDT.

The most important component is the transducer. Gericke uses lead–zirconate–niobate transducers that produce a pulse, which contains various frequency components covering a band of approximately 3–10 MHz. The spectrum of the initial pulse is determined by the frequency response of the transducer and by the length of the rectangular voltage pulse used to excite the transducer. Gericke states, "If the excitation pulse length is properly matched to the transducer response curve, a sufficiently uniform ultrasonic spectrum can be generated to permit successful ultrasonic spectroscopy."

Commercially available transducers may exhibit complex spectral response, in both the time and the frequency domains, which make it important that their characteristics are evaluated, and known by users. An example of the time domain waveform and spectrum for a commercial 5 MHz compression wave transducer, with short pulse in the time domain and smooth spectrum, is shown in Figure 7.5. Such a transducer acts as a band-pass filter, where the sensitivity and achievable signal-to-noise ratio are strongly frequency dependent. The spectral analysis shows a band of

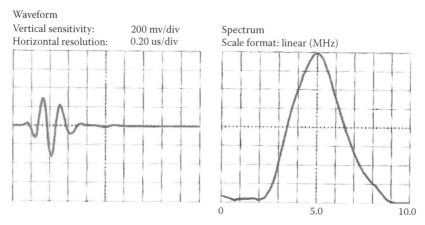

FIGURE 7.5 Example of the time-domain response (waveform) and spectrum for a commercial 5 MHz compression wave transducer. (Courtesy of Olympus NDT.)

frequencies with good sensitivity; in this case, the −6 dB down bandwidth is between about 3.5 and 6.5 MHz. For transducers to be used for ultrasonic spectroscopy, a wide bandwidth is required, and these transducers use special designs, including, in some cases, thin polymer films as the transducer element. The desirable characteristics are very low Q, no significant resonances in the operating band, and good sensitivity.

Whaley and Cook [42] developed a means of ultrasonic frequency analysis, which employs commercially available electronic equipment and can be used with conventional ultrasonic instruments operating in either a contact or an immersion mode. The application goes beyond analyzing test material to include the evaluation of the characteristics of transducers, instrument types, transducer positioning, tuning devices, and collimation. All these parameters are important, especially in comparing inspection data between instruments and also at various intervals of time. Differences in response and setup can result in serious problems of interpretation of inspection data. For their work, Whaley and Cook preferred an immersion setup.

Serabian [43] used spectrum analysis to show that as an ultrasonic pulse propagates through highly attenuating, nondispersive materials, preferential attenuation of high-frequency components causes the pulse to change in shape continuously as it propagates through the medium.

The center frequency shifts toward the low-frequency end of the pulse, which also may be lower in intensity than the center frequency. As a result, sensitivity to defects decreases with distance in such materials not only because of absorption of energy but also because of the lower sensitivity to defects at lower frequencies. Thus, spectral changes in pulses propagating through such materials (for instance, graphite) impose a limitation on the use of ultrasonics for their inspection.

Ultrasonic spectroscopy is seeing application for the characterization of colloidal systems, for slurries, and at higher frequencies for material characterization in both fluids and gases where molecular absorption modes are excited [44–46]. It is discussed further in several later chapters in this book.

7.4.5 CRITICAL ANGLE ANALYSIS

Solids are capable of sustaining several wave modes. When a solid is immersed in a liquid and an ultrasonic wave impinges on the solid from the liquid, several wave modes may be generated in the solid by mode conversion. The critical angle for a given mode is the angle of incidence, which produces an angle of refraction of 90° for that mode. According to Snell's law, the refraction angle is a function of the angle of incidence and of the relationship between the velocity of sound in the liquid and in the solid. The principle offers a sensitive means of analysis, which has been used to a limited extent.

Critical angle analysis makes use of phenomena that are easily observed. A reflected wave at a liquid/solid interface undergoes a shift along the interface, which is dependent upon the angle of incidence. The amplitude distribution across the reflected beam differs from the distribution across a focused incident beam when the axis of the incident beam coincides with the Rayleigh critical angle. The amount of amplitude redistribution depends upon the density and the elastic properties of the solid. The shift of the position of the reflected wave is the major factor in the redistribution of the amplitude in the reflected beam. The principle is illustrated in Figure 7.6.

When a focused beam impinges on a liquid/solid interface so that the ray along the axis of the beam is at the Rayleigh critical angle, the beam also contains rays at incident angles, which are either greater or less than the critical angle. The rays at angles exceeding the critical angle are totally reflected. Those at angles less than the critical angle are partially reflected and partially refracted into the solid. As a result, the reflected rays in the latter case are less intense than they would be if they were reflected at an angle greater than the critical angle. In addition, the varying amount of beam shift due to the multiple angles of incidence in a focused beam causes variation in the interference pattern of the reflected beam, according to Huygens' principle.

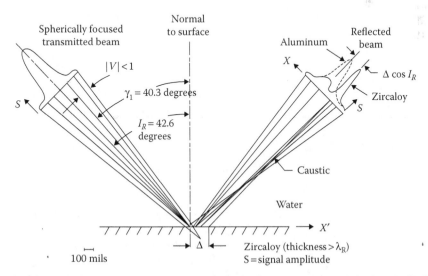

FIGURE 7.6 Ray diagram illustrating amplitude distribution and beam shift of a spherically focused beam reflected from a liquid–solid interface at the Rayleigh critical angle.

In the application of the critical angle technique, the focused beam is adjusted so that the ray along the axis is at the Rayleigh critical angle. The beam shift and intensity distribution of the reflected wave are then characteristic of the material under inspection. If the beam is caused to scan the surface of the metal while maintaining a constant angular relationship, the conditions within the reflected wave remain constant unless a change occurs in the elastic properties or the density of the solid (assuming a perfectly smooth surface). Since changes in elastic properties or density cause corresponding changes in the propagation velocity of ultrasonic waves, the former changes also cause a corresponding change in critical angle. Rays other than those on the axis of the beam are located at the new critical angle. Consequently, a change in amplitude distribution occurs, which can be related to the new properties of the material.

The lateral shift of the reflected beam is given by

$$\Delta = -\frac{\lambda}{2\pi}\left(\frac{\partial\phi}{\partial\sin\alpha_1}\right) \tag{7.8}$$

where
λ is the wavelength of sound in the liquid
ϕ is the phase of the reflection coefficient
α_1 is the angle of incidence and the angle of reflection when α_1 is greater than the angle of total internal reflection

Equation 7.8 is valid when the width of the incident beam is much larger than wavelength. The phase shift is approximately 2π when the angle of incidence is nearly equal to the Rayleigh critical angle. The lateral shift for this condition is

$$\Delta = \lambda\left(\frac{2}{\pi}\right)\frac{\rho_m}{\rho_f}\sqrt{\frac{r\,(r-s)}{s\,(s-1)}}\left[\frac{1+6s^2(1-q)-2s(3-2q)}{s-q}\right] \tag{7.9}$$

where
ρ_f is the density of the fluid
ρ_m is the density of the metal
$r = (c_s/c_f)^2$; $s = (c_s/c_R)^2$; $q = (c_s/c_L)^2$
c_R is the velocity of Rayleigh waves in the metal
c_s is the shear-wave velocity in the metal
c_f is the velocity of sound in the liquid
c_L is the velocity of longitudinal waves in the metal

In practice, a focused transducer is oriented with respect to the solid surface in an immersion system at the Rayleigh critical angle (see Figure 7.7). A miniature receiving transducer is mounted on a support, which is adjusted by micrometer. The scan of the receiver passes through a null region caused by destructive interference between reflected rays. The sensitivity of the miniature receiving transducer, which is approximately 2.0 mm in diameter, is enhanced by a collimator.

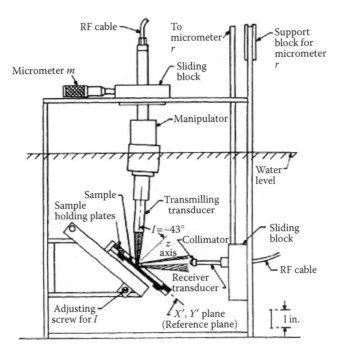

FIGURE 7.7   Schematic of ultrasonic mechanical rig. (Courtesy of Bettis Atomic Power Laboratory.)

The technique thus described has been used by Willard [47] to detect hydrides in Zircaloy. The sensitivity was 600 ppm, being limited by the surface roughness of the specimens. The presence of hydrides in Zircaloy affects both the density and the elastic properties. Frequencies used by Willard were 1.0, 2.25, and 5.0 MHz in flat plate specimens 0.355 mm thick.

Hunter [48] has used the critical angle method to determine irradiation damage. Both shear and longitudinal waves are slowed by such damage, but, in some cases, specimens irradiated at high temperature exhibited increasing velocities of sound prior to the decreasing trend. Radiation damage in low-alloy steels appears as impact embrittlement.

The critical angle method of analysis is potentially more useful than its limited application would indicate. The observed effects are attributable to boundary wave generation, and these are potentially useful for determining near-surface conditions, such as measuring residual stresses, and detecting variations in elastic moduli, degree of cold work, grain orientation, composition, and surface treatment. Replacing the miniature receiving transducer with an electronic–acoustic image converter might enhance the sensitivity of the method and permit more accurate evaluations. In addition, the electronic–acoustic image converter permits faster scanning than is possible by mechanical scanning.

Techniques have been developed that combine critical angle analysis and both pulse-echo and mode-conversion. The measured signals with these systems typically give a very good signal-to-noise ratio.

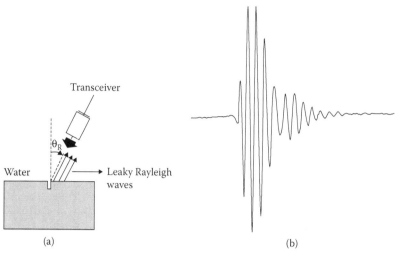

FIGURE 7.8 Leaky Rayleigh wave pulse-echo system. (a) Basic configuration. (b) Signal obtained at 1 MHz with a 0.5-mm-deep defect in a Dural test block ($\theta_R = 30°$). (Reprinted from Bond, L. J., and N. Saffari, Mode conversion ultrasonic testing, In *Research Techniques in Non-Destructive Testing*, ed. R. S. Sharpe, Vol. 7, Chap. 5, 145–89, Academic Press, London, 1984. With permission.)

The pulse-echo technique using leaky Rayleigh waves, as shown in Figure 7.8a [49], has been applied in various forms, including for ceramic inspection at 50 MHz by Khuri-Yakub et al. [50]. It has been implemented at frequencies between 10 and 50 MHz by Johar et al. [51] and Fahr et al. [52]. The technique has also been applied at lower frequencies to inspect metals by Bond and Saffari [53]. An example of the signal obtained for a 1 MHz pulse incident on a 0.5-mm-deep defect is shown in Figure 7.8b. A through-transmission technique, also detecting leaky Rayleigh waves, has also been demonstrated by Bond and Saffari [53,54], and an example of the configuration and resulting signal, using the same test block as was employed to give data in Figure 7.8, is shown in Figure 7.9. These techniques are found to have good signal-to-noise properties, and defect characterization seems to be possible from a combination of arrival time, radiation pattern, and frequency content data. This technique is reported as applied as part of the U.S. Air Force (USAF) Retirement for Cause (RFC) program [55].

7.5 INSTRUMENTATION

Each type of ultrasonic NDT instrumentation forms a *system* comprised of several elements. These are a specific form of measurement system that is designed to detect particular classes of material anomalies. High levels of performance are required in applications where "quality" is important and/or it is necessary to ensure operational safety and fitness for service. In many applications, systems are designed to detect particular types of cracks, in a particular size range, or particular classes of manufacturing defects.

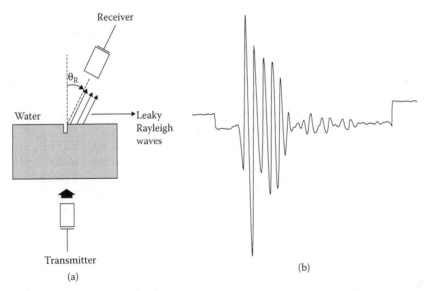

FIGURE 7.9 Leaky Rayleigh wave through-transmission system. (a) Basic configuration. (b) Signal obtained for test block used for Figure 7.8 at 1 MHz. (Reprinted from Bond, L. J., and N. Saffari, Mode conversion ultrasonic testing, In *Research Techniques in Non-Destructive Testing*, ed. R. S. Sharpe, Vol. 7, Chap. 5, 145–89, Academic Press, London, 1984. With permission.)

7.5.1 COUPLING ENERGY TO THE TEST OBJECT

A means must be provided for transferring ultrasonic energy between the piezoelectric transducers generally used in NDT and the test object. If an ultrasonic transducer is operated into air, the great difference between the acoustic impedances of air and solids causes nearly 100% of the ultrasonic energy impinging on air/solid boundaries to be reflected. Maximum energy is transferred across an interface when the acoustic impedances of the two media are equal, or a matching layer is employed. Although *common liquids* seldom meet the conditions for maximum energy transfer, they are most generally used as the coupling medium for NDT because they do provide adequate energy-transfer capability and usually they are convenient to use.

Basically, three common methods of coupling energy to a specimen material are used: (1) direct *contact* between the transducer and the test piece with a liquid or grease couplant between, (2) *immersion* of the transducer and the test piece in a liquid bath and coupling through the intermediate liquid path, and (3) flooding a space separating the transducer and the test piece with a liquid (column or jet). Most other methods of coupling—liquid-filled tire, liquid-filled boot, and so on—are modifications or combinations of these techniques. If liquids are not permissible, sometimes thin films of soft pliable polymers or high pressure gas can be used as coupling materials.

Noncontact coupling, that is, coupling across an air gap, offers advantages in several ways: (1) it avoids contamination of the test object by fluid couplant material, (2) it avoids wear on the face of the transducer over hard and rough surfaces, (3) it permits testing at elevated temperatures, and (4) in some cases, it leads to faster inspection [56].

The eddy sonic method [33] was developed several years ago with the objective of generating ultrasonic energy in a test object across an air gap. Here the test piece actually becomes a part of the transducer; that is, the electrical driving energy is actually transformed to mechanical energy within the test piece.

Botsco's eddy sonic method was used originally to test materials at frequencies in the sonic range. Today, noncontacting coupling into electrically conductive materials is accomplished routinely with electromagnetic acoustic transducers (EMATs) operating in the low-megahertz range of frequencies typically used in ultrasonic NDT [57,58].

Ultrasonic waves are produced electromagnetically by Lorentz forces. These are forces generated in an electrically conductive material carrying high-frequency eddy currents and simultaneously being subjected to a steady magnetic (bias) field. As one would expect at frequencies of 1.0–10 MHz, the penetration depth of the eddy current is very shallow, that is, only a few micrometers. Therefore, compared with the wavelengths of the ultrasonic waves, the effect is that of a very thin transducer in the surface of the test object. The type of wave, longitudinal or shear, is determined by the direction of the bias magnetic field compared with the direction of eddy current flow. A bias field normal to the direction of current flow and parallel to the surface causes a longitudinal wave that propagates in a direction normal to the surface. A bias field under similar circumstances but directed normal to the surface generates shear waves that propagate in the same direction. The same principle can be applied for receiving ultrasonic waves.

Efficiency of the EMAT is low, when compared with PZT-based transducers, it being determined by the electromechanical conversion effectiveness of the electromagnetic fields and also by the spacing between the transducer assembly and the test object. Although improvements have been made in recent years for a pulse-echo system, the received signals are from 40 to 100 dB below those of conventional probes [57].

An EMAT transducer may be round, elongated rectangular, and so on, to produce predetermined beam profiles. EMAT transducers are marketed by several companies including for testing for discontinuities in centerlines on bar and rod in continuous casting and for more general use regardless of material or purpose. Powerful pulser/receiver units capable of outputs up to 8 kW are available for driving these low-sensitivity units.

Capacitance transducers offer another means of coupling ultrasonic energy into a test object across an air gap and similarly of detecting acoustic waves [48,56]. A capacitor can be formed by positioning an electrode close to and parallel to a grounded, electroded surface of a test object. An attractive electrostatic force is produced between the two electrodes whenever a voltage appears across them. By imposing an alternating voltage and a very much higher bias voltage across the gap, the attractive force is modulated at the driving frequency to produce longitudinal waves in the test object. The chief advantage of this technique lies in its very broadened capability. Its drawbacks are the need to maintain a very high bias voltage, the need for a constant and extremely short gap distance, and the extremely low energy conversion efficiency.

Laser techniques of generating [60] and detecting [61] ultrasonic waves have been studied for possible use in NDT. Their advantages include broadband pulse characteristics, coupling through any transparent medium such as air or vacuum, operation on rough or coated surfaces, and high sensitivity as a receiver through a high-precision interferometer. The equipment is expensive and bulky. However, the techniques show considerable promise in the laboratory for flaw characterization, that is, interaction of ultrasonic waves with defects in a test material, and for transducer calibration.

Laser pulses applied to a surface of a test material generate ultrasonic waves by rapid thermal expansion. Rise time is fast. Cooling is much slower. Lasers can be operated in two regimes, depending on power: (1) thermo-elastic and (2) ablation. These are the phenomena that produce the wideband characteristics.

In direct, contact, gap, or immersion methods, nearly all modes of ultrasonic waves used in NDT, except longitudinal, are produced by mode conversion at the boundaries between the couplant and the test piece. The incident wave is generated by a transducer operating in a longitudinal mode. If testing is to be done using a longitudinal wave, the transducer is oriented so that the wave impinges at normal incidence on the surface of the test piece. If a shear-wave technique is to be used, the incidence angle is made larger than the critical angle for longitudinal waves. When direct contact coupling is used with shear-wave testing, the transducer is mounted on a solid wedge (plastic) in which the velocity of sound is lower than that in the test piece. The angle of incidence is designed to produce the desired refraction angle for the shear wave in the test piece. When immersion coupling is used, the inspector can exercise greater freedom in choosing the incidence angle and, therefore, the refraction angle within the test piece.

In immersion testing, the distance between the transducer and the surface of the test piece is chosen to avoid confusing the surface echo with reflections from within the test piece. At normal incidence, the pulse will bounce back and forth between the transducer and the surface of the test piece. Following the first return, time is allowed to receive all necessary information from the test piece before the second reflection returns to the transducer. The spacing may be determined by the relative velocities in the two media. For example, ultrasonic pulses travel nearly four times as fast in steel as they do in water. If the ratio were 4:1, the recommended separation between the transducer and test material in a water bath would be at least 2.5 cm for every 10 cm the pulse travels in one direction in the test material.

7.5.2 RESONANCE METHODS

Resonance methods of measuring the thickness of materials were among the earliest ultrasonic methods to be implemented and placed on the NDT market. These early basic systems are reviewed here in recognition of the fact that resonance methods continue to offer unique possibilities in certain limited areas of application such as testing very thin metal sections, locating nonbond or delamination areas between rubber or plastic on the back side of a metal plate, and evaluating multilayered structures.

Generators designed for resonance testing consisted of a continuous wave oscillator, an amplifier, and a means, either manual or automatic, of modulating the frequency. Frequency modulation was obtained by varying either the inductance or the capacitance in the oscillator circuit.

Inductance modulation was obtained by an electrically variable saturable-reactor type of inductor in the circuit of the oscillator. The capacitance in the circuit was small. The upper limit on the frequency range usually was on the order of 30 MHz, which corresponds to the fundamental resonance of a steel or an aluminum plate of 0.1 mm thick. Modern pulse-type instruments are capable of measuring similar plates to the same resolution.

When the oscillator frequency swept through a point of resonance in the material being inspected (i.e., the fundamental frequency or its harmonics), the amplitude of vibration of the material, under excitation of the transducer, increased characteristically. The energy dissipation increased, and this was accompanied by an increase in driving current of the oscillator, which was easily detected.

The information display was synchronized with the sweep of the frequency range. If an oscilloscope was used, the horizontal sweep was triggered to coincide with the start of the frequency sweep (starting at the high-frequency end of the range). The sweep rate was correlated with the frequency range. The vertical input was proportional to the current of the oscillator. Thus, resonance peaks occurred on the oscilloscope screen at locations corresponding to resonant frequencies of the test material. Calibrated scales placed on the screen indicated the thickness directly in case of the fundamental frequency, or the thickness could be determined by means of Equation 7.3 in case harmonics were used.

Historically, correlation between frequency and indicator was ensured in a stroboscopic type of thickness tester in which a disk with a neon lamp mounted on its outer edge rotated with the capacitor in the tank circuit of the oscillator. The increased plate current at resonance caused the neon lamp to flash, giving the appearance of a stationary light at positions corresponding to the fundamental frequency (or actual thickness) or the harmonics. In the case of harmonic indications, the thickness is determined by Equation 7.3.

Some early, small, portable instruments were battery operated, and frequency was controlled manually. A plate-current meter indicated resonance. Thickness readings were obtained from the position of a pointer over a calibrated dial at resonance positions. Headphones or loudspeakers were often used, and resonance was indicated by a rise in the sound level in the devices. Kits are also available containing software for transferring grids, waveforms and custom interface setups to and from a computer. Other types including large graphic displays, data loggers, scrolling B-scans, single and dual point calibration, and multichannel systems are available [56].

7.5.2.1 Transducers

Resonance measurements are dependent upon the change in impedance reflected into the oscillator circuit when the test material resonates. For this reason, the transducer must have a flat response over the frequency range of the test in order to avoid false indications. This is accomplished by using a transducer that resonates fundamentally at a frequency above the tuning range (usually 10%–20% higher). Proper coupling to the test object provides some damping. Piezoelectric elements that resonate at a fundamental frequency of 30 MHz are from 0.051 to 0.076 mm thick. The fragility of such thin elements is a limiting factor on the upper frequency that can be used for resonance testing.

The transducer consists of an undamped piezoelectric element. It may be flat for use on flat plates or contoured to match specific geometries.

Coupling between the transducer and the test piece should be neither too tight nor too loose.

7.5.2.2   Data Recording

Using resonance methods, automatic recording of thickness on a go/no-go basis is accomplished by gating circuits or in software on a computer. The location and width of the gate were set to coincide with the thickness and allowable variations in thickness of the material being inspected. When indications occurred outside the gate limits, the recorder received no signal and, therefore, showed a zero reading. A number of gates can be used, each one supplying a signal to a separate recorder channel.

Actual thickness variations were recorded by means of an electronic circuit that produced a voltage, which was proportional to the distance the indication appeared from the start of the gate. A similar technique was used to produce a meter reading of thickness.

7.5.3   Pulse Methods

Considering the numerous wave modes, associated characteristics of these waves, and effects of the media on the propagation characteristics (velocity and attenuation), there probably is no other energy form that can provide as much information about a material nondestructively as ultrasonic energy. Pulse methods are the most versatile of the ultrasonic diagnostic systems used both in NDT and in medicine.

The versatility of pulse methods of NDT is demonstrated by a variety of instruments available for generating ultrasonic pulses and recording test information. These instruments reflect the tremendous advances of recent years in electronics—transistors, integrated circuits (ICs), microprocessors, digital functions, and computers—in their compactness and in their data storage and processing capabilities. A basic system consists of a pulse generator, a pulse amplifier, a transducer for generating ultrasonic waves, a transducer for receiving the ultrasonic energy (which may or may not be the same one used for the transmitter), an amplifier for amplifying the received signal, and a means of displaying or storing the test information.

Broadband operation is desirable for most pulse-type applications. The pulse generator and amplifier must provide a "sharp" electrical pulse to excite the transmitter. The receiving transducer and amplifier should have similar broadband characteristics to avoid distorting the received signal.

The initial electrical impulse is produced by a pulse generator at the desired pulse repetition rate (0.5–5 kHz is typical).

7.5.3.1   Transducers

Modern design improvements have led to transducers with excellent properties for controlling pulse shape and characteristics, such as the very short pulses needed for good resolution and accuracy in measuring thickness particularly of thin sections.

In contrast to transducers used for resonance measurements, which have undamped, high-Q piezoelectric elements, broadband pulse-type transducers are low-Q transducers. Spatial resolution for indications improves as the Q decreases.

Low-Q broadband characteristics are obtained by loading the nonradiating surface with a lossy material. The damping material preferably has the same characteristic impedance as the piezoelectric element. Sometimes the backing material is shaped to enhance its ability to absorb energy from the piezoelectric element by reflecting the energy in directions that increase its path length.

The piezoelectric element is protected from the surface of the test material by a faceplate. In contrast with the backing material, the faceplate is carefully matched acoustically to the piezoelectric material for maximum transfer of energy into the test material. Bonding uniformity and careful positioning and alignment of the piezoelectric element on the faceplate are very important for good performance characteristics of the transducer.

Commonly used piezoelectric materials are the various forms of lead–zirconate–titanate, (PZT) and lithium niobate.

Transducers for pulse-type measurements in NDT may be classified according to their methods of coupling to the test piece—immersion, direct contact, gap, or coupling member (tire or sliding boot). They may also be classified according to their principle of transduction—piezoelectric, EMAT, capacitance, or laser—and, within each class, according to the wave modes for which they are intended—longitudinal, shear, surface, and so on. Laser and capacitance techniques have yet to be developed for a general industrial-scale market.

Contact types of transducers are available for producing the various types of waves used in NDT, that is, longitudinal waves, shear waves, surface waves (Rayleigh and Love), and Lamb waves. Longitudinal waves are produced by a transducer element vibrating in its thickness mode and in a direction normal to the surface of the test piece. Other vibrational modes are produced by mode conversion. To generate modes other than longitudinal, a piezoelectric element vibrating in the thickness mode is bonded to a surface of a wedge of material in which the velocity of sound is low compared with that in the test material. The surface to which the element is bonded is inclined at such an angle to the surface of the test piece that the angle of incidence exceeds the critical angle of all modes, which travel in the test material at velocities higher than the velocity of the desired mode.

Immersion types of transducers for NDT are sealed liquid-tight. The beam may be plane or focused. Focusing is done through a lens attached to the face of the piezoelectric element. The transducers are mounted on immersion tubes, the mounts sometimes containing means for adjusting the incidence angle of the ultrasonic beam.

The immersion tubes are attached to manipulators located on a traversing mechanism, which in turn is mounted on a bridge. The bridge travels the full length of an immersion tank, and the traversing mechanism is designed to scan the width of the tank. The manipulator also contains angle adjustments to facilitate obtaining the proper incidence angle and inspection coverage.

Sometimes collimators are attached to immersion-type transducers (1) to improve the sensitivity to and location of defects, (2) to decrease the effects of surface roughness that tend to increase the length of the pulse reflected from the interface and thus reduce the resolving power, and (3) to produce a beam of limited size or controlled cross-sectional dimensions to prevent generating several modes in materials with irregular surfaces. The cross section of the beam may be round or rectangular. The walls of collimators are designed to absorb and reflect energy that does not pass through the opening so that it does not return to interfere with the received signal.

Two or more transducer elements may be located in one head. One example is a dual-element transducer used for thickness measurements. Two identical transducers are mounted on separate acoustical delay lines separated by an absorbent layer to avoid cross talk. The transducers operate in a pitch–catch fashion, one being a transmitter and the other a receiver. The delay lines prevent echoes from returning to the receiver before the initial pulse has damped out, thus enhancing resolution. Thicknesses ranging from 0.1 mm to 30 cm with 0.003 mm resolution in steel are measured by typical instruments on the market.

EMATs, as described in Section 7.5.1, are used for "conventional" pulse-echo and through-transmission types of inspection of metals in which, for one reason or another, contact methods are undesirable.

There are several companies specializing in manufacturing ultrasonic NDT transducers. Hundreds of types of different dual focuses angle beam transducers, automated scanning transducers, tandem duals, contacts, immersion, array-type transducers, and more are available. There are companies manufacturing transducers for NDT, thickness gaging, and medical and research applications. There are many other types including ultrasonic systems for manual and automated NDT, electromagnetic types, rapid rates C-scans, magnetostrictive types for testing concrete, rods, cables, and similar structures, ultrasonic scanning and measurement systems, stand-alone and computer-based ultrasonic plug-in cards, portable workstations and Windows-based control and data acquisition software, large gantry and robotic systems, large and small immersion systems, and high-frequency acoustic microscopes. The microscope is capable of, at bright frequency, detecting submicron defects. There are also systems for measuring velocity, thickness, heat-treat quality, and cracks. Some companies provide testing and engineering services either on the customer's job site or in their own laboratories. As one looks over the equipment and services available, he or she must be impressed with the large scope of useful ultrasonic NDT.

7.5.3.2   Data Recording

Data from nondestructive tests using pulse methods may be presented in many ways. A few of the basic techniques will be discussed here. Applications presented in Chapters 8 through 10 and, for medical applications, in Chapter 14 will demonstrate how and where these techniques are used.

The most familiar and basic data presentation is the A-scan, shown in Figure 7.10, which displays the various significant signals on an oscilloscope screen. The sweep is triggered by the impulse that energizes the transmitting transducer. Thus, the two events are synchronized. The sweep may be delayed and the sweep rate controlled to enable thorough examination of a desired region. This latter feature is especially useful in immersion testing. The settings may be adjusted so that front-surface reflection is indicated at the left of the oscilloscope or computer screen, a back-surface reflection occurs at the right side of the oscilloscope screen, and reflections from discontinuities within the material occur between the front- and back-surface reflections.

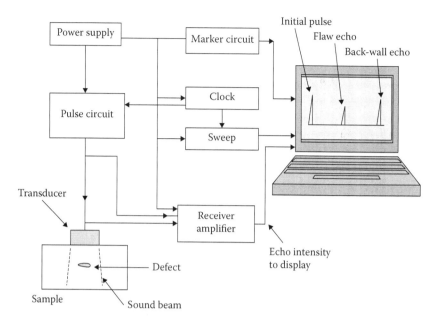

FIGURE 7.10 Block diagram for an A-scan system, for basic pulse-echo ultrasonic inspection (electronics now commonly fabricated as cards that fit inside a computer with A-to-D conversion).

B-scan presentations are pictorial displays of the cross section of a test piece. The schematic of a typical B-scan system is shown in Figure 7.11. They are more commonly used in medical diagnosis than in commercial NDT applications. In modern systems, the mechanical scanning, system settings, and data recording are computer controlled, and data are displayed on a computer screen.

Pulse-type NDT instruments may be equipped with gating circuits and recorder amplifiers, which permit the retention of a test record usually now using a computer system. This type of display is especially useful in high-speed automatic scanning. Digital systems, such as counters that sum the amount of time the received signal is either above or below a reject level, and warning devices may also be employed.

A plan-view record of an ultrasonic NDT is called C-scan. The schematic of a typical C-scan system is shown in Figure 7.12. The presence, but not the depth, of discontinuities is indicated. If indications of discontinuities within certain depth limits are desired, the gate settings may be adjusted to these limits and the signals from the discontinuities obtained by pulse-echo means. Multiple gates would permit obtaining plan views at various depths.

In modern C-scan systems, the mechanical scanning, system settings, and data recording are now routinely controlled using a computer, which will also provide software for advanced data display and analysis, and alarms for automated defect

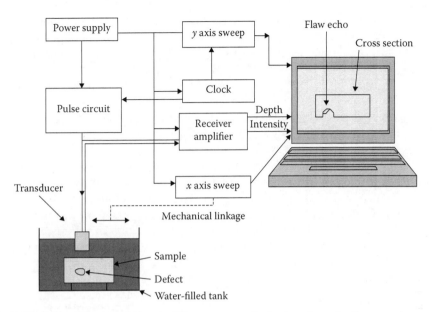

FIGURE 7.11 Schematic for typical B-scan system for basic pulse-echo line-scan immersion inspection (electronics now commonly fabricated as cards that interface with a digital system).

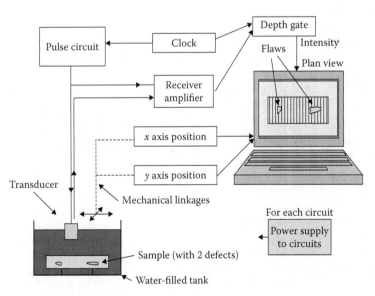

FIGURE 7.12 Schematic for typical C-scan system used for basic pulse-echo immersion inspection.

detection. In some system, the full "RF" data record, the unrectified pulse-echo signal, is recorded for every index point in the scan. Such data files are large, but they do enable a wide range of post-processing, off-line review and data achieving to be performed.

The spectrum of an ultrasonic pulse often may provide valuable information regarding a test specimen. Techniques have been developed for analyzing pulses before and after passing through a specimen, and by special data-processing methods, it is possible to determine specific qualities that are not readily recognized in the signal amplitude alone.

The types of information available during pulse methods of ultrasonic inspection are ideal for use with digital systems and computers. Many instruments on the market are equipped to interface with computers and printers. Each of the types of images, B-scan and C-scan, can be enhanced by off-line computer processing.

Many small, portable pulse-type thickness gauges have built-in data loggers capable of storing as many as 1000 thickness measurements with identification codes with added capabilities to review stored data in the field without off-loading. Many other microprocessor and computer functions are available in commercial instruments, and some of which are discussed in Chapters 8 through 10 and 14.

7.5.4 ACOUSTIC EMISSION METHODS

The instrumentation used in acoustic emission depends upon the parameters to be measured. A minimum system includes a sensor, amplifier, filter, signal processor, and display recording system. Such a simple system with only one sensor might be used to monitor developments in an isolated source area such as stress-wave emissions produced during laser welding [20]. For locating the source (actually approximate location) of acoustic emissions, the system becomes more complicated. Such a system may consist of four or more transducers located strategically on the surface of the test material. Source location is determined by triangulation based on relative positions of the transducers and the difference in time of arrival, which translate into differences in path length from source to detectors if the velocity of sound is the same in all directions. A very interesting source location system has been developed by Barsky and Hsu [62], which utilizes rather inexpensive transducers and inexpensive IC amplifiers for the output of these transducers. Their system includes four transducers and IC preamplifiers, four input latches, four counters, a time control, error detection logic, a clock, T buffer lathes, a microprocessor, and an output buffer through which the output data are presented to a suitable display or data processor.

When the parameters to be measured or the analyses to be performed increase in complexity, the instrumentation chosen must exhibit suitable characteristics. For example, the system may be required not only to detect and locate defects, as mentioned previously, but also to indicate the type of damage occurring by amplitude and frequency analyses and other information processing. The choice of the transducer is as critical in this case as is the choice of the data-processing equipment.

Equipment is available that contains several channels, which acquire acoustic emission events independently and report signal amplitudes (dB), counts, and duration for each event. Facilities are included for storing complete field data including date and time of test, transducer configurations, description of part tested, names of test personnel, and so on.

Some companies specialize in acoustic emission equipment and systems. Among such systems, some are now in the form of battery-operated units utilizing note-book personal computers or customized hardware and displays for the user interface. The more powerful computer capabilities that have become available are being utilized in more complex digital acoustic emission measurement systems that perform source analysis [21].

7.5.5 PHASED ARRAYS SYSTEMS

Several companies advertise phased arrays. These arrays consist of a grouping of transducers such that by controlling the timing of the energizing of each element in the array, the direction of the beam axis, the focusing of the beam, or other effects may be controlled.

One such system is a battery-operated ultrasound phased-array with 32 active elements for high volume resolution "to precisely map and size discontinuities." It supports up to 2048 beam angles.

One unique unit is a rotating ultrasonic probe, which provides a "radar map" of its surrounding for corrosion surveys. Another unit is a miniature system for portable, automated time of flight diffraction, pulse-echo and corrosion mapping with up to 32 channels. The TD-Focus Scan is a phased array system providing up to 256 elements, 8 conventional and 1–64 active channels.

Phased array systems are also now being used in NDT with various forms of data processing, including synthetic aperture focusing techniques [63].

7.5.6 SOME SPECIALIZED EQUIPMENT

There are companies who specialize in making components that a customer can incorporate into a test facility. Some important types of equipment are portable ultrasonic test systems, automated and manual scanners, ultrasonic testing boards and custom cables and connectors, and plus high-speed boards for "on board" signal processing.

Some companies make data acquisition software of signal and image processors that users can cascade together to form more complex processors with no programming required.

Another company produces industrial 3-D computed tomography scanners and real-time industrial visualization workstations. Computed tomography refers to the reconstruction "by computer of a tomographic plane or slice of a test object." The company utilizes "advanced 3-D, cone beam computed tomography scanners, fast image reconstruction computers, and high-resolution industrial viewer workstations" to provide a high-resolution test for industrial applications."

There are also companies who specialize in NDT Standards and Test Blocks.

7.5.7 Commonly Used Specifications and Standards

7.5.7.1 Standards for Ultrasonic Inspection

Standard reference blocks are included as accessories with commercially available ultrasonic equipment used in NDT. These blocks are used for standardizing the ultrasonic equipment and for a reference in evaluating indications from the part being inspected. The types of standards available with the equipment depend upon the use for which the equipment is designed. The standards are made from carefully selected material, which must meet predetermined standards of ultrasonic attenuation, grain size, and heat-treat condition.

Examples of types of standard reference blocks include area/amplitude blocks (see Figure 7.13), distance/amplitude blocks (see Figure 7.14), an ASTM basic set of blocks that includes both area/amplitude and distance/ amplitude, shear-wave testing blocks (see Figure 7.15), and thickness-testing reference blocks (Figure 7.16), which are used for calibrating resonance and pitch–catch (or pulse) types of thickness-measuring

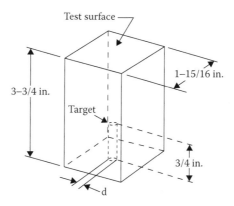

FIGURE 7.13   Area/amplitude standard reference block.

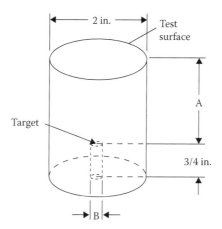

FIGURE 7.14   Distance/amplitude standard reference block.

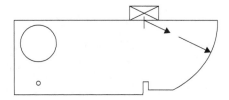

FIGURE 7.15   Shear-wave testing reference blocks. (a) IIW—Type 1 Block. (b) Miniature angle-beam block.

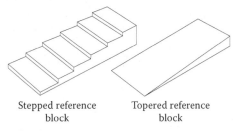

Stepped reference   Topered reference
block   block

FIGURE 7.16   Thickness-testing reference blocks.

equipment. The dimensions for ASTM reference blocks are given in ASTM E 127 and recommended practice for fabrication is found in ASTM E 428. The International Institute of Welding (IIW) has also developed a set of commonly used ultrasonic reference blocks.

The commercially available standard reference blocks fulfill many inspection needs. However, metallurgical and geometric differences between the test block and the part to be inspected often require that special test blocks be fabricated from the same material as that to be inspected. Artificial defects, which best simulate the types and sizes of defects to be detected, are introduced into the test block.

Comparisons between the results of two different ultrasonic inspections carried out presumably under similar conditions may show differences. The reasons for these differences often may be traced to differences in the equipment rather than in the personnel operating the equipment. In cases where a baseline inspection is to be performed, it is good to file the characteristics of the inspection equipment with the test data. Such equipment data include make and model of the instrumentation, characteristics of the probe including directionality, sensitivity, and frequency response, and the coupling conditions. As an example, data for an immersion, pulse-type probe may be obtained in a small immersion-type anechoic tank in which a rod for reflecting a signal is located. The probe is mounted to operate in a horizontal direction. The directionality of the probe is determined by recording the amplitude of a reflected pulse in each of several angular positions relative to the positions of the probe and reflector.

Additional information of importance includes an analysis of the frequency response of the probe and equipment. Often, the rated frequency of a commercial probe differs from the actual center frequency of the pulse. In order to compare test results with one another, it is essential that the equipment used in the individual tests have identical characteristics.

Characterizing ultrasonic transducers used for NDT and for medical diagnostic purposes is important to ensuring that the transducer meets the expected performance criteria. Almost all those ultrasound practitioners who have used any reasonable number of ultrasonic transducers for NDT have found significant variability in units provided with nominally identical specification and fabricated by the same manufacturing process [64]. Ultrasonic transducer fabrication is a combination of both art and science, and assembly is typically not automated. To address the issues of calibration, examples of ways to evaluate transducer tolerances have been reported [65] and evaluation/calibration standards developed. One standard commonly used for NDT transducers is the ASTM E 1065 [66]. High-frequency focused transducers used for special systems, such as acoustic microscopes, can require additional characterization, and parameters analogous to those used in optics including the line spread function and modulation transfer function [67,68] have been applied to ultrasonic imaging systems.

For a complete description of transducer performance and characteristics quantification, it is necessary to consider the effects on measured signals for all parts of a system. This includes the electronics used for excitation, any matching networks, the transducer (usually modeled as an electrical equivalent circuit), and both theoretical prediction and experimental characterization of the wave fields in both fluids and/or solids [69, 70]. Typically four parameters measured are (1) beam characteristics by an imaging technique, (2) transducer electrical impedance, (3) insertion loss, and (4) impulse and frequency response. Beam imaging provides information about beam angle, beam shape and symmetry, and distance–amplitude characteristics. Transducer electrical impedance is related to matching the transducer to pulsing equipment and to transducer construction. Insertion loss is a relative measure of transducer sensitivity and penetration capability. The insertion loss together with the frequency response provides transducer spectral characteristics including bandwidth, pulse shape, and damping characteristics.

Imaging and computer modeling techniques [71] can be used to evaluate and visualize transducer wave fields and wave-field–target interactions. Early work with ray theory models is fully described by Marsh [72] and Slater et al. [73], and further examples of the model results are presented in a series of elegant conference publications, including that by Baborovsky et al. [74,75].

The similarities between ultrasonic transducer fields and optical waves can be seen in the visualization of an ultrasonic pulse traveling through a direct ultrasonic visualization system, which uses a stroboscopic Schlieren system to visualize wave fields [76].

Differences in transducers may be attributed to variability in piezoelectric characteristics, damping, mounting and alignment of elements, and quality of faceplates. Equipment variations include bandwidth characteristics of the amplifiers and deterioration of components in the instrument.

Couplants are important elements in NDT. These can be bought in many places that supply oils, greases, or gels, which seem to fit the problem at hand. There are companies that supply couplants especially for NDT. One company that specializes in NDT couplants advertises that it produces 45 couplants in 7 categories and also formulates custom products and packages couplants and similar fluids, gels, and greases to meet specialized customer requirements.

7.5.7.2 Methods Used to Determine Flaw Size

Various methods are used for the determination of the actual flaw size during ultrasonic inspection. A few methods are presented here. Problems unique to specific applications of general interest will be discussed more fully in later sections. The factors that influence the nature of the indications include orientation, location, type of defect (void or inclusion), ratio of wavelength to dimensions of the defect, size of the defect compared with the dimensions of the ultrasonic beam, grain size and acoustic scattering properties (anisotropy and boundary conditions), and geometry of the defect.

Defects may change the wave front of a beam on reflection and thus affect the amplitude of the defect indication. For example, a plane wave striking a large-plane reflecting surface will be reflected as a plane wave. However, a plane wave incident on a spherical surface is transformed into a spherical wave on reflection (or refraction). Diffraction also occurs, and if the beam is larger than the defect, the defect casts a shadow that may disappear within a distance, which depends upon the wavelength and the dimensions of the defect. Nevertheless, the total energy of the beam is decreased by the amount of energy reflected out of the beam.

It is possible to calculate theoretically the energy that might be reflected and the energy lost from an ultrasonic beam due to a reflector if the scattering cross section is known. The scattering cross section, therefore, may form the basis for a determination of defect size. Here frequency or wavelength of the ultrasound and dimensions of the reflector are factors. Since nearly all pulse-echo inspections use broadband pulses, spectral analysis is a useful tool for evaluating flaw sizes [77].

A laminar defect lying in a known plane, which has a projected area in a plane that is normal to the direction of propagation and which is large compared with the cross section of the beam, provides a simple illustration. An example of this condition is large-area lack of bond between core and cladding in flat nuclear fuel elements. In this case, inspection by an immersion through-transmission method using a collimated beam can be used to map the extent of the flaw quite accurately. The accuracy depends upon how well the position at which the defect influences the signal amplitude can be determined. To give a size for a target, a standard approach is to scan a transducer over a feature, as illustrated in Figure 7.17, and to use the drop in signal by 6 dB (half the maximum response) as the definition of defect edge, hence the name of the method, "6 dB drop." A variation of this method is to use the 20 dB drop contour, and these can both be mapped, giving a projection of the defect in 2-D.

The sizes of defects that are smaller than the cross section of the ultrasonic beam are more difficult to determine. Such defects may occur in an infinite number of configurations and sizes. They may be voids or they may contain inclusions, which are capable of conducting energy away from the interface. An analysis of the energy reflected by

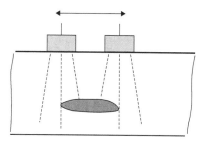

FIGURE 7.17   Schematic showing transducer scanning for "6 dB drop method," used for larger defect sizing.

defects of irregular shape can be simplified by the assumption that the defect approximates a simple geometry: cylinder, sphere, plane, or rectangular parallelepiped.

The frequency and time domain response for a spherical void can be predicted by the Ying and Truell [78] scattering theory. The normalized responses for a spherical target are shown in Figure 7.18 [79]. For simple spherical and elliptical targets illuminated with a short pulse of ultrasonic energy, response models can predict the observed on-axis separations seen in the pulse-echo reflected signals. This observation is illustrated with the data for a spherical void shown in Figure 7.19 [79]. The time T is the separation between the front surface impulse response and the creeping wave component, which travels around the surface of a void.

The normalized response for a rigid sphere in a liquid is shown in Figure 7.20, together with the specular reflection, for a flat surface, and that using a Rayleigh scattering approximation. In general for features much smaller than a wavelength, a Rayleigh scattering approximation can predict response, and for large flat targets, with beam diameter smaller than the target area, a near-specular reflection is observed. It is for targets with dimensions of the order of a wavelength that a more complex, and typically oscillatory, response is seen.

If the geometry of a void-type defect approximates a cylinder, the total power scattered by the cylinder may be calculated by [80]

$$P_s \begin{cases} \left(\dfrac{6\pi^5 a^4}{\lambda^3}\right)I_0 & (\lambda \gg 2\pi a) \\ 2aI_0 & (\lambda_1 \leq 2\pi a \leq \lambda_2) \\ 4aI_0 & (\lambda \ll 2\pi a) \end{cases} \tag{7.10}$$

where
I_0 is the intensity of the wave incident on the defect (W/cm^2)
P_s is the total power scattered per unit length (W)
a is the radius of the cylinder (cm)

The limiting value of total scattered power for very short wavelengths is the power contained in a beam twice as wide as the cylinder.

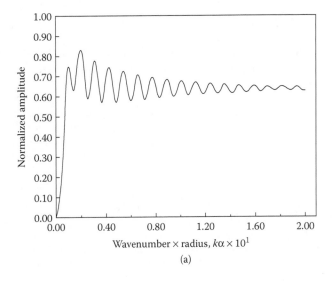

(a)

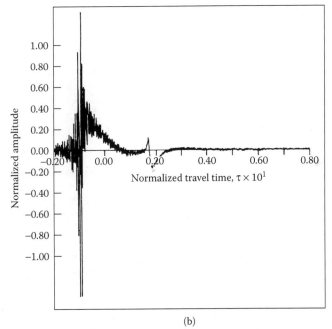

(b)

FIGURE 7.18 Response of a spherical void, with normalized radius. (a) Frequency domain predicted using Ying and Truell theory (k is wave number, α is radius). (b) Corresponding time domain response with top hat window function applied to limit high frequency content (time units: $1 = 2a/c$, where c is velocity of compression wave). (Reprinted from Zhang, H., and L. J. Bond, *Ultrasonics*, 27(2):116–9, 1989. With permission.)

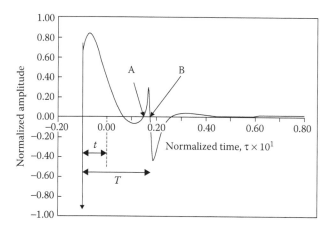

FIGURE 7.19 Pulse-echo response of a void in a solid. Time T is the separation between the front surface impulse response and the creeping wave component ($1 = 2a/c$). (Reprinted from Zhang, H., and L. J. Bond, *Ultrasonics*, 27(2):116–9, 1989. With permission.)

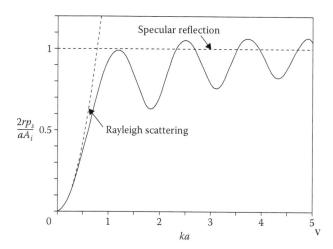

FIGURE 7.20 Normalized response of a rigid sphere in a liquid. Solid line shows analytical solution and dashed lines are plots for specular reflection and Rayleigh scattering. (From Kino, G. S., *Acoustic Waves*. Prentice-Hall, Englewood Cliffs, NJ, 1987. With permission.)

For a spherical cavity in which $\lambda \ll 2\pi a$, an approximate value of the scattered power is

$$P = 2\pi a^2 I_0 \tag{7.11}$$

where a is the radius of the sphere.

In terms of intensity, the ratio of transmitted average intensity (I_t) across a defect to that which would exist (I_i) were the defect not present and the corresponding voltage ratios (V_d/V_0) measured across the terminal of a receiving transducer in a through-transmission system are given by

$$\frac{I_t}{I_i} = \left(\frac{V_d}{V_0}\right)^2 = 1 - \left(\frac{D_1}{D_2}\right)^2 \qquad (7.12)$$

for circular delaminations, or

$$\frac{I_t}{I_i} = \left(\frac{V_d}{V_0}\right)^2 = 1 - \frac{4A_1}{\pi D_2^2} \qquad (7.13)$$

for any geometry, where
D_1 is the diameter of the circular delamination
D_2 is the diameter of the cross section of the beam
A_1 is the area of a delamination

The corresponding equation for transmission past a cylindrical void for which the long-wavelength case applies is

$$\frac{I_t}{I_i} = \left(\frac{V_d}{V_0}\right)^2 = 1 - \frac{24\pi^4 a^4 W}{D_2^2 \lambda^3} \qquad (7.14)$$

For the intermediate-wavelength case

$$\frac{I_t}{I_i} = \left(\frac{V_d}{V_0}\right)^2 = 1 - \frac{8aW}{\pi D_2^2} \qquad (7.15)$$

where W is the length of the cylinder ($0 \le W \le D_2$).

Theoretical and experimental intensity ratios and voltage ratios are given in Table 7.1, showing effects of defects in thin plates on receiver voltage in a through-transmission test in which collimators were used on both transducers. The agreement of experimental values with calculated values is typical of that found in several materials. If the flaw is an inclusion, the value of V_d/V_0 may increase by an amount dependent upon the material of the inclusion and the manner in which it is coupled to the base material.

In pulse-echo inspection, sometimes a comparison of the echo indication from a defect with the amplitude of a back echo provides a more reliable indication of flaw size than does the defect echo alone. A second receiver on the opposite side of the defect (through transmission) can provide a signal for comparison with the two echoes and thus enhance the flaw size evaluation. If the orientation is such that only a small amount of the energy reflected from the defect is returned to the receiver, the amplitude of the signal corresponding to the defect will be correspondingly small and give a false indication of the size of the defect. The amplitude of the back echo will decrease by an amount corresponding to the projected area of the defect.

Two characteristics of the ultrasonic beam are of importance in determining flaw size. These characteristics are (1) the effective beam form, that is, the effective distribution of energy across the wave front, and (2) the effective bandwidth or the effective frequency content of the transmitted and received pulses.

TABLE 7.1

Effects of Defects in Thin Plates on Receiver Voltage

Type of Defect	Size of Defect (mm)	Theoretical		Experimental	Remarks
		$I_t/I_i{}^a$	V_d/V_0	$V_d/V_0{}^a$	
Circular Delamination	1.6 mm	0.75	0.866	0.80	
Circular Delamination	0.8 mm	0.4375	0.662	0.65	
Circular Delamination	4.35 mm	0.75	0.866	0.84	Defect opposite transmitter
Circular Delamination	2.4 mm	0.75	0.866	0.82	Defect on transmitter side of plate
Rectangular Void	2.4×3.23 mm	0.364	0.603	0.60	
Long Cylindrical Void[b]	0.25 mm	0.996	0.998	—	
Long Cylindrical Void[b]	0.51 mm	0.934	0.966	—	
Long Cylindrical Void[b]	0.76 mm	0.668	0.818	—	Long-wavelength formula
Long Cylindrical Void[b]	0.76 mm	0.694	0.832	—	Intermediate-wavelength formula
Long Cylindrical Void[b]	1.02 mm	0.593	0.77	—	Intermediate-wavelength formula
Short Cylindrical Void	0.76×1.52 mm	0.841	0.917	—	Long-wavelength formula
Short Cylindrical Void		0.853	0.924	—	Intermediate-wavelength formula

[a] Transmitter collimator: 6.4 mm; I_t, transmitted intensity; I_i, incident intensity. Receiver collimator: 3.2 mm; V_d, receiver voltage measured as result of defect; V_0, receiver voltage measured as result of nondefective area of plate.

[b] Cylinder longer than incident beam. Axis intersects axis of incident beam at right angle.

In the far-field, the energy distribution across the beam typically peaks at the center and drops off rapidly at the edges. Therefore, the maximum sensitivity coincides with location of the defect at the center of the beam. In the near-field, the intensity varies according to near-field conditions.

The frequency effect is related to the relationship between the wavelength and the dimensions of the flaw. Equations 7.10 and 7.15 are examples of frequency effects. A defect in the path of a beam may be treated as a new source of ultrasonic energy for analytical purposes. Huygens' principle may be used to analyze the "radiation" pattern from the defect. Reflected energy corresponds to radiation of the same frequencies from a source of the size and location of the defect. Energy bypassing the defect corresponds to energy radiated from a transducer in the same location as the defect but with an area equal to the cross section of the beam at that location and the energy distribution in the plane corresponding to the exit side of the defect.

When ultrasound of short wavelengths impinges on a plane surface, energy is reflected in a plane wave. If the wavelength is large compared to the area of the surface, the surface simulates a point source and the reflected wave is spherical. If a plane wave impinges on a spherical surface, the reflected wave is spherical regardless of the wavelength; however, the amount of energy reflected is a function of the scattering cross section.

Since an ultrasonic pulse used in NDT consists of a spectrum of frequencies, frequency analysis of pulse-echo indications from defects provides a possible means of determining flaw size; however, applications of frequency analysis to determining flaw size have been only marginally successful. Calibration of the system is performed using suitable standards.

In surface wave inspection, just as with volumetric targets, there are three distinct interaction regimes.

For features much less than a wavelength, when the impinging beam is at normal incidence, for constant depth the reflected signal amplitude is proportional to the surface breaking length of the defect. For both down-steps and open cracks of constant surface length (width across the beam) or where the feature is wider than the beam, the reflected amplitude response as a function of depth exhibits characteristics somewhat similar to those for a sphere. To illustrate the form of the response commonly seen, a compilation of then available theoretical and experimental data for a Rayleigh wave and down-step interaction is shown in Figure 7.21 [81,82].

In shear-wave inspection by the angle beam method, the amplitude of echoes from cracks open to the surface is proportional to the depth of the cracks to depths of approximately two wavelengths [83]. Interferences from reflections such as cylindrical waves from edges or corners prevent continued increase at a proportional rate as the cracks become deeper. However, depths of deep cracks may be determined geometrically. If the actual angle at which the beam is refracted within the test piece is known, reflections corresponding to the top and bottom of the defect can be used to determine the approximate depth. The probe is moved to a position corresponding to the first indication of a defect (see Figure 7.22). The transducer is then moved away from the defect area until the defect indication

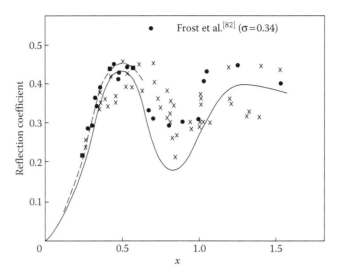

FIGURE 7.21 Rayleigh wave and down-step interaction. Reflection coefficient as a function of depth [82,83]. (Reprinted from Bond, L. J., *Ultrasonics*, 17:71–7, 1979. With permission.)

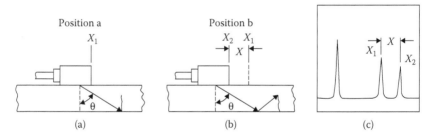

FIGURE 7.22 Measuring crack depth by a shear-wave technique. (a) Pulse-echo shear wave to crack root. (b) Scan on surface (X) and reflection from crack tip. (c) Differences in pulse arrival times shown in A-scan. Defect size estimated from geometry (b) or X shown in (c).

disappears. At this point, the transducer has moved a distance corresponding to the depth of the crack plus the effective beamwidth. The depth of the crack is

$$h = x \cos \theta \qquad (7.16)$$

where x is the total distance that the transducer moved during the measurement less the effective width and θ is the refraction angle [84].

In addition to amplitude-based methods, a variety of techniques, which developed from the geometrical theory of diffraction and employ the relative arrival times of the various pulses for sizing, are being utilized, and these are collectively called "time-of-flight" methods [85].

One method of determining flaw size and configurations combines B-scan and C-scan displays to present a 3-D view of the defect on an oscilloscope screen [86].

A single transducer scans the test piece in an immersion-test setup. The transducer oscillates so that a complete isometric view is obtained.

An amplitude correlation and differencing method for monitoring flaws for possible growth in repeat ultrasonic inspection, particularly of large and expensive components, has been reported by Burch [87]. The method was developed for use on digital, computer-based systems for the recording and analysis of data accumulated by the pulse-echo techniques. It can be applied if hardware is available for digitizing the peak pulse amplitude at each position of a transducer within an automated scan. Each inspection includes repeated scanning to provide an effective method for setting a reporting level for the amplitude difference values computed by the method, but a separate calibration experiment is needed to relate the reporting level to the dimensions of the minimum detectable growth.

REFERENCES

1. Thompson, D. O., and D. E. Chimenti, eds. 1980–2009. *Review of Progress in Quantitative Nondestructive Evaluation*. Vol. 1–28. Plenum: AIP Conference Proceedings Series.
2. Bond, L. J. 1988. Review of existing NDT technologies and their capabilities AGARD/ SMP Review of Damage Tolerance for Engine Structures, 1. Non-Destructive Evaluation. AGARD Report 768. Paper 2. 16. Luxembourg: AGARD.
3. Krautkramer, J., and H. Krautkramer. 1990. *Ultrasonic Testing of Materials*. 4th ed. Berlin: Springer-Verlag.
4. *American Society for Testing and Materials—Annual Book of Standards*. 2009. (New editions issues annually). Vol. 03.03, Nondestructive testing. Philadelphia, PA: ASTM
5. IAEA Report, TECH-DOC 462, Ultrasonic testing of materials at Level 2. Vienna, AU: IAEA.
6. Bray, D. E., and R. K. Stanley. 1989. *Nondestructive Evaluation*. New York: McGraw-Hill.
7. ASM International. 1989. *Metals Handbook*. 9th ed., Vol. 17. Metals Park, OH: Nondestructive Evaluation and Quality Contriol, ASM International.
8. Crawley, P. 2001. *Proc Inst Mech Eng L J Mater Des Appl* 215(L4):213–23.
9. Mohan, J. R. 2004. Nondestructive Test Equipment Equipment Market—Consolidating toward a digital future—Part 2. Frost & Sullivan. (May, 17, 2004) Briefing Paper.
10. Rosales, J. 2006. Nondestructive Test Equipment Segments—an overview. Frost & Sullivan. (Setpember 6, 2006) Briefing Paper.
11. Szilard, J., ed. 1982. *Ultrasonic Testing, Non-Conventional Testing Techniques*. Chichester: Wiley.
12. Migkiori, A., and J. L. Sarrao. 1997. *Resonant Ultrasound Spectroscopy*. New York: Wiley.
13. Zadler, B. J., J. H. L. Le Rousseau, J. A. Scales, and M. L. Smith. 2004. *Geophys J Int* 156:154–69.
14. McSkimmin, H. J. 1964. Ultrasonic methods for measuring the mechanical properties of liquids and solids. In *Physical Acioustics*, ed. W. P. Mason, Vol. 1A, Chap. 4, 271–335. New York: Acadeic Press.
15. Visscher, W. M., A. Migliori, T. M. Bell, and R. A. Reinert. 1991. *J Acoust Soc Am* 90:2154–62.
16. Frederick, J. R. 1970. *Mater Eval* 28(2):43–7.
17. Scofield, B. H. 1966. Technical Report No. AFML-TR-66-22, Air Force Materials Laboratory Research and Technology Division, Air Force Systems Command, Wright-Patterson Air Force Base, Ohio.
18. Martin, R. L. 1970. *Instrum Control Syst* 43(12):79–82.

19. Hutton, P. H. 1968. *Mater Eval* 26(7):125–9.
20. Awerbuch, J., M. R. Gorman, and M. Madhukar. 1985. *Mater Eval* 43:754–64, 770.
21. Gorman, M. R. 1990. *J Acoust Soc Am* 90(1):358–64.
22. Goranson, U. G. 1997. Jet transport structures performance monitoring. In *Structural Health Monitoring*, ed. F.-K. Chang, 3–17. Boca Raton, FL: CRC Press.
23. Romrell, D. M., and L. R. Bunnell. 1970. *Mater Eval* 28(12):267–70, 276.
24. Ogilvy, J. A. 1991. *Theory of Wave Scattering from Randon Rough Surfaces*. Bristol: Inst. Of Physics.
25. Goebbles, K. 1981. Structural analysis by scattered radiation. In *Research Techniques in NDT*, ed. R. S. Sharpe, Vol. 4, 109–57. London: Academic Press.
26. Goebbels, K. 1994. *Materials Characterization for Process Control and Product Conformity*. Baco Raton, FL: CRC Press.
27. Papadakis, E. P. 1981. Scattering in polycrystalline media. In *Ultrasonics, Methods of Experimental Physics*, ed. P. D. Edmonds, Vol. 19, Chap. 5, 337–98. New York: Academic Press.
28. Papadakis, E. P. 1968. Ultrasonic attenuation caused by scattering in polysrystaline media. In *Physical Acoustics*, ed. W. P. Mason, Vol. 4B, 269–29. New York: Academic Press.
29. Pao, Y.-H. 1983. *Trans ASME* 50(12):1152–64.
30. Kino, G. S. 1987. *Acoustic Waves*. Englewood Cliffs, NJ: Prentice Hall.
31. Urick, R. J. 1983. *Principles of Underwater Sound*. 3rd ed. New York: McGraw-Hill.
32. Rose, J. H. 1989. *PAGEOPH* 131(4):716–39.
33. Botsco, R. 1968. *Mater Eval* 26(2):21–6.
34. Moore, J. F. 1967. *Mater Eval* 25(2):25–32.
35. Generazio, E. R., and D. J. Roth. 1986. *Mater Eval* 44:863–70.
36. Godfrey, H. L. 1965. *Combustion* 36(8):19–20.
37. Schroeer, R., R. Rowand, and H. Kamm. 1970. *Mater Eval* 28(11):237–43.
38. Sansalone, M. J., and W. B. Streett. 1997. *Impact-Echo: Nondestructive Evaluation of Concrete and Masonry*. Jersey Shore: Bullbrier Press.
39. Fitting, D., and L. Adler. 1981. *Ultrasonic Spectral Analysis for Nondestructive Evaluation*. New York: Plenum Press.
40. Gericke, O. R. 1966. Technical Report No. AMRA-TR-66-38, U.S. Army Materials Research Agency, Watertown, Massachusetts, December 1966.
41. Gericke, O. R. 1968. Ultrasonic Pulse-Echo Spectroscopy, Technical Report, Materials Testing Laboratory, Army Materials and Mechanics Research Center, Watertown, Massachusetts.
42. Whaley, H. L., and K. V. Cook. 1970. *Mater Eval* 28(3):61–6.
43. Serabian, S. 1968. *Mater Eval* 26(9):173–9.
44. Challis, R. E., M. J. W. Povey, M. L. Mather, and A. K. Holmes. 2005. *Rep Prog Phys* 68(7):1541–637.
45. Pappas, R. A., L. J. Bond, M. S. Greenwood, and C. J. Hostick. 2007. On-line physical property measurements for nuclear fuel recycling, Proceedings GLOBAL 2007, Advanced Nuclear Fuel Cycles and Systems, Boise, ID, September 9–13, 2007, Am Nuclear Soc 1808–16.
46. Matherson, A. J. 1971. *Molecular Acoustics*. London, New York: Wiley.
47. Willard, H. J. Report No. WAPD-TM-833, Bettis Atomic Power Laboratory, Pittsburgh, Pennsylvania, October 1969; available from Clearinghouse for Federal Scientific and Technical Information, National Bureau of Standards, U.S. Department of Commerce, Springfield, Virginia 22151.
48. Hunter, D. O. 1969. *Ultrasonic Velocities and Critical-Angle-Method Changes in Irradiated A302-B and A542-B Steels*. Richland, Washington: Battelle Memorial Institute, Pacific Northwest Laboratories.

49. Bond, L. J., and N. Saffari. 1984. Mode conversion ultrasonic testing. In *Research Techniques in Non-Destructive Testing*, ed. R. S. Sharpe, Vol. 7, Chap. 5, 145–89. London: Academic Press.

50. Khuri-Yakub, B. T., Y. Shui, G. S. Kino, D. B. Marshall, and A. G. Evans. 1984. Measurement of surface machining damage in ceramics. In *Proc Review of Progress in Quantitative Nondestructive Evaluation (QNDE)*, ed. D. O. Thompson, and D. E. Chimenti, August 7–12, 1983, Vol. 3A, 229–37. Santa Cruz, CA: Plenum Publishing Corp. (New York).

51. Johar, J., A. Fahr, M. J. Murthy, and W. R. Sturrock. 1983. Detection of surface flaws in high performance ceramics by surface acoustic wave technique. In *Proc. 2nd IMRI-NRC Conf. Advanced NDE Technology (1982)*, ed. J. F. Bussiere, 15–20. Canada: NRC.

52. Johar, J., A. Fahr, M. J. Murthy, and W. R. Sturrock. 1984. Surface acoustic wave studies of surface cracks in ceramics. In *Proceedings, Review of Progress in Quantitative Nondestructive Evaluation (QNDE)*, ed. D. O. Thompson, and D. E. Chimenti, August 7–12, 1983, Vol. 3A, 239–49. Santa Cruz, CA: Plenum Publishing Corp. (New York).

53. Bond, L. J., and N. Saffari. 1984. Defect characterization in turbine discs. In *Proc. Review of Progress in Quantitative Nondestructive Evaluation (QNDE)*, ed. D. O. Thompson, and D. E. Chimenti, August 7–12, 1983, Vol. 3A, 251–62. Santa Cruz, CA: Plenum Publishing Corp. (New York).

54. Bond, L. J., and N. S. Saffari. 1983. UK Patent Application N0 8313118.

55. Chern, E. J., and J. M. Raney. 1983. *Keynote Paper, Review of Progress in Quantitative Nondestructive Evaluation (QNDE)*. Book of meeting abstracts, Center for NDE, Iowa State University, Ames, Iowa, August 7–12, 1983.

56. Szilard, J., ed. 1982. *Ultrasonic Testing, Non-Conventional Testing Techniques*. 381–99. New York: Wiley.

57. Maxfield, B. W., A. Kuramoto, and J. K. Hulbert. 1987. *Mat Eval* 45(10):1166–83.

58. Thompson, R. B. 1990. *Phys Acoust* 19:157–200.

59. Exhibitors Gallery, Materials Evaluation. 2003. *Mater Eval* 61(10):1087–110.

60. Scruby, C. B., and L. E. Drain. 1990. *Laser Ultrasonics*. Bristol: Adam Hilger.

61. Dewhurst, R. J., and Q. Shan. 1999. *Meas Sci Technol* 10(11):R139–68.

62. Barsky, M., and N. N. Hsu. 1985. *Mater Eval* 43:108–10.

63. Schuster, G. J., S. R. Doctor, and L. J. Bond. 2004. *IEEE Trans Instrum Meas Technol* 53(6):1526–32.

64. Bond, L. J., N. Jayasundere, A. D. Sinclair, and I. R. Smith. 1982. Investigation of ultrasonic transducers, as used for NDT. In *Proc. Review of Progress in Quantitative Nondestructive Evaluation (QNDE)*, ed. D. O. Thompson, and D. E. Chimenti, August 2–7, 1981, Vol. 1, 701–12. Boulder, CO: Plenum Publishing Corp. (New York).

65. Hotchkiss, F. H. C., D. R. Patch, and M. E. Stanton. 1987. *Mat Eval* 45(10):1195–202.

66. ASTM E 1065-08. 2008. *Standard Guide for Evaluating Characteristics of Ultrasonic Search Units*. Philadelphia, PA: Am Soc Testing and Materials (ASTM).

67. Shiloh, K., A. K. Som, and L. J. Bond. 1991. *IEE Proc A* 138(4):205–12.

68. Lee, U. W., and L. J. Bond. 1994. *IEE Proc Sci Meas Technol* 141(1):48–56.

69. Ilan, A., and J. P. Weight. 1990. *J Acoust Soc Am* 88(2):1142–51.

70. Good, M. S., and L. G. Van Fleet. 1988. Mapping ultrasonic fields in solids. In *Proc. Review of Progress in Quantitative Nondestructive Evaluation (QNDE)*, ed. D. O. Thompson, and D. E. Chimenti, Williamsburg, Virginia, June 22–26, 1987, Vol. 7A, 637–46. New York: Plenum Publishing Corp.

71. Bond, L. J. 1982. Methods for the Computer Modelling of Ultrasonic Waves in Solids. In *Research Techniques in Non–Destructive Testing*, ed. R. S. Sharpe, Vol. 6, Chap.3, 107–50. London: Academic Press.

72. Marsh, D. M. 1973. In *Research Techniques in Nondestructive Testing*, ed. R. S. Sharpe, Vol. 2, 317–67. London: Academic Press.

73. Slater, E. A., V. M. Baborovsky, and D. D. Marsh. 1974. A computer model of the interaction between an ultrasonic pulse and a surface defect, TIRL Report 369 A: TI Research Labs (Hinxton Hall, Saffron Walden, Essex, UK).

74. Baborovsky, V. M., D. Marsh, and E. Slater. 1973. *Nondestr Test* 6(4):200–7.

75. Baborovsky, V. M., E. A. Slater, and D. M. Marsh. 1975. The response of ultrasound to defects. In *Proc. Ultrasonics Int. 1975*, Imperial College, London, March 24–26, 46–53. Guildford: IPC Press.

76. Heyman, A. J. 1977. *Schlieren Visualization of Focused Ultrasonic Images*, Ph.D. Thesis. London: The City University and Hayman, A. J., and P. D. Hanstead. 1979. *Ultrasonics* 17(3):105–12.

77. Sinclair, A. 1985. *Mater Eval* 43:105–7.

78. Ying, C. F., and R. Truell. 1956. *J Appl Phys* 27(9):1086–97.

79. Zhang, H., and L. J. Bond. 1989. *Ultrasonics* 27(2):116–9.

80. Morse, P. M. 1948. *Vibration and Sound.* 2nd ed. New York: McGraw-Hill.

81. Bond, L. J. 1979. *Ultrasonics* 17(2):71–7.

82. Frost, H. M., J. C. Sethares, and T. L. Szabo. 1975. Applications for new electromagnetic SAW transducers. In *Proc. 1975 Ultrasonics Symposium*, ed. J. de Klerk, Los Angeles, CA, Sept. 22–24, 604–7. New York: IEEE.

83. Frielinghaus, R., J. Krautkramer, and U. Schlengermann. 1970. *Non Destr Test* 3(2):125–7.

84. Hagemaier, D. 1968. *Proc. 1968 Symposium on the NDT of Welds and Materials Joining,* American Soc Nondestructive Testing, Evanston, IL: ASNT.

85. Charlesworth, J. P., and J. A. G. Temple. 2001. *Engineering Applications of Ultrasonic Time-of-Flight Diffraction.* 2nd ed. Baldock: Research Studies Press.

86. Anderson, L. J., M. R. Wick, and G. J. Posakony. 1963. Technical Documentary Report No. ASD-TR-61-205, Part II, Directorate of Materials and Processes, Aeronautical Systems Division, Air Force Systems Command, Wright-Patterson Air Force Base, Ohio, November 1963.

87. Burch, S. F. 1985. *Ultrasonics* 23(6):246–52.

8 Use of Ultrasonics in the Nondestructive Testing and Evaluation of Metals

8.1 INTRODUCTION

Ultrasonic methods have been applied to nondestructive testing (NDT) and material characterization in numerous ways and are among the most important of the techniques used for inspecting metal products. The applications presented in this chapter are only representative. A more detailed presentation is beyond the scope of this book. However, applications and methods are presented in a manner intended to show what can be done using instrumentation that is readily available. For a more detailed approach to specific areas, the reader should consult the literature cited in and references at the end of this chapter. A book written by Krautkramer and Krautkramer [1], one edited by Szilard [2], and several others are devoted entirely to the applications of ultrasonics in NDT. There are also specialized tests that consider specific applications and testing approaches (e.g., [3,4]). The journals *Materials Evaluation*, published by the American Society for Nondestructive Testing, and *Ultrasonics and NDT & E International*, now published by Elsevier, are good sources of information on applications of ultrasonics in this field. Many codes and standards include ultrasonic methods as being recommended or required for inspecting materials (e.g., ASTM standards [5]). Ultrasonic inspection is safe, versatile, and economical.

Recent years have seen fundamental changes in the philosophical approach that is used to define the requirements for, and implementation of, many ultrasonic inspections. Early standards and methods used NDT as a check on "workmanship" and fabrication acceptance criteria. For example, defects that were detected in welds were assumed to need to be repaired: a zero acceptable (detectable) defects approach was applied. There is now increasing use and acceptance of fitness-for-service (FFS) methods, where NDT provides defect size that is then used for a fracture mechanics assessment. It is this assessment of defect significance that then becomes the basis for accept–reject decisions. Such analysis combines required design/service life, data for material toughness, estimates for expected loadings during service life, and cracks growth rates, and these factors combine to give criteria for maximum acceptable initial defect size, in the context of component life. At the same time, there has been a move to statistical assessments of the probability of detection, and this combines defects in probabilistic methods for risk assessment. These changes have required the development of new and improved methods for defect sizing, particularly when applied to evaluation of welds used in high-value or safety critical components.

Effective use of ultrasonic NDT requires skilled operators. It also requires an understanding of the theory of wave propagation and of the characteristics of available equipment on the part of the person responsible for prescribing the testing procedure. Many short courses are available to those interested in developing such an understanding and skill.

Presented in this chapter are common applications of ultrasonics in the inspection of metal products. The use of ultrasonics for evaluating nonmetallic products is discussed in Chapter 9.

8.2 INTERNAL STRUCTURE OF METALS

Measuring the propagation characteristics of ultrasonic energy is a means of determining internal (or metallurgical) structures. Evaluation may be based on attenuation characteristics of wave components at various frequencies, on the magnitude of echoes produced by internal structures, or on the velocity of sound in the material. When performing inspections, a good reference standard is essential to a reliable interpretation of the data. The ultrasonic techniques have been used successfully in assessing the structure of heat-treated steel castings, the form and amount of graphite in iron castings, the tensile strength of materials, the hardening depth or quality of chilled iron rolls, and the grain size and orientation in uranium, brass, and other alloys.

In common methods of indicating average grain size, use is made of (1) decrease (attenuation) in amplitude of the first back-echo pulse, (2) decrease in amplitude of the received pulse in a through-transmission arrangement of transducers, or (3) rate of decrease of successive back echoes. The first two methods are most easily adapted to recording. Recording when using the third method requires at least two gates and a means of comparing the second back echo with the first—simple requirements for modern instruments and methods of data processing. Standard, commercially available transducers are used for all the three methods. They are useful in separating materials with grain sizes larger than an acceptable limit and at the same time respond to other factors that also affect the attenuation. As grain size indicators, their main weakness lies in the fact that properties other than grain size contribute to attenuation. Some studies have aimed at evaluating methods of spectral analysis for grain size determination. Given the high quality and repeatable characteristics of modern transducers, recent studies (e.g., [6,7]) have shown that spectral analysis of pulsed CW techniques can provide sufficient indication of differences in grain size and grain size distribution or orientation in similar materials for some practical applications. Comparing ultrasonic pulse data obtained from similar metals but with varying histories known to cause grain growth or variations in grain orientation allows evaluation of grain properties on the basis of change in pulse amplitude and shape.

The relationship between attenuation (due to scattering at grain boundaries) and the ratio of wavelength to grain dimensions may vary among materials because of variations in anisotropy and elastic properties at interfaces, but generally, when the ratio between wavelength and grain size exceeds 30:1, little change in attenuation with change in wavelength occurs. When the ratio falls below 10:1, attenuation increases rapidly, reaching a maximum within the range 3:1 to 1:3.

Aisin [8] reported using the pulse-echo method to determine the macrostructure of bars of ball-bearing steel 100–160 mm in diameter. Hafemeister [9] used ultrasonic methods to detect grossly coarsened grain structure in finish-machined steel parts.

Dunegan [10] developed the following equation for describing the attenuation of sound in uranium as a function of average grain diameter:

$$\alpha = Ae^{kD} \tag{8.1}$$

where

α is the attenuation (dB/μs)
A is a frequency-dependent constant
k is 46×10^3 dB/μm
D is the average grain diameter in micrometer

Sharpe and Aveyard [11] demonstrated the facsimile recording of coarse-grain areas in cast uranium bars using the ultrasonic pulse through-transmission technique. Ultrasonic transmission through coarse-grain regions is poor. Sections of the facsimile recording corresponding to these coarse-grain regions are shown as zero-intensity areas, thus producing a plot of the exact location of the coarse grains.

Nemkina et al. [12] evaluated the use of ultrasonic attenuation in determining the grain orientation of transformer steel. Using a pulse-echo method and a frequency of 5 MHz on samples 0.5 and 0.35 mm thick, they concluded that the position of the amplitude minimum of the reflected signal near the direction of rolling uniquely determines the orientation of the (001) axis and can be used to determine the deviations of the (001) orientation of the grains from the ideal orientation.

There has been a growing interest in the use of scattering and attenuation data due to grain morphology for materials characterization. The earlier work is reviewed in book chapters by Goebbels [6] and Papadakis [7] and several other review papers (e.g., [13,14]).

As early as 1941, Sokolov [15] reported a method of measuring the depth of hardening in steel by ultrasonic means. Yegorov [16,17] and Koppelman [18] also claimed good accuracies using similar methods. These measurements are based upon the backscattering of ultrasonic pulses from grain boundaries using high-sensitivity detectors. Yegorov [16,17] used a pitch–catch arrangement of transducers and determined the depth of hardening by the angle of incidence of the beam on the non-tempered metal layer and the distance between transducers. Since the structure of the hardened layer differs from that of the nonhardened layer, reflections occur at the boundary between the two layers. Koppelman [18] used a pulse-echo method to arrive at the statistical backscattering of ultrasonic waves (i.e., noise caused by scattering from the grain boundaries). By averaging over an area, he obtained sufficient accuracy for practical measurements. Measurements on hardened-steel rolls are done at frequencies ranging from 30 to 60 MHz.

Fenkner [19] found a reproducible relationship between heat treatment and the velocity of sound in ball-bearing steel by which he was able to determine the amount of residual austenite in the steel. He measured the velocity of sound in the material by a pulse-echo method using an ultrasonic material tester with a calibrated transmit-time base. The abrupt changes of elastic constants occurring with polymorphous transformations of solids formed the basis of the measurements. Fenkner obtained the following formula for the retained austenite:

$$Y_g = RY_A + (1 - R)Y_M \qquad (8.2)$$

where

Y_g is the ultrasonically determined Young's modulus
Y_A is the Young's modulus of pure austenite
Y_M is the Young's modulus of pure martensite
R is the retained austenite

Since Y_A cannot be measured because pure austenite is unstable at room temperature, Fenkner extrapolated its value from a retained-austenite value obtained by other methods. He claimed an approximately linear relationship between retained austenite and longitudinal wave velocity. The velocity decreased with increasing amounts of austenite, dropping from slightly over 5900 m/s at 5% austenite to slightly under 5850 m/s at 25% austenite.

The anisotropic properties and coarse grain structure associated with austenitic steels present problems in the ultrasonic examination of these steels. Their crystallographic structure is dendritic. Grain boundary scattering in these materials leads to excessive attenuation and confusion in identifying defects (see Chapter 2, Section 2.10.2.6). The previous discussion (Chapter 2) is related to Rayleigh scattering, which occurs when $\lambda \gg D$, that is, the wavelength is much larger than the average grain diameter. At higher frequencies for which the wavelength is of the same order of magnitude as the average grain diameter, scattering becomes random, or stochastic, and attenuation due to scattering increases considerably. At still higher frequencies, $\lambda \ll D$, scattering is diffuse. These three conditions describe the three scattering regimes for which expressions for the ultrasonic attenuation coefficient, α, have been developed, that is, the Rayleigh scattering regime, the stochastic scattering regime, and the diffusion scattering regime. Papadakis [20] presented equations for Rayleigh and stochastic scattering and tabulated Rayleigh scattering factors for a number of cubic and hexagonal materials and stochastic scattering factors for cubic materials. He suggested that more experimental work is needed to fully confirm extant scattering theory.

Hecht et al. [21] have developed a method based upon backscatter of ultrasonic energy from sheets of austenitic steel. The sheets are scanned at the critical Rayleigh angle at the center frequency of 25 MHz in an immersion system. Both Rayleigh surface waves and bulk longitudinal waves are produced so that the signal received back into the transducer consists of components of both wave modes. Their results showed that the amplitude of backscattered waves increases with grain size. The slope of the backscattering curves as a function of distance traveled in the media is a function of sheet thickness rather than grain size; however, the amplitude is a function of average grain size. The curves in materials of equal thickness but with differing grain sizes are parallel. They applied the technique to a 1-mm austenitic strip 1761 m long to detect grain size changes caused by changes in production speed in a strip-sheet mill with fairly good correlation with metallographic measurements.

Baligand et al. [22] have studied the problem of inspecting coarse-grained austenitic steel products by an ultrasonic signal processing method. They use a broadband, nonfocused transducer with a center frequency of 2.25 MHz, 38 mm in diameter, in an immersion system. The test specimen is a cast stainless steel test block that represents the coarse-grained metallurgical texture of the cast bends of the primary piping in nuclear power stations. The received signal is filtered in narrow frequency bands, and then searched for anomalies associated with defects by processing the envelope of each filter output signal. The technique is similar to that used in radar systems to increase the signal-to-noise ratio. Their technique appears to show promise as a means of testing austenitic steel products.

Propagation characteristics of ultrasonic energy are influenced even by factors operating on a microscopic scale. Merkulov et al. [23] claimed to have shown good agreement in the determination of trace impurities in metals by ultrasonic and other means. The total number of impurity atoms determines the degree of absorption of ultrasound and the frequency at which maximum absorption occurs.

The work by Goebbels [6] has been developed and expanded, particularly as modern computers have enabled major advances to be made in data analysis and visualization. Scattering and attenuation are being investigated to enable characterization of material damage and aging and prediction of remaining service life [14]. Examples of such work now include characterization of void swelling in pressurized water reactor core internals [24].

The scattering of ultrasonic signals that occurs when conventional NDT methods are used to try and inspect coarse grain materials, such as cast austenitic stainless steels, remains a challenge, and the "noise" due to grain structure and beam bending due to anisotropy can mask responses from defects. A novel low-frequency phased array system that employs synthetic aperture focusing (SAFT) to process data has provided a high-resolution ultrasonic imaging capability that has been demonstrated to be able to image weld grains and small defects that are otherwise undetectable in welds [25,26].

8.2.1 MATERIAL EVALUATION BASED ON VELOCITY AND ATTENUATION OF ULTRASOUND

In general, an increase in the attenuation or a decrease in the velocity of ultrasound in a material is an indication of possible degradation or loss of strength of the material. Measurements of these properties can be used to detect (1) changes in the properties of materials that occur during such processes as nuclear irradiation, deformation, stress cycling and fatigue, creep, temperature fluctuations, and pressure fluctuations; (2) changes in purity or changes in state (such as aging, hardening, and formation of precipitates); or (3) phase changes. These conditions influence the propagation of all modes. The choice of method depends upon which one is most affected by the condition to be monitored and the ability to implement the technique. When a study involves both ultrasonic attenuation and velocity measurements, these should be performed concurrently.

The potential use of velocity measurements to determine preferred grain orientation in polycrystalline metal plates is discussed by Thompson et al. [27]. This is done by comparing velocities of symmetrical and of horizontally polarized shear plate waves with those of their plane-wave counterparts—longitudinal and horizontally polarized shear waves.

The quality of certain geometric structures can be ascertained by resonating the parts. This approach has been developed into resonant ultrasound spectroscopy (RUS), which has now been extensively used to study material properties as well as used as a nondestructive evaluation (NDE) tool [28]. RUS can determine the full elastic tensor in a single measurement with high accuracy [29]. This is achieved by obtaining a spectrum of resonant peaks for the material specimen of interest and then performing an inversion that utilizes an appropriate forward mathematical model or results from reference standards. In addition to the usage of RUS for material characterization, RUS-based techniques are becoming increasingly attractive to structural and machine component manufacturers. In these situations, a component is excited ultrasonically and defects are detected on the basis of changes seen in the pattern of vibrational resonances, when compared with computed or experimentally measured spectra of known flawless components [28]. In dealing with simple component geometries, the mode type and frequency can be calculated and selection of diagnostic modes can be based on these results. A mathematical model using the computationally efficient "XYZ Algorithm" of Visscher et al. [30] can be used to investigate results for the effects of the influence of the presence of embedded flaws/defects in a test specimen having simple geometrical shape [31]. When cracks or other defects form in a vibrating part, the resonant frequency drops and frictional losses at the new mating surfaces cause an increase in the damping factor. Resonance methods have been used to inspect grinding wheels, crankshafts, honeycomb panels, laminated materials, and corrugated materials. Types of defects located by this method include lack of bond, cracks, crushed cores, and voids.

Attenuation measurements have been used to determine average grain size in metals. As previously implied, grain size is a determining factor in the size of a defect that can be detected ultrasonically.

Measurement of attenuation also is useful for basic studies of the properties of materials. Bratina et al. [32] explored the use of the attenuation of ultrasonic energy in the megahertz range associated with vibration of free dislocation segments to study mechanical characteristics of metals over a wide temperature range (−150°C to +150°C). Attenuation associated with vibration of free dislocation segments is called dislocation damping. The studies of these investigators include the strain aging and ductile–brittle transition in iron and carbon steels. A measure of the depinning of dislocations from interstitial atoms was indicated by the amount of dislocation damping caused when an external load was applied to the specimen.

Tietz [33] used ultrasonic loss measurement at 4 MHz to determine the yield point of materials during elastic–plastic deformation. Both shear and longitudinal waves propagating in a direction normal to the applied tensile stress were used during these studies. An abrupt increase in the value of the loss coefficient for structural steel occurs at the yield point. Tietz attributed the increased attenuation to the action of dislocations, which are formed in certain grains at the upper yield point, are torn from the cloud of foreign atoms by the critical stress, and pile up at grain boundaries to form stress fields in which the separation of the dislocations in adjacent grains occurs.

In more recent studies, Generazio [34] has measured ultrasonic attenuation in polycrystalline samples of nickel and copper with various grain size distributions produced by heat treatment. His objective was to develop a means of determining the mean grain size in materials instead of the usual time-consuming, expensive, and destructive metallographic methods. His studies resulted in an approach in which he measured attenuation as a function of frequency in a sample having a known grain diameter $\bar{D}$. This function can be scaled to determine the mean grain size of other samples of the same materials but with different mean grain diameters. His approach is not to attribute attenuation to a particular scattering mode, since, in general, Rayleigh, stochastic, and diffusive scatterings may occur simultaneously because structural materials generally exhibit complex microstructures.

To develop Generazio's theory, one may note that the attenuation coefficient, α_R, for Rayleigh scattering as given in Chapter 2, Section 2.10.2.6, may be written in the form

$$\alpha_R = C_R \bar{D}^3 f^4 \text{nepers/cm} \tag{8.3}$$

where C_R is a constant that depends upon the material parameters other than mean grain diameter as given in Chapter 2. The equation for α_R is of the same form regardless of whether longitudinal or shear waves are involved and whether the crystalline structure is hexagonal or cubic; however, these factors influence the value of C_R. Similarly, expressions can be written for the stochastic scattering regime and the diffusion scattering regime. The corresponding relationships are, for the stochastic scattering regime

$$\alpha_S = C_S \bar{D} f^2 \text{nepers/cm} \tag{8.4}$$

and for the diffusion scattering regime

$$\alpha_d = \frac{C_d}{\bar{D}} \text{nepers/cm} \tag{8.5}$$

Each of these Equations 8.3 through 8.5, contains a coefficient that is dependent on $\bar{D}$. If the relationships given in Chapter 2 for C_R are correct, C_R should not be affected by changes in mean grain diameter. Likewise, C_S and C_d should remain the same if *mean grain size* is the only difference between the materials. Also, by assuming that the *shape* of the grain size distribution function and grain morphology are unaltered by the mean grain size, Generazio is able to derive a scale factor relationship,

$$\sigma = \frac{\lambda_n}{\lambda_0} = \frac{\overline{D}_n}{\overline{D}_0} = N^{th} \text{ scale factor} \tag{8.6}$$

where

λ_0 is an arbitrary wavelength for a system having mean grain diameter $\overline{D}_0$
λ_n is the wavelength for a system having mean grain diameter $\overline{D}_n$ for which the attenuation constant α is the same as that for λ_0

He proved his theory successfully using high-purity nickel and copper samples, making attenuation measurements in the frequency range 25–100 MHz. Microstructures were modified by appropriate heat-treating and annealing procedures. These results are very interesting; however, Generazio recognizes that, in general practice, theories and empirical relationships for attenuation due to many other factors appearing in more complicated structures are still insufficient to describe mean grain size, grain size distribution, and grain-boundary topology.

The value of attenuation, although it is influenced by grain size, is not limited to grain size determination in NDE of materials.

Mechanical properties may vary from one steel ingot to the next. Tight customer specifications are creating a need to develop means for accurately measuring 100% of steel materials shipped. Obviously, destructive metallurgical and mechanical testing methods are inadequate to meet such demands. Reliable nondestructive methods for on-line measurement of microstructure and mechanical properties are needed. Bussierre [35] has reviewed several methods that might be applied to on-line evaluation of the properties of steel and concluded that a combination of complementary techniques, such as ultrasonic methods for determining grain size and anisotropy with magnetic measurements to narrow the uncertainty in yield strength predictions obtained from ultrasonic attenuation, probably would be best for on-line evaluation. This is a good suggestion but much work appears yet to be done before reliable systems can be implemented on-line.

Bussierre refers to studies conducted at Bethlehem Steel aimed at practical on-line ultrasonic methods for predicting properties such as yield strength, tensile strength, and fracture toughness of steel. These investigators base their approach on the Hall-Petch relations

$$M = M_0 + M_1 \overline{D}^{-1/2} \tag{8.7}$$

which relate mechanical property, M, to grain size, $\overline{D}$, where

M_0 is a constant dependent on composition, phase morphology, etc.
M_1 is a constant

They estimate grain size nondestructively from ultrasonic attenuation measurements. The Rayleigh scattering regime is defined by Bussierre. He reports, "by a judicious choice of frequency (typically 5–10 MHz) the attenuation due to grain scattering can be used as a proxy for the metallographically measured grain size in the estimation of yield strength, tensile strength, and transition temperature." Whether or not grain size alone is the controlling factor in the relationship between these mechanical and metallurgical properties, it is not unreasonable to expect a correlation between them and attenuation. Plots of calculated and measured quantities presented appear to support this conclusion.

Nadeau et al. [36] studied the relationship between ultrasonic attenuation and fracture toughness in type 403 stainless steel. Rather than duplicating the previous practice of varying fracture toughness by changing the microstructure or composition of specimens, these researchers used a single sample of constant microstructure in which fracture toughness was changed by a factor of nearly three by varying the temperature. Attenuation changes were less than the 6% change expected from grain-boundary scattering theory.

The use of ultrasonic technology has matured, and systems have been developed and deployed in plants to provide correlations between case depth and hardness in steels. A laboratory system that uses megahertz frequency ultrasound to analyze microstructure and determine heated-treated or case depth measurement is shown in Figure 8.1 [37]. This same measurement concept has also been applied in several forms including to the evaluation of heat treatment efficacy, ultrasonic fingerprinting of materials, and evaluation of formed metal parts.

FIGURE 8.1   (See color insert.) Ultrasonic case depth and hardness measurement system. (Courtesy of Pacific Northwest National Laboratory.)

8.2.2 Surface Hardness Measurements

There are several commercial ultrasonic systems that can be used to measure the surface hardness of a wide range of materials (e.g., [38]). The hardness can be registered directly on a dial, or the process can be automated in a go/no-go form on a production line. It is claimed that the equipment can be calibrated in Rockwell C units in the range of 20–70 with an accuracy of ±1 unit. This technique produces an instantaneous electronic output that is a function of the contact area of the indenter. Surface indentation is as small as 0.0046 mm (4.6 micron). Commercial equipment is available in a portable version and in industrial versions offering variable loads. A laboratory microhardness tester is also available. Readings can be made every 3–10 seconds, depending upon the model used.

Kleesattel developed the ultrasonic contact impedance (UCI) hardness test method in 1961. The hardness tester is a portable electronic unit, which uses a Vickers indenter at the end of a vibrating rod with a load of about 5 kg force applied. For best results, the surface finish should not exceed 50 μm rms, but, by averaging, surfaces of 125 μm rms can be measured. The depth of penetration is only a few micrometers and, although primarily designed for steel, the instrument can be calibrated for most other materials. The advantages of the ultrasonic test are speed, ease of automation, accuracy, and nondestructiveness. A schematic showing the elements in an UCI unit is given in Figure 8.2 [39].

To measure hardness, a vibrating rod is applied lightly to the surface. The rod is mounted on a heavy mass at one end. The other end is free and supports the Vickers diamond indenter. Under no-load, the resonant rod is a half-wavelength long. As the load is applied and the area of the indentation increases, the impedance of the material being tested causes the point of contact to move toward a nodal condition, thus shortening the wavelength. The resonant frequency is forced upward by an amount, which depends inversely upon the hardness of the material. The accuracy

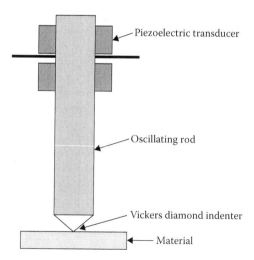

FIGURE 8.2 Schematic showing elements in an ultrasonic contact impedance unit.

of the measurement is dependent upon having both the mass to which the rod is permanently clamped and that of the load appear as infinite masses.

The methodology used in measurement of the UCI is now the basis of an ASTM standard (A 1038).

8.2.3 EVALUATION OF SINTERED PRODUCTS

NDT of powdered metals is in many respects similar to NDT of concrete. The velocity of sound in a material is a function of the elastic moduli, the density, and Poisson's ratio. In sintered materials, the elastic moduli are related to the density of compaction of the materials. The moduli increase at a greater rate than does the density of the material; therefore, the velocity of sound increases with increased density. Measurements have been made to correlate the velocity of sound in sintered products with density.

Brockelman [40,41] claimed a linear relationship between the tensile strength of powder-iron compacts and both the resonances of specimens and the velocity of sound in them. The strength of critical regions in complex shapes can be evaluated by using a narrow ultrasonic beam. The velocity of sound provides a better nondestructive test of tensile strength than of the sintered density. Both the velocity of sound and the tensile strength are influenced in the same way by pore size, shape, and distribution.

Abbe and Korytoski [42] used an immersion through-transmission technique for evaluating powder-metal materials. Information was stored on magnetic tape, and the data were compared with data from a standard test specimen. A technique similar to this has been used to test brake friction disks, in which the material is evaluated on the basis of attenuation of a signal through the disk. Poor bonding between individual particles within the powdered-metal friction layers increases the scattering from such particles and thus contributes to attenuation. Techniques have been developed for semiautomatic inspection on production lines.

Examples of applications now include measurements on iron compacts, refractory materials, and sintered uranium dioxide fuel pellets [43]. The variation of longitudinal ultrasonic velocity in sintered uranium dioxide pellets in the range of 89%–96% of the theoretical density was initially studied using an ultrasonic thickness/velocity meter [43]. The approach was subsequently developed to demonstrate that longitudinal velocity can be used to determine correlations with both elastic moduli and density, and is found to be near-linear over a change in Young's moduli value of ~50% [44].

8.2.4 ELASTIC AND ANELASTIC ASYMMETRY OF METALS AND
ACOUSTIC MEASUREMENT OF RESIDUAL STRESS

Both elastic and anelastic characteristics of materials are nonlinear at relatively low stress amplitudes [45]. Since these are conditions that affect the propagation of ultrasonic waves, it is natural to look to ultrasonics as a means of indicating stress levels in materials—particularly the residual stress.

The theoretical background for ultrasonic stress measurement is presented in several papers and summarized in a text by Bray and Stanley [46] and also in an earlier book chapter by Allen et al. [47]. The Chapter by Allen et al. includes useful appendices, which provide in one place the explicit expressions for velocity in deformed media, relationships that relate to harmonic generation both with and without static strain, data on grain size in steels, and relationships for ultrasonic velocity and attenuation dispersion. Both references cite further sources that include data for examples of practical studies. Although apparently simple in concept, significant challenges are faced both in providing the needed experimental accuracy and in relating measured travel times to stress/strain values, in large part due to the effects of underlying microstructure and temperature on the measured transit times.

The velocity of a transverse wave propagating in a direction that is perpendicular to the direction of stress depends upon the direction of its polarization relative to that of the stress. The velocity of ultrasonic waves polarized perpendicular to the stress direction differs from the velocity of those polarized parallel to the stress direction by a quantity that increases linearly with the stress [48].

Several investigators have studied the effects of stress-induced anisotropy on the propagation of ultrasonic waves. Most of the proposed techniques are in the development stage and have been used primarily in laboratories. Change in velocity with change in stress in metals is small, and many other factors also affect the velocity by amounts that may obscure stress-induced changes. Some of these factors are grain orientation, variations in mechanical treatment, and varying concentrations of alloying elements. In the laboratory, these factors can be identified, but in the field, identification may not be possible. Progress has been made in recent years toward developing methods of separating these factors from stress-induced changes in velocity of sound. These methods could lead to field measurements in certain limited, but important applications, especially with recent advances and the precision available for measuring velocity of sound.

Conventional pulse-echo methods of indicating velocity changes in metals due to stress usually are not practical. Various early investigators tried different techniques for indicating velocity changes due to stress. Benson and Realson [49] investigated the birefringent effect of stress on polarized shear waves. By orienting the wave at 45° to the stress orientation, the ultrasonic wave is broken down into two components. One component is polarized parallel to the direction of the stress, and the other is polarized normal to the direction of stress. Since these two components propagate at different velocities, the wave undergoes a change in polarization as it propagates through the medium. Benson and Realson used two identical shear-wave transducers aligned at 45° to the direction of stress on opposite and parallel sides of the specimen. The amplitude of the first signal was plotted as a function of applied stress. As the stress increased, the phase relationships between the two wave components shifted accordingly. As a result, the received-signal amplitude went through a series of maxima and minima, maxima occurring when the velocity difference produced a phase difference of $2n\pi$, where n is an integer (including 0 when the medium is isotropic and the components are in phase).

An objection to this method of measuring residual stress is that it gives an integrated average and, except in structures of certain geometries, the algebraic sum of the residual stresses through a test section is likely to be zero. Thus, on ideal rectangular specimens, the polarized-beam method of Benson could measure applied stress; however, its use in measuring local residual stresses is limited.

Benson and others also investigated the effects of stresses on the velocity of surface waves to detect residual stresses in the surfaces of materials. Benson [50] used a phase-comparator method. McGonagle and Yun [51] referred to the sing-around method of measuring surface-wave velocity as a possible means of measuring surface residual stresses. (In the sing-around method, each received pulse is used to trigger the succeeding ultrasonic pulse. Velocity is determined as a function of pulse repetition rate.)

Rollins et al. [52] attempted to develop a technique of measuring residual stresses at any location within a material by analyzing the interaction of two intersecting plane waves. Such an interaction results, in theory, in a scattered wave with a frequency corresponding to either the sum of or the difference between the frequencies of the intersecting waves. Although Rollins et al. apparently were successful in generating such waves, the principle has not been applied successfully to measurement of local stresses.

In the practical area, an ultrasonic method of measuring changes in bolt resonances attributable to elongation and changes in velocity of sound in bolts under stress is the basis of an instrument called the reflection oscillator ultrasonic spectrometer [53]. The instrument is frequency-locked to the peak of a mechanical resonance frequency of the bolt. It is capable of indicating changes in bolt strain to better than one part in 10^4. Accuracy requires certain precautions such as

- Grinding the ends of the bolt flat and parallel
- Applying the transducer to the bolt with considerable care and consistency to minimize coupling-related frequency shifts
- Recording the ambient temperature when tightening the bolt and when making subsequent measurements in order to compensate for thermal expansion

Pulsed radio-frequency (rf) or tone-burst methods have been used successfully in the laboratory to measure stresses in metals and to develop acoustic stress constants for different alloys [54]. An immersion method employing longitudinal modes was used. (The same reference describes effects of hardening on velocity of sound in a low-alloy heat-treatable steel and proposes this method for measuring hardness in these materials.) For the case of a longitudinal wave propagating in the z direction at normal incidence to a plane specimen in plane stress, the relationship between acoustic wave velocity and stress is

$$\frac{\Delta c_\ell}{c_\ell} = \frac{c_\sigma - c_\ell}{c_\ell} = C_\ell(\sigma_1 + \sigma_2) \qquad (8.8)$$

where

c_σ is the velocity of the wave in the stressed material
c_ℓ is the velocity of the wave in the unstressed state
C_ℓ is an acoustic stress constant for longitudinal waves that depends upon the elastic constants of the material
$(\sigma_1 + \sigma_2)$ is the planar first stress invariant

A similar equation for shear waves polarized along the principal stress directions is

$$\frac{c_{s1} - c_{s2}}{1/2(c_{s1}+c_{s2})} = C_S(\sigma_1 - \sigma_2) = B \tag{8.9}$$

where

c_{s1} is the velocity of the wave polarized along the σ_1 direction
c_{s2} is the velocity of a wave polarized in the σ_2 direction
σ_1 is the stress in the 1 direction
σ_2 is the stress in the 2 direction
B is the acoustic birefringence
C_S is the stress constant for shear waves

When acoustic birefringence measurements are made using piezoelectric shear-wave transducers, variations in thickness of coupling fluid cause differences of arrival time that are sources of error. This problem can be overcome by using electromagnetic acoustic transducers (EMATs) rather than piezoelectric transducers [55]. The full range of EMATs now commercially available is illustrated with the various systems shown in Figure 8.3. The use of EMATs offers several advantages in addition to overcoming the coupling error problem. They can cause any of the wave modes previously used with piezoelectric crystals to propagate in directions normal to the surface typical of the direction used in most previous studies, and they can generate shear waves that are polarized parallel to the plate surfaces (S-H waves) and propagate at an angle to the plate normal. Perhaps the most important advantage to be gained by using EMATs is the capability they afford to remove the unstressed birefringence (that are attributable to the material properties). This is done by measuring arrival times at two or more angles of incidence at a given location.

Kwun [56] measured the distribution of residual bulk stresses in a rectangular, butt-welded ASTM A-588 steel specimen and a T-shaped, full penetration-welded ASTM A-514 steel specimen using a 10-MHz longitudinal-mode transducer, a 5-MHz

FIGURE 8.3   (See color insert.) Some examples of commercial electromagnetic acoustic transducers on pipes, on plates, and with rollers. (Courtesy of Sonic Sensors.)

shear-wave transducer, and magnetic fields applied either parallel or perpendicular to the weld lines. The ultrasonic waves were propagated in the thickness direction of the specimens. Kwun's results showed good agreement with destructive stress measurements; therefore, the procedure is deemed a practical means of nondestructively measuring bulk stresses in ferromagnetic structural steels. Magnetically induced velocity changes observed under various uniaxial stress conditions were used for determining the unknown residual bulk stresses. The method makes use of the fact that the magnitude of magnetically induced velocity changes and the way the changes vary as a function of the applied magnetic field are characteristically different depending on (1) the magnitude of the stress, (2) the sign of the stress (i.e., tensile or compressive), and (3) the relative orientation of the stress, the magnetic field, and the propagation direction (or polarization for shear waves) of the ultrasonic waves.

There remains interest in developing and using EMATs to give stress, including through the measurement of changes seen in acoustic birefringence. An article by Kristan and Garcia summarizes work performed at the National Institute of Standards and Technology (NIST) with the support of the Association of American Railroads. Good correlations were reported to have been obtained between experimental data and results from numerical models [57]. This approach is also being considered for use on pipelines (e.g., [58]).

8.2.5   Fatigue, Aging, and Monitoring for Metals

Ultrasonics has been used to study and detect fatigue in metals. High-intensity ultrasonic waves often induce fatigue cracks in materials, and this phenomenon has sometimes been applied to accelerate fatigue tests.

NDT methods of monitoring or detecting the presence of fatigue in metals are based on either the attenuation of the amplitude of the waves or the reflection of energy from the crack as a discontinuity in the material. The methods have been applied successfully to follow the growth of cracks on specimens in fatigue-testing machines. In other studies, attempts have been made to detect the initiation of cracks and to predict the remaining life of parts.

Lamb waves and Rayleigh surface waves are especially sensitive to the development of fatigue cracks. The effect on surface waves can be visualized by the fact that the effective depth of penetration of such waves is approximately one wavelength. For example, the velocity of a Rayleigh wave in steel is approximately 2996.0 m/s. The effective depth of penetration of a surface wave at a frequency of 5 MHz in steel is 0.559 mm. The wave is influenced most by defects nearest the surface. A 5-MHz Rayleigh wave in steel can be used to detect fatigue cracks less than 0.0508 mm in depth, about one tenth of a wavelength.

According to Rasmussen [59], the first indication of a change in the properties of the material under dynamic stress is given by the formation of slip lines, which may be observed using electron microscopy. These slip lines may form as early as at 1% of the fatigue life of the part. Fatigue fracture begins as microcracks forming along active slip planes at an early stage. In some cases, extrusions, intrusions, or pits are formed in the surface at an early stage. At about 60%–70% of the fatigue life, the microcracks start joining to form larger cracks. Some of these growing cracks cause the final failure. Fatigue life of a part depends upon the manner in which the cracks form under fatigue-stress conditions. Prediction of fatigue life, then, is based on locating points on a characteristic curve corresponding to the developing cracks under known conditions of cyclic stress. Rasmussen claimed to have detected fatigue damage in the form of microcracks and other discontinuities mentioned previously in 2024ST3 aluminum alloy plate, which was electropolished at 39%–44% of the fatigue life during dynamic loading. The primary variable, which influences how early detection can occur, is the surface roughness of the specimen. Rasmussen used surface waves and claimed very reproducible results and claimed further to be able to extrapolate from the given indications when total failure would occur. The method was based on measuring the attenuation of surface waves due to the formation of the microcracks.

Critical angle tests have been used to detect fatigue in metals [60]. These tests involve the use of surface waves generated when the transmitted wave is incident on a test specimen at or near the Rayleigh critical angle. The results were similar to those of Rasmussen.

Klima and Freche [61] applied a standard longitudinal-mode pulse-echo technique to detect and measure the growth of fatigue cracks in notched cylindrical specimens of 2014-T6 aluminum, 5Al-2.5Sn titanium, a maraging steel, and a cobalt-base alloy, L-605. The investigators were able to detect cracks, which generally extended around the specimen circumference, by the time they reached a depth between 0.0127 and 0.1016 mm. Initially, detectable cracks were formed within approximately 10%–40% of life-to-fracture, depending on material and cyclic stress.

In a similar method using 10-MHz shear waves, Clark and Ceschini [62] moved the transducer along the surface of a specimen relative to a crack tip so that the crack length could be interpreted in terms of transducer location.

Norris et al. [63] combined surface-wave and longitudinal-mode techniques to study fatigue-crack initiation and growth in low-cycle fatigue testing. The surface waves were used to determine crack initiation and the longitudinal waves for monitoring crack propagation.

The attenuation of through-transmission ultrasonic energy has been applied by Cole [64] to determine the location and criticality of fatigue-induced damage in filament-wound fiberglass cylinders. The application was for the determination of residual life remaining in composite, deep submergence, multi-dive vehicles.

Adoption of an ultrasonic monitoring technique to estimate fatigue-limited serviceability of aircraft structures has been proposed by Zirinsky [65]. The proposed technique would involve the use of a single instrument to monitor several permanently mounted shear-wave transducers, which can be manually operated or used with a recorder. Recent years have seen research focused to develop techniques that can inspect aging aircraft and other civil infrastructure. There is a growing interest in the use of ultrasonic sensors to provide on-line monitoring and in detecting the effects of aging and degradation on materials, and then using the combination of the NDE data with models that predict remaining life [66,67]. There is also work to demonstrate the potential for structural health monitoring to applications such as advanced designs of nuclear power plants [68].

8.3   INSPECTION OF BASIC STRUCTURES AND PRODUCTS

Many structures are readily adaptable to ultrasonic inspection. Shafts and bar stock of simple geometries with flat, parallel ends can easily be inspected using longitudinal modes. If the material has low attenuation (e.g., aluminum or low-alloy steel with small grain size), fairly accurate and rapid inspection can be performed. Wall effects, which include mode conversion as the beam spreads and impinges on the walls, will cause trailing signals of noise in long, relatively slender bars. However, it is usually possible to isolate the back-echo and defect indications from these noise signals. Bar-type structures typically inspected by ultrasonics include ordinary steel round stock, engine crankshafts, railroad axles, large bolts, steel billets used in the fabrication of landing-gear pistons (AISI 4340 steel), cold-extruded transmission shafts (SAE 4027 steel) and axle shafts (SAE 1052 steel) with hot-forged flanges, generator rotors and armature shafts, rolls of cast and forged Cr–Mo steel alloys used in rolling mills (inspected for macrodefects), reconditioned or built-up drill bits, and steel trunnions for steel casting ladles.

Pulse-echo longitudinal- and shear-wave techniques have been used to detect defects in high-pressure vessels, high-pressure gas cylinders, riveted drums, and many other similar structures.

Surface-wave inspection has been useful in locating defects such as fatigue cracks in helicopter and propeller blades and turbine and compressor blades. The advantages of surface waves over other types in such applications are that they follow the contour of the pieces without loss due to direction changes.

The details of specific methods for the inspection of items are to be found in numerous quality assurance plans, which, in many cases, are derived directly from accepted codes and standards, for example, the ASTM standards [5]. More general information on ultrasonic inspection is provided in NDT texts, such as text by Krautkramer and Krautkramer [1], and in the training manuals for inspectors provided by large companies, professional societies, and international organizations, such as the International Atomic Energy Agency (IAEA) [69].

8.3.1   WELDS

Ultrasonics is one of the most widely used techniques for the nondestructive inspection of welds, and consequently much has been written on the subject. The most commonly used method is angle-beam testing in which shear waves are transmitted through the part, as shown in Figure 8.4. The shear wave is generated by mode conversion from longitudinal waves incident on the surface of the test material. The longitudinal wave is eliminated from the test material by using an angle of incidence, which exceeds the critical angle for total reflection. Shear waves may also be generated directly using EMAT transducers. Methods of inspecting T-sections using longitudinal waves are illustrated in Figure 8.5.

Any of the three common methods of coupling energy from piezoelectric transducers to the part are used in weld inspection: (1) direct contact, with the transducer mounted on a plastic wedge; (2) immersion; and (3) gap, using a liquid stream, liquid-filled wheel, or liquid-filled boot.

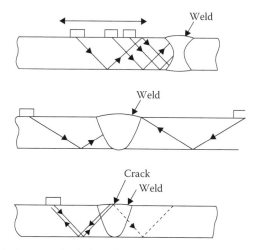

FIGURE 8.4   Angle-beam method of weld inspection by ultrasonics.

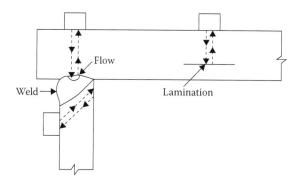

FIGURE 8.5   Methods of inspecting T-joints.

The proper design of any ultrasonic inspection method takes into account the possible generation of unwanted wave modes confusion by the fact that they propagate at different velocities. Figure 8.6 illustrates what can occur when using the same ultrasonic beam dimensions to inspect welds in tubing of two different sizes. In the large-diameter tubing, the variation in the angle of incidence across the beam is small, and the shear wave that is generated is close to being similar to that which would occur if the tube were a flat plate. In the small-diameter tubing, the incidence angle varies from front to back by a large amount. As a result, longitudinal waves, shear waves, surface waves, and sometimes Lamb waves may travel in the material at the same time and in both directions around the circumference. A thinner beam in the circumferential direction is necessary to eliminate this problem.

Typical weld defects include cracks from various causes (crater, hot cracking, cold cracking, etc.), incomplete penetration, pickup, blowholes, porosity, slag inclusions, lack of fusion, undercutting, and pipes.

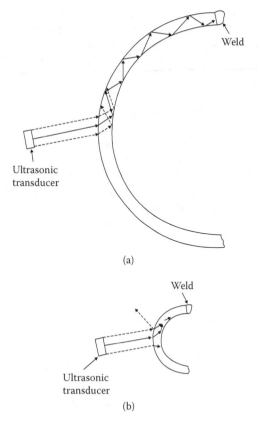

(a)

(b)

FIGURE 8.6   Problem of applying same wide beam to large- and small-diameter tubes.

Crater cracks are formed while a weld cools. The crater solidifies from the outside toward the center, a condition which favors the production of shrinkage cracks.

Hot cracking refers to intergranular cracks, which form as a weld metal cools and solidifies. When these cracks are opened at the surface of the metal, oxidation occurs on the walls of the crack because of the temperature of the metal and exposure to the atmosphere.

Cold cracking refers to cracks that form after freezing has been completed. These cracks usually cut across grains. There are many causes of this type of defect, all of which can be related to high stress concentrations and a brittle characteristic of the material at the site of the fault.

Penetration is a term used to refer to any of several conditions. In fillet welds, it may refer to fusing completely through a layer of mill scale on the surface of the welded parts. It may also refer to the depth below the surface of the base metal to which the fused zone penetrates. In a butt joint, it may refer to the depth of the joint to which the weld metal is able to flow.

Pickup refers to the intermixing of a filler rod and base metal where these materials are of different compositions.

Blowholes are rounded gas bubbles with smooth, bright surfaces. They are caused by gases either being released from solution in the base metal or being produced by fluxes that dissociate at high temperatures. The gas bubbles are prevented from rising to the surface by the slowing effect of the viscosity of the melt and by the cooling rate of the weld metal.

Porosity is related to the formation of blowholes. Gas bubbles liberated during crystallization of the weld metal are entrapped by the growing crystals, which results in clusters of pores.

Slag inclusions are nonmetallic materials other than graphite that are embedded in a metal. They may originate in slag or flux on the surface of the metal or they may originate in the metal as products of a deoxidation reaction.

Lack of fusion refers to a lack of coalescence between filler metal and base metal, or of base metal only, during welding.

Undercutting is illustrated in Figure 8.7. This refers to a groove melted into the base metal adjacent to the toe of a weld that is left unfilled by weld metal.

Pipe may occur in fillet welds as a result of solidification starting at both sides of the weld. As freezing progresses, the solid, being more dense than the liquid, occupies less space than it did when it was liquid, and the level of the remaining liquid falls. Thus, when solidification is complete, there remains a deep cavity, or pipe.

Sometimes the type of the defect can be ascertained by knowledge of the materials being inspected and the nature of the ultrasonic indication. More often, the identification of defect types requires a knowledge of the location of the defects within the weld since certain types may occur only in certain parts of the weld.

One may understand from the foregoing the importance of using good test equipment, good standards for calibrating the equipment and assessing the nature of defects, and adequate training of the inspector.

A complete discussion of weld inspection by ultrasonic means would be a nebulous task. Those interested in more detailed descriptions of procedures and problems are directed to some excellent literature on the subject.

Ostrofsky [70] discusses the control of beam spread, calibration procedures, weld-inspection methods, and characterization of defects and difficulties in interpretation. He also includes a table of helpful information for identifying weld defects from ultrasonic indications.

An excellent treatise on the fundamentals of weld inspection has been presented by Hagemaier [71]. He discusses various aspects of instrument calibration and transducer characteristics as well as factors involved in the test setup, identification of

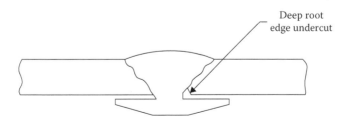

FIGURE 8.7   Deep root edge undercut sometimes encountered in backing type of pipe weld.

defect location, and interpretation of test results. Others, Harris-Maddox [72], for example, provide useful discussion and tables on the identification of weld defects.

The training handbooks issued by IAEA [69] and a number of companies and NDT societies are of considerable value to those interested in developing skill in ultrasonic inspection of welds.

Starting in the 1970s, new methods for defect sizing were developed. This was accompanied, in many high-technology industries, by the replacement of ultrasonic NDT as a "workmanship" standard, with its use as the basis for FFS criterion. The FFS accept–reject criteria, when applied to weld defects, require accurate measurement of defect dimensions that are then used in fracture mechanics assessments. The literature that now discusses ultrasonic methods that determine defect dimensions and locations is extensive (e.g., [71–79]). An overview of crack sizing was provided by Silk in 1977 [80], and there have been a number of subsequent papers and books on this topic.

The newer ultrasonic techniques that are being introduced can measure defect height (depth), at least in principle. Initially, ultrasonic amplitude methods were used for height measurement, but these proved unreliable. Now diffraction methods, especially Time-Of-Flight-Diffraction (TOFD), are being used in conjunction with amplitude methods. An assessment of the current state of the art for pipeline weld inspection and defect characterization is provided by Moles [81]. This article reviews data from several larger nuclear industry studies, including the European Community, Joint Research Center Programme for the Inspection of Steel Components, Phase II (PISC II) and some pipeline studies. The best methods used in the nuclear industry for defect sizing are those based on tip-diffraction phenomena. These methods give flaw measurements that have been shown through destructive examination to give characteristics that are within a few millimeter of true size. Current best practice gives accuracies for defects that are typically ±1 mm, which correlate with the beam spot size and typical weld pass dimensions.

Weld beads and certain metallurgical characteristics of some types of weld metals may cause such severe scattering of ultrasonic energy that effective inspection for internal weld-metal defects is precluded. Grinding weld beads flush with the base metal can eliminate scattering from this source, but scattering from internal structures, such as large grain boundaries, makes other inspection methods necessary.

Specific areas of application of ultrasonic inspection of welds are pipe and tubing welds, welds in rocket-engine hardware (aluminum, steel, and stainless alloys), structural-steel welds in ship and submarine structures, welds in pressure vessels, and end-cap welds of nuclear containment capsules. Hagemaier and Posakony [82] describe the use of a 25-MHz transducer and high-resolution equipment to inspect welded couplers in 6–25 mm outer diameter (OD) hydraulic lines. They found ultrasonics to be the only feasible method of inspecting melt through fusion welds joining the sleeve or adaptor to the tube.

Center and Roches [83] propose combining ultrasonics and radiography for inspecting heavy-filled pressure-vessel weldments, an area in which, until recently, there has been a reluctance to use ultrasonics. Their tests show that ultrasonics could make a valuable contribution, particularly to the detection of planar-type defects such as cracks, lack of fusion, and lack of penetration. Regarding lack of fusion in any weld, it is possible to have such defects under so much compression that little energy is reflected from the defect and it goes undetected by ultrasonics.

Lamb waves are often used for the inspection of welds in thin-walled tubing [81]. Stainless steel or other thin-walled tubing 10–70 mm in diameter can be inspected and recorded at the rate of 3 m/min.

Automatic and semiautomatic systems are being used for ultrasonic NDT of welds and these have been discussed in various papers [81,85–87]. The problem in the development of such systems involved incorporating or simulating the versatility of manual operation. Certain types of welds in which the defects are oriented along a known direction or plane are easily adapted to automated ultrasonic tests.

Methods of inspecting spot welds include measuring the transmission of an ultrasonic beam through the weld as it is being formed [88] or observing differences in the pattern of multiple echoes from different welds [89–90]. To facilitate inspection of the weld as it is being formed, ultrasonic transducers are built into the electrodes.

Most normal contact and delay line transducers have difficulty coupling into the geometry of spot welds, which are typically cup-shaped and 3–6 mm in diameter. Because they are often very thin, best inspection frequencies range from 15 to 30 MHz. These problems are met by commercially available captive water-column transducers with a pliable rubber member enclosing a column of water to conform to the depression in the weld and thus to couple sound into it. A straight beam pulse-echo technique is used with the water column acting as a delay line. The reflected signals have characteristics identifiable with good or poor welds.

In a study aimed at determining whether acoustic emission (AE) from laser spot welding would be capable of providing a power-controlling feedback system to the welder, Weeter and Albright [91] determined that such a feedback system should be feasible because spot weld emissions differ significantly between full penetration, minimal penetration, and no penetration. This in a sense is applying AE nondestructive testing to power feedback control.

The orientation of a defect with respect to the direction of propagation of the ultrasonic beam determines whether the reflected pulse will return to the receiver. Cross et al. [92] developed a method of detecting weld defects regardless of orientation. This method, called the delta method, which was originally reported by Posakony [93], is illustrated in Figure 8.8. The transmitter produces an ultrasonic wave at an angle to the surface of the base metal. The energy propagates into the weld metal and strikes a discontinuity. If the defect is a surface oriented so that energy is reflected upward, the receiver, which is located directly over the weld, receives the reflected pulse. If the defect is oriented normal to the surface, a surface wave is generated by mode conversion on the surface of the defect. The wave propagates downward, converts into a longitudinal wave at the edge of the defect, and reflects from the bottom. The receiver then receives "reradiated" energy from the defect. Another orientation may reflect energy toward the underside of the weld, where it is reflected upward and into the receiver.

Over the years, a number of techniques that utilize mode-conversion phenomena, in many cases derivatives or extensions of the Delta technique, have been proposed and deployed in both laboratory studies and for practical applications [94].

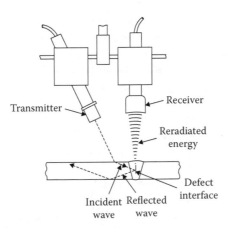

FIGURE 8.8   Delta configuration technique.

8.3.2   TUBES, PIPES, AND SHELLS

Ultrasonic inspection of tubes, pipes, and shells is closely related to and may include weld inspection, but it also includes the location of defects in any part of the base metal. As illustrated in Section 8.3.1 (Figure 8.6), the geometry of the test specimen enters into the inspection system design. It is important not only to control the relationship of the beam dimensions to the tube diameter but also to maintain a fixed alignment between the transducers and the surface of the item being tested. A slight shift in the position of the tubing axis may result in a large shift in the angle of incidence of the ultrasonic beam if the transducers are not moved in synchronism with the tube.

Most wave modes used in other types of ultrasonic inspection are used for inspecting pipe and tubing. Shear waves are used most often, but longitudinal waves for thick-wall tubing and Lamb waves for thin-wall tubing may be used occasionally.

Tubing as small as 5-mm OD with walls as thin as 0.20 mm has been inspected satisfactorily with sensitivities to defects of 3% of the wall thickness.

Sometimes pipe and tubing are inspected manually using direct contact methods and contoured plastic shoes. Production speeds require automated systems. Coupling in automated systems is done by immersion, gap, sliding boot, or wheel filled with liquid. The positions of the transducers relative to the tubing under inspection are rigidly maintained. For external application, control of the transducer position may be maintained by guides on the transducer holder, which may consist of guide sleeves matching and concentric with the tubing or of wheels, which support the transducer carriage. Either the lateral motion of the tubing must be completely restrained or the transducer carriage should be free-floating to follow the motion of the tubing as it passes through the inspection system.

The ultrasonic beam is introduced into the tubing in such a manner as to minimize the number of wave modes present in the material. The critical angle for longitudinal waves in steel corresponds to a shear-wave angle of refraction, which is slightly less than 36°. Therefore, to eliminate the longitudinal wave from a shear-wave test in steel requires a shear-wave refraction angle greater than 36°. With shear-mode inspection, the angle of refraction determines the wall thickness that can be inspected. For instance, the maximum wall thickness that can be inspected using a 45° shear wave is 20% of the inner diameter.

The ultrasonic beam may be collimated or focused to minimize the spread of the incidence angle. Focusing to a "point" provides the greater concentration of energy and, therefore, greater sensitivity to defects. However, a "line" focus is more practical for production speeds because it covers more of the tubing per rotation of the tube. True point focusing and line focusing are impossible because the wavelengths of the ultrasound are finite. By making the circumferential dimension of the focused beam on the intercepting surface of the test material small compared with the diameter, the incidence angles vary little from those that would occur if the surface were a flat plate.

There are many approaches to inspecting pipe and tubing ultrasonically. These may involve only one transducer or a number of transducers. In one arrangement, one transducer is used to transmit a wave in a circumferential direction and another to transmit a wave in the axial direction. This arrangement increases the possibility of detecting defects regardless of their orientation. Additional protection is provided by using a third probe, which also transmits in a circumferential direction but opposite to that of the first probe, so that waves are propagated in both directions simultaneously around the tube. Thus, certain types of defects, such as lap-type seams, that might be missed in one direction may be detected in the other. Figure 8.9 illustrates several transducer arrangements.

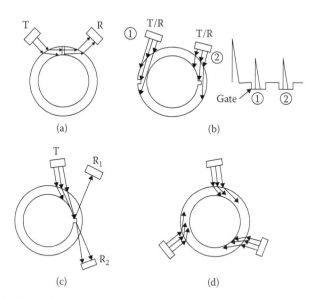

FIGURE 8.9 Examples of transducer arrangements for inspecting pipe and tubing. (a) Transmit–receive methods of inspecting electric–resistance–welded tubing, etc. (b) Two probe method of distinguishing internal and external flows. Probe 1 is adjusted to obtain maximum echo from external defect. Probe 2 is adjusted to obtain maximum echo from internal defects. (c) Three probe echo and transmit–receive methods. (d) Multiple probe method of convering entire circumference of tubing without rotation.

When an angled shear wave is used to inspect pipe or tubing in which the OD and internal diameter (ID) are not concentric, false echoes may be received when the wave propagates in the direction of decreasing thickness. The angle of incidence decreases for each reflection from the surfaces of the tube for a few reflections, after which the wave reverses itself and returns to the point of entry. This type of reflection can be distinguished from a true defect by the fact that, as the tubing rotates, the indication moves across the screen as does a defect but dies out as it approaches the transducer, whereas the true-indication remains. For automatic testing, the gate can be set so that the false echo disappears before approaching the gated area. Another method of reducing or eliminating such false indications is to cause the beam to propagate in a helical pattern around the tube.

As in all ultrasonic inspection applications, suitable standards are essential for adjusting and calibrating the equipment. Standards are produced usually from the same material as the tubing to be inspected. Since the geometry of the test specimen is an important factor in determining the inspection modes, the standard may be made from a piece of tubing taken from the production lot. This tubing is then carefully inspected for defects. After thorough inspection ensures that the tubing to be used for a standard is free of defects, artificial defects are punched, drilled, or electromachined into the tubing to be used for calibration purposes.

The echo method of inspecting tubing is more sensitive to defects than the transmit–receive method.

Ultrasonic inspection of tubing often involves rotating the tubing in order to obtain total inspection. In some cases, it is practically impossible to rotate the tubing. The Krautkramer Company developed a method that involves rotating the ultrasonic transducers about the tubing instead of rotating the tubing [95]. Rotation is at speeds up to 3000 rpm. At least three probes are used, two of which scan circumferentially in opposite directions to detect longitudinal defects while the third scans in an axial direction to detect circumferential defects. Electrical connection between the test probes and the electronic components is by inductive coupling. Pulse repetition rate is 3000/s. Coupling is provided through water pumped under pressure into the rotary test block. Alignment is maintained by guides and rollers. This equipment is reportedly capable of increasing the inspection speed by a factor of 10 over previous methods.

The speed at which a tube can be inspected by pulse methods depends upon the effective dimensions of the ultrasonic beam and the pulse repetition rate. The defect should be interrogated by a sufficient number of individual pulses to provide the maximum signal possible from that defect. Speeds as high as 61 m/min have been claimed for the Krautkramer equipment.

As mentioned previously, controlling the orientation of the ultrasonic beam with the surface is important in proper ultrasonic inspection of tubing. Obviously, tapered tubing presents a problem in controlling the angle of incidence because of the varying diameters. McClung and Cook [96] developed a method of inspecting tapered tubing of 0.23 and 1.27 mm wall thicknesses used in nuclear heat exchangers. They claimed reproducible sensitivity to discontinuities only 0.127 mm in depth and 3.175 mm long. Their method involves scanning the tube in an immersion tank by rotating the tubing and translating the transducers in an axial direction. The ultrasonic beam is transmitted circumferentially. Alignment of the transducers with the tubing surface is maintained constant by rotating the tubes in Teflon bushings machined to fit the maximum diameter of the tubing and located on either side of the transducers. V-blocks mounted alongside each bushing apply mechanical pressure to the tubing and maintain a constant angle of incidence.

Inspection of silver-brazed pipe joints has been accomplished using an automated pulse-echo system [97]. A single probe on a plastic delay line is caused to rotate about the circumference of the brazed section. Longitudinal waves at 5 MHz are transmitted into the joint. If the bond is good, the energy reflects from the ID of the pipe. If the bond is poor, energy is reflected from the ID of the fitting. The acoustic impedances of the fitting, the silver braze, and the pipe are sufficiently similar that little energy is reflected from the bonded interface.

Measurement of thickness is also an important application of ultrasonics in tubing inspection. At one time, boiler tubes were inspected in situ by manual contact methods from the outside. A more recent development makes use of an immersion method to inspect boiler tubes for corrosion from within [98]. The inspection unit is carried by six wheels mounted on retractable legs that permit the assembly to pass through IDs and to expand into a larger ID. The mechanism maintains the probe to produce normal incidence. It can be rotated to inspect the circumference at any desired location. The electrical cables to the transducer are located inside a 14 meters long rubber-coated flexible shaft, which is used to manipulate the probes. A pulse-echo technique is used.

The method is not effective in areas having a substantial amount of scale. Nor is it suited to finding localized pitting since this would require full coverage of an area, which is time-consuming.

Resonance measurements of thickness are possible in tubing if the surfaces are regular and parallel. The transducer, however, should be contoured to match the surface of the tubing.

The dual-probe, pulse method of measuring thickness also is widely used (see Section 8.6).

The ultrasonic micrometer, based on the interferometric cancellation of an ultrasonic wave in a quarter-wave thickness and minimum loss in a half-wave thickness, has been used to measure wall-thickness variations of 0.0001 mm in 0.0057-mm-wall tubing [99].

Chretien and Huard [100] discuss a Lamb-wave technique of inspecting tubes intended for use as fuel-element cladding. The method is based on differences in the damping of the Lamb waves at the air–metal and metal–water interfaces, which allow separate recordings to distinguish between internal and external defects. The instrument employs a variable gain along the time base. It has been used to inspect zirconium–copper cladding with thicknesses in the range of 0.7–1.2 mm. It has also been used for stainless steel tubes 6.3- and 0.35-mm thick.

Typical applications of ultrasonic inspection of tubing and pipe include the inspection of seamless tubes, arc-welded pipe, resistance welded pipe, high-pressure couplings and tubing, Zircaloy-2 cladding tubes, thin-walled, aluminum-clad, uranium–aluminum tubes, aluminum tubes, stainless steel tubes, welded pipes as large as 1.07 m or more, and many others. Automatic systems capable of inspecting welds in pipes at speeds exceeding 100 m/min are possible.

Along with all other areas of NDT, the inspection of pipes and tubing has benefited by the remarkable developments in electronics of recent years. Digital processing of received data has led to wide usage of imaging as a means of determining size, shape, and position of defects and to measuring accurately changes in condition of a component. These benefits are demonstrated by a method developed by the U.K. Atomic Energy Authority Research Laboratory–Harwell in an ultrasonic testing system for inspecting piping and pressure-retaining components [101]. The method relies on signals diffracted from the tips of defects or cracks in material. Two transducers are used, a transmitter and a receiver, located on the outside circumferential surface on a line parallel to the pipe axis. The test area lies between the two probes. Sound waves from the transmitter are diffracted from the tips of any crack that may obstruct these waves. The remainder of the beam is reflected from the outside diameter and the back wall. Time of flight of sound waves subjected to diffraction by a defect is compared with time of flight of the portions of the beam subjected to reflection from the outside surface and the back wall. From these data, the size and location of the defect can be determined accurately by developing an image while scanning the test object. The system eliminates the need for costly surface grinding of weld surfaces, and defect orientation presents no problem because the beam spreads at a wide angle.

A permanent digital record permits comparison of repetitive inspection data to monitor changes in the condition of the component. The images can be further enhanced by off-line computer processing.

Mahmoud et al. [102] describe a use of a time-of-flight method in pulse-echo sizing of planar defects. The method was intended to improve on the often poor correlation between echo amplitude and defect size by methods described in Chapter 7. Estimates of defect depth are independent of echo amplitude.

Testing was performed in an immersion tank. Time of travel includes the passage of the pulse through the liquid couplant as well as all paths inside the test object. Experiments were run at two probe angles. To measure the times of flight to obtain accurate defect measurements, it is necessary to relate the beam path length to the measured probe angles and other geometric measurements. Using their method with repeated scans and using maximum rather than mean predictions of defect depths gave results that are within 2–3 mm of the actual depth. The time-of-flight technique of Mahmoud et al. was not very suitable for defining defect length, especially when the curvatures at the end of the defects were steep. They resorted to an echo-dynamic technique in which the actual crack length was determined by drop in echo amplitude. The actual end points of the defects were found to lie between positions corresponding to drops of 14 and 20 dB in reflected signal amplitude.

Mosfeghi [103] describes a method of using caustics in testing cylindrical objects for radial and surface defects. A caustic is a focal region along which neighboring rays in a beam touch. The method has been used for imaging radial cracks and surface defects in round rods and pipes. The experimental procedure includes locating the test object in an immersion tank, interrogating the part with pulses from a 1.3-cm unfocused wideband 5-MHz transducer, and rotating the sample with the probe fixed in position. In some cases, part of the beam was masked to produce a smaller rectangular beam.

When the axis of the incident beam is directed to one side of the axis of the test object so that all rays of the beam are refracted in the same general direction, the beam will spread within the tube or rod according to the range of incident angles corresponding to rays in the incident beam. The internal reflections of these rays between the two circumferential extremes cause caustics to form within the material. A caustic effectively acts as an acoustic microscope system, which probes the surfaces of the defects as they intersect it. The returned pulse-echo data can be used to determine the size, shape, and location of the defects.

Mosfeghi used shear-wave scanning incidence angles. A simple calculation shows that there is a maximum depth of rods or thickness, T, of pipes, as stated earlier, that can be insonified by shear waves if no longitudinal waves are to be produced. The ratio of thickness, T, to outside diameter, D, is given by

$$\frac{T}{D} = 0.5\left(1 - \frac{c_S}{c_L}\right) \tag{8.10}$$

where

c_S is the velocity of shear waves in the material
c_L is the velocity of longitudinal waves in the material

The dimensions and positions of the probe and the sample were fed to a plotting program, which also calculates the first internal reflection of the rays inside the sample.

According to the author, shear-wave scanning of cylindrical samples produces two distinct acoustic wavefield regimes inside the samples. The beam diverges cylindrically in the first regime. Pulse-echo data in this region can be processed, using tomographic image reconstruction algorithms for cylindrically diverging beams, to produce acoustic reflection images. These algorithms involve backprojection of the pulse-echo data over circular arcs. By using a wideband pulse and a large synthesized aperture, these algorithms effectively perform a synthetic focusing function to produce high-resolution images in all directions.

Another aspect of tube and pipe inspection is related to obstructions inside the components. Such obstructions can lead to serious consequences. One method of locating obstructions in pipes or tubing in which an open end is available is echo ranging through the air in the tube. An acoustic ranger is an instrument that emits audio frequency sound waves, which propagate through the pipe [104]. At any position where the wave encounters a change in acoustic impedance, for example, at the surface of an obstruction, it will be partially or entirely reflected, depending upon the nature of the impedance. Location of the problem area is determined by sonar techniques based on the velocity of sound in air and time of travel.

Ultrasonic methods of locating obstructions in pipe are effective provided the pipe can be filled with a suitable liquid [105]. Burkle's [105] ultrasonic technique is a variation of pulse-echo liquid-level detection. The method is based on transmitting an ultrasonic pulse along a diameter of the pipe filled with water or other suitable liquid. In the absence of an obstruction, the ultrasonic pulse is transmitted through the pipe wall, across the fluid inside, and through the wall on the opposite side with only the echoes corresponding to the various interfaces being observed. Loss of back echoes indicates that the pipe is not filled with liquid at the point of test or that sound transmission is being blocked by obstructing matter.

8.3.2.1   Acoustic Emission Monitoring of Structural Integrity of Underground Pipelines

One of the promising applications of AE techniques for NDT is monitoring the structural integrity of underground pipelines. Detection and location of AEs at the first signs of failure in pipelines will allow remedial actions to be initiated to ensure safe operation.

The first signs of failure are bursts of acoustic energy associated with the formation of stress-induced microcracks. As the tensile stresses in a region exceed the tensile strength of the material, microcracks form suddenly. The amplitude of the stress across a crack suddenly drops, and its load is just as suddenly transferred to the remaining intact material, thus creating an impulse that propagates along the pipeline for many feet. Because of the very manner in which it is generated, the acoustic impulse is broadband.

Researchers at Battelle's Columbus Laboratories [106] have reported a computer model, for the gas industry, to simulate transient acoustic wave propagation through buried, coated, high-pressure pipeline. This model treats the AE signal as a transient elastic wave propagating in a pipeline. Conventional AE monitoring has been shown to oversimplify the nature of the AE signal as a simple pulse traveling at a constant speed, which is not the case. This conventional approach has led to erroneous and often conflicting results. The Battelle model predicts the propagation characteristics of the most energetic mode that can be detected at a distance of several hundred feet from the source. The computer model eliminates irrelevant data and identifies waveforms with their several packets of energy in terms of the predominant mode, group velocity, frequency range, and phase between the source and the transducer. This process makes it possible to locate a defect by AE pulses detected over several hundred feet from the source. The system in which transducers are permanently attached to a pipeline at logical intervals provides a means of continuously and passively monitoring pipelines for defects that could lead to hazardous leaks.

Theoretical models of the AE process have been verified in the field, and equipment, including transducers and digital processing systems, is being developed for field units tailored to monitoring buried pipelines, chemical plant piping, and other structures.

8.3.3   PLATES AND STRIPS

Conventional ultrasonic methods are used to inspect plates and strips for defects and for thickness variations. The geometry of plates provides a fairly constant surface for maintaining a constant angle of incidence, and this is an advantage to the inspector.

Inspection of plates and strips may be divided into methods applicable to thin sheets and those applicable to thick plates. Surface waves and shear waves are used to inspect thick plates for defects such as cracks or inclusions at positions remote from the transducer. Longitudinal waves are useful for detecting defects such as laminations and inclusions in a plane parallel to the major surfaces of the plates. If the conditions of the edges of the plate permit, longitudinal waves also can detect cracks normal to the major surfaces.

Lamb-wave techniques are widely used to inspect thin plates. There are numerous discrete values of velocities of Lamb waves in a plate, which depend upon the plate thickness and test frequency. Lamb waves affect the entire thickness of the plate and travel parallel to its surface. In accordance with Snell's law, a particular Lamb-wave mode will be excited most easily at discrete values of incidence angle corresponding to the velocity of the particular wave mode. The velocity of Lamb waves, in addition to being a function of the product of the frequency times the plate thickness, is a function of the elastic properties and density of the materials of the plate, as are all other types of ultrasonic waves. Figure 8.10 shows the incident angle required to produce given Lamb-wave modes in aluminum sheets in water [107].

Techniques for inspecting plates and sheets vary from manually scanning the materials with direct contact probes to fairly rapid automatic systems. Manual inspection with longitudinal waves is slow and impractical for production use except, perhaps, where the sides of the plates are suitable for coupling.

Thickness measurements of plates are made using longitudinal waves and either resonance or pulse methods. The probes must scan the entire surface of the plate for complete coverage, if this is desired. To make such an inspection less time-consuming and more adaptable to production speeds, complete scanning can be accomplished by employing an array or assembly of transducers, which extend the width of the area covered by a given movement of the apparatus. The transducers are "sampled" by switching from one to the other and correlating the output signal

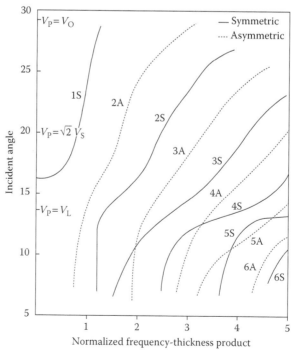

FIGURE 8.10 Incident angle required to produce various Lamb-wave modes in aluminum in water, normalized frequency-thickness product (in 2.54×10^5 cm Hz).

with the position of the probe sampled. Each of the probes must have its own associated electronic system for generating and detecting the signal so that they may be calibrated to similar references.

Harris [108] describes one multiple-unit device for measuring the thickness of sheets of metal moving at 1.09 m/s. The system uses the resonance principle to cover the thickness range 0.025–0.25 cm. Five identical transducers are water-coupled to the sheet of metal. Each transducer operates through its own electronic system. The system provides both visual displays and permanent recordings.

An automatic system described by Silber and Ganglibauer [109] is used to test slabs in the thickness range of 80–250 mm at a speed of 6 m/min. The transducers operate through a 13% chromium steel wear plate.

A manually operated instrument for inspecting plates ultrasonically using multiple transducers was developed by Homer Research Institute, Bethlehem Steel Corporation [110]. The inspection device is mounted on a cart that is equipped with push-type handles so that a person can manipulate the unit in an erect position. The transducers operate through a water column and generate longitudinal waves in the thickness direction in the plate. The transducers are located in direct contact with one another to provide continuous inspection of the full width of the probe assembly. Each transducer section is separately calibrated so that all sections are identically sensitive to defects.

A wheel type of inspection unit has been used to continuously inspect 5.7-cm-wide by 1.27-mm-thick 1050 steel strip for laminar defects [111]. The liquid-filled wheel contains two transducers transmitting Lamb waves in a cross pattern, which provides 100% coverage of the entire width of the strip. The direction of propagation is opposite to the direction in which the strip moves through the test area. The strip is uncoiled and passed through a straightener to eliminate a crown, or curvature, perpendicular to the length of the strip before it passes under the wheel. The wheel is oil-coupled to the strip. The testing speed is limited to 12.2 m/min by the straightener.

A method that has been used to eliminate the need to readjust the incident angle when continuously inspecting a strip when the thickness of the strip changes involves the use of a wide angle of spreading of the incident wave as it propagates from the source transducer [112]. With this type of probe, the proper angle of incidence for a specific Lamb-wave mode is always available.

Lamb waves, shear waves, and surface waves are used for automatic inspection of wide strip. In most cases, immersion testing is not practical, and coupling of the energy to the test strip is best accomplished by water jets or bubblers, which produce a water film between the transducer and the strip. Grasshoff and Tobolski [113] have used Lamb waves in a pulse-echo manner to inspect strip and sheet up to 8 mm thick and 600 mm wide. Their apparatus is used in the pickling line between the scale breaker and the pickling bath and has a good sensitivity to shrinkage cavities and other defects.

Figure 8.11 illustrates a side-looking ultrasonic wheel for generating 45° shear waves in a steel plate to be inspected. The wheel contains a 12 by 24.5 mm 2.25-MHz transducer. Table 8.1 shows the effectiveness of the transducer in detecting defects in 2.1-mm-thick steel strip at various speeds. The strip was 30 cm wide and contained artificially milled defects 0.127, 0.254, 0.635 mm deep, each 2.54 cm long. The wheel was located at one edge of the strip and water-coupled to the strip.

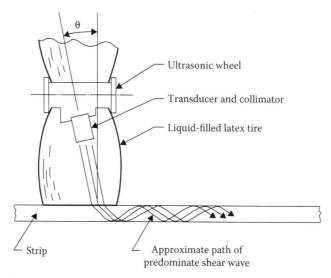

FIGURE 8.11   Ultrasonic wheel assembly used for inspecting plates and pipes.

TABLE 8.1
Amplitude of Indications from 2.54-cm-long Defects in 1055 Steel Strip Using Ultrasonic Wheel and 2.25-MHz Shear Wave[a]

Plate Speed (m/min)	0.127 mm defect	0.254 mm defect	0.635 mm defect
2.15	12	22	Off scale
8.1	5	17	Off scale
35.4	4	14	Off scale
59.6	3	7	Off scale
97.5	–	5	35

[a] Defects were on the same side of the strip as the transducer.

The pulse repetition rate was 500/s. The gate was adjusted to correspond to the width of the strip from slightly past the edge of the wheel to slightly short of the opposite edge (to prevent false alarms from back echoes).

Figure 8.12 illustrates another inspection method in which a collimator with a 3.175 by 12.7 mm aperture terminating in a liquid-filled latex boot was used. The assembly into which the transducer and collimator were mounted could be angularly adjusted to optimize the incidence conditions. Strip chart recordings obtained by using the transducer and collimator shown in Figure 8.13 to inspect the same material used to obtain the data of Table 8.1 are given in Figures 8.13 and 8.14. In Figure 8.13, the defects are on the same side of the plate as the transducer, and in Figure 8.14, they are on the opposite side of the plate.

The same factors affect the attenuation of Lamb waves used for inspecting plate and strip that affect the attenuation of other modes. Curvature in sheet materials has

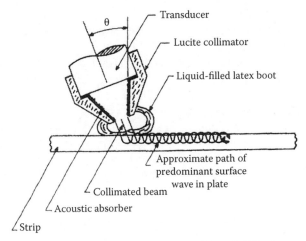

FIGURE 8.12   Ultrasonic rectangular collimator assembly and sliding boot used for inspecting plates and pipes.

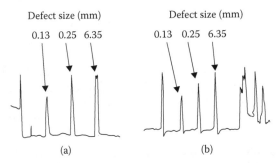

FIGURE 8.13   Indications of defects obtained with collimated ultrasonic probe at a speed of (a) 0.12 m/s and (b) 1.57 m/s (defects and transducer on same side of plate, incidence angle 18°, plate 3A (1055 steel), plate thickness 0.21 cm, and frequency 5 MHz).

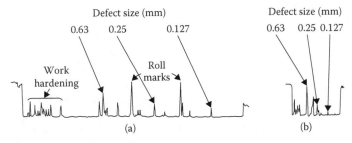

FIGURE 8.14   Indications of defects obtained with collimated ultrasonic probe at speed of (a) 0.02 m/s and (b) 1.53 m/s (defects and transducer on opposite sides of plate, incidence angle 20°, plate 3A (1055 steel), plate thickness 0.21 cm, and frequency 5 MHz).

no appreciable effect. The frequency spectra of Lamb waves influence the attenuation by factors related to the dependence of the velocity of such waves on the product of frequency and thickness.

Isono et al. [114] claimed to have discovered an extraordinary attenuation of Lamb waves in mild-steel sheets up to 500°C, which they attributed to damping associated with either carbon or nitrogen, or both, dissolved in the steel. They claimed that measuring the group velocity of Lamb waves is a suitable means of obtaining a measure of the internal friction associated with the concentration of solute carbon and nitrogen atoms.

Progress toward improved plate defect evaluation is being made in studies involving (1) time-of-flight techniques, (2) spectral analyses, and (3) amplitude methods using guided waves.

8.3.4 FORGINGS

Materials suitable for forgings are usually suitable for ultrasonic inspection both before and after forming. For this reason, ultrasonic NDT has been used successfully to inspect forgings such as large turbine-generator rotors, aluminum-alloy die forgings, and many others.

Typical defects found in forgings and for which ultrasonic inspection is recommended are bursts, cracks, hydrogen flakes, inclusions, laminations, laps and seams, and unfused porosity [115].

Forging bursts are ruptures, usually running parallel to the grain, which can occur either at the surface or internally. They are attributed to excessive working or movement of the metal or to forging at too low a temperature. Bursts vary in size and are often large. Ultrasonic pulse-echo methods are perhaps the best means of detecting bursts.

Hydrogen flakes are very thin internal fissures with bright silvery surfaces usually found in heavy steel forgings, billets, and bars. They usually run parallel to the grain. They are caused by high localized stresses caused by hydrogen coming out of solution as hot-worked material cools. Ultrasonic immersion or contact testing methods are used to detect hydrogen flakes. Flakes usually present a number of scattering (or reflecting) sites that cause a loss of back echo and considerable "hash" on the oscilloscope screen with the A-scan presentation.

Inclusions may consist of many types, but they may be categorized in two groups: (1) nonmetallic, tightly adhering particles, which are elongated during processing parallel to the flow lines and (2) non-plastic materials, which appear as randomly oriented masses. Inclusions appear as definite interfaces within the metal; however, the amplitude of reflections from the interfaces depends upon the coupling at the interface and the material of the inclusions. The maximum signal corresponding to echoes from the interfaces occurs when the beam is incident at 90° to the reflecting surface. For this reason, the preferred direction of the beam during ultrasonic testing usually is normal to the direction of grain flow. Pulse-echo techniques generally are used for detecting inclusions coupling by either contact or immersion methods. Indications may range from loss of back echo to numerous echoes corresponding to scattered inclusions.

Laminations are impurities in metals that have become very thin during processing. They form layers parallel to the work surface and alternate between layers of the metal. The choice of ultrasonic inspection method and wave mode depends upon the dimensions of the material to be inspected.

Laps are characterized by shallow wavy lines on the surface of a forging, which usually result from improper forging operation. Corners may be folded over, or excess material may be smeared and pressed into the forging without fusing.

Seams are similar to laps but are caused by defects introduced into the material during a prior operation. They are elongated fissures parallel to the direction of forging and are often deep and sometimes very tight.

Ultrasonic testing generally is used to evaluate wrought material prior to forming. The surface-wave technique is used to locate and determine the dimensions of laps and seams. Laps and seams produce clear, definite indications of the interfaces on an A-scan presentation.

Unfused porosity consists of very thin fissures aligned with the grain flow existing in wrought aluminum forgings. It is the result of porosity existing in the cast ingot, which is flattened without fusion during the forging operation. Pulse-echo methods are capable of locating and establishing the extent of these defects if the geometry of the part permits testing in more than one direction.

There is a growing interest in the use of phased arrays for the inspection of materials. A number of computer-based inspection systems have now been developed for specific classes of application. Portable commercial phased array control and display units, such as that shown in Figure 8.15, are seeing increasing use. This type of unit is used with transducers such as that shown in Figure 8.16, which is a 1.0-MHz transmit–receive unit originally designed for evaluating inspection effectiveness of phased array methods on components with inlays, onlays, and overlays. It consists of two 5-element by 10-element matrix arrays. One array is used for transmitting, the other for receiving ultrasonic signals. An example of an array in use is shown in Figure 8.17, which shows a mechanical scanner mounted on a ring section in the background with the scanner arm and probe extending to the foreground (left) and a side view of the 1.5-MHz two

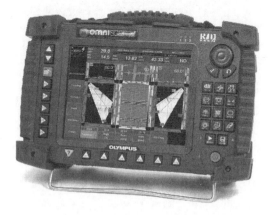

FIGURE 8.15   **(See color insert.)** Example of a commercial phased array control and display unit.

FIGURE 8.16   Custom phased array transducer, set on a wedge for use in pipe and weld inspection. (Courtesy of Pacific Northwest National Laboratory.)

FIGURE 8.17   **(See color insert.)** Phased array with mechanical scanning being set onto a composite pipe, with image display on computer monitor. (Courtesy of Pacific Northwest National Laboratory.)

array unit with separate transmit (T) and receive (R) arrays operating at low-frequency (L), know as a TRL probe, probe next to a weld (right). Another example of such a unit is a phased array, ultrasonic inspection system, the LOGIQ 9NDT, from GE Inspection Technologies [116]. This type of system significantly reduces inspection times. They incorporate medical system technology and claim to be the most versatile and productive engine forgings inspection system currently available. It can drive up to 1024 element probes and can accommodate four phased array probes, with one active at any particular time. The launch application (2006) used a 10-MHz linear probe having 768 pairs of elements, which can be scanned and focused electronically to provide very high resolution and sensitivity. The unit can inspect to greater depths than other units, which typically only use 256 pairs of elements. It has been demonstrated to work well to depths of up to 14 cm, in metallic alloys including Titanium, Nickel, Aluminum, and Steel. Data can be presented as A-, B-, and C-scan displays [116].

8.3.5   BEARINGS AND BEARING MATERIALS

Conventional methods of ultrasonic inspection have been used to detect defects in bearings and bearing races. Ball bearings as small as 4.7 mm in diameter have been inspected successfully for cracks. Brown and Dunegan [117] used a focused, 15-MHz lithium sulfate transducer in conjunction with a collimator with a terminating inside diameter of 4 mm to inspect ball bearings. The end of the collimator was dished out to accept the bearings and to hold them in the proper position for testing. Water was pumped through the collimator to provide coupling. The focal region was adjusted to the bearing surface to minimize mode conversion. The appearance of a crack in the bearing produced a reflection, and when the defect was properly oriented with respect to the incident beam, the defect indication on the oscilloscope screen appeared between those corresponding to the front reflection and the back reflection. If the crack is large enough, it will completely eliminate the true back reflection, although reflections from the surface returning to the transducer by reflection from the defect may give the appearance of a back echo. The existence of a crack, however, is easily ascertained by observing the effects on amplitude and position of indications corresponding to internal reflections when the orientation of the ball with respect to the incident beam is varied.

Accurate measurement of the velocity of sound provides a means of detecting polymorphous transformations of solids by the corresponding changes in elastic constants accompanying these changes. Equipment that provides a capability for measuring velocity of sound in metals is available commercially. Fenkner [19] used such equipment to determine a correlation between the velocity of sound and the amount of residual austenite in roller-bearing steel. Fenkner's data indicated a linear drop in velocity with increased retained austenite in chromium-steel bearing alloy. Between 4% and 28% retained austenite, the velocity dropped from approximately 5905 m/s to approximately 5840 m/s.

Through-transmission and pulse-echo methods are effective in locating unbonded regions between the liner and the base metal of bearings, as well as inclusions and other defects within these materials [118]. Diesel-engine bearings and other large bearings are especially adaptable to ultrasonic inspection.

Lamb waves have been used to detect defects that may lead to fatigue failure in aircraft bearing races, employing a spherically focused transmitter and receiver. The test reveals lack of bond between lining and base material, cracks, and inclusions.

8.3.6 CASTINGS

The metallurgical and physical characteristics of many types of castings preclude their inspection by conventional ultrasonic means, precluding primarily rough surfaces, large grain size, and unfavorable geometries. In the present context, conventional means refers to the typical pulse-echo and through-transmission techniques involving either immersion or direct contact coupling to the part and generally testing at frequencies exceeding 1 MHz. Not all castings are excluded from ultrasonic testing, as attested by a number of such applications presently being used. Certain types of castings can be inspected by the conventional ultrasonic means, and many of those that cannot be inspected by high-frequency ultrasonic methods can be inspected at lower frequencies for certain factors that affect quality.

In general, ordinary cast-iron parts are not inspected ultrasonically, except for those with fairly simple geometries, which are inspected by resonance methods. Low-frequency resonance methods can be used to detect cracks, voids, porosity, mottle, shrink, and inclusions in both cast-iron and cast-steel parts if the extent and location of the defects are of such a nature that they affect the resonance characteristics. For example, voids at high-stress locations within a resonant structure lower the resonance frequency of the part, whereas similar voids located at a stress node may have little effect on the resonance.

Various resonances are associated with complex parts. If these resonances can be identified for a standard good part, it is possible to evaluate similar parts for defects by shifts in the points of resonance if the parts can be excited through a range of frequencies.

Inspection of cast-iron using ultrasound was first performed in the 1950s, and relationships have been established between ultrasound properties and metallurgical characteristics. Much of the modern work is based on a study by Ziegler and Gestnet, 1957, [119], and a useful review of early Russian and European work is provided by Voronkova and Bauman [120].

An example of inspection of cast-iron that can be accomplished ultrasonically is that of cast-iron rolls. Sasaki and Ono [121] employed immersion, pulse-echo ultrasonics to estimate the thickness of the hardened layer. They found that the most suitable frequency for this measurement was always less than 3 MHz. To detect internal defects, Sasaki and Ono used frequencies of 0.4 and 0.75 MHz.

Typical types of defects that can be located in castings and billets, but only if the conditions are favorable, are shrinkage defects, hot tears, cracks, gas pockets and porosity, pipe, and coarse nonmetallic inclusions such as slag and sand. Typical castings that have been inspected successfully ultrasonically include cast-iron brake drums, cast-iron rolls (as mentioned previously), cast-iron crankshafts containing globoidal graphite, cast-aluminum and aluminum alloys, cast piston rings, and ductile-iron parts.

The quality of cast-iron products is often related to the velocity of sound or to the relative attenuation through the material. These parameters are affected by the size, shape, and distribution of graphite and alloy constituents, by the microstructure, and by the presence of defects.

The velocity of sound is also used to measure the degree of nodularity in cast-iron, distinguish between nodular iron and gray iron, and detect the presence of gray iron inclusions in nodular iron castings. Equipment is commercially available for measuring velocity of sound in materials. Typical velocities are, in pure iron, approximately 5.9×10^5 cm/s; in nodular iron, approximately 5.6×10^5 cm/s; and in gray iron, 4.8×10^5 cm/s.

Concentrations of graphite will produce ultrasonic echoes similar to those from poorly defined voids. Regions of graphite precipitation will show a distinct rise in reflection amplitude over reflections from grain boundaries obtained from good castings. Severe cases will produce distinct defect indications. Equipment selection and settings are determined by the details of the application.

Microshrinkage or shrink porosity can be located ultrasonically. Concentration of many voids too small to be detected individually will cause flaw-type echoes and severe attenuation of ultrasonic energy.

Figures 8.18 through 8.26 illustrate various steps that have been used successfully by the first author to inspect ductile-iron industrial wheels. These figures illustrate typical defects, methods of locating and identifying these defects, and the types of

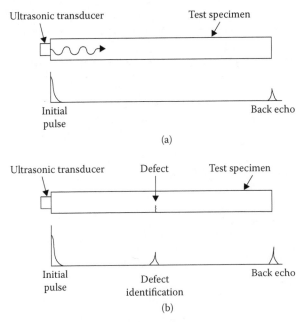

FIGURE 8.18   Pulse-echo method of detecting defects in the rim of industrial ductile-iron wheels. (a) Good region. (b) Crack indication.

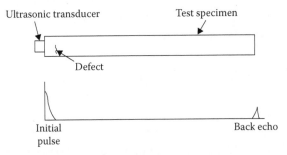

FIGURE 8.19   Pulse-echo technique with defect near the transducer (reflection from defect obscured by initial pulse and back echo diminished as a result of a shadow from the defect).

indications observed on an A-scan presentation. Both contact and immersion pulse-echo inspections were used; however, the latter method is preferred because of the roughness and irregularity of the surface. Both longitudinal- and shear-wave methods were necessary to obtain the required coverage using both 2.25- and 5.0-MHz nominal test frequencies.

The inspection and characterization of cast-iron remains of industrial importance. Computer-based instruments are now available for rapid characterization of samples. Research is ongoing to investigate the potential for the use of ultrasound for the characterization of damages, including fatigue damage, in various types of nodular cast-iron [122].

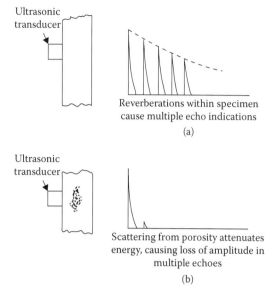

Reverberations within specimen
cause multiple echo indications

(a)

Scattering from porosity attenuates
energy, causing loss of amplitude in
multiple echoes

(b)

FIGURE 8.20 Pulse-echo techniques applied in radial direction to detect porosity in the rim of industrial ductile-iron wheels. a) Multiple echoes within specimen indicating reverberation (limited attenuation). b) Scattering from porosity adds attenuation and causes loss of multiple echoes.

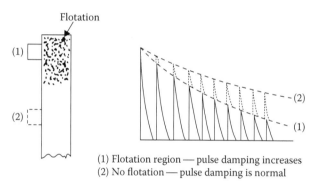

(1) Flotation region — pulse damping increases
(2) No flotation — pulse damping is normal

FIGURE 8.21 Pulse-echo techniques of detecting flotation in the rim of large industrial ductile-iron wheels.

8.3.7 RAILS

The railroads provided one of the earliest commercial uses of ultrasonics in NDT. Both resonance [123] and pulse-echo [124] methods were used to inspect rails.

The resonance type of rail-flaw detector consisted of a piezoelectric quartz-crystal type of transducer mounted in the end of a cane. The operator carried the oscillator and a battery pack on his back and inspected the rails in a walking position. The transducer signal was coupled to the rail through a film of oil. The electrical signal

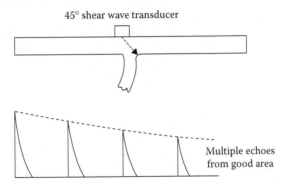

FIGURE 8.22 Pulse-echo techniques of locating defects in the web-rim region of large industrial ductile-iron wheels. Indication typical of no defect.

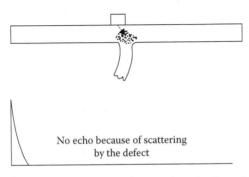

FIGURE 8.23 Pulse-echo techniques of locating defects in the web-rim region of large industrial ductile-iron wheels. Shrink porosity causes scattering of energy and loss of back echo.

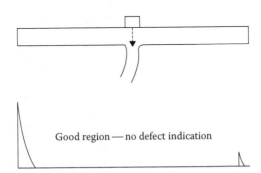

FIGURE 8.24 Pulse-echo techniques of locating defects in the web-rim region of large industrial ductile-iron wheels. Longitudinal pulse encounters no defect no energy is reflected back to the transducer, which indicates a good region.

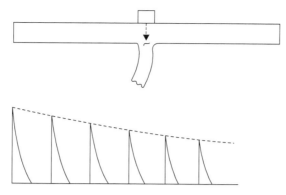

FIGURE 8.25   Pulse-echo techniques of locating defects in the web-rim region of large industrial ductile-iron wheels. Compressional pulse encounters defect; energy is reflected back to the transducer from the defect. Discontinuity below the rim will give multiple echoes if discrete properly oriented. Attenuation of multiple echoes depends upon type of defect. Spacing between echoes depends on distance to defect.

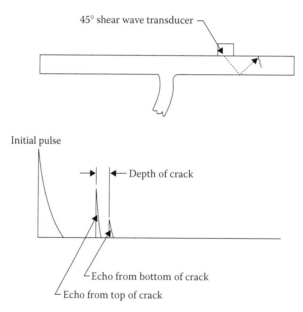

FIGURE 8.26   Angle-beam shear-wave technique of locating cracks in the rim of large industrial ductile-iron wheels.

to the transducer was frequency modulated. The modulation width included several harmonics of the resonant frequency of the rail.

The operator would hear a tone through a set of earphones at a frequency equal to the product of the number of resonance peaks contained in one sweep times the number of sweeps per second. The resonance frequency of a shallower depth corresponding to the distance to a bolt hole or to a crack that lay parallel to the surface would be higher than that of sound rail. Such a discontinuity would allow fewer harmonic peaks within a sweep, and the audible tone generated would drop accordingly. If the crack was not parallel to the running surface, the wave would be reflected away from the crystal and the audible tone would disappear, except for noise associated with the electronics of the system.

The pulse-echo equipment was mounted on a small test car. The operator worked from a low hanging seat at the rear of the car. The signal was coupled into the rail through an oil film. Most of the inspection was done, using a 45° shear wave to detect cracks of various orientations to the surface. Occasionally, a longitudinal wave normal to the surface was used to aid in interpreting certain types of defects. Typical A-scan indications were obtained on an oscilloscope.

Cowan [125] devised a method of inspecting railroad rails automatically at high speed using the pulse-echo technique. Differentially oriented beams of ultrasound simultaneously probe different regions of the rail. The characteristics of the targets are determined by counting the returning echoes. Logic systems are incorporated so that response characteristics are adjusted to correlate with the types of target being inspected, leading to an automatic decision as to the type of target and whether it is a defect.

Rail inspection cars have been used for years to monitor the condition of rails. These cars are equipped with conventional ultrasonic pulse-echo equipment often in cooperation with magnetic inspection. Transducers are housed in either liquid-filled rubber wheels or sliding runners. Both shear and longitudinal waves are transmitted into the rail to provide full coverage of the critical areas of the rail. A water spray ahead of the transducer assembly produces a coupling film between the wheel, or runner, and the rail.

Both A-scan and B-scan presentations indicate the condition of the rail to an observer. When indications of defects of a serious nature appear on the screen, the vehicle is stopped and the defect is marked for immediate repair. Test speeds range from 11 to 21 km/h, depending on the type of display and the ability of the observer to identify indications.

Studies have been conducted, which indicate that a maximum speed of inspection for critical flaws can approach 80 km/h. Higher pulse repetition rates and computer data acquisition and processing methods are necessary for meaningful testing at these higher speeds.

Occasionally, rails have broken from head defects apparently missed by a flaw-detector car. These failures have led to some interesting studies of the effects of cold-working in the railhead on the velocity of shear waves. A change due to cold-working in the velocity of vertically polarized shear waves (SV waves), which are the type used in the angle beams of rail inspection cars, could have a deleterious effect on the detection of flaws from the railhead surface. Using a variable-angle transducer and comparing time of flight at 100-mm incremental spacings of the probes, Najm and Bray [126] measured the velocity of SV, longitudinal, and Rayleigh waves in the heads of new and used rails. They discovered that the effect of cold-working on the velocity of SV waves is most significant, with effects on Rayleigh- and longitudinal-wave velocities being less significant. Their data analysis yielded a constant multiplier that is required to correctly describe the refraction pattern in the used rail with Snell's law as it applies to isotropic materials, that is,

$$c_S = KA\left(\frac{1}{\sin\theta_S}\right) \tag{8.11}$$

where

A is the longitudinal velocity of sound in the medium of the incident wave
c_S is the velocity of the SV wave in the railhead
θ_S is the refracted angle

For a typical used rail, the actual distance to a reflector is about 14% closer to the detector than that predicted by Snell's law for isotropic material.

Stresses in continuously welded railroad rails vary considerably between extremes of temperature. Measurement of the absolute stress in rails is important in preventing buckling under these changes in condition. Bray and Leon-Salamanca [127] have derived a theoretical formula for measuring bulk stresses in rail, which also applies to other steel and aluminum shapes such as beams and plates. The applicable equation is

$$\sigma = \frac{E}{2.45t_0}(t - t_0 - \Delta t_T) \tag{8.12}$$

where

E is Young's modulus of elasticity
t is measured travel time
t_0 is travel time in the absence of stress
Δt_T is travel-time effect of temperature difference at time of measurement from a standard temperature

The technique measures bulk stresses rather than localized or surface stresses by measuring the velocity of bulk longitudinal waves in the member. It is based on an averaging technique that is aimed at establishing zero-force travel time.

Bulk waves are produced by critical refraction, that is, impinging an incident wave at the critical angle, or angle refracted at 90° and parallel to the incident surface. These waves are little affected by surface irregularities such as minor corrosion or letter stampings. They also are readily accessible from the surface for ease in detection for velocity measurements. The experimental apparatus includes a transmitter on a 28° wedge and two receivers on identical 28° wedges spaced 216 mm apart. Bray and Leon-Salamanca concluded that the particular nature of head waves makes them uniquely suited for measurement of longitudinal loads in steel and aluminum structural shapes such as rails, beams, and plates.

A review of the state of the art in rail flaw detection, together with an assessment of needs for future development, was provided by Clark [128]. There remains a requirement to improve sensitivity and enable inspections to be performed at higher inspection train speeds. Development work continues to seek to demonstrate and deploy both EMAT and laser-based ultrasound on rails (e.g., [129]).

8.3.8   WIRES

Ultrasonic inspection of wires has received some attention for several years, but to date use in industry is limited, if not nonexistent. Techniques of generating wire waves include (1) using the magnetostriction characteristics of (steel and nickel) wire, (2) setting up flexural vibrations by drawing the wire across a metallic shoe, (3) drawing the wire through a mating groove in the shoe of an angle probe used conventionally to generate pulses of shear or surface waves and coupling through a thin film of oil, and (4) passing the wire through an immersion bath for pulse-echo or transmit–receive testing.

The magnetostriction technique has the advantage that energy is coupled to the wire electromagnetically so that the driving coil does not touch the wire. The decrease in depth of flux penetration with increasing frequencies limits the method to inspection at low frequencies. Wires 2 mm in diameter and larger can be inspected by this means at approximately 4 m/s. Defects can be detected at depths of 0.05 mm or more.

Wires may sustain torsional (shear) waves, longitudinal waves, or flexural waves. Like Lamb waves, wire waves may be produced in many modes, and the velocities of propagation of these modes depend upon the mode and frequency. Therefore, ultrasonic pulses in wires become somewhat broadened as a result of velocity dispersion of the various frequencies contained in the pulse.

8.3.9   RIVETS

The geometry of rivets is amenable to ultrasonic testing. Wiseman and Garcia [130] were successful in detecting corrosion at the roots of the flush and formed heads of aluminum rivets in a riveted aluminum diesel cruiser. The rivets were flush-head 2S aluminum, 4.76 mm in diameter and 12.7-mm long. By using a 2.25-MHz, 6.35-mm-diameter longitudinal-wave transducer, a pulse repetition rate of 13.5 kHz, and a Polaroid camera, these investigators were able to detect hairline cracks as well as larger defects extending over 20% of the total cross-sectional area of the rivet.

Inspection of smaller rivets and greater defect sensitivity should be possible, especially in materials with as good ultrasonic transmission characteristics as those of aluminum. Higher frequencies would increase resolution and defect sensitivity.

The present-day inspection procedure for rivets is similar to that used for inspecting bars and bolts, which is done routinely using commercially available equipment.

8.4 INSPECTION OF HOT METALS

8.4.1 HOT STEEL

Ultrasonic inspection of materials, in general, has been limited to temperatures only slightly above room temperature by the low Curie points of commonly used transducer materials. Inspection at higher temperatures requires special considerations. Beaujard et al. [131] used water for the dual purpose of cooling transducers employed for inspecting hot steel and of coupling the energy into the steel. A stream of water is forced between the transducer and the hot surface, and the area beneath the transducer is cooled locally below the boiling point. Coupling into hot metals through a water jet continues to be practiced today. The water serves the additional purpose of keeping the transducer cool. Either longitudinal or shear waves may be coupled momentarily into the part by this method. Steel slabs and billets have been inspected by the water jet method at temperatures as high as 1100°C. The method has been recommended primarily for use with soft or extra soft steels at high temperatures.

Another method of coupling into hot metals includes a water-cooled buffer rod between the transducer and the test piece, which in brief contact time (0.1–1.0 second) has been used to test bulk solids up to 2500°C. Fairly high coupling pressure is required. The buffer rod technique has significant limitations. High losses at the interfaces between the buffer rod and the couplant and between the couplant and the test piece weaken the signal considerably. Also, mode conversion in the rod causes secondary echoes. Mahmoud [132] has used low-melting alloys, such as Cerrobend, as ultrasonic couplants for high-temperature testing with good success. The sound velocity does not change significantly with temperature, and the improved acoustic impedance match leads to better signal transfer across the couplant into and out of the test material. EMATs have been used successfully in testing hot metals.

A typical application of ultrasonic testing of hot steel is in determining the extent of pipe in billets prior to cropping. Atkins and Druce [133] proposed that transducers be located inside rolls, which could be pressed against the surfaces of billets. Oil under pressure was used as a couplant, lubricant, and coolant.

In addition to detecting pipe, Lynnworth and Carnevale [134] used both longitudinal- and shear-wave ultrasonics to measure thickness and temperature during the continuous casting of steel. Average temperature along a chosen path was determined by measuring the longitudinal- and shear-wave velocities. The temperature dependence of these velocities was predetermined in similar materials.

The velocity of sound has been measured at elevated temperatures to determine the physical properties of hot metals, including shear and Young's moduli and Poisson's ratio [135].

The high Curie point of lithium niobate (1210°C) makes it potentially useful for high-temperature studies. Care must be taken to prevent thermal shock to the piezoelectric element.

There remains interest in providing sensors to provide real-time data on parameters such as grain size during hot rolling of steel. As discussed in earlier parts of this chapter, ultrasonic techniques have been demonstrated that can measure grain size. A study at NIST reported on a novel sensor that used a transducer coupled to

FIGURE 8.27 High temperature grain-size measurement system, showing buffer-rod transducer in open furnace. (With kind permission from Springer Science+Business Media: *Proceedings 6th International Symposium on NDE of Materials*, Sensor system for intelligent processing of hot-rolled steel, 1993, Clark, A. V., M. Lozev, B. J. Filla, and L. J. Bond, Plenum Press, New York.)

the part using a loaded buffer rod to perform measurements at temperatures up to 1000°C. The transducer from this study is shown in a furnace during tests at elevated temperature in Figure 8.27 [136].

8.4.2 FOLLOWING THE LIQUID–SOLID INTERFACE DURING COOLING OF INGOTS

The velocity of sound and, therefore, the acoustic impedance of a material differ considerably between the solid and the liquid states. Ultrasonic energy is reflected at a liquid–solid boundary. Ultrasonic pulse-echo methods have been used to monitor the location of solid–liquid interfaces during the freezing of various materials. The accuracy of the measurement is affected by the nature of the freezing interface. Dendritic formations or other roughness characteristics cause excessive scattering and a corresponding loss of signal. A distinct interfacial surface may not be formed so that accurate measurements are impossible.

Bailey and Dule [137] used miniature probes with pulse-echo methods to follow the motion of the solid–liquid interface during freezing of water, where the heat transfer is unidirectional. Kurz and Lux [138] used ultrasonic-reflection methods to study the freezing of Wood's metal (melting point 80°C), a Pb–Sn alloy containing 20% tin, and an alloyed steel solidifying between 1530°C and 1430°C. An advantage of the ultrasonic method of monitoring the interface is that no foreign substances are introduced into the system. The interfaces between the liquid and the solid phases of these materials are rough and cause severe scattering of reflected energy.

Kurz and Lux [139] reported that casting conditions in the continuous casting of steel are more easily controlled by continuous measurement of the thickness of the solidified portion. In actual tests, they found that reflections are ideal only during the initial process of solidification. As the thickness of the solidified portion increases, dendritic formation inhibits return of reflected energy and the measurement becomes less reliable. However, they claimed that the method is suitable for locating the solid–liquid boundary in steel during solidification and remelting and for determining the depth of the first "bubble-ring" during casting of rimmed steel.

Work has continued to develop ultrasonic methods to use in the monitoring of the solidification of metals (as well as other materials). Time-of-flight methods have been used to provide reconstructions of solid–liquid interfaces [140]. Laser–ultrasound systems have been investigated to achieve coupling, so as to enable measurement to be performed at high temperatures [141]. In support of such work, a range of more sensitive laser-based detection tools have been also developed [142]: laser ultrasonics is discussed further in Chapter 5.

8.5 DETERMINATION OF BOND INTEGRITY

Bonded materials may be inspected for either cohesive strength or adhesion by ultrasonic means, with certain limitations. There is no certain way of determining the actual strength of adhesion. However, if the bonding material is sufficiently matched acoustically to the base material, as in silver brazing of steel to steel, nonbonded areas can be detected by the size of the echo received from the interface, by the amplitude of an indication corresponding to the depth of the bond line from the surface obtained by a resonance-type thickness gauge, and by the reduced signal received in a through-transmission test.

Lamb waves have been used to detect nonbonds between laminations when at least one is an outer layer. The incidence angle of the transmitted wave is set to generate a Lamb wave of a specific mode corresponding to the material and thickness of the outer layer. The Lamb wave will be generated only when the interface is not damped by a bond to the adjacent surface.

The sensitivity of techniques based on through-transmission, pulse-echo, or resonance methods to nonbond defects depends upon the relationship between the wavelength of the ultrasound used and the area of the defect. The sensitivity of through-transmission and resonance methods is also a function of the ratio of beam diameter to defect size. Sensitivity decreases with decreasing frequency or increasing wavelength.

Ultrasonic methods have been used to examine multiple thin-layered diffusion-bonded composites for nonbonds and interface locations and for inspection of brazed and diffusion-bonded honeycombs.

Interference phenomena occurring in thin plates may affect the amplitude of the received signal during inspection for delaminations in thin sections by the through-transmission method. The effects can be minimized by tilting the specimen a small amount with respect to the axis of the incident beam or by avoiding frequencies at which destructive interference may occur. When an acoustic plane wave of finite cross section traveling in a medium encounters an extremely thin flexible plate in which the wavelength is very large compared with the thickness, the plate flexes with the medium. Very little energy is lost in the flexural wave thus produced in the plate, and the pressures on the two faces of the plate are approximately equal in magnitude. As the plate thickness increases, multiple reflections from the interfaces interfere with the primary transmitted wave. The extent of interference depends upon the ratio of thickness of the plate to wavelength, the coupling to the surrounding medium, the pulse length and shape (when pulses are used), and the internal-absorption losses. When the ratio of thickness to wavelength is an odd multiple of quarter-wavelengths, destructive interference occurs within the plate, which may completely prevent transmission through the plate. Resonance occurs when the plate thickness is a multiple of half-wavelengths and the energy is transferred through the plate with minimum loss. Material thickness is much less important when the specimen is several wavelengths long.

If a nonbond region in a thin section is large compared with the cross section of the incident beam in through-transmission inspection, a barrier to transmission results and no signal is received by the receiving transducer. When the collimated beam is larger than the defect, the total energy transferred past the defect is

$$E_t = a_2^{-2}(a_2^2 - a_1^2)E_i \qquad (8.13)$$

where

E_i is incident energy; E_t is the total energy that would be transferred where the
  defect is not present
a_2 is the radius of a collimated beam concentric with a delamination of radius a_1

Cohesive strength of adhesive bonds is related to the acoustic impedance of the adhesive. The response of a joint to ultrasonic excitation is determined primarily by the properties of the adhesive layer beneath the transducer. Viscoelastic properties, porosity, and mass are the primary factors that determine the cohesive strength of an adhesive. These same factors affect the velocity of sound and density, that is, acoustic impedance. Inspection of adhesive bonds is based on this effect on acoustic impedance. The load that the adhesive imposes on the top face, or lamination, influences the resonance frequencies of the layer. Resonance-type instruments, which have been calibrated against a free plate of the same material that is used for the top lamination, have been employed for the inspection. Nonbonds, voids in the glue line, and degree of cure are determined by the shift in resonance indication.

Smith and Cagle [143] were able to correlate bond strength determined by a tedious procedure of destructive measurements with ultrasonic resonance measurements.

Nonbonds are often assessed by means of pulse-echo methods, by comparing the echo from the bond line with the back echo.

Solid state bonding of metals, diffusion bonds, and both friction and inertia weldings are all increasingly being considered for use in advanced systems, particularly within the areospace community. If parts fabricated using these jointing techniques are to be utilized, it is necessary to have inspection technologies, which can reliably ensure that the required quality/strength is achieved. A summary of work that investigated using novel ultrasonic inspection configurations to measure bond quality was provided by Thompson et al. [144]. Bond lines were considered as an array of "flaws," and some correlations between quality-strength and ultrasonic responses were demonstrated. Challenges however remained for identifying partially bonded conditions: major delaminations are detectable and good bonds, where the bond line becomes transparent to ultrasound, can be identified. To address this challenge, work has also been performed to use an acoustic microscope to characterize bond lines. An example of a composite compressional wave image, measured at 50 MHz, for a diffusion bonded titanium bonded plate, with various bond-line conditions, is shown in Figure 8.28 [145]. The general differentiation between acceptable and unacceptable is seen, but there are underlying issues that require attention if the interaction is to be well understood and used for characterization. One approach used line-scan data, and those for three classes of bond are shown in Figure 8.28b [146]. It is seen that for any individual region within the point response data, there is overlap between the signals from acceptable and unacceptable regions. In analyzing the data,

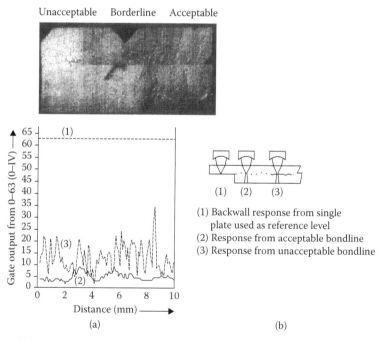

FIGURE 8.28 Bond-line response from various regions of diffusion bonded titanium plate using compressional waves: C-scan image of plate at 50 MHz. (a) Bond-line response from regions (1), (2), and (3), identified in (b). (With kind permission from Springer Science+Business Media: *Rev Prog Quant Nondestruct Eval*, Characterization of diffusion bonds using an acoustic microscope, Vol. 10B, 1991, pages 1391–8, Som, A. K., L. J. Bond, and K. J. Taylor, Figs. 5 and 6, Plenum Press, New York.)

the response is found to be a complex function of frequency, spot size (area of bond illuminated), the ratio between spot-size and the size of the grains, and "roughness" within the grain structure in the bonded zone. In general, it was found to be necessary to provide an integrated or average measure of the bond-line response that covers a scale, which can be related to that for macromechanical properties. On the microscale, fractal concepts and chaos may be needed to relate the microstructure that develops across the bond to measures of bond strength. For the inspection of solid-state bonds, the general problem of bond-line characterization remains a challenge, and this remains an active research area.

8.6 THICKNESS MEASUREMENTS

The thickness of materials may be determined by the time of travel of an ultrasonic pulse between opposing surfaces or by the thickness resonance frequencies. Portable instruments of both types were once commercially available.

The principle of a common pulse type is illustrated in Figure 8.29. In Figure 8.29a, the transmitting and receiving piezoelectric elements are separated from the surface of the specimen to recover from the effects of the initial pulse before the received pulse returns. The triggering of the oscilloscope is delayed by a corresponding amount of time so that the left side of the oscilloscope screen corresponds to the top surface of the specimen. Little error is caused by the angle between the incident wave and the reflected wave from the back surface if the specimen is thick compared with the spacing between transducer elements. Errors in measurement in thinner specimens may be caused by multiple reflections within the specimen. Most commercially available instruments of this type are suitable for measuring thicknesses down to 1.27 mm; however, instruments are available that can measure thickness down to 0.10 mm.

A combination of pulse and resonance methods has been used to produce an "ultrasonic micrometer" [146]. An oscillator produces long pulses of carrier driving an ultrasonic transducer. The transducer is coupled to the material to be measured through a water column. The frequency of the carrier is adjusted so that the electrical output signal of the transducer exhibits a null in the center of the pulse, which is an indication that the sheet is very nearly in half-wave resonance or resonating at one

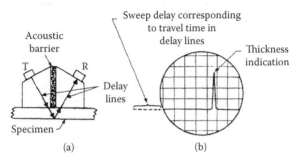

FIGURE 8.29 Measurements of thickness by pitch–catch method. (a) Transducer arrangement. (b) Oscilloscope trace.

of the overtones. If the acoustic-impedance mismatch between the sheet to be measured and the water column is sufficiently great and if the beam width and frequency are chosen properly, the receiving transducer will exhibit a null output. The null can be made to occur close enough to a half-wave dilatational resonance to be used for measuring thicknesses.

Arave [147] reported the use of an ultrasonic system to plot the thickness profile of a fuel-element coolant channel. Two 8-MHz lead–zirconate–titanate piezoelectric crystals are mounted one on each wall of the channel. One transducer emits a signal, which is detected by the second transducer, and the received signal is used to trigger a time-to-analog converter that is connected to the y-axis of a recorder. Longitudinal distance in the channel is converted to an analog output to drive the x-axis of the same recorder. The claimed accuracy is 0.0127 mm, and the resolution is 0.0025 mm with a range of 1.27–4.445 mm.

A method of inspecting copper, silver, brass, and bronze platings on steel [148] has as its basis the relationship between the thickness of the metallic layers and the phase velocity of ultrasonic waves. The method is sensitive both to defects in the metal and plate and to the size of the particles in the plate. Interference patterns developed as a result of the difference in velocities of sound between the plating and the substrate are used to assess the character of the metal/plating interface.

Pitch–catch types of thickness-measuring devices with digital readout are manufactured by several companies. The tremendous improvements in transducer and electronics technologies of recent years have resulted in systems of high resolution and accuracy in metals from 0.1 mm and up. Because of the resolution and accuracy of these instruments, they have replaced resonance-type thickness gauges on the market.

Both continuous-wave and pulse-type ultrasonic micrometers have been used in nuclear applications to measure wall thickness. Corrosion inside forced-circulation piping in reactor pressure systems is one application.

Resonance methods have been used in the past to assess the degree of corrosion in harbor structural elements located underwater. The corrosion rate of unprotected steel wharf and pier pilings and bulkheads in harbors has been monitored by means of ultrasonic probes.

Typical thickness measurements made by ultrasonic means, in addition to those mentioned previously, include measurements of thickness of hollow helicopter shafts, detection of corrosion in aircraft fuel tanks, detection of corrosion and erosion in pipelines and closed vessels, particularly pressure vessels, determination of the thickness of ship steelwork, measurement of wall thickness in diesel cylinder liners, and testing for corrosion and wall thickness reduction in petroleum refineries.

Corrosion is a general term referring to the destruction of metals by chemicals or electrochemical action [149]. Direct chemical corrosion occurs only under unusual conditions involving highly corrosive environments or high temperatures or both; for example, metals in contact with strong acids corrode rapidly. Overheated metallic components, such as boiler tubes, will cause dissociation of water in contact with them and consequent corrosion.

Most common forms of corrosion of metals in contact with water—that is, metals with atmospheric films of water or metals submerged in or containing water—are caused by electrochemical activity. This activity may occur between dissimilar metals. The most common cause is the heterogeneous polarity, which is characteristic of corrodable materials. The process may be accelerated by various factors in addition to heat, such as stress, microbes, contact with various foreign particles, and irregular distribution of oxygen in the water.

Aircraft materials are subjected to high stresses, and corrosion in aircraft components is a major concern. Ultrasonic and AE techniques are among those being used to detect corrosion in aircraft parts.

Modern ultrasonic pulse-echo thickness-measuring systems are used for many diverse applications such as measuring the critical dimensions of soft contact lenses, measuring plating thickness (such as nickel plating on steel), measuring wall thickness of turbine blades, and measuring the thickness of hot cannon tubes immediately following forging at temperatures as high as 1100°C.

Pulse-echo methods of measuring wall thickness of turbine blades, such as those used in aircraft engines, are determined by the geometric and acoustical properties of the blade to be measured. Delay line or immersion methods are used. For the thicker sections, measurements are based on the time separating the arrivals of the front interface reflection from the back interface echo. If the points to be measured are located where multiple echoes between the front and back walls are possible, thinner sections can be measured.

Rumbold and Krupski [150] used a standard ultrasonic instrument and a 10-cm-long beryllium copper delay line pressure-coupled at approximately 910 kg to the test surface to measure wall thickness of hot cannon tubes. The delay line was water-cooled. Instead of using automatic electronic means of compensating for the shift in the velocity of sound with temperature (approximately −17% between room temperature and 1000°C), only the uniformity of the thickness was measured. This fulfilled the objective, which was to determine the eccentricity rather than the absolute thickness, and this was accomplished well within the required accuracy.

Figure 8.30 illustrates the compactness and versatility of equipment for measuring thickness that is typically available on the market. Thicknesses measured by such instruments range from 0.1 mm to 50 cm (0.10–500 mm) with an accuracy of ±1 count of the least significant digit. An internal data logger stores more than 1000 thickness measurements with identification codes.

FIGURE 8.30 Example of family of commercial hand-held ultrasonic digital thickness gage, showing examples of transducers and samples on which they are used. (Courtesy of Olympus NDT.)

Another thickness measurement technique is based on the propagation of Lamb waves in thin plates [151]. Pulsed ultrasonic waves having a wide frequency spectrum are caused to travel through a coupling medium and to strike the test sample at a predetermined angle of incidence. Lamb-wave modes are excited within the test sample by discrete frequencies within the incident ultrasonic waves. The types of Lamb-wave modes generated depend upon the thickness of the plate and the velocity of the ultrasonic waves along the surface of the test sample. The excited Lamb waves radiate pulsed ultrasonic waves from the plate at an angle of exit equal to the incidence angle. The waves radiated contain only the discrete frequencies of the Lamb waves, and detection of these discrete frequencies gives a measurement of the thickness of the metal plate.

Stress corrosion cannot be detected by means of thickness measurements. However, cracks formed by stress corrosion can be detected by shear-wave techniques, and the degree of damage can be assessed by the magnitude of echoes obtained from the cracks or by the attenuation of surface waves propagating through stress-corroded regions.

8.7   INSPECTION OF SOLDER JOINTS

Solder joints between lead wires, terminal conductor pads, and computer components or between two wires may temporarily provide good conduction but lack the mechanical strength to survive movement, physical shock, and fatigue in their intended use. For example, discrete elements on printed circuit boards may initially have a continuous electrical contact through a solder joint to the printed conductor, but when used in mobile systems, they may be subjected to vibration that causes cracks or physical discontinuities in the solder joints resulting in current interruption. Electrical quality-control testing would be insufficient to detect physical defects in the solder joint, which may cause the joint to fail in use. Such defects might include poor adherence of the solder and wire, inclusions, porosity, cold solder, a break, or other similar flaws.

Two methods have been developed for inspecting solder joints ultrasonically. The first includes two transducers, one located on each side of the solder joint. One transducer introduces an acoustic signal into the "upstream" lead, which passes through the solder joint and is detected by the "downstream" transducer [152]. The system may or may not include simultaneously introducing an electric current through the joint to enhance sensitivity to defects. The joint reaction to the acoustic stimulus modulates the acoustic signal and the electric current flow, if electric current is used, and the nature of these modulations identifies the conditions within the joint. In the presence of a good solder joint, the received acoustic signal corresponds to the initial signal although it may be somewhat attenuated depending upon the nature of the solder joint. In a good solder joint, the electrical signal is not modulated; voltage or current flow remains constant.

A defective solder joint may permit minute relative motion between segments of the joint, which produce a detectable noise in the acoustic signal. If an electric current is applied simultaneously, the relative motion will cause a variation in electrical impedance ranging from minor changes in resistance to a complete break in electrical continuity, again depending upon the type of joint interrogated and its condition. This first technique may be used with most solder joints of any size where there is easy access to both sides of the joints.

A very important class of solder joints includes those on a conductor pad on a printed circuit board (flat packs) wherein only one side of the joint is accessible. The second method includes an active and a passive transducer combination, which contacts only one element of the joint. Evaluation is based on the response of the joint to an acoustic stimulus of constant force magnitude and contact pressure. The response of the joint depends upon its acoustic impedance or stiffness [153]. A good joint shows high stiffness and damps the vibrations of the active transducer throughout the active band of frequencies. Several frequency ranges have been used, including 150 to 650 kHz, 200 to 950 kHz, and 500 kHz to 1.5 MHz. Any defect in the solder joint affects its impedance, which is reflected in the response of the joint to the acoustic stimulus. Even voids, inclusions, and other perhaps nonserious defects cause effects in the spectra that are identifiable. The most serious defects are large cracks in the heel, dewetted joints, and joints that are barely tacked at some position. Figure 8.31 presents tracings of the spectra obtained from (a) a good solder joint, (b) a solder joint that is barely tacked at the heel and toe, and (c) a solder joint that contains a 15% crack at the heel but otherwise is solid.

Figure 8.32 is a prototype laboratory system for evaluating soldered joints on printed circuit boards. The transducers are typically separated by 0.5 mm or less. Each transducer is spring-loaded independently with forces ranging from 4–16 g. The performance of the transducers is independent of force over this range of pressures. The dimensions of the tips of the transducers are approximately 0.127–0.762 mm. Often the differences between spectra from good joints and defective joints are

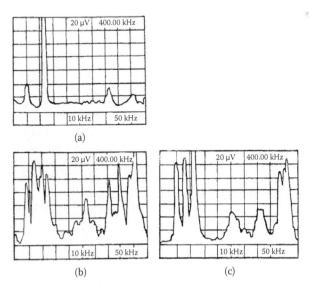

(a)

(b)   (c)

FIGURE 8.31 Spectrum of (a) a good solder joint, (b) a solder joint that is barely tacked at the heel and the toe, and (c) a solder joint containing a 15% crack but otherwise solid.

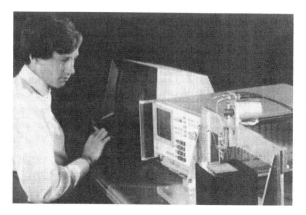

FIGURE 8.32 Prototype laboratory system for evaluating solder joints on printed circuit boards. (Courtesy of Battelle Columbus Laboratories.)

subtle. An experienced operator can usually reliably identify the spectra associated with defective joints on careful examination; however, the system uses a computer analysis, which has proved to be effective in classifying the joints as good or faulty. The algorithm used integrates the amplitude of the received signal over the inspection frequency range. A reduction in acoustic impedance corresponding to a faulty solder condition causes an increase in the received signal amplitude over the chosen range of frequencies. Integrating the signal over frequency takes these increases into account regardless of the frequency at which they occur. The magnitude of the integrated signal is compared with an empirically determined threshold level. If the received signal is greater than the threshold, the joint is declared faulty. If it is below the threshold, the joint is declared good [154].

8.7.1 Acoustic Microscopy

In addition to the use of ultrasonic spectroscopy, there has been a growing interest in the application of acoustic microscopy to inspect the initial quality of multilayered assemblies, particular those found in IC assemblies in modern computer systems, and to diagnose issues in units that fail in service.

Acoustic microscopy systems have now been developed to provide imaging at frequencies up to about 2 GHz, which give submicron resolution. Reviews of industrial applications, including some in the electronics industries, were provided in articles by Nikoonahad [155] and Gilmore et al. [156]. The second article considers systems covering the frequency range from 10 to 100 MHz. Acoustic microscopy is now recognized to be a powerful tool for the inspection of microstructures in electronic devices, and it is being employed to inspect chips, integrated circuits, flip-chip assemblies, ball-grid arrays, and layered assemblies for a wide variety of delaminations including die attachment and poor heat-sink attachment, for bond-line integrity, and to find voids and cracks that could cause in-service hotspots and failures [157]. An example of an image obtained with a commercial 50 MHz imaging system used to inspect for delaminations in an IC package is shown in Figure 8.33.

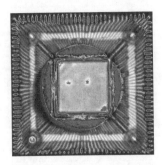

FIGURE 8.33 (See color insert.) Example of an acoustic microscope image showing delamination (red) in an IC. (Courtesy of Sonoscan Inc.)

8.8 IN-SERVICE INSPECTION OF NUCLEAR REACTORS

Ultrasonic methods of nondestructive inspection are used widely in the nuclear reactor field, where a real concern for safety exists. Inspection procedures are applied at many stages of reactor development—when materials are selected, while the components are being fabricated, and after the elements have been in service.

Ultrasonic methods are used to indicate size and homogeneity of grains by attenuation; to measure cladding thickness; to locate lack of bond between cladding and core; to detect pipe, shrink porosity, inclusions, and microinclusions in uranium-alloy castings; to inspect welds for wall penetration, porosity, laminar voids, unfused joint interfaces, cavities at weld root, intergranular cracking, and cracks; to locate wall and core defects; and to indicate density and porosity of metallic-uranium core. Density and porosity are determined by ultrasonic velocity and attenuation measurements.

Ultrasonic methods of inspecting reactor materials and components include resonance methods for thickness measurements (sometimes), through-transmission pulse methods, and pulse reflection methods.

In addition, ultrasonics may be used to indicate the position of control rods in a nuclear reactor, to indicate the steam–water interface in a boiling-water reactor, and to detect boiling in pressurized-water systems.

The immersion method of coupling is commonly used for ultrasonic inspection of fuel elements (Figure 8.34). Automatic scanning and recording devices have been developed. The system design may vary between laboratories, but, in general, a C-scan recording of data is made. Plate-type elements are scanned linearly, and cylindrical elements are scanned in a helical pattern.

A portable imaging system has also been developed and used to characterize microstructure signatures in fuel assemblies. The Ultrasonic Intrinsic Tag (UIT) provides a local scan and images microstructure at 25 MHz. The transducer is coupled to the part of interest using a liquid-filled bladder and a thin film of couplant. This system is shown in use on a fuel assembly in Figure 8.35 [158].

Years of development and experience have produced reliable inspection methods for use during fabrication of certain components used in nuclear power plants.

Nondestructive testing of metals

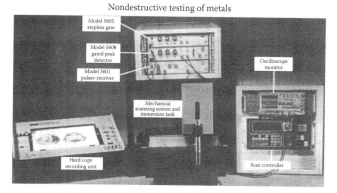

FIGURE 8.34   Bench-top ultrasonic immersion tester and data processing unit. (Courtesy of Panametrics.)

FIGURE 8.35   **(See color insert.)** Ultrasonic Intrinsic Tag, shown in use on a fuel rod assembly. (Courtesy of Pacific Northwest National Laboratory.)

However, after a reactor goes into service, several of these methods are no longer useful. Access to many areas of the reactor is difficult. Radiation levels prevent an individual's remaining in an area for an extended length of time, and thus remote methods of inspection are preferred. Present methods of in-service inspection require a complete shutdown of the reactor until the required inspection can be completed.

The American Society of Mechanical Engineers (ASME) code for in-service inspection calls for volumetric examination of a large percentage of the components of nuclear power plant. This means that a penetrating form of energy is required. Only two nondestructive methods are capable of practical examination of thick steel structures. These are radiography and ultrasonics. Radiography can be used before the plant is placed in service. However, after the reactor goes into service and becomes radioactive, ultrasonics is about the only method that can be used where volumetric inspection is required.

Automatically operating ultrasonic systems are used for remote inspection. The waves are insensitive to gamma rays (although the materials of the ultrasonic system might be affected). The problem is not exactly and completely solved, however, because the materials used in certain components of the reactor cause severe scattering of ultrasonic waves generated by conventional inspection equipment. Problems exist in the use of ultrasonics even when the energy is coupled through a somewhat ideal surface material such as the outer shell of the reactor vessel. Undulations at the cladding/shell interface cause serious scattering, and detecting defects in clad materials requires a systematic comparison between initial baseline data and later inspection data. This process of data comparison with the transmit–receive method of inspection is susceptible to the introduction of a number of errors. Slight changes in incidence angle, coupling, or indexing between inspections could result in errors unless their effects are detected and adaptively corrected in an accurate fashion.

In tests during fabrication of certain reactor vessels [159], inspection has been performed through cladding at normal inspection frequencies of 2.25 MHz. The cladding was weld-deposited stainless steel, and longitudinal waves at 2.25 MHz might be expected to penetrate the cladding satisfactorily. The cladding was inspected for quality of fusion to the base metal, using a wheel-type search unit.

Shear waves at 2.25 MHz, using a standard 45° angle probe, have revealed cracks during inspection from the cladding side. However, the wavelength of shear waves in steel is only approximately half that of longitudinal waves at the same frequency. Scattering is so excessive that the gain of the instrument must be increased by an order of magnitude above what might be considered normal; that is, the gain setting must be increased to at least 10 times the calibration sensitivity. This practice would not be acceptable in a remote, automatic device, and more reliable means of inspection should be developed.

Certain components of the reactor system are especially difficult to inspect ultrasonically. For instance, cast stainless pump castings and valve bodies consist of large grain structures that are almost impervious to ultrasonic energy using conventional NDT methods.

The current state of the art for inspection of cast stainless steels has been reviewed in an Electric Power Research Institute (EPRI) report [160], which shows that significant progress has been made, but challenges remain. Recent developments include using ultrasonic phased array technology which, with data processing, has demonstrated the ability to detect small features in at least some coarse grain materials [26].

The development of nuclear power, using mostly light-water reactor technology, has continued outside the United States of America, and there are currently (2009) 439 nuclear power plants in the global fleet, with about another 30 under construction, mostly in Asia. The cost of oil, the need for electric power to enable global sustainable development, and the desire to deliver electricity without carbon emissions are all combining to increase global interest in nuclear power. Construction of new nuclear plants is again being considered in the United States of America and other western countries, which currently have aging nuclear plants for which life extension is now also being considered. In the current light-water reactor fleet, new degradation processes have appeared on average at a rate of one mechanism every 7 years. Operators need information to better manage power-plant life holistically, adjusting operating conditions to reduce the impact of stressors. The result is that there is a growing need to provide new and improved in-service inspection technologies to support life-extension for the legacy fleet, both in the United States of America and globally, and to provide improved inspection technologies to support the building of new nuclear power plants, where the initial design life is being increased from 40–60 years [161,162].

Looking to the longer term, there is also interest in new advanced gas reactors and other advanced designs (Generation IV concepts) for nuclear power plants [161]. Such plants are expected to operate with high capacity factor (90%+), for longer fueling cycles (4–6 years) and to have necessary inspections and maintenance performed during shorter outages. One challenge to be faced is limited knowledge of material performance for next-generation designs, including balance of plant and secondary units for process heat or hydrogen production. Generation IV nuclear power plants will operate at higher temperatures (potentially 510°C–1000°C), and operation in this temperature range brings to the forefront the potential for many new degradation processes that have not been experienced in current reactors and thus are not well understood or accounted for in the design. Since periodic inspections, which typically occur during refueling outages, cannot be assumed to be adequate to help ensure fitness for service for critical safety systems and components or help ensure optimal plant life management, developing methodology and designing systems for on-line continuous monitoring becomes critical. It is needed to provide operators with better plant situational awareness and reliable predictions of remaining service life of critical systems and components. Particular challenges are faced in some designs for fast reactors that use liquid metals, such as sodium, as the reactor coolant presents inspection challenges. For example in optically opaque sodium systems, ultrasonic under sodium viewing is needed and is being developed to address some critical inspection needs [163,164].

As monitoring and inspection for legacy reactors is increasing and improved inspection is being deployed to ensure safe and reliable operation for new reactors in high radiation areas, methods for remote inspection are being investigated. Systems can include permanent tracks located in appropriate regions of the reactor, such as in the space between the reactor vessel and the surrounding structure [165]. The tracks guide a device fitted with such items as a closed-circuit television camera, ultrasonic inspection equipment, and other tools used on the reactor vessel. The needs for enhanced inspection, on-line monitoring, and diagnostics are also being driven by the need to move life-management from where the operator responds to the findings from an inspection, by being reactive, to rather deployment of technologies that enable the proactive management of materials degradation [166]. The role for ultrasound is still evolving but AE and guided waves are among the technologies being considered for use to support longer term operation.

REFERENCES

1. Krautkramer, J., and H. Krautkramer. 1999. *Ultrasonic Testing of Materials.* 4th ed. Berlin: Springer-Verlag.
2. Szilard, J., ed. 1982. *Ultrasonic Testing: Non-Conventional Testing Techniques.* Chichester: Wiley.
3. Charlesworth, J. P., and J. A. G. Temple. 2001. *Engineering Applications of Ultrasonic Time-of-Flight Diffraction.* 2nd ed. Baldock, UK: Research Studies Press LTD.
4. Holler, P., ed. 1983. *New Procedures in Nondestructive Testing.* Berlin: Springer-Verlag.
5. ASTM Annual. 2008. *ASTM Book of Standards: Section 3, Metals Test Methods and Analytical Procedures Vol. 03.03 – Nondestructive Testing.* Philadelphia, PA: Am. Soc. for Testing and Materials.
6. Goebbels, K. 1981. Structure analysis by scattered ultrasonic radiation. In *Research Techniques in Nondestructuve Testing*, ed. R. S. Sharpe, Vol. 4, 87–157. London: Academic Press.
7. Papadakis, E. P. 1981. Scattering in polycrystalline media, In *Ultrasonics, Methods of Experimental Physics*, ed. P. D. Edmonds, Vol. 19, 337–298. New York: Academic Press.
8. Aisin, R. G. 1967. *Zavod Lab* 33(8):1038–9.
9. Hafemeister, R. N. 1955. *Iron Age* 176:95–117.
10. Dunegan, H. L. 1962. *Grain Size Measurement in Uranium by Use of Ultrasonics.* Report dated July 2, 1962. Livermore, California: Lawrence Radiation Laboratory.
11. Sharpe, R. S., and S. Aveyard. 1962. *Appl Mater Res* 1(3):170–6.
12. Nemkina, E. D., G. D. Radaev, and A. A. Zborovskii. 1966. *Zavod Lab* 32(1):51–52.
13. Goebbels, K. 1986. *Phil Trans R Soc Lond A* 320(1554):161–9.
14. Thompson, R. B. 1998. Determination of texture and grain size in metals: An example of materials characterization. In *Topics on Nondestructive Evaluation*, ed. G. Birnbaum and B. A. Auld, Vol. 1, 23–45. Columbus, OH: Sensing for materials characterization, processing and manufacturing, *Am Soc NDT.*
15. Sokolov, S. Ya. 1941. *J Tech Phys USSR* 7:160.
16. Yegorov, N. N. 1958. *Sb. Primeneniye Ul'traakust. Issled. Veshchestva* (7):169–83.
17. Yegorov, N. N. 1959. *Zavod Lab* 25(7):829–33.
18. Koppelman, J. 1967. *Materialpruefung* 9(11):401–5.
19. Fenkner, M. 1969. *Mater Eval* 27(1):11–15, 22.
20. Papadakis, E. P. 1965. *J Acoust Soc Am* 37(4):703–10.
21. Hecht, A., R. Thiel, E. Neumann, and E. Mundry. 1981. *Mater Eval* 39:934–8.
22. Baligand, B., M. Grozellier, and D. Romy. 1986. *Mater Eval* 44:577–81.
23. Merkulov, L. G., L. A. Yakovlev, and E. K. Guseva. 1967. *Defekto-Skopiya* (6):33–42.
24. Garner, F. A., P. D. Pannetta, II Balachov and A. V. Kozlov. 2006. In-situ measurement of void swelling of austenitic steel used in PWR core internals. In *4th International Conference Nuclear and Radiation Physics.* Almaty, Kazakhstan, PNNL-SA-6136.
25. Doctor, S. R., G. J. Schuster, L. D. Reid, and T. E. Hall. 1996. *Real-Time 3-D SAFT-UT System Evaluation and Validation.* WA: Pacific Northwest National Laboratory, PNNL-10571 and US Nuclear Regulatory Commission NUREG/CR-6344.
26. Schuster, G. J., S. R. Doctor, and L. J. Bond. 2004. *IEEE Trans Instrum Meas Technol* 53(6):1526–32.
27. Thompson, R. B., S. S. Lee, and J. F. Smith. 1987. *Ultrasonics* 25(3):133–7.
28. Migliori, A., and J. L. Sarrao. 1997. *Resonant Ultrasound Spectroscopy.* New York: John Wiley & Sons, Inc.
29. Migliori, A., J. L. Sarrao, W. M. Visscher, T. M. Bell, M. Lei, Z. Fisk, and R. G. Leisure. 1993. *Physica B* 183(1–2):1–24.
30. Visscher, W. M., A. Migliori, T. M. Bell, and R. A. Reinert. 1991. *J Acoust Soc Am* 90(1):2154–62.

31. Ahmed, S., and L. J. Bond. 2007. Modeling of resonant ultrasound spectroscopy based nondestructive evaluation using the "xyz-algorithm." In *Proceedings, International Congress on Ultrasonics*, eds. E. Benes and S. Radel, Technical University Vienna, Vienna, Austria, April, 9–13, 2007, Paper # 1322, 4pp.

32. Bratina, W. J., J. T. McGrath, R. I. Moore, O. M. Mracek Mitchell, and R. F. Love. 1967. In *Proc. 5th International Conference on Nondestructive Testing*, Canadian Institute for NDE, 140–5. Montreal, Canada.

33. Tietz, D. 1966. *Wiss Z Tech Hochsch Otto Guericke* 10(b):SOS–SOS.

34. Generazio, E. R. 1986. *Mater Eval* 44:198–202, 208.

35. Bussierre, J. F. 1986. *Mater Eval* 44:560–7.

36. Nadeau, F., J. F. Bussierre, and G. Van Duenen. 1985. *Mater Eval* 43:101–4, 7.

37. Good, M.S., and J. L. Rose. 1984. Measurement of Thin Case Depth in Hardened Steel by Ultrasonic Pulse-Echo Angulation Techniques. In *Nondestructive Methods for Material Property Determination*, ed. C. O. Ruud and R. E. Green Jr., 189–203. New York: Plenum Press.

38. Krautkramer, H. 1990. *Echo J* 35:9.

39. Frank, S. 2001. Innovations in portable hardness testing. *NDT.net* 6(09):6.

40. Brockelman, R. H. (1966). *Met Prog* 90(1):95.

41. Brockelman, R. H., and K. A. Fowler. 1966. *Int J Powder Metall* 2(4):45–53.

42. Abbe, E. H., and R. D. Korytoski. 1961. In *Proc. 16th Meeting of the Metal Powder Industries Federation*, 5–18. Chicago, 1960. Princeton, NJ: Metal Powder Industries Federation.

43. Panakkal, J. P., J. K. Ghosh, and P. R. Roy. 1984. *J Phys D Appl Phys* 17(9):1791–5.

44. Panakkal, J. P. 1991. *Trans IEEE UFFC* 38(3):161–5.

45. Benson, R. W. 1963. In *Proceedings of Symposium on Physics and Nondestructive Testing*, 92–103. Southwest Research Institute, San Antonio, TX.

46. Bray, D. E., and R. K. Stanley. 1989. *Nondestructive Evaluation*, 143–64. New York: McGraw-Hill.

47. Allen, D. R. W. H. B. Cooper, C. M. Sayers, and M. G. Silk. 1984. The use of ultrasonics to measure residual stresses. In *Research Techniques in Non-destructive Testing*, Vol. VI, ed. R. S. Sharpe, 152–209. London: Academic Press.

48. Charlet, G. 1967. *Rev Fr Mecan* Vol 21. pp 57–86.

49. Benson, R. W., and V. T. Realson. 1959. *Prod Eng* 30:29, 56.

50. Benson, R. W., and Associates, Inc. 1967. *Monthly Progress Report No. 24 (Quarterly Report)*, Contract No. NAS8–20208 for the George C. Marshall Space-Flight Center of the National Aeronautics and Space Administration, June 30, 1967.

51. McGonagle, W. J., and S. S. Yun. 1967. In *Proc. 5th International Conference on Nondestructive Testing*, Canadian Institute for NDE, 159–64. Montreal, Canada.

52. Rollins Jr., F. R., D. R. Kobett, and J. L. Jones. 1963. *Technical Documentary Report No. WADD-TR-61–42, Pt. 2*, January 1963; Available from Armed Forces Technical Information Agency, Arlington, Virginia.

53. Heyman, J. S. 1977. *Exp Mech* 17:183–7.

54. Kino, G. S., D. M. Barnett, N. Grayeli, G. Hermann, J. B. Hunter, D. B. Ilic, G. C. Johnson et al. 1980. *J Nondestr Eval* 1(1):67–77.

55. Clark Jr., A. V., and J. C. Moulder. 1985. *Ultrasonics* 23(6):253–259.

56. Kwun, H. 1986. *Mater Eval* 44:1560–6.

57. Kristan, J., and G. Garcia. 1998. *Adv Mater Processes* 154(5):25–7.

58. Hirao, M., and H. Ogi. 1999. *NDT&E Int* 32(3):127–32.

59. Rasmussen, J. G. 1962. *Nondestr Test* (Chicago) 20:103–10.

60. Dixon, N. E. 1967. In *Proc. 6th Symposium on Nondestructive Evaluation of Components and Materials in Aerospace Weapons and Nuclear Applications*, Southwest Research Institute and American Soc for NDT, 16–35. SwRI, San Antonio, Texas, April 17–19.

61. Klima, S. J., and J. C. Freche. 1968. *NASA Tech Memo* NASA TMX- 52421, October 28-November 1.
62. Clark Jr., W. G., and L. J. Ceschini. 1969. *Mater Eval* 27:180–4.
63. Norris, E. B., S. A. Viaclovsky, and A. R. Whitney. 1969. In *Proc. 7th Symposium on Nondestructive Evaluation of Components and Materials in Aerospace Weapons Systems and Nuclear Applications*, Southwest Research Institute and American Soc for NDT, 32–6. SwRI, San Antonio, Texas, April 23–25.
64. Cole, C. K. 1969. In *Proceedings of 15th National Symposium and Exhibition, Society of Aerospace Material and Process Engineers, Materials and Processes for the 70's*, Society of Aerospace and Process Engineers (SAMPE), 923–42. Los Angeles, April 29–May 1.
65. Zirinsky, S. 1967. In *Paper 67–793 presented at the American Institute of Aeronautics and Astronautics Fourth Annual Meeting and Technical Display*, Anaheim, California, October 23–27, AIAA, Reston, VA.
66. Dobmann, G., M. Kroning, W. Theiner, W. Herbert, and U. Fiedler. 1992. *Nucl Eng Des* 157(1–2):137–58.
67. Dobmann, G., N. Meyendorf, and E. Schneider. 1997. *Nucl Eng Des* 171(1–3):95–112.
68. Nakagawa, N., F. Inanc, A. Frishman, R. B. Thompson, W. R. Junker, F. H. Ruddy, A. R. Dulloo, J. M. Beatty, and N. G. Arlia. 2006. On-line NDE and structural health monitoring for advanced reactors. In *Advanced Nondestructive Evaluation I, Parts 1 and 2, Proc. Key Engineering Materials, On-line NDE and Structural Health Monitoring for Advanced Reactors*, 321–3 and 234–9.
69. International Atomic Energy Agency. 1988. IAEA Technical Document (TECH-DOC 462). IAEA, Vienna, Austria.
70. Ostrofsky, B. 1965. *Weld J Res Suppl* 44:97s–106s.
71. Hagemaier, D. 1968. In *Proc. 1968 Symposium on the Nondestructive Testing of Welds and Materials Joining*. Los Angeles, March 11–13, Am. Soc. Nondestructive Testing, ASNT, Columbus, OH.
72. Harris-Maddox, B. 1963. *Ultrasonics* 1(4):189–95.
73. Raikhman, A. Z. 1963. *Zavod Lab* 29(10):1186–8.
74. Scott, E. 1966. *Ultrasonics* 4(3):152–6.
75. Di Giacomo, G., J. R. Crisci, and S. Goldspiel. 1970. *Mater Eval* 28:189–93, 204.
76. Dianov, V. F. 1969. *Defektoskopiya* (5):522–4.
77. Feoktistov, V. A., and V. V. Grebennikov. 1967. *Defektoskopiya* Vol. 4:12–15.
78. De Sterke, A. 1967. *Br Weld J* 14(4):183–90.
79. Grebennik, I. L., M. F. Krakovyak, V. V. Rakhmanov, I. A. Vyatskov, and I. N. Ermolov. 1967. *Sov J Nondestr Test* (2):184–7; Translated from Defektoskopiya.
80. Silk, M. G. 1977. Sizing crack like defects by ultrasonic means. In *Research Techniques in Non-destructive Testing*, Vol. III, ed. R. S. Sharpe, 51–99. London: Academic Press.
81. Moles, M. 2004. Defect sizing in pipeline welds—What can we really achieve? In *Proceedings of ASME PVP Conference: July 2004*. San Diego, California; Paper # PVP2004–2811, Am. Soc. Mechanical Engineers.
82. Hagemaier, D., and G. J. Posakony. 1968. *Mater Eval* 26(11):221–6.
83. Center, D. E., and R. J. Roches. 1969. *Mater Eval* 27(5):107–17.
84. Anikejev, Ya. F. 1963. *Zavod Lab* 29(10):1191.
85. Deutsch, W. A. K., P., Schulte, M. Joswig, and R. Kattwinkel. 2006. Automated inspection of welded piles. *Proc ECNDT* 2006, European Conference NDT (ECNDT) organized by the European Federation for Non-Destructive Testing, Brussels, Belgium, (Paper 2.3.1.).
86. Heckhauser, H., and S. Schultz. 1995. *Insight* 37(6):440–3.
87. Lawson, S.W., and G. A. Parker. 1996. Automatic detection of defects in industrial ultrasound images using a neural network. *Proc SPIE* 2786:37–47.

88. Murray, E. E. 1967. *Mater Eval* 25(10):226–30.
89. Anon. 1966. *Ultrasonics* 4(2):84–7.
90. Mundry, E. 1967. *Schweissen Schneiden* 19(4):165–71.
91. Weeter, L., and C. Albright. 1987. *Mater Eval* 45:353–7.
92. Cross, B. T., K. J. Hannah, and W. M. Tooley. 1968. In *Proceedings of 1968 Symposium on the Nondestructive Testing of Welds and Materials Joining*, 482–511. Los Angeles, March 11–13, Am. Soc. Nondestructive Testing (ASNT), Columbus, OH.
93. Posakony, G. J. 1966. *The Delta Technique*. Report No. TR66–24. Boulder, CO. USA: Automation Industries Inc.
94. Bond, L. J., and N. Saffari. 1984. Mode conversion ultrasonic testing. In *Research Techniques in Non-Destructive Testing*, Vol. 7, ed. R. S. Sharpe, 145–89. London: Academic Press.
95. Hohl, C. E. 1967. In *Proceedings of 5th International Conference on Nondestructive Testing*, 347–51. Montreal, Canada, Canadian Institute for NDT, Hamilton, ON, Canada.
96. McClung, R. W., and K. V. Cook. 1966. *Mater Eval* 24(10):573–6.
97. McFarlan, T. W. 1965. *Mater Eval* 23(11):553–7.
98. Ostrofsky, B., and C. B. Parrish. 1968. *Mater Eval* 26(6):106–108.
99. Sharpe, R. S., and S. Aveyard. 1963. *J Iron Steel Inst* 201(10):856.
100. Chretien, J. F., and J. Huard. 1969. *Bull Inf Sci Tech* (Paris) Vol. 135, 35–42.
101. Bernard, L. 1987. *Mater Eval* 45:507.
102. Mahmoud, M. A., G. A. Hewitt, and D. J. Burns. 1986. *Mater Eval* 44:1125–31.
103. Mosfeghi, M. 1986. *Ultrasonics* 24(1):19–24.
104. Morgan, E. S. 1981. *Mater Eval* 39:926–30.
105. Burkle, W. S. 1981. *Mater Eval* 39:931–3.
106. Anon. 1986. *B-TIP Technology Sensor*, No. 3, p. 2, B-TIP Program Office, Battelle Columbus Laboratories. Battelle Memorial Institute, Columbus, OH
107. Schmitz, G., A. Wieczorek, and M. Levine. 1964. Technical Documentary Report N. ML-TDR-64–278, Air Force Materials Laboratory Research and Technology Division, Wright-Patterson Air Force Base, Ohio.
108. Harris, R. V. 1963. *Automation* 10(8):71.
109. Silber, F. A., and O. Ganglibauer. 1966. Berg Huettemaenn. *Monatsh* 11(9):393–8.
110. Nusbickel, E. M., and R. N. Cressman. 1970. *Mater Eval* 28(1):1–7.
111. Hatch, H. P., and K. A. Fowler. 1965. *Mater Eval* 23(10):496–9.
112. Kopineck, H. J., G. Sommerkorn, and W. Bottcher. 1967. *Stahl Eisen* 87(20):1198–201.
113. Grasshoff, H. W., and J. Tobolski. 1967. *Baender Bleche Rohre* 8(11):763–43.
114. Isono, E., T. Ueno, and T. Udagawa. 1968. *Nippon Kinzoku Gakkaishi* 32(10):976–81.
115. ASM. 1989. *Nondestructive Evaluation and Quality Control*, Vol. 17, *Metals Handbook*. 9th ed. Metals Park OH: ASM International.
116. Howard, P., R. Klaassen, N. Kurkcu, J. Barshinger, C. Chalek, E. Nieters, Zongqi Sun, and F. deFromont. 2007. Phased array ultrasonic inspection of titanium forgings. In *Review of Progress in QNDE, AIP Conf. Proc.*, eds. D. O. Thompson and D.E. Chimenti. Vol. 894, 854–861.
117. Brown, A. E., and H. L. Dunegan. 1965. *Mater Eval* 23(1):46–7.
118. Matting, A., and A. Heideman. 1966. *Materialpruefung* 8(5):175.
119. Ziegler, R., and R. Gertner. 1958. Die scallgereschwindigkeit als keenzeichnendl grobe fur die beirteilung von gußeisen. *Gißerei* 45(10):185–93 (In German).
120. Voronkova, L. V., and N. E. Bauman. 2006. Ultrasonic testing possibilities of cast iron ingots. *Proc. ECNDT 2006*, European Conference NDT (ECNDT) organized by the European Federation for Non-Destructive Testing, Brussels, Belgium. Paper 2.2.3.
121. Sasaki, S., and K. Ono. 1967. In *Proc. 5th International Conference on Nondestructive Testing*, 341–4. Montreal, Canada, Am. Institute NDE, Hamilton, ON, Canada.

122. Liu, J. H., L. G. Li, X. Y. Hao, D. B. Zeng, and Zh. H. Sun. 2001. *Mater Lett* 50(4):194–8.
123. Bloch, P. K. 1952. *Nondestr Test* (Chicago) 10(3):16–8.
124. Drake, H. C. 1950. *Railway Age* 128(9):55–8.
125. Cowan, G. de G. 1968. U.S. Pat. 3,415,110 (December 10, 1968).
126. Najm, M., and D. E. Bray. 1986. *Mater Eval* 44:463–70.
127. Bray, D. E., and T. Leon-Salamanca, 1985. *Mater Eval* 43:854–8, 63.
128. Clark, R. 2004. *NDT & E Int* 37(2):111–8.
129. Nielsen, S.A., A. L. Bardenshtein, A. M. Thommesen, and B. Stenum. 2004. Automated laser ultrasonics for rail inspection. In *Proceedings 16th WCNDT*, 8. August 30–September 3, Montreal, Canada, Paper # 377.
130. Wiseman, H. A. B., and E. V. Garcia. 1968. *IEEE Trans Sonics Ultra-son* SU-15(1):57–8.
131. Beaujard, L., J. Mondet, and J. Kapluzak. 1965. *Rev Met* 62(12):1187–91.
132. Mahmoud, M. A. 1985. *Mater Eval* 43:196–200, 205.
133. Atkins, M., and M. Druce. 1966. *J Iron Steel Inst* (London) 204(6):607–8.
134. Lynnworth, L. C., and E. H. Carnevale. 1967. In *Proceedings of 5th International Conference on Nondestructive Testing*, 300–7. Montreal, Canada, Canadian Institute for NDE, Hamilton, ON, Canada..
135. Morgner, W., K. H. Schiebold, and H. Krause. 1987. *Mater Eval* 45:569–71.
136. Clark, A. V., M. Lozev, B. J. Filla, and L. J. Bond. 1993. Sensor system for intelligent processing of hot-rolled steel. In *Proceedings 6th International Symposium on NDE of Materials*, eds. R. E. Green, K. J. Kozaczek and C. O. Rudd, Oahu, Hawaii, June 7–11, 1993, 29–36. New York: Plenum Press.
137. Bailey, J. A., and A. Dule. 1967. *Rev Sci Instrum* 38(4):535.
138. Kurz, W., and B. Lux. 1966. *Z Metallkd* 57(1):70–3.
139. Kurz, W., and B. Lux. 1968. *Arch Eisenhuettenwes* 39(7):521–30.
140. Queheillalt, D. T., and H. N. G. Wadley. 1997. *J Acoust Soc Am* 102(2):2146–57.
141. Queheillalt, D. T., Y. Lu, and H. N. G. Wadley. 1997. *J Acoust Soc Am* 101(2):843–53.
142. Dewhurst, R. J., and Q. S. Shan. 1999. *Meas Sci Technol* 10(11):R139–68.
143. Smith, D. F., and C. V. Cagle. 1966. *Mater Eval* 24:362–70.
144. Thompson, R. B., O. Buck, D. K. Rebhein, F. J. Margetan, and T. A. Gray. 1989. Ultrasonic nondestructive evaluation of solid state bonds. In *Proceedings, IEEE Ultrasonics Symposium*, Vol. 2, 1117–23. Montreal, Canada, November 3–6, 1989. New York: IEEE.
145. Som, A. K., L. J. Bond, and K. J. Taylor. 1991. Characterization of diffusion bonds using an acoustic microscope. In *Proceedings, Review of Progress in Quantitative Nondestructive Evaluation*, (QNDE), ed. D. O. Thompson and D. E. Chimenti, Vol. 10B, 1391–8. New York: Plenum. La Jolla, CA, July 15–20, 1990.
146. Aldridge, E. A. 1967. *IEEE Trans Sonics Ultrason* SU-14(2):89.
147. Arave, A. E. 1969. *Nucl Appl* 6:332–5.
148. Bobrov, V. A., and N. V. Khimchenke. 1967. *Defektoskopiya* (6):50–4.
149. Hagemaier, D. J., A. H. Wendelbo Jr., and Y. Bar-Cohen. 1985. *Mater Eval* 43:426–37.
150. Rumbold, J. G., and S. Krupski. 1981. *Mater Eval* 39:939–42.
151. Worlton, D. C. 1980. U.S. Pat. 3, 120,120 (February 4, 1964). D. Ensminger, U.S. Pat. 4,218,922 (August 26, 1980).
152. Ensminger, D. 1980. U.S. Pat. 4,218,922 (August 26, 1980).
153. Ensminger, D. 1981. U.S. Pat. 4,287,766 (September 8, 1981).
154. Shilliday, T. S., T. L. Fletcher, F. B. Stulen, and D. Ensminger. 1985. In *Paper presented at the 9th Annual Soldering Technology Seminar*. China Lake, California: Naval Surface Weapons Center, February 19–20, 1985.
155. Nikoonahad, M. 1984. Reflection Acoustic Microscopy for industrial NDE. In *Research Techniques in Non-Destructive Testing*, Vol. 7, ed. R. S. Sharpe, 217–58. London: Academic Press.

156. Gilmore, R. S., K. C. Tam, J. D. Young, and D. R. Howard. 1986. *DR Phil Trans R Soc Lond A* 320(1554):215–35. International society for optics and photonics, Bellingham, WA.

157. Fassbender, S. U., and K. Kraemer. 2003. Acoustic microscopy: a powerful tool to inspect microstructures of electronic devices, Testing, Reliability, and Application of Micro- and Nano-Material Systems. In *Proceedings of the SPIE*, Vol. 5045, ed. N. Meyendorf, G. Y. Baaklini, and B. Michel, 112–21.

158. Good, M. S., B. E. Simpkins, L. J. Kirihara, J. R. Skorpik, and J. A. Willett. 2003. *Ultrasonic Intrinsic Tagging for Nuclear Disarmament: A Proof-of-Concept Test.* Pacific Northwest National Laboratory, Report PNNL-14462. PNNL, Richland, WA.

159. Proctor, E. S. 1969. In *Proc. 7th Symposium on Nondestructive Evaluation of Components and Materials in Aerospace Weapons Systems and Nuclear Applications*, 180–7. San Antonio, Texas, April 23–25, 1969.

160. O'Regan, P. 2005. Assessment of cast stainless steel inspection. EPRI Report # 1011600 (November 2005). Electric Power Research Institute, Palo Alto, CA.

161. Bond, L. J., and S. R. Doctor. 2007. From NDE to prognostics: a revolution in asset management for Generation IV nuclear power plants. In *Proceedings, SMIRT* 19, Toronto, Canada, August 13–18, *International Association for Structural Mechanics in Reactor Technology*, 7. Paper # O-03-3. IASMiRT, Rayleigh, NC.

162. Bond, L. J., S. R. Doctor, D. B. Jarrell, and J. W. D. Bond. 2008. In *Improved economics of nuclear plant life management Second International Symposium on Nuclear Power Plant Life Management to be held in Shanghai, China*, 26. from 15 to 18 October 2007, IAEA Paper IAEA-CN-155–008KS. IAEA, Vienna, Austria.

163. Bond, L. J., S. R. Doctor, K. J. Bunch, M. Good, and A. E. Waltar. 2007. Instrumentation, monitoring and NDE for new fast reactors. In *Proceedings, GLOBAL (2007), Advanced Nuclear Fuel Cycles and Systems*, 1274–9. Boise, ID, September 9–13, 2007, *Am Nuclear Soc.* Am Nuclear Society, La Grange Park, Il.

164. Karasawa, H., M. Izumi, T. Suzuki, S. Nagai, M. Tamura, and S. Fujimori. 2000. *J Nucl Sci Technol* 37(9):769–79.

165. Ikegami, T., T. Shimura, and M. Koike. 2001. *Hitachi Rev* 50(3):84–8.

166. Bond, L. J., T. T. Taylor, S. R. Doctor, A. B. Hull, and S. N. Malik. 2008. Proactive management of materials degradation for nuclear power plant systems. In *Proceedings, Int. Conf. Prognostics and Health Management 2008*, 9. Denver CO, 6–9 October *IEEE Reliability Soc* (10.1109/PHM.2008.4711466). New York: IEEE.

9 Use of Ultrasonics in the Inspection and Characterization of Nonmetals

9.1 INTRODUCTION

Ultrasonics is used extensively for inspecting many nonmetallic products. The principles of ultrasonic inspection of metals and nonmetals are in general similar and often similar or even, in some cases, the same equipment is used for both. However, nonmetallic products include a wide range of materials that have a diverse range of properties, and their inspection often requires special consideration. In particular, the frequencies and transducers employed and the methodology for coupling ultrasound into the material can be challenging and may require methods that are different from those commonly used with metals. For example, water or a gel couplant cannot be used on porous materials. Velocity and attenuation of sound in these materials range in values from far below to much greater than those commonly observed in metals. For this reason, inspection often has to be conducted at low-kilohertz frequencies. Concrete in large civil structures, such as dams, polymers, including rubber used in tires, thick section fiber-reinforced composites, and wood and many wood products are examples of materials and products that require low-frequency examination. Certain ceramic products have excellent transmission properties and may be inspected at frequencies commonly used in inspecting metals or in many cases at higher frequencies. As with metals inspection the wavelength employed, and hence frequency, need to be related to propagation characteristics, the dimensions of the item under examination and the size of defect which requires detection. For example, for high-strength engineering ceramics, critical defects can be a few micron long, requiring use of 100 MHz, or higher frequencies, and for structural concrete, defects can be many centimeter (or even more than a meter) long, requiring inspection frequencies at 20 kHz, or even lower, in large structures.

This chapter discusses some typical examples of the use of ultrasonics in the inspection of nonmetallic materials and products, that is, products made of concrete, ceramics, wood and wood composites, selected polymers, plastics, and polymer-composite materials. Ultrasonic testing of livestock, food, and vegetables is becoming widely accepted: some applications of ultrasound for animals in food production and veterinary medicine use methods that are considered in Chapter 14, and other food characterization and food processing uses are included in the discussion in Chapters 10 through 13, respectively.

9.2 CONCRETE

Initial quality, composition, and processing, which give initial strength, as well as subsequent degradation or damage, all combine to determine the ultrasound transmission characteristics through concrete [1]. Ultrasonic testing of concrete has been undertaken in the laboratory and onsite since the 1950s [2]. According to Deltuva [3], the strength of concrete can be determined within 10% by measuring the velocity of sound. The accuracy depends upon the aggregate used. Silicate products provide good uniformity and are easily adapted to ultrasonic testing. The velocity of ultrasonic waves in concrete increases with cure (or hardness). Cracks can be detected through their effect on energy losses and scattering.

According to Gemesi [4], measurement of the velocity of sound is the only nondestructive means for monitoring the strength of concrete during steam curing. The velocity increases with steam-curing time, passes through a maximum, and then drops to about 80% of the maximum value, where it remains constant. There is some evidence that the optimum time of steam curing corresponds to the maximum in ultrasonic velocity.

Ultrasonics has been used to examine the homogeneity of concrete in nuclear reactor shielding prior to operation [5]. The conditions of density and strength were related to ultrasonic propagation through the concrete wall.

Organic matter in sand–cement mixes can slow down the curing rate to the extent that hardening may be delayed by weeks. Jones [6] studied the effects of soil in sand–cement mixtures by ultrasonic means. He claimed a relationship between compressive strength (C) and pulse velocity (c) of the form $C = ac^3$, where a is a constant equal to 0.35 for the soil–cement tested. Savchuk and Filichinski [7] gave the following empirical formulas relating the compressive strength (C) of concrete to ultrasonic velocity (c):

$$C = c^{3.75} \quad \text{to} \quad C = c^{3.85} \tag{9.1}$$

Jones [8] also used ultrasonics to measure the thickness of concrete slabs (under favorable conditions) and to determine to some extent the severity of visible cracks in concrete.

Mikhailov and Radin [9] gave the following values for determining the quality of concrete: (1) over 4500 m/sec indicates excellent quality, (2) 4000–4500 m/s indicates good quality, (3) 2300–4000 m/s indicates acceptable quality, and (4) lower than 2300 m/s indicates very poor quality.

Simeonov [10] found a similar correlation between compressive strength and velocity of sound. He reported that the velocity of sound in concrete cured under controlled laboratory conditions increased steadily with time. The velocity in specimens cured under atmospheric conditions varied irregularly, depending on the average air temperature. The velocity of sound in specimens cured under either condition corresponded well to the increase of strength measured mechanically.

These results correspond to what would be expected from a consideration of the dependence of velocity of sound on density and elastic properties. However, as Pawlowski and Raniszewski [11] warned, evaluation of the quality of concrete based upon velocity of sound must be related to the type of concrete. The velocity depends on composition including aggregate material type and size, age, porosity, loading, method of reinforcing, and microcracks. Kordina [12] obtained consistent results with 27 different concretes and concluded that the ultrasonic technique of evaluating concrete is suitable for on-site quality control in the building industry. Shum [13] also recommended ultrasonic nondestructive testing for continuous production-line quality control in the manufacture of concrete products.

The relationship between velocity and strength for an isotropic material, which is used for material characterization, is shown in Equation 9.1. The velocity–strength relationships for concrete materials, of the type shown as Figure 9.1, become the foundation for concrete monitoring using ultrasound. The demonstration that the velocity of ultrasound can be used to monitor the degree of cure, as well as degradation and damage, resulted in the development of commercial, and relatively inexpensive, instruments in the 1970s which all measure ultrasonic pulse velocity (UPV). Three basic UPV measurement configurations are shown in Figure 9.2. The attenuation of ultrasound is higher in concrete than for most metals and most common materials. For example, attenuation coefficients of −0.7 dB/mm at 200 kHz and −2.7 dB/mm at 800 kHz have been measured in concrete [14]. The effect of the higher attenuation, when compared with metals at typical inspection frequencies, is that the testing of concrete is usually performed at frequencies ranging from 25 to 100 kHz.

Since the introduction of UPV, there has been a growing interest in the monitoring of aging civil infrastructure. The UPV meter has been adopted as the basis for a standard and has became widely used in testing of civil structures, including to determine the strength of concrete beams [15,16]. Equipment to perform such tests is now available in the form of a family of commercial instruments, known in Europe as PUNDIT the UPV meter now made by Proceq SA, which is shown in its latest form in Figure 9.3 and in the United States as the James V-meter. An example of

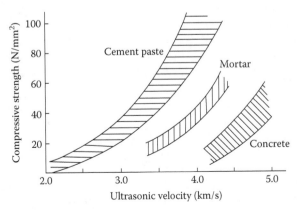

FIGURE 9.1   Ultrasonic velocity–strength relationships for cement materials.

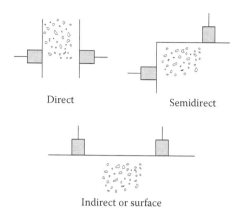

Direct   Semidirect

Indirect or surface

FIGURE 9.2   Transducer arrangements for ultrasonic pulse-velocity testing of concrete.

FIGURE 9.3   (**See color insert.**) Commercial ultrasonic pulse-velocity meter. (Courtesy of Proceq SA.)

such a commercial instrument is shown in use on a concrete sample as Figure 9.4. These instruments provide travel time measurements, and they can detect cracks and voids, thickness, modulus and measure uniformity and deterioration.

In recent years, numerous studies that use ultrasound to investigate concrete have been reported in the literature [1]. Four broad categories of nondestructive testing methods are now used for concrete characterization, and these are (1) methods to estimate strength, (2) methods to estimate elastic properties, (3) methods which locate and characterize voids and cracks within the structure, and (4) methods used to evaluate reinforcing steel [17]. Ultrasound is receiving increasing attention and use for the characterization of the effects of aging and degradation, as well as impact and other forms of damage in concrete. This, like several other specialist application areas within ultrasonics, now has its own literature, and community that hold their own meetings. This community encompasses on both the fabricators, and their professional societies (civil engineers), and engages wider user communities (e.g., nuclear power plant and hydraulic structure owners and operators).

A useful introduction to ultrasound and other methods used for concrete non-destructive testing (NDT), basic manufacturing processes and defects, and wider aspects of quality control is given in an IAEA *Guidebook on NDT of Concrete*

FIGURE 9.4 Commercial pulse-velocity meter in use on a concrete sample. (Courtesy of James Instruments Inc.)

Structures [18]. This report covers active ultrasonic NDT, monitoring using acoustic emission (AE), strain sensing, a summary of the various correlations between velocity and compressive strength, methods used to perform a survey of a structure, case studies, and listings for a range of both international and national standards, including those for the United States, Great Britain, Germany, and Japan. The compilation of information provided by the report is complemented by a good bibliography that list more comprehensive documents, which cover basic and current inspection methods applied to concrete.

The concrete inspection research literature continues to be developed. Papers are being published in both inspection related and civil engineering, as well as presented at specialty meetings, such as those that consider aging concrete in civilian nuclear power plants, integrity of defense structures, hydraulic structures, and bridges. Examples of studies that employ specific methods illustrate the wide use of the following: (1) UPV [19], (2) impact-echo [20], (3) surface waves [21], and (4) ultrasonic tomography [1,22], applied to mass concrete structures including dams.

Studies are also addressing the issue of damage evolution under load [23], the use of backscattered and diffuse fields [24], and novel methods for coupling energy in and sensing fields propagating in a structure, for example, using air-coupled ultrasonics [25]. Modern pattern recognition techniques are also being employed that use artificial intelligence to extract features in data and classify concrete samples [26]. Fully automated robotic systems for specific applications, such as concrete sewer inspection, are also being reported [27].

9.3   CERAMICS AND CERAMIC COATINGS

Ceramic materials such as porcelain, glass, and dense aluminum oxide have good ultrasonic transmission characteristics. Aluminum oxide or ceramics used in making spark plugs have a velocity of sound, which is considerably higher than that in steel. An array of engineering ceramics with strength, hardness, wear, and corrosion resistance, as well as high temperature stability are being developed and deployed in advanced engineering systems. Such materials present inspection challenges, not least because in some cases, critical defects are small and failure is rapid through brittle fracture. Materials being investigated and deployed include silicone nitride, silicone carbide, zirconia, and alumina. The applications for these materials can be as diverse as turbine blades and coatings in tristructural-isotropic (TRISO) particle fuel pellets, used in advanced designs of nuclear power reactors. Structural glass, fixed sheet glass, and polished plate glass of standard types have sound velocities that are approximately equal to that of steel. Ultrasonics is applied commercially for measuring the thickness of glass and porcelain enamels.

The velocity of ultrasonic waves in porcelain is a function of the density/porosity. Measurements of velocity of sound have been used to determine porosity-related mechanical strength of high-voltage ceramic porcelain insulators [28,29]. Ultrasonic pulse-echo methods also have been used for detecting internal defects in these materials. Inspection of porcelain is usually done within the frequency range of 1–5 MHz [30,31].

Techniques similar to those for testing concrete have been used to evaluate materials used in fabricating steel-mill casting ladles (stoppers and ladle bricks). The evaluation is based upon the relationship between the velocity of sound, porosity, and compressive strength [32].

Cracks in a ceramic alter the elastic properties of the material. As a result, resonances associated with a given geometry shift in accordance with the extent of cracking. This resonance shift has been used as a basis for detecting flaws in spark-plug ceramics. A doughnut-shaped transducer formed to match the tip of the ceramic is coupled to the ceramic by oil. Resonant frequencies are located by sweeping the frequency of the driver oscillator over a range, which includes at least one resonance of the specimen. Voltage peaks associated with impedance changes at resonance are displayed on an oscilloscope.

Evaluation of ceramic coatings on metallic substrates by ultrasonic methods has been investigated. Ceramic coatings may vary widely in their acoustic properties. Glassy materials exhibit good transmission properties, whereas porous, coarse-grain structures may cause high acoustic attenuation. These differences influence the method of inspection and in some cases may rule out the use of ultrasonics.

Two types of information may be of interest in inspecting ceramic coatings: (1) adhesion between the coating and the substrate and (2) the quality of the coating itself.

Imperfect bonds between coatings and substrate have been detected through the use of the nonlinear interaction between high- and low-frequency ultrasonic waves [33]. At an unbonded region, a low-frequency wave encounters a stress discontinuity that results in a perturbation of the coating–substrate interface. The intermittent separation of the materials at this interface creates a periodic barrier to the transmission of high-frequency waves. High-frequency waves transmitted across the defective region are thus modulated by the low-frequency wave. Acoustical imaging methods, including acoustic microscopy, have been used to detect such discontinuities, but other methods based upon through-transmission of ultrasonic waves could be equally effective.

Rayleigh wave techniques have been applied to the inspection of ceramic coatings to detect defects such as cracks. This method can be used on a limited number of coatings, namely those that are at or near their theoretical density, have fairly low attenuation, and are thick enough to permit the use of ultrasonic frequencies below 50 MHz. (The effective penetration depth of Rayleigh waves is approximately one wavelength.) Attenuation of Rayleigh waves in many coatings is high so that their use as a means of inspection is best implemented using a large-aperture, relatively low-frequency scanning acoustic microscope, operating typically below 100 MHz [34,35,36]. This method generates a ring of Rayleigh waves in the surface. Recent work, using Rayleigh waves, has included detection of subsurface damage, assessing coatings and visualizing residual stress in ceramic and semiconductor coatings [37].

The relative strength of fairly thick ceramic coatings sometimes has been determined by measuring the velocity of either shear or longitudinal waves in the coating or by the relative attenuation of such waves. Weak coherence of the materials lowers the elastic moduli and may greatly increase the attenuation due to scattering.

Jones et al. [38] have measured the velocity of longitudinal and shear waves in aluminum oxide samples in both the green and sintered states. Dry coupling techniques were developed for use with both longitudinal and shear waves. Longitudinal waves were coupled through a surgical glove type of natural latex elastomer under a pressure of 50 kPa. Shear waves were dry coupled into the specimen by means of a proprietary elastomer material at a pressure of less than 10 kPa.

The dry coupling technique makes it possible to evaluate ceramic materials from the green to the fully dense and sintered states without damage or contamination of the product. As expected, velocities of both longitudinal and shear wave modes increased by nearly an order of magnitude from the green to the dense state.

Stockman et al. [39,40] have evaluated scattering of focused ultrasonic pulses with a center frequency of 30 MHz from voids in glass of less than 100 μm diameter and from microspherical oxide (zirconia and magnesia at diameters less than 200 μm) inclusions in glass at a center frequency of 25 MHz. Conducting the studies in glass allows direct optical characterization of the defects so that calculations can be directly compared with experimental results. The scattered signal spectra depend on sound velocities and incoming pulse-frequency distribution. Changes in defect composition produced less changes in backscattered signals than particle size and host matrix composition, as theory predicts. These studies demonstrate the feasibility of characterizing microdefects in vitreous ceramics by ultrasonic means.

Cielo et al. [41] used three laser-based techniques for characterizing zirconia samples. One of these techniques was ultrasonic for evaluating mechanical properties, grain size, density, and distribution of pores in fired ceramics. Each method showed good correlation with independently measured density of the zirconia samples. The ultrasonic converging-wave technique used is fast and noninvasive and takes advantage of well-documented literature on ultrasonic properties for material evaluation. However, the laser technique used requires high-power lasers operated in tightly enclosed areas for safety, and it is difficult to implement with rough-surface materials and in the presence of strong environmental vibration noise.

A scanning laser acoustic microscope (SLAM) is used for evaluating certain ceramic products. The acoustic microscope reveals localized changes in elastic structural properties of a material by measuring the point-to-point interactions of the specimen with an ultrasonic wave. Ultrasonic waves within 30–1000 MHz frequency range are transmitted through the specimen. The ultrasound reaching the surface opposite the source is detected by a rapidly scanning laser beam in synchronism with a television monitor, thus producing an image of the ultrasonic field at that surface. The technique has been used, for example, to evaluate electronic components such as ceramic capacitors [42] and structural ceramic elements, such as silicon nitride (Si_3N_4) and silicon carbide (SiC), considered for use as hot-section component materials in advanced heat engines [43]. Polishing the surface of the carbide and nitride materials permitted reliable detection of surface voids as small as 100 μm in diameter. Polishing or smooth grinding the surfaces also increased the inspection rate by a factor of 10 over the rate at which as-fired surfaces could be inspected.

Scanning acoustic microscopy has been used to examine the sintered ceramic materials that are used in ultrasonic transducers. An example of a transmission scanning acoustic microscope image of 320 μm thick PZT5 (PZT5 is a specific form of PZT, Lead Zirconate Titanate) transducer material imaged at 50 MHz is shown in Figure 9.5. The density and/or velocity variations in the individual grains contribute to the fine structure seen in the image [44].

The possibility of monitoring tensile strength relationships by AE has been evaluated for porcelain and various ceramic materials. Two methods for determining tensile strength using AE measurements were substantiated. One method involves determining and using the relationship between the strength during subsequent loading and the stress at which during proof testing the AE activity changes from continuous to discrete as the load diminishes. The second method involves two loading–unloading cycles and determining the relative change in pulse amplitude during the unloading periods.

There is ongoing interest in deploying acoustic microscopy for inspecting ceramics and ceramic coatings. A range of signal processing, including ultrasonic spectroscopy, has been used for microstructural characterization [45,46].

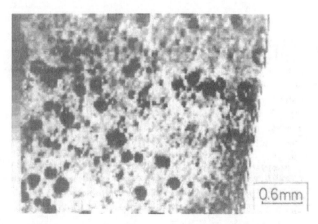

FIGURE 9.5 Acoustic microscope image of PZT5 transducer material. (With kind permission from Springer Science+Business Media: *Proceedings, Review of Progress in Quantitative Nondestructive Evaluation*, Investigation of ultrasonic transducers as used for non-destructive testing, Vol. 1, 1982, 697, Bond, L. J., N. Jayasundere, A. D. Sinclair, and I. R. Smith, Fig. 7, Plenum Press, New York.)

9.4 TIMBER, WOOD, AND WOOD COMPOSITES

Wood and its various processed forms are commonly used as building materials, particularly in much of the United States. It is commonly used in small bridges, piers, and jetties, and in both commercial and residential buildings. It is also to be found in many historic structures and in numerous works of art, including painted wooden panels, which are found throughout the world.

Simple tapping has long been used to determine the soundness of wood. The "Death Watch" beetle, common in historic European church structures, is known for its characteristic AEs, which are audible, as it bores it way through wood beams. Such active natural AE signatures, which in some cases extend into the ultrasonic range, are produced by a diverse range of insects [47], and monitoring these sounds is one method which enables insect infestation to be detected and the treatment and repair areas localized.

The use of ultrasound to measure wood properties, and in its passive listening form, AE, has a long history. AEs, in the audio range, are well known to anyone who breaks a stick. AEs in wood are also generated at higher frequencies (in the ultrasonic range) by a variety of loading processing. Such signals are known to occur when timbers in mines are loaded and measuring these sounds dates back to the 1930s. More recent studies have also considered the emission (at low levels) due to phenomena such as water transport in trees and plants [48]. A good review of AE in wood and wood products is provided by Bucur [49].

Although wood has been used extensively in construction for centuries, including for bridges, quantification of properties and grading has tended to be based on visual, rather than mechanical properties. Mechanical properties in wood can be measured using ultrasound over a range of frequencies [50,51].

Decay affects the propagation of ultrasonic energy through timber [52]. The velocity of sound decreases and the attenuation increases as a function of the amount of decay. In making such measurements in the field, the bark on opposite sides of the tree, standing or fallen, is removed and the transducers are coupled to the tree through a liquid at these points. Evaluation is based on comparison of the attenuation and velocity measurements in a standard table. The quality of timber, even in the absence of decay, can be evaluated by the longitudinal ultrasonic waves directed along the diameter of the logs [53].

However, until relatively recently (~1990) the characterization of wood using ultrasound had only been investigated by a small number of researchers, including Bucur, who over the years published a series of papers and a book [49,54]. Significant growth in the use of ultrasound for the NDT on wood and wood materials was spurred in the United States by symposia, starting in the 1970s, and organized by Pellerin and McDonald [55], and these events were held in Pullman, Washington. These meetings subsequently evolved into an ongoing series with published proceedings.

The imaging of the structure of wood, on various scales, has been investigated using several imaging methodologies, including ultrasound. A review of the approaches that can be deployed is given by Bucur [56]. Properties including those for green sap wood and their change through the drying process, and the effects of temperature and moisture content on velocity during curing have been investigated [57,58]. Such data can become a database and be used for process monitoring. The resulting cured construction timber can also be graded using ultrasonic velocity [59], and the wood screened for quality. For example, ultrasonic transmission has been used to screen marine timber for knots [60].

The application of ultrasound to wood and wood products was in part facilitated by improvements in transducers, which have increased system dynamic range and transmission efficiency since about 1990. Air-coupled ultrasonic inspection for wood has moved from the laboratory to become a viable online NDT technique, which can be implemented in several configurations. Wood, wood panel, plywood, and wood-composite material inspection are being performed using air-coupled ultrasounds in transmission, in angled transmission (to generate shear waves), and using plate waves, all in the frequency range between 50 and 400 kHz [61].

Ultrasonic through-transmission techniques have been used to detect blisters in plywood with thickness ranging from 5 mm to 2 cm. The equipment for screening will operate at speeds up to 152 m/min. The data are processed in near real time and automated grading is accomplished at production-line speeds, with the voids being counted and marked automatically [62]. Ultrasound has now been demonstrated as a QA/QC tool for wood-composite panels, using both transmission and low-frequency plate waves (~15 kHz) and measurements reported across the range from 5 to 50 kHz [63].

In the United States, interest in methods to check wooden bridges has been spurred since 1988 by two legislative acts, which have resulted in research and development activities [64]. The U.S. Department of Agriculture, Forrest Products Laboratory, has developed a number of publications, including a guide for inspection of timber bridges using stress-wave methods that was published in 1999, and which includes tabulated data for ultrasonic properties for a wide range of tree species [65]. A review of AE and acoustoultrasonic techniques for wood and wood-based products was published by Kawamoto and Williams in 2002 [66]. There is also a report on an evaluation of several stress wave (ultrasonic) methods for condition assessment of timber structures [67]. There are also studies that report specific investigations into the condition of particular bridges [e.g., 68]. The inspection of wood using stress-wave (acoustoultrasonic) methods has now become the basis for several ASTM standards [69]. Some of the simplest measurements involve the use of the pulse-velocity meters, which were originally developed for concrete and that can also, subsequently, have been applied to timber and timber structures.

Air-coupled ultrasound has also been used for some more exotic applications involving wood, for example, the detection of flaws in historic works of art that are painted on wooden panels [70]. The use of air-coupled ultrasound was extended and developed, through the use of pulse compression and signal processing, including advanced Fourier-based methods, to use a pitch–catch and plate wave scanning method to investigate the state of conservation of ancient wooden panel paintings [71].

9.5 PAPER

The use of ultrasound to characterize paper and inspect paper products goes back at least into the 1960s.

Researchers at the Swedish Wood Center in Stockholm, Sweden, have applied ultrasonics in an interesting way to measure the thermal softening of paper products and to determine the influence of thermal autocrosslinking reactions [72]. To determine the moduli of elasticity for five different papers in the temperature range −80°C to +420°C, they first heated or cooled dry samples in the nip between precooled or preheated platens for 2–30 seconds. This was followed by measurements of the velocity of ultrasonic waves in the material. From these measurements and the apparent density, they were able to determine Young's modulus of elasticity. For all samples tested, they claimed that the modulus of elasticity decreased rapidly with increasing temperature in the ranges of 0°C–50°C, 100°C–240°C, and greater than 300°C. Between 240 and 300°C, the modulus of elasticity increased or remained constant. The results also indicated the second-order transitions at approximately 25°C–210°C. The latter was attributed to a glass transition in the amorphous part of the cellulose. Autocrosslinking took place at greater than 240°C in spite of short heating times, giving a thermal hardening effect in the range 240°C–300°C. Thermal hardening effects were separated from physical softening effects, which were less time dependent, by varying resistance time in the heated nip.

The relatively low acoustic impedance contrast between paper and air enabled some of the earliest implementations of air-coupled ultrasound, for online inspection, to be developed. Luukkala et al. [73] in 1971 reported the demonstration of the use of air-coupled ultrasonic plate waves to provide contactless measurement of properties in a paper web. This innovative work was subsequently extended to the inspection of thin metal plates [74]. A number of studies have followed from this early work that have sought to give online web property measurement on running paper-webs [75].

Work on developing inspections for wood resulted in advances in transducer materials, and in particular new matching materials, which have enabled much improved transducer dynamic range and efficiencies to be achieved. These advances in transducers have opened up the possibility of new applications, including inspecting inside paper containers. McIntyre et al. [76] demonstrated air-coupled ultrasonics on paper and extended the application to cardboard, packages and some examples of package contents. They used improved capacitive transducers to give better matching to air and hence increase the S:N and dynamic range for the inspections. Other studies have investigated the effects of paper wetting and paper surface roughness on scattering. Further work has used air-coupled ultrasound, together with ultrasonic spectroscopy, to investigate water flux and filtration properties through filtration membranes [77].

There has also been work performed to develop ultrasonic methods for use during wood-pulp/paper manufacturing to provide needed process measurement, monitoring, and potentially control [78]. Such methods are considered further in Chapter 10.

9.6 LEATHER

The quality of leather can be ascertained by measuring the speed of sound, which varies with changes in chemical and physical structure and in fiber orientation in the leather. The velocity may be indicative of modifications of the fibrous order produced by strain, aging, and impregnating material [79]. If a material produces appreciable acoustic attenuation, the velocity of sound is a function of the modulus of elasticity, the density, and Poisson's ratio, and also of attenuation. Velocity measurements are sufficient to determine the deviation of the quality of a sample from a control. The attenuation and frequency are of only secondary importance.

In recent years, the interest in ultrasound and leather has mostly been through the application of ultrasound at high power to perform processing, such as to clean hides and to enhance dye penetration. Such phenomena are considered further in Chapters 11 through 13.

9.7 PLASTICS, POLYMERS, AND COMPOSITES

Plastics, polymers, and composites have become a ubiquitous part of modern life and ultrasound is now used extensively in manufacturing QA/QC and the in-service inspection process for these materials. Plastic parts cover a size range with dimensions from millimeter, and submillimeter, in plastic encapsulation for integrated circuit packages, through the centimeter to meter range, with numerous domestic parts and components, including the cases for innumerable electronic products and car parts, and sizes up to many meters, and tens of meters, for ship hulls, piping, aircraft components and wind turbine blades.

Ultrasound has a long history of application to polymers, and polymer processes, in both fluid and solid forms. Ultrasound was first reported applied to monitor a polymerization reaction in 1946 [80]. Ultrasound is rich in information that can be used for material characterization, process monitoring, inspection measurements, and the resulting data that can be used for process control. In this context, ultrasound has been described as the least understood of the chemometric (and inspection) tools that can be employed in process analytical chemistry [81]. Workman et al. provide an extensive review of ultrasonic analysis, applied to polymer and other chemical processes, in a review, which is a part of a wider review of process analytical measurements [81].

There are observed differences between the static and dynamic properties of polymers. It is important to note that for many polymers the assumptions of linear elasticity are inadequate. The absorption mechanisms are different from metals and even with small amplitude wave propagation absorption commonly becomes nonlinear. The response of filled polymers and many composites to ultrasound can become increasingly complex. Ultrasonic attenuation in many plastics is significantly higher than that typically found with metals at the same frequency. Some examples of data for the attenuation–frequency relationships for some common plastics are shown in Figure 9.6 [after 82]. These data give attenuation coefficients of the order of 1 dB/mm, at 2 MHz, as compared with, 5.6 dB/m for compression waves in a pearlitic steel at 2.25 MHz.

In addition to the significant differences in attenuation between metals and plastics, the velocity of ultrasound in plastics is typically lower than that for metals (e.g., Plexi glass has a compression wave velocity of about 2730 m/s and for many steel alloys, velocities are close to 5800 m/s). In many composite materials, there can be significant anisotropy and velocity can have significant directional dependence. In addition to differences in propagation characteristics, the nature and response of defects are different from those commonly found in metal parts. Some examples of polymer degradation and damage are given in Table 9.1.

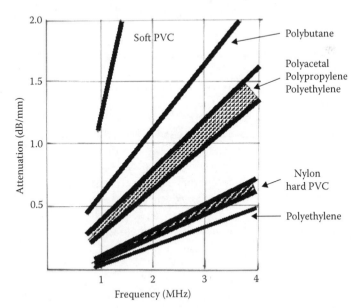

FIGURE 9.6 Frequency–attenuation response shown for some typical plastics. (After Frielinghaus, R., *The Ultrasonic Inspection of Plastics*, Int Conf NDT, Paper J 18, Warsaw, Poland, 1973.)

TABLE 9.1

Examples of Degradation and Damage in Polymers Detectable by Ultrasound

Discrete "Macro-Scale" Defects	Bulk Property Changes "Micro-Scale" Changes	Damage Accumulation
Delamination	Cross-link polymer density	Chemical changes
Porosity	Crystallite formation	Physical "damage"
Cracks	Phase-separation	
	Micro cracks	

Source: Pacific Northwest National Laboratory.

NDE, including methods using ultrasound, has been developed to specifically address the challenges presented by polymers and polymer matrix composites and including application to the geometries in which they are commonly used. This topic is included within the scope of major NDE conference and technical literature [83,84], and in addition, as with some many specializations, within ultrasonics, it has developed its own series of dedicated meetings and conferences [e.g., 85].

Inspection of plastics, polymers, and composites is discussed in several parts of the *ASTM Annual Books of Standards* [86]. Specific ultrasonic methods have been implemented on extruders, to monitor properties, and for the detection of defects, such as bubbles (voids), inhomogeneities, and inclusions. Techniques have also been developed to inspect joints and welds in polymer pipes in a range of geometries that use several different jointing processes, which do commonly involve heating and application of pressure to bring soft/molten material together to form a joint. The evaluation of defect significance and the quantification of ultrasonic techniques for determination of probability of detection with these increasingly used materials remains an area of ongoing research.

This section in this chapter focuses on inspection of some selected polymers, plastics, and composites. The use of ultrasound to provide data for process measurement monitoring and control, including density, viscosity, and particle size are all discussed in Chapter 10.

Many polymers and polymer composites are manufactured in a fluid form and are placed or injected into molds or extruders during fabrication. In addition to formed parts, larger billets, in a range of cross sections, and sheets, that will subsequently be machined, or are used as building materials are produced. For such production, it is necessary, in some cases, in addition to identifying manufacturing defects, and during manufacture to monitor the progression and completeness of solidification and cure.

Ultrasonic techniques operating in pulse-echo have been used to monitor the position of interfaces during solidification. Early work by Bailey and Davila [87] considered both low-melting-point metals and N-paraffin's. Subsequent studies by Parker and Manning have monitored the solid–liquid interface during both solidification and melting [88]. Such measurements were performed by Parker et al. in the 1–5 MHz frequency range and included investigation of the properties of the "mushy zone," the solid–liquid transition zone, which is of finite thickness, and related interfacial structures [89]. The same basic approaches have been used by McDonough and Faghri [90] to follow the interface positions during solidification in various model systems, including aqueous sodium carbonate. The use of simple pulse-echo measurements can be extended to include the reconstruction of solid–liquid interfaces, as was shown by Mauer et al. [91].

Ultrasonic thickness measurements are used to assist operators in setting up extrusion equipment and in monitoring performance by periodic checks during a production run. Production measurements are usually made with immersion types of transducers, which focus a sound beam through a water column, provided by either a squirter or bubbler, to avoid the need for direct contact with hot plastic materials.

Typical products to which ultrasonic measurements are applied include pipe, milk bottles, lotion bottles and similar structures, tubing, sheet, wire insulation, and cable jacketing.

Multilayer plastics are used extensively for packaging food products and carbonated beverages. Typically, these products are made of two layers of a common and inexpensive plastic, such as polyethylene, with one or more barrier layers of another material between them. The manufacturers need to measure both the thickness of the barrier layer and its position within the multilayer structure. These measurements are made at high ultrasonic frequencies in order to obtain sufficient resolution.

Barrier thicknesses are typically 0.025–0.125 mm, too thin to be measured by conventional ultrasonic thickness gauges. Layer thickness is measured based upon the velocity of sound in the material and the round-trip transit time of broadband pulses (at least 300 MHz bandwidth) through the thickness of the laminated structure. Velocity is measured using calibration samples. In the absence of good calibration samples, an estimated velocity of 0.25 cm/μs is usually used. Transducer coupling and alignment are critical to accurate evaluation.

A water path delay may be necessary to measure the thinnest sections (0.25 mm). A similar method may be used to measure thicknesses of plastic coating, for example, on glass lamps. A manufacturer of ultrasonic thickness gauges claims that an accuracy of ±0.005 mm should be possible under good conditions, that is, with proper gauge calibration, alignment of sample, and good quality of echoes.

9.7.1 INSPECTION OF FIBROUS-BONDED COMPOSITES

Many of the basic ultrasonic measurement methods and approaches developed for metals have been adapted and then deployed to inspect composites. Many composites require thickness measurement, which can be performed using pulse-echo instruments similar to those used on metals. However, an inspection based on scanning a hand-held transducer and recording point measurements, which can be used for many crack-sizing measurements in welds in metals, may not be adequate to give a record of the inspection and for making an assessment of condition. The nature of many defects in composites that require characterization are different: they include lack of homogeneity (on various scales), delaminations, and porosity. These defects may occur in the form of a volume with anomalous properties. To investigate composites and provide adequate characterization, there is, in many cases, a need for a recorded image of the inspection.

A heterogeneous distribution of elastic properties is characteristic of certain materials such as graphite-fiber materials, filament-wound structures, and epoxy-glass laminates. Variations in the degree of cure or in the distribution of the binding agent lead to variations in the velocity of sound and in the attenuation of sound by scattering. The strength of the bond between adjacent fibers also affects the velocity and scattering. Ultrasonic measurements of thickness of reasonable accuracy may not be possible without the aid of a secondary method since such measurements are based upon the velocity of sound. Resolution and defect sensitivity also are affected. The failure mechanism is more complicated in composites than in metals [92]. However, degree of cure and quality of bonds are important and their influence on the propagation of ultrasonic waves makes ultrasonics of considerable value in the nondestructive inspection of composite materials.

Measurement of the velocity of sound generally is a good method of determining variations in the strength of composite materials. Attenuation can be correlated with delaminations, porosity, and resin content. Defects can be located by through-transmission or pulse-echo methods.

In discussing problems related to inspection of composites, Bar-Cohen [92] states, "One of the most applicable NDE parameters is ultrasonic attenuation. It has served in studying many nonrelated phenomena, including fatigue damage, weathering degradation, and resin-to-fiber volume ratio." He also mentions an ultrasonic backscatter technique combined with C-scanning, which allows imaging of matrix cracking, as a significant contribution to prevention of in-service failure of composite structures. The cracks relative to each fiber orientation are visualized by proper control of incidence angle. Cracks at 45° and 90° are located by separate scans and corresponding proper incidence angles with no significant interference between signals from the crack orientations.

Williams and Lee [93] illustrate the effectiveness of ultrasonic through-thickness attenuation as an indicator of residual static and fatigue strengths of specimens and structures with plots of (1) ultrasonic attenuation versus fracture strength of flawed and unflawed fiberglass-reinforced polyester composites, (2) initial ultrasonic attenuation versus cycles to failure for transfiber compression-compression fatigue of unidirectional carbon fiber-reinforced epoxy composites, and (3) increase of ultrasonic attenuation relative to initial attenuation at 4 MHz versus residual tensile strength for impact-damaged carbon fiber-reinforced epoxy composites and various hybrids of carbon and Kevlar fiber-reinforced epoxy. Additional theoretical analyses by these and other investigators are progressing toward more reliable inspection of composite materials and structures, and ultrasonic methods are expected to play a major role in this important development [94].

Rogovsky [95] has investigated ultrasonic and other methods of inspecting composite tubular parts used in the aerospace industry. His concern was composites other than the fibrous materials and included "multilayer, bonded structures comprising graphite-epoxy and glass-epoxy laminates, metals, rubber, and other dissimilar materials." Defects of interest are voids and disbonds including defect sizes.

These structures are simpler from an ultrasonic inspection point of view than the fibrous composites. A system of multiple ultrasonic reflections was used to evaluate bonds between layers. Good bonds permit maximum transfer of energy across the interface between materials and fewer echoes occur due to energy dissipation in both material layers. Poor bonds cause larger and more echoes from the interface. Hand-held contact ultrasonic C-scan and imaging techniques were used to determine defect size and locations. Ultrasonic imaging was the most efficient for providing three-dimensional and amplitude information on tested composite joints.

Hagemeier and Fassbender [96] include both carbon-epoxy composite laminates and honeycomb structures in a discussion of methods of these materials used in the aerospace industry. They also refer to the fact that ultrasonic inspection is the most favorable method for detection of defects generated in carbon-epoxy composites. The defect types listed include delaminations, interlaminar porosity, voids, foreign materials, and voids or porosity in adhesive-bonded joints. They describe reference standards, with built-in defects, used in practice. Design of these standards and the size of built-in defects must be appropriate to the structural requirements. Acceptance/rejection criteria are developed in advance from design considerations. Usually, the highest quality will be required at the area of maximum stress or strain.

One ultrasonic technique of evaluating anomalies is ultrasonic feature mapping. Nestleroth et al. [97] describe their procedure for ultrasonic feature mapping as including the traditional C-scan analysis and noting changes in signal amplitude or noting changes in the feature value of a back-wall echo or in a through-transmission signal. They experimented with laminated graphite-epoxy composite panels. Anomalies consisted of (1) small Teflon wafers inserted between plies at one-quarter of the thickness, midplane, and three-quarters of the thickness during layup of the panels; (2) multiple voids or porosity that formed in the matrix during the manufacturing process; and (3) true delamination or separation of the interface between layers. The mapping procedure was successful in identifying the anomalies. They conclude that a classification algorithm will always have to be fine-tuned as material systems are changed or as a new group of anomaly geometric characteristics are identified. The feature selection process for producing a meaningful anomaly-identification feature map is important. This requires an estimate of the signal characteristics from an anomaly in order to properly select suitable signal features, test frequencies, frequency bandwidths, and numbers and locations of receiving transducers.

A polar scan described by van Dreumel and Speijer [98] might be useful in the nondestructive determination of layer orientation in composite materials. In the polar scan, two transducers are operated in the through-transmission mode. The incidence angle is varied continuously and the attenuation as a function of angle of incidence is simultaneously recorded by imaging. This method identifies the critical angle at which energy is totally reflected from an interface. Critical angle is a function of relative velocities in the media on opposite sides of an interface and, therefore, is related to stiffness and density.

To obtain information about the directional elastic properties of a material, a polar scan could be made covering the angles of a half-sphere. Rotation about two axes is considered sufficient. The authors' results indicate that the method gives a reasonable qualitative impression of the directional elastic properties. They believe that a library of polar patterns stored as "fingerprints" might be useful in laminate identification by pattern recognition.

Several computer-controlled ultrasonic scanning and data collection systems have been developed to reduce inspection costs, minimize operator dependence, and increase inspection reliability of composites [99,100]. These systems are in operation online, providing data for characterizing and classifying many types of flaws without the need for repeated scanning of parts.

Acoustic emission has been used to locate areas of damage progression and to measure damage severity in composite rocket motor cases [101,102].

Modern aircraft and wind turbines manufacturing are two industries, where developments are being driven by both energy and greenhouse gas concerns. Inspection needs in these areas will be used to illustrate some of the developments in composite inspections. For example, the new Boeing 787, the Dreamliner, has about half of its fuselage and wings constructed of composite materials. This makes the aircraft about 18,000 kg lighter than one of similar size fabricated using conventional construction. This weight difference is reported to give an increase of about 20% in fuel efficiency; however, ensuring parts have needed strength requires advanced inspections. Composites, including those formed using resin transfer molding (RTM), are also being increasingly used in military aircraft. The development of large wind turbine blades involves composites that consist of multiple layers and skins and is requiring large area scanning, both at the time of manufacture, and at periodic in-service inspections. In both cases, ultrasound can be used to meet some of the inspection needs.

To provide one-sided inspection of the large number of composite parts being used in aircraft, Dassault Aviation, and other companies, has invested in new inspection techniques, which include automated scanning, acoustical holography, acoustical microscopy, and laser-based ultrasound. In the case of Dassault, the latter being developed in cooperation with Aerospatiale. Dassault Aviation is also investigating some emerging acoustical techniques, which include air-coupled ultrasonics and acousto-ultrasonic inspection [103].

An assessment of conventional and some advanced methods for the inspection of composites used for aircraft was recently reported [104]. This activity included an assessment of performance of traditional tap tests and both ultrasonic and other NDE methodologies on a diverse range of samples that included honeycomb parts and multilayer carbon-epoxy modules. Probability of detection curves were developed for the various methodologies investigated: for many applications, the MAUS scanner, an ultrasonic scanner originally developed for the U.S. Air Force, performed well, when operating in a resonance mode.

Technologies for inspection of composites continue to be investigated, and this includes the use of ultrasonic spectroscopy and plate waves in a variety of forms. An example of a commercial phased array with water coupling on a section of tube is shown in Figure 9.7. The complexity of shape and large size, together with need for a record of inspection, is encouraging use of inspection techniques that combine robotics and noncontact coupling, such as ultrasonic methods that use water jets, lasers, and air-coupled transducers.

FIGURE 9.7   **(See color insert.)** Example of automated ultrasonic inspection of a tube using a phased array system with shoe and water coupling. (Courtesy of Olympus.)

One approach used for inspection of large composite parts is acoustic holography. A real-time image of a region, that can be 30 cm in diameter, is given. The first group to develop ultrasonic holography was at Pacific Northwest National Laboratory (PNNL), when it was named Pacific Northwest Laboratory/Battelle-Northwest Division (Richland, WA). Work commenced in 1962, but it was not disclosed until 1967 [105,106]. This group developed "RTUIS," the real-time ultrasonic imaging system. A description of this work and a comparison of various acoustical holography methods were provided by Brenden [107] and a schematic showing the elements in RTUIS is shown in Figure 9.8. Since that time, the various forms of acoustic imaging, including both acoustic holography and acoustic microscopy, have developed by several groups for a diverse range of both industrial and medical application. This topic is discussed further in Chapter 10.

To meet aerospace industry needs, Dassault Aviation, working in cooperation with Battelle developed, optimized, and delivered a custom RTUIS for the inspection of some composite parts to be used in the RAFALE fighter [108]. RTUIS (shown in Figure 9.8) is an ultrasonic camera that uses a large ultrasonic transducer (in some cases more than 20 cm in diameter) to generate plane waves, which provide the ultrasonic illumination, and then employs liquid surface holography, through the interaction with a beam from a reference transducer, to convert the sonic image into a wave pattern, on a liquid surface. A pulsed laser is used to capture the diffraction image of the liquid surface, giving an image of the part with a video rate frame of 50–60 per second. In the system delivered to Dassault, specimens to be inspected were placed in front of a 10 cm diameter ultrasonic transducer, which can give discrete frequencies from 3 to 5 MHz in a tone burst mode creating wave trains of 100 microsecond duration. Parts could then be scanned to give larger area inspection.

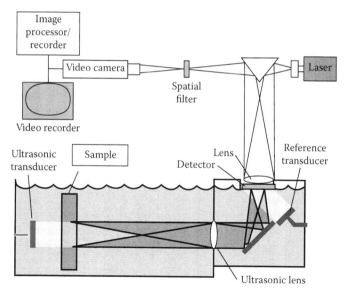

FIGURE 9.8   **(See color insert.)** Schematic showing elements in RTUIS. (Courtesy of Pacific Northwest National Laboratory.)

The technique has been shown to offer significant advantages for composite inspection. It provides the following:

- Rapid inspection (10–30 m²/min) with a single transducer
- High spatial resolution, when expressed in terms of the diameter of a resolution element, is in the order of 0.025 cm²
- Relative easy automation for inspecting large structures
- Good sensitivity and resolution for 30 mm thick composite materials
- Better tolerance for beam incidence (±7°) enabling the inspection of fairly complex geometry part with easy positioning

Applications have been shown to cover a wide range of parts as diverse as spars and stiffeners with radius, curved panels, and self-stiffened composite boxes. An example of an image for a self-stiffened box with simulated delaminations is shown in Figure 9.9 [103].

As the range of applications of composites has increased, the diversity of inspection techniques adapted or developed to meet requirements has expanded. The growing use of plastic and composite pipes in utility industries and nuclear power plant applications, for service water system use, is requiring validated inspection methods.

A recent study has performed an assessment of butt fusion piping joints, for lack of fusion, using ultrasonic and other NDE technologies [109]. The ultrasonic methods investigated included phased arrays and time-of-flight diffraction (TOFD), both adapted from metal pipe inspection, and the TOFD system is shown as Figure 9.10. The thick wall plastic pipe weld contained simulated defects and results for four

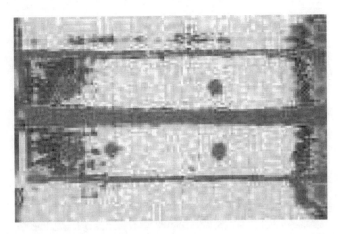

FIGURE 9.9   **(See color insert.)** Image of a RAFALE CFRP self-stiffened box with simulated delaminations. (Courtesy of NDT Source.)

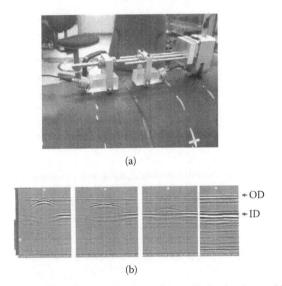

FIGURE 9.10   Time-of-flight diffraction on a thick wall plastic pipe weld with simulated defects. (a) TOFD probe assembly. (b) TOFD results from four ID-connected saw cuts from left-to-right through-wall depths are 75%, 50%, 25%, and 10%. (Courtesy of Pacific Northwest National Laboratory.)

ID-connected saw cuts are shown in Figure 9.10b. A phased array unit using a 1.5 MHz probe applied to a similar polymer pipe is shown in Figure 9.11, with the B-scan, showing defects on a computer screen behind the operator.

    There is also a growing body of work that is investigating damage initiation and development and evolution in composites, including hidden (or subsurface) damage, resulting from impacts. A review of one project, which employed 1.9 MHz ultrasound

FIGURE 9.11 (**See color insert.**) Phased array in use on a thick-walled polymer pipe, inspecting for lack of joint fusion, showing B-scan on computer screen behind operator. (Courtesy of Pacific Northwest National Laboratory.)

to scan simulated aircraft assemblies was reported by Liu [110]. Ultrasound is a very effective tool for detection and imaging of subsurface features, including hidden damage, where there is only access to an outer skin. The linkage between inspection and materials science, including degradation and damage models for composites, remains an area of ongoing research.

9.7.2 Tires

Ultrasonics has been used successfully in inspecting tires used on passenger cars, trucks, and aircraft. The technique is capable of detecting porosity and delaminations, measuring tread uniformity, and determining the state of the cure of the rubber [111]. The test is based upon an ultrasonic through-transmission technique [112]. Halsey [113] claimed to have developed an ultrasonic technique for measuring tread thickness and tread uniformity, and equipment is commercially available for measuring the distance to steel or fiber ply cords or belting and measuring total wall thickness. Because of the high attenuation due to the scattering from the fabric within the tire, the frequencies used are usually much lower than those normally used for NDT in metals. A technique employed by Patton and Hatfield [114] involved placing a crystal transmitting a 50-kHz beam from inside the tire to a number of receiving crystals outside the tire, which was immersed in water and slowly rotated. Faulty areas caused a drop in signal from one or more of the receivers. Defects smaller than 3×9 mm could be detected at a frequency of 180 kHz.

Thickness gauges for use on rubber tires are available operating between 0.5 and 2.25 MHz. The instrument is calibrated by measuring the velocity of sound in the rubber using a sample of known thickness. The velocity of sound in rubber is very sensitive to temperature; therefore, calibration must be done at a temperature closely corresponding to measurement conditions.

FIGURE 9.12   Tire scanning tank and electronic controller rack. (From Downes, J., P. Zhang, and M. L. Peterson. 1999. *IEEE/ASME Trans Mechatron* 4(3):301–11. With permission.)

Rieckmann [115] used a liquid-levitation technique for visualizing defect areas in tires. He placed a 200-kHz ultrasonic transmitter inside the tire, which was immersed in a reflection-free trough. He used a narrow parallel beam of light, the diameter of which was small in comparision with the ultrasonic wavelength, to make the interference patterns visible. He also described a barium titanate hydrophone or microphone for picking up the ultrasonic wave after it passed through the tire. Rieckmann claimed that the sensitivity of the method was affected if the tire had a marked tread pattern.

Ultrasound inspection of tires has been performed using both air and water coupling. A 1999 study reported a high-speed high-resolution multichannel ultrasonic tire inspection machine, operating at 500 kHz using through-transmission. The scanning tank and system electronics for that system are shown in Figure 9.12 [116].

9.7.3 POLYMER MEMBRANES

Porous polymer membranes are of increasing significance in a number of application areas, particularly in chemical engineering separation science. These membranes, which are typically about 150 μm thick, are comprised of several sublayers, and they are used to process both liquid and gas systems. To enable properties in and near the membranes to be studied and performance investigated, it is necessary to provide online, real-time, noninvasive measurements for process monitoring and control. Ultrasonic methods have now been developed and applied to meet this need and used for the following: (1) fouling characterization, (2) compaction measurements, and (3) solidification monitor, during membrane formation.

The initial work that applied ultrasonic measurements in membrane science used a pulse-echo method, "time-domain reflectometry" (TDR), named by analogy to optical methods, to give noninvasive characterization of fouling during a separation process, at elevated pressure [117]. The method was developed and first applied to a reverse osmosis (RO) desalination membrane, to give through-wall measurements, from outside the separation cell using normal incidence 5 and 10 MHz compression waves [118]. The ultrasonic TDR used is an extension of pulse-echo ultrasound, as shown in schematic form in Figure 9.13, where the amplitude and arrival times of the reflections from the various interfaces are tracked. Interpretation is then related to the appropriate process model, for example, in the case of fouling, signal amplitude was related to the developmental stages in fouling formation.

The success achieved with fouling characterization spurred the investigation of the use of acoustic (ultrasonic) TDR for other measurement challenges in membrane science. The method was developed and used for assessing structure–property characteristics of polymeric thin films [119], for real-time measurement of thickness changes during evaporative casting of polymeric films [120], measurement of compressive strain during membrane compaction [121], and simultaneous real-time measurement of membrane compaction and performance during exposure to high-pressure gasses [122]. As the diverse range of processes was investigated, material properties, including compaction and strain, were correlated with ultrasonic response. This class of measurements has been seen to have the ability to provide data from a wide range of pressurized fluid and gas-membrane systems.

In gas-separation systems, the ultrasonic velocity, which is typically lower than for a fluid, effectively increases resolution for measurement of surface displacement and the application of pressure increases density. The increase in acoustic impedance reduces the impedance contrast at the separation-cell and the gas-membrane boundaries. The high pressure also causes drop in the attenuation in the gas at higher

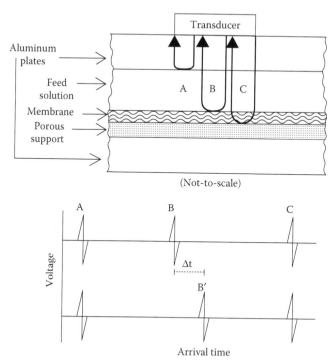

FIGURE 9.13 Schematic showing primary interface responses for acoustic time-domain reflectometry in a reverse osmosis separation cell.

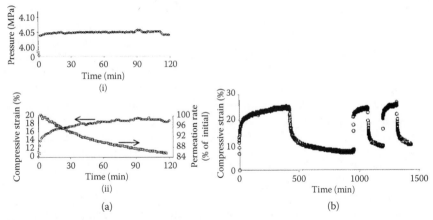

FIGURE 9.14 Example of membrane strain and recovery for a cellulose acetate membrane in a gas-separation system. (a) Compaction data for a Dow BW-30 reverse osmosis membrane: (i) The top trace shows upstream pressure and (ii) bottom trace shows compressive strain and permeation rate. (b) Compressive strain data for both creep and recovery membrane with 36% porosity and an initial thickness of 180 μm. (Reprinted from *Desalination*, 116(2–3), Peterson, R. A., A. R. Greenberg, L. J. Bond, and W. B. Krantz, pp. 115–22, 1998, with permission from Elsevier.)

frequencies, as had been reported previously and in that case used to enable in gas-coupled acoustic microscopy [123].

The ability to monitor membranes in gas-separation systems, including investigation of compressive strain in an online separation cell is illustrated with data shown in Figure 9.14 for a Dow BW-30 RO membrane operating under pressure. The material is a cellulose acetate (CA) membrane with 36% porosity and an initial thickness of 180 μm. The top trace shows upstream pressure and the bottom compressive strain and permeation rate. Compressive strain data for both creep and recovery phases are shown in Figure 9.14b [121].

9.7.4 ENERGETIC MATERIALS AND SOLID ROCKET MOTORS

Ultrasonic techniques have been implemented on large rocket motors, ordnance items, and explosives since at least the late 1960s. Such measurements are becoming an integral part of enhanced manufacturing quality assurance, process control and provide data for safe life, stockpile surveillance, and life extension programs. Such life-related aging programs traditionally focused on obtaining data from accelerated aging, periodic sampling, firing, and destructive examination.

A review of developments in the use of ultrasonic assessment techniques, during the last 20 years, as they pertain to explosives, propellants, and solid rocket motors was published in 2001 [124]. There is a long history for the application of ultrasonics to large solid rocket motors [125–128]. In the early 1970s, a system was developed for the U.S. Navy to inspect a prototype charge for use in a 5-in. (127 mm) shell [129,130]. More complex acoustic imaging methods were also developed for inspection of cylindrical billets [131]. Measurement techniques were developed for the assessment of various forms of damage, including cumulative damage due to mechanical loading [132–135] and for the determination of mechanical properties. The early work identified the potential and also some fundamental limitations of these forms of ultrasonic inspection. However, the then available instrumentation severely limited the sophistication of the systems deployed at that time [e.g., 136,137]. The adoption of ultrasonic inspection proceeded at a slow pace. One factor was that only limited details were available in the open literature. Additionally, there has been a relative unavailability of expertise in this rather specialized field of application. Finally, within the United States, the standard form of NDT applied to ordnance items continues to be radiography, and acceptance standards are most commonly based on such inspections.

A number of imaging systems have been developed to meet the needs of particular rocket motors using ultrasonic systems operating between about 500 kHz and 1–5 MHz. For example, an automated, multitransducer automated inspection and sorting system was developed to inspect parts of the 2.75-in. (Hydra-70) solid rocket motor [138] using the combination of configurations shown in Figure 9.15. To meet the need to inspect case-bonded solid rocket motors, a field deployable system, based on a pipe inspection unit, working with a custom transducer head, was developed and tested for the Sparrow motor and used for stockpile screening [124].

Advanced ultrasonic defect characterization tools, which do more than simple imaging, have been developed and demonstrated for the characterization of both fluid- and gas-filled voids in large rocket motors [139]. One family of studies used Born inversion [140]. This method involves use of a pulse-echo technique and then application of an inversion algorithm to provide void characteristics: size and type (gas or liquid filled). The Born area signature, derived from processing pulse-echo data, has also been demonstrated to be able to characterize inclusions [141] but it has not been reported as used on propellants or explosives.

Impediography is another signature obtain from pulse-echo data and has been used. It is again applied to pulse-echo data, to reconstruct the acoustic impedance profile looking through a case-liner-propellant systems, to determine the degree of bonding and interface conditions [142]. Attenuation and scattering data have been shown to be able to provide characterization of both aging and cumulative damage for rocket propellant materials, which are in essence filled polymers.

The work with ultrasound for stockpile screening has shown that ultrasonic inspections can be a cost-effective tool [138] and can provide valuable data. It is also

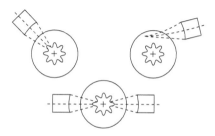

FIGURE 9.15 Basic ultrasonic immersion inspection transducer configurations for the Hydra-70 rocket motor grain.

receiving ongoing attention for characterization of energetic materials [143]. That said it must be noted that the ultrasonic contrast mechanisms are fundamentally different from those found with radiography and understanding the nature of defects and damage of interest needs analysis, prior to the use of ultrasound for applications, such as stockpile screening.

9.7.5   LOW DENSITY FOAMS (AEROGEL)

Aerogels are low-density foams, which are highly porous and can have bulk densities in the range 0.004–0.5 g/cm^3. These materials have large open pores, high specific surface areas, and they form a distinct family of materials that are used in very specific applications. The synthesis, properties, and some applications are discussed by Husing and Schubert [144].

For ultrasonic applications, there is interest in the use of these materials in matching layers, as they have low and variable acoustic impedance, making them of interest for air-coupled ultrasonic transducer matching layers. This use has required the development of air-coupled methods to characterize quarter-wavelength thick silica aerogels in the low MHz frequency range [e.g., 145].

Aerogels and related low-density foams are also used in a range of novel science and engineering applications including in the fabrication of the targets used in inertial confinement fusion experiments, such as those being performed at the National Ignition Facility (NIF), in California. The materials must be homogeneous and can be required to support inclusions held in the bulk material. Inspecting these materials presents particular challenges as in many, if not all, cases they are hydroscopic, highly fragile and loading causes the foam prestructure to collapse. One solution to these challenges is the use of gas-coupled ultrasound, and inspections have been performed at megahertz frequencies using both air and high-pressure gas as the coupling medium [146].

9.8   ADHESIVE BOND INTEGRITY

Adhesive bond integrity may refer to (1) the quality of adhesion between surfaces, (2) the presence or extent of nonbond areas, or (3) the strength and quality of the adhesive material at the bond line. Unless some other mechanical force is holding the joint under compression with adhesive acting as coupling agent, conditions (1) and (2) are fairly easy to identify. If the adhesive layer is very thin and a good bond exists, some energy may be reflected from the interface and most transferred to the second medium, especially if the two materials have similar acoustic impedances. If an area lacks adhesive or if there is lack of adhesion with adhesive present, a larger portion of energy will be reflected from the interface, especially if the two media have similar acoustic properties.

A second condition involves bonding of two materials of very different acoustic impedances, for example, plastic to metal. In this case, considerable echo will return from the interface whether it is well bonded or not. In this case, bond integrity can be assessed by observing the effect of the interface on the phase of the echo. The phase of the echo is reversed when the acoustic impedance of the medium of the incident wave is greater than that of the medium of the opposite side of the interface, but it is not reversed when the impedance of the second medium is greater than that of the first. For example, when a pulse impinges on an interface from a low-impedance plastic bonded well to a high-impedance steel, its echo back into the plastic exhibits positive polarity, that is, no 180° phase shift. If the bond separates from the steel, the polarity reverses to negative—a shift of 180° in phase.

Evaluating the adhesion state between layers in a composite structure is a more complex problem for which there is no satisfactory solution. Most production tests are based upon detecting areas of bond or nonbond, as described previously. There may be either through-transmission or pulse-echo scans of bonded areas.

In principle, the strength or quality of the adhesive layer might be evaluated on the basis of propagation characteristics, since velocity of sound and attenuation are determined by the elastic properties and loss mechanisms (viscosity, scattering, etc.) of a medium. As an adhesive cures, these properties change.

The acoustic impedance of an adhesive differs in most cases from that of any of the components to which it is attached. This property affects the wave propagation characteristics through the assembly to the extent of its effect on the overall volume of the wave path. Various ultrasonic methods have been studied in efforts to evaluate adhesive quality but with only limited success. The "ideal" wave would be one that travels primarily through the adhesive layer. In a laminated structure, such waves are called interface waves, or Stoneley waves. However, the impedance relationships between metals and polymers prevent their existence in metal/polymer interfaces.

Lamb waves are sensitive to delamination and voids and have been evaluated for use in pulse-echo inspection of laminated structures. Surface or Lamb waves can be generated at an interface by critical angle techniques. A sample is placed in an immersion tank on a rotatable mounting [147]. Ultrasonic pulses emitted by the transducer are reflected from the outer surface and from the inner surface (opposite side). Adhesion conditions at the interface are reflected in velocity-related effects. As the orientation of the surface of the specimen relative to the axis of the incident beam is changed from normal incidence, the position of the reflected beam shifts by an amount controlled by the velocity of sound in the bond material according to Snell's law. In a second method, using only one probe in the pulse-echo mode, echoes from both surfaces return to the transducers. The angles at which maximum reflection amplitude occurs from the liquid/solid interface are noted. Change in angle is related to change in velocity. Higher phase velocity usually indicates greater degree of adhesion.

A comprehensive review of early work that investigated NDT of bonded structures, including ultrasonic imaging, AE, and tap testing, is given by Segal and Rose [148]. Ultrasonic methods can measure bond-line thickness and monitor homogeneity of the adhesive layer.

A variety of imaging systems, including methods that use Rayleigh and plate (Lamb) waves, have been investigated. In addition, some investigators have considered the use of ultrasonic spectroscopy. Rose and Meyer [149] claimed that empirical differences in spectra could be correlated to bond strength. A number of studies in the 1970s investigated spectral characteristics for adhesive-bonded lap-joints between aluminum strips [150,151] and when scanning across bonded honeycomb-filled sections with aluminum plates glued to seal, the honeycomb fill and gross defects were able to be identified.

A diverse range of models for inspections of adhesive-bond integrity have been reported. For example, finite element methods have been used to solve basic elastodynamic equations and good agreement obtained between the model and experimental data [152].

C-scan imaging can, in many cases, detect the presence or lack of presence for a glue-line and has been able to image some forms of anomalies in glue lines. Acoustic microscopy is being used for characterization of interfaces in IC packages [153], as illustrated with Figure 8.34. Recent studies have continued to investigate characterization of adhesively bonded joints. Quantification of adhesion, including detecting degradation has been claimed when using angled beam ultrasonic spectroscopy [154]. A further approach that is receiving attention is the use of nonlinear ultrasound. It has been applied to structural epoxy-metal bonds, and a review of this approach was provided by Bockenheimer et al. [155].

The detection of regions with the "kissing-bond" where there is intimate physical contact and no (or minimal) mechanical adhesion and the prediction of bond gradation in bond strength remains challenging. For bonded joints, the average, as well as local adhesion, are important in prediction of joint strength. Higher frequency imaging, the use of EMATs (when metal layers are involved), and laser ultrasound, including generation of shear and interface waves have all been reported applied to this problem.

REFERENCES

1. Bond, L. J., W. F. Kepler, and D. M. Frangopol. 2000. *Constr Build Mater* 14(3):133–46.
2. Jones, R. 1970. A review of the NDT of Concrete. In *Proceedings of ICE Symposium on NDT of concrete and Timber*, 1–7. London, UK: Inst. Civil Engineers.
3. Deltuva, J. 1962. *Statyba ir Arch* 1:159–64.
4. Gemesi, J. 1967. *Epitoanyaq* 19(1):31–4.
5. Honig, A., and V. Zapletal. 1966. *Inz Stavby* 4:145–53.
6. Jones, R. 1956. *J Appl Chem* (London) 6(5):226–30.
7. Savchuk, A., and A. Filipchinski. 1958. *Beton Zhelezobeton* 2:45–52.
8. Jones, R. 1963. *Ultrasonics* 1(1):78–82.
9. Mikhailov, A. V., and A. N. Radin. 1956. *Beton Zhelezobeton* 2:266–8.
10. Simeonov, J. 1965. *Pol Akad Nauk Zeszyty Probl Nauki Polskiej* (26):410–3. (in English).
11. Pawlowski, Z., and J. Raniszewski. 1965. Pol. Akad. Nauk Zeszyty Probl. *Nauki Polskief* 26:373–289.
12. Kordina, K. 1967. *Materialpruefung* 9(3):81.
13. Shum, Y. F. 1966. *Beton Zhelezobeton* 12(6):15.
14. Carleton, H. R., and J. F. Muratore. 1986. Ultrasonic evaluation of concrete. In *Proceedings of IEEE Ultrasonics Symposium*, November 17–19, Williamsburg, VA, ed. B. R. McAvoy. 1017–20. New York: Ultrasonics, Ferroelectrics and Frequency Control Society, IEEE Institute of Electrical and Electronics Engineers.
15. Malhtotra, V. M., and N. J. Carion. 1991. *CRC handbook on NDT of concrete*. Boca Raton, FL: CRC Press.
16. ACI. 1998. NDT Methods for evaluation of concrete in structures. Report ACI Committee 228, Report No. 228 2R-98, ACI International, Am Concrete Inst Farmington Hills, MI.
17. Nazarian, S., and L. D. Olson, eds. 1995. SPIE NDE of aging structures and dams. *Proceedings No.* 2457. Bellingham, WA: SPIE.
18. IAEA. 2002. *Guidebook on NDT of Concrete Structures: Training Course Series No. 17*, 242. Vienna: International Atomic Energy Agency.

19. Trtnik, G., F. Kavcic, and G. Turk. 2009. *Ultrasonics* 49(1):53–60.
20. Sansalone, M., and N. J. Carino. 1991. *Impact-echo: A Method for Flaw Detection in Concrete Using Transient Stress Waves. Report NBSIR 86–34452.* Washington, DC: National Bureau of Standards.
21. Piwakowski, B., A. Fnine, M. Goueygou, and F. Buyle-Bodin. 2004. *Ultrasonics* 42(1–9):395–402.
22. Kepler, W. F., L. J. Bond, and D. M. Frangopol. 2000. *Constr Build Mater* 14(3):147–56.
23. Radakovic, Z., W. Kepler, and L. J. Bond. 1996. Ultrasonic assessment of concrete in uniaxial compression. In *Proceedings 3rd Conference NDE of Civil Structures and Materials, Boulder, CO*, 455–69. Sept 9–11 1996. Boulder, CO: Express Press.
24. Chaix, J.-F., V. Garnier, and G. Corneloup. 2003. Quantitative ultrasonic characterisation of concrete damage using a multiple scattering model. In *Proceedings, World Congress Ultrasonics*, 451–4. Sept. 7–10, 2003. Paris: French Acoustical Soc.
25. Berriman, J. R., D. A. Hutchins, A. Neild, T. H. Gan, and P. Purnell. 2006. *IEEE Trans UFFC* 53(4):768–76.
26. Kim, S.-D., D.-H. Shin, L.-M. Lim, J. Lee, and S.-H. Kim. 2005. *IEEE Trans UFFC* 52(7):1145–51.
27. Elkmann, N., S. Kutzner, J. Saenz, B. Reimann, F. Schultke, and H. Althoff. 2006. Fully automated inspection system for large underground concrete pipes partially filled with wastewater, IEEE Intelligent Robots and Systems. In *Proc. 2006 IEEE RSJ, Int. Conf. October*. 4234–8. New York: IEEE.
28. Bosek, M., and J. Ranachowski. 1965. *Pol Akad Nauk Zeszyty Probl Nauki Polskiej* (26):326–33.
29. Stuber, C. 1967. U.K. Atomic Energy Authority, AERE-Trans.-1077. Report AERE (Atomic Energy Research Establishment) Transactions 1077. Harwell, Oxfordshire, UK: AERE.
30. Haro, L., and S. Hyvrylainen. 1968. *Bull Assoc Suisse Electr* 59(11):486–90.
31. Weyl, D. 1957. *Arch Tech Mess* 257:121–4.
32. Zabinska, T., and R. Pampuch. 1965. *Pol Akad Nauk Zeszyty Probl Nauki Polskiej* 26:439–48.
33. Whymark, R. R., and W. E. Lawrie. 1960. *Materials Analysis and Evaluation Techniques*, Report for 1 April 1959–29 February 1960, Armour Research Foundation, Chicago, May 1960.
34. Som, A. K., and L. J. Bond. 1988. Ultrasonic inspection of engineering ceramics. In *Reliability in Non-Destructive Testing*, ed. C. Brook and P. D. Hanstead, September 12–15, 1988. Portsmouth, UK. 235–48. Oxford, UK: Pergamon.
35. Gilmore, R. S., K. C. Tam, J. D. Young, and D. R. Howard. 1986. *Phil Trans Roy Soc Lond A* 320(1554):215–325.
36. P. Kauppinen, H. Jeskanen and J. Pitkanen. 2000. Characterization of ceramic coatings with large-aperture low-frequency transducers. In *Proc. 15th WCNDT, Roma, Italy, 15–21 Oct 2000*. Paper No. 041. Brescia, Italy: Italian Soc NDT and Monitoring Diagnostics.
37. Q. U. Jun and P. J. Blau, *Ceramic Eng Sci Proc* 27(2), 267–274 (2007).
38. M. T. Jones, G. V. Blessing, and C. R. Robbins. 1986. *Mater Eval* 44:859–62.
39. Stockman, A., and P. S. Nicholson. 1986. *Mater Eval* 44:756–61.
40. Stockman, A., P. Mathieu, and P. S. Nicholson. 1987. *Mater Eval* 45:736–42.
41. Cielo, P. X. Maldaque, S. Johan, and B. Lauzon. 1986. *Mater Eval.*
42. Kessler, F. A. 1986. *Mater Eval.* 44:626–7.
43. Roth, D. J., S. J. Klima, J. D. Kiser and G. Y. Baaklini. 1986. *Mater Eval.* 44: 726–68.
44. Bond, L. J., N. Jayasundere, A. D. Sinclair, and I. R. Smith. 1982. Investigation of ultrasonic transducers, as used for NDT. In *Proceedings, Review of Progress in Quantitative Nondestructive Evaluation (QNDE)*, Vol. 1, ed. D. O. Thompson and D. E. Chimenti, 701–712. New York: Plenum Publishing Corp., Boulder, CO, Aug 2–7, 1981.
45. Kulkarin, N., B. Mudgil, and M. Bhardwaj. 1994. *Am Ceramic Soc Bull* 73(6):146–53.

46. Kulkarin, N., B. Mudgil, and M. Bhardwaj. 1994. *Am Ceramic Soc Bull* 73(7):83–5.
47. Ganchev, T., I. Potamitus, and N. Fakotakis. 2007. Acoustic monitoring of singing insects. Acoustics, Speech and Signal Processing, 2007. In *IEEE Int. Conf., 15–20 April 2007*, Vol. 4, IV 721–4. New York: IEEE.
48. Haack, R. A., and R. W. Blank. 1990. *J Acoust Emission* 9(3):181–7.
49. Bucur, V. 1995. *Acoustics of Wood*. New York: CRC Press.
50. Ouis, D. 2002. *Wood Sci Technol* 36(4):335–46.
51. Bucur, V., and F. Feeney. 1992. *Ultrasonics* 30(2):76–81.
52. The Echo 23, 319–20, published by J. Krautkramer and H. Krautkramer, Cologne, Germany, April 1970.
53. Bucur, V. 1985. *Ultrasonics* 23(6):269–75.
54. Bucur, V. 2003. Nondestructive characterization and imaging of wood. In *Springer Series in Wood Science*, ed. T. E. Timell and R. Wimmer. Berlin: Springer.
55. Falk, R. H., M. Patton-Mallory, and K. A. McDonald. 1990. NDT of wood products and structures: State of the art. In *NDT and evaluation for manufacturing and construction. Proc.*, ed. L. M. Henrique, 137–47. Aug 9–12, 1988 Champaign, IL. New York: Hemisphere Pub. Co.
56. Bucur, V. 2003. *Meas Sci Technol* 14(12):R91–8.
57. Beall, F. C. 2002. *Wood Sci Technol* 36(3):197–212.
58. Kang, H., and R. E. Booker. 2002. *Wood Sci Technol* 36(1):41–54.
59. Sandoz, J. L. 1989. *Wood Sci Technol* 23(1):95–108.
60. Machado, J. S., R. A. Sardinha, H. P. Cruz. *Wood Sci Technol* 38(4):277–84.
61. Buckley, J. 2000. Air-coupled ultrasound—A millennial review. In *Proc. 15th WCNDT*, 15–21 October, Roma, Italy, Paper No. 507. Brescia, Italy: Italian Soc NDT and Monitoring Diagnostics.
62. Collins, J. T. 1967. *Instrum Soc Am Trans* 6(4):303–6.
63. Tucker, B. J., and D. A. Bender. 2003. *Forrest Products J* 53(6):27–32.
64. Ritter, M. A., S. R. Duwadi, and E. Cesa. 2000. *Prog Struct Eng Mater* 2(3):290–5.
65. Ross, R. J., R. F. Pellerin, N. Volny, W. W. Salsig, and R. H. Falk. 1999. *Inspection of Timber Bridges Using Stress Wave Timing Nondestructive Evaluation Tools*. U.S. Dept. Agriculture: Forest Products Laboratory, Report FPL-GTR-114. Madison, WI: USDA.
66. Kawamoto, S., and R. S. Williams. 2002. *Acoustic Emission and Acousto-Ultrasonic Techniques for Wood and Wood-Based Composites: A Review*. U.S. Dept. Agriculture: Forest Products Laboratory, Report FPL-GTR-134. Madison, WI: USDA.
67. Brashaw, B. K., R. J. Vatalaro, J. P. Wacker, and R. J. Ross. 2005. *Condition Assessment of Timber Bridges: 2. Evaluation of Several Stress-Wave Tools*. U.S. Dept. Agriculture: Forest Products Laboratory, Report FPL-GTR-160. Madison, WI: USDA.
68. Emerson, R. N., D. G. Pollock, D. I. McLean, K. J. Fridley, R. J. Ross, and R. E. Pellerin. 1999. Nondestructive testing of large timber bridges. In *Proc. 11th Int. Symp. NDT of Wood, Madison, Wisconsin, Sept 9–11, 1998*, 175–84. Madison: Forest Products Soc.
69. ASTM International. 2008. *Wood Stand* 4.10, Annual.
70. Murray, A., F. Mecklenburg, C. M. Fortunko, and R. E. Green. 1996. *J Am Inst Conserv* 35(2):145–62.
71. Siffiolo, A. M., L. D'Acquisto, A. R. Maeva, and R. G. Maev. 2007. *IEEE Trans UFFC* 54(4):836–46.
72. Black, E. L., M. T. Htun, M. Jackson, and F. Johanson. 1967. *TAPPI J* 50(11 Pt. 1):542–7. (in English)
73. Luukkala, M., P. Heikkila, and J. Surakka. 1971. *Ultrasonics* 9(4):201–8.
74. Luukkala, M., and P. Merilainen. 1973. *Ultrasonics* 11(5):218–21.
75. Kopkin, B. 1999. *TAPPI J* 82(5):137–40.
76. McIntyre, C. S., D. A. Hutchins, D. R. Billson, and J. Stor-Pellinen. 2001. *IEEE Trans UFFC* 48(3):717–27.

77. Gomez, T. E. 2003. *IEEE Trans UFFC* 50(6):676–85.
78. Greenwood, M. S., P. D. Panetta, L. J. Bond, and M. W. McCaw. 2006. *Ultrasonics* 44(Supp 1):e1123–6.
79. Anon. 1957. *Ultrasonic News* 1:26.
80. Sokolov, S. I. 1946. *Zh Tekh Fiz* 16:283.
81. Workman, J., D. J. Veltkamp, S. Doherty, B. B. Anderson, K. E. Creasy, M. Koch, J.F. Tatera et al. 1999. *Anal Chem* 71(12):121R–80R.
82. Frielinghaus, R. 1973. The ultrasonic inspection of plastics. *Int Conf NDT Warsaw Poland*, 7th World Congress NDT Warsaw, Poland, Organized by committee "ICNDT" International Committee NDT (secretariat British Inst. NDT - Northampton, UK). Paper # J18.
83. Thompson, D. O., and D. E. Chimenti, eds. 1982–2010. Review of progress in quantitative nondestructive evaluation. In Vols 1–39. New York: Plenum. now *Am. Inst. Phys, Conf. Series* (1982-2010): *Annual Review of Progress in QNDE, AIP Conf. Proceedings*. 1096.
84. *NDT&E International*. Amsterdam: Elsevier.
85. Ashbee, K. H. G., ed. 1986. *Polymer NDE, Proc. European Workshop on NDE of polymers and polymer matrix composites, Termar do Vimeiro, Portugal, Sept. 4–5, 1984*. Lancaster: Technomic Press.
86. ASTM International. 2008. *Plastics and Rubber*, Sections 08 and 09, Annual.
87. Bailey, J. A., and J. R. Davila. 1971. *Appl Sci Res* 25(3–4):245–61.
88. Parker, R. L., and J. R. Manning. 1986. *J Cryst Growth* 79(1–3):341–53.
89. Parker, R. L., J. R. Manning, and N. C. Peterson. 1985. *J Appl Phys* 58(11):4150–64.
90. McDonough, M. W., and A. Faghri. 1993. *Exp Heat Transf* 115(4):1075–8.
91. Mauer, F. A., S. J. Norton, Y. Grinberg, D. Pitchure, and H. N. G. Wadley. 1991. *Metall Trans B* 22B(4):467–73.
92. Bar-Cohen, Y. 1986. *Mater Eval* 44:446–54.
93. Williams Jr., J. H., and S. S. Lee. 1985. *Mater Eval* 43:561–5.
94. Henecke II, E. G., and J. C. Duke Jr. 1985. *Mater Eval* 43:740–5, 53.
95. Rogovsky, A. J. 1985. *Mater Eval* 43:547–55.
96. Hagemeier, D. J., and R. H. Fassbender. 1985. *Mater Eval* 43:556–60.
97. Nestleroth, J. B., J. L. Rose, M. Bashyam, and K. Subramanian. 1985. *Mater Eval* 43:556–60.
98. van Dreumel, W. H. M., and J. L. Speijer. 1981. *Mater Eval* 39:922–5.
99. Jones, T. S. 1985. *Mater Eval* 43:746–53.
100. Chang, F. H., J. R. Bell, J. L. Brown, and R. W. Haile. 1985. *Mater Eval* 43:1117–23.
101. McNally, D. J. 1985. *Mater Eval* 43:728–32.
102. Hill, E. V. K., and T. J. Lewis. 1985. *Mater Eval* 43:859–63.
103. Tretout, H. 1998. Review of advanced ultrasonic techniques for aerospace structures. In *Proc. 7th European Conference NDT, Copenhagen, 26–29 May*, Vol. 3(9). Paper No. 46, NDT.net.
104. Roach, D. 2008. Assessing conventional and advanced NDI for composite aircraft. *Composites World*, (Web published), posted 6/30/2008. 7 pages. Reprinted from *High Performance Composites*, July 2008. Gardner Publications Inc.
105. Aldridge, E. E. 1970. Ultrasonic holography. In *Research Techniques in NDT*, ed. R. S. Sharpe, Vol. 1, 133–54. London, UK: Academic Press.
106. Aldridge, E. E., and M. J. -M. Clement. 1982. Ultrasonic holography. In *Ultrasonic testing*, ed. J. Szilard, 103–166. Chichester, UK: Wiley.
107. Brendon, B. B. 1969. A comparison of acoustic holography methods. In *Acoustical Holographys. Proc. 1st Int. Symp. Acoustical Holography, Huntington Beach, CA, Dec 14–15, 1967*, 57–71. New York: Plenum Press.
108. Durruty, A., G. Lefebvre, and J. Bernardi. 1994. New acoustic holography applied to composite materials. In *Proc. 6th ECNDT, Nice France*, 24–28 October, 1994.

109. Crawford, S. L., S. E. Cumblidge, S. R. Doctor, T. E. Hall, and M. T. Anderson. 2008. *Preliminary assessment of NDE methods of inspection of HDPE butt fusion piping joints for lack of fusion*, May, 2008 PNNL Report-17584. Richland, WA. Pacific Northwest National Laboratory.

110. Liu, C. T. 2000. *Comput struct* 76(1–3):57–65.

111. Halsey, G. H. 1969. *Rubber Age* 101(2):70–5.

112. Kirk, G. R., and L. U. Rostrelli. 1968. *Nondestructive Inspection Techniques for Aircraft Tires, Final Technical Report*, Southwest Research Institute, San Antonio, Texas, April 1968.

113. Halsey, G. H. 1968. *Mater Eval* 26(7):137–42.

114. Patton, G. R., and P. Hatfield. 1952. *Electron Eng* 24:522–5.

115. Riechmann, P. 1956. *Z Angew Phys* 8(8):386–91.

116. Downes, J., P. Zhang, and M. L. Peterson. 1999. *IEEE/ASME Trans Mechatron* 4(3):301–11.

117. Bond, L. J., A. R. Greenberg, A. P. Mairal, G. Lest, J. H. Brewster, and W. B. Krantz. 1995. Real-time nondestructive characterization of membrane compaction and fouling. In *Proc. Review of Progress in Quantitative nondestructive Evaluation (QNDE)*, Vol. 14, ed. D. O. Thompson and D. E. Chimenti, 1167–1173. New York: Plenum, Snowmass, CO, July 31-Aug 5, 1994.

118. Mairal, A. P., A. R. Greenberg, L. J. Bond, and W. B. Krantz. 1999. *J Membrane Sci* 159(1–2):185–96.

119. Konagurthu, S., R. Peterson, L. J. Bond, A. R. Greenberg, W. B. Krantz, D. Kishoni, D. Kuhns, L. Burgess, and A. M. Brodsky. 1996. Development of an acoustic technique for assessing structure–property characteristics of polymeric thin films. In *Proc. IEEE Ultrasonic Symposium*, eds. M. Levy, S. C. Schneider, and B. R. McAvoy, Vol. 1, 413–7. San Antonio, TX: IEEE, Nov. 3–6, 1996.

120. Kools, W. F. C., S. Kongagurthu, A. R. Greenberg, L. J. Bond, W. B. Krantz, Th. van den Boomgaard, and H. Strathmann. 1998. *J Appl Polym Sci* 69(10):2013–9.

121. Peterson, R. A., A. R. Greenberg, L. J. Bond, and W. B. Krantz. 1998. *Desalination* 116(2–3):115–22.

122. Reinsch, V. E., A. R. Greenberg, S. S. Kelly, R. Peterson, and L. J. Bond. 2000. *J Membrane Sci* 171(2):217–28.

123. Wickramasinghe, H. K., and C. R. Petts. 1980. Acoustic microscopy in high pressure gases. In *Proc. 1980 IEEE Ultrasonics symposium*, ed. B. R. McAvoy, 5–7 Nov 1980, Boston MA, 668–72. New York: IEEE.

124. Bond, L. J. 2001. Inspection of solid rocket motors and munitions using ultrasound. In *Proc. 50th JANNAF Propulsion Meeting*, July 11–13, 2001. CD Publication No. 705. Boston, MA: Chemical Propulsion Information Agency.

125. Dean, D. S., and D. Young. 1973. The reduction and display of ultrasonic data. In *Research Techniques in NDT*, Vol. 2, ed. R. S. Sharpe. London, UK: Academic Press.

126. Dean, D. S., and D. T. Green. 1979. Ultrasonic imaging applied to NDT of rocket propellants. In *AGARD Conference Proceedings 256*, 1–13. AGARD/NATO, Paper 33. Advanced Fabrication Processes, Report # AGARD-CR-256. January 1979.

127. Rooney, M. 1990. *NDE of Structural Bonds for Large Solid Rocket Motor Applications*. M. S. Thesis. Center for NDE, Johns Hopkins University, MD.

128. Nici, R. 1993. *Ultrasonic Wave Propagation Model for Nondestructive Evaluation of Solid Rocket Motor Propellant*. PhD Thesis. Boulder, CO: University of Colorado.

129. Springer, C. E. 1972. *Ultrasonic imaging of cylindrical explosive loads*. Dahlgreen, VA: NSWC. TR-2737.

130. Blessing, G. V., and J. M. Warren. 1975. *Mater Eval* 35(9):69–75.

131. Knollman, G. C., J. L. Weaver, J. J. Hartog, and J. L. Bellin. 1975. *J Acoust Soc Am* 58(2):455–70.

132. Elban, W. L., A. L. Bertram, G. V. Blessing, and R. R. Bernecker. 1979. *Use of Ultrasonics for the NDE of Damage in Mechanically Deformed Propellants*. Silver Springs VA: Naval Surface Warfare Center Report.

133. Knollman, G. C., R. H. Martinson, and J. L. Bellin. 1979. *J Appl Phys* 50(1):111–20.

134. Knollman, G. C., R. H. Martinson, and J. L. Bellin. 1980. *J Appl Phys* 51(5):3164–70.

135. Knollman, G. C., and R. H. Martinson. 1979. *J Appl Phys* 50(12):8034–7.

136. Kent, D. A. 1967. *Non-destr. Test* 1:108–12.

137. Blessing, G. V., and J. M. Warren. 1977. *Mater Eval* 35(9):69–75.

138. Morris, E., L. M. Glowacki, and L. J. Bond. 1997. Improved Quality Assurance Program for the Hydra'70 Solid Rocket Motor. In *Proc. Review of Progress in Quantitative Nondestructive Evaluation (QNDE), San Diego, CA, July 27–Aug 1, 1996*, ed. D. O. Thompson and D. E. Chimenti, Vol. 17, 1973–80. New York: Plenum.

139. Chaloner, C. A., and L. J. Bond. 1987. *IEE Proc Part A* 134(3):257–65.

140. Chaloner, C and L. J. Bond. 1986. *Electronics Letters* 22(3):171–3.

141. Bond, L. J., C. A. Chaloner, S. J. Wormley, S. P. Neal, and J. H. Rose. 1988. Recent advances in Born inversion (weak scatterers). In *Proc. Review of Progress in Quantitative Nondestructive Evaluation (QNDE), Williamsburg, Virginia, June 22–26, 1987*, Vol. 7, ed. D. O. Thompson and D. E. Chimenti, 437–44. New York: Plenum.

142. Smith, P. R. 1987. *Ultrasonics* 25(3):138–40.

143. Cobb, W. N. 2007. Ultrasound characterization of energetic materials. In *Proc. 19th Int. Congress Acoustics, Madrid, Spain, 2–7 Sept., 2007*. Paper No. ULT-17–009. Sociedad Espanola de Acustica (SEA).

144. Husing, N., and U. Schubert. 2005. Aerogels. In *Synthesis of inorganic material*, 1–27. Weinheim Germany: Wiley–VCH Verlag GmbH & Co.

145. Gomez, T. E., F. M. de Espinosa, E. Rodriguez, A. Roig, and E. Molins. 2002. Fabrication and characterization of silica aerogels for air-coupled piezoelectric. In *Proc. 2002 IEEE Ultrasonics Symposium*, eds. D. E. Yuhas, and S. C. Schneider, October 8–11, 2002, Munich, Germany, 1107–10. New York: IEEE.

146. Good, M. S., L. J. Bond, S. N. Schlahta, and B. E. Johnson. 2004. *Ultrasonic Characterization of Low Density Foam Material: Phase 2 Report, December 2004*. Richland WA: Pacific Northwest National Laboratory.

147. Pilarski, A. 1985. *Mater Eval* 43:765–7.

148. Segal, E., and J. L. Rose. 1980. Nondestructive testing techniques for adhesive bonded joints. In *Res. Tech. in NDT*, Vol. 4, ed. R. S. Sharpe, 275–316. London, UK: Academic Press.

149. Rose, J. L., and P. A. Meyer. 1973. *Mat Eval* 31(6):109–114.

150. Brown, A. F. 1982. Ultrasonic spectroscopy. In *Ultrasonic Testing*, ed. J. Szilard, 167–215. Chichester, UK: John Wiley.

151. Brown, A. F. 1973. *Ultrasonics* 11(5):202–10.

152. Sullivan, J. M., R. Ludwig, and Y. Geng. 1990. Numerical simulation of ultrasound NDE for adhesive bond integrity. In *Proc. IEEE Ultrasonics Symposium*, Dec. 4–7, Waikiki, HI, ed. B. R. McAvoy. Vol. 2, 1095–8. New York: IEEE.

153. Ong, S. H., S. H. Tan, and K. T. Tan. 1997. Acoustic microscopy reveals IC packaging hidden defects. *Electronic Packaging Technology Conf. Oct 8–10, 1997*, Singapore, (EPTC'97), 297–303. New York: IEEE.

154. Adler, L., S. Rokhlin, and A. Baltazar. 2001. Determination of material properties of thin layers using angle beam ultrasound spectroscopy. In *Proc. IEEE Ultrasonics Symposium*, 701–4.

155. Bockenheimer, C., D. Fata, W. Possart, M. Rothenfusser, U. Netzelmann, and H. Schaefer. 2002. *Int J Adhes Adhes* 22(3):227–33.

10 Imaging, Process Control, and Miscellaneous Low-Intensity Applications

10.1 INTRODUCTION

Although nondestructive testing (NDT) provides one of the larger and better known industrial markets for ultrasonic equipment, it is only one of many areas in which low-intensity ultrasonic energy is applied. Several additional areas of applications are discussed in this chapter. The development of numerous industrial sensing or metrology applications of ultrasound has occurred in parallel with the development of medical applications. In many cases, similar or related technology has been used in the underlying research and development activities, particularly in applications that involve optically opaque media.

Other applications discussed in this chapter are process monitoring, measurement, and control devices that measure an array of properties including density, viscosity, and particle size; sound navigation and ranging (SONAR) and fish locators; underwater-communication devices; surface acoustic wave (SAW) devices and sensors; delay lines; flowmeters; anemometers; pressure- and temperature-measuring devices; leak detectors; and intrusion detectors.

10.2 ULTRASONIC IMAGING

Modern ultrasonic imaging is, in most peoples' minds, now almost invariably and nearly automatically associated with its numerous uses in medicine. The development of this technology can be seen to have been spurred in the United States by a comprehensive assessment of the then state-of-the-art that was provided by the National Science Foundation Blue Ribbon Task Force on Ultrasonic Imaging (1973–1974). The potential for medical applications was recognized, and it was seen that focused activities were needed to move the science and technology forward. Major advances were then developing using ultrasonic forms of holography, which leveraged concepts based in the work of Dennis Gabor, for which he received the 1971 Nobel Prize for Physics. The various forms of early acoustic microscopy and phased array technology were also being developed in the same time frame. Much of the literature that reported developments in imaging was reported in the conference series that started as Acoustical Holography (1967) and was subsequently retitled Acoustical Imaging [1]. The 30th International Symposium on Acoustical Imaging was held in Monterey, California in 2009. Advances in digitization and modern computer data processing and display have enabled a diverse array of systems to be developed and deployed,

Acoustic Field Visualization

• Particle visualization (in fluids – liquid + gas)	• Thermal techniques
• Dye visualization	• Sonogram technique
• Moire fringe	• Ultrasonic scanning
• Interferometer techniques	• Acoustic microscopy
• Shadow photography	• Holography
• Schlieren	• Computer models (field simulation)

FIGURE 10.1   Visualization of ultrasonic fields. (After Stephens, R. W. B., An historical review of ultrasonics. In *Proc. Ultrasonics International '75*, 9–19, ICP Press, Guildford, UK, 1975.)

particularly in medical ultrasonics. This chapter is focused on industrial ultrasonics, with medical ultrasonics considered further in Chapter 14.

Industrial applications and process imaging are less numerous than those within medicine, not least due to the need for customization of implementations and systems: human morphology and composition is simply more uniform than that found in the industrial process streams to which ultrasound could be applied. The majority of ultrasonic sensors used in process monitoring, measurement, and control do not currently use imaging [2–4]. However, as the costs of imaging systems reduce, particularly those using phased arrays, the numbers of industrial applications of ultrasonic imaging are increasing. The applications of ultrasonics, which are now used for particular material characterization and process characterization implementations, are forming a bridge between traditional ultrasound used for NDT and materials science, and on into the measurement technologies used for a diverse array of process measurements.

The various forms of visualization applied to ultrasonic fields are summarized in Figure 10.1. The history for several of these areas is discussed in Section 10.2.1 of this chapter.

10.2.1   HISTORICAL BACKGROUND

Ultrasonic imaging is generally defined as any technique that forms a visible display of the intensity and, in some systems, phase distributions in an acoustic field. This definition covers (1) the electronic–acoustic image camera; (2) B-scan systems, which are increasingly used in NDT and are commonly used in medical diagnosis; (3) C-scan systems, which are widely used in NDT, generally for inspecting flat materials or cylinders, and their high-frequency forms "acoustic microscopy," which is being routinely used in the semiconductor industry and investigated for both industrial and medical research; (4) liquid surface levitation presentations; (5) liquid crystal, photographic, and similar presentations; (6) light-refraction methods; (7) mechanical methods; and (8) acoustical holography.

The origin of ultrasonic imaging has been attributed to various scientists, and the dates of invention for the different technologies vary widely. The similarities between acoustical "optics" and light optics were known to Lord Rayleigh in the nineteenth century.

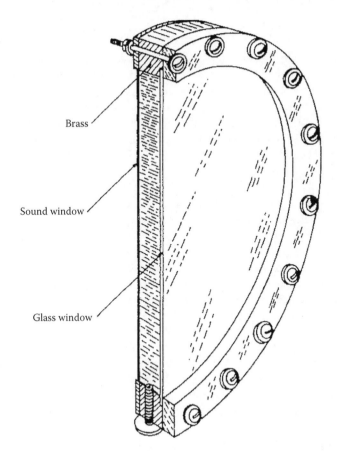

Brass

Sound window

Glass window

FIGURE 10.2   Pohlman-cell construction.

In 1937, Pohlman [6] initiated ultrasonic optical imaging by demonstrating that small, nonspherical particles have a tendency to align themselves so that the normal to their major surfaces becomes parallel to the direction of wave propagation. Pohlman produced a cell for imaging in which he placed a large number of disk-shaped aluminum particles suspended in toluene. The diameter of the aluminum particles was much smaller than the wavelength in toluene, and the thickness of the particles was much smaller than the diameter. Figure 10.2 shows the details of construction of a cell based on the Pohlman principle. Pohlman used the cell for testing materials during World War II.

According to Suckling [7], the first published demonstration of an actual sonic image was by Barbieri [8] in 1942.

Imaging using schlieren visualization started with the work of Robert Hooke (1635–1703) who used it to visualize thermals in air caused by a candle. In the nineteenth century, August Toepler reinvented the schlieren technique between 1859 and 1864 [9]. Toepler published papers on methods of making visible sound waves produced by a spark discharge in 1866 and 1867. The schlieren technique is based on refraction of light as it passes through a transparent medium in which a pressure gradient exists.

The Soviet scientist Sokolov is recognized as having invented the electron ultrasonic image device, the "ultrasonic microscope" [10]. There are indications that Sokolov's actual reduction of the device to practice came in 1947, although he claims that the 1947 ultrasonic microscope is a refinement of one used by him prior to World War II [11]. All subsequent electron acoustic image converters use the principles he originally proposed and thus verify the soundness of his method; a review of devices produced up to 1975 is provided by Brown et al. [12]. In recent years, although there has been a transition to the use of phased arrays for imaging, several novel ultrasonic cameras have been reported (e.g., Lasser et al. [13]).

Various investigators have studied photographic techniques of one form or another, such as those involving the use of sound-sensitive photolayers, starch–potassium iodide layers on paper, thermosensitive paper, phosphorescent screens, and liquid crystals. These methods, however, require intensities that are too high for most practical applications.

The liquid levitation method was developed by Schuster, according to Freitag [14]. In this case, the plane of the sound image is made to coincide with the surface of a liquid so that the sound radiation pressure interacts with the force of gravity in such a manner as to cause the fluid surface to rise, in dependence upon the acoustic intensity. The sound image appears as a relief image on the fluid surface and is then converted into a light image by means of schlieren optics and made visible on a screen or ground glass plate. The sensitivity, which is quite acceptable for practical requirements, depends on the curvature at the edges of the bulge rather than on the total height of the bulge.

Recent developments in computer science and digital electronics have led to routine use of B-scan and C-scan ultrasonic images in medical diagnosis and materials evaluation. These images can be stored, enhanced, and displayed in any conventional form digital image display. Factors involved in developing an image are transducer position and angle relative to the object, pulse travel time and echo amplitude, and known acoustical properties of the coupling fluid and the object, such as velocity of sound, attenuation, scattering phenomena, phase, diffraction, and acoustic impedance, which determine reflection. Most of these factors (and all factors in the case of velocity dispersion) are frequency dependent. These frequency relationships are important for developing images with good resolution.

Many of the historical methods leading to the present-day imaging procedures are discussed in the following sections.

10.2.2   Electron Acoustic Image Converter

Electron acoustic image converters are modified from the original Sokolov disclosures. Sokolov envisioned a television-camera tube in which the light-sensitive element is replaced by a pressure-sensitive element—a piezoelectric plate. Secondary electrons are given off when the plate is struck by a scanning beam of electrons. Under the influence of an ultrasonic field, electrical potentials are produced on the faces of the piezoelectric plate that are proportional to the impressed acoustic pressure, and these electrical potentials modulate the secondary emission of electrons. Sokolov used quartz plates. Figure 10.3 is a simplified diagram of Sokolov's ultrasonic microscope, which is also known as the Sokolov camera.

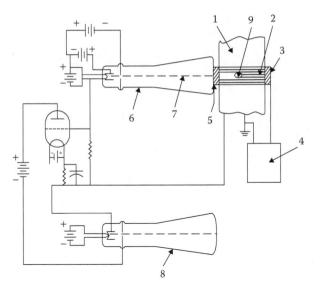

FIGURE 10.3 Electron ultrasound camera by Sokolov. (1) object under examination, (2) ultrasound beam, (3) source transducer, (4) source oscillator, (5) receiving transducer, (6) television iconoscope as used by Zworykin, (7) scanning electron beam, (8) picture-reproducing television tube scanned in sympathy with the camera 6, and (9) defect.

Secondary emission resulting from the impingement of an electron beam on the surface of a piezoelectric plate is a function of the velocity of the primary electrons and depends on the plate material. A secondary-emission ratio is defined as the ratio of the number of electrons leaving the plate surface to the number of electrons arriving at the surface. When the number of electrons leaving the plate equals the number of those arriving at the plate, the secondary-emission ratio is unity. If the primary-electron velocity is increased from zero, the secondary-emission ratio also increases from zero, passes through a maximum, and then decreases. For some piezoelectric materials, the maximum ratio is greater than unity and there are two velocities, one above and one below the point of maximum secondary emission, at which the ratio becomes unity. For other materials, the maximum ratio never exceeds unity.

If the secondary-emission ratio is less than unity, the excess charge accumulates on the surface of the plate, building up a potential in opposition to the scanning beam of electrons and reaching a limit when the incident beam is completely repelled.

At secondary-emission ratios greater than unity, a space charge accumulates in front of the plate surface. The space charge retards the scanning beam and leads to a loss of sensitivity toward the edges of the picture.

Starting with the basic design of Sokolov, Smyth et al. [15] developed an ultrasonic-imaging camera (Figure 10.4) that could be operated in either an amplitude-sensitive mode or a phase-sensitive mode.

For the amplitude-sensitive mode, a low-voltage (160–200 V) scanning electron beam was used. At this low voltage, the secondary-emission ratio was less than unity, which caused the electrical potential of the surface of the piezoelectric plate, which in

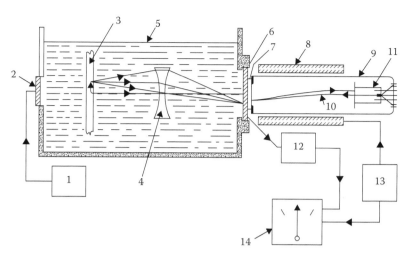

FIGURE 10.4 Ultrasound image camera of Smyth. (1) 4-MHz oscillator, (2) ultrasound transmitter, (3) material under inspection, (4) acoustic lens, (5) water, (6) signal plate, (7) quartz plate, (8) deflection and focus coils, (9) ultrasound camera tube, (10) ion-trap mesh, (11) electron beam, (12) electron gun, (13) video frequency amplifier, and (14) deflection generator.

Smyth's camera was quartz, to be brought down to that of the cathode of the electron gun in the absence of an ultrasonic field. After scanning in the absence of ultrasound, the ultrasound was turned on and the plate was scanned again. Electrons collected over the surface of the plate in proportion to the piezoelectric voltage present at each part of the surface. A corresponding image charge was formed on an anode located outside the tube, and the video signal was derived from the anode charges. The ultrasound was turned off, and the negative charge on the piezoelectric plate was removed by flooding with gas ions. Time required to remove the charge was 0.02 second. Smyth et al. [15] also recommended the use of high-voltage scanning to remove the negative charge, followed by low-voltage scanning to bring the surface back to cathode potential, both with the ultrasound turned off. Then, with the ultrasound on, the plate was scanned again at low voltage to form the image as before. This procedure avoided the need for ion recharging and thus allowed the use of a high-vacuum tube.

For the phase-sensitive mode, higher voltage beams (600 V) were used along with an auxiliary grid on which the secondary-emission current was collected. The inner quartz surface was stabilized at about the voltage of the auxiliary grid. The piezoelectric charge on the quartz plate tended to become neutralized and, as a result, the current to the quartz plate and to the collector grid fluctuated at the frequency of the ultrasonic waves. The output signal at each picture point was, therefore, proportional to the incident ultrasonic pressure.

Alternative means for reproducing the image on the piezoelectric surface have been devised to avoid some of the difficulties inherent in high-velocity scanning-beam stabilization of an insulating surface. One proposed method involves the use of a photoemissive surface deposited on the piezoelectric surface, scanning then being accomplished by a flying spot of light [16]. Mechanical scanning methods also have been used and are considerably more sensitive than scanning with an electron beam. The surface of a piezoelectric plate located in an ultrasonic field was scanned mechanically using a capacitive, noncontacting electrode or by sliding a small electrode in contact with the surface. The speed of operation of the electron acoustic image converter more than offset the disadvantages of lower sensitivity.

Smyth [17] also considered pulsed operation, which avoids cancellation effects due to reflected-wave interferences. However, the scanning beam and the arrival of the pulse must be synchronized and the pulse repetition rate must be low enough to avoid interferences between pulses. Only one point per pulse is obtained, and scanning the ultrasonic field to obtain one complete image may take several seconds.

The quartz plate of Smyth's ultrasound image camera was bonded by an epoxy cement to an annular plate of nonmagnetic steel. The steel plate was in turn cemented to one end of a glass tube 4.5 cm in diameter. The mating bonded surfaces were lapped flat so that very little resin was exposed to the vacuum. An electron gun and a zirconium getter were fitted to the other end of the glass tube. An ion-trap gauze was also located in the tube near the quartz plate.

Jacobs [18] developed ultrasound image converters also based on Sokolov's method but with an electron multiplier built into the camera tube (Figure 10.5). This construction provided shielding for the low-level circuits and virtually eliminated ground loops. The effects of stray electromagnetic radiation fields from the ultrasound generator were minimized. The electron multiplier amplified the signal currents to approximately 10^5 times greater than the threshold values before they were introduced into the associated amplifiers. In Jacobs' camera, the secondary-emission electrons produced by the scanning beam were attracted toward a positively charged electrode in the electron-multiplier structure. The electron multiplier then served as a wideband amplifier with good noise characteristics. Its output constituted the video signal, which was processed through a conventional closed-circuit television system. Jacobs [18] used both sealed ultrasonic camera tubes with quartz crystals permanently sealed to the end of the tube and tubes with interchangeable crystals that permit operation over the frequency range of 1 to 15 MHz.

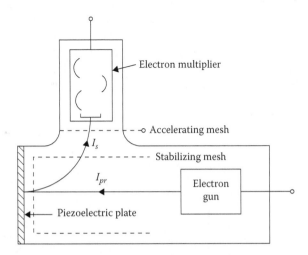

FIGURE 10.5 A schematic diagram of Jacobs' ultrasonic camera tube.

Jacobs et al. [19] introduced the use of color to enhance the sensitivity of the ultrasonic image converter. The method used the National Television System Committee system of video signal processing, which is the standard for color television broadcasting in the United States. The operating frequency was 3.58 MHz. The advantage of the color display is the greater sensitivity of the human eye to differential changes in color than to differential changes in brightness. The color display indicated both relative amplitude and phase of the received ultrasonic signal. Jacobs et al. [19] claimed that at a frequency of 3.58 MHz, the minimum resolvable detail is approximately 0.6 mm without lenses and approximately 1 mm with lenses.

The minimum resolvable detail in an electron acoustic image converter depended on the thickness of the piezoelectric plate, which, at resonance, is one-half wavelength. The thickness of the plate also is the limiting factor on the size of the image. The diameter-to-thickness ratio was necessarily small enough for the crystal to withstand the vacuum. Decreased sensitivity and diffraction effects within the crystal at the harmonics of the crystal limited the usefulness of operation at these frequencies.

In using the electron acoustic image converter, it was often desirable to make use of acoustic lenses. There are at least two reasons for this: (1) the sensitivity of the systems may be increased by the higher concentration of energy and (2) observations may be made over a larger area by passing the image through a lens and concentrating it on the receiving piezoelectric element of the image converter. This would be particularly advantageous at higher frequencies, at which the resolution is best but small-diameter receiving elements are required because of the increased fragility of the piezoelectric unit, as mentioned previously. The greatest problem with acoustic lenses is image distortion due to the wavelengths involved.

Acoustic lenses used for focusing ultrasound or for imaging purposes must satisfy two conditions. The velocity of sound in the lens material must differ significantly from that of the surrounding medium and the reflection of energy at the boundary between the lens and the surrounding medium must be a minimum. The latter condition is automatically fulfilled when the acoustic impedance of the lens material exactly equals that of the surrounding medium. Liquid lenses of chloroform or carbon tetrachloride, which have acoustic impedances similar to that of water, have been used in the laboratory, but their use in industry poses a problem. Plastic lenses also have been used. The velocity of sound in these materials exceeds that in water, and the materials present a mismatch in impedance between the lens and the water, but not to the extent of rendering them useless. Their relatively high absorption limits their use to frequencies below 15 MHz. Aluminum and similar metallic lenses have lower absorption characteristics, so they could be useful at frequencies far above 15 MHz; however, the impedance mismatch between these materials and water prevents their effective use.

Since the velocity of sound in solids is higher than that in liquids, solid concave lenses are convergent and convex lenses are divergent.

In practice, the object and the image camera were located in the far field of the transmitting transducer. The field intensity was, thus, made fairly uniform over the object. Diffraction made it necessary to locate the camera as close as possible to the object. The shadow that formed the image could be carried in the sound beam for only a short distance. The discontinuity within the plane in which it appeared may be considered analytically as a new source of sound with its own near field and far field.

Use of lenses on the receiver side made it possible to move the object farther from the image camera. The lens also produced diffraction rings. Over 80% of the acoustic energy is contained within a central lobe, and it is this central lobe that was used in producing the image. However, the distance that a shadow can be carried depends on the relationship between the wavelength and the dimensions of the object casting the shadow and on the cross-section of the irradiating beam.

The sensitivity of the electron acoustic image converter was as high as 3×10^{-9} W/cm^2, with 10^{-7} W/cm^2 being typical. The signal-to-noise ratio in the camera was a function of the transconductance of the camera, noise resistance of the tube, coefficient of secondary-electron emission, frequency, electron beam current, collector capacitance, stray capacitance, input capacitance of the amplifier, capacitance of an element of resolution of the piezoelectric plate, and passband required for undistorted transmission of the signal. The passband was determined by the number of frames per second and the number of image lines that were scanned.

Within the receiving crystal, there was a source of image interference, which resulted from mode conversion. At certain frequencies, plate modes of relatively high intensity were generated and these showed up as waves progressing laterally across the image. The result was that the maximum resolution obtainable was a function of frequency. The range of usable frequencies was approximately 5% below the half-wave resonance (thickness mode) of the crystal and 10% above the half-wave resonance, with operation being best at resonance [20].

Electron acoustic image converters usually were operated using continuous waves (CW). CW operation is conducive to standing waves between successive reflecting surfaces. The standing waves tended to obscure the image, making it necessary to eliminate them if possible. These waves were eliminated by (1) tilting the object, if it was thin and flat, to cause reflections to deflect away from the source and the receiver; (2) inserting transition layers to produce a gradual acoustic match between adjacent media (a difficult procedure that usually degraded the image); and (3) inserting an attenuator between the reflecting surfaces to attenuate the reflected signal. The third method was particularly useful between the lens and the receiving crystal as there was little variation in the relationship between the positions and orientations of the lens and the receiving crystal.

Almost 40 years after the original work by Sokolov, a 210- $\times$ 150-mm image converter operating at 2 MHz was developed in the early 1970s. The faceplate construction and electrode layout were generally similar to earlier units, but the unit was significantly larger than units produced previously [12]. When compared with phased arrays, such units tend to lack ruggedness and sensitivity, and in recent years, these types of devices have tended to be largely replaced by various forms of piezoelectric material–based phased arrays. There is also a growing interest in the potential use of silicon micromachined ultrasonic transducers in applications in both air and water [21].

10.2.3 Schlieren Imaging

Schlieren imaging has been used for many years to visualize sound fields, and it has been a useful tool for studying the propagation of ultrasonic waves. Pressure gradients in an ultrasonic wave cause density gradients. When these gradients lie in the path of a light beam, the beam is refracted. In schlieren devices, the refracted light is used to produce the image of the sound field. This is done by one of the following methods:

1. Interrupting the refracted part of the beam to remove it from the field and focusing the remainder of the field on an image detector, that is, a ground glass screen or camera film. The image produced is dark on a light background.
2. Focusing the refracted rays on the image plane and eliminating the remainder of the beam of light from the image. The result is a light image on a dark background.

Figure 10.6 illustrates a schlieren system that has been used for direct viewing of ultrasonic waves used in NDT. The image may be viewed directly using a ground plate glass screen.

In Figure 10.6, the lamp, condenser lenses, filters, first knife edge, and first collimator are mounted on a single 0.5-m optical bench. The second collimator, second knife edge, camera lens, and camera back are mounted on a second 0.5-m optical bench. These two benches are mounted on a 25-cm-wide, 2.0-m-long steel channel. The water tank is mounted completely independent of the optical system so that changes made in the tank will not misalign the optical system. The adjustment of a high-sensitivity schlieren system can be upset by alignment changes of only a few micrometers. Hence, the optical system must be mechanically isolated from all other parts of the apparatus.

In operation, the image of the light source is centered on the first knife edge, after which the collimating lenses, tank windows, second knife edge, camera lens, and viewing screen are centered in the resulting light beam. Maximum sensitivity and uniform field result only when optics are properly centered and knife edges are placed accurately at the focal points of the corresponding collimating lenses. When properly aligned and focused, the field can be made to go from light through gray to dark uniformly by adjusting the second knife edge farther into the light beam until it intercepts it completely. In the range of gray settings, convection currents, both in the water and in the air, are clearly visible. The ultrasonic field is made most easily visible by setting the

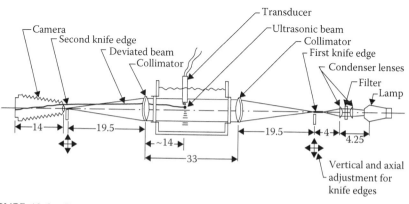

FIGURE 10.6 Sketch of schlieren apparatus for viewing ultrasonic fields (dimensions in centimeters).

second knife edge to intercept all of the main beam, permitting only the light refracted in the ultrasonic field to pass and thus provide illumination in the viewing plane.

Great care must be taken in selecting optics that are free of defects, such as bubbles, pits, scratches, surface haze, and dust, as the defects will scatter light to such a degree as to be evident in the image. Similarly, the water in the tank must be free of particles, such as dust, corrosion products, and algae, which also can scatter light and show in the image. The windows on the tank and the collimating lenses must be accurately surfaced so that light is not deviated from the main beam except by the ultrasonic field.

The optical system also may be adjusted for bright-field operation, in which the ultrasonic disturbance appears as dark shadows. Although not generally so sensitive as dark field, the bright-field adjustment can, in some cases, show details not evident in the dark field.

A zirconium arc lamp makes a good light source for studying continuous ultrasonic waves. For making pulsed waves visible, it is necessary to use a stroboscopic light source synchronized with the ultrasonic source. Xenon flash lamps or commercially available stroboscopes are useful for this purpose. The flash of the lamps must be synchronized with the ultrasonic pulse. This is done by using an electrical signal from the pulse generator and a circuit containing a controllable time delay to facilitate photographing pulses at any point in their travel through the optical field. Movies of pulses propagating through water have been made while varying the delay on the stroboscopic light.

Color schlieren photography has been used for many years and can be quite useful in the study of ultrasonic waves. In particular, color is useful for showing the various pressure levels in an ultrasonic field. One method makes use of a spectroscopic prism between a slit (located in the position of the first knife edge; Figure 10.6) and the first collimating lens. This method produces colors ranging from red or from blue to green. Waddell and Waddell [22] developed a method of producing a complete color spectrum. They eliminated the spectroscopic prism and used a color-filter matrix in place of the second knife edge and a vertical slit in place of the first knife edge. They used a high-pressure mercury arc for a light source.

The color matrix used by the Waddells consists of three gelatin color filters mounted between 5- × 5-cm glass-slide covers. The three filters are the primary colors—red, green, and blue. The green filter, which is the width of the projected image of the slit (approximately 0.31 cm.), is located in the center, and the red and blue filters (approximately 23.8 mm wide) are located adjacent to the green filter in such a manner that no white light can pass through the matrix.

The work of the Waddells was shock wave oriented. The system can be used for ultrasonics and where, instead of vertical orientation, horizontal positioning of the optical slit and the filter might be preferable to correspond to the orientation of the wave fronts (if the wave propagation is in the vertical direction).

The quality of the color image will depend on the correlation between the transmission properties of filters used and the light source used. The Waddells used Wratten 25 (red), 57 (green), and 47 (blue) filters manufactured by Eastman-Kodak with a BH6 mercury arc lamp.

Schlieren systems are now commonly used to investigate a variety of phenomena, including ballistics, aerodynamics, and acoustic-ultrasonic wave phenomena [23]. Such systems are used in ultrasonics to visualize transducer wave fields and wave–target interactions, and are particularly useful when teaching students about wave field phenomena. Enhanced systems do continue to be developed and they remain a valuable research tool (e.g., Neumann and Ermert [9]). Over the years, a number of large schlieren systems have been used in NDT research programs [24]. In recent years, such systems have tended to be replaced by computer modeling capabilities; however, such visualization is still used by some researchers to validate data given from numerical simulations.

Light-optical techniques are used to make other types of ultrasonic images visible, including liquid levitation methods and those using nematic liquid crystals.

10.2.4 LIQUID LEVITATION IMAGING

Liquid levitation imaging refers to the production of an ultrasonic image on the surface of a liquid. The sound radiation pressure interacts with the force of gravity to cause the fluid surface to rise in dependence upon the acoustic intensity. The sound image appears as a relief image on the fluid surface, is then converted into a light image by means of schlieren optics, and is made visible on a ground glass plate or photographic film. The sensitivity, which is quite acceptable for practical requirements, depends on the curvature at the edges of the bulge rather than on the total height of the bulge.

Gericke and Grubinskas [25] developed a cell based on the liquid levitation principle for use in the nondestructive inspection of materials. The cell could be coupled to an object for inspection without the need for immersing the object. Their results included good images of holes 8 and 16 mm in diameter drilled into the side of aluminum cylinders 75 mm in diameter and 50 mm in height. Attenuation in aluminum is low, which produces a good situation for this application. The technique offered several advantages: potentially unlimited area of field, good image resolution, inexpensive equipment, and satisfactory sensitivity for NDT.

In most cases, pulsed or frequency-modulated ultrasound produced the best images. CW ultrasonic energy produced the best detail resolution but suffered from ambiguities caused by standing waves.

The liquid surface levitation converter had an inherent time constant of about 0.01–0.1 seconds, depending on the viscosity of the conversion liquid.

10.2.5 ULTRASONIC IMAGING WITH LIQUID CRYSTALS

Various investigators have experimented with the use of liquid crystals for imaging ultrasonic fields. Liquid crystals are liquids having certain physical properties shown by crystalline solids but not by ordinary liquids. Sproat and Cohen [26] used a cell composed of an acoustical absorber and cholesteric liquid crystals. These crystals change color at various transition temperatures, and their sensitivity to temperature depends on the particular chemical composition. The purpose of the acoustical absorber was to increase the rate of absorption at a location near the crystals and thus to increase the sensitivity of the liquid crystal cells. Liquid crystals are subject to rapid oxidation, and encapsulation retards this action.

In practice, the object to be imaged was placed in water. A beam of CW ultrasonic energy was passed through the object and impinged on the imaging cell. The ambient temperature of the water path was kept slightly below the transition temperature. The higher amounts of energy absorbed at the high-intensity portions of the beam produced local heating above the transition temperature and thus the image was formed.

With this cell, very high intensities (approximately 100 W/cm^2 in early experiments) were necessary to produce color change. Such an intensity is much too high for practical use in any imaging system.

A second class of liquid crystals, known as nematic crystals, were used in which the ultrasonic energy produced a disturbance within the crystals, which could be detected optically with sensitivity equivalent to that of schlieren systems or liquid levitation systems. Nematic crystal cells provided high sensitivity, high resolution, low cost, large-area capability, and handling convenience.

Developments in nematic liquid crystal image converters have increased the range of applications. Dion and Jacob [27] reported a system for intensity and phase imaging operating at 3.6 MHz, which was developed for use in direct viewing of ultrasonic holography images. Gerdt et al. [28] used an underwater imager to directly produce the image. Research into developments into liquid crystals has continued. Some novel applications of liquid crystal usage have already been developed.

In addition to imaging devices, ultrasonic methods have been used extensively to investigate properties of the liquid crystal materials. A review of some of the earlier studies on liquid crystals using ultrasonic was provided by Natale [29]. The ongoing interest in the use of ultrasonics for the characterization of liquid crystal materials can be illustrated with a recent paper by Sukhovich et al. [30], who investigated refraction and focusing phenomena in two-dimensional (2D) phononic crystals.

10.2.6 PHOTOGRAPHIC METHODS OF IMAGING BY ULTRASONICS

One of the earliest approaches to imaging ultrasonic fields used the effects of ultrasonic energy on photographic emulsions. These early experiments led to other similar methods involving thermal and chemical processes to produce a permanent record of the intensity distribution in an ultrasonic field.

One proponent of ultrasonic imaging by photographic methods was Arkhangelskii [31]. One approach taken by this investigator involved exposing a photographic film (photodiffusion method) to light before developing the film under the influence of ultrasonic irradiation. The ultrasonic intensity variations due to an object located in the ultrasonic field affected the rate of development accordingly. Arkhangelskii [31] concluded that the main factor giving rise to acceleration of the development of an exposed film in an ultrasonic field was the acoustic pressure and the second most important factor was a local increase in temperature of the sensitive layer. The ultrasound promoted penetration of the liquid into the "pores" of the gel during the compression half of the cycle, but during the rarefaction half of the cycle, the liquid remained in the cell because of absorption or other forces.

Arkhangelskii [31] found a correlation between the concentration of the developer, the exposure time, and the contrast of the image. Exposure time required increased with decreasing concentration of developer, but the corresponding contrast increased considerably with lower concentration of developer.

Arkhangelskii and Pinus [32] investigated the influence of ultrasound on the process of diffusion of a colored electrolyte in a gelatin film. Exposing the entire diffusion system to ultrasound at frequencies ranging from 500 kHz to 9 MHz at an intensity of 0.3 W/cm^2, these investigators were able to show a 15% increase in diffusion rate in a gel containing 75% water. The cause of the increased diffusion rate was traced to the acoustic pressure.

Berger and Kraska [33] concluded that the direct action of ultrasonic waves on a photographic emulsion was influenced to a great extent by the softness of the emulsion. The emulsion was softened by soaking in water before exposure, thereby making it more sensitive. Film was exposed to ultrasonic irradiation in an iodine–water solution without darkroom techniques. The iodine affected color changes in the irradiated areas, allowing the ultrasonic image to be observed as it formed.

Schilb et al. [34] described a "bleach-out" method of producing visible images of sound fields. The basic materials required for the method were a dye, a bleach, and gelatin-coated glass plates. The plates were outdated photographic plates, which were fixed, hardened, and washed. They were then soaked in the dye solution for 24 hours and exposed before they had an opportunity to dry. The action of the bleach on the dye at any point occurred at a rate depending on the intensity of the ultrasound—the lighter areas correspond to the positions of higher intensity in the field.

Eckardt and Fintelman [35] produced sound images using pigment papers that consisted of a nonhardening gelatin layer with a mixture of insoluble color pigments. Such films could be used as transfer agents in the copper deprint method. They could be hardened in a 0.001% potassium permanganate solution under the influence of ultrasound, so that instead of light refinement there was sound refinement. Suitable coloring of the pigment paper could be obtained with a 5-minute exposure at a frequency of 1.4 MHz and a sound intensity of 0.25 W/cm^2. Higher intensities (1.5 W/cm^2) produced detrimental heat effects.

Although it is not a photographic technique, the influence of ultrasonics on the luminescence of phosphors has been used to produce ultrasonic images in much the same manner as it has been used to produce images photographically. An object in the path of ultrasonic rays incident on a luminescent screen casts an acoustic shadow, which is transformed into a visible image by the effects of ultrasonic energy on the luminescence of the phosphors of the screen [36]. The action of ultrasound is explained in part by the heating effect of the ultrasonic energy.

All photographic methods and the method based on the luminescence of phosphors under the influence of ultrasound exhibited extremely low sensitivity, that is, they required high intensities, and were slow. For these reasons, they have found no practical industrial applications.

10.2.7   ULTRASONIC HOLOGRAPHY

Holography is a form of three-dimensional (3D) imaging conceived and named by Gabor. Gabor revealed his discovery in a paper [37] written in 1948 and published in 1949. It was first used in electron microscopy. Gabor's [37] objective appears to have been one of overcoming problems associated with correcting for spherical aberration of electron lenses. In the new method, the problem was shifted to light optics, where refracting surfaces can be figured to any shape without the limitations associated with electron optics imposed by laws of the electromagnetic field. It is necessary only to imitate the aberrations to an accuracy corresponding to the accuracy required for a given resolution.

The principle of holography, as presented by Gabor [37], is that if a diffraction diagram of an object is taken with coherent illumination, and a coherent background is added to the diffracted wave, the photograph will contain the full information on the modifications suffered by the illuminating wave in traversing the object. The object can be reconstructed from this diagram by removing the object and illuminating the photograph by the coherent background alone. The wave emerging from the photograph will contain as a component a reconstruction of the original wave that appears to issue from the object. Thus, a hologram may be defined as a recording (or photograph) of two or more coherent waves. If one recorded wave is from an illuminated object and another is a reference wave, illuminating the hologram with the reference wave reconstructs twin images of the original object in three dimensions.

Gabor [37] used a monochromatic light source placed behind the object to provide the coherent light waves. His technique was, thus, limited to microscopic use. The development of the laser has made available large and much more powerful coherent light sources and has been responsible for renewed interest in this technique.

Gabor [37] explained his discovery mathematically. A coherent monochromatic wave U with a complex amplitude striking a photographic plate may be described by

$$U = Ae^{j\theta} = U_0 + U_1 = A_0 e^{j\theta_0} + A_1 e^{j\theta_1} \tag{10.1}$$

where A and θ are real, U_0 is a background wave, U_1 is a remainder, and the absolute value of amplitude is

$$|A| = [A_0^2 + A_1^2 + 2A_0 A_1 \cos(\theta_1 - \theta_0)]^{1/2}$$

During the reconstruction process, illuminating the positive hologram with the background wave, U_0, alone produces a new wave, U_s, such that

$$U_s = A_0 e^{j\theta_0} [A_0^2 + A_1^2 + 2A_0 A_1 \cos(\theta_1 - \theta_0)]^{\gamma/2} \tag{10.2}$$

where γ is a power relating transmission intensity to the exposure. The best choice of γ is 2. When $\gamma = 2$

$$U_s = A_0^2 e^{j\theta_0} \left[A_0 + \frac{A_1^2}{A_0} + A_1 e^{j(\theta_1 - \theta_0)} + A_1 e^{-j(\theta_1 - \theta_0)} \right] \tag{10.3}$$

Thus, if the background is uniform, the new wave, U_s, contains a component proportional to the original U, which is

$$A_0^2 e^{j\theta_0} \left[A_0 + A_1 e^{j(\theta_1 - \theta_0)} \right]$$

It also contains a spurious component

$$A_0^2 e^{j\theta_0} \left[\frac{A_1^2}{A_0} + A_1 e^{j(\theta_1 - \theta_0)} \right]$$

These two components constitute the twin images, the twin waves carrying equal energies.

Each part of the hologram contains the total information about an object. Therefore, it is possible to reconstruct the object from a section cut from the original photograph. The result is similar to viewing the object through a window.

An interesting fact is that the wave used in the reconstruction need not be the original, and this is the factor that permits the use of ultrasonic waves and the subsequent reconstruction of the image using coherent light from a laser. However, the size of the image is changed in proportion to the ratio of the wavelength of the reconstructing wave to the wavelength of the original illuminating wave.

Several methods of making acoustical holograms have been used. One method (Figure 10.7) makes use of liquid levitation, and Figure 10.7b. shows a detailed schematic for the surface deformation and light interaction. The reference beam may be obtained either by reflecting a portion of the irradiating beam onto the surface or by generating a separate wave using a second transducer. The height of the bulge of the surface is critical. The height of the bulge should be small relative to the wavelength of the light. Sometimes, a wetting agent is added to improve the image formation.

Other methods of recording holograms include electron acoustic image converters and mechanical scanning.

Mechanical scanning consists of an acoustic detector (microphone or hydrophone), which produces a signal amplitude proportional to the intensity of the sound field at a given position. The amplified signal from the detector controls the output

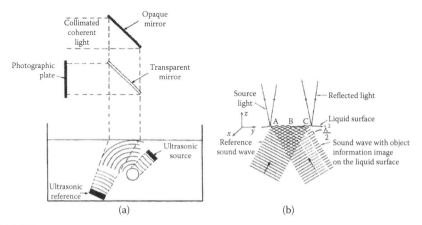

(a)   (b)

FIGURE 10.7 Liquid levitation ultrasonic holography. (a) System schematic. (b) Interference between light and liquid surface in detail.

of a light. The acoustic detector scans the sound field, and the light moves in correspondence with the position of the source, producing a record on a photographic film. This is a slow method but can be used in air and liquid. In the place of the light, the corresponding data, including detector position, intensity, and time, can be stored and processed by computer.

The electron acoustic image converter was a good means of producing ultrasonic holographs but suffered from limitations related to frequency or the size of piezo-electric crystal that could be used in the camera.

Young and Wolfe [38] proposed the use of deformable films on a solid substrate to produce holograms. Thermoplastic coatings were placed on suitable solid substrates, and these were located at the surface of the liquid bath. The materials were either soft at the ambient temperature or softened by electrical heating. After the hologram was formed, it was preserved by solidification of the film. The image was then reconstructed by conventional coherent light-optical techniques, the depth of the deformations being sufficiently shallow for suitable light reproduction. Young and Wolfe [38] claimed to obtain adequate deformations at frequencies ranging from 5 to 30 MHz and intensities between 0.1 and 1.0 W/cm^2. These intensities are too high for many practical applications, particularly medical diagnostic applications. The fundamentals and early applications have now been discussed in various review papers (e.g., Aldridge and Clement [39]). Acoustic holography is of research interest; however, there are currently few commercial systems deployed. Deployment has been limited by conceived system complexity, challenges in providing larger fields of view, in particular relating to size limits for uniform transducer, system signal to noise, and sensitivity. Although there are advantages with acoustic holography, particularly in that the illuminated field is viewed in real time, C-scan systems and phased array systems are tending to dominate the market, particularly for medical applications.

Ultrasonic holographic systems are receiving attention for some specific medical and industrial applications. This technology does have some particular advantages: an example is in the inspection of composite parts, where defects may not be in the form of single isolated features, but regions with anomalous properties. Brenden [40] developed a real-time ultrasonic imaging system that used a 100-μs pulse that was transformed into an optical image by a pulse of coherent light of 6-μs duration, giving images at video frame rates. The system has been applied to visualize multiphase flows such as fluidized beds, bubbly flows, thermal gradients, and mixing phenomena in real time through vessel walls as shown by Shekarriz and Sheen [41]. Figure 10.8 shows real-time bubble formation, rise, and interaction and particle motion in a fluidized bed around a flat plate.

The same technology has also been developed to meet the need for one-sided inspection of a large number of composite parts used in aircraft (Dassault Aviation). The technology was developed in cooperation with Aerospatiale, and this work is discussed in Chapter 9 [42].

Research on medical application of acoustic holography for real-time breast imaging, including monitoring biopsy needle manipulation [43], and other aspects of medical imaging are discussed in Chapter 14.

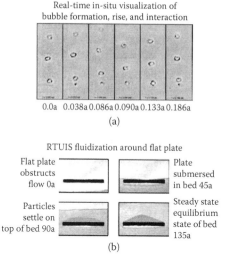

FIGURE 10.8   (**See color insert.**) Examples of acoustic holography visualization. (a) Real-time in situ visualization of bubble formation, rise, and interaction. (b) Fluidization of particles around a flat plate. (Courtesy of Pacific Northwest National Laboratory.)

10.2.8 Acoustic Microscopy

Acoustic microscopes are a well-established tool in research laboratories around the world and are seeing increasing application in manufacturing for process monitoring and control. For industrial applications, these instruments typically operate in the range of 10 to 300 MHz, and the highest operating frequency for a research system reported to date is 15 GHz.

These microscopes exist in two general families. The first are the scanning laser acoustic microscopes, which were first discussed in 1970 [44]. These instruments illuminate an object with continuous ultrasonic waves and detect surface displacements using a scanning laser system. The second, and more common, are the scanning acoustic microscopes (SAMs). This type of instrument was created by Lemons and Quate at Stanford University in 1973 [45,46]. Its scanning was mechanically driven, and it operated in the transmission mode. Since then, mechanical and electronic circuit improvements have been made and image recording has been automated. In general, acoustic microscopes now work in the reflection mode. Instruments have been developed to operate in a variety of liquids, including water and mercury, with pressurized gases and gases in the form of liquids at cryogenic temperatures.

The lateral resolution of SAM is dependent on the frequency of the acoustic waves and the couplant used: the best units have achieved about 0.75 μm resolution. This submicron resolution was achieved with a 15-GHz unit operating at cryogenic temperatures [47]. A reflection high pressure gas–coupled SAM was reported by Petts and Wickaramasinghe [48], which operated at 500 MHz and 147 Pascal. Improved transducers and electronics subsequently enabled the demonstration of a transmission high pressure gas–coupled acoustic microscope, operating between 10 and 30 MHz in argon [49,50]. Important advances have also been made in the direction of imaging subsurface structures at high temperatures. Ihara et al. [51] developed a sound imaging technique to see a small steel object immerged in molten zinc at 600°C. Recently, a new high-frequency (1 GHz) time-resolved acoustic microscope was developed at the Fraunhofer Institute for Biomedical Engineering [52]. It is based on an optical microscope from Olympus and operates in a reflection mode. The new microscope permits fluorescence microscopy and scanning acoustic microscopy at the same time. This gives the ability to perform synchronous optical and acoustical investigation of cell components with changing cellular stiffness.

Several reviews of acoustic microscopy have now been published. Early industrial application using frequencies between 10 and 100 MHz was reviewed by Gilmore et al. [53]. Books by Briggs [54,55] and Briggs and Arnold [56] provide a good introduction to the subject, and a paper by Khuri-Yakub [57] provides a brief history and identifies the companies that were producing commercial SAM instruments in 1993. Commercial instruments are available from several vendors that operate up to 300 MHz, and units have been sold operating at frequencies up to 2 GHz. These instruments are now increasingly being used within the materials science community to study local properties.

Variation of the mechanical properties with depth can be studied by scanning at various defocus values. C-scan acoustical images obtained at different defocus positions are used for the detection of subsurface voids and cracks. Collecting images obtained at various defocus positions allows a 3D image to be reconstructed, representing the volume of the entire microstructure of the investigated sample. Time-resolved acoustic microscopy adds an additional degree of freedom for quantitative measurement. In time-resolved acoustic microscopy, a short sound pulse is sent toward a sample. For layered materials, the reflected signal represents a train of pulses (a time varying or radio frequency (RF) signal or an A-scan). The first pulse is attributed to the reflection from the liquid–specimen interface. The second pulse appears as a result of reflection from the first internal interface. The time delay of the pulses and their amplitudes provides information about the elastic properties and attenuation of sound in the layer. The velocity of the wave can be determined by measuring the time delay of the corresponding pulse. Time-resolved images obtained by mechanical scanning along a line are presented as B-scans. When a wide-aperture lens is defocused, the interaction between the direct reflection and a surface wave can be used to give a $V(Z)$ characteristic, which is used for material elastic property characterization.

Measurement of the elastic properties of solids and thin films, measurement and visualization of adhesion in layered structures, subsurface imaging of defects in coatings and crack visualization, visualization of stress inside solid materials, characterization of carbon fiber–reinforced composites, and characterization of cells and biological tissues are some of the recent developments in acoustic microscopy.

For acoustic microscopes, the largest impediments to wider use are the challenges faced by manufactures in lens design, particularly above 300 MHz. Issues include the problems relating to the fabrication of the matching layers needed on lenses to improve acoustic transmission and hence sensitivity, and the electrical matching circuits used to improve both transmission and detection. Applications are also limited by the relatively small size of the community that understands the complexities of the propagation and contrast mechanisms, and relating observations to fundamental material properties.

Within the research community, the majority of interest has now moved from instrument development to focusing on understanding and utilizing the insights obtained through applications of acoustic microscopy. Subspecialties are now emerging within the microscopy research community, such as ultrasonic "biomicroscopy," and this is discussed further in Chapter 14.

10.2.9   ULTRASONIC ARRAYS

An ultrasonic array is any transducer that is composed of a number of individually connected elements. Such arrays occur in a diverse range of geometries as illustrated (Figure 10.9). Array transducers are commonly used in medical ultrasound, and such arrays are considered in Chapter 14. Arrays are also used in SONAR and geophysical exploration. There is growing use of arrays in both NDT and industrial process imaging. A review of arrays used in NDT is provided by Drinkwater and Wilcox [58], and similar geometries are also used in industrial process imaging and process tomography.

To further enhance material visualization, a number of data processing schemes are being developed and deployed. An example of both the hardware and software used with one scheme is that for synthetic aperture focusing, which is used for the inspection of metals, and this is reported by Schuster et al. [59]. Systems have been developed for a number of challenging environments, including for sodium viewing in fast-spectrum nuclear power reactors, where transducers operate at 260°C, and this technology was reviewed by Bond et al. [60]. B-scan systems have been used to image processes, including mixing, crystallization, and dissolution [61] and two-phase flow, all imaged using linear arrays operating at 7.5 MHz [62]. Arrays have also been developed into handheld units for NDT and for a portable underwater camera [62]. A further example of signal processing to give novel images is found in time-reversal imaging, which is being considered for some interesting imaging

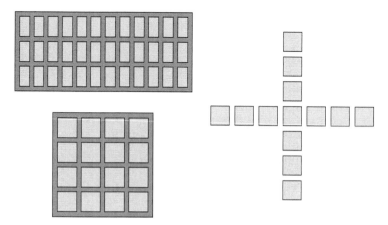

FIGURE 10.9   Schematic showing examples of configurations for some two-dimensional ultrasonic arrays.

challenges, in particular, as a tool to remove artifacts, in medical applications such as those due to the skull in transcranial images [63,64].

10.2.10 Applications of Ultrasonic Imaging

The nature of ultrasonic waves and their reaction to various discontinuities of acoustic impedance made acoustic imaging of particular interest in two technical areas. These are NDT and medical diagnosis. Using the earlier techniques, the applications were restricted to areas in which the image could be maintained in the ultrasonic beam. Diffraction effects limit the distance to which a shadow can be maintained in a beam with a cross-section that is large compared with the size of the obstacle. A typical distance in these previous methods was 8–10 diameters of the obstacle. One area of application of acoustical imaging by early methods was the inspection of flat plates such as flat fuel elements for lack of bond between core and cladding.

Acoustical holography offers an additional advantage of providing a 3D display of the defect in NDT or medical diagnosis, thus enhancing interpretation. Neither the electron acoustic image camera nor holography received widespread acceptance in industry or medicine.

After many years of experimentation, all of the earlier methods were abandoned for various reasons. However, interest has continued or revived in some areas so that more modern methods, including holographic methods using liquid surface and film deformation techniques, have been investigated. Various technical and economic factors have caused a move away from the holographic techniques and toward the more practical methods of building images by B-scan and C-scan techniques. Developments in electronics and computer science have contributed significantly toward obtaining and processing B-scan and C-scan data both in medicine and in NDT.

The diverse range of ultrasonic arrays used in medical 2D, 3D, and now four-dimensional (4D) imaging is discussed further in Chapter 14.

10.3 PROCESS MONITORING, MEASUREMENT, AND CONTROL

Ultrasonic properties of materials can be measured and used to control various processes. In one type of process, as a material crystallizes from a liquid or syrupy state, the attenuation of ultrasonic energy increases significantly. Thus, if ultrasonic waves are caused to pass through such a medium, the decrease in signal amplitude can be used to control the energization of a relay to initiate follow-up action.

Similarly, methods based on either the transmission properties of the materials or the receipt of echoes from inclusions have been used to indicate the solid content of liquid slurries or the presence of bubbles or other foreign particles in enclosed systems.

As shown in Chapter 2, the primary factors that influence the velocity of sound in a fluid are the density, compressibility, and ratio of specific heats. These factors and viscosity, or attenuation, are functions of temperature. By correlating mole fraction, temperature, and velocity of sound in a known mixture, it is possible to monitor the ratios of its constituents. Instruments have been developed for such purposes, particularly for monitoring the buildup of a component that is potentially dangerous above a certain level.

To enable better process monitoring, measurement, and control, and to improve product quality, reduce variability, reduce energy utilization, and reduce waste, there has been a growing need to add new and improved process sensors. Online, continuous, real-time ultrasound–based physical property measurement technologies are increasingly being deployed on a wide range of fluid and slurries found in process industry plants. These ultrasonic sensor technologies are noninvasive, are online, require low maintenance, and can provide real-time data for large-volume samples. To meet these needs, ultrasound is one of the sensor technologies that has received increasing attention in recent decades.

The potential offered through the use of ultrasonics has long been recognized [2,3]. The number of types of ultrasonic sensors and the diverse forms for their implementation have also been well recognized [4,65,66]. In general terms, the process industries encompass chemical, petrochemical, pharmaceutical, food and beverage, biotechnology, water and waste water, and some classes of power generation industries. Although the specific nature of the materials and the process conditions, in particular temperature and the volumes of the processes involved, vary, fluid streams are involved that may be a single phase or multiphase. Ultrasound can be used to measure a range of properties. A schematic showing some of the fundamental sensor configurations is given in Figure 10.10. These include (a) pitch-catch, (b) and (c) pulse-echo, with (b) using a reflector and in some cases standing waves, (d) transmission, (e) reception (for detection of acoustic emissions [AE]), and (f) a sensor such as a SAW device: such units are connected to a source in the form of a pulse generator or oscillator, signal acquisition, and some form of interface, now commonly analogue-to-digital conversion to computer-based analysis and display. These sensors can investigate arrival time, amplitude variability, and frequency sensitivities resulting from changes in velocity (c), density (ρ), attenuation (α), and scattering responses. Several of these configurations have similarities to geometries used in NDT, passive monitoring that is analogous to AE for NDT, and surface sensors that are in many cases based on SAW technology.

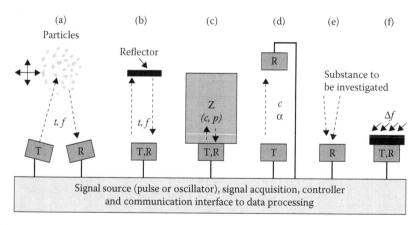

FIGURE 10.10 Schematic illustrating ultrasonic transducer configurations for sensor systems. (After Hauptmann, P., N. Hoppe, and A. Puttmer, *Measurement Science and Technology*, 13(8):R73–R83, 2002.)

In addition to the low-power sensing applications on various process streams, there are also a growing number of high-power applications, where ultrasound is used to treat and modify materials. The various aspects of high-power ultrasound, including sonochemistry, are considered in Chapters 11 through 13 and medical applications are discussed in Chapters 14.

10.3.1 ULTRASOUND IN PROCESS INDUSTRIES

In addition to some studies [2,3], there are studies that discuss applications of ultrasound in process industries, for example, as a novel process analytical tool for chemometrics [67–69].

Ultrasonic measurements have been performed on solids, liquids, gases, and various types of mixtures and multiphase systems for over 50 years [70]. The first observations of ultrasound applied to a reacting chemical system were made in 1946, when the extent of polymerization in a condensation or radical process was measured [71]. Since that time, a diverse range of measurements have been made that consider ultrasonic interactions in terms of the effect of materials or processes and changes on four ultrasound metrics:

- Velocity
- Attenuation
- Acoustic impedance
- Absorption and scattering

These parameters are in general functions of frequency, composition, process reaction or phase (time/rate), and temperature and relate to fundamental thermophysical parameters. Their form factors are inherently robust, and they have been deployed to provide data and material signatures/properties at both low (cryogenic) and elevated temperatures in highly corrosive (high and low pH) and high-radiation environments. However, the various ultrasound methods are probably the least understood of all the tools used in what can generally be defined as process analytical metrology or, by some, chemometrics.

Recent years have seen a resurgence of interest in ultrasound technologies and its application for material characterization, process measurement, monitoring, and control. This has, to a significant extent, been facilitated by the formation of interdisciplinary teams that work on process problems and by the major advances in the development of new and improved ultrasound transducers and in hardware and software elements in computers, display systems, and interface and digitization cards. This section considers ultrasonic technology, with focus on its application to chemical and other process industry plants.

10.3.2 ULTRASONIC SYSTEMS AND MEASUREMENTS

Many of the fundamental aspects of the physics of ultrasonic measurements are discussed in earlier chapters and in classic texts, such as that by Truell et al. [72]. Aspects of measurement systems are also discussed in earlier chapters, with a further good reference that considers ultrasonic transducers, velocity and attenuation measurements, viscosity measurement, and ultrasonic chemical relaxation spectroscopy being the contributed chapters in the book, *Ultrasonics*, in the series *Methods of Experimental Physics*, edited by Edmonds [73]. The relationships between ultrasound interaction with materials and molecular structure were of considerable interest in the 1970s (e.g, Matheson [74]), and these topics have recently started to receive some increasing attention, particularly in the process chemistry community (e.g., Capote and Lugue de Castro [69] and Pethrick [70]). Ultrasonic broadband spectroscopy remains a topic for current research.

Much of the early literature that applies ultrasonic measurements for process measurement is reviewed and cited in the text by Lynnworth [2]. This book focuses on applications and considers ultrasonic methods for flowmeter, thermometer, density, porosity, and property measurement, and interface sensing, together with a review of selected topics from more traditional NDT. The fluid metrology literature has been supplemented by two significant texts: *Ultrasonic Techniques for Fluids Characterization*, by Povey [75], and *Ultrasonic Sensors for Chemical and Process Plant*, by Asher [3]. These books include extensive bibliographies for the earlier literature, including data for numerous examples of applications of ultrasound technology to chemical measurements and industrial processes.

10.3.3 Velocity and Attenuation Measurement to Characterize Media and Monitor Processes

Numerous transmission measurements that give the velocity and attenuation properties of a fluid or mixture between a pair of transducers have been made by Povey [75], and others. Except for simple systems, these data do not directly give thermophysical properties and the interpretation of the velocity and attenuation data as a function of process parameters, and frequency can, however, present challenges. The properties of single-phase media, in general, follow well-behaved compositional and temperature dependent, ultrasound parameter relationships; however, a priori such relationships may not follow expected trends. An appendix in the text by Asher [3] provides a good discussion and summary of the properties and ultrasonic responses of many mixtures.

The theory for the velocity and attenuation of ultrasound in suspensions of particles in fluids was reviewed and extended by Harker and Temple [76]. Subsequent work by Harker et al. [77] further extended these approaches to the analysis of ultrasonic propagation in slurries, demonstrating the analysis for systems consisting mainly of silicon carbide in water and ethylene glycol. Tsouris and Tavlarides [78] used the speed of sound coupled with a linear model to estimate the volume fraction of water in oil.

Ultrasonic systems that give the density from ultrasonic data have been devised by Hale [79], which can be used for process control. This devise has been developed into an ultrasonic pulse-echo reflectometer, for both pulse-echo and swept frequency (0.3–6 MHz) operations, by McClements and Fairley [80,81]. Such measurements can be used to investigate the properties of solutions, emulsions, and colloidal systems, including monitoring both melting and crystallization as shown by McClements et al. [82].

There have been studies that consider acoustic and electroacoustic spectroscopy for characterizing emulsions and colloids (e.g., Dukhin and Goetz [83]).

The fundamentals of acoustics in liquids, including the theory for scattering and absorption in colloidal systems, together with a review of ultrasonic methods for characterizing colloids was presented in a 2002 text by Dukhin and Goetz [84]. A comprehensive 96-page review, which cites 348 references, discussing ultrasound techniques for the characterization of colloidal dispersions has been written by Challis et al. [85]. This review covers the phenomenology of ultrasound propagation in colloidal mixtures, evolution of comprehensive theories, and effects of concentration on response, giving some examples of ultrasonic instruments.

Ultrasonic spectrometry for liquids is also increasingly being seen as a tool in chemistry and chemical physics, with ultrasonic absorption data being correlated to fundamental structural resonances and absorption [86,87]. The review by Kaatze et al. [86] presents a comprehensive overview and cites over 300 papers. There is ongoing research and publication in this area that analyze increasing range of materials.

Ultrasound has been extensively used by Lavallee et al. [88] to analyze the thermal and viscoelastic behavior of polyurethane–poly(vinyl chloride) blends. Morphology of polymer blends has been analyzed by Piau and Verdier [89] and Verdier and Piau [90] using ultrasound. These approaches have been extended and applied to the study of the temperature-dependent properties of epoxy prepolymers, and the ultrasound data have been compared with that from Brillouin scattering as demonstrated by Matsukawa et al. [91]. The use of ultrasound velocity and attenuation measurements for online measurements has been integrated into extrusion systems by Gendron et al. [92] for online real-time monitoring.

Measurement of the properties of metal–powder–viscous liquid suspensions used in slurry casting synthesis has been investigated by using ultrasound by Schulitz [93].

10.3.4   Monitoring Solidification (Interface Sensing)

Ultrasonic techniques operating in pulse-echo have been used to monitor the position of interfaces during solidification. Early work by Bailey and Davila [94] considered both metals with low melting point and N-paraffin. Subsequent studies by Parker and Manning [95] have monitored the solid–liquid interface during both solidification and melting. Such measurements have been performed by Parker et al. [96] in the 1- to 5-MHz frequency range and included investigation of the "mushy zone," which forms between the solid and fluid phases and these interfacial structures. These basic approaches have been applied more recently by McDonough and Faghri [97] to follow the interface positions during solidification of an aqueous sodium carbonate. The use of simple pulse-echo measurements can be extended to include the reconstruction of solid–liquid interfaces in solidifying bodies, as was shown by Mauer et al. [98].

10.3.5   Acoustic Time Domain Reflectometry

This pulse-echo technique, acoustic time domain reflectometry (A-TDR), is named by analogy to optical TDR and can be used to provide backscatter data, including data on interface sensing. The use of scattered ultrasound energy for structural analysis was initially developed for material characterization of solids, including grain size analysis (considered in Chapter 2). It was subsequently extended to the analysis of two-phase and multiple scattering. A review of the fundamental aspects of structure analysis is provided by Goebbels [99]. The development of ultrasonic measurements for online real-time monitoring of processes in membrane science was started in 1994 by Bond et al. [100], and this topic is discussed in Chapter 9, Section 9.7.3. Other applications of this approach include monitoring of sedimentation, solidification, membrane treatments, and characterization of inorganic, biological, and particulate fouling in a variety of processes.

One specific implementation of A-TDR has been for determination of liquid composition, in what is called the "acoustic inspection device," a portable, battery-operated, handheld system for real-time, sealed-container inspection and content (liquid/material) identification [101]. This unit was commercialized as the Product Acoustic Signature System (PASS). The technology uses sound waves to acoustically detect and identify substances and materials such as explosives and hazardous and flammable liquids (e.g., liquid explosives) at security checkpoints. PASS is also suitable for checking source and feedstock liquid-based materials in manufacturing environments to ensure the correct labeling of products and to detect foreign objects inside fluid-filled containers. It is a versatile tool that can rapidly screen bulk containers in the field ranging from a 55-gallon oil drum to test tubes. PASS significantly reduces the amount of time required to screen such containers while at the same time increasing the reliability and accuracy of the screening. PASS uses advanced acoustic signal processing methods to consistently and accurately measure the temperature-corrected acoustic velocity (speed of sound) and relative acoustic attenuation in a fluid or material to

- Sort and classify liquid types into groups of like and unlike
- Identify liquids and bulk solids as a function of temperature
- Determine the fill level in liquid-filled containers

PASS technology has been successfully used in national security applications, including the identification of chemical warfare agents. It has proven useful in the augmentation of on-site material characterization efforts that use the expensive and time-consuming processes of direct sampling and laboratory analysis techniques. An example of a unit in use is shown in Figure 10.11.

10.3.6 THREE-PHASE REACTORS

Ultrasonic techniques have been used to investigate properties of both saturated and partially saturated rock systems, those involving hydrocarbons, gases, and water. In some laboratory studies, measurements are made at very high temperatures and pressures, and in others, three-phase systems including glass beads are used to simulate sandstone or catalysis beds. Such studies provide many insights that are directly relevant to the measurement of multiphase process beds at elevated temperatures.

The ultrasonic characteristics of the three-phase material mix in catalysis units have been investigated. Preliminary ultrasonic velocity and attenuation data have been reported by Soong et al. [102] for slurries consisting of water, glass beads, and nitrogen bubbles. This work was then extended to characterize slurries consisting of

FIGURE 10.11   (**See color insert.**) The Product Acoustic Signature System (PASS) in use. (Courtesy of Pacific Northwest National Laboratory.)

molten paraffin wax, glass beads, and nitrogen bubbles at 198°C by Soong et al. [103]. These studies demonstrated that ultrasonic techniques have potential for online, real-time monitoring of, at least some, two- and three-phase process reactors.

10.3.7 PROCESS TOMOGRAPHY USING ULTRASONIC METHODS

Recent years have seen a growing interest in the implementation of various ultrasonic measurement schemes that provide imaging for industrial flows and processes, and a family of methods, now known as process tomography, has been developed and implemented. An interesting review of the concept of image analysis providing information useful in process chemistry was provided by Geladi and Esbensen [104].

A review of the various technologies for the noninvasive tomographic and velocimetric monitoring of multiphase flows was provided by Chaouki et al. [105]. This covers all measurement modalities that have been applied to give such data. The ultrasonic portion of the review is less comprehensive than other parts of the article. However, it provides much useful data and citations. A book by Plaskowski et al. [106] provides the fundamentals of imaging industrial flows. It also includes a short section on applications of ultrasound, but it is only superficial in its coverage and gives a negative view on the assessment of the advantages of ultrasound. It should be noted that these comments were written at a time when ultrasonic process tomography was still in its infancy.

A good review of the background, advantages, and difficulties of ultrasound for process tomography is given by Hoyle [107]. More recent literature has reported a number of successful applications of process tomography using ultrasonic sensors.

Martin et al. [108] report on the application of ultrasonic imaging to pilot plant-scale gas and liquid processing vessels. The use of ultrasonic tomography for monitoring gas–liquid flow is also reported by Xu et al. [109], and the use of air-coupled ultrasound transducers for imaging temperature and flow fields in gases was demonstrated by Wright et al. [110]. Hoyle et al. [111] report the use of ultrasound technology in a multimodal measurement system that combines sensing modalities and fuses data.

10.3.8 ULTRASONIC TRANSDUCERS: PROCESS INDUSTRY APPLICATIONS

Ultrasonic measurements can be implemented with a diverse range of types of transducers and include the use of designs and materials specifically developed to meet the needs for operation in hazardous environments. Conventional ultrasonic transducers, used for NDT, are typically designed to operate up to 50°C. Clamp-on and waveguide transducers operate up to at least 300°C, and these have been discussed by Lynnworth et al. [112]. Custom and semicustom transducers, which are practical, have been shown by Dreacher-Krasicka et al. [113] to operate up to 600°C.

Waveguide/buffer rods have been used to further extend the ranges of process temperatures where measurements can be made: these have included use with molten magnesium at 690°C [114] and molten aluminum at 780°C [115]. A team at the National Institute of Standards and Technology demonstrated a waveguide–/buffer rod–based sensor used to measure grain size in steel samples during hot rolling to operate with samples at temperatures up to 1000°C [116].

The specific challenges of transducers for use in liquid metal–cooled nuclear reactors are discussed in a review of high-temperature transducers by Kazys et al. [117]. The state-of-the-art for the ultrasonic instrumentation, including NDT, imaging, and process sensors, for fast-spectrum nuclear reactors has been reviewed by Bond et al. [60]. Several programs are currently developing sensors for deployment in liquid metal–cooled fast-spectrum nuclear reactors (e.g., Karasawa et al. [118]).

Alternate ultrasound generation and detection that use lasers or electromagnetic acoustic transducers (EMATs) are also of interest for on-process applications [119]. For example, both laser systems and EMATs have been investigated for online measurements near the outlet of a steel manufacturing continuous casting unit. Both these modalities have particular advantages for specific types of application, and these transduction technologies are discussed further in Chapter 5.

10.3.9 DENSITY MEASUREMENT

Several manufacturers produce instruments that use ultrasound to measure density online and at-line, as well as supply laboratory instruments. Anton Paar [120] has developed instruments that measure the speed of sound through a liquid as a function of temperature. Previous calibration measurements are used to determine the density of a given liquid for online measurements. Instruments using the frequency of oscillation of a U-shaped tube are also used to determine the density. Other manufacturers who produce instruments based on measuring the speed of sound through a liquid, according to Povey [121], are Cygnus Ltd. [122], Canongate [123], and Nusonics [124]. In addition, Fuji Ultrasonics [125] is another manufacturer to use the speed of sound technique for online measurements.

An online ultrasonic viscometer, designed by Sheen et al. [126–128], measures the density and viscosity of a liquid or slurry flowing through a pipeline. It consists of two transducer wedges mounted on a pipe opposite one another and flush with the inner surface of the pipe. A longitudinal-wave transducer is mounted on one wedge and measures the reflection of ultrasound from the wedge–liquid interface; a shear-wave transducer on the second wedge measures a similar quantity. Each wedge has an offset surface so that a portion of the ultrasound is reflected from a wedge–air interface to provide a continuous reference signal for self-calibration. The density is determined by measuring the acoustic impedance of the liquid and the speed of sound through the liquid. The shear reflection coefficient is used to calculate the density–viscosity product. The typical error in the density measurement is 1% and in the viscosity measurement is 5%. The minimum value of the viscosity measurement is 50 cP. Different shear-wave wedge designs and materials offer the potential for measurement of the properties of lower viscosity fluids. One advantage of this instrument is that it measures both the density and viscosity online. Another advantage is that no calibration measurements are needed to interpret the results of the measurements. The details of such instruments are discussed further in Chapter 6.

Kline [129] describes a method to measure the density of a liquid such as aircraft fuel. The densitometer uses a transducer to transmit an ultrasonic pulse through the liquid to a block of reference material. The pulse is reflected from the near side and far side of the block. There is also an additional signal the properties of which are reflecting the condition of the block, and this appears between the near and far side reflections as well as an internally reflected portion of the pulse. The density is determined as a function of the amplitudes of the three return pulses. The velocity of sound in the liquid is determined from its measured time of flight. This system seems to be more complicated than the system of Sheen et al. [126–128] that also measures the time of flight.

McClements and Fairley [80] have designed an ultrasonic pulse-echo reflectometer for use in the laboratory that measures the ultrasonic velocity, attenuation coefficient, and characteristic impedance of liquid materials. The density is obtained from the impedance and velocity measurements. This instrument is similar to that described by Hale [79]. The error in the density measurement was 0.5%. The instrument was immersed in a thermostatically controlled water bath to ensure constant temperature, and results were averaged over 2000 signals.

Greenwood et al. [130] used ultrasonic attenuation measurements to determine the concentration of a slurry. This method was used to determine the concentration of components in a slurry in experiments carried out in a 1/12-scale model of a double-shell tank on the Department of Energy (DOE) Hanford reservation [131]. These were online measurements. However, in order to interpret the online attenuation measurements, calibration measurements of attenuation as a function of volume fraction of the slurry were carried out beforehand.

Greenwood et al. [132,133] also developed an online ultrasonic density sensor in which six transducers (five longitudinal-wave transducers and one shear-wave transducer) are mounted on a plastic wedge. The base of the wedge is in contact with the liquid or slurry. Longitudinal ultrasonic beams strike the base of the wedge at angles of 0°, 45°, and 60° with respect to the normal to the base. A shear wave at 0° also strikes the base of the wedge. The reflection at the plastic–liquid interface depends on the incident angle, density of the liquid or slurry, speed of sound in the liquid or slurry, and wedge parameters. The voltages on the receive transducers are compared with those when the base is immersed in water, and the reflection coefficients are determined. Using the reflection coefficient at two angles, the density of the liquid and the speed of sound can then be calculated. The error in the density measurement is 0.5% for liquids and about 1%–2% for slurries. There are several advantages to this instrument. The first is that the sensor consists of only one component. Since the measurement depends on the reflection at the plastic–liquid interface, and not on the passage of ultrasound through a liquid, the density of a liquid that severely attenuates ultrasound can be determined. For implementation, the wedge base can be placed in a cutout section of a pipeline wall, and the compact design allows deployment in short pipe spool pieces. The sensor is not affected by electromagnetic noise and can be located in harsh environments and other areas crowded with machinery.

The need for a compact, noninvasive, real-time measurement of the density led to the development of a second ultrasonic sensing technique based on the reflection of ultrasound at the solid–fluid interface (Figure 10.12). The transducer is mounted on the outside of the pipeline wall, and an ultrasonic pulse is reflected at the solid–fluid interface and again reflected at the solid–transducer interface and makes over 15 so-called echoes within the pipeline wall. These echoes are recorded by the same transducer. Each time the ultrasound strikes the solid–fluid interface, some fraction is reflected and some fraction is transmitted into the fluid. These fractions are dependent on the properties of the slurry, its density, and velocity of sound. The many echoes act as an amplifier of this effect, and observation of the echo yields the product of the density of the slurry and the velocity of sound in the slurry. Observation of the time-of-flight across the pipeline leads to a measurement of the velocity. The combination of the two measurements gives the density of the slurry. The uncertainty in the density measurement is about 0.2%–0.5%. The ultrasonic sensor unit consists of transducers with a 5-mm-thick disk of solid material cemented to the front face of each transducer (Figure 10.13). The disk can be made from stainless steel, aluminum, or quartz; but the less dense material, such as aluminum or quartz,

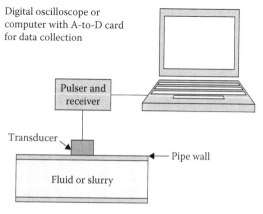

FIGURE 10.12 Conceptual drawing of ultrasonic density meter.

FIGURE 10.13   (See color insert.) Field deployment configuration of density meter. (Courtesy of Pacific Northwest National Laboratory.)

is preferred. By measuring the diminishing amplitudes of successive echoes within the solid disk and well-known theoretical reflection coefficients, the acoustic impedance (defined as the product of the density of the slurry and the velocity of sound in the slurry) is determined. By measuring the time-of-flight through the slurry, the density of the slurry is determined, which can be converted to a volume percentage or a weight percentage. This methodology is discussed further in Chapter 6.

10.3.10   Ultrasonic Characterization of Multiphase Fluids and Flow

Ultrasonic signals have attributes that are well suited for characterization of multiphase fluids and flows. The signals have the ability to interrogate fluids and dense opaque suspensions and penetrate vessel and process plant walls, and they are not degraded by noisy conditions because the signal frequencies used in ultrasound measurement differ from that due to machinery. Also, measurements are not affected by most practical flow rates.

Ultrasonic sensors can be designed to provide real-time in situ measurement or visualization of process characteristics; the sensors and sensing systems are compact, rugged, and inexpensive. Ultrasonic sensors can be designed to measure fluid density, viscosity, and velocity; slurry density, particle size, weight or volume percent solid concentration, stratification, and rheology; and colloid gelation and particle size and to quantify multiphase flow interfaces, state of mixing, homogeneity, and slurry transport.

10.3.10.1   Slurry Particle Size and Concentration

Scott and Paul [134] provided a summary of instruments available for online particle characterization. The sensing techniques reviewed include laser-light scattering or diffraction, laser Doppler and phase Doppler interferometry, electrical impedance or capacitance tomography, in-process video, focused-beam or dual-beam reflectance, electroacoustic spectroscopy, and ultrasonic attenuation or extinction. The majority of the instruments operate on dry (gas–solid) mixtures. For measuring the particle size distribution or concentration of particulate-laden slurries, ultrasonic attenuation has significant advantages. Sympatec Inc., System-Partikel-Technik [135], markets an in-line system to measure particle size covering the range of 0.1 to 3000 μm at particle concentrations up to 70 volume percent.

Ultrasonic attenuation spectra from ultrasound signals provide information about slurry particle size and concentration, and this is reported by Bamberger et al. [136]. As the particle size and the acoustic frequency are changed, the relative importance of attenuation mechanisms changes, and the acoustic attenuation can be dominated by different effects. Three regimes are identified for slurries consisting of solids in a liquid such as fine sand in water, and these are the viscous regime, inertial regime, and Rayleigh scattering regime, which are discussed by Kytömaa [137]. Researchers have used transitions between two regimes, in particular the viscous and inertial regime, to quantify particle size and concentration in real time as discussed by Boxman et al. [138].

Ultrasonic spectroscopy to measure the particle size of industrial slurries and colloids has been developed [139–142]; an acoustic spectrometer for laboratory-scale measurements based on this technique is marketed by Dispersion Technology, Inc., Bedford Hills, NY. Pendse and Han [142] have extended this method to address the challenge of characterization of nonspherical particles. Pendse and Sharma [140,141] applied this method on a titanium dioxide slip stream. The unit operated at a low flow rate (2 L/minute) and used frequencies up to 100 MHz. The configuration is limited to small transducer separation distances, with 24.5 mm being the largest distance investigated (H. Pendse, pers. comm., January 28, 1998).

For characterizing diverse slurries, ultrasonic systems that capture transitions between three regimes are more robust, as was shown by Kytömaa [137]. This extended frequency regime requires additional theory. In the viscous regime, Biot [143], Urick [144], Hampton [145], Atkinson and Kytömaa [146], and Allegra and Hawley [147] have shown that the viscous boundary layer thickness is larger than the particle radius. As frequency is increased, the viscous boundary layer becomes thinner than the particle radius; losses occur in the thin boundary layer surrounding the particles as presented by Biot [143], Johnson et al. [148], Sheng and Zhou [149], Zhou and Sheng [150], Atkinson and Kytömaa [151], and Allegra and Hawley [147]. Atkinson and Kytömaa [152] described the nonlinear behavior of attenuation at concentrations above 10%–20% for both regimes. Kytömaa and Corrington [153] carried out the first documented experiments in the inertial regime to derive the nonlinear behavior of the added mass coefficient with concentration. This theory spans the two regimes, and the representation for the complex wave number was developed in detail by Atkinson and Kytömaa [152]. Derksen and Kytömaa [154] used ultrasonics to determine the added mass coefficient. At sufficiently high frequencies, Rayleigh scattering initiates, where energy from coherent incident sound is scattered by the random distribution of particles attenuating the coherent waves as discussed by Allegra and Hawley [147]. Experiments conducted by Salin and Schön [155], near-maximum packing concentrations of glass spheres, showed that the transition in the response can well be described by an additive combination of the Atkinson and Kytömaa theory to the Rayleigh scattering theory. This was later confirmed for slurries by tests conducted by Greenwood et al. [156]. Theoretical studies have also been performed by Spelt et al. [157]. For a fixed particle size, the transition between each of these regimes occurs at a frequency defined by the particle size and the mechanical properties of the solid and the fluid. The viscous regime exhibits a quadratic scaling with frequency; the inertial regime scales with second power of frequency, while the Rayleigh scattering regime scales with the fourth power of frequency. Each regime exhibits a constant slope on a log–log plot. The first transition from the viscous to the inertial regime is gradual and occurs over two decades or more, while Rayleigh scattering sets in over a narrower range in frequency. The frequency or range of frequencies over which transition occurs is strongly dependent on particle size. For larger particles, the transitions will tend to occur at lower frequencies and for smaller particles, conversely, at higher frequencies.

A real-time in situ sensor to measure slurry mean particle size, distribution width, and concentration has been developed and tested with slurries composed of a range of particle types. The approach has been demonstrated for specific slurry types. Accuracy for mean particle diameter determination was within 1 µm, and accuracy for solid fraction was 1% as described by Bamberger et al. [136].

10.3.10.2 Ultrasonic Device for Empirical Measurements of Slurry Concentration

Greenwood and coworkers [101,156] have developed a method for slurry concentration measurement using the multiple echoes that are observed as the ultrasound reflects between two transducer faces (Figure 10.14). The multiple echoes are recorded and plotted similarly to the density measurement. The multiple reflections (about 20 echoes) extend the path length to many times the distance between the plates, making this attenuation method very accurate and acutely sensitive to small changes in the volume percentage of the slurry. In a previous application, 0.025 weight percent silica in water was accurately measured, the equivalent of 0.25 g of silica in 1 L of

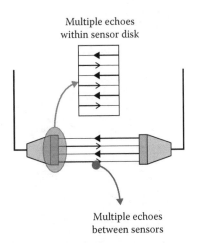

FIGURE 10.14 Concept drawing of ultrasonic concentration meter. (From Pappas, R. A., L. J. Bond, M. S. Greenwood, and C. J. Hostick, On-line physical property process measurements for nuclear fuel recycling, In *Proc GLOBAL 2007, Advanced Nuclear Fuel Cycles and Systems*, Boise, ID, 1808–16, © 2007 by the American Nuclear Society, La Grange Park, IL. With permission.)

water. This method is ideal for measurement of very dilute slurries that is often not possible with current commercial instruments.

As with the density measurement, this method is self-calibrating and provides an uncertainty metric. To achieve a quantitative measurement, representative samples must be available beforehand for empirical laboratory calibration measurements. In recent development activities, this method has been integrated with the multiple reflection technique discussed earlier.

10.3.10.3   Measurement of Viscosity Using Ultrasonic Reflection Techniques

Greenwood and coworkers [101,156] have demonstrated an online computer-controlled sensor to measure the product of the viscosity (η) and density (ρ), $\rho\eta$, of a liquid or slurry for Newtonian fluids and the shear impedance of the liquid for non-Newtonian fluids. Measurements have been carried out for a quartz wedge having a base angle of 70°, as shown in Figure 10.15, and for one with a base angle of 45°. This method is discussed further in Chapter 6.

10.3.10.4   Ultrasonic Diffraction Grating Spectroscopy for Particle Size and Viscosity in Slurries

A variety of techniques for measuring the particle size and viscosity of a slurry have been demonstrated in a laboratory setting, but few methods are available for online and real-time measurement [101]. Ultrasonic diffraction grating spectroscopy has been investigated by Greenwood et al [156]. The results demonstrate a new technique for the measurement of particle size and show effects of the viscosity of the slurry [156], and the velocity of sound in the slurry is also measured by this technique. This method is discussed further in Chapter 6.

10.3.10.5   Ultrasonic Backscatter Measurement for Slurry Concentration and Phase Changes

Greenwood et al. [156] are also developing an ultrasonic measurement methodology to characterize solid–liquid suspensions at high concentrations based on novel backscatter methodologies. A schematic of the measurement system is shown in

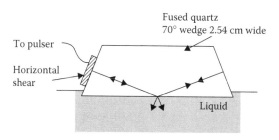

FIGURE 10.15   Diagram of fused quartz wedge and illustration of multiple reflections of a horizontal shear wave. (From Pappas, R. A., L. J. Bond, M. S. Greenwood, and C. J. Hostick, On-line physical property process measurements for nuclear fuel recycling, In *Proc GLOBAL 2007, Advanced Nuclear Fuel Cycles and Systems*, Boise, ID, 1808–16, © 2007 by the American Nuclear Society, La Grange Park, IL. With permission.)

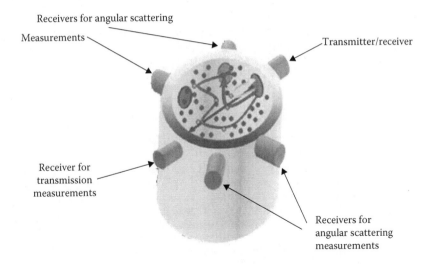

Receivers for angular scattering

Measurements

Transmitter/receiver

Receiver for
transmission
measurements

Receivers for
angular scattering
measurements

FIGURE 10.16   (See color insert.) Schematic of experimental configuration. (From Pappas, R. A., L. J. Bond, M. S. Greenwood, and C. J. Hostick, On-line physical property process measurements for nuclear fuel recycling, In *Proc GLOBAL 2007, Advanced Nuclear Fuel Cycles and Systems*, Boise, ID, 1808–16, © 2007 by the American Nuclear Society, La Grange Park, IL. With permission.)

Figure 10.16. It is designed to acquire and analyze ultrasonic signals to be used for attenuation, speed of sound backscattering, and the diffuse field. The measurement cell is composed of a ring with opposing transducers to send and receive the ultrasonic fields. The speed of sound is determined from a measurement of the transit time across the chamber, and the attenuation is a measurement of the energy lost due to scattering and absorption as the sound traverses the chamber. The backscatter is a measurement of the energy that is returned to the transmitting transducer. The diffuse field develops after the ultrasonic field has undergone many scattering events with the particles in the fluid and the walls of the chamber. For the attenuation, backscattering, and diffuse field, the Fourier amplitude is calculated to produce an ultrasonic response as a function of frequency. The effects from the measurement system and the transducer are removed from the received signal by calibration using an appropriate reference signal.

Preliminary investigations using 35 and 70-μm glass spheres are promising, as shown in Figure 10.17, and the results suggest that the direct and off-angle scatter signals are especially useful for characterizing high-concentration slurries. [101]. Work continues on backscatter methodologies in the deconvolution of particle sizes and concentrations in slurries. This method should be particularly useful in detecting the onset of some system phase changes such as precipitation and agglomeration.

10.3.11   Fluid Flow Measurement, Velocity Profiles, and Rheology

Ultrasonic methods can provide flow rate and rheological information on the contents of a process stream that can be invaluable in the monitoring and control of

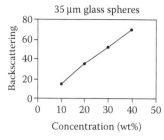
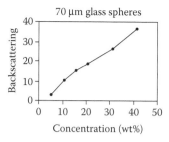

FIGURE 10.17 Preliminary results for 35- and 75-μm glass spheres. (From Pappas, R. A., L. J. Bond, M. S. Greenwood, and C. J. Hostick, On-line physical property process measurements for nuclear fuel recycling, In *Proc GLOBAL 2007, Advanced Nuclear Fuel Cycles and Systems*, Boise, ID, 1808–16, © 2007 by the American Nuclear Society, La Grange Park, IL. With permission.)

product quality. There are numerous commercial devices for monitoring average flow rates based on ultrasonic transit time measurements in both liquid and gas flow rates. All are based on one of two principles: (1) the effect of the moving fluid on the time of travel of sound between two points or (2) the effect of the moving stream on the deflection of an ultrasonic beam that is propagating normal to the direction of flow. Ultrasonic flowmeters are capable of measuring fluid flow without altering the flow pattern and thus they introduce little or no pressure loss.

A flowmeter developed by the British Scientific Instrument Research Association [158] is described as one that provides an electrical output consisting of pulses of constant amplitude and repetition rate (25V and 800/second, respectively). The width of the output pulse is proportional to the rate of flow. The pulse is averaged by a resistor–capacitor (RC) network, giving a direct voltage with an amplitude that varies with pulse width. The voltage from the RC circuit is applied to drive an integrator motor, which is used to integrate the flow rate with respect to time.

In one method, upstream pulse travel times are compared with downstream travel times. Figure 10.18 illustrates the principle of the design. In practice, the transducers may be located side by side. The distance between each transmitting transducer and its corresponding receiving transducer is a constant d. If the travel time of a pulse from an upstream transmitter to a downstream receiver is T_1 and the travel time upstream is T_2, then

$$T_1 = \frac{d}{c' + v_f \cos \theta}$$

$$T_2 = \frac{d}{c' + v_f \cos \theta}$$

(10.4)

where c' is an equivalent velocity of sound in the fluid and includes an increment corresponding to a distance the pulse travels between the transducers and the boundary of the flowing fluid and v_f is the fluid velocity. When the fluid is stationary, $T = d/c'$

In a variation of the meter shown in Figure 10.18, only two identical transducers are used. Each alternates between transmitting and receiving.

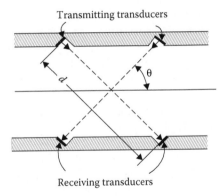

Transmitting transducers

Receiving transducers

FIGURE 10.18 Flowmeter based on comparing upstream and downstream pulse propagation rates.

Various techniques have been used to compare the upstream and downstream pulse propagation rates. One method involves the use of a "sing-around" principle in which a pulse generator produces a short train of ultrasonic waves. The received signal is amplified and used to retrigger the pulse generator. Neglecting delay times due to the electronic system and the distance the pulse travels outside the fluid stream, the difference between the downstream and upstream pulse repetition rates is

$$f_1 - f_2 = \frac{2v_f \cos\theta}{d} \qquad (10.5)$$

Flowmeters that are based on comparison of upstream and downstream transit times between two cylindrical transducers located around the blood vessel have been developed for measuring blood flow [159]. It is claimed that this instrument can measure flow rates of 1–100 cm/second. The lower limit represents a difference in transit time of 2×10^{-10} seconds, or a phase difference of 0.03° (operating frequency between 300 and 400 kHz).

Acoustic anemometers that use the same principles illustrated in Figure 10.18 have been developed.

Further examination of Equation 10.4 will show that

$$f_1 + f_2 = \frac{2c'}{d} \approx \frac{2c}{d} \qquad (10.5a)$$

where c is the velocity of sound in the fluid. By inserting an additional transducer that senses the load impedance, providing the characteristic impedance of the liquid (ρc), it is possible to obtain a measure of the mass flow rate with the same meter. A computer determines the velocity of sound, c, from the sum frequency, divides this quantity into impedance to obtain the density, ρ, and, from the average flow rate obtained from the difference frequencies, determines the mass flow rate.

The most significant errors in meters of the type shown in Figure 10.18 are caused by the velocity profile. The meter measures the average flow velocity along the path of the beam. For laminar flow, the meter reading is two-thirds of the maximum velocity at the center of the stream. The true average is given by one-half of the maximum velocity. For turbulent flow, the relationship depends on the Reynolds number for the pipe or channel.

In a flowmeter developed by Westinghouse, an upstream-traveling pulse and a downstream-traveling pulse are initiated simultaneously [160]. Each transmitter then becomes a receiver and detects the received pulses. The meter then measures the difference in travel time and relates this to the average flow velocity, where, for beam direction parallel to flow from Equation 10.4 the relationship becomes:

$$T_1 = \frac{d}{c + v_f} \quad \text{and} \quad T_2 = \frac{d}{c - v_f} \tag{10.6}$$

Then,

$$T_2 - T_1 = \frac{2dv_f}{c^2 - v_f^2} = \frac{2dv_f}{c^2} \quad v_f \ll c \tag{10.7}$$

The second type of ultrasonic flowmeter is based on the deflection of an ultrasonic beam by the fluid flow [161]. A transmitting transducer located on one side of the pipe emits a CW into the fluid stream. A split transducer on the opposite side of the pipe (Figure 10.19) determines the amount of beam deflection. The outputs of the two transducers are compared by use of a sensitive differential amplifier. If there is no flow, the beam strikes midway between the two receiver sections, the two sections generate equal voltages, and the output from the differential amplifier is zero. When the fluid is flowing, the beam shifts in the direction of the flow by an amount corresponding to the flow rate, and the outputs from the two sections differ. The difference voltage corresponds to the flow rate. If the flow rate were constant across the pipe, the beam deflection, ϕ, would be given by

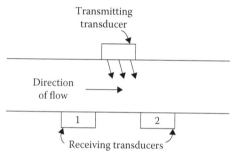

FIGURE 10.19   Beam-deflection ultrasonic flowmeter.

$$\phi = \tan^{-1}\left(\frac{v_f}{c}\right) \tag{10.8}$$

Applications of ultrasonic flowmeters have included measurements of flow rates of rivers, reactor heat-exchanger fluids, gas-containing emulsions, slurries (sewage, paper pulp, etc.), corrosive liquids of varying viscosities, and wind velocities.

One advantage of ultrasonic flowmeters is the fact that they introduce negligible pressure loss in a system. Also, they are inexpensive to operate and can handle a large variation in flow rates in a wide range of pipe diameters and over a wide range of pressures.

Quantities of large particles, including gas bubbles and solids, carried in a liquid stream place a restriction on the effective use of an ultrasonic flowmeter. When the size of these particles approaches one wavelength or larger, scattering produces maximum attenuation and therefore loss of signal.

In a related area, leak detection devices are available that operate on AE caused by escaping fluids. Aboveground storage tanks are monitored for leaks using AE technology and advanced computer analysis. This technique is used while the tanks are in service, using relatively inexpensive equipment and providing a very economical monitoring system.

Other AE systems are used to detect gas leaks. As gas moves from a high-pressure volume to a low pressure, as from the atmosphere into a vacuum or out of a high-pressure cylinder or line to the atmosphere, it causes turbulence that produces a sound spectrum. Certain systems sense the ultrasound in the spectrum created by the turbulence and use this information to locate the leaks.

The Doppler principle has been explained on pages 67–70. Flowmeters have been developed based on the Doppler principle. As an ultrasonic beam is directed upstream of fluid flow, the frequency of the received signal increases. As the beam travels downstream, the frequency of the received signal decreases. Flow rate is based on the shift in frequency and direction of the beam compared with the direction of flow.

10.3.11.1   Velocity Profiles and Rheology

Handa et al. [162] obtained velocity profiles of rotating flow of a magnetic fluid using an ultrasound velocity profile monitor; in addition, the 3D flow structures were clarified. Takega et al. [163] used ultrasonic Doppler velocimetry to measure mercury flow in the target of a neutron source. The technique provides spatiotemporal information about the flow field and is used for flow mapping. Shekarriz et al. [164] used ultrasonic Doppler velocimetry and time-of-flight measurements, and from these measurements, the local shear rate was determined using the local velocity in the pipe.

In response to DOE waste remediation needs, an advanced ultrasonic device has been developed for flow characterization [41,101,165]. The device provides rheological information based on velocity profiles generated with range-gated Doppler measurements. It also integrates interface detection and concentration measurements with the rheology determination capability into a single ultrasonic monitoring device. The performance of this device has been tested in a 5-cm-diameter flow loop using shear thinning and pseudoplastic materials (Figure 10.20).

In the monostatic operation mode, range-gated Doppler measurements are made at two frequencies with two dual-frequency probes located at opposite sides of the pipe. The ultrasonic Doppler velocimetry system measures the velocity profile in a pipe flow system by transmitting and receiving a wide-bandwidth coherent burst signal. This signal is typically 4–20 cycles in length. The bandwidth is then from 5% to 25% of the center frequency. The expected Doppler frequency, D_f, is given by

$$2D_f = \frac{2v_f}{c} \tag{10.9}$$

where v is the fluid/particle velocity, c is the speed of sound, and f is the ultrasonic frequency. The Doppler frequency is on the order of kilohertz, whereas the bandwidth of the transmitted signal is on the order of megahertz. For this reason, the Doppler shift cannot be detected in a single signal transmit–receive sequence but is determined by sampling the transmitted and received signals at a fixed-pulse repetition frequency. The range and Doppler information are then used to build a flow velocity profile. Shear rate information is then determined from this profile and is combined with pressure drop measurements to provide a rheogram, an example of which is given in Figure 10.21.

For use with slurries, the Doppler attribute of the backscatter signals measures flow and can identify solid stratification and fouling that can be present in some dense slurries and precursors to plugging. Recent work has shown that the blending of Doppler-derived measurements and acoustic impedance boundary reflections provides an exceptionally robust method for monitoring setting and fouling in vessels

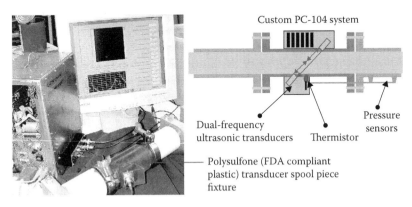

FIGURE 10.20 **(See color insert.)** Ultrasonic Doppler–based device for flow characterization. (From Pappas, R. A., L. J. Bond, M. S. Greenwood, and C. J. Hostick, On-line physical property process measurements for nuclear fuel recycling, In *Proc GLOBAL 2007, Advanced Nuclear Fuel Cycles and Systems*, Boise, ID, 1808–16, © 2007 by the American Nuclear Society, La Grange Park, IL. With permission.)

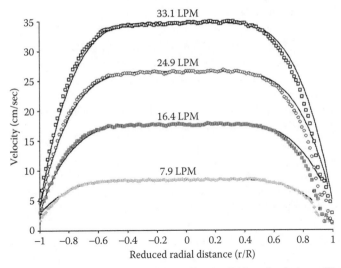

FIGURE 10.21   (See color insert.) Velocity profiles for Carbopol solutions. (From Pappas, R. A., L. J. Bond, M. S. Greenwood, and C. J. Hostick, On-line physical property process measurements for nuclear fuel recycling, In *Proc GLOBAL 2007, Advanced Nuclear Fuel Cycles and Systems*, Boise, ID, 1808–16, © 2007 by the American Nuclear Society, La Grange Park, IL. With permission.)

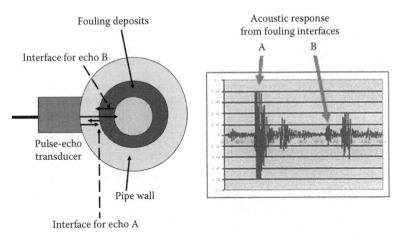

FIGURE 10.22   (See color insert.) Schematic showing detection of impedance boundaries to provide a means to track pipe inside pipes. (From Pappas, R. A., L. J. Bond, M. S. Greenwood, and C. J. Hostick, On-line physical property process measurements for nuclear fuel recycling, In *Proc GLOBAL 2007, Advanced Nuclear Fuel Cycles and Systems*, Boise, ID, 1808–16, © 2007 by the American Nuclear Society, La Grange Park, IL. With permission.)

and piping. The capability to detect impedance boundaries provides means to track pipe fouling inside pipes, and this is illustrated with the transducer pipe schematic in Figure 10.22. This implementation is essentially another implementation of A-TDR, which is discussed in Chapter 9, applied to polymer membranes.

10.3.11.2   Ultrasonic Liquid-Level Methodology

Tremendous economic savings can be realized if feedstock inventories are carefully controlled. With judicious timing of the delivery of feedstock inventory, initial process costs are reduced and daily operational costs are optimized. Robust liquid-level monitoring is an important component of this concept. A wide range of ultrasound-based, liquid-level monitors are on the market [3], and these use a diverse range of types of transducer configurations and sensor locations on tanks and vessels.

Researchers at Pacific Northwest National Laboratory have developed and validated a novel concept for an ultrasonic liquid-level monitor that is responsive to the market need for a low-cost, noninvasive, ultrasonic liquid-level monitoring device that can be easily installed, maintained, and operated [101]. The device comprises a single transducer mounted onto the outside surface of the side of a tank. The transducer generates acoustic burst signals with transverse and oblique propagation paths. The time-of-flights of the different echoes depend on the liquid composition, liquid temperature, and level of the liquid surface. The monitor continuously self-calibrates to account for temperature and composition changes, assuring accurate fill-level determinations. The liquid level is estimated from these time-of-flights using a physical model and rigorous mathematical algorithm. The estimation algorithm relies on the distinctively large echoes that return from the corner reflectors formed at the intersection of the liquid surface and the tank sidewalls. This monitor has several important features that distinguish it from other available monitors:

- Corner-shot physics model uses oblique angle operation versus normal to the surface operation
- External single sensor is side mounted rather than on the bottom, top, or inside of the tank
- Cost-effective, potential, single-transducer pulse-echo operation is possible versus multiple-transducer pitch-catch/pulse-echo operation
- Self-calibrating mechanism accounts for differences in the densities and temperature of the various liquids a tank may contain over time

10.3.11.3   Multiphase Flow "Visualization"

For multiphase flows, Kytömaa and Corrington [153] used clinical ultrasound backscatter to distinguish liquefied from settled states in transient events. The liquefied regions were identified by characteristic small-scale random fluctuations consistent with particle diffusion in suspensions. Good et al. [166] used acoustics to monitor the curing of grout by measuring the elastic properties. Clark et al. [167] investigated ultrasonic methods for mapping radioactive waste stored in million gallon tanks using attenuation and speed of sound, making measurements over the kilohertz to megahertz frequency ranges.

Most recently, ultrasonic methods have been demonstrated for the online real-time characterization of state of mixing. This work includes transducers that operate in the low-megahertz frequency range in an A-scan (pulse-echo or A-TDR) mode and medical-style B-scan (2D sector) imaging that provides real-time video images of mixing processes, as shown by Bond and coworkers [61].

10.3.12   PRESSURE AND TEMPERATURE

Temperature measurements by ultrasonic means are based on factors affecting the velocity of sound in the material. If the elastic properties of a material as a function of temperature are known, a measurement of the velocity of sound in the material can be related to the temperature. In some solids, the velocity decreases at a fairly constant rate as temperature increases until it reaches a point that is slightly below the melting point. Water-cooled buffer rods are used between the transducer and the heated section.

Lynnworth [168] has measured temperatures to above 2,775°C by a technique based on a pulse-echo type of velocity measurement. The instrument contains a thin wire probe, which is inserted into the medium (fluid or gas). An ultrasonic pulse is transmitted into the probe element, partially reflecting at a point at the beginning of the thin wire section and again at the end of the wire. The difference in time between the two echoes is a measure of the velocity of sound and hence of the temperature.

Lynnworth [168] also claims that measuring the velocity of sound in gases directly enables measurement of temperature to above 16,666°C. Obviously, various problems would be associated with making such measurements.

Pressure also can be measured on the basis of the velocity of sound in gases, liquids, and some solids. Temperature compensation is necessary to eliminate errors from this source.

10.4   UNDERWATER APPLICATIONS

SONAR [169] is a form of high-frequency sound that is used in underwater acoustic beacons that mark a spot in the sea, in detecting enemy submarines hundreds of kilometers away, in sounding of the depth of water beneath the ship, in transponders (devices that receive signals from one device and transmit the corresponding signals to another), in speedometers, which measure the components of ship speed over the bottom, in acoustical navigational aids, and in locating and tracking of fish and submarines. Side-scan SONARs and bottom profilers are used to probe the ocean floor to the side of the vessel and to explore what lies beneath the ocean floor. The maximum practical intensity of the source located near the surface of the sea is 0.36 W/cm²; however, the practical upper limit of intensity increases with increasing depth.

Underwater sound signals and echoes are detected by a variety of audio and visual displays. Earphones and loudspeakers are still useful accessories to more complex presentations. Computers are commonly used in many SONAR systems for recording and display of characteristics such as the bearing-time history of the target [144].

There are noises of various sources in the sea that interfere with the wanted noise from the target. These unwanted noises come from distant ship traffic; biological noisemakers such as fish, porpoises, and other kinds of oceanic life, which also cause noise scattering; rough sea surface; reverberation; and seismic activity. Any motion of the target causes a slight shift in frequency of the echo from the background of reverberation in which it occurs, so that the echo can often be separated from the reverberation by filtering.

There are two general types of SONAR devices: active and passive.

Active SONAR transmits a signal. The signal is reflected from the target and detected as an echo by the transducer that sent it. It is converted back to an electrical signal. This signal is amplified and altered by various techniques known as signal processing and presented to the observer on a display where it may be seen or heard.

Passive SONAR uses the underwater noise made by the target, such as a distant ship or a school of fish. This noise is received on an array of transducers and is again processed and displayed to the observer. One important type of passive SONAR, called SOSUS, is an array of transducers laid on the deep ocean floor for tracking noisy submarines to long distances and for listening to explosive sources such as underwater volcanoes and missiles that fall into the sea. It is also used to study the effects of climate changes on the temperature in the ocean. Such arrays are located in many oceans [170,171].

Sound underwater is also used for communication by voice and by code. Such communications can be made to distances of approximately 10 km if water temperatures are favorable. Practical audio frequency ranges are approximately 150–3000 Hz using carrier frequencies in the range of 8–10 kHz. Devices using these frequencies have been used not only for underwater communication between ships and submarines at sea but also by swimmers in lakes and ponds. This wide range of application of ultrasonic energy provides many commercial outlets for underwater sound equipment.

An overview, in the form of a brief tutorial paper, covering acoustic telemetry was provided by Baggeroer [172] in 1984. The late 1980s and early 1990s saw an increase in interest in acoustic communication underwater. This area has, like several aspects of ultrasonics, become its own specialized area for both research and deployment. An assessment of the state-of-the art, focused on developments since 1982, in underwater acoustic telemetry was provided by Kilfoyle and Baggeroer [173] in 2000. A major challenge is developing high-rate communication systems. Freitag et al. [174] include a perspective on this topic and future challenges in this area.

In addition to more modest person-to-person communication technologies for use by divers, there is increasing interest in acoustic networks to provide increased local area network coverage, analogues to the cell phone networks now common on land [175]. The potential for use of underwater ultrasonic measurements is increasing with application to measurement that includes thermometry and mapping of sea properties.

10.5 SURFACE ACOUSTIC WAVE SENSORS AND DELAY LINES

Ultrasonic bulk and the SAW devices were used in televisions and early computers in a variety of forms. These devices use both longitudinal and surface waves on quartz, silicon, and semiconductor substrates. Some of the bulk wave devices operated at frequencies of 20 and 100 MHz. A comprehensive treatment of bulk and SAW devices is provided by Kino [176].

Ultrasonic delay lines are being used in radar and computer signal processing. A signal is stored in the form of a short ultrasonic pulse for a brief period of time and retrieved at a proper time for display or computer use. Two general types are used: (1) fixed delay and (2) variable delay.

A typical fixed delay line consists of a many-sided (flat) low-loss material such as quartz with a transmitting transducer attached to one side and a receiving transducer attached to another side. The pulse of longitudinal-wave ultrasonic energy reflects alternately from each of the flat sides until it finally impinges on the receiving transducer.

Techniques that provide variable acoustic delay lines include (1) the use of high-frequency surface waves and (2) the deflection of laser beams. With surface waves, which travel at a much slower velocity than the longitudinal waves, the time is varied by varying the position of the receiving transducer.

With the laser beam technique, the input transducer is attached to a transparent acoustic delay bar [177]. A laser beam is directed onto a rotatable mirror, which directs the laser beam by way of a converging lens through the delay bar, then through a second converging lens, and finally onto a photodetector. As the sound wave encounters the laser beam, it acts as a moving diffraction grating on the beam and frequency-shifts the laser beam by an amount equal to the ultrasonic frequency. The undisturbed portion of the light beam and the diffracted portion are heterodyned together to produce an output signal at the photodetector at the input frequency. The delay is accomplished by changing the rotation of the mirror. By scanning the light beam along the bar, it is possible to generate other effects. For example, if the laser beam sweeps the bar at a rate higher than the rate at which ultrasonic wave propagates, the output signal is reversed in time. If the scanning rate falls between twice the velocity of sound and the velocity of sound, the output signal will be expanded. If the velocity of sound is greater than the scanning rate, the output is compressed and upshifted. When the scanning rate is exactly twice the velocity of sound, the output is a perfect time-reversed replica of the input pulse.

According to Smitz and Sittig [178], ultrasonic delay lines for digital storage, designed for bit rates near or beyond 100 MHz and capabilities around 1000 bits, compete favorably with integrated circuit storage in terms of economy, ease of fabrication, and speed.

SAW devices that operate at frequencies as high as 1.1 GHz have been developed, and in addition to their use as delay lines, they are increasingly being modified into devices to perform signal processing and as sensors.

Electronic thin film fabrication technologies have been used to move beyond simple mass sensing. A number of chemical and biochemical sensors [179] now use active coatings. These technologies are reviewed in a series of papers by Grate and his coworkers [180–183]: the devices use sorbent polymer coatings and have been used with both liquid and gas phase sensing. Other sensors based on SAW devices (mass sensing) are used for applications such as dew point sensing [184], and yet others, for more novel applications, include a SAW sensor array for wine discrimination [185]. These classes of SAW sensors are also being used in wireless modules. A review of wireless SAW sensors, which provides a comprehensive survey of the state-of-the-art, is given by Pohl [186]. Such sensors offer new devices for a wide range of applications, including remote monitoring and control of moving parts in harsh environments.

10.6   APPLICATION IN GASES

Several practical applications are included in this category of low-intensity ultrasonics. They include vehicle detection and speed measurement, television control,
garage-door openers, intrusion detection (burglar alarms), level detectors (liquids
and solids), conveyor-type monitoring, gas-pressure measurement, guidance for the
blind, and air flow measurements (anemometry).

There are also a growing number of air-coupled measurements in the low-megahertz range being made for both NDT and real-time online process measurements.

Ultrasonic sensors for use in air operate on principles involving transmission and
reception of echo pulses from a target, breaking an ultrasonic beam (pulse or CW),
sensing a given frequency to trigger a relay, or sensing Doppler shifts corresponding
to the movement of a target.

Many patents have been issued on inventions involving the detection of vehicles by ultrasonic means. The approach of a vehicle can be detected (1) by beaming
pulsed ultrasonics in a direction that intercepts the anticipated path of the vehicle
and detecting echoes, (2) by causing the vehicle to break an ultrasonic beam (CW
or pulsed) between a transmitting and receiving transducer, or (3) by measuring the
Doppler frequency caused by the motion of the vehicle into an ultrasonic beam. The
first and second methods are useful for counting vehicles. The third method can
be used to determine the velocity of approach by applying the appropriate Doppler
equations. A practical range for monostatic units operating in open quiet air is about
10 m at a frequency of 26 kHz. The range increases as the frequency decreases.
High-intensity broadband pulses also make possible much greater distances. High-
intensity pulse-echo equipment has been used to detect an automobile in an otherwise empty parking lot at distance of 53 m. Air turbulence (wind) places a practical
upper limit on the effective distance of these detectors. Noise due to turbulence is
controlled by appropriate filters and signal-cancellation techniques, but the range of
velocities is limited by the degree of such filtering.

An interesting passive device that senses the approach of another vehicle from the
rear has been developed by the company Sylvania [187]. The device responds only
to sounds emanating from vehicles entering a precisely controlled surveillance area
by sensing engine and tire sounds. It discriminates between moving and stationary
objects. It detects at a distance of 8 m from the receiving transducer in a target area
greater than a single traffic lane and at target speeds not less than 40 km/hour. The
warning is given either audibly or by dashboard display lamp.

Some early television controls for switching channels from across a room were
based on generating sympathetic resonances in elements within the set in response
to sound waves emitted by the control box. Pushing a button causes a striker to
resonate one of the bars. The pickup unit within the television set generates a signal to operate the switching mechanism. In another method, a resonant disk that
vibrates in flexure as it is struck is used [188]. The resonant frequency is controlled
by constraining dampers and selective location of a striking hammer. Channel
selection, volume control, and color adjustment can be accomplished ultrasonically
with at least eight channels being accommodated in the frequency range from 34 to
45 kHz [189].

Similar techniques are used to open garage doors and to switch on or off lights and appliances. Some interesting interactions have occurred in households using multiple units, such as having a coffee percolator energized when it was only intended that the garage door be closed. Sometimes, low-flying jet aircraft have caused garage doors to open.

Intrusion-detection devices (burglar alarms) work either on amplitude variations of an ultrasonic signal caused by an intruder or, commonly, on sensing Doppler signals caused by the intruder. The latter devices are very effective and sensitive. Air turbulence, heat sources, and vibrations or motions of objects (walls, drapes, etc.) in the room are major contributors of noise in installations using these units. However, modern equipment contains compensating circuitry for noises due to these sources and, if it is properly installed, they will generate few false alarms.

Measurement of liquid level can be accomplished from either the gas phase or the liquid phase, depending on which appears to be the more practical approach. The measurement usually is based on the time required for a pulse of ultrasound energy to travel from a transmitter to a free surface, reflect from the surface, and return to a receiver, which may be the same transducer used to generate the pulse. Some installations make use of the change in load impedance that occurs when the liquid level falls below or rises above a fixed position marked by the location of the ultrasonic transducer. The same method can be used to discriminate between liquid level, foam, and air.

Conveyor-type monitoring includes a variety of applications. Ultrasonic switches are available that count, sort, position, or control other functions associated with a conveyor-line operation. Ultrasonic beams in the 30- to 40-kHz frequency range are broken or reflected from an object to be detected, such as a box or carton. Either loss of signal in a broken beam or receipt of an echo by a receiving transducer is used to operate a relay, counter, or other device as the operation requires. Detection range is up to 12 m. Such systems have been used to detect breaks in paper webs and to control automatic production-line operations. An example of the latter is the control of the application of cream and jam to sponge cakes [190]. The cakes are split and the jam and cream are applied to the lower half. Ultrasonic sensors are used to detect the presence or absence of a cake in a pan. This is done by locating ultrasonic transducers above and below the conveyor and transmitting ultrasonic energy between them. The cake tray has a hole in the bottom through which the beam passes if no cake is present. When the receiver detects a signal indicating that no cake is present in the tray, its output is amplified and it operates a relay to stop the cream and jam applicator to avoid waste.

Similar systems have been used to monitor the flow of coal in mines by counting cars, indicating blocked chutes, and measuring the height of coal in bins. Measuring the height of material in bins is not restricted to coal mines but can be used wherever bin storage is required.

The U.S. Bureau of Mines has shown the feasibility of using pulse-echo techniques for evaluating coal mine roof strata [191]. Defects in the roof are indicated by echoes. One problem is the availability of compact, explosion-proof equipment.

Applications of NDT for metals were discussed in Chapter 8, as were measurements in high-pressure gas, which now include acoustic microscopy. New transducers, such as those employing polymers, aero-gels or micro-machined silicon, are achieving better matching to gases, as they exhibit reduced acoustic impedance contrast. This improved matching increases energy transmission and is opening up many new uses for ultrasound in gases [192].

The velocity and absorption of ultrasound in air has been measured to 20 MHz [193] at local atmospheric pressure. Operation of ultrasound in a noncontact mode in gases has significant advantages in terms of the resolution that is achievable at a particular frequency; this is due to the lower wave velocity in the gas. Ultrasonic parameters in gases have also been measured in both rarefied and highly compressed gases to investigate fundamental thermodynamic properties of many gases. The theory for the use of ultrasound to measure both composition and specific gas properties in binary and three-component mixtures has been developed and reported (e.g., Phillips et al. [194] and Zipser et al. [195]).

The developments in both transducers and systems are enabling an increasing number of novel measurements to be made using air-coupled ultrasonics. For example, Gan and Hutchins [196] have reported tomographic imaging of high-temperature flames using 1-MHz ultrasound, and the same group has also characterized capacitive micromachined ultrasonic transducers in rarefied gases [197].

Ultrasonic detection systems for use in guiding the blind have been investigated for many years. In most cases, these operate on echo principles and provide an audible signal to the user by means of earphones. One particular method rectifies the received pulses and produces an audible signal with a pitch that corresponds to the time between successive echoes and therefore to the distance to the obstacle. The effectiveness of the unit is, thus, dependent on the user's ability to estimate pitch.

REFERENCES

1. Methereli, A. F., and Larmore, eds. 1967–2009. Acoustical holography. In *Proceedings V 1-29: Continued as Acoustical Imaging*. New York: Plenum, and now Berlin: Springer.
2. Lynnworth, L. L. 1989. *Ultrasonic Measurements for Process Control*. Boston: Academic Press.
3. Asher, R. C. 1997. *Ultrasonic Sensors for Chemical and Process Plant*. Bristol: Institute of Physics Publishing.
4. Hauptmann, P., N. Hoppe, and A. Puttmer. 2002. *Meas Sci Technol* 13(8):R73–83.
5. Stephens, R. W. B. 1975. An historical review of ultrasonics. In *Proc. Ultrasonics International, 1975*. 9–19. Guildford: ICP Press.
6. Pohlman, R. 1937. *Z Phys* 107(7–8):497.
7. Suckling, E. E. 1957. *J Acoust Soc Am* 29(1):146–8.
8. Barbieri, O. 1942. *Alta Freq* 11:383.
9. Neumann, T., and H. Ermert. 2006. A new designed Schlieren system for the visualization of ultrasonic pulsed wave fields with high spatial and temporal resolution. In *Proc, Ultrasonics Symposium*, ed. M. P. Yuhas, B. C. Vancouver, Canada, October 2–6. 244–7. New York: IEEE.
10. Ya, S. 1937. Sokolov U.S. Pat. 2,164,185.
11. Ya, S. 1949. *Sokolov Zavod Lab* 14(11):1328–35.
12. Brown, P. H., R. P. Randall, R. F. Sivyer, and J. Wadley. 1975. A high resolution, sensitive ultrasonic image converter. In *Proc. Ultrasonics International*, 73–9. Guildford: IPC.
13. Lasser, R., M. Lasser, J. Gurney, J. Kula, and D. Rich. 2005. Multi-angle low cost ultrasound camera for NDT field applications, Pre-print ASNT Fall Conference. http://www.imperiuminc.com/PDF/ASNT-Paper-Summary-2005.pdf
14. Freitag, W. 1958. *Jenaer Jahrbuch* Pt. I:228–74.

15. Smyth, C. N., F. Y. Poynton, and J. F. Sayers. 1963. *Proc. I EE* 110(1):16–28.

16. Fetland, R. A. 1960. U.S. Pat. 2,919,574.

17. Smyth, C. N. 1965. In *Proc. Symposium on Ultrasonic Imaging*, 39–58. Washington, DC: Mine Advisory Committee, National Academy of Sciences – National Research Council.

18. Jacobs, J. E. 1962. In *Proc. Symposium on Physics and Nondestructive Testing*, 59–74. San Antonio, Texas: Southwest Research Institute.

19. Jacobs, J. E., K. Reimann, and L. Buss. 1968. *Mater Eval* 26(8):155–8.

20. Jacobs, J. E. 1965. In *Proc. Symposium on Ultrasonic Imaging*, 59–75. Washington, DC: Mine Advisory Committee, National Academy of Sciences – National Research Council.

21. Khuri – Yakub, B. T., F. L. Degertekin, X.-C. Jin, S. Calmes, I. Ladabaum, S. Hansen, and X. J. Zhang. 1998. Silicon Micromachined ultrasonic transducers. In *Proc. IEEE Ultrasonic Symposium*, eds. S. C. Schneider, M. Levy, and B. R. McAvoy, Oct. 5–8, Sendi, Miyagi, Japan, 985–91. New York: IEEE.

22. Waddell, J. H., and J. W. Waddell. 1970. *Res Dev* 30:32.

23. Settles, G. S. 2001. *Schlieren and Shadowgraph Techniques*. Berlin: Springer.

24. Baborovsky, V., D. Marsh, and E. Slater. 1973. *Nondestr Test* 6(4):200–7.

25. Gericke, O. R., and R. C. Grubinskas. 1969. *J Acoust Soc Am* 45(4):872–80.

26. Sproat, W. H., and S. E. Cohen. 1970. *Mater Eval* 28(4):73–6.

27. Dion, J.-L., and A. D. Jacob. 1987. *IEEE Trans UFFC* 34(5):550–7.

28. Gerdt, D. W., M. C. Baruch, and C. M. Adkins. 1999. Ultrasonic liquid crystal based underwater acoustic imaging. In *Proc. Liquid Crystal Materials, Devices and Applications VII*, Vol. 3635, 58–65. San Jose, CA: SPIE Proceedings.

29. Natale, G. G. 1977. Recent ultrasonic studies of liquid crystals. In *Proc. IEEE Ultrasonic Symposium*, eds. J. de Klerk and B. R. McAvoy, Oct 26–28, Phoenix, AZ, 366–73. New York: IEEE.

30. Sukhovich, A., L. Jing, and J. H. Page. 2008. *Phys Rev B* 77(1):014301.

31. Arkhangelskii, M. Ye. 1964. *Sov Phys Acoust* 9(3):301–2.

32. Arkhangelskii, M. Ye., and G. N. Pinus. 1960. *Sov Phys Acoust* 6(3):276–81.

33. Berger, H., and I. R. Kraska. 1962. *J Acoust Soc Am* 34(4):518–9.

34. Schilb, T., C. Bennett, and C. E. Adams. 1957. *J Acoust Soc Am* 29(1):145–6.

35. Eckardt, A., and D. Fintelman. 1955. *Naturwissenschaften* 42(20):555–6.

36. Kudryavtsev, B. B., A. N. Medbedev, and A. P. Ponomarev. 1959. Sb. Primeneniya Ultraakust Issled. *Veshchestva* Vol. 9:139–45.

37. Gabor, D. 1949. *Proc R Soc (London) Ser A* 197:454–87.

38. Young, J. D., and J. E. Wolfe. 1967. *Appl Phys Lett* 11(9):294–6.

39. Aldridge, E. E., and M. J.-M. Clement. 1982. Ultrasonic holography. In *Ultrasonic Testing*, ed. J. Szilard, 103–66. Chichester: Wiley.

40. Brenden, B. B. 1994. Ultrasonic holography using a liquid surface sensor. In *International Advances in Nondestructive Testing*, ed. W. J. McGonnagle, Vol. 17, 31–62. Langhorne, PA: Gordon and Breach.

41. Shekarriz, A., and D. M. Sheen. 1998. Slurry pipe flow measurements using tomographic ultrasonic velocimetry and densitometry. In *Proc. Fluids Engineering Division Summer Meeting FEDSM'98*, Paper # 5076, Am. Soc. Mech. Eng., June 21–25, Washington, DC. New York: ASME.

42. Tretout, H. 1998. Review of advanced ultrasonic techniques for aerospace structures. In *Proc. ECNDT'98*, May 26–29, Copenhagen, Denmark, Vol. 3, Paper No. 9. Proceedings now online at www.ndt.net.

43. Lehman, C. D., M. P. André, B. A. Fecht, J. M. Johansen, R. L. Shelby, and J. O. Shelby. 2000. *Acad Radiol* 7(2):100–7.

44. Korpel, A., and L. Kessler. 1971. Comparison of methods of acoustic microscopy. In *Acoustical Holography*, ed. A. F. Metherell, Vol. 3, 23–43. New York: Plenum.

45. Lemons, R. A., and C. F. Quate. 1974. *Appl Phys Lett* 24(4):163–5.

46. Quate, C. F., A. Atalar, and H. K. Wickramasinghe. 1979. *Proc IEEE* 67(8):1092–114.

47. Heiserman, J., D. Rugar, and C. F. Quate. 1980. *J Acoust Soc Am* 67(5):1629–37.

48. Petts, C. R., and H. K. Wickaramasinghe. 1980. *Electron Lett* 16(1):9–11.

49. Chiang, C.-H., L. J. Bond, and W. Dube. 1994. Non-linear attenuation of ultrasonic waves in pressurized gases in a gas-coupled transmission-type acoustic microscope. In *Proc. 1994 Far East Conf. NDT (FENDT'94) and 9th ROCSNT Annual Conference Taipei, ROC*, November, 449–56. Taipei, Taiwan, R.O.C.: Chinese Society for Materials Science.

50. Bond, L. J. 1992. Through transmission gas and pulsed water-coupled microscopy of electronic packaging and composite materials, August 1990-July 1992, Final Report to Materials Reliability Division, NIST. Boulder, CO: Department Mechanical Engineering, University of Colorado at Boulder.

51. Ihara, I., C.-K. Jen, and D. Ramos França. 2000. *Rev Sci Instrum* 71(9):3579–86.

52. Lemor, R. M., E. C. Weiss, G. Pilarczyk, and P. V. Zinin. 2003. *Measurements of Elastic Properties of Cells Using High-Frequency Time-Resolved Acoustic Microscopy, 2003 IEEE Ultrasonic Symposium.*, ed. D. E. Yuhas, Honolulu, HI, Oct. 5–8, 752–6. New York: IEEE.

53. Gilmore, R. S., K. C. Tam, J. D. Young, and D. R. Howard. 1986. *Philos Trans R Soc Lond A* 320(1554):215–35.

54. Briggs, A. 1992. *Acoustic Microscopy*. Oxford: Oxford University Press.

55. Briggs, A., ed. 1995. *Advances in Acoustic Microscopy*. Vol. 1. New York: Plenum.

56. Briggs, A., and W. Arnold. 1996. *Advances in Acoustic Microscopy*. Vol. 2. Berlin: Springer.

57. Khuri-Yakub, B. T. 1993. *Ultrasonics* 31(5):361–72.

58. Drinkwater, B. W., and P. D. Wilcox. 2006. *NDT&E Int* 39(7):525–41.

59. Schuster, G. J., S. R. Doctor, and L. J. Bond. 2004. *IEEE Trans Instrum Meas Technol* 53(6):1526–32.

60. Bond, L. J., S. R. Doctor, K. J. Bunch, M. Good, and A. E. Waltar. 2007. Instrumentation, monitoring and NDE for new fast reactors. In *Proc. GLOBAL 2007, Advanced Nuclear Fuel Cycles and Systems*, 1274–9. Boise, ID, Sept. 9–13. La Grange Park, IL: Am. Nuclear Soc.

61. Bond, L. J., M. Meenaksh, and H. O. Matthiesen. 1998. Ultrasonic methods for the on-line real-time characterization of state of mixing. In *Proc. 16th ICA and 135th Meeting Acoustical Society of America*, ed. P. K. Kuhl, and L. A. Crum, 1161–2. Seattle, WA, June 20–26. Woodberry, NY: Acoustical Soc Am.

62. Roh, Y., and B. T. Khuri-Yakub. 2002. *IEEE Trans UFFC* 49(3):293–8.

63. Fink, M. 1992. *IEEE Trans UFFC* 39(5):555–66.

64. Wu, F., J.-L. Thomas, and M. Fink. 1992. *IEEE Trans UFFC* 39(5):567–78.

65. Hauptmann, P., R. Lucklum, A. Püttmer, and B. Henning. 1998. *Sens Actuators A Phys* 67(1–3):32–48.

66. Hauptmann, P., R. Lucklum, and B. Henning. 2001. *Sensors Update* 3(1):163–207.

67. Workman, J., D. J. Veltkamp, S. Doherty, B. B. Anderson, K. E. Creasy, M. Koch, J. F. Tatera et al. 1999. *Anal Chem* 71(12):121R–80R.

68. Workman, J., K. Creasy, S. Doherty, L. J. Bond, M. Koch, A. Ullman, and D. J. Veltkamp. 2001. *Anal Chem* 73(15):27–2718.

69. Capote, F., and M. D. Lugue de Castro. 2007. *Analytical Applications of Ultrasound, Vol. 26. Techniques and Instrumentation in Analytical Chemistry*. Amsterdam: Elsevier.

70. Pethrick, R. A. 1989. Chapter 17, In *Comprehensive Polymer Science*, ed. G. Allan, and J. C. Bevington, Vol. 2, 571–600. New York: Pergamon Press.

71. Sokolov, S. I. 1946. *Zh Tekh Fiz* 16:283.

72. Truell, R., C. Elbaum, and B. B. Chick. 1969. *Ultrasonic Methods in Solid State Physics*. Boston: Academic Press.

73. Edmonds, P. D., ed. 1981. Ultrasonics. In *Methods of Experimental Physics*, ed. P. D. Edmonds, Vol. 19. New York: Academic Press.

74. Matheson, A. J. 1971. *Molecular Acoustics*. New York: John Wiley & Sons.

75. Povey, M. J. W. 1997. *Ultrasonic Techniques for Fluids Characterization*. Boston: Academic Press.

76. Harker, A. H., and J. A. G. Temple. 1988. *J Phys D Appl Phys* 21(11):1576–88.

77. Harker, A. H., P. Schofield, B. P. Stimpson, R. G. Taylor, and J. A. G. Temple. 1991. *Ultrasonics* 29(6):427–38.

78. Tsouris, C., and L. L. Tavlarides. 1993. *Ind Eng Chem Res* 32(5):998–1002.

79. Hale, J. M. 1988. *Ultrasonics* 26(6):356–7.

80. McClements, D. J., and P. Fairley. 1991. *Ultrasonics* 29(1):58–62.

81. McClements, D. J., and P. Fairley. 1992. *Ultrasonics* 30(6):403–5.

82. McClements, D. J., M. J. W. Povey, and E. Dickerson. 1993. *Ultrasonics* 31(6):433–7.

83. Dukhin, A. S., and P. J. Goetz. 1998. *Colloids Surf A* 144(1–3):49–58.

84. Dukhin, A. S., and P. J. Goetz. 2002. Ultrasound for characterizing colloids. In *Studies in Interface Science*, ed. D. Mobius, and R. Miller, Vol. 15. Amsterdam: Elsevier.

85. Challis, R. E., M. J. W. Povey, M. L. Mather, and A. K. Holmes. 2005. *Rep Prog Phys* 68(7):1541–637.

86. Kaatze, U., T. O. Hushcha, and F. Eggers. 2000. *J Solution Chem* 29(4):299–368.

87. Kaatze, U., F. Eggers, and K. Lautscham. 2008. *Meas Sci Technol* 19(6):1–21.

88. Lavallee, C., M. Carmel, L. A. Utracki, J. P. Szabo, I. A. Keough, and B. D. Favis. 1992. *Polym Eng Sci* 32(22):1716–26.

89. Piau, M., and C. Verdier. 1993. A feasibility study of the on-line characterization of polymer blends using ultrasound. In *Proc. Ultrasonics International*, Vienna, Austria, July 6–8, 423–6. Guildford: Butterworth.

90. Verdier, C., and M. J. Piau. 1996. *J Phys D Appl Phys* 29(6):1454–61.

91. Matsukawa, M., N. Ohtori, I. Nagai, K. Bohn, and J. K. Kruger. 1997. *Jpn J Appl Phys* 36(5B):2676–980.

92. Gendron, R., R. J. Tatibouet, J. Guevremont, M. M. Dumouloin, and L. Piche. 1995. *Polym Eng Sci* 35(1):79–91.

93. Schulitz, F. T., Y. Lu, and H. N. G. Wadley. 1998. *J Acoust Soc Am* 103(3):1361–9.

94. Bailey, J. A., and J. R. Davila. 1971. *Appl Sci Res* 25(3–4):245–61.

95. Parker, R. L., and J. R. Manning. 1986. *J Crystal Growth* 79(1–3):341–53.

96. Parker, R. L., J. R. Manning, and N. C. Peterson. 1985. *J Appl Phys* 58(11):4150–64.

97. McDonough, M. W., and A. Faghri. 1993. *Exp Heat Transfer* 115(4):1075–8.

98. Mauer, F. A., S. J. Norton, Y. Grinberg, D. Pitchure, and H. N. Wadley. 1991. *Metall Trans B* 22B(4):467–73.

99. Goebbels, K. 1981. Structure analysis by scattered ultrasonic radiation. In *Research Techniques in Nondestructuve Testing*, ed. R. S. Sharpe, Vol. 4, 87–157. London: Academic Press.

100. Bond, L. J., A. R. Greenberg, A. P. Mairal, G. Loest, J. H. Brewster, and W. B. Krantz. 1995. Real-time nondestructive characterization of membrane compaction and fouling. In *Review of Progress in Quantitative Nondestructive Evaluation*, ed. D. O. Thompson, and D. E. Chimenti, Snowmass, CO, Jul. 31–Aug. 5, 1994, Vol. 14, 1167–73. New York: Plenum.

101. Pappas, R. A., L. J. Bond, M. S. Greenwood, and C. J. Hostick. 2007. On-line physical property measurements for nuclear fuel recycling. In *GLOBAL 2007, Advanced Nuclear Fuel Cycles and Systems*, Boise, ID, 1808–16. La Grange Park, IL: Am. Nuclear Soc.

102. Soong, Y., I. K. Gamwo, A. G. Blackwell, R. R. Schehl, and M. F. Zarochak. 1995. *Chem Eng J* 60(1–3):161–7.

103. Soong, Y., I. K. Gamwo, A. G. Blackwell, F. W. Harke, R. R. Schehl, and M. F. Zarochak. 1996. *Ind Eng Chem Res* 35(6):1807–12.

104. Geladi, and K. J. Esbensen. 1989. *J Chemometrics* 3(2):419–29.

105. Chaouki, J., F. Larachi, and M. P. Dudukovic. 1997. *Ind Eng Chem Res* 36(11):4476–503.
106. Plaskowski, A., M. S. Beck, R. Thorn, and T. Dyakowski. 1995. *Imaging Industrial Flows*. Bristol: Institute of Physics Publishing.
107. Hoyle, B. S. 1996. *Meas Sci Technol* 7(3):272–80.
108. Martin, P. D., M. Beesley, and P. E. Myers. 1995. *Chem Eng J* 56(3):183–5.
109. Xu, L., Y. Han, L.-A. Xu, and J. Yang. 1997. *Chem Eng Sci* 52(13):2171–83.
110. Wright, W. M. D., D. W. Schindel, D. A. Hutchins, P. W. Carpenter, and D. P. Jansen. 1998. *J Acoust Soc Am* 104(6):3446–55.
111. Hoyle, B. S., X. Jia, F. J. W. Podd, H. I. Schlaberg, H. S. Tan, M. Wang, R. M. West, R. A. Williams, and T. A. York. 2001. *Meas Sci Technol* 12(8):1157–65.
112. Lynnworth, L. L., G. Jossinet, and E. Cherifi. 1996. 300°C clamp-on ultrasonic transducers for measuring water flow and level. In *Proc. Ultrasonics Symposium*, Vol. 1, 407–12. New York: IEEE.
113. Dreacher-Krasicka, E., H. T. Yolken, W. C. Carter, and D. de Fontaine. 1992. Ultrasonic monitoring of phase transformations during processing of YBa2Cu3Oz. In *Review of Progress in QNDE*, ed. D. O. Thompson, and D. E. Chimenti, Brunswick, ME, Jul. 28–Aug. 2, 1991, Vol. 11, 1837–44. New York: Plenum.
114. Ono, Y., J.-F. Moisan, and C.-K. Jen. 2003. *IEEE Trans UFFC* 50(12):1711–21.
115. Ono, Y., J.-F. Moisan, and C.-K. Jen. 2004. *Meas Sci Technol* 15(2):N25–9.
116. Clark, A. V., M. Lozev, B. J. Filla, and L. J. Bond. 1994. Sensor System for intelligent processing of hot-rolled steel. In *Proceedings 6th International Symposium on NDE of Materials*. eds. R. E. Green, K. J. Kozacker, and C. O. Ruud, Oahu, HI, June 7–11, 29–36. New York: Plenum Press.
117. Kazys, R. A., A. Volesis, and B. Voleistene. 2008. *Ultragarsas (Ultrasound)* 63(2):7–17.
118. Karasawa, H., M. Izumi, T. Suzuki, S. Nagai, M. Tamura, and S. Fujimori. 2000. *J Nuclear Sci Technol* 37(9):769–79.
119. Silk, M. G. 1984. *Ultrasonic Transducers for Nondestructive Testing*. Bristol: Adam Hilger.
120. Density & sound velocity meter (trade literature). 1998. Ashland, VA: Anton Paar USA.
121. Povey, M. J. W. 1997. *Ultrasonic Techniques for Fluids Characterization*. 48. New York: Academic Press.
122. Ultrasound velocity meter (trade literature). 1998. Dorchester, England: Cygnus Ltd.
123. DensiCheck (trade literature). 1998. Edinburgh, Scotland: Canongate.
124. Concentration monitors. (technical literature). 1998. Tulsa, OK: Nusonics.
125. Ultrasonic flowmeter (technical literature). 2010. Tokyo, Japan: Fuji Ultrasonics.
126. Sheen, S.-H., W. P. Lawrence, H. T. Chien, and A. C. Raptis. 1994. U.S. Patent 5,365,778.
127. Sheen, S. H., H. T. Chien, and A. C. Raptis. 1995. An on-line ultrasonic viscometer. In *Proc. Rev. Prog. Quant. Nondestr. Eval*, ed. D. O. Thompson, and D. E. Chimenti, Snowmass, CO, Jul. 31-Aug. 5, 1994, Vol. 14A, 1151–65. New York: Plenum Press.
128. Sheen, S. H., H. T. Chien, and A. C. Raptis. 1996. Measurement of shear impedances of viscous fluids. In *Proc. IEEE Ultrason. Symp*. eds. M. Levy, S. C. Schneider and B. R. McAvoy, Nov. 3–6, San Antonio, TX, Nov. 3-6, Vol. 1, 453–7. New York: IEEE.
129. Kline, B. R. 1991. U.S. Patent 4,991,124.
130. Greenwood, M. S., J. L. Mai, and M. S. Good. 1993. *J Acoust Soc Am* 94(2):908–16.
131. Fort, J. A., J. A. Bamberger, J. M. Bates, C. W. Enderlin, and M. R. Elmore. 1993. Pacific Northwest National Laboratory Report, PNL-8476. Richland, WA: PNNL.
132. Greenwood, M. S., J. R. Skorpik, J. A. Bamberger, and R. V. Harris. 1999. *Ultrasonics* 37(2):159–71.
133. Greenwood, M. S., J. R. Skorpik, and J. A. Bamberger. 1998. On-line Sensor to Measure Density of a Liquid or Slurry. In *Science and Technology for Disposal of Radioactive Tank Waste*, ed. W. W. Schulz, and N. J. Lombardo, 479. New York: Plenum Press.
134. Scott, D. M., and B. O. Paul. 1999. *Chemical Process,* Vol 61.

135. Sympatec GmbH. 1999. Ultrasonic extinction sensor (OPUS). Clausthal-Zellerfeld, Germany: System-Partikel-Technik, Sympatec GmbH.
136. Bamberger, J. A., M. S. Greenwood, and H. K. Kytömaa. 1998. *Proc. Ultrasonic characterization of slurry density and particle size Fluids Engineering Division Summer Meeting FEDSM'98*, Washington, DC, June 21–25, Paper # 5075, New York: ASME.
137. Kytömaa, H. K. 1995. *Powder Technol* 82(1):115–21.
138. Boxman, A., D. M. Scott, and C. E. Jochen. 1995. *PARTEC*. Germany: Nurnberg.
139. Pendse, H. 1991. DOE/Industry Advanced Sensors Technical Conference, U.S. Department of Energy, 9.
140. Pendse, H., and A. Sharma. 1993. *Part Part Syst Charact* 10(5):229–33.
141. Pendse, H., and A. Sharma. 1994. On-line Instrumentation for Size Distribution Analysis of Industrial Colloidal Slurries Using Ultrasound," First International Particle Technology Forum Proceedings. Part 1, 136–41. New York: AIChE Press.
142. Pendse, H., and W. Han. 1994. *Ultrasound Characterization of Concentrated Colloidal Suspensions Containing Nonspherical Particles, First International Particle Technology Forum Proceedings*. Part 1, 142–7. New York: AIChE Press.
143. Biot, M. J. 1956. *Acoust Soc Am* 28(2):179–91.
144. Urick, R. J. 1947. *J Appl Phys* 18(11):983–7.
145. Hampton, L. 1967. *J Acoust Soc Am* 42(4):882–90.
146. Atkinson, C. M., and H. K. Kytömaa. 1991. In *Liquid Solid Flows*, ed. M. C. Roco, and T. Masuyama, Fluids Engineering Division FED-Vol-118. 145. New York: American Society of Mechanical Engineers.
147. Allegra, J. R., and S. A. Hawley. 1972. *J Acoust Soc Am* 51(5):1545–64.
148. Johnson, D. L., K. Koplik, and R. Dashen. 1987. *J Fluid Mech* 176:379–402.
149. Sheng, P., and M. Zhou. 1988. *Phys Rev Lett* 61(14):1591–4.
150. Zhou, M., and P. Sheng. 1989. *Phys Rev B* 39(16):12027–39.
151. Atkinson, C. M., and H. K. Kytömaa. 1992. *Int J Multiphase Flow* 18(4):577–92.
152. Atkinson, C. M., and H. K. Kytömaa. 1993. *J Fluids Eng* 115(4):665–75.
153. Kytömaa, H. K., and S. W. Corrington. 1994. *Int J Multiphase Flow* 20(5):915–26.
154. Derksen, J. S., and H. K. Kytömaa. 1994. *Acoustic Properties of Solid-Liquid Mixtures in the Inertial Regime: Determination of the Added Mass Coefficient In Solid Liquid Flows*. 75–81. New York: American Society of Mechanical Engineers. FED-vol. 189, Fluids Engineering Division.
155. Salin, D., and W. Schön. 1981. *J de Phys Lett* 42(22):L477–80.
156. Greenwood, M. S., J. L. Mai, and M. S. Good. 1993. *J Acoust Soc Am* 94(2):908–16.
157. Spelt, P. D. M., M. A. Norato, A. S. Sangani, and L. L. Tavlarides. 1999. *Phys Fluids* 11(5):1065–80.
158. Bertova, H. C. 1960. *Electron Eng* 32(389):442–3.
159. Haugen, M. G., W. R. Farrall, J. F. Herrick, and E. J. Baldes. 1955. *Proc Natl Electron Conf* 11:465–75.
160. Anon. 1967. *J Acoust Soc Am* 41(2):535–6.
161. Dalke, H. E., and W. Welkowitz. 1960. *Instrum Soc Am J* 7(10):60–3.
162. Handa, T., T. Sawada, and G. Yamanaka. 1998. Proc. Fluids Engineering Division Summer Meeting FEDSM'98, Washington, DC, June 21–25, Paper # 5073. New York: ASME.
163. Takeda, Y., and H. Kikura. 1998. Proc. Fluids Engineering Division Summer Meeting FEDSM'98, Washington, DC, June 21–25, Paper # 5074. New York: ASME.
164. Shekarriz, A., B. B. Brenden, and H. K. Kytomaa. 1998. Planar ultrasonic technique for real-time visualization and concentration Fluids Engineering Division Summer Meeting FEDSM98, Washington, DC, June 21–25, Paper # 5226. New York: ASME.
165. Pfund, D. M., M. S. Greenwood, J. A. Bamberger, and R. A. Pappas. 2006. *Ultrasonics* 44(Suppl. 1):e477–82.

166. Good, M. S., G. J. Schuster, and H. L. Benny. 1990. *Mater Eval* 48(4):456–60.
167. Clark, M. A., D. M. Martin, and J. Gajda. 1993. Acoustic properties of underground storage tank stimulant waster. In *Review of Progress in Quantitative Nondestructive Evaluation*, ed. D. E. Chimenti, and D. O. Thompson, La Jolla, CA, July 19–24,1992, Vol. 12B, 2273–80. New York: Plenum Press.
168. Lynnworth, L. C. 1969. *Mater Eval* 27(3):60–6.
169. Kinsler, L., and A. Frey. 1950. *Fundamentals of Acoustics*. Chap. 15–6. New York: Wiley.
170. Lerner, R. G., and G. L. Trigg, eds. 1991. *Encyclopedia of Physics*. 2nd ed. New York: VCH Publishers Inc.
171. Spiesberger, J. L. 2003. *J Acoust Soc Am* 144(5):25557–60.
172. Baggeroer, A. B. 1984. *IEEE J Oceanic Eng* OE-9(4):229–35.
173. Kilfoyle, D. B., and A. B. Baggeroer. 2000. *IEEE J Oceanic Eng* 25(1):4–27.
174. Freitag, L., M. Stajanovic, D. Kilfoyle, and J. Preisig. 2004. High-rate phase coherent acoustic communication: A review of a decade of research and a perspective on future challenges. In *Proc. 7th European Conf. Underwater Acoustics, ECUA 2004*. The Netherlands: Delft. The Hague: TNO Physics and Electronics Laboratory.
175. Proakis, J. G., E. M. Sozer, J. A. Rice, and M. Stajanovic. 2001. *IEEE Commun Mag* 39(1):114–9. The Hague: TNO Physics and Electronics Laboratory.
176. Kino, G. S. 1987. *Acoustic Waves*. Englewood Cliffs, NJ: Prentice-Hall.
177. Riezenman, M. J. 1969. *Electron Des* 12:28.
178. Smitz, F. M., and E. K. Sittig. 1969. *Ultrasonics* 7(3):167–70.
179. Lucklum, R., and P. Hauptmann. 2003. *Meas Sci Technol* 14(11):1854–64.
180. Grate, J. W., M. H. Abraham, and R. A. McGill. 1996. Sorbent polymer coatings for chemical sensors and arrays. In *Handbook of Biosensors: Medicine, Food, and the Environment*, ed. E. Kress-Rogers, and S. Nicklin, 593–612. Boca Raton, FL: CRC Press.
181. Grate, J. W., S. J. Martin, and R. M. White. 1993. *Anal Chem* 65(21):940A–8A and 65(22):987A–96A.
182. Grate, J. W., and G. C. Frye. 1996. Acoustic wave sensors. In *Sensors Update*, ed. H. Baltes, W. Goepel, and J. Hesse, Vol. 2, 37–83. Weinheim: VSH.
183. Grate, J. W. 2000. *Chem Rev* 100(7):2627–48.
184. Hoummady, M., C. Bonjour, J. Collins, F. Lardet-Vieudrin, and G. Martin. 1995. *Sens Actuators B* 27(1–3):315–7.
185. Santos, J. P., M. J. Fernández, J. L. Fontecha, J. Lozano, M. Aleixandre, M. Garcia, J. Gutiérrez, and M. C. Morrillo. 2005. *Sens Actuators B* 107(1):291–5.
186. Pohl, A. 2000. *IEEE Trans UFFC* 47(2):317–32.
187. Anon. 1970. *Electromech Des* 14(11):4–5.
188. Rossteutscher, G. 1969. *Radio Mentor (Germany)* 35(5):329–33.
189. Rossteutscher, G. 1969. *Radio Mentor (Germany)* 35(5):329–33.
190. Anon. 1967. *Ultrasonics* 5(4):198.
191. Mongan Jr., C. E., and T. C. Miller. 1960. *U S Bur Mines Rep Invest* 5617.
192. Mágori, V. 1994.Ultrasonic sensors in air. In *Proc. 1994 IEEE Ultrasonics Symposium*, eds. M. Levy, S. C. Schneider, and B. R. McAvoy, Cannes, France, Nov. 1–4. 471–81. New York: IEEE.
193. Bond, L. J., C.-H. Chiang, and C. M. Fortunko. 1992. *J Acoust Soc Am* 92(4):2006–15.
194. Phillips, S., Y. Dain, and R. W. Lueptow. 2003. *Meas Sci Technol* 14(1):70–5.
195. Zipser, L., H. Franke, and W.-D. Bretschneider. 2006. *IEEE Sensors J* 6(3):536–41.
196. Gan, T. H., and D. A. Hutchins. 2003. *IEEE Trans UFFC* 50(9):1214–8.
197. Davis, A. J., D. A. Hutchins, and R. A. Noble. 2007. *IEEE Trans UFFC* 54(5):1065–71.

11 Applications of High-Intensity Ultrasonics

Basic Mechanisms and Effects

11.1 INTRODUCTION

High-intensity applications of ultrasonics are those that produce changes in or effects on the media, or the contents of the media, through which the waves propagate. Various mechanisms may be activated by the ultrasonic energy to promote the effects, but the mechanisms involved are not always well known or understood. The phenomena that occur with high-intensity ultrasound interactions are in most cases nonlinear [1]. The complexity of these phenomena can best be illustrated with an example: energy is concentrated and transformed in cavitation. In cavitational collapse, temperatures of more than 5000 K and pressures greater than 50 MPa (above 500 atmospheres) are reported, and under some circumstances there are emissions of picosecond duration flashes of visible light, called sonoluminescence.

The phenomenon of sonoluminescence has been known since 1934. This conversion of energy from sound to light was first interpreted as blackbody radiation from the hot compressed bubbles, but it is now known that the radiation can exhibit a discrete spectrum due to excited atoms and molecules, superposed on a broad background. As Putterman points out, somehow the diffuse energy densities of about 10^{-11} eV/atom in a sound field are being concentrated into the energies of a few electron volts per atom that are needed to produce visible radiation. Sound, heat, and light are all engaged in these energy transformations. Exactly how this energy concentration occurs has never been fully explained, although various theories have been proposed and it certainly involves complex processes [2,3].

Cavitation and its numerous related interactions and energy transformations that are encountered in high-power ultrasound interactions are now used in a range of industrial processes, including sonochemistry, which remains a field of active research. Most of the effects can probably be related to the following:

1. Heat: As ultrasound progresses through a solid or fluid medium, energy is lost to that medium in the form of heat. Losses vary according to the nature of the medium. At certain interfaces, absorption may be high because of factors such as shear (friction) across the interface, and viscous effects in fluids increase heating.

2. Stirring (acoustic streaming): Intense ultrasound will produce violent agitation in a liquid medium of low viscosity and disperse material by resulting streaming currents of liquid or accelerations imparted to the particles.

3. Cavitation: Many of the effects associated with ultrasonics occur in the presence of cavitation. Emulsification of otherwise immiscible liquids occurs under this condition.

4. Chemical effects: Chemical activity, "sonochemistry," especially oxidation reactions, may be accelerated, sometimes manyfold, under the influence of ultrasonically produced cavitation. The effects have been variously attributed to heat and to mechanical rupture of chemical bonds. In some cases, the effect is the result of mechanical mixing or of dispersion of saturated layers that ordinarily form at an interface between the participants in the reaction.

5. Mechanical effects: Stresses developed in an ultrasonic field can cause ruptures to occur in materials. They may also cause relative motion between surfaces, which produces selective absorption at these surfaces, as in ultrasonic bonding of materials and the phenomena of softening or acoustoplasticity in metals, such as gold. Stresses developed in cavitation bubble walls can cause severe erosion of surfaces.

6. Electrolytic effects: It has been shown that when two metals separated on the electrolytic scale by even a small amount are exposed to intense ultrasonic irradiation in water, an accelerated galvanic action may be induced, which causes electrolytic corrosion.

7. Diffusion: Ultrasonic energy promotes diffusion through cell walls, into gels, and through porous membranes.

8. Vacuum effects: During the low-pressure phase of each cycle, boiling in liquids may be induced and fluids may be drawn into tiny pores.

9. Cleansing: Sometimes, the observed effects may result from acoustically eroding a protective coating from a surface so that reactions between two materials may occur that would not be possible otherwise.

10. Particle motion: Standing waves are used to manipulate and aggregate particles in separators, and ultrasonic fields are also used for particle manipulation, for example, as tweezers.

Although the mechanisms may not be fully understood and unexpected effects are still sometimes produced, all effects produced have logical explanations. Significant progress has been, and is being, made toward understanding the underlying phenomena used in high-power ultrasound (e.g., Beyer [1], Leighton [3], Brenner et al. [4], Suslick [5], Blake [6]). Basic mechanisms involved in high-intensity applications and effects produced are discussed in this chapter. Cavitation, which plays an important part in many reactions, is discussed more thoroughly in Chapter 2.

11.2 GENERAL DISCUSSION

High-intensity ultrasonic energy has produced many very interesting effects in the laboratory. Some of these effects are discussed in this chapter and in Chapters 12 and 13. An increasing number of these phenomena observed in the laboratory have been developed into industrial-scale processes. A greater number of these phenomena are yet to be developed beyond the laboratory stage. The most consistent commercial areas of application based on high-intensity ultrasound are welding and staking of plastic parts and cleaning. There is growing use of ultrasonics in the electronics industry for jointing "flip-chip" assemblies. There is a diverse range of ongoing research into various applications that are sonochemical processes, and most of these applications are restricted to production of high-value products. High-power and sonochemistry applications also include food processing, waste water treatment, water and food product sterilization, and an increasing range of applications in medical therapy and surgery, of which the best known are probably phacoemulsification (cataract disruption and removal) and ultrasound-assisted lipoplasty ("liquefaction" of fatty tissue in liposuction). Ultrasonic homogenizers have been available commercially for many years and are extensively used as laboratory devices, most often used for sample preparation requiring dispersing and lysis of biological cells and tissues. Ultrasonic soldering baths are used in some commercial installations. Home humidifiers based on ultrasonically atomizing water without heat can be purchased. There are special job shops where ultrasonic machining of ceramics and hard brittle materials is performed. A limited number of industrial installations have been using ultrasonics on bar and tube-forming dyes for many years. Studies combining ultrasound with other forms of energy synergistically appear to be leading to some practical large-scale processes. A notable example is electroacoustic dewatering—a process that combines the principles of ultrasonic energy with those associated with electric fields such as electroosmosis and electrophoresis. Although ultrasonic atomization of fuels has been studied at various levels for many years, its general, commercialization remains in the future. Many of the medical applications of high-power ultrasound are discussed in Chapter 14 and industrial applications are discussed in Chapter 12.

Large-scale acceptance of high-intensity ultrasonic processes in industry depends on

1. Design simplicity—for example, welders and cleaners.
2. Unique capabilities offered.
3. Utilization of localized reaction zones. In welding, drawing, and forming, the reaction zones are at or near the interfaces between the tool and the work surface or between mating surfaces of the materials to be joined.
4. Operational simplicity.

Factors that appear to hinder large-scale applications of high-intensity ultrasonic energy include

1. Competition from other less expensive methods.
2. Scale-up problems—particularly with volumetric effects requiring cavitation.
   a. Cavitation is energy consuming.
   b. Associated energy loss limits the total volume through which cavitation can be effective.
   c. Total energy required for an effective process is a function of volume to be treated.
3. General lack of information, training, and equipment developed for the proposed application.

The last issue of the availability of information, training, and equipment has changed significantly over the past 20 years. There has been growth in both research and applications in sonochemistry, and there are now a growing number of industrial processing and high-power medical systems available and in routine use.

Developing ultrasonic methods into practical applications is a multifaceted process. Identifying a need that might be satisfied is only the first step. High-intensity ultrasonic energy accelerates chemical and mechanical reactions, causes the formation of new chemical compounds that are produced by no other method, pulverizes and disperses materials into a medium, cleans surfaces to provide electrical continuity between an electrolyte and electrode, compresses solid particles into higher density compacts, atomizes liquids into fine mists, produces droplets of uniform size in a controlled flow system, produces emulsions, produces agglomeration and deagglomeration depending on conditions in the ultrasonic field, reduces drawing forces in wire and bar drawing, enhances surface coating on small particles or large areas, overcomes friction between surfaces in order to promote relative motion, enhances plastic flow, breaks through oxide coatings to permit operations such as soldering of aluminum, produces localized frictional effects that cause welding of metals and plastics, and so on. Ultrasonic energy is also a means of producing "deep heat" for medical therapy and in near-surface surgery. There is a selective absorption that has been used beneficially in medicine. For example, cancerous tissue absorbs ultrasonic energy at a higher rate than does normal tissue, and this property has been utilized effectively in treating cancer. The last few years have seen a resurgence of interest in therapeutic ultrasound at low power (~30 mW) and use at intermediate and high powers for surgery [7]. It is now commonly used in several specialized procedures including phacoemulsification (cataract removal) and lipoplasty (ultrasound-assisted liposuction). These and other high-power medical topics are discussed further in Chapter 14.

It is not difficult to recognize the benefits these effects can offer industry and medicine. However, the big second step is the transition between the laboratory and practice. This step is guided by basic principles that include not only propagation, absorption, stress distribution, and other factors associated with ultrasonics but also the nature of the materials and media affected, scale-up factors, and other implementation and personnel factors. It is often advantageous in large installations to combine forms of energy that work synergistically. Each form of energy performs what it can best accomplish. For example, combining ultrasonics with electrophoresis is a proven method for dewatering certain slurries and sludges. There are several identifiable mechanisms associated with the ultrasonic part of the process, such as cleaning screens to promote electrical continuity into the slurry and compaction, which causes similar electrical continuity through the slurry. Thus, the ultrasonic energy operates primarily in a localized region but assists the electrical energy in its operation on the bulk product.

Many illustrations could be given to show the importance of ultrasonic factors and of the nature of the products being treated in deciding whether and how to use high-intensity ultrasonic energy in a process. These principles will be demonstrated in the applications that are described in this and the following chapters.

11.2.1 ENERGY AND ENERGY CONVERSION

High-power ultrasound phenomena are enabled by the delivery of ultrasonic energy to the interaction zone. The two simplest high-power ultrasound configurations are the horn and the bath, to which transducers are attached. These basic configurations are shown in schematic form in Figure 11.1. In recent years, there are also a growing number of designs that use multiple transducers in various types in flow-through and batch-tank geometries, including those that use transducers operating at multiple frequencies [8,9].

Many high-power processes in fluids employ multibubble cavitation, in which a zone or volume of cavitation bubbles is generated. For industrial application, a central aspect of ultrasound performance evaluation, including comparisons with alternate methodologies, is the efficiency and effectiveness or yield given. A number of studies have sought to measured energy efficiency. Among these are several studies by Gogate et al. [9]. Although apparently simple, the analysis of the phenomena, and particularly the energy partition, proves to be quite challenging.

In evaluating ultrasound processing, it is found that the effectiveness of the interaction is particularly difficult to quantify. Results obtained with different systems appear to be both process and geometry dependent. There are also strong indications that there are cases where frequency, and the use of multiple frequencies, gives different efficiencies. The efficiency is also a function of applied power: including cases where higher powers may give lower efficiency. At present, although there is still only limited reliable data available, some trends do seem to be developing. Multifrequency cells have higher energy efficiency than either single horns or baths, but in some studies, an orifice plate (used in hydrodynamic reactors) gives the highest cavitational yield [9], although this in itself is the most important phenomenon. An example of a comparison of efficiency with different types of cavitation equipment is shown in Figure 11.2.

Vichare et al. [10] provide an energy analysis for acoustic cavitation. Starting from the assumptions of Flynn [11], they investigated parameters that included

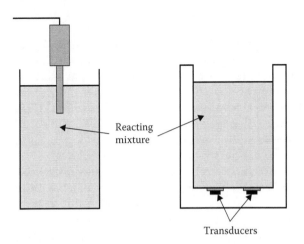

FIGURE 11.1 Schematic representations of ultrasonic horn and bath.

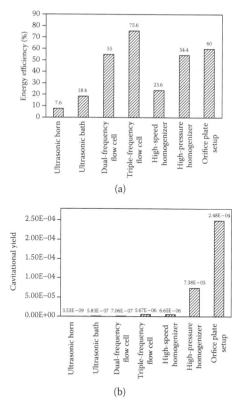

FIGURE 11.2 An example of a comparison of efficiency of different cavitational equipment: (a) energy efficiency and (b) cavitational yield. (Reproduced from Gogate, P. R., R. K. Tayal, and A. B. Pundit, *Curr Sci*, 91(1):35–46, 2006. With permission.)

frequency and bubble collapse radius and provided an estimate for a cavitational efficiency parameter. The data given show that for the various examples considered, efficiencies range between 1% and 20% when estimated calorimetrically.

Loning et al. [12] used electrical power measurements and calorimetry to investigate energy transformation in sonochemical processes for a horn-based system. Efficiency is shown to be dependent on both the system and the fluid properties. Energy efficiency of two ultrasonic devices, a horn and a custom bath, has been investigated by Wu and Ondruschka [13]. Performance at 850 kHz is compared with that at 24 and 30 kHz. In this study, oxidation of potassium iodide (KI) was used as a standard reaction to measure efficiency or effectiveness of interaction. In general, lower power intensity or higher power density is found to be generally favorable to KI oxidation. However, for benzenes, degradation energy efficiency is enhanced with increase in power intensity and density at lower power and decreased with increase in power intensity and density at higher input power.

In addition to the challenges of comparing different systems, it is well known that sonochemistry is, in general terms, less efficient at higher intensities. This has been attributed to decoupling and shielding, which involve a small high-intensity interaction zone that is found immediately under a horn tip and where there is a complex bubble cloud that moves away from the horn under the forces of acoustic streaming [14].

The various energy analyses have shown that effectiveness of enabled phenomena (e.g., chemical yield) and cavitational efficiency are functions of excitation frequency, intensity applied, fluids, reactions induced, and device geometry (e.g., a horn as opposed to a disk applied to a vessel wall or an enclosed pipe geometry). The yield of particular interactions also depends on a particular mechanism that induces the desired sonochemical effect: heating, cavitational phenomena (high pressures–temperatures), and/or streaming/mixing and fluid-catalysis interactions. Different system parameters are required to be optimized so as to give the optimal interaction/yield, depending on which of the different underlying phenomena is of greatest importance.

11.2.2 INTERACTION ZONES

High-power ultrasonic fields can vary significantly in terms of their three-dimensional form, temporal geometry, and space–time evolution. In general terms, there are three fundamental configurations used, which are shown in schematic form in Figure 11.3. The geometries are

- Horn near or applied to a surface (this can be either a horn and a trapped thin layer or a jackhammer, if in contact with the surface of a material)
- Standing wave (e.g., as used in separation devices or resonant interaction cell)
- Semiinfinite volume (a horn in a container)

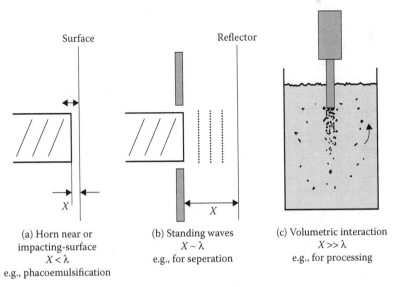

(a) Horn near or impacting-surface
$X < \lambda$
e.g., phacoemulsification

(b) Standing waves
$X \sim \lambda$
e.g., for seperation

(c) Volumetric interaction
$X \gg \lambda$
e.g., for processing

FIGURE 11.3 Interaction volumes for a horn with different horn–surface separations.

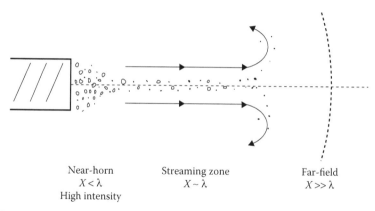

Near-horn	Streaming zone	Far-field
$X < \lambda$	$X \sim \lambda$	$X \gg \lambda$
High intensity		

FIGURE 11.4   Interaction zones for an ultrasonic horn.

The classes of interaction and the interaction volumes involved vary significantly for each geometry.

For the standing wave system, separation of a multiple number of half-wavelengths is needed, together with parallel surfaces. Also, there are upper and lower bounds for the energy that can be delivered if well separated static bands or flow is to be achieved. An example of the application of such a system is a flow-through cell or nanoparticle separation system.

For a horn (or transducer applied to a baffle), as shown in Figure 11.3c., there are distinct interaction zones. These have been visualized in several classic papers using ingenious approaches and high-speed photography. A good summary of the phenomena and illustrations, with references to the early papers, are given in the book by Beyer [1]. The basic phenomena are illustrated in schematic in Figure 11.4. The first zone is immediately under the horn: *highest intensity zone*, where there are multiple bubbles and under the correct conditions sonoluminescence is commonly seen; *streaming field of bubbles*, this is a volume of fluid moving out from the horn with return fluid flow that can be visualized (by injecting dye using a hollow needle); and *far field*, where there are no macrostreaming effects and where energy is absorbed through heating.

The size and shape of the interaction zones depend on the horn parameters, amplitude/power, and the fluid properties including temperature and viscosity. The nature of the cavitation phenomena, the duration or life of an individual bubble, also depends on system experimental conditions. Those topics, and some of the implications, are discussed further in parts of this chapter and in Chapters 12 and 13. Those particularly interested in investigating these phenomena further should start with a text such as that by Beyer [1].

11.3   MECHANICAL EFFECTS

11.3.1   CAVITATION

Some of the principles of cavitation are discussed in Chapter 2. Many of the effects produced by ultrasonics in liquids are associated with the production of cavitation.

Ultrasonic cavitation is the formation of bubbles or cavities in liquids during the low-pressure portion of a pressure wave cycle. The bubbles may be filled with gas coming out of the solution in the liquid under reduced pressure or with vapors of the liquid itself. Cavitation is a transitory phenomenon: if the bubble collapses within a single pressure wave cycle, it is termed "inertial cavitation" (formerly transient cavitation), and if the bubble oscillates through several cycles before collapse, it is termed "stable cavitation." In the limit, it is possible for a single bubble to be driven, at say 20 kHz, and remain oscillating and stable in terms of diameter and emitting light via sonoluminescence for long periods (many minutes). A photograph of stable single-bubble sonoluminescence is shown in Figure 11.5. The flask has a 20-kHz ring transducer attached, and to maximize light emissions, the water is precooled to about 4°C. The single blue "dot" can remain with a constant diameter when oscillating over many minutes, commonly until fluid temperature increases significantly, which quenches the phenomena.

For a horn immersed into a fluid, when operating above the cavitation threshold, multibubble systems are typically seen streaming away from the transducer: this geometry is shown in Figure 11.4. One issue that is encountered with all horn-based ultrasonic systems is cavitational erosion of the horn tip. Some examples of the effect of erosion on replaceable titanium horn tips are shown in Figure 11.6. Such erosion reduces efficiency of energy transmission and can detune the horn away from its designed resonant frequency. In addition, the erosion introduces metallic contamination into the fluid being sonicated. Instantaneous stresses and temperatures associated

FIGURE 11.5   (**See color insert.**) Photograph showing sonoluminescence for single-bubble cavitation. (Courtesy of L. R. Greenwood and Pacific Northwest National Laboratory.)

FIGURE 11.6   Examples of cavitational erosion seen in replaceable titanium ultrasonic horn tips. (Courtesy of Pacific Northwest National Laboratory.)

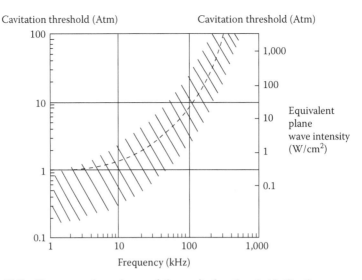

FIGURE 11.7   Frequency dependence of the cavitation threshold. Continuous wave data, fresh water at atmospheric pressure. (After Urick, R. J., *Principles of Underwater Sound*, 3rd ed., p. 76, McGraw-Hill, New York, 1983.)

with the formation and collapse of cavitation bubbles have been estimated at various levels, with 1000 MPa 10,000 atmospheres, and 2000–3000°C being typical. At the same time, free chemical radicals are produced, and even very tough metals are eroded.

There is a strong frequency dependence for the cavitation threshold for continuous waves in freshwater at atmospheric pressure, which is shown with the data summarized by Urick [15] in Figure 11.7. This threshold shifts with impurities in water and can be suppressed through the application of a hydrostatic pressure. Cavitation is of interest in both sonochemistry and sonar. Numerous papers, book chapters, and books have been published on the topic (e.g., Leighton [3]). A collection of papers that discuss most aspects of acoustic cavitation and sonoluminescence is given in a journal special issue edited by Blake [6].

Significant progress has been made in developing theories for cavitation. Single-bubble cavitation is discussed in detail by Brenner et al. [4], and multibubble sonoluminescence/cavitation, including hot-spot conditions, is considered by Suslick et al. [16]. As recently as 2006, Margulis [17] commented in a paper, discussing the mechanisms of multibubble sonoluminescence, that "the mechanism of luminescence induced by ultrasound waves in a liquid is an unsolved problem of acoustics and modern physical chemistry." He addresses the contradictions based on assumptions made for theories based on thermal recombination and chemiluminescent mechanisms for multibubble sonoluminescence.

11.3.2   DISPERSIONS, HOMOGENIZATION, AND EMULSIFICATION

The production of various types of solid–liquid and liquid–liquid dispersions by ultrasonic means has many applications both in the laboratory and industry. The basic mechanism appears to be cavitation. The energy released as the bubbles collapse imparts high stresses and accelerations to particles in close proximity to the bubbles. The particles are, thus, propelled into the fluid medium, and at extremely high intensities, the particles may be fractured or shattered.

The choice of equipment is based on two fundamental considerations: the difficulty of the task and the quantity to be processed.

The earliest experiments using ultrasonics to emulsify immiscible liquids were performed with transducers of X-cut quartz crystals. Energy was coupled from the radiating surface of the transducer through transformer oil to the specimen. Quartz crystals are capable of producing high intensities at frequencies ranging from 100 kHz to approximately 8 MHz. Large quartz crystals are expensive. The high operating voltages associated with high ultrasonic intensities often lead to arc-over between electrodes and concurrent damage to the crystal.

The development of ceramic piezoelectric materials has made possible high-intensity ultrasonic energy at frequencies above 100 kHz at lower voltages. These materials have the advantage of being formable into shapes for focusing to obtain intensities that will produce dispersions and emulsions. Bowl shapes have been used for treating small quantities at high intensities, and cylindrical shapes for continuous flow systems. However, these types of transducer have found limited use as dispersers and emulsifiers compared with systems using stepped mechanical amplifiers and "liquid whistles."

Very high intensities may be produced in small volumes for analytical purposes, using transducers equipped with stepped mechanical amplifiers. Systems using the stepped horn are sold extensively for biological investigations and industrial research, particularly for sonochemistry. Dispersions that require several minutes for suitable preparation in ball mills and high-speed blenders may be prepared within a few seconds by ultrasonic means. Applications range from dispersion of bull semen for evaluation to disruption and homogenization of cells from various organs of the body.

The so-called liquid whistle is adapted to large-volume production of emulsions and dispersions [18]. The whistle is based on a principle proposed and developed by Pohlmann. A high-pressure, flat jet of liquid impinges on the edge of a thin blade. The blade is supported at the nodal points and located in a resonant cavity. The unstable condition created by the jet on the edge of the blade causes the blade to vibrate in resonance. The intensity of the vibrations and other fluid dynamic forces are sufficient to cause localized cavitation and thus produce dispersions or emulsions at a high rate. These systems are adaptable to production installations, a typical system handling from 1136 to 1325 L/hour. Systems have been designed for handling abrasive materials. Liquid whistles should not be expected to break down large solid particles other than agglomerates.

The liquid whistle is claimed to have a long operating life. However, failure of the blade by fatigue has been a problem. This was especially true when treating viscous materials if the system "starved" for fluid supply.

A typical commercial use of the liquid whistle was in the preparation of resin-bonded solid-film lubricants [19]. These lubricants consisted of molybdenum disulfide, graphite, and epoxy resins. Production rates with a 2.2-kW motor powering the liquid whistle system were claimed to be two times those with a 8.9-kW colloid mill.

Dispersions and abrasive emulsions have been prepared with liquid whistles during the manufacture of waxes and polishes. Whistles for dispersing abrasive materials were designed with abrasion-resistant parts. One whistle reportedly handled 1136–1325 L/hour of waxes and polishes. The whistle also has been used to produce DDT emulsions [20] and margarine emulsions [21–23]. A sulfide liquor has been found to be a good emulsifying agent for use in the preparation of stable DDT emulsions.

Liquid whistles are useful for removing ink in the recovery of cellulose fibers from slurries of waste papers of copiers, printers, and fax machines [24]. Ultrasonics can also be used to remove xerographic ink particles from waste paper by focusing high-frequency pulses on a small spot on the paper. The energy required is sufficiently high to remove the particles of ink but low enough not to disturb the paper. The pulse width is 10 μs, the pulse repetition rate is 1 kHz, and the sonic input pressure is 45 atm peak negative. The delinked spot is approximately 4 mm in diameter.

The permanence of an emulsion prepared by ultrasonic means depends on the relative characteristics of the liquids. Some oil–water emulsions may remain stable for weeks without the addition of an emulsifier. Mercury in water without an emulsifier usually separates within a few minutes, but if it is carefully handled, it may last several hours.

Marboe and Weyl [25] showed that the dispersion of mercury (Hg) in a liquid by ultrasonic means requires the presence of strong forces between the metal and the medium. For example, mercury cannot be dispersed in paraffin oil at 25°C under the influence of ultrasonic energy without strengthening the forces between the two liquids. When these forces are strengthened by adding stearic acid to the paraffin oil, the mercury disperses in the paraffin oil, resulting in a stable emulsion. In the latter case, the incompatible liquids are linked together by a molecular film of mercury stearate.

Similar results are obtained with metallophilic groups. It is difficult to produce an emulsion between mercury and deaerated (by boiling) distilled water. However, mercury–distilled water emulsions are readily produced after the water is aerated. The air oxidizes the mercury under the influence of ultrasonic energy, thus producing highly polarizable cations of Hg^+ and Hg^{2+}. The necessary binding forces between the mercury and water are provided by the resulting ionic character of the mercury surface. The Hg^{2+} ion in deaerated distilled water can be produced by adding a small quantity of mercury(II) chloride or mercuric chloride (formerly corrosive sublimate; $HgCl_2$) solution to the water; then, when ultrasonic energy is applied to the combination, a mercury–water emulsion results. Emulsification of mercury in water can be hindered or prevented by adding *potassium chloride* (KCl) solution to the water. The KCl changes the Hg^{2+} ion into a complex anion of the type $(HgCl_4)^{2-}$.

Marboe and Weyl [25] claimed that polar molecules exert image forces upon mercury so that it can be dispersed in polar liquids such as nitrobenzene but not in nonpolar liquids such as carbon tetrachloride.

A summary of the science and applications of emulsification and aggregate dispersal, including mechanisms and efficiency, is provided by Rooney [26]. Applications in food processing are increasing, and these are considered by Mason [27]. An overview of recent developments in ultrasonic emulsification is also provided by Canselier et al. [28] and Priego Capote and Lugue de Castro [29].

11.3.3   AGGLOMERATION AND FLOCCULATION

Intense sonic and ultrasonic energy can cause agglomeration and precipitation of some particles in fluid media. Thus, several proposals have been made to use ultrasonics to precipitate smoke from industrial chimneys and fog from airports. In general, high-intensity sirens operating at frequencies between 1 and 30 kHz have been considered, and these were claimed to induce the desired agglomeration and precipitation. The mechanisms of smoke agglomeration are discussed more extensively by Bergmann [30] and Hueter and Bolt [31]. With improvements in efficiency of the coupling of ultrasound into air, the use of airborne ultrasound for precipitation of smokes, fine dusts, and powders and the destruction of foams in both industrial and food processes is receiving renewed attention. Developments and examples of applications have been reviewed and described by Riera et al. [32].

High-intensity ultrasound applied to liquid suspensions of fibrous materials, such as paper pulp, causes the particles to become entangled and agglomerate.

Ultrasound, for example, is being investigated as an aid to dewater biosludge [33]. It is also being investigated for applications in food processing [34], and several application areas within process chemistry are discussed by Priego Capote and Lugue de Castro [29].

11.3.4   PRECIPITATES AND SOLS

Kukoz et al. [35] investigated the effects of temperature, ultrasonic energy, and rate of mechanical mixing on the fractional composition and particle size of nickel hydroxide precipitated from aqueous solutions of nickel sulfate ($NiSO_4$) and sodium hydroxide (NaOH). Particle size was influenced more by mixing rate than by temperature. The finest particles were produced by a combination of ultrasonic irradiation and intense mixing while saturating the medium with gas. The ultrasonic dispersion was greatest at the moment of precipitation.

Some Indian scientists have investigated the effects of ultrasonic energy on the stability of cobalt ferrocyanide [36] and nickel ferrocyanide sols [37]. In both cases, ultrasonic energy increased the stability of the sols, with or without oxygen. The increased stability was attributed to attrition of the particles due to cavitation in the case of nickel ferrocyanide and to an increase in hydroxide *ion* (OH) ions in the case of cobalt ferrocyanide. Both sols were negatively charged.

The growing interest in nanotechnology is seeing new applications of ultrasound to provide precipitates and sols in nanoparticle preparation. For example, Okitsu et al. [38] investigated ultrasound frequency effects on synthesis of gold nanoparticles. Optical active materials are also being obtained by using ultrasonic spray pyrolysis [39]. Yet other applications are emerging in food processing [29,34].

11.3.5   ENHANCEMENT OF HEAT TRANSFER

Ultrasonic energy can be beneficial in increasing heat transfer, but the effects are not limited to ultrasonic frequencies. Lemlich [40] noted an improvement in the coefficient of heat transfer from wires vibrated in transverse modes at frequencies between

39 and 122 Hz. Lemlich used electrically heated wires of diameter 6.4, 10.0, and 20.6 mm and obtained heat-transfer coefficients with vibration that were as much as four times the coefficients obtained without vibration. He noted an increase in coefficient with both amplitude and frequency, but the effects were independent of the direction of vibration. He proposed a "stretched-film" theory to explain his observations. Lemlich arrived at the following empirical equation, which is applicable to higher than 10% increase in heat transfer in air and other diatomic gases:

$$\frac{h}{h'} = 0.75 + 0.0031 \frac{\overline{R}_e^{2.05}(\beta \Delta t)^{0.33}}{(G_r)^{0.41}} \tag{11.1}$$

where h is the heat-transfer coefficient with vibration (Btu/hr-ft-°F), h is the heat-transfer coefficient without vibration (Btu/hr-ft-°F), R_e is Reynolds number computed from the diameter of the wire and the average vibrational velocity (dimensionless), β is the thermal coefficient of volumetric expansion (°F⁻¹), Δt is the difference between the surface temperature of the wire and the ambient temperature (°F), and G_r is the Grashof number (dimensionless).

Lemlich made an approximate extension to include other fluids, such as air, by the following equation:

$$N_u = \left[0.75 + 0.0022 \frac{\overline{R}_e^{2.05}(\beta \Delta t)^{0.33}}{P_r^{1.54} G_r^{0.41}} \right] \left[0.63 + 0.35(G_r \times P_r)^{0.17} \right]^2 \tag{11.2}$$

where N_u is Nusselt's number ($=h/h'$) (dimensionless) and P_r is the Prandtl number (dimensionless)

Fand [41] concluded that the increase in heat-transfer rate from a heated horizontal cylinder subjected to transverse horizontal vibrations is caused by thermoacoustic streaming. Both Fand [41] and Westervelt and coworkers [42–44] stated that there is a critical intensity below which vibrations applied to a cylinder in a fluid produce no effect on heat transfer. According to Fand [41], this critical intensity is approximately 9.1 cm/second for both vertical and horizontal transverse vibrations of a heated horizontal cylinder in air. The intensity is the product ξf, where ξ is the sinusoidal displacement amplitude of vibration and f is the frequency of vibration. According to Fand [41], the ratio of displacement amplitude to cylinder diameter, ξ/D, is an important parameter in all fluid dynamical problems involving transversely vibrating cylinders. This ratio is included in the Reynolds number (Equations 11.1 and 11.2) and therefore is taken into account in all equations involving the Reynolds number.

Westervelt [43,44] proposed that modifications in heat transfer by sound are caused by modifications of the convective flow in the inner-streaming boundary layer. The critical intensity is determined by the ratio of particle displacement amplitude to the acoustic boundary layer thickness, ξ/δ_{ac}, and it is reached when $\xi/\delta_{ac} = 1$. In terms of critical sound pressure level, N_{spl}, in air,

$$N_{spl} = (136 - 10 \log f) \quad db \frac{\xi}{\delta_{ac}} > 1 \tag{11.3}$$

where f is frequency in kilohertz. At a higher critical pressure level, the inner-streaming boundary layer collapses and the action near the surface becomes vigorous and chaotic. Westervelt [43,44] believed that this activity tends to greatly increase the mixing of hot fluid near the cylinder with cooler fluid away from the cylinder and thus causes an additional and significant increase in heat transfer.

Each of the different hypotheses advanced by the previously mentioned investigators indicates that whether the heated surface is vibrated or is stationary and located in an intense acoustic field, heat transfer should improve by virtue of the acoustic vibration when the intensity of vibration exceeds a critical value. In studying heat transfer in forced convection, steam condensation, and nucleate boiling, the former U.S. Department of Interior Office of Saline Water [45] found that of the two methods, vibrating the heated surface in water is the more effective. The difference in the results obtained may be due to the difficulty of controlling the experimental variables, for if the same relative conditions are obtained in and about the inner-streaming boundary layer, the end result by either method should be the same. The Office of Saline Water attributed the improved heat transfer to combined axial and oscillating cross-flows.

Several investigators have studied the effects of ultrasonic vibrations on heat transfer. Any difference between the heat-transfer mechanisms at ultrasonic frequencies and those at low frequencies, if such a difference exists, might be related to the wavelengths involved. At ultrasonic frequencies, the stretched-film theory of Lemlich [40] may be less applicable than the modification of the convective flow in the inner-streaming boundary layer proposed by Westervelt [43,44]. The latter theory should apply regardless of frequency.

Bergles and Newell [46] investigated the influence of ultrasonic energy on heat transfer to water flowing in annuli consisting of two concentric stainless steel tubes. The inner tube was heated electrically. Ultrasonic energy at 70 and 80 kHz was directed through the water toward the heated tube from cylindrical transducers mounted on the outer tube. An improvement of 40% in heat transfer was obtained and was attributed to cavitation near the surface of the heated tube.

As the temperature of the liquid increases, the intensity of ultrasonic energy at which vaporous cavitation occurs decreases. Since the liquid is hottest near the source of heat, cavitation is quickest to form in this region. However, when the temperature reaches the boiling point, the additional stimulus from ultrasonic energy becomes insignificant. Bubbles formed during the boiling reflect and scatter the ultrasound and thus prevent its acting at the heated surface. It would be impractical to overcome the effects of boiling by increasing the intensity of the ultrasound because cavitation would occur nearer the source and thus cause even worse scattering. An alternative is to vibrate the heated surface and thus avoid having to transmit energy through a medium in which the attenuation is so affected by temperature. The results obtained by Bergles and Newell [46] confirm this. They showed that ultrasonic agitation was most effective in improving heat transfer at low rates and at temperatures below which surface boiling became well established. Low flow rates are more conducive to formation of cavitation because cavitation is time dependent.

In studies on effects of ultrasonics in a nucleate boiling system, Schmidt et al. [47] obtained results that can be explained in a similar manner. They applied ultrasonic energy at 20 kHz to a flash boiling system and concluded that the major contribution of the ultrasonic energy in increasing heat-transfer rate was to increase the turbulence in the system by cavitation. When bubble density in the fluid from other sources is high, the effectiveness of cavitation is overshadowed and actually degraded by scatter and attenuation of the ultrasonic energy. Therefore, Schmidt et al. [47] found that at large values of parameters that control bubble density—properties of the fluid, degree of superheat, fluid velocity, and system pressure—the effect of ultrasonics on heat transfer was not detectable.

However, at low values of superheat (approximately 1°C) and low flow rates (0.3–1.0 kg/minute), heat transfer increased 50%–60%.

McCormick and Walsh [48] claimed that ultrasonic energy gives a threefold improvement in heat transfer and suggested its use in cooling high-power tubes.

Cheng [49] claimed to obtain an average increase of 80% in the "coefficient of thermal conductivity" between hot and cold water. The increase between water and lubricating oil was 20%. He believed these values were far below the theoretically possible maximum increase.

Ultrasonic and acoustic vibrations and the effects on both heating and cooling continue to receive attention. Fairbanks [50] summarized three exploratory investigations, including radiant heat into water, heat conduction through metals, and melting. He observed significant effects, including up to a 30% increase in melting. A study by Loh et al. [51] included the use of computational fluid dynamics codes, validated with experiments, to investigate acoustic streaming induced by flexural vibration in a beam and the associated enhancement of convective heat transfer.

There is a growing interest in thermal conduction effects, and the effects ultrasound has, for applications such as nanotube systems. Amrollahi et al. [52] have investigated the effects of temperature and vibration time on the thermophysical properties of a carbon nanofluid (a carbon nanotube suspension). The size of agglomerated particles was decreased and the thermal conductivity increased with elapsed sonication time. Improved thermal performance has also been reported for nanofluids containing diamond nanoparticles [53]. Acoustic streaming induced by a 30-kHz longitudinal vibration is also being investigated for possible applications to improve cooling in electronics packaging [54].

11.3.6 DIFFUSION THROUGH MEMBRANES

Diffusion of ions and liquids through membranes and cell walls can be accelerated ultrasonically. Woodford and Morrison [55] attributed the effect to disruption of the boundary layer at the phase interface. They found that ultrasonic irradiation increased the diffusion rate of sennoside A through a cellulose membrane into water irrespective of temperature conditions under which diffusion took place. Gilius [56] also found a correspondence between the rate of diffusion of sugar and water through sugar beet membranes under the influence of ultrasound. The increased transfer rate continued even after irradiation was stopped. Water passed more easily through plasmolyzed membranes, while sugar passed more easily through fresh membranes. These observations are in agreement with those of other investigators [57,58].

Fairbanks and Chen [59] obtained substantial increases in the amount of flow of crude oil through porous sandstone and in the amount of water through porous stainless steel filters under the influence of ultrasound at 20 kHz.

The diffusion of hydrogen, generated in a cell during electrolysis of 0.1 N sulfuric acid solution, through a 0.17-cm-thick Armco iron membrane to a compartment filled with glycerol was enhanced under the influence of ultrasound at 24.5 MHz and 3 W/cm^2 when irradiation was applied from the glycerol compartment [60]. However, diffusion was hindered when irradiation was from the opposite side of the membrane.

The rate of diffusion of salt from 5% saline solution through a cellophane membrane into distilled water was increased by as much as 100% when ultrasound was applied in the direction of diffusion [61]. Diffusion was slower when irradiation was in a direction opposite that of diffusion. The increased rate was proved to be not a result of damage to the membrane or of temperature rise in the medium. Increased diffusion was greatest when irradiation was in the direction of the force of gravity. Frenzel et al. [62] obtained similar results with ultrasound promoting diffusion of potassium oxalate through a parchment membrane. The increase in diffusion over that without ultrasound was 69% with irradiation in the same direction as diffusion; when the two directions were in opposition, the increase was only 35%.

The effect of ultrasonics and its mechanisms for diffusion of electrolytes through cellophane membranes as a function of intensity were considered by Lenart and Ausländer [63]. They performed experiments at 1 MHz with intensities in the range of 1 to 6 W cm^2 for 5–30 minutes at 20°C. Significant enhancement of material diffused was reported, and it was attributed to the effect of acoustic microcurrents.

Julian and Zentner [64] investigated the effects of ultrasound and enhanced permeation through polymer membranes. A 23% increase in permeability of hydrocortisone in a cellulose film was reported. A subsequent paper, by the same authors, focused on mechanisms [65] that were attributed to the effects of a decrease in the measured activation energy necessary for diffusion.

High-power ultrasound has been investigated with regard to its application to give enhanced membrane-based filtration. Kyllönen et al. [66] attributed the effects to microjets, acoustic streaming, and microstreaming (associated with cavitation). A number of medical applications have also been investigated. Ultrasonically mediated transdermal drug delivery (phonophoresis/sonophoresis), increased permeability of the blood–brain barrier, and some ultrasonically mediated controlled-release drug delivery systems have all been investigated, and these topics are discussed further in Chapter 14.

11.4 CHEMICAL EFFECTS: SONOCHEMISTRY

The chemical effects of ultrasound have a long history. Alfred L. Loomis is credited with having noticed the first chemical effects of ultrasound in 1927; however, the field of sonochemistry lay fallow for nearly 60 years. The renaissance of sonochemistry started in the 1980s. This increase in research activity followed soon after the availability of inexpensive and reliable high-intensity ultrasound generators for laboratory use [5,9,67]. An example of such a commercial bench-top high-power

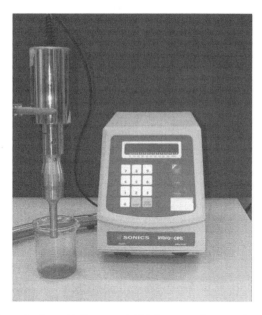

FIGURE 11.8 (**See color insert.**) Commercial high-power ultrasound unit with horn of type commonly used in laboratory sonochemistry.

ultrasound system is shown in Figure 11.8. Processing "cells" that can operate with such horns have now been developed into a diverse array of forms. Units are available that enable flow-through processing, provide temperature control and monitoring of temperatures and pressures, and use of a range of on-process instrumentation.

Sonochemistry has become its own field of endeavor. A good review of the science and engineering aspects of sonochemistry is provided by Thompson and Doraiswamy [8]. This paper includes a categorization of many of the families of chemical reactions and discusses the various ultrasonic systems that are used in sonochemical processing.

Scientists now know that the chemical effects of ultrasound are diverse and, for some chemical reactions, include substantial improvements in yield for both stoichiometric and catalytic chemical reactions. In some cases, ultrasonic irradiation can increase reactivity by nearly a million-fold. In seeking to analyze these activities, the chemical effects of ultrasound are considered to fall into three areas:

1. Homogeneous sonochemistry of liquids
2. Heterogeneous sonochemistry of liquid–liquid or liquid–solid systems
3. Sonocatalysis (which overlaps the first two)

Because cavitation can take place only in liquids, chemical reactions do not generally occur during the ultrasonic irradiation of solids or solid–gas systems.

In addition to the literature that has discussed the chemistry, there are a number of studies that are focused on the acoustic bubble [3] and the phenomena of sonoluminescence, the emission of light from collapse of cavitation bubbles.

The chemical activity promoted by ultrasonic energy may be the result of many different mechanisms. These may include intimate mixing of reactive substances, production of localized heat particularly in association with cavitation, production of free ions associated with the production of cavitation, and stress-associated molecular breakdown. Among other effects, ultrasound promotes oxidation reactions, promotes polymerization or depolymerization depending on the nature of the molecule being treated, affects the potential of polarized electrodes, increases the rate of crystallization, and facilitates extraction of oils.

The mechanisms involved in the chemical effects of ultrasonics have been widely studied, and many interesting applications of these effects have been made. Suslick et al. [68–70] attribute the major, or more important, chemical effects to cavitation. Boudjouk [71] also states, "One of the points on which there is general agreement is that, in order to produce a chemical effect in liquids using ultrasonic waves, sufficient energy must be imparted to the liquid to cause cavitation, i.e., the formation and collapse of bubbles in the solvent medium and the consequent release of energy." Suslick et al. [68–70] emphasize the importance of avoiding the complete degassing of solutions before sonolysis, that is, the application of ultrasonic energy to promote chemical activity in solutions, and the part played by the vapor pressure in the rates of sonochemical reactions. The rates of sonochemical reactions decrease with increasing solvent vapor pressure and increase with decreasing solvent vapor pressure. Suslick et al. [68–70] emphasize that solvent control exerted on sonochemical activity can be extreme, with changes in vapor pressure responsible for relative product yields differing by a hundredfold for reaction pathways that differ significantly in their activation energies.

Not all beneficial chemical effects are due to cavitation. For example, the curing of rubber is a thermal process. The thermal conductivity of rubber is very low. Rubber is cured conventionally by conducting heat from an outside source into the volume of material. This process results in nonuniform heating and curing. The volume of uncured rubber can be heated uniformly by applying ultrasonic energy at a wavelength that is long compared with the thickness of the volume of material to be cured. Heating in this case is not dependent on the thermal conductivity of the material but on the absorption of ultrasonic energy within the rubber. The result is a more uniformly and more rapidly cured rubber product.

New efforts are in progress toward obtaining a better understanding of mechanisms of sonochemistry and toward applying the effects on a commercial scale [9]. Chemical effects may be involved in many industrial applications of ultrasonics. Ultrasonics may increase the rate of a reaction occurring under normal circumstances or may induce reactions that would not occur under normal circumstances.

11.4.1 DEPOLYMERIZATION

Depolymerization of certain types of high–molecular weight polymers occurs under the influence of ultrasonic energy. The initial structure of the molecule, rather than its size, determines whether it will be broken by this means. Degradation of the molecules continues to a limiting value and cannot be carried further. The mechanism appears to be cavitation. Cavitation is difficult to achieve in viscous materials corresponding to high–molecular weight polymers. For this reason, dilute solutions of the polymers have been used for studying ultrasonic depolymerization.

Newton and Kissel [72] fragmented tobacco mosaic virus at 7 MHz and observed the results using electron microscopy. At an intensity implied to be below cavitation level, polymeric forms of the virus were fractured into predominantly monomeric lengths, and this was accompanied by partial fragmentation of a small percentage of the monomers, the amount of fragmentation being a function of time. Breakage of the monomer occurred most frequently at a distance of 175–205 μm from one end, which appears to indicate a fundamental weakness at this location in the monomer. At high intensities and correspondingly high cavitation activity, the monomers were extremely fragmented. The extent of fragmentation was relatively independent of time. A large number of the fragments were spherical in appearance. The virus had lost 95% of its infectivity during intense treatment.

Brett and Jellinek [73] and Weissler [74] demonstrated the role of cavitation by comparing ultrasonic depolymerization of polystyrene under conditions that were conducive to cavitation and conditions that inhibited cavitation. Brett and Jellinek [73] ultrasonically irradiated 1% solutions of polystyrene in benzene at constant intensity and temperature and at a frequency of 500 kHz. Cavitation activity was controlled by pressurizing the samples either with nitrogen or with mercury. Increasing pressure decreases the tendency of a liquid to cavitate under the influence of ultrasonics. When pressure was applied through nitrogen, depolymerization occurred even at 3.1 MPa at a rate only slightly lower than that obtained at atmospheric pressure. When the pressure was applied through a mercury column, the rate of depolymerization decreased as the pressure increased in accordance with the effect of pressure on cavitation. At 3.1 MPa, very little depolymerization was evident. The upper limiting pressure, of course, depends on the intensity of the ultrasonic field.

Weissler [74] ultrasonically irradiated a 1% solution of polystyrene in toluene at 400 kHz. When the solution was treated without prior degassing, cavitation activity was apparent and depolymerization occurred as expected. When the solution was degassed by boiling prior to treatment, there was no visible cavitation during irradiation and depolymerization did not occur.

Weissler [74] obtained similar results with hydroxyethyl cellulose dissolved in ordinary water and in degassed water. He further attempted to prove that degradation is not due to oxidants produced during ultrasonic irradiation of solutions containing nitrogen or oxygen. Depolymerization occurs at approximately the same rate even when purified helium is the only gas present.

If cavitation alone is responsible for depolymerization under the influence of ultrasonic energy, varying frequency would affect depolymerization only as it affects cavitation. Langton and Vaughan [75] compared the results of depolymerizing polystyrene at 1 MHz with the results obtained at lower frequencies by other investigators. The results were similar. Langton and Vaughan [75] proposed an equation to describe the rate of degradation as follows:

$$\frac{dP_t}{dt} = -K(P_t - P_\infty) \tag{11.4}$$

where P_t is the degree of polymerization after t seconds of irradiation, P_∞ is final limiting degree of polymerization, and K is a constant.

Ultrasonic energy is equally effective in depolymerizing dextrin at 20 and 250 kHz.

Results of depolymerization studies conducted by Head and Lauter [76] illustrate the importance of molecular structure to the susceptibility of the molecule to rupture under the influence of ultrasonic energy. They treated 1% and 2% aqueous solutions of acacia (gum arabic), 0.5% aqueous solutions of Viscarin, 0.25% aqueous solutions of agar, 0.5% aqueous solutions of locust bean gum, 0.5% aqueous solutions of tragacanth, 1% solutions of karaya gum, and a 0.5% aqueous solution of Methocel (4 Pascal second) (Dow methylcellulose) at 275 kHz. Viscarin, agar, locust bean gum, and Methocel (4 Pascal second) are linear polymers with molecular weights ranging from 140,600 to 310,000. These materials depolymerized at comparable rates. Acacia (gum arabic) has a molecular weight of 250,000; however, it is not a linear polymer and did not depolymerize. Karaya, having a much larger and more complex molecule (molecular weight 9,500,000), apparently did not depolymerize. On the other hand, tragacanth, with a molecular weight of only 5000, which has a complex structure, appeared to polymerize (characterized by an increase in viscosity) to some maximum and then to depolymerize to less than half its initial molecular weight.

According to Hanglein [77], when poly(methyl methacrylate) in benzene solution is degraded ultrasonically in the presence of iodine, long-chain free radicals are produced that react rapidly with the iodine in the absence of oxygen. Spectroscopic studies with I^{131} show that for each ruptured bond, two atoms of iodine are used and these are incorporated into the fragments. The rate of depolymerization is somewhat greater in iodine-containing solutions than in iodine-free solutions. In the absence of iodine, 20% of the polymer radicals either recombine or combine with radicals from other macromolecules. In the presence of oxygen, the reaction is inhibited.

Kossler and Novobilsky [78] studied the ultrasonic degradation of polychloroprene both fresh and aged in air. They concluded that cross-linking occurs by the formation of labile peroxide bonds that prevail, in the first stages of aging, over cross-links of another kind. Infrared spectra showed an increased content of C=O groups in the first fraction. The number of bonds broken by ultrasound per unit time decreases with increasing concentration. The addition of gases retards the depolymerization according to the sequence, air, nitrogen, oxygen, and carbon dioxide.

Degradation of yeast ribonucleic acid (RNA) was observed at a frequency of 0.8 MHz under cavitational conditions when a 0.005% solution of RNA was treated [79]. No degradation occurred at 2.4 and 4.0 MHz. The degradation of RNA was attributed to the formation of hydrogen peroxide and nitrous acid in the presence of cavitation, which occurred at 0.8 MHz but did not occur at 2.4 and 4.0 MHz. Higher concentrations of RNA (0.025% and 0.2%) retarded degradation. The pH, which drops to 3.3 in distilled water in 0.005% RNA solutions, does not drop below 6.5 in 0.2% solutions.

Decomposition of the benzene ring was observed when phenol and benzene in aqueous solutions were treated at 0.8 MHz [80]. The decomposition was attributed to two different reactions: (1) the benzene ring is oxidized, which leads to successive formation of phenol, monosubstituted benzenes, quinones, and unsaturated acids (muconic, maleic), and (2) direct splitting off of parts of benzene molecules occurs with formation of formaldehyde; under proper conditions, the latter condenses with phenol to form resins. In treating phenol solutions at a concentration of 48 mg/L ultrasonically, decomposition of the ring occurs only as a result of oxidation, according to the reference. The process occurs in acid, neutral, or weakly alkaline media. In a strongly alkaline medium, the process is severely inhibited. The rate of decomposition of phenol depends on its initial concentration in the solution; the amount of oxidized phenol increases with increase in concentration to a certain point, above which an increase in concentration has no effect on reaction rate. With increase in sound intensity and temperature, the reaction rate increases. Prolonged ultrasonic treatment of a phenol solution in a strongly acid medium (in contact with air) results in formation of small quantities of mononitrophenols.

Inhibition of bromine release from bromoform under the influence of ultrasonics was observed when water, chloroform, diethylene glycol, butyl alcohol, glycerol, ethyl alcohol, ethylene glycol, diacetone alcohol, methyl alcohol, dioxan, xylene, acetone, and benzene were added to aqueous solutions of the bromoform [81]. The inhibition was attributed to certain physical properties such as surface tension, viscosity, vapor pressure, and internal pressure, which were affected by these nonelectrolytes. All these factors affect the intensity of cavitation.

Shaw and Rodriguez [82] have shown that samples of polysiloxane solutions with different initial distributions approach a uniform distribution after 120 minutes of ultrasonic degradation. Dilution of the polymer to 1.0–0.01 g/dL more than triples the degradation per gram of polymer, but it requires 10 times as much energy per gram to degrade from an average molecular weight of 10^6 to 10^5. However, doubling the power input (36–72 W) nearly doubles the degradation rate per gram of polymer.

Degradation of agar-agar solutions in water and methyl sulfoxide at concentrations ranging from 0.1 to 2 g/L showed that degradation is more extensive at the higher concentrations [83]. A reversible viscosity decrease was observed at concentrations greater than 2 g/L and was attributed to the degradation of the gel. At concentrations of less than 0.3 g/L, only slight degradation of agar-agar occurred. The viscosity of agar-agar decreased very rapidly during the first stages of ultrasonic treatment and later converged toward a limiting value. The gas in contact with the solution influenced the rate of degradation. Air produced the most extensive degradation; in nitrogen, degradation was slower and not so extensive; for sulfur dioxide, the quantity $(t_u - t_v)$ was linear and an increase in the sulfur content of agar-agar was noted. Here, t_u is the ultrasonic treatment time and t_v is the time of flow through an Ostwald viscometer. Argon was the most suitable gas for study of degradation. Temperature (16, 20, and 36°C) had no effect on variation in viscosity at concentrations of 0.1–0.3 g/L. After 100 minutes of ultrasonic radiation, the pH decreased from 6.98 to 3.96 at a concentration of 0.8 g/L. These authors proposed that degradation proceeds according to a four-step mechanism: (1) cleavage of the 1,3- and 1,4-glycosicidic bonds in agar-agar, yielding macroradicals; (2) recombination of these macroradicals; (3) reaction of water with the macroradicals OH + agar − CO → agar − COOH Carboxylic acids; and (4) transfer between agar-agar and radicals of water.

Witekowa [84,85] explained the chemical action of ultrasonics in various quantities. In explaining the kinetics of the reduction of acidic potassium permanganate solutions in an ultrasonic field, Witekowa [84,85] claimed that acidic permanganate

MnO_4^- solutions are reduced to Mn^{2+} in several minutes, whereas in the absence of the field, they remain unchanged for days. The reaction occurs according to the equation $-d[KMnO_4]/dt = k_0$, where k_0 is a constant. The value of k_0 increases with pH and temperature, and at $10°C$–$30°C$, the average energy of activation is found to be 10.6 kcal/mole. Witekowa [84,85] explained the reaction as being dependent on the concentration of excited OH^- ions. The latter depends on the initial OH^- concentration (i.e., on pH), frequency, and power of the ultrasonic field. At a suitable frequency, the OH^- radicals may be partly liberated from the surrounding water molecules and become more active.

In explaining the kinetics of oxidation of aqueous potassium ferrocyanide, Potassium ferrocyanide, also known as potassium prussiate or yellow prussiate of potash or potassium hexacyanidoferrate(II), Potassium ferrocyanide, also known as potassium prussiate or yellow prussiate of potash or potassium hexacyanidoferrate(II), $K_4[Fe(CN)_6]$, Witekowa [85] claimed that these solutions are oxidized rapidly in an ultrasonic field to potassium ferricyanide, $K_3[Fe(CN)_6]$. The rate constant increases with temperature, hydrogen ion concentration, and the square root of ionic strength. The activation energy at $10°C$–$30°C$ is 6.88 kcal/mole. The reaction $[Fe(CN)_6]^{4-} + (H+) \rightarrow [Fe(CN)_6]^{3-} + 1/2H_2$ is supposed to be the rate-determining reaction. The mechanism is complicated by the possible formation of nitrous acid and nitric acid in an ultrasonic field.

Zaskal'ko et al. [86] studied the degradation of 7% polyisobutylene solution in oil at $100°C$. They showed that there is more degradation of long-chain polymer chains than the short chains. After 5-hour exposure to ultrasound, 81.5% of the polyisobutylene of molecular weight 85,000 was degraded, but there was no loss of polyisobutylene of molecular weight 2000.

Ultrasound-enhanced polymer degradation remains of interest [67], and it has also become a topic of interest to the medical community. This phenomenon is of interest and importance in controlled release of incorporated or enclosed materials. Ultrasound has been shown to give a 5-fold, and in some cases a 10-fold, increase in degradation and the release rate of incorporated molecules [87].

11.4.2 POLYMERIZATION

Ultrasonics can be used not only to depolymerize materials but also to produce the opposite effect, that is, polymerization. Hastings and Bolten [88] showed that applying ultrasonics during the polymerization of styrene produced a sharper distribution of the molecular weight than that obtained by degradation of the polymer in toluene, and less polymer of low molecular weight was present. An increase in molecular weight and sharpening of distribution was achieved under some conditions compared with photoinitiated polymer.

Graft copolymerization of cellulose with vinyl monomers in the presence of salts of group VIII metals, cerium or Cu^{2+}, was accelerated by ultrasonic treatment, as observed by Okamura et al. [89]. Treating a mixture of 438.1 mg of viscose sponge, 8 cm^3 of methyl methacrylate, and 2 cm^3 of 0.1 M cerium nitrate ultrasonically at 420 kHz and 200 W at $40°C$ for 2 hours produced 1525 mg of graft copolymer extracted with benzene. A similar run without ultrasonics yielded 4309.5 mg of copolymer from 454.3 mg of viscose.

The intrinsic viscosity of polymers obtained by the polymerization of methyl methacrylate with Grignard reagents and benzoyl peroxide dropped markedly upon ultrasonic irradiation at 500 kHz [90]. Osawa and Igarashi [90] believed that in the transition state in polymerization, the growing polymer end and the catalyst form bonds that are stronger than the C–C covalent bonds in the main chain and that are not broken by ultrasonic energy. The stereoregularity of the polymers is not affected. Polymers polymerized with Grignard reagent and benzoyl peroxide are predominantly isotactic and syndiotactic in structure, respectively.

Romanov and Lazar [91] studied the modification of a copolymer of atactic polypropylene in an ultrasonic field. They prepared a 15% solution of atactic polypropylene in styrene and exposed the mixture to ultrasonic radiation at 12 W/cm^2 in a nitrogen atmosphere for 11 hours to produce a highly viscous solution. They prepared another copolymer chemically by initiating polymerization of styrene with ozonized atactic polypropylene. After the two copolymers had been isolated and precipitated out, they discovered that the chemically prepared copolymer had more of the character of atactic polypropylene, whereas the copolymer prepared by ultrasonic irradiation had more of the character of polystyrene, although from the elemental analysis, the opposite would be expected. They concluded that the chemically prepared sample is a graft copolymer and the ultrasonically prepared sample is a block copolymer.

Shiota et al. [92] showed that ultrasonics increases the rate of polymerization and the molecular weight in the production of polyamide, an effect they attributed to an increase in terephthalyl chloride, $p\text{-}C_6H_4(COCl)_2$, and ethylenediamine, $H_2NCH_2CH_2NH_2$, in the presence of cavitation. Interfacial polycondensation occurs at ultrasonic frequencies such as 200–2000 kHz. They observed that the optimum concentration of the reagents appears to be independent of the ultrasonic irradiation.

There are now several books and reviews available that discuss polymerization. This literature includes a paper by Price [93], and the topic is covered in the review paper by Thompson and Doraiswamy [8] and book by Mason and Lorimer [67]. The role of ultrasound in a range of polymerization reactions is also considered in the wider review of applications of ultrasound to materials chemistry by Suslick and Price [94].

11.4.3 Catalysis

Ultrasonic energy acts in conjunction with a catalyst to increase the yield or accelerate chemical reactions. When a chemical reaction occurs in the presence of both a catalyst and ultrasonic irradiation, the activity usually increases, but some reactions may be inhibited by the effect of the ultrasonics on the catalyst.

The application of ultrasound in hydrolysis of metallic peroxyacetate, using hydrochloric acid as catalyst, produces an increase in the reaction rate with increasing intensity [95]. Ultrasound has no effect on activation energy and produces no extra hydrogen ions.

The rate of oxidation of cyclohexanone in air in the presence of magnesium stearate and benzyl alcohol decreases with ultrasonic irradiation [96]. Ultrasonic irradiation does not influence the rate of oxidation in the presence of adipic acid alone, or if no acid is added. With adipic acid–benzyl alcohol mixtures, the effect is immediate. Apparently, the active catalyst is a benzyl alcohol–magnesium stearate complex, which is decomposed by ultrasound, favoring a less active but more stable adipic acid–magnesium stearate catalyst.

Menzhulin and Ponomareva [97] compared the reduction of selenium in sulfuric acid solutions with the reduction of copper, copper–zinc couple, zinc, iron, aluminum, and magnesium with and without ultrasound. Without ultrasound, reduction of copper and copper–zinc was relatively rapid. The other metals were rather inactive, with no reduction in aluminum being detected after 3 hours. Ultrasonics accelerated the rate of reduction in all cases, with the reduction of copper and copper–zinc being the most pronounced. Ultrasonics also caused a finer crystallization of the deposits. The authors attributed the effects of ultrasound on the reduction of selenium to the destruction of the deposited film and to the acceleration of anodic and cathodic processes.

Parthasarathy et al. [98] showed that ultrasonics promoted esterification of ethyl alcohol–acetol without the use of a catalyst.

Prakash and Prakash [99] reviewed the decolorization of aqueous solutions of rhodamine B, bromocresol green, Congo red, aniline blue, chrysoidine, and brilliant green by ultrasound at 1 MHz. Decolorization is a unimolecular process, and there appears to be no single explanation; oxidation, reduction, and decomposition reactions are all involved.

There are now numerous examples of ultrasound combined with catalytic reactions in the literature. Such reactions involving catalysis cover both decomposition and formation reactions, of a wide variety of types [5,8,67]. A review of "Sonocatalysis" is provided by Suslick and Skrabalak in the *Handbook of Heterogeneous Catalysis* [100]. There are also now numerous papers in the literature that report reactions for specific chemical systems: for example, a comparison between conventional and ultrasound-mediated heterogeneous catalysis for hydrogenation of 3-buten-1-ol aqueous solutions [101]. Given the shear number of papers on specific sonochemical reactions, the reader with interests in a particular chemical reaction is best advised to search computer databases for the relevant papers.

11.4.4   PRECIPITATION

Ultrasound has been found to cause precipitates of dissolved or suspended materials in liquids. Examples are calcium carbonate (in the form of aragonite crystals) [102] and certain metallic compounds such as nickelous oxide hydrate [103]. Under the influence of ultrasound, nickelous oxide hydrate precipitates at room temperature. When this process occurs in an alkaline battery, the electrical characteristics of the battery remain unimpaired. High-purity nickelous oxide hydrate is produced at low temperatures without the use of excess base in the reactor.

11.4.5   METALS

Certain chemical and chemically related effects of ultrasound are beneficial in certain metal-processing applications. Ultrasound accelerates nitriding in a molten bath with ammonia bubbling through it. The depth of the nitriding layer and its microhardness are increased under such treatment [104]. The ultrasound treatment (1) promotes mixing of the liquid; (2) removes reaction products from the surface of the metal; (3) cleans the surface of the metal and activates it, permitting surface-active substances to be adsorbed; and (4) promotes diffusion of nitrogen into the metal as a result of cavitation.

The application of intense ultrasound accelerates the dissolution of the steel in dilute sulfuric acid. The accelerated activity probably is due to the removal of hydrogen bubbles, which screen the surface from the acid [105].

11.4.6 CRYSTALLIZATION

Two categories of crystallization are recognized: (1) crystallization from a solution and (2) crystallization from a melt.

Crystallization from solution involves the formation and growth of solid crystals of a solute in a supersaturated liquid solution. Supersaturation occurs when solvent is removed or the temperature is lowered so that the solubility limit of the solute in the liquid is exceeded and solid solute separates spontaneously from the liquid.

Crystallization from the melt occurs when the temperature of a liquid is lowered below its freezing point and solid crystals of the substance are allowed to form.

The steps in a process of forming individual crystals, such as sugar, from solution include (1) causing supersaturation either by lowering the temperature or by removing solvent or both; (2) forming or adding crystals (nucleation); (3) growing the crystals to a suitable size, which involves diffusion of the solute from the solution to the solid surface; and (4) separating the crystals from the remaining liquid.

Ultrasonic energy initiates crystallization processes and produces effects related to size and character. The mechanisms of crystallization under the influence of ultrasonic energy are similar whether crystallization is from a melt or from a solution. A major contribution comes from cavitation. In gaseous cavitation, during the expansion phase, crystal nucleation may occur by an instantaneous local cooling and solvent evaporation sufficient to produce microcrystals. Ultrasonic energy also promotes microstreaming, which carries these microcrystals into the melt volume. Crystallization centers such as oxides, nonmetallic particles, and intermetallic compounds may be stirred into the melt by the ultrasonic energy. Crystallization growth may be aided by the ultrasonic field by aiding diffusion of the molecules of the crystallizing solute to the surface of the crystal; however, ultrasonics also promotes a uniform grain refinement by the increased number of nucleation centers and stirring effects that distribute and sometimes break the crystals in the melt or solution.

Abramov [106] confirmed the effectiveness of ultrasound in generating crystallization centers in transparent organic substances such as betol, naphthalene, azobenzene, salol, and thymol. Ultrasonics produced crystallization centers at lower supercooling temperatures and in shorter times than are required with the usual processes; however, the increase in the threshold of metastability differed with different materials. Depending on the nature of the material and the conditions for crystallization, dispersion of growing crystals and increased growth rate were effected by ultrasound, with either dispersion or increased growth rate being the predominant process. Cavitation was a factor in both processes.

Ultrasonics applied to solidifying metals also results in uniform grain refinement, as shown by several investigators.

Small crystals of hydrocortisone have been obtained by ultrasonic treatment of saturated solutions of isopropyl alcohol, ethyl alcohol, and a 50% by volume blend of isopropyl alcohol and ethyl alcohol [107]. Particle size was independent of treatment time, and distribution was similar at various coolant temperatures. Crystals obtained from saturated solutions in ethyl alcohol were smaller than those obtained from isopropyl alcohol or the mixture of isopropyl alcohol and ethyl alcohol. The degree of saturation affected the particle size and distribution. It also affected the yield. Smaller crystals and greater yields generally were obtained from higher levels of saturation. The extent of crystallization increased with decreasing temperature. Weak cavitation produced larger crystals, and low pressures increased yields.

Kapustin [108] suggested that ultrasonics induces crystallization by changing the density of crystal nuclei in the melt. He found that in crystallizing o-chloronitro-benzene by means of ultrasonic energy, no crystallization was induced at temperatures of 20°C–28°C, but crystallization centers appeared either immediately or within a few seconds of the application of ultrasound near the limit of the metastable region (18°C–20°C). High-intensity ultrasound seldom produced crystallization.

In general, it has been found that exposure of supersaturated solutions to an ultrasonic field speeds up crystallization and results in smaller crystals. de Castro and Priego-Capoto [109] reported dramatic reductions in induction time with supersaturated conditions.

This approach has been used in large instillations in the pharmaceutical industry to provide crystals of small and controlled size [110]. Li et al. [111] reported experiments to investigate crystallization of 7-amino-3-desacetoxy cephalosporanic acid and showed that mixing was more efficient and uniform and both the induction time and agglomeration were reduced. They also show that ultrasonic parameters can be used to modulate both crystal size and the resulting size distribution. Chen and Huang [112] used ultrasound to obtain high purity (98.9% by weight) 2,4-dinitrotoluene from spent acid.

11.4.7 SONOLUMINESCENCE

The phenomenon of sonoluminescence was discussed briefly earlier in this chapter in the context of energy partition and related phenomena. This phenomenon is the weak emission of light that sometimes accompanies cavitation in an ultrasonic field. Several theories of the origin of sonoluminescence have been proposed, but the validity of none of the theories has been proved satisfactorily. The two most common theories are (1) that electrostatic discharges occur as a result of the formation of electrical charges on the walls of the cavitation bubbles and (2) that temperatures sufficiently high to cause the vapors of the liquid to become incandescent occur as a result of microshocks associated with the collapse of cavitation bubbles. There appears to be some justification for both theories. Grandchamp and Prudhomme [113] showed a drastic reduction in sonoluminescence and oxidation reactions in the presence of hydrogen and in the presence of ether. They attributed the reduction in these reactions in the presence of hydrogen to inhibition of the formation of electrical charges on the walls of cavitation bubbles. The reduction in the presence of ether was attributed to the fact that ether accepts free radicals and increases the size of the cavitation bubbles. Both materials lower the intensity of cavitation bubbles, which results in a correspondingly lower number of free radicals being produced. The proven appearance of free radicals in the presence of cavitation and experiences with induced electrolytic erosion in the presence of high-intensity cavitation seem to make the electrostatic discharge theory a plausible one. The effect is similar to the phenomenon of lightning in nature.

Hickling [114] claimed that sonoluminescence is largely thermal in origin, with the spectra showing a direct relation between luminous intensity and the temperature generated inside the collapsing bubbles. He explained the strong dependence of the luminous intensity on the nature of the gas dissolved in terms of thermal conduction. On this basis, if the bubbles are small, loss of heat from the bubbles into the liquid can significantly reduce the temperatures attained during the collapse, with a corresponding reduction in luminous intensity. Small cavitation bubbles are produced under two conditions: (1) the intensity is low, in which case luminescence is minimal or nonexistent, and (2) no gas is present in the liquid, the liquid has a low vapor pressure, and the surface tension of the liquid is high so that it is difficult to produce cavitation. On the other hand, dissolved gases and liquids with high vapor pressures lower the cavitation threshold and weaken the intensity of the cavitation. In the latter case, luminescence will not occur. This appears to contradict the thermal conductivity explanation of luminous intensity.

Parke and Duncan [115] observed a broad band in the visible spectrum while ultrasonically irradiating aerated solutions of bromine in water and aerated Carbon disulfide (CS_2) in water. Intensity of sonoluminescence increased with time of irradiation up to equilibrium values, which varied with ultrasonic intensity. Oxidation increased linearly with time in solutions of oxygen in water, oxygen and KI in water, and air, KI, and carbon tetrachloride in water. The oxidation rate in these three materials increased linearly with both ultrasonic intensity and intensity of sonoluminescence. Oxidation also increased linearly with time in solutions of argon in water and nitrogen in water. Neither the addition of salts, except KI, nor the addition of sulfuric acid affected luminescence or oxidation. However, lowering the pH of solutions of O_2-KI-H_2O resulted in a threshold increase in the rate of iodine liberation without changing luminescence.

Both single- and multibubble sonoluminescence have been studied [3,4,11] and remain the subject of active research. An example of the sonoluminescent glow seen below an ultrasonic horn driven at 20 kHz is shown in Figure 11.9.

A series of studies have investigated the size of the light-emitting region in an individual sonoluminescent bubble [116], which is seen as key to understanding this process. The theory and experimental measurements of spectra for sonoluminescent bubbles, including the cases for rare gas-filled bubbles [117], are found to be in agreement. The shape of bubbles during acoustic cavitation and the form of the sonoluminescent field are also being studied [118].

In addition to using a horn, sonoluminescence can be induced in several other configurations, including using a ring transducer applied to the outside of a pipe,

FIGURE 11.9 **(See color insert.)** Photograph of sonoluminescence in water below a 20-kHz ultrasonic horn. (Courtesy of Pacific Northwest National Laboratory.)

FIGURE 11.10 **(See color insert.)** Piezoelectric element used as a ring transducer. (Courtesy of Pacific Northwest National Laboratory.)

as given in Figure 11.10. The light emissions that result when such a transducer is attached to glass tube and driven at a resonant frequency are shown in Figure 11.11. This transducer configuration has several advantages, not least that the light emissions (and potentially sonochemistry) occur in a larger volume than with the horn and there is no metallic contamination of the fluid resulting from cavitational erosion of a transducer horn tip. The ring transducer has now been developed for pilot-scale applications on a 5-cm-diameter stainless steel pipe, and the transducer configuration used in the flow loop is shown in Figure 11.12.

A related and, for some, a contentious topic has been the feasibility of using ultrasound for achieving the conditions for neutron emission from hydrogen or deuterium, or other isotope, in sonoluminescent bubbles. Although discussed for many years, a U.S. patent, filed by Hugh Flynn in 1978, appears to be the earliest document to propose methods for neutron production during cavitation [119]. Since this time, there have been several papers and proposals discussing this basic concept [120].

FIGURE 11.11 (See color insert.) Example of volumetric sonoluminescence produced with a ring transducer applied to a 5-cm-diameter glass cylinder. (Courtesy of L.R. Greenwood and Pacific Northwest National Laboratory.)

FIGURE 11.12 (See color insert.) In-line sonochemistry processing unit with a series of cylindrical ring transducers applied to a stainless steel pipe. (Courtesy of Pacific Northwest National Laboratory.)

Taleyarkhan et al. [121–123] reported thermonuclear fusion generated by cavitation in deuterated acetone in 2002, 2004, and 2006. These studies generated significant discussion and controversy. There have also now been four other studies, by others, that investigated and sought to reproduce the phenomena and that report null results [124–126]. The most recent reported study is discussed by Camara et al. [127] who presented an unsuccessful attempt to reproduce the apparatus and claimed effect reported by Taleyarkhan and his colleagues. The paper by Camara et al. [127] does, at the very least, define an upper bound for the possibility of neutron emissions in the case of deuterated acetone.

The phenomenon of "sonofusion" remains an intriguing and tantalizing possibility, which can be expected to remain as such until it is conclusively demonstrated since the proving of the phenomenon's nonfeasibility remains equally challenging (and expensive).

11.5 METALLURGICAL EFFECTS

Certain beneficial effects have been demonstrated by applying ultrasonic and low-frequency vibratory energy to molten and solidifying metals. Degassing, reduction of grain size, and uniform mixing of different metals are some of the effects.

Degassing occurs by the same mechanisms that cause degassing in other liquids. During the low-pressure phase of each cycle, gas is released from solution. The effect is only partially reversible, so that only part of the gas returns to solution during the high-pressure phase of the cycle within the time available. As the process continues, small bubbles coalesce and rise to the surface. Combined vacuum and ultrasonic degassing results in considerable improvement over the rate and effectiveness of either method used separately.

Reduction of grain size is obtained under certain conditions, and it is also known to give faster solidification from melts [128]. Ultrasonic and low-frequency vibratory energy effect the formation of crystallization centers in supercooled liquids. If the distribution of energy throughout the melt is such that the entire melt is subjected to intensities above a certain threshold level, uniform and refined grain sizes will result. Refinement is independent of intensity above the threshold level. In molten metals, refinement appears to be most effective in alloys that have a low alloy content and are predominantly solid solutions in structure. Grain refinement is best when the melt is cooled slowly with the irradiation continuing until solidification is nearly complete.

Ultrasonics tends to emulsify or mix liquids and disperse particulate matter uniformly throughout the affected volume. This phenomenon has been demonstrated in molten metals by applying ultrasonics to disperse oxides, graphite, or other metals through a melt. Reportedly, a better distribution of lead in aluminum is obtained by applying vibratory energy at 500 Hz than is obtained by the chemical lead chloride process [129].

Although many beneficial metallurgical effects have been shown, the commercial application of ultrasonics in foundry processes has not been realized. However, a renewed interest among scientists in the fields of metallurgy and ultrasonics engendered by present-day needs plus modern developments in both fields shows encouraging prospects that this condition is about to change. Hindrances to this application have been inherent in the methods and materials involved in generating the energy in the melt and in containing the molten metals. Various means have been used to generate energy in moderately large batches. These include (1) vibrating the mold using eccentrics, aerodynamic, or other mechanical means of producing vibrations and (2) generating vibrations within the melt by electromagnetic induction. Neither of these methods, vibrating the mold or vibrating the metal by electromagnetic induction, is likely to produce the total effects observed in a small-crucible laboratory-scale experiment because of the limitations on intensity and beneficial mechanisms produced by these methods. Cavitation, fluid movement such as "sonic wind," and radiation pressure on bubbles (for degassing) appear to be major factors in producing desirable metallurgical effects. Cavitation would be limited in a low-frequency vibration of the mold by the dynamic forces that the mold itself could withstand. Vibrating the melt by induction is inefficient, so that cavitation would also be minimal under this type of excitation. However, induction can be used to promote flow and mechanical stirring. Radiation pressure is a function of intensity, and since intensity in either process is low, radiation pressure from either process probably would have little influence on removing gas bubbles from a melt. However, shaking the mold or melt should help release gas bubbles from solution, allowing them to coalesce and rise to the top surface of the melt. Sonic wind is produced best by irradiating a portion of a cross-section of a fluid so that radiation pressure will produce a flow of fluid along the radiation axis away from the radiation surface. The fluid is recycled by returning to the radiating surface along an outer path away from the sonic field. The effect is a stirring action. If the entire melt is irradiated uniformly, sonic wind is inhibited or prevented. On a small scale, ultrasonic vibrations may be introduced through compatible metal rods, ceramic-tipped or ceramic-coated metal rods, or ceramic rods by which cavitation and sonic wind and local eddying are produced easily. Low–melting temperature alloys can be treated in Pyrex beakers placed in a pan to which a suitable transducer is coupled. Coupling to the bottom of the beaker is accomplished through a molten layer of metal.

The greatest problem with most materials used to contain molten metals or to couple energy to the metal is erosion and contamination of the melt. Kudel'kin et al. [130] found that in treating various steels and alloys, the percentage of non-metallic inclusions was higher in the materials treated ultrasonically than in the controls. Otherwise, ultrasonically treated materials showed superior properties in uniformity of deformation and plasticity indexes, while other mechanical properties were similar to those of the controls. The effects of ultrasound on the deformation characteristics of metals are discussed by Langenecker [131,132]. The phenomena in solids include heating, and changes in the internal structure and internal friction that contribute to acoustic softening that has the ability to aid in forming processes and applications to wire drawing, deep drawing, and bending have all been investigated [110].

The application of ultrasound to melts has now resulted in the development of several novel processes. For example, Ichitsubo et al. [133] reported experiments and models for ultrasonically accelerated crystallization of Pb-based metallic glasses.

Present-day interest and developments in laboratory studies aimed at implementing ultrasonics in metallurgical processes are encouraging. Metallic couplings are now recognized, and ultrasonics can be applied at stages where the actual volume being treated at any instant is relatively small. The complaint that the high temperatures of the molten metals cause excessive shifts in resonance frequency of the ultrasonic coupling device from the frequency at room temperature can be met by proper design and choice of transducer. The allowable frequency shift for a typical piezoelectric power-type transducer is much smaller than it is for a magnetostrictive transducer. Therefore, magnetostrictive transducers are preferred for high-frequency applications of ultrasonic energy to metallurgy. They have the additional property of being water cooled.

In summary, the beneficial mechanisms of ultrasonics in metallurgy are [134]

1. Cavitation, which promotes crystallization and grain refinement
2. Degassing
3. Microstreaming, which disperses crystallization centers, significantly diminishes the thickness of a diffusion layer, thus intensifying processes where diffusion through a boundary layer is a limiting factor, and increases the homogeneity of the ingot

These mechanisms are responsible for eliminating columnar structure, forming fine equiaxed grains, breaking up precipitates of excessive phases, and reducing zonal and dendritic segregation.

REFERENCES

1. Beyer, T. 1997. *Non-linear Acoustics*. New York: Acoustical Society of America.
2. Levi, B. G. 1991. *Phys Today* 44(11):17–8.
3. Leighton, T. G. 1994. *The Acoustic Bubble*. San Diego: Academic Press.
4. Brenner, M. P., S. Hilgenfeldt, and D. Lohse. 2002. *Rev Mod Phys* 74(2):425–91.
5. Suslick, K. S. 1988. *Ultrasound: Its Chemical, Physical and Biological Effects*. New York: VCH Publishers.
6. Blake, J. R. ed. 1999. Acoustic cavitation and sonoluminescence. *Phil Trans R Soc Lond A* 357(1751):199–369.
7. ter Haar, G. 2008. *Ultrason* 48(4):233.
8. Thompson, L. H., and L. K. Doraiswamy. 1999. *Ind Eng Chem Res* 38(4):1215–49.
9. Gogate, P. R., R. K. Tayal, and A. B. Pandit. *Curr Sci* 91(1):35–46.
10. Vichare, N. P., P. Senthilkumar, V. S. Moholkar, P. R. Gogate, and A. B. Pandit. 2000. *Ind Eng Chem Res* 39(5):1480–6.
11. Flynn, H. G. 1964. Physics of acoustic cavitation in liquids. In *Physical Acoustics*, ed. W. P. Mason, 57–172. New York: Academic Press.
12. Loning, J.-M., C. Horst, and U. Hoffermann. 2002. *Ultrason Sonochem* 9(3):169–79.
13. Wu, Z.-L., J. Lifka, and B. Ondruschka. 2006. *Chem Eng Technol* 29(5):610–5.
14. van Iersel, M. M., N. E. Benes, and J. T. F. Keurentjes. *Ultrason Sonochem* 15(4):294–300.
15. Urick, R. J. 1983. *Principles of Underwater Sound*. 3rd ed. New York: McGraw-Hill.
16. Suslick, K. S., W. B. Mcnamara, Y. Didenko. 1999. Hot spot conditions during multi-bubble cavitation. In *Sonochemistry and Sonoluminescence*, eds. L. A. Crum, T. J. Mason, J. Reisse, and K. S. Suslick, 191–204. Doedrecht, Netherlands: Kluwer Publishers.
17. Margulis, M. A. 2006. *Russian J Phys Chem* 80(10):1698–702.
18. Alexander, P. 1951. *Paint Manuf* 21:157–61, 175.
19. Anonymous. 1963. *Soap Chem Spec* 39(10):175.
20. Wang, H. C. 1965. *Yao Hsueh Hsueh Pao* 12(5):346–8.
21. Witkowski, S. 1966. *Tluszcze Jadalne* 10(2–3):56–60.
22. Witkowski, S. 1966. *Tluszcze Jadalne* 10(4):129–35.
23. Witkowski, S., and M. Otowski. *Tluszcze Jadalne* 10(5):192–203.
24. Ramasubramanian, M. K., and S. L. Madansetty. 2000. In *Proceedings, 2000 TAPPI Recycling Symposium; March 5–8, 2000; Washington, DC*, 21. Atlanta, GA: TAPPI Press.
25. Marboe, E. C., and W. A. Weyl. 1950. *J Appl Phys* 21:937–8.
26. Rooney, J. A. 1988. Other nonlinear acoustic phenomena. In *Ultrasound: Its Chemical, Physical and Biological Effects*, ed. K. S. Suslick, 65–96. New York: VCH Publishers.
27. Mason, T. J. 1998. Power ultrasound in food processing—the way forward. In *Ultrasound in Food Processing*, eds. M. J. W. Povey, and T. J. Mason, 105–26. London: Blackie Academic & Professional.
28. Canselier, J. P., H. Delmas, A. M. Wihelm, and B. Abismail. 2002. *J Dispers Sci Technol* 23(1):333–49.
29. Priego Capote, F., and M. D. Lugue de Castro. 2007. *Analytical applications of ultrasound*. Vol. 26. *Techniques and instrumentation in analytical chemistry series*. Amsterdam: Elsevier.
30. Bergmann, L. 1954. *Der Ultraschall*. Stuttgart: Hirzel Verlag.
31. Hueter, T. F., and R. H. Bolt. 1955. *Sonics* New York: Wiley.

32. Riera, E., J. A. Gallego-Juárez, and T. J. Mason. 2006. *Ultrason Sonochem* 13(2):107–16.
33. Yin, X., P. Han, X. Lu, and Y. Wang. 2004. *Ultrason Sonochem* 11(6):337–48.
34. Povey, M. J. W., and T. J. Mason, eds. 1998. *Ultrasound in Food Processing*. London: Blackie Academic & Professional.
35. Kukoz, F. I., E. M. Feigina, and N. Ya. Avdeev. 1966. *Issled Obi Khim Istochnikov Toka* 35–45.
36. Prakash, S., O. Prakash, and S. Prakash. 1965. *Bull Chem Soc Jpn* 38(9):1426–30.
37. Prakash, S., O. Prakash, and S. Prakash. 1967. *J Indian Chem Soc* 44(6):508–12.
38. Okitsu, K., M. Ashkkumar, and F. Grieser. 2005. *Phys Chem B: Lett* 109(44):20673–5.
39. Jokanović, V., M. D. Dramićanin, Ž. Andrić, T. Dramićanin, M. Plavšić, and M. Miljković. 2008. *Opt Mater* 30(7):1168–72.
40. Lemlich, R. 1957. In *Proc. 2nd ICA (International Congress on Acoustics) Congress— Sound and Man*, Boston/Cambridge, Massachusetts, 17–23 June 1956. Abstract JD6 (1956). New York: American Institute of Physics.
41. Fand, R. M. 1962. *J Acoust Soc Am* 34(12):1887–94.
42. Raney, W. P., J. C. Corelli, and P. J. Westervelt. 1954. *J Acoust Soc Am* 26(6):1006–14.
43. Westervelt, P. J. 1960. *J Acoust Soc Am* 32(3):337–8.
44. Westervelt, P. J. 1963. *J Acoust Soc Am* 35(4):618.
45. Anon. 1962. U.S. Office of Saline Water—Research and Development Progress Report 65.
46. Bergles, A. E., and P. H. Newell, Jr. 1965. *Int J Heat Mass Transf* 8(10):1273–80.
47. Schmidt, F. W., D. F. Torok, and G. E. Robinson. 1967. *J Heat Transf* 89(4):289–94.
48. McCormick, J. E., and T. W. Walsh. 1965. *Res Technol Briefs* 3(3):7–14.
49. Cheng, H. 1964. English translation of Wu Li Hsueh Pao (Chinese People's Republic), 20(4):368–73. AD-639161.
50. Fairbanks, H. V. 1979. Influence of ultrasound upon heat transfer systems. In *Proc. 1979 Ultrasonics Symposium*, eds. J. deKlerk and B. R. McAvoy, New Orleans, LA, Sept. 26–28, 384–7. New York: IEEE, Sonics and Ultrasonics Group.
51. Loh, B. G., S. Hyun, P. L. Ro, and C. Kleinstreuer. 2002. *J Acoust Soc Am* 111(2):875–83.
52. Amrollahi, A., A. A. Hamidi, and A. M. Rashidi. 2008. *Nanotechnol* 19(31):Article # 315701.
53. Xie, H. Q., W. Yu, and Y. Li. 2009. *J Phys D—Appl Phys* 42(9):Article # 095413.
54. Lee, D. R., and B. G. Loh. 2007. *IEEE Trans Compon Packag Technol* 30(4):691–9.
55. Woodford, R., and J. C. Morrison. 1969. *J Pharm Pharmacol* 21(9):602–6.
56. Gilius, I. 1969. *Liet TSR Aukst Mokyklu Mokslo Darb Chem Chem Technol* 9:199–205.
57. Tamas, G., and T. Tarnoczy. 1955. *Magy Fiz Foly* 3:543–52.
58. Tarnoczy, T. 1956. *Magy Fiz Foly* 4:67–74.
59. Fairbanks, H. V., and W. I. Chen. 1969. *Ultrason* 7(3):195–6.
60. Kuznetsov, V. V., and N. I. Subbotina. 1965. *Elektrokhimiya* 1(9):1096–8.
61. Baumgarte, F. 1949. *Arztliche Forsch* 3(21):525–30.
62. Frenzel, H., K. Hinsberg, and H. Schultes. 1935. *Z Gesamte Exp Med* 96:811–816.
63. Lenart, I., and D. Ausländer. 1980. *Ultrason* 18(5):216–8.
64. Julian, T. N., and G. M. Zentner. 1986. *J Pharma Pharmacol* 38(12):871–7.
65. Julian, T. N., and G. M. Zentner. 1990. *J Control Release* 12(1):77–85.
66. Kyllönen, H. M., P. Pirkonen, and M. Nyström. 2005. *Desalination* 181(1–3):319–335.
67. Mason, T. J., and J. P. Lorimer. 1988. *Sonochemistry*. New York: Ellis Horwood Ltd (Chichester) a division of Wiley.
68. Suslick, K. S., P. F. Schubert, and J. W. Goodale. 1981. In *Proceedings IEEE Ultrasonics Symposium, Chicago, October 14–16, 1981*. 2 vol. ed. B. R. McAvoy, 612–6, New York: IEEE.
69. Suslick, K. S., P. F. Schubert, and J. W. Goodale. 1981. *J Am Chem Soc* 103:7342–4.
70. Suslick, K. S., J. J. Gawienowski, P. F. Schubert, and H. H. Wang. 1984. *Ultrason* (1):33–6.

71. Boudjouk, P. 1986. *J Chem Educ* 63:427–9.
72. Newton, N., and J. W. Kissel. 1953. *Arch Biochem Biophys* 47(2):424–37.
73. Brett, H. W. W., and H. H. G. Jellinek. 1956. *J Polym Sci* 21:535–45.
74. Weissler, A. 1950. *J Appl Phys* 21:171–3.
75. Langton, N. H., and P. Vaughan. 1957. *Br J Appl Phys* 8:289–92.
76. Head Jr., W. F., and W. M. Lauter. 1957. *J Am Pharm Assoc* 46:617–21.
77. Hanglein, A. 1955. *Z Naturforsch Teil B* 10:616–22.
78. Kossler, I., and V. Novobilsky. 1963. *Collect Czech Chem Commun* 28:578–84.
79. Levinson, M. S., and V. P. Nefedov. 1964. *Izv Sib Otd Akad Nauk SSSR Ser Biol Med Nauk* (1):145–7.
80. Lur'e, Yu. Yu., P. F. Kanzas, and A. A. Mokina. 1961. *Vestn Tekh Ekon Inf Nauchno Issled Inst Tekh Ekon Issled Gos Kom Khim Prom Gosplane SSSR* (6–7):54–6.
81. Pandey, J. D., S. Prakash, and S. Prakash. 1965. *J Phys Chem (Leipzig)* 228(3–4):272–6.
82. Shaw, M. T., and F. Rodriguez. 1967. *J Appl Polym Sci* 11:991–9.
83. Simionescu, Cr., and V. Rusan. 1966. *Celul Hirtie (Bucharest)* 15(10):369–73.
84. Witekowa, S. 1962. *Rocz Chem* 36:693–702.
85. Witekowa, S. 1962. *Rocz Chem* 36:703–11.
86. Zaskal'ko, P. P., G. I. Kichkin, and V. L. Lashkhi. 1969. *Khim Tekhnol Topi Masel* 14(6):46–9.
87. Kost, J., K. Leong, and R. Langer. *Proc Natl Acad Sci* 86(20):7663–6.
88. Hastings, G. W., and R. F. E. Bolten. 1966. *Soc Chem Ind London Mongr* (20):274–81.
89. Okamura, S., K. Hayashi, T. Sasaki, and T. Shioda. 1965. *Japanese Pat* 16(July):155.
90. Osawa, Z., and N. Igarashi. 1965. *J Appl Polym Sci* 9(9):3171–6.
91. Romanov, A., and M. Lazar. 1963. *Plaste Kautsch* 10(8):470–1.
92. Shiota, T., Y. Goto, S. Tazuke, K. Hayashi, and S. Okamura. 1965. *Kobunshi Kaqaku* 22(239):186–92.
93. Price, G. J. 1996. *Ultrason Sonochem* 3(3):S229–38.
94. Suslick, K. S., and G. J. Price. 1999. *Annu Rev Mater Sci* 29:295–326.
95. Chen, J. W., and W. M. Kalback. 1967. *Ind Eng Chem Fundam* 6(2):175–8.
96. Chervinskii, K. A., and V. N. Mal'tsev. 1966. *Ukr Khim Zh (Russ Ed)* 32(1):69–71.
97. Menzhulin, Yu. N., and E. I. Ponomareva. 1962. *Tr Inst Metall Obo—gashch Akad Nauk Kaz SSR* 5:24–8.
98. Parthasarathy, S., M. Pancholy, and T. K. Saksena. 1962. *Curr Sci (India)* 31:500–2.
99. Prakash, S., and S. Prakash. 1965. *J Sci Ind Res (India)* 24(12):629–30.
100. Suslick, K. S., and S. E. Skrabalak. 2008. Sonocatalysis. In *Handbook of Heterogeneous Catalysis*, vol. 4, eds. G. Ertl, H. Knözinger, F. Schüth, J. Weitkamp, 2006–17. Weinheim: Wiley-VCH.
101. Disselkamp, R. S., K. M. Judd, T. R. Hart, C. H. F. Peden, G. J. Posakony, and L. J. Bond. 2004. *J Catalysis* 221(2):347–53.
102. Cypies, R. 1957. *Bull Cent Beige Etud Document, Eaux (Liege)* Vol. 35, 52–55.
103. Gaivoronskaya, N. P., L. A. Kukoz, F. I. Kukoz, M. F. Skalo—zubov, and S. F. Sorokina. 1966. *Akust. Uagn. Obrab. Veshchestv Novocherkassk* 51–7.
104. Zemskov, G. Z., E. N. Dombrovskaya, V. T. Yarkina, L. K. Gushchin, and A. K. Parfenov. 1964. *Metalloved. Term. Obrab. Met.* (1):52–5.
105. Kapustin, A. P., and M. A. Fomina. 1952. *Dokl Akad Nauk SSSR* 83(6):847–849.
106. Abramov, O. V. 1965/1966. *Rost Nesoversh. Metal. Krist. Dokl. Vses. So—veshch Kiev* 326–337.
107. Cohn, R. M., and D. M. Skauen. 1964. *J Pharm Sci* 53(9):1040–5.
108. Kapustin, A. P. 1954. *Kafedra Chshchei Fiz* 88:53–6.
109. de Castro, L., and F. Priego-Capote. 2007. *Ultrason Sonochem* 14(6):717–24.
110. Shoh, A. 1988. Industrial applications of ultrasound. In *Ultrasound: Its Chemical, Physical and Biological Effects*, ed. K. S. Suslick, 97–122, New York: VCH Publishers.

111. Li, H., H. Li, Z. Guo, and Y. Liu. 2006. *Ultrason Sonochem* 13(4):359–63.

112. Chen, W.-S., and G.-C. Huang. 2008. *Ultrason Sonochem* 15(5):909–15.

113. Grandchamp, CI., and R. O. Prudhomme. 1967. *Acta Polytech (Prague)* 4(1):21–30.

114. Hickling, R. 1963. *J Acoust Soc Am* 35(7):967–74.

115. Parke, A. V. M., and T. Duncan. 1956. *J Chem Soc (London)*, Paper #855, (Pt. 4):4442–50.

116. Dam, J. S., and M. T. Levinsen. 2004. *Phys Rev Lett* 92(14):114301.

117. Hammer, D., and L. Frommhold. 2000. *Phys Rev Lett* 85(6):1326–9.

118. Lauterborn, W., T. Kurz, R. Geisler, D. Schanz, and O. Lindau. 2007. *Ultrason Sonochem* 14(4):484–91.

119. Flynn, H. G. 1982. US 4333796.

120. Crum, A. 1998. *J Acoust Soc Am* 103:3012.

121. Taleyarkhan, R., C. West, J. Cho, R. Lahey Jr., R. Nigmatulin, and R. Block. 2002. *Science* 295(5561):1868–73.

122. Taleyarkhan, R., J. Cho, C. West, R. Lahey Jr., R. Nigmatulin, and R. Block. 2004. *Phys Rev E*, Paper #4036109, 69(3 Pt 2):036109.

123. Taleyarkhan, R., C. D. West, R. T. Lahey, R. I. Nigmatulin, R. C. Block, and Y. Xu. 2006. *Phys Rev Lett* 96(3):034301.

124. Shapira, D., and M. Saltmarsh. 2002. *Phys Rev Lett* 89(10):104302.

125. Tsoukalas, L., F. Clikeman, M. Bertadano, T. Jevremovic, J. Walter, A. Bougaev, and E. Merritt. 2006. *Nucl Technol* 155(2):248–51.

126. Camara, C. G., S. D. Hopkins, K. S. Suslick, and S. J. Putterman. 2007. *Phys Rev Lett* 98(6):064301.

127. Saglime III, F. 2004. MS Thesis Rennsselaer Polytechnic Inst.

128. Jian, X., H. Xu, T. T. Meek, and Q. Han. 2005. *Mater Lett* 59(2–3):109–193.

129. Hiedemann, E. A. 1954. *J Acoust Soc Am* 26(5):831–42.

130. Kudel'kin, V. P., M. M. Klyuev, S. I. Filipov, E. N. Milenin, and I. I. Teumin. 1969. *Izv Vyssh Uchebn Zaved Chern Metall* 12(11):64–70.

131. Langenecker, B. 1966. *IEEE Trans Sonics Ultrason* SU-13(1):1–8.

132. Langenecker, B., and O. Vodep. 1975. Metal plasticity in macrosoniic fields. In *Proc. Ultrasonics International, 1975*, ed. Z. Novak, 202–5, Guildford, UK: IPC.

133. Ichitsubo, T., E. Matsubara, T. Yamamoto, H. S. Chen, N. Nishiyama, J. Saida, and K. Anazawa. 2005. *Phys Rev Lett* 95:24551.

134. Abramov, O. V. 1987. *Ultrason* 25(2):73–82.

12 Applications of High-Intensity Ultrasonics Based on Mechanical Effects

12.1 INTRODUCTION

In the strictest sense, all high-intensity applications of ultrasonic energy are based upon mechanical effects. All of these result from particle motions. However, as a result of the particle (or mechanical) motion, certain secondary effects are produced. Some of these effects are strictly mechanical, such as both the dispersion and the separation of particles in a liquid. Some others are combinations of mechanical effects working in tandem with certain secondary effects, such as the mechanical eroding of an oxide coating from a metal surface, exposing the surface to chemical attack, which the ultrasonic energy also accelerates.

The applications discussed in this chapter are those in which mechanical effects are the primary factors in the process, although other factors may also be active. Chapter 13 presents discussions of applications in which chemical factors (i.e., sono-chemical processes) are of primary importance in the processes, although it is recognized that mechanical effects are always present. In Chapter 14, applications for a range of medical and biomedical processes are discussed, including certain applications that use predominantly mechanical, and others that use chemical or biochemical, effects.

There are many high-intensity applications and potential applications of ultrasonic energy [1,2]; these include cleaning, soldering, plastic bonding, spot welding of metals, liquid atomization, materials forming, foam control, emulsification of liquids, drying and dewatering (particularly in combination with other forms of energy such as electrophoresis), compaction of powdered metals and plastics, acceleration of chemical activity, accelerated diffusion, alleviation of frictional effects, depolymerization, fragmentation and dispersion of particles, separation, enhanced filtration, control of grain size in solidifying melts, dust and smoke precipitation, and many others. Among these, cleaning and plastic bonding are probably best known and offer the most generalized market for high-intensity ultrasonics. However, soldering equipment is used in many industrial installations, machining service shops are available, and ultrasonic atomization equipment is used both in home and hospital humidifiers and has been considered for use in atomizing fuels for various applications as well as for preparing various powder products.

12.2 CLEANING

The market for ultrasonic cleaners has benefited from the development of ceramic transducer materials and power electronics. The equipment required is of relatively simple design and ranges in capacity from a liter to several hundred liters, and it is used by military, industrial, medical, and domestic personnel.

12.2.1 PRINCIPLES OF ULTRASONIC CLEANING

The effectiveness of ultrasonic energy in cleaning materials may be attributed to phenomena that accompany cavitation. These include (1) development of stresses between the cleaning fluid and the contaminated surface, (2) agitation and dispersion of contaminant throughout the cleaning fluid, (3) increasing of the attractive forces between contaminant and cleaning fluid, (4) promotion of chemical reactions at the contaminated surfaces in some cases, and (5) effective penetration of pores and crevices.

Cavitation forming at a liquid–solid interface imposes severe stresses on the solid surface such that, in time, the surface may itself become severely eroded. When a surface containing a contaminant is exposed to cavitation, these stresses operate to disperse the contaminant. The intensity of the stress when cavitation occurs is a function of the vapor pressure of the liquid, the gas content of the liquid, and the adhesive force between the liquid and the surface [3].

Particles of contaminant that are removed from the surface are immediately subjected to the violent activity of cavitation bubbles. These particles are, thus, propelled with high initial accelerations and dispersed throughout the solvent.

Further, the agitation provides a scrubbing action which promotes the removal of contaminant. Such contaminants may be loose, solid particles or materials that dissolve or emulsify in the cleaning fluid.

Ions are produced in and near the walls of cavitation bubbles. When ions from the cleaning fluid show a preferential attraction for ions from the contaminant, the net attractive forces are greater than those that would exist between the liquid and the contaminant in the absence of ultrasonic agitation. As a result, the contaminant is more readily removed by the cleaning fluid.

In some cases, ultrasonic cleaning may be attributed to the promotion of chemical reaction. When the contamination is a part of the material to be cleaned, cavitation erosion alone is insufficient to perform the cleaning operation at a practical rate. This situation is typical of heat-treated specimens. The heat-treat scale is often removed in a hot, acid-pickling operation. It may be removed at a much higher rate in a lukewarm, ultrasonically agitated pickling bath.

When a contaminated surface is located in an ultrasonic cleaning bath such that it is perpendicular to the direction of the ultrasonic stress, contaminants in pores, cracks, and crevices extending from this surface are withdrawn and replaced with solvent. Other orientations that provide suitable components of stress that are directed into the pores are also effective. Formation of cavitation bubbles over the pores provides a strong "vacuum cleaner" action.

12.2.2   FACTORS THAT AFFECT THE CLEANING OPERATION

The effectiveness of an ultrasonic cleaner depends upon the cleaning liquid and the ability to provide sufficient energy at the contaminated surface to promote the desired cleansing. This section is devoted to those factors that affect the distribution of power in the cleaning tank. The choice of solvent will be considered in a later section.

The amount of power required to remove contaminants from a surface varies with the contaminant and the solvent used. Because cavitation is associated with the cleaning rates for which ultrasonics is noted, the intensity at the surface must exceed the threshold of cavitation. The threshold varies with the solvent, but intensities within the range 0.5–6 W/cm^2 will include the cavitation threshold of solvents most generally used. Commercial equipment is usually designed to draw approximately 200 W of power from the line for each gallon of tank capacity. This is true irrespective of whether piezoelectric or magnetostrictive materials are used in the transducer.

The conversion efficiency of the electronic generator and of the transducer determines the amount of power available to the cleaning solution. Several factors must be considered in order to make the best use of the available power.

The ideal ultrasonic cleaner would maintain a high-intensity field over all surfaces to be cleaned under all load conditions. This ideal, best fulfilled in a diffuse field, is generally impractical because of the nature of acoustic waves, in particular, the reflection of energy from boundaries between media of different acoustic impedances. Certain methods are used in attempting to approach the diffuse field conditions, for example: (1) placing the radiating surfaces at an angle to the surface of the tank and to any other large reflecting surface in order to avoid establishing standing waves, and (2) introducing several frequencies simultaneously into the cleaning tank. In general, tank-mounted transducers in cleaners are suspended from the bottom of the tank and operate at discrete frequencies. Their effectiveness is influenced by the presence of standing waves resulting from surface reflections. Such cleaners operate best when their solvent depths are multiples of half-wavelengths.

The operation of ultrasonic cleaners at discrete frequencies has some advantages. High-intensity cavitation forms in local regions rather than throughout the cleaning tank. This concentration of cavitation permits operation at relatively low power input per unit volume of solvent. The disadvantage is that parts to be cleaned must be located in the high-stress region of the solvent. Intensities in other regions may not be sufficient for satisfactory cleaning. Materials closest to the radiating surface will receive the greatest benefit from the cleaning activity. These surfaces may shield materials located behind them with respect to the radiating surface. Diffuse fields produced by placing the radiating surfaces at angles to large surfaces distribute the high-stress regions through more of the tank. The effect is to increase the active volume, and the power required to obtain cleaning action is correspondingly higher.

The multifrequency system represents an attempt at approximating uniformity of field. With a multiplicity of wavelengths, high-stress regions are brought into proximity. Destructive interference by reflections from the surface of the solvent or from large surfaces placed in the tank is less of a problem than it is with single-frequency operation. If one wavelength is canceled by destructive interference, activity continues at the remaining frequencies. The intensity of the energy at each frequency must be higher than the cavitation threshold. The cavitation threshold increases as with higher frequency. The total energy consumed is necessarily higher than that for single-frequency units, but the additional power of the multifrequency units is generally better distributed.

As with any other application of ultrasonic energy, basic principles govern the effectiveness of an ultrasonic cleaning operation. These principles involve the choice of cleaner, the choice of cleaning solvent, and the distribution of energy to the surfaces to be cleaned. Some cleaning solvents will react unfavorably with certain types of soils and, in fact, make it more difficult to disperse the contaminants ultrasonically. Certain manufacturers have developed data on the effectiveness of solvents for cleaning various classes of materials, and can provide not only these data but also prepared solutions that are often sold under trade names that indicate the applications for which they are suited. Solvents for ultrasonic cleaning are discussed more elaborately in a later section.

Hindrances to the proper distribution of energy to surfaces to be cleaned may sometimes be subtle. Any coating or obstacle that prevents the ultrasonic energy from acting on a surface to be cleaned through a suitable solvent is a detriment to effective cleaning. Piling too many items into a cleaning tank is, obviously, poor practice.

12.2.3   TYPES OF ULTRASONIC CLEANERS

The basic equipment required for ultrasonic cleaning includes the transducer, tank, and electronic generator. Applications may vary from simple use of the basic components to large automated installations complete with conveyors, additional rinses, heaters for the solvents, dryers, mechanical pre-washers, and solvent-recovery units. Integrated systems are available that, for example, combine the benefits of ultrasonic cleaning and vapor degreasing. Products are ultrasonically cleaned in one tank and vapor rinsed and dried in a second tank. Figure 12.1 depicts a view of a typical commercial-type ultrasonic cleaning tank.

FIGURE 12.1   Bench-top ultrasonic cleaning tank. (Courtesy of Sonic System, Inc.)

Both magnetostrictive and piezoelectric transducers lead zirconate titanate, also called PZT, have been used widely in the past. These transducers may be obtained either in immersion form or attached directly to the cleaning tank.

Immersion transducers are separate, completely sealed units designed such that they can be placed in any suitable tank in any desired position. A number of immersion transducers can be arranged to focus energy for increased intensity.

Most ultrasonic cleaners are constructed with the transducers attached to the tank [2]. In early piezoelectric models, the transducer was a piezoelectric slab that was epoxy-bonded to the bottom of the tank. This construction has certain drawbacks. The epoxy bond may deteriorate with age, with an accompanying loss of acoustic coupling to the cleaning bath. The total power-handling capability of the epoxy-bonded, freely suspended slab is limited by the strength of the slab and by that of the bond.

Improvements in coupling, power-handling capabilities, and efficiency are obtained with PZTs by relatively simple design considerations. A discrete frequency system can be designed using a sandwich construction. In this design, the transducer plus the bottom of the tank constitutes a half-wave vibrator. The piezoelectric slab (or pair of slabs) is located at the node sandwiched between two metal plates, preferably with the node on the central plane of the slab or slabs. The metal plate nearest the bottom of the tank is welded or silver-brazed to the tank. The welding or brazing must be free of unbonded areas or voids, which hinder the coupling of energy to the solvent in the tank. The metal backup plate is bolted to the mounting plate, which has been welded to the tank and torqued to produce a bias compressive stress across the piezoelectric element and corresponds to the stress required for maximum operating efficiency recommended by the manufacturer of the element. If the elements of the transducer, including the bottom of the tank, have been appropriately matched acoustically, the transducer will operate effectively when the cleaning solution is an integral number of half-wavelengths deep.

Magnetostrictive transducers can be silver-brazed or welded to the bottom of the cleaning tank. As with piezoelectric systems, the weld should be free of voids and the transducer should operate as a half-wave system.

Magnetostrictive transducers often appear as a laminated stack suspended from the bottom of the tank. The stack is surrounded by a coil, and a separate iron yoke completes the magnetic system. Although this design has the advantage of low-cost construction, it is less efficient than other magnetostrictive designs. The air gap inherent in this design produces the major magnetomotive force (mmf) drop in the magnetic circuit.

A second and more efficient magnetostrictive transducer consists of a laminated stack with a window punched through the center. The coil is wrapped around the two legs on either side of the window. Best operation occurs when the coil is centered over the velocity nodal plane.

Magnetostrictive transducers require a bias magnetic field. Otherwise, they vibrate at twice the frequency of the driving voltage but at low efficiency. The amount of bias affects the efficiency.

The bias flux may be supplied by a permanent magnet or by passing direct current through the coils. The advantages of permanent-magnet biasing are that (1) no separate power supply is necessary and (2) no additional provision for isolating direct current from the output transformer of the power oscillator and alternating current from the power supply is required. The magnet must be capable of enduring high-frequency mmf fluctuations to which it might be subjected. Care must be exercised to prevent overbiasing.

Advantages of biasing with direct current in the coils are that (1) the bias field can be adjusted for optimum performance of the transducer and (2) the bias magnetic path is identical to that of the alternating field. The alternating current must be electrically isolated from the direct-current source, usually by means of a high inductance in the direct-current branch, and the direct current is isolated from the output transformer of the high-frequency source by means of blocking condensers in the alternating current branch.

12.2.4 ELECTRONIC GENERATORS FOR ULTRASONIC CLEANERS

Recent developments in electronics, and specifically in solid-state devices and printed circuitry, have benefited the ultrasonics industry. Generators are more compact, more rugged, and can be more reliable than their electron-tube predecessors. The electronic generator basically consists of single oscillator, amplifier, and output stages.

Good impedance matching among all components of the system is important for maximum power transfer. Matching must include consideration of the acoustic impedance provided by the cleaning bath as well as consideration of each element of the transducer and of the electronic generator. The acoustic impedance of the load reflected into the electronic system as electrical impedance depends upon the type and density of the cleaning fluid, the temperature of the bath, the type of cleaning load, the depth of the fluid, and the cross-sectional area of the radiating surface. All of these factors are variable with the exception of the cross-sectional area of the radiating surface of the transducer. Their influence on the reactive component of the impedance reflected into the transducer has its effect on the resonance frequency of the transducer. The oscillator is tuned to the resonant frequency of the transducer–tank combination, either manually or by feedback control of the oscillator frequency. Tuning becomes less critical with decrease in mechanical Q. A low-Q installation will operate effectively across a range of frequencies with little loss in efficiency.

12.2.5 CHOICE OF ULTRASONIC CLEANING FLUIDS

Two obvious considerations in the choice of cleaning fluids for use in ultrasonic cleaners are (1) chemical compatibility of the cleanser with the materials to be cleaned and the materials of the container and (2) the effectiveness of the cleaner in removing the soil. There are several important factors that determine compatibility and effectiveness in the presence of intense ultrasonic fields. Stresses in the walls of cavitation bubbles are concentrated and can be extremely severe. These stresses are controlled, in part, by the vapor pressure of the liquid. As vapor pressure increases and surface tension decreases, the maximum stress associated with cavitation decreases. Higher stresses are associated with smaller bubbles. The intensity of the stress waves associated with the collapse of cavitating bubbles is a function of the ratio of maximum bubble size to minimum bubble size. Bubbles in liquids with high vapor pressure or in liquids containing a large amount of dissolved gases tend to grow under the influence of the ultrasonic field, and the ratio of expansion to contraction under the alternating stresses diminishes rapidly.

There are two fundamental types of cavitation, defined in terms of the bubble lifetime and these are "stable"—involving long-lived bubbles that exist for many cycles of the driving ultrasonic field—and "inertial or transient," where the life cycle of the bubble occurs within one cycle of the applied ultrasonic field (at 20 kHz—within 1/20,000 of a second).

In addition to the life time of the bubble, relative to that of the period of the applied ultrasonic field, the properties of the cavitation bubble depend on the fluid properties, applied pressure, and temperature. In this context, there are four fundamental phenomena associated with driving bubble growth:

1. For a gas-filled bubble, pressure reduction or increase in temperature. This is called gaseous cavitation.
2. For a vapor-filled bubble, pressure reduction. This is called vaporous cavitation.
3. For a gas-filled bubble, diffusion. This is called degassing as gas comes out of the liquid.
4. For a vapor-filled bubble, sufficient temperature rise. This is called boiling.

In a fluid, the properties of the dissolved gases, vapor pressure of the liquid, and the cohesive strength of the liquid all interact with temperature, applied static/hydrostatic pressure, and the compression and rarefaction associated with the applied ultrasonic pressure field. In particular, it is the rate of change of pressure which is critical in determining which of the cavitation-linked phenomena and mechanisms, including sonoluminencense, will occur.

Dispersed materials of solid particles or gas bubbles, small point-protrusions, and rough surfaces form nuclei for the formation of cavitation bubbles. Therefore, the stresses that might be associated with cavitation in a given solvent are a major consideration in the choice of solvent. This is important with regard to the possible effects produced on the contaminated surface. For instance, high-intensity cavitation will erode plated, and many coated, surfaces. The effect can be minimized or eliminated by the choice of high-vapor-pressure solvents. Sometimes, the desired effect is accomplished by blending solvents to obtain the desired properties of solubility and nondestructive stress levels. For example, high cavitation intensities are possible in water. Dissolving a small quantity of high-vapor-pressure material, such as ethyl or methyl alcohol, in water can decrease the intensity of the cavitation stresses considerably. A blend of the two materials makes it possible to take advantage of the solvent properties of the water without the destructive intensities being applied to the contaminated surface. One might consider the alcohol as being representative of a mechanical fuse.

The chemical classification of the solvents and contaminants is a major consideration in the choice of solvent—that is, whether they are polar or nonpolar. In general, polar solvents are chosen to remove polar contaminants and nonpolar solvents to remove nonpolar contaminants. According to the dictionary, a polar chemical is one that is capable of being ionized as sodium chloride (NaCl), hydrogen chloride (HCl), or sodium hydroxide (NaOH). Further, the polar group includes materials, such as alcohol, acetone, acids, aldehydes, water, nitriles, and amines, that are chemically active. A distinguishing characteristic of polar materials is a high dielectric constant. Nonpolar substances are characterized by a low dielectric constant and include materials such as hydrocarbons and their derivatives—benzene, carbon disulfide, gasoline, kerosene, and chloroform, which are comparatively inert. For a given polarity of solvent, the best penetration of the contaminants is obtained at lower molecular weights.

The possible effect of solvents, particularly acids, on the material to be cleaned may be a factor. Acids may cause hydrogen embrittlement, and the effect is accelerated under ultrasonic excitation. In addition, the cavitation may produce a chemical reaction between the solvent and a base metal. This author has twice had the experience of having used acidic solvents recommended for cleaning certain stainless steel items, which were supposedly unaffected by these chemicals, only to have a rapid chemical action occur that dissolved the specimen. Commercially available cleaners of the strong-mineral-acids group contain wetting agents and inhibitors. The purpose of the inhibitor is to minimize hydrogen embrittlement and base-metal attack. Such inhibitors usually are organic materials that may break down at high temperatures.

Other effects caused by solvents include dissolving of base material, swelling of materials such as elastomeric coatings, cracking and distorting of certain plastics, and coagulation of certain materials (e.g., proteinaceous materials at high temperatures) that inhibit their removal and dispersion by ultrasonic energy. These considerations are necessary when choosing a solvent that has material compatibility.

Other considerations in the choice of solvents for ultrasonic cleaning include the ability to contain fumes (e.g., the fuming of an acid, such as hydrochloric acid, is accelerated under the influence of ultrasonically induced cavitation) and toxicity, ability to dissolve the contaminant, as well as its ultrasonic transmission properties. Materials that cavitate at a very low intensity (low cavitation threshold) inhibit the production of high-intensity cavitation and limit the transmission of ultrasonics through the liquid to low intensities.

Cleansers used most frequently for ultrasonic cleaning include aqueous materials (including household detergents), acidic chemicals, alkaline chemicals, and hydrocarbons. Commonly used solvents and their recommended applications are listed in Table 12.1. Most manufacturers of ultrasonic cleaners have prepared literature on recommended solvents, and these can be obtained upon request. These manufacturers often offer proprietary trade-name formulations of their own for use with specific classes of soils and other classes of materials.

12.2.6 PROCEDURES FOR ULTRASONIC CLEANING

The operation of bottom-mounted transducers is affected by the depth of solvent in the tank and by loading in the tank. When the resonance in the tank coincides with the resonance of the transducer, maximum agitation occurs (i.e., when the depth

TABLE 12.1
Commonly Used Solvents for Ultrasonic Cleaning

Solvents[a]	Base Materials	Contaminants	Remarks
Hydrochloric acid (HCl) with inhibitor	Ferrous alloys	Oxides, heat-treat scale	Plastic containers preferred
Sulfuric acid (H_2SO_4) with inhibitor	Ferrous alloys	Oxides, heat-treat scale	Plastic containers preferred
Nitric acid (HNO_3) with inhibitor	Stainless steel, some aluminum	Oxides, heat-treat scale	316 stainless and plastic are acceptable materials
Phosphoric acid with inhibitor	Iron, steel, brass, zinc, aluminum	Light scale and oxides, some drawing compounds, oil, grease	316 stainless is an acceptable container
Water	Any material not damaged by water	Loose soils, water-soluble soils	Stainless steel containers
Detergent and water	All metals and many other materials	Oils, greases, loose soils, etc.	Stainless steel containers
Soap and water	All metals and many other materials	Oils, greases, loose soils, etc.	Stainless steel containers
Trichloroethylene	All metals	Oil, grease, buffing compounds, etc.	Stainless steel containers
Tetrachloroethylene	All metals	Oil, grease, buffing compounds, etc.	Stainless steel containers
Acetone	All metals	Oil, some plastic-base cements (polar)	Stainless steel containers
Benzene	All metals	Oil, grease	Stainless steel containers
Xylene	All metals	Oil, grease, (slightly polar)	Stainless steel containers
Chlorothene	Most materials (bearing materials, meters, circuit boards, electronic components, etc.)	Grease, oils, dust, lint, flux, oxides, pigments, inks	Stainless steel containers
Freons[b]			
Trichlorotrifluoroethene (Freon TF)	Most materials (circuit boards, etc.)	Fluxes, loose soils, oils, other organic compounds	Stainless steel, aluminum, or plastic containers can be used; easily distilled and recycled
Tetrachlorodifluorethane	Most materials (circuit boards, etc.)	Fluxes, loose soils, oils, other organic compounds	Stainless steel, aluminum, or plastic containers can be used; can be used up to 150°F (66°C) but expensive
1,1,1-Trichloroethane	Electrical and electronic assemblies	Flux, dust, etc.	Stainless steel containers

a Wetting agents are recommended for most solvents used in ultrasonic cleaners
b Du Pont trademark.

of the solvent is a multiple of half-wavelengths at the driven frequency). When the depth is an odd multiple of quarter-wavelengths, the reflected wave is 180° out of phase with the transmitted wave, and, in such case, maximum destructive interference occurs. When the tank is loaded in such a manner that the standing-wave pattern is destroyed, the depth of loading becomes less significant and the acoustic impedance of the load becomes, primarily, a function of the characteristic impedance of the fluid in the tank. The effective cleaning volume is, then, a function of the total energy that is converted to acoustic energy in the fluid.

Sometimes, it is desirable to clean specimens in a basket or in a container filled with solvent that is different from that in the tank. In either case, care should be taken to obtain good transmission into the container. Wire-mesh baskets may be composed of wire of such a scattering cross section that the transmission of energy to the parts to be cleaned is inhibited. Considerable energy may be lost during transmission through glass beakers with thick bottoms. Sometimes, it is advantageous to use plastic bags, such as polyethylene, to contain the fluid and the part to be cleaned, while being careful not to puncture the bag. A container consisting of a stainless steel cylinder with a thin bottom of metal or plastic can be created such that transmission losses through the bottom are negligible. The buildup of gas bubbles on the bottom of a container will form a barrier to the transmission of ultrasonic energy and this should be prevented.

Soft-glass containers and thick thermoplastic materials should not be used in the presence of strong cavitation. The glass will fracture under such conditions. The plastic will be damaged and, perhaps, punctured by intense cavitation.

The arrangement of parts in a cleaning bath should be such that all areas to be cleaned are accessible to the cavitation. If parts are acoustically shielded by other heavier parts, they should be rearranged or removed for cleaning at a later time. Large parts may have to be reoriented occasionally to obtain full coverage.

The choice of frequency depends upon the size of parts that the unit is designed to clean. Low frequencies are less subject to scattering and losses due to loading than are high frequencies, whereas high frequencies produce a greater number of stress maxima in a given volume.

When properly used, ultrasonic energy provides an effective means of accelerating cleaning rates and of cleaning materials that are, otherwise, difficult to clean. It is particularly valuable in removing debris from pits and crevices.

The L&R Manufacturing Company of Kearny, New Jersey, offers the following DOs and DON'Ts with their cleaners, and these rules can be applied to the equipment from other manufacturers as well.

DOs
1. Do use the proper cleaning solution for the job.
2. Do use beakers and positioning cover and/or auxiliary pan for additional solutions.
3. Do rinse thoroughly after each cleaning application.
4. Do keep tank bottom free of silt and dirt accumulation.

DON'Ts
1. Don't place items directly on tank bottom. Use a basket or suspend into tank.
2. Don't run machine without solution in the main tank.
3. Don't use acidic or strong alkaline solutions in stainless steel tanks (use beakers).
4. Don't allow skin contact with chemicals.

Two novel cleaners are shown in Figures 12.2 and 12.3. Figure 12.2 depicts a strip-cleaner for cleaning components such as assemblies of soldered detonator wires. The strip is fed through a flooded cavity, where it is exposed to ultrasonic waves and is cleaned. Solvent is fed into the cavity through a manifold and is discharged through slits into a reservoir, where it is reclaimed and recirculated. This cleaner utilizes a magnetostrictive transducer.

Figure 12.3 presents a continuous dry cleaner used for cleaning belts, such as sanding or grinding belts, which move over the output surface of the transducer assembly. It is claimed that this device increases belt life by up to three or four times.

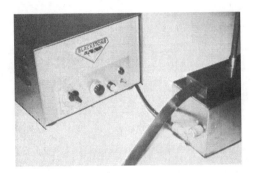

FIGURE 12.2   Ultrasonic strip cleaner. (Courtesy of Blackstone Corp.)

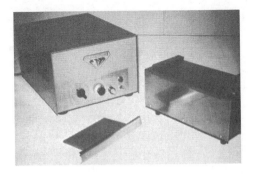

FIGURE 12.3   Ultrasonic continuous dry cleaner. (Courtesy of Blackstone Corp.)

12.2.7 Methods of Evaluating Ultrasonic Cleaners

The rapid growth of the market for ultrasonic cleaners induced the recognition of a need for techniques suitable for effectively evaluating these cleaners. The Ultrasound Industries Association (UIA; formerly the Ultrasonic Manufacturers Association [UMA]) attempted to develop a standard for this purpose. Ultrasonic cleaners are usually rated according to a power expressed in watts, for example, a 500-W system has a 500 W average power delivered by the electrical power supply. However, this does not mean that 500 W of cleaning power is delivered. The power rating actually refers to the line power rather than to the power density in the cleaning tank. The actual measurement of power density and total power within a cavitating ultrasonic bath is difficult. Table 12.2 lists a number of techniques that

TABLE 12.2
Methods of Measuring Ultrasonic Activity in Treatment Systems

Probe Methods	Sonochemical	Calorimetric
Acoustic pressure sensitive	Chlorine release	Substitution
Velocity sensitive	Bromine release	Thermal isolation
Acoustic absorption	Sonoluminescence	Constant flow
Displacement sensitive	Chemical oxidation	Acoustic dilatometer
Feel	Hydrogen peroxide (H_2O_2)	
Bouncing beads	formation	**Other**
Calorimetric expansion	Depolymerization	Visual
Noise sensitive		Complex transducer electrical
Rayleigh disc	**Surface Properties of**	impedance
Radiometer	**Cleaned Sample**	Reaction of radiation pressure
Hot-wire microphone	Water break	Fountain or surface distortion
Thermocouple in lossy matrix	Coefficient of friction	Volume flow rate
Acoustic fluxmeter	Surface electrical conductivity	Heat transfer
Signal to noise	Adhesion of evaporated or	Complex mechanical impedance
Total noise	sprayed-on metal film to	Light scattering
Subharmonic measurement	nonmetallic surface	Volume change
Semiconductor thermo-resistors	Thermally induced oxide film	Change in conductance
Photoelastic	(440°C) (metals)	Attenuation by scattering
Emulsion probe	Plating effectiveness (metal)	Printing
Electrolytic probe		Biological cell breakage
	Standard Soil	
Erosion	Radioactive tracer	
Foil (heavy)	Reflectometer	
Soil removal	Light transmission	
Strips or rods	Chemical tracer	
Foil (light)	Fluorescent dye	
Dispersion	Chemical dye	
	Graphite-coated ceramic	

have been used for measuring an ultrasonic field or for evaluating an ultrasonic system. Some of these techniques are applicable to intensities less than the cavitation threshold only. Few of the other techniques can fulfill the requirements of a general standard.

Effective cleaning and processing in liquids at medium and high intensities are usually associated with the occurrence of cavitation. Cavitation bubbles disrupt the normal propagation of high-intensity waves such that a true measurement of the power density in a bath can be obtained by few methods. If one can assume that all of the energy is absorbed within the liquid, calorimetric methods can sometimes be used to determine the power radiated by the transducer. However, calorimetric measurements are only limited to systems of certain designs.

All reactions or ultrasonic effects, in some way, relate to the frequency and intensity of the impressed field. The most popular methods of evaluating ultrasonic cleaners are based on one or more of these effects, and may be classified according to whether they operate by (1) sonochemical activity, (2) erosion effects, or (3) probe reactions.

Cavitation tends to produce free chemical radicals at and near the bubble walls. The quantity of these radicals in a given chemical mixture depends upon the intensity of the stresses involved and the total wall areas of all radical-producing bubbles. Previously, this phenomenon was investigated by the UIA as a possible basis for a standard [4]. In the UIA proposed standard, 200 ml of a filtered 0.1 M potassium iodide solution saturated with carbon tetrachloride is placed in a cell—a 10×15-cm polyethylene bag with a 0.1-mm wall thickness.

The solution in the tank to be evaluated consists of tap water, which has been aged for 12 h at room temperature or ultrasonically agitated for 30 min in order to stabilize the gas content of the liquid. A wetting agent is added to the water at concentrations that range from 0.005% to 1%. The evaluation is performed at temperatures between 25.5°C and 27°C.

The cell is placed in the tank such that the liquid level in the cell is at the same level as that in the tank. The cell may be suspended from a scanning mechanism for an integrated evaluation of the activity in the tank, or it may be held stationary to indicate the activity at any desired location. (The UIA-proposed standard gave the design of a traversing mechanism for scanning a tank.) The final reading is reduced to a units-per-minute basis (mg Cl_2/L·min).

The UIA technique has been called a chlorine-release method because, when the test solution is subjected to intense ultrasonic radiation, free radicals of chlorine are released from the carbon tetrachloride at or near the surface of cavitation bubbles, as described previously. The chlorine replaces the iodine released from potassium iodide (KI) to form potassium chloride (KCl) and I_2. The iodine thus released is soluble in the KI solution and can be measured either by titration or by spectrophotometric means. The measurement must be performed very carefully. If the titration method of analysis is used, the titration solution must be standardized on each day that it is used. (The UIA-recommended titration solution is a 0.0028 N sodium thiosulfate solution prepared with distilled water. A starch-indicator solution can be prepared by dissolving 0.1 g of potato starch, by boiling, in 100 ml of water.)

Titration or spectrophotometric measurements should be completed within a few minutes after exposure.

The linearity with time of the chlorine release in a constant-intensity field has been proved empirically [5]. Similarly, the rate of chlorine release appears to be proportional to cavitation "intensity," that is, the actual power density within the cell.

To overcome the objection that little, if any, ultrasonic cleaning occurs at 27°C, Jacke [6] obtained results similar to those obtainable with the chlorine release by using CBr_4 instead of CCl_4 at 65°C. Bromine release occurs slower than the chlorine release, but this approach does hold promise as an evaluation technique at common cleaning temperatures.

A popular and simple method of evaluating the activity in ultrasonic cleaning tanks consists of immersing a strip of aluminum foil mounted on a wire frame in a vertical position in the tank. The evaluation is based either upon the amount of pitting and erosion that occurs within a specified duration of time or upon the time required to erode holes in the foil. The foil-erosion method provides a good indication of the distribution of the cavitation within the tank. However, there are problems associated with its use that prevent it from becoming a suitable standard. However, foils of the same brand may exhibit different degrees of hardness and toughness to cavitation erosion. Particles of the foil resulting from cavitation erosion contaminate the solvent and must be removed. Further, the method of mounting the foil on the wires and the wire sizes and types themselves may affect the formation of cavitation pits in the foil.

Several types of probe techniques have been developed. These may include pressure probes (most of which are useless in a cavitating field), particle-velocity probes, emulsion probes, weight-loss probes (an erosion technique), noise-measurement probes, thermoelectric or heat-sensitive probes, and electrolytic probes.

Hydrophones can be calibrated to measure actual acoustic pressure at intensities less than the cavitation level. At intensities greater than the cavitation threshold, pressure-sensitive probes can be used to measure noise associated with cavitation. Such probes are easy to use. The noise measurements correlate fairly well with the chlorine-release technique to an intensity that corresponds to the formation of a continuous cavitation bubble between the bottom of the tank and the cleaning fluid. At this point, the noise-measurement values continue to increase but the chlorine release decreases. Gas films forming on the surface of the probe and spurious noises from other sources in the unit that are not related to cavitation adversely affect the meter indications.

Relative to the tank, the intensity of the ultrasonic field and its distribution at the operating conditions are, perhaps, the most important factors controlling the effectiveness of ultrasonic cleaning. The intensity within the tank is controlled primarily by the radiation from the transducer. The most effective transfer of energy occurs when maximum displacement occurs at the transducer–solvent interface, which is usually at the bottom of the tank. The intensity at this surface is a function of the displacement times of the frequency, and it is primarily responsible for the cavitation activity within the liquid load.

Velocity-sensitive, or displacement-sensitive, probes are of little value in measuring acoustic fields at intensities greater than the cavitation threshold. However, they can be used to determine the displacement of the radiating surface and, thus, to evaluate the performance of the transducer. Probes of this type may operate on magnetic principles, capacitance principles, or piezoelectric principles. Methods based on magnetic principles involve variations in the length of gap in a magnetic circuit caused by the ultrasonic vibrations. Similarly, methods based on capacitance principles involve change in capacitance between plates of a condenser (one of which may be the radiating surface) due to vibrations induced in one of the plates. Piezoelectric displacement probes may be bimorphic, and the charges developed on the faces of the piezoelectric elements are functions of the displacement of the elements.

A probe for indicating the amplitude of vibration of the radiating surface has been used for evaluating transducers [7]. It is a resonant accelerometer with a bandwidth of approximately 1 kHz, which must be brought into contact with the surface that is to be measured. The amplitudes at various points on the bottom of the tank are measured and recorded. Any later measurements must be made at the same points. Comparative evaluations between cleaners are limited to those of exactly the same frequency. The reaction of the probe on certain types of tank designs and the problem of maintaining a satisfactory acoustic coupling to the surface are detriments.

Emulsion probes consist of two immiscible liquids. The cavitation activity is evaluated by the rate at which the liquids are emulsified. The probe may be a small plastic bag containing distilled water and a piece of absorbent material saturated with oil. Alternatively, the oil-saturated material may be immersed directly in the cleaning fluid. Although this method does not provide an absolute measurement of "cavitation activity," it does provide a rapid and visual indication of the potential effectiveness of the cleaner.

Weight-loss probes are in a category similar to the foil-erosion technique. A probe of material is located in an ultrasonic field, and the cavitation activity is determined on the basis of the amount of material lost by erosion during a specified period of time. Various materials have been used, with lead being the most common.

Thermoelectric probes consist of thermocouples embedded in materials that are good absorbers of ultrasonic energy. The acoustic intensity is determined by the heating rate in the absorber.

Regardless of the type of probe used, the effect that inserting it into the field has upon the field itself will have a corresponding effect upon the accuracy of the measurement. The ideal probe would be small compared with the wavelength of ultrasound and the dimensions of the tank. Large probes that are acoustically mismatched to the surrounding liquid may cause undesirable reflections, which, in turn, may cause either destructive or constructive interferences as well as consequential errors.

A UK National Physical Laboratory (NPL) study has investigated the issues relating to standardization of measurement methods for calibrating high-power ultrasonic fields, including those used in sonochemistry. Hodnett and Zeqiri [8] provided a review of the literature in 1997 relating to calibration for industrial, medical, and sonochemical applications. This study was followed by the development of a novel sensor for monitoring acoustic cavitation [9,10]. Later, two papers reported developments made in moving toward developing a "reference 25 kHz ultrasonic cavitation vessel" [11,12]. All of these activities are part of an ongoing project to establish a reference facility for acoustic cavitation at the NPL.

In spite of what has been discussed above, it should be noted, however, that current ultrasonic cleaners are not readily calibrated; their stated acoustic output is not, in general, accurate and the conditions in the tank, even without a load, tend to be variable. When the variations caused by the mass and geometry of the load, the surface conditions, and chemistry are included, further variability is observed when measurements are made. In addition, there are wild variations in local acoustic/cavitation power that are attributed to the effects caused by minute changes in positioning of samples and the arrangement of sample supports or baskets in the tank. All of these issues can result in highly variable local performance and power.

None of the previously mentioned techniques is entirely acceptable to all manufacturers and consumers of ultrasonic cleaners. A method that is gaining acceptance, however, is based upon the removal of graphite stains from aluminum oxide blocks. The face of the block is coated with a dark layer of pencil-type graphite and located within the cleaning tank. The tank is filled with water and operated for a predetermined length of time. Then, the block is removed, allowed to dry, and, thereafter, compared with a set of blocks rated from 1 to 10 according to the amount of light reflected from the surfaces (i.e., according to the amount of graphite removed).

The increasing regulation of the storage and disposal of hydrocarbon-based cleaning solvents by many federal, state, and local organizations has increased interest in the use of ultrasonic cleaning. As a result, a range of alternate-cleaner performance-validation tests are being developed [13].

12.3   MACHINING, FORMING, AND JOINING

12.3.1   MACHINING

Ultrasonic machining is one of the earliest of the commercial high-intensity applications of ultrasonic energy. The earliest method, and one that continues to be used in special job shop operations, was based upon techniques described by Farrer [14] and Balamuth [15]. In this method, a slurry consisting of abrasive particles in a low-viscosity liquid is washed over the end of a tool shaped to conform to the desired geometry of the impression intended. The axial motion of the tool tip and the resulting cavitation impart high accelerations to the abrasive particles and, thus, erode the work away. The method is particularly useful for machining brittle ceramic materials, but it has also been used for machining some powder-metal parts. In the latter application, evidences of highly accelerated electrolytic erosion have been observed on occasion. The effect may be adverse in that the tool, rather than the work, is "eaten" away.

Shown in Figure 12.4 is a typical early model of ultrasonic-machining assembly consisting of a half-wave magnetostrictive transducer silver-brazed to a half-wave monel coupling horn. Figure 12.5 presents a photograph of a 2-kW magnetostrictive transducer attached to a rectangular, wedge-shaped horn of stainless steel. A second half-wave horn is attached, by means of a fine-thread stud, to the coupling horn; this

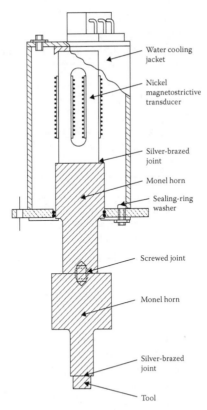

Water cooling
jacket

Nickel
magnetostrictive
transducer

Silver-brazed
joint

Monel horn

Sealing-ring
washer

Screwed joint

Monel horn

Silver-brazed
joint

Tool

FIGURE 12.4   Typical early-model ultrasonic machining assembly consisting of three half-wavelength wave resonant sections.

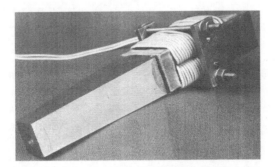

FIGURE 12.5   Two-kilowatt magnetostrictive transducer.

second horn is used as a toolholder. Toolholders, routinely, are made of Monel or stainless steel and are designed to resonate at the transducer frequency when the tool is attached. The tool is usually designed to be a fraction of an inch longer than the length required to match the transducer and it is removed when it wears to a fraction of an inch shorter than the resonant length. Equations for designing horns are given in Chapter 4.

Materials that are typically used for making horns include Monel (66 nickel (Ni), 28 copper (Cu), 5 iron (Fe), plus manganese (Mn), silicon (Si), and carbon (C)), titanium 318A (6 aluminium (Al), 4 vanadium (V), and 90 titanium (Ti), Admiro 5 (mainly copper and zinc, with small amounts of Ni, Mn, and Al), 304 stainless steel, and other stainless materials consisting of 1.3 C, 12.5 tungsten (W), 1 molybdenum (Mo), 4 V, 5 Columbium (also called Niobium) (Co), 4.3 cromium (Cr), and remainder comprising of Fe.

For best operation, the combination of toolholder and attached tool should resonate at the resonant frequency of the transducer. To accomplish this, the length of the horn should be such that the acoustic impedance at the junction is equal in magnitude in both the tool and the toolholder at the resonant frequency. The diameter of the selected horn at the junction should preferably be only slightly larger than that of the tool. The length of the tool usually is less than 2.54 cm at 20 KHz, although holes with diameter of 0.076 cm and depth of 3.8 cm have been drilled into glass at low driving power. The reason for the limitation in length is that the stress at the junction increases as the tool length increases, and, when this stress exceeds the strength of the bond, the tool falls off. Therefore, joining at a point of maximum stress in a horn should be avoided. In tapered horns, maximum stress occurs near the small end, which is the end to which the tool is attached. In stepped horns (with two quarter-wavelength sections) or cylinders, the point of maximum stress is located at the center of the length of the horn. Brazed joints should be free of defects and preferably filleted. A rule of thumb for the selection of lateral dimensions of the tool is 0.2 (or lesser) of the wavelength of the material of the toolholder. A drill rod is used most often as the tool material.

Some deep holes are drilled using the slurry method with half-wave tools attached to the half-wave toolholder. Such drilling is done only occasionally because of the difficulty involved in pumping slurry to the tool tip after the tool has penetrated to only a shallow depth.

Best machining is accomplished when the driving frequency is at the resonant frequency of the transducer and tool assembly. Tuning was done manually in older models. Some attempts at feedback from the transducer to control the oscillator frequency have been made with magnetostrictive transducers, but this has not been entirely satisfactory under load conditions. Later models use ceramic elements in a sandwich-type construction instead of the magnetostrictive transducer, and these are more readily adapted to feedback control of the driver frequency.

Abrasive slurry is frequently applied by means of a recirculating pump; however, it is sometimes advisable to use an eyedropper or other means of application. Satisfactory slurries comprise abrasives of boron carbide, aluminum oxide, garnet, silicon carbide, and diamond dust in liquids such as water, kerosene, or other low-viscosity liquids that are compatible with the work and present no health hazard. Abrasive-grit sizes are usually of 220 mesh or smaller, with the smaller sizes producing better surface finishes and the larger sizes producing faster machining.

A slurry consisting of one part of water to one part of abrasive, by volume, is best for shallow holes. Slightly thinner slurries are better for deep holes because they penetrate the hole better. In thicker slurries, the mobility of the particles is restricted and the machining is hindered. Thin slurries are less effective because of the smaller number of cutting edges available. The abrasive should be harder than the work to be machined.

Conventional ultrasonic machining units operate with weights, or other means of controlling the pressure at the tip of the tool, applied to force the tool into the hole. The speed of drilling is affected by the applied force. Too much force can be as detrimental to the machining process as can too little force. It may prevent free flow of abrasive and change the reflected impedance, causing the unit to operate at less than optimum frequency. In machining difficult materials, it is often advisable to relieve the pressure periodically to permit free flow of abrasive. Best pressures are, most often, determined empirically.

Tools containing cavities that trap air limit the depth to which holes can be drilled unless the air pocket is relieved by venting, for example, by a small hole open to the atmosphere at the top and entering the cavity at its highest position. The pressure-release hole can be used to advantage in drilling deep through holes. Hollow tools drill faster than solid tools because they have less material to remove per given depth of hole and, in addition, because they permit greater slurry mobility across the work face.

Tool wear is a function of the heat treatment of the tool, the abrasive slurry used, and the relationship between hardness and toughness of the tool and the work. Martensitic-type tools show good wear resistance, as do hard alloy tools.

Small handheld ultrasonic drills have been constructed for machining small holes. A skilled person can drill holes of 0.025 mm diameter to depths of up to 1.58 mm. A drop of slurry is placed on the surface of the material to be drilled, and the operator feeds the tool into the hole as it is being drilled. This operation is extremely difficult in opaque materials. The fact that the eye can follow the progress of the drill into the work makes it much easier to drill very small holes into transparent material.

A laboratory model of a handheld drill with bits and attachments for drilling deep holes is shown in Figure 12.6. Tools and toolholders use threaded attachments. The transducer consists of lead-zirconate-titanate plates in a sandwich construction designed to resonate at 28 kHz.

Dry ultrasonic machining has been accomplished without slurry by using chisel-type tool tips. The tool can become embedded in the work, and, therefore, it is necessary to continually change its position to present new shearing surfaces. This operation requires fairly soft work material and specially designed tool tips that are hard, tough, and sharp. Energy concentration at the cutting edges can quickly heat and melt the tip to a dull bead.

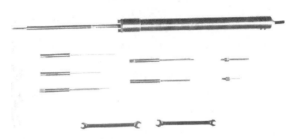

FIGURE 12.6   Handheld ultrasonic drill with tools, toolholders, and extensions for deep holes.

Materials most often machined by conventional ultrasonic machining techniques that use abrasive slurries are glass, ceramics, stones, occasionally brittle metals, and sintered materials.

Combining ultrasonics with electrolytic or spark-erosion machining of conducting materials results in faster work removal than is accomplished using either electrolytic or spark-erosion machining alone. However, this combination is seldom used for reasons of economic feasibility.

Further, ultrasonics has been used with chemical etching to machine materials into controlled shapes. The application of ultrasound increases the etching rate by a large factor which depends upon the material to be machined and the chemical being used.

Other methods include (1) revolving a workpiece and ultrasonically energizing a conventional drill bit to drill materials that are, otherwise, difficult to machine, (2) energizing a lathe tool to obtain improved surfaces and reduce chatter in materials that have a tendency to chatter normally, and (3) energizing reamers to increase feed rates and improve surface finish. There are reports of improved grinding effects under the influence of ultrasonic energy.

Ultrasonic rotary drills with conventional diamond tips are commercially available. These units are capable of drilling holes in ceramic materials as well as in metals. A commercially available rotary-type ultrasonic drill is shown in Figure 12.7. Several typical specimens machined with the rotary drill are shown in Figure 12.8. The bit is water-cooled, but no slurry is required. This equipment is, primarily, of value for machining ceramic-type materials, but it shows promise for improved machining of certain metals as well. The effectiveness of the imposed ultrasonic forces in improving the machining effectiveness is a function of the ratio of maximum vibratory-tip velocity to the peripheral velocity of the tool relative to the work surface. The effectiveness is diminished as this ratio decreases to less than unity.

FIGURE 12.7   Commercially available rotary-type ultrasonic drill press. (Courtesy of Branson Sonic Power Co.)

FIGURE 12.8 Typical parts machined with a rotary-type ultrasonic drill press. (Courtesy of Branson Sonic Power Co.)

Increase of the ratio to greater than unity should be beneficial if the cutting edges of the tool remain in good condition. At very high ratios, heating is a factor and careful cooling is required to prevent damage to the tool.

In addition to the traditional horn designs, which are discussed in Chapter 4, several new designs have been investigated as researchers seek to improve efficiency and performance [16]. Ultrasonic drills are being developed for certain novel applications. For example, the National Aeronautics and Space Administration (NASA) has developed an ultrasonic rock drill/corer [17]. This unit employs a PZT stack, which drives a horn. The researchers report the equivalent electrical circuit, an analysis of the mass–horn interaction, and an integrated computer model used for design and development evaluations.

12.3.2 Metal Forming

A plate of soft-annealed copper, with thickness measuring 3.2 mm, can be reduced to a thin foil by forcing it between the flat tip of an ultrasonically vibrating tool and an anvil when the spacing between the anvil and the tool is fixed. The force necessary to maintain a continuous feed rate can be applied easily by hand. This example illustrates the reduction in drawing force that can be accomplished with the application of ultrasonic energy to drawing dies and other forming operations.

The drawing force necessary for certain forming processes can be reduced by more than 50% using ultrasound. Langenecker [18] claims that drawing forces can be reduced by 80% at drawing speeds of 49 cm/s. Two factors believed to improve formability of metals with the application of ultrasonic energy are (1) the softening effect on the crystals (volume effect) and (2) the reduction of frictional forces between the workpiece and the tool. The potential advantages include the increase in metal formability and the beneficial influence of ultrasonic energy in forming difficult-to-form materials such as case-hardened steel and tungsten. Other benefits that have been claimed, particularly for drawing operations, are increased area reduction per pass, better size control, improved inside-diameter surface finish (tubing), greater tool life, and minimized chatter.

Balamuth [19] proposed a thermal-equivalent theory to explain the volume effect relative to drop in force required to draw materials. The theory is based upon the internal thermal energy of a solid being present in the form of incoherent atomic vibrations; the combination corresponds to an effectively increased kinetic energy content per atom. This argument suggests that ultrasonic energy introduced into a solid should have the same effect on the behavior of the metal as elevating the temperature of the material by an equivalent amount. Balamuth then claimed to show experimental evidence of a drop in drawing force under ultrasonic excitation that exceeds the dynamic force applied. Reduction in apparent static yield stress occurs under the influence of high-intensity ultrasonic energy for strain rates up to 100/s.

Prolonged exposure of materials to intense ultrasound before, during, and after forming can produce undesirable and irreversible effects.

Further, Langenecker [20] explained plastic deformation with the aid of ultrasonics from the total-energy viewpoint. Acoustic strain energy interacts with materials in several time-dependent ways. Any resonance absorbs energy in the form of higher dynamic strains. Absorption due to internal friction or absorption at lattice defects or grain boundaries raises the total thermal energy of the material. Some acoustic energy is converted to residual strain energy through plastic deformation.

Lehfeldt [21] stated that the lowering of the yield point under the influence of ultrasound is a spurious effect rather than a specific change in the material.

Pozen et al. [22] investigated the metallurgical effects of high-intensity ultrasound on a specimen of carbon steel. The microstructure of sections cut from the maximal stress zone revealed grain boundaries between pearlite and ferrite that were less sharp than those in untreated materials.

Effects of superimposed ultrasonic vibrations on compressive deformation were investigated by Izumi et al. [23]. The flow stress of compression can be reduced ultrasonically, but the effectiveness is dependent upon the acoustic impedance, Young's modulus, melting point, work hardenability, and stacking-fault energy. In addition, temperature rise in the material with increased intensity is material dependent, as one would expect. Ultrasonics influences the hardness distribution of the material in accordance with the distribution of the stresses in the field.

Kralik and Weiss [24] found that work hardening and tensile strength of wires of face-centered cubic metals increased, particularly when these materials (copper, gold, silver, and nickel) were treated at 20 kHz at various temperatures. However, the work-hardening response to ultrasonic treatment decreased for solid solutions of face-centered cubic metals. They suggested that the work-hardening effect might be related to some as-yet-unexplained mechanism involving dislocation loops and vacancy agglomerations in the materials.

Langenecker [25] claimed that yield strengths of metals such as zinc and aluminum are greatly increased by application of ultrasonic energy for a few minutes at an acoustic pressure exceeding 25,592 gm/cm^2. The ultrasound oscillates and distributes the crystal-lattice dislocations similarly to work hardening, but without change in shape. He claimed that yield strength in shear is increased to 50% by radiation at 4500 kN/cm^2, and is as high as 150% at 5000 kN/cm^2. Fracture along grain boundaries occurs at stresses exceeding 5500 kN/cm^2.

Multiplication of dislocations in nickel, copper, and aluminum (as well as lithium fluoride and sodium chloride crystals) was found, by Pines and Omel'yanenko [26], to begin at a certain threshold amplitude of vibrations, and this threshold decreases with increase in temperature. Above the threshold amplitude, the dislocation density increases with the ultrasonic intensity. In addition, the dislocation density increases linearly with time, initially, but tends toward saturation later. The saturation level increases both with increasing temperature and with increasing amplitude of vibration.

Chachin and Skripnichenko [27] found that the use of ultrasonics during the tempering of cutting blades increased their wear resistance more than that obtained with the normal treatment. Specimens of an alloy steel containing 0.87% carbon, 4% chromium, 6.3% tungsten, 2.5% molybdenum, and 1.7% vanadium were subjected to intense ultrasonics in a salt bath at 560°C for durations ranging from 15 to 60 min. Comparisons of hardness, impact and bending strength, red hardness, and wear resistance of these specimens with specimens that were similarly heat treated, but in the absence of ultrasonics, showed that (1) hardness increased to more than that of usual tempering with a 15-min application of ultrasonics; (2) impact strength and bending strength after a 60-min application of ultrasonics corresponded to values obtained after a 120-min normal tempering period; and (3) ultrasonics applied during tempering increased the red hardness and wear resistance of this alloy. These effects are probably associated both with internal mechanisms and with mechanisms associated with heat transfer.

Schmid [28] stated that the application of ultrasonic energy during a static test causes softening of a metal; however, if the insonation is superimposed prior to the static deformation, the sample is usually hardened by an amount which depends upon the intensity of the ultrasound, temperature, and amount of pre-strain. Increase in pre-strain is accomplished by decreasing the hardening effect of ultrasonics until, at a certain level of pre-strain, softening occurs. Schmid believed that the decrease in hardening may be attributed to recovery and re-crystallization. Recovery occurs for polycrystalline materials if the insonation of the samples is stopped before fracture occurs. Increases in hardening are less when insonation continues to fracture due to re-crystallization. According to Schmid, ultrasonics induces types of dislocation structure that are the same as those induced by alternating stress at low frequencies.

Ultrasonics has been investigated as a possible means of enhancing many types of forming operations, including wire and bar drawing, tubing drawing, sheet-metal rolling, densification of powder-metal parts, and grinding.

Experiments, conducted by Winsper and Sansome [29], with 18.7-kHz ultrasonic vibrations applied to a drawing die for mild-steel wire, with oscillations in the axial direction, indicated that the only reduction in force was attributable to the superposition of ultrasonic force on the drawing force rather than to heat and reduction of friction. No reduction in drawing force was observed for drawing speeds equal to or greater than the peak velocity of the oscillations. Specimens drawn under the influence of ultrasound revealed no changes in surface finish or mechanical properties.

Pohlman and Lehfeldt [30] claimed that internal friction can be reduced by ultrasound and that considerable reduction of the external friction between the die and the workpiece can be achieved in wire-and tube-drawing operations. Their experiments included materials such as copper, spring steel, and other steel alloys. In addition, their work showed that the surface conditions and the micro-hardness of copper wires were similar to those of wires drawn without application of ultrasound.

Kralik [31] reported that, if wires are drawn through orifices located at the node of a rod vibrating at 20 kHz, with the vibration direction being perpendicular to the direction of drawing, then, the rotational symmetry of the drawing texture is destroyed.

Severdenko and Clubovich [32] found that vibration amplitudes in the range of 0.012–0.02 mm at 23 kHz, when applied to a wire-drawing die for copper wire, not only caused a 50% reduction in drawing force but also increased the elongation with a corresponding reduction in quality, estimated to be approximately 15%. The diameter of standard copper wire was reduced from 1.57 to 1.25 mm.

In a previously mentioned wire-drawing work, the transducers were attached to the die and the die vibrated. Lorant [33] adopted a different approach to draw wires of high-purity copper, aluminum, and nickel to a diameter of 18 μm under the influence of ultrasound. The wire-drawing machine, using 11 dies, was immersed in a small ultrasonically agitated bath of liquid. Reduction was recorded as 30% per die at the rate of 300 m/min. He claimed that the procedure has the following benefits: (1) reduction in wire breakages, (2) increases in the life span of drawing dies, and (3) closer dimensional control.

Maropis and Clement [34] reported substantial reductions in bar-drawing forces for rods of 6061–0 aluminum alloy, AISI 4340 alloy steel, and 6A1–4V-Ti alloy, which they attributed to the reduction of coefficient of friction between the die and the material. Power input to the transducers was as high as 7 kW at a frequency of 15 kHz, and vibrated the die in the axial direction. Stock of 1.75-cm initial diameter was reduced in area by 15% per pass at the rate of 18 m/min. At 4.5-kW input to the transducers, the drawing force was reduced to between 15% and 20% for the aluminum alloys and 4.5% and 6.5% for the steel alloys. Surface finish was reportedly improved and micro-hardness and microstructure were not affected. Therefore, increasing of the ultrasonic intensity was claimed to result in corresponding increases in the magnitude of effectiveness.

In tube drawing, a reduction by 36% in the time required to draw 316 stainless steel tubing through the application of ultrasonic energy was claimed by researchers [35]. Thin-walled tubing was drawn to 0.26 cm outside diameter (OD), and a wall thickness of 0.254 mm at a tolerance of 0.025 mm using a fixed-plug mandrel was reported. Longer mandrel life and improved surface finish were also claimed in the report. Additional claims include (1) drawing 1.27 cm OD 25% silver-alloy tubing with a wall thickness of 0.025 mm held to 0.3 μm inside surface have been made for ultrasonic plug drawing [36], (2) 321 stainless steel bellows tubing being held to a tolerance of 7.62 micron on a 76-μm wall, and (3) an Fe–Ni–Co tube with a 381-μm wall being drawn to a minimum inside corner radius of 381 μm. Reduction of drawing forces was claimed to be as high as 10%.

Increased area reduction with ultrasonics reduces the number of drawing steps. Alloys, in addition to those previously mentioned, that have been formed with the aid of ultrasonics include 7075 aluminum alloy, 6061-T6 alloy tubing, 1100 aluminum (impact extruded), 3003-H14 alloy covers, 304 stainless steel, phosphor bronze, titanium and beryllium copper, Inconel 718, Kovar, brass, and nickel-plated steel. In addition, forming operations include flaring required to form stainless steel-tubing connections and production of brass cartridge cases [37].

Further, the application of ultrasonic energy to conventional lathe tools and to grinding operations was reported to be beneficial [38]. The poor surface quality often observed when cutting light-metal alloys at low speeds can be eliminated by ultrasonically agitating the tool. The effect drops sharply with increase in cutting speed, with the ratio of maximum vibratory velocity to cutting speed, again, being important. Quality improvement is dependent upon the vibratory motion of the tool in the direction of the main cutting forces. Nothing was gained by vibrating the tool in the direction of traverse.

According to Balamuth [39], grinding with ultrasonic assistance nearly always produces results superior to those obtained with regular grinding. Fatigue properties are not impaired, burning is almost totally absent, and, with light cuts, vibrating the wheel does not materially affect the surface finish. Vibration-assisted grinding with heavy cuts, in general, produces a superior surface finish. Chatter is reduced, resulting in a better surface finish. It is further claimed that ultrasonically assisted grinding takes about one-third as much power on the spindle as does an identical, but unaided, grind.

Ultrasonically assisted rolling has been more difficult to implement. However, Cunningham and Lanyi [40] reported that a 5% load reduction was attained by attaching a magnetostrictive transducer to the workpiece. A similar load reduction was attained when the transducer was designed to be integral with the roll; however, the system was plagued with short roll life and electromechanical difficulties.

The topic of acoustoplasticity or thermosonic bonding has been receiving increasing attention [41], particularly in the electronics community, for flip-chip bonding of electronic assemblies [42,43].

In all applications of ultrasonic energy, whether they are forming operations or any other types, the design of the system must conform to good, basic ultrasonic principles to ensure that the ultrasonic energy is delivered into the appropriate regions to be treated in order for the ultrasound to be effective.

12.3.3   ACCELERATED FATIGUE TESTING

Fatigue failure is the most important factor in applying ultrasonic energy to any process at high intensities. Fundamental equations employed in ultrasonic design are discussed in Chapter 3, and design equations for ultrasonic horns for processing applications are presented in Chapter 4. Stress-distribution equations are provided for each design. For longitudinally vibrating horns, these equations are based upon the "average" stresses in planes normal to the axes of the horns. This is justified by the assumption that the lateral dimensions are small enough to allow bar acoustic properties to prevail.

In applications such as ultrasonic abrasive slurry machining, a tool is attached to a horn usually by silver soldering. The horn and the tool are a half-wave assembly.

Therefore, neither component is a full half-wavelength long. The length of tool and the length of the horn are chosen such that the stress at the junction does not exceed the strength of a good silver-solder joint at the displacement amplitude at the tip of the tool under operating conditions. Ensminger and coworkers have developed tool-design charts for various horn designs and, from these, a slide rule which is used by ultrasonic machining job shops. Their use has been proved to be reliable; if the tool length or the displacement amplitude exceeds the limit calculated to be safe, fatigue failure occurs very rapidly.

If the density, elastic properties, frequency, and displacement amplitude at a displacement antinode are known accurately and vibration is restricted to a single mode, stress distribution in a vibrating system can be calculated, with longitudinal bar resonance being the simplest. Stress-distribution calculations become increasingly complicated when complex modes occur or when configurations other than the simple geometric configurations are vibrated; such configurations may complicate both stress patterns and vibratory modes.

Stress concentrations due to scratches, machine marks, and pits increase the stress level at their locations, and, if they are located at points of high stress level, stress concentrators can cause very rapid fatigue failure at ultrasonic frequencies. Stress concentrations in fillets can be calculated using equations available, for example, in the book by Roark and Young [44], if the fillet radius is very small when compared with a wavelength.

Because most failures in dynamic structural elements are traceable to fatigue, considerable effort is expended in fatigue-life measurements, or the development of S–N curves for various materials or structures. The S–N curve relates stress level (S) to number of cycles to failure (N). Standard fatigue testing is done at frequencies ranging from a fraction of a cycle per second to 200 Hz. Therefore, developing a complete S–N curve by conventional methods for any given material occupies months of testing. For example, it takes more than 1.6 years to expose a test material to 10^{10} cycles at the rate of 200 Hz. By comparison, at 40 kHz, materials are subjected to 3.456×10^9 cycles per day and to 10^{11} cycles in less than 30 days.

The high-frequency cyclic stresses to which several modern machine components, such as high-intensity ultrasonic devices and high-speed engine components, are subjected introduce a need to expand the range of frequencies normally covered by standard fatigue testing of materials. Not only is it important to know whether transferring low-frequency fatigue data to components subjected to ultrasonic frequencies is valid but also the converse of the question is pertinent. Are fatigue data obtained at ultrasonic frequencies applicable to low-frequency engineering needs? There is certainly a need for high-frequency fatigue test data and, if the high-frequency test data are applicable to low-frequency needs, the shorter test periods required offer many advantages. Studies of the effects of temperature, environmental media, mechanical conditions such as surface finish and geometry, metallurgical conditions, and so on can be conducted within relatively short times.

Puskar [45] states that "Currently, research is being carried out on the problems of high-frequency fatigue of various materials, from engineering, technological, physico-metallurgical, and testing viewpoints." He then proceeds to discuss common aspects of high-frequency testing, results of comparative studies of fatigue-limit values obtained at different frequencies, the influence of various factors on the high-frequency fatigue limit, and the accumulation of damage and formation of fracture areas.

There are significant differences between ultrasonic fatigue testing and low-frequency testing. The most significant differences involve the rate of cycling and the methods of determining stress in the test piece. Conventional low-frequency fatigue testing is performed according to either of two modes: (1) constant maximum displacement amplitude (strain) or (2) constant maximum force amplitude (load). Applied force is monitored continuously in either mode. At constant strain amplitude, some materials soften with time and applied load decreases accordingly. At constant load amplitude, the strain amplitude increases with softening. Therefore, at ultrasonic frequencies, it is easier to maintain constant strain amplitude.

At ultrasonic frequencies, the frequency, modal characteristics, and displacement amplitude are the primary variables that are monitored. Stress level is determined by the displacement amplitude and calculated using the elastic constants of the specimen material. One may use published values for elastic constants; however, for greater accuracy, monitoring frequency and modal characteristics of thin, longitudinally vibrating bars yields bar velocity of sound at any moment and, by having weighed the specimen before starting the test, yields Young's modulus of elasticity from the equation for bar velocity of sound, that is, $c = \sqrt{E / p} = \lambda f$.

Accuracy in measuring wavelength, frequency, and displacement is essential to a reliable S–N curve at ultrasonic frequencies.

Figure 12.9 depicts an S–N curve for SNC-631, a Japanese steel alloy, obtained at 38.6 kHz. The curves were developed using stepped thin bars in longitudinal half-wave resonance in air. The step occurred at a velocity node located between two cylindrical sections that were 1.58 and 0.79 cm in diameter, respectively. The

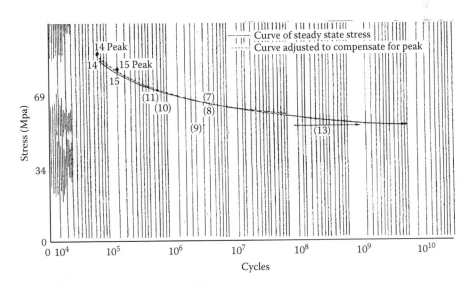

FIGURE 12.9 S–N curve for steel alloy.

step was filleted and the stress-concentration factor in the fillet was calculated to be 2.8535, using the method of Roark and Young. Displacement amplitude was monitored using capacitance-type probes and very light accelerometers. Only two points, (9) and (10), are significantly removed from the curve. Specimens corresponding to these points contained additional stress enhancers, in the form of deep machine marks in the fillet area, which resulted in early failure of these specimens. Tests were run at constant room temperature with only air cooling on the specimen. Tests were suspended immediately at the point that resonance frequency began to shift rapidly and unidirectionally—a certain indication that failure was in progress.

Well-recognized factors influencing fatigue life at ultrasonic frequencies include specimen temperature, surface condition, and environment or surrounding medium (e.g., atmosphere, liquids, and abrasives).

An appropriate high-frequency fatigue-testing facility includes

1. Suitable electronic power oscillator with feedback control to the electronic oscillator to maintain resonance operation throughout the test
2. An accurate, preferably non-contacting, displacement-measuring device for monitoring displacement at the free end of a resonant test specimen
3. Accurate means of locating the first displacement node from the free end of the specimen and, thus, determining a quarter-wavelength of sound in the bar
4. An accurate frequency meter for monitoring frequency and providing a basis for determining a count of the number of cycles of stress applied
5. An accurate timer
6. Means of stopping the test and indicating the time that the test was stopped at definite signs of failure
7. Strip chart recorder for continuously monitoring displacement amplitude and time
8. Appropriate cooling means
9. Thermocouple and associated instrumentation for monitoring specimen temperature

Further, Puskar mentions the value of accelerated fatigue testing to the theoretical aspects and the significance of these tests for basic research. He notes that "Knowledge on the mechanisms of damage cumulation, the origination of crack nuclei and their propagation, and a knowledge of the character of fracture area of different materials under various conditions can contribute to precision improvement and development of the theory of fatigue ... Recently, we have seen that the method of securing samples and their stressing during high-frequency tests give very good possibility for direct observation of formation and growth of fatigue zones; their interaction; the formation of submicroscopic cracks; and finally of the fracture area, making great use of the scanning electron microscope."

Results reported by different investigators vary. In many cases, fatigue limit at high frequency exceeded those at low frequencies by various amounts. Under certain circumstances, fatigue limits at ultrasonic frequencies were lower than those obtained at low frequency. Often, low-and high-frequency data were similar.

There are many possible explanations for these differences, including (1) differences in experimental facilities and procedures, (2) differences in specimen mechanical and metallurgical properties (e.g., surface condition, dimensions, and grain size), and (3) internal friction, dislocation dynamics, and other factors associated with the structure and history of the specimens.

The fact that fatigue data can be developed at such high rates at ultrasonic frequencies offers interesting possibilities for studying many important parameters related to specimen environment, material properties, and temperature. Amplitudes of vibrations (strains) are very low at ultrasonic frequencies, and, therefore, this parameter must be measured very carefully and accurately. Accuracy is dependent both upon careful and accurate measurements and definite identification of modal conditions (preferably vibration in a single mode, the longitudinal mode being the simplest to follow and calculate). In addition, it is dependent on specimen properties including both metallurgical properties and surface contour and conditions.

The development of *S–N* curve data for a range of metals and composite materials remains of importance, not least with regard to aerospace and wind-turbine applications. Several accelerated life models have been developed to address differences that are found between low-cycle and high-cycle responses. One such approach to address the observed differences is the Bairnbaum–Sanders model reported by Owen and Padgett [46]. In addition, there are, at present, piezoelectric-driven fatigue machines that are capable of providing 10^{10} cycles in less than a week [47]. Further, data are being accumulated using accelerated fatigue, at cryogenic temperatures, which can reduce run times by 400–500 times and give significant savings in cost through reduction in the volume of cryogenic gases required [48].

12.3.4 DEBURRING

Ultrasonics has been used to remove burrs created by forming or machining. The technique can be used on precision parts where fine deburring is required if tolerances equal to or more than 2.54 μm. This method is based upon the fact that ultrasonic energy accelerates particular reaction rates, particularly in the presence of cavitation. When cavitation erodes a metallic surface, the pits formed may be fairly rough at the opening. If the erosion is due to accelerated chemical attack, the pits may be smooth at the entrance.

Small burrs not only form nuclei to encourage the formation of cavitation but also present maximum area per unit of volume to the attacking fluid. For this reason, ultrasonics applied to an acid will accelerate the dissolution of small metallic burrs from such items as gear wheels. Ultrasonic deburring has been used on materials such as stainless steels, mild steels, and copper and nickel alloys. Its use requires careful formulation of the deburring bath and good control of temperature and time to avoid excessive erosion of the part.

The items to be deburred are placed in a cleaning-type tank prepared with an acid-resistant lining. The tank contains an acidic slurry formulated specifically for the type of metal to be deburred. The unit is then turned on for a predetermined time, with the temperature of the bath set at a predetermined level. The deburring cycle is followed by ultrasonic cleaning in an alkaline bath to neutralize the acid, and this is followed by a clean-water rinse.

Ultrasonic deburring does not remove heavy burrs without excessively eroding the part. It should be used only on parts where the economics of the operation justify its use and not on inexpensive, non-precision parts.

The ultrasonic deburring process has been used and developed by Yeo et al. [49] with thick metallic and non-metallic burrs that previously could not be deburred. The ability of ultrasound to propagate through elastic media and the fact that it does not require any tooling makes this technique very useful to treat surfaces that are generic in nature and difficult to access.

12.3.5 COMPACTION OF POWDER METALS AND SIMILAR MATERIALS

The use of ultrasonics in compacting metal powders makes higher densities with greater dimensional stability possible. The feasible application of ultrasonics in the compaction of powder materials is limited to small parts with thicknesses of only a fraction of a wavelength (~0.03λ) in the material of the compact.

Two methods of applying the ultrasonics are available. For compacting loose or porous metals and plastics (thermoplastic), in which the displacement between particles is relatively free, the compact is best located at the end of a half-wave tool—a position corresponding to minimum acoustic impedance. This method is acceptable for low-melting-point materials; however, protection against oxidation is usually necessary. The pressure required is generally low. In addition, 304 stainless steel specimens measuring $1.27 \times 5.7 \times 0.64$ cm^3 pressed in low-thermal-conductivity dies reach red-hot temperature within a few seconds at 1-kW input to the transducer. When subjected to similar conditions, silver will melt within a few seconds.

Transverse strengths of various iron compacts prepared in this manner, without pre-sintering or reduction of oxides and using only dead-load clamping pressure, showed improvements ranging from 7.5 to 15 times the green fiber-stress values for the materials. The corresponding increases in density were between 8% and 10%. Higher strengths were associated with higher intensities. Massive oxides were present near the pores and on the outside surfaces. These oxides may have been partially produced during ultrasonic excitation, but there is some evidence that these particles result from agglomerations of oxides that are already present in the compact. The frequency of the applied ultrasonic energy was 20 kHz, the specimen dimensions were $0.64 \times 1.27 \times 0.64$ cm^3, and the power to the transducer was in the range 500–1000 W.

On the other hand, if the specimens have been pre-compacted to equal or greater than 75% theoretical density, particle motion within the compact is restricted, corresponding to a high acoustic impedance. The specimen is then located at a high-impedance point in the vibratory system. In a uniform bar, this corresponds to the velocity node and stress anti-node. Figure 12.10 shows a basic design of an ultrasonic system for producing powder-metallurgy parts. The arrows indicate the points at which the external force may be applied. This particular design was employed for a 0.75 MN laboratory press. With the pressure applied between nodal

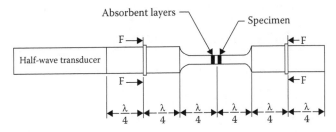

FIGURE 12.10 Basic design of ultrasonic system for producing powder-metallurgy parts.

points, little energy is lost to the press. The energy loss within the specimen causes considerable heating and, again, a protective atmosphere may be necessary to inhibit oxidation. Compactions of various alloys to 100% theoretical density have been produced with the system, with power input to the transducer ranging from 1200 to 1650 W.

Compacts of iron specimens which were hydrogen annealed before green pressing to remove the inherent oxides and, in addition, pre-sintered in a hydrogen atmosphere showed improved characteristics after ultrasonic compaction in the system shown in Figure 12.10. Transverse strengths of these specimens were 1.5 to 2.5 times greater than those of samples prepared from the as-received powders. Increases in strength were obtained with both increased intensity and increased clamping pressure. In one instance, doubling the pressure resulted in a 66% increase in strength. Metallographic examination indicated that only a normal amount of oxides was present.

Lehfeldt and Pohlman [50] indicated some improvement in density for a given punch load at low densities and some reduction of temperature for equivalent density in hot pressing of metal powders with superimposed ultrasonic vibrations.

Pokryshev and Marchenko [51] found an initial intensive shrinkage in the first 5–10 min of sintering hot-extruded iron powder in the presence of ultrasonics, after which time the shrinkage rate decreased. This was attributed to acceleration, by the ultrasound, of the transition from nonstationary to stationary flow in the powder. They, too, found that ultrasonic irradiation effectively reduces the temperature of hot extrusion by up to 20%.

Sonic vibrations in the range 5–15 kHz at displacements of 3–6 micron (μm) were found by Pines et al. [52] to activate sintering and inter-particle diffusion in various copper powders when the amplitude of vibration exceeded a critical threshold for each type. Powder compacts of 25%–55% porosity were used for these studies. Sintering of compacts comprising 50% Cu–50% Ni powder showed that sonic vibrations stimulate the process of heterodiffusion.

Hoffmann and coworkers [53,54] applied ultrasonics to precision casting, as used in dentistry and jewelry making. Ultrasonic energy was applied to promote flow and improve properties of castings from molds prepared by the lost-wax process. Ultrasonics promotes flow by reducing the apparent surface tensions and by driving the molten metal into various regions of the mold to completely fill the mold. In addition, it eliminates dissolved gases and, in general, improves the quality of the product.

Kostin et al. [55] studied the effects of ultrasonics on porosity and hardness of drill bits formed by pressing powdered materials. They obtained an average decrease in porosity of 10%–12% and average increase in hardness of 20%–30%.

Ultrasonic methods have been found to be useful in the compaction of PZT ceramics [56] intended for use as ultrasonic transducers. In addition, it is being increasingly applied for the compaction and processing of nanopowders.

12.3.6 Soldering

One of the first commercial applications of high-intensity ultrasonics was in soldering difficult-to-solder materials, namely aluminum [57]. In the earliest model, a soldering tip was incorporated in the working end of a transducer. The transducer was a magnetostrictive Permendur stack. The assembly was housed in the form of a conventional soldering gun and operated in a similar manner.

In later models, the transducer and soldering tip were separate, and outside sources of heat, such as a hot plate, were relied upon to melt the solder.

Soldering pots and soldering baths of various sizes can be obtained that contain a built-in heater and an attached transducer (see Figure 12.11). When soldering aluminum, the joints are immersed in a puddle of solder such that surfaces to be bonded are completely protected from the atmosphere. The soldering tip is then brought into contact with the workpiece. The ultrasonic vibrations erode the protective oxide coating, bringing the solder into direct contact with the base metal, to which it becomes attached [58].

The ultrasonic energy not only disperses the oxide coatings but also promotes flow into crevices and cavities by reducing the apparent surface tension of the solder and by the imposed stresses that promote flow. For this reason, it is also applied to other soldering applications in production lines to ensure better coverage.

Ultrasonic soldering has received attention as a flux-free method for use in the electronics industry. [59] and in soldering hard-to-join material combinations such as copper–aluminum pipe elements [60].

FIGURE 12.11 Commercial soldering pot for fluxless soldering of aluminum sheets and tubes such as those used in the manufacture of aluminum heat exchangers. (Courtesy of Blackstone Corp.)

12.3.7 WELDING

12.3.7.1 Welding Metals

The earliest studies in ultrasonic welding were conducted on metals, and commercial equipment was introduced later in the 1950s [61]. When two materials are placed in intimate contact and pressure is applied normal to the interface, ultrasonic vibrations producing shearing stresses at the interface cause heating as a result of local absorption. This phenomenon has been used to produce spot-and-seam welds in metals such as aluminum, brass, 304 and 321 stainless steel, copper, zirconium, titanium, niobium, commercial iron, chromel, nickel, molybdenum 0.5%, titanium, tantalum, gold, and platinum. Applications include bonding of both similar and dissimilar metals, in areas ranging from very small wires used in the fabrication of microelements to the seam welding of metal plates that are up to 5.0 mm in thickness (25 kW). All materials that can be pressure welded can probably be ultrasonically welded.

The primary mechanism of bonding appears to be associated with the formation of a thin molten film in the interface [62]. However, there is some indication of a solid-state bonding. A galling effect has been observed in experiments with similar materials in which the ultrasonic intensity was believed to be too low to cause melting. According to Ginzburg et al. [63], the welding process occurs by diffusion in the welding zone at low amplitudes and high applied pressure.

Optimum conditions for formation of welded joints by ultrasonics depend on the amplitude of vibration of the welding tip and applied static pressure. Joints can be made either at high vibration amplitudes and low applied static pressure or at low amplitudes and high applied pressure. Welding at high amplitudes and low pressures is similar to friction welding. The choice of method depends primarily upon the materials to be joined. Very ductile materials will yield under ultrasonic strain without interfacial sliding, even at low pressures. Hard, brittle materials will tend to fatigue under strains and pressures required for welding.

Typical ultrasonic welder arrangements are shown in Figure 12.12. Figure 12.12a is a diagram of a seam welder with a sliding or rotating tip. Either spherical or wheel types of tips are used. Figure 12.12b is a diagram of the typical spot welder in which

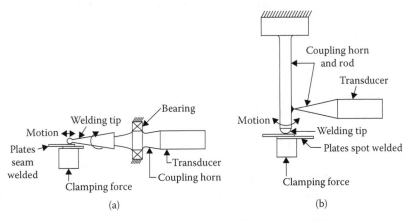

FIGURE 12.12 Ultrasonic welder arrangements. (a) Seam welder. (b) Spot welder.

the longitudinal motion of the coupling horn is converted to flexural motion in the welding tip. In both cases, shearing stresses are generated in the specimen in a direction parallel to the transducer axis. In a third type of arrangement, a torsional type welding tip is used to apply shear in a plane normal to the axis of the welding tip. Torsional types of ultrasonic welders have been used to produce hermetic seals in propellant containers [64]. The cover material comprised of 127-μm-thick 3003-H19 aluminum alloy. Welding time was 0.5 s, with an input power of 5400 W and 680-kg clamping force, to produce hermetic seals. Other applications of torsional types of bonders are described in a patent issued to Jones and DePrisco [65].

Removing the oxide film prior to welding aluminum does not affect the strength of welds in this material. The oxide layers are broken up and, often, forced to the outer perimeter of the weld by the ultrasonic energy.

Cracking frequently occurs at the edge of spot welds made in various combinations of heat-resistive alloys. These cracks are probably attributed to fatigue failure caused by high cyclic stresses occurring in these areas. Fatigue cracking is minimized by restricting weld times to a minimum. Another problem encountered occasionally is the welding of the tip to the material to be welded.

Some additional typical applications of ultrasonic welding that have been reported include the bonding of 0.2-mm anodized aluminum to 0.381-mm-thick brass within 0.35–0.45 s [66]. Pressure and power applied in the procedure were not reported. However, this reference does claim that the microstructure of the weld showed a good quality of contact, which was free of fragments of the anodized layer. Shin and Gencsoy [67] claimed good results in bonding low-carbon steel to polypropylene, acrylics, and glass by using aluminum foil interleafs. The presence of oxygen, aluminum, silicon, or magnesium as a main constituent in the nonmetallic materials enhanced the strengths of the welds. These materials easily form atomic or ion bonds with aluminum-foil interleaf.

Varga [68] recommended aluminum sheet thicknesses ranging from 0.01 to 0.8 mm for spot welding and 0.009 to 0.6 mm for seam welding using ultrasonics.

Pfaelzer and Frisch [68] compared results of welding copper in air and in a vacuum at 1.33×10^{-8} Pascal. Welds produced under vacuum were much stronger than those produced in air. According to these authors, the strength of welds depends on interfacial areas that have not slipped relative to each other.

In a related area, several studies have been performed with regard to control of grain size and quality of welds by applying ultrasonics and low-frequency vibrations to the welding bath during arc welding of aluminum and other alloys. Russo [70] applied low-frequency vibrations to 18-mm-thick plates of low-alloy Cr–Ni–Mo steel being welded manually using an eccentric device at frequencies ranging from 30 to 50 Hz and amplitudes ranging from 0.01 to 0.1 mm. The grain size of the welds was considerably reduced. Treatment at 20 kHz during crystallization of an Al–Mg-alloy ingot that was 18 mm in diameter and 10-mm high produced appreciable reduction in grain size of the initial structural grain and a more regular distribution of the second phase along the grain boundaries. Similar results have been obtained by applying ultrasonic energy to the weld pool by using tungsten, molybdenum, or ceramic rods to transfer the energy from the transducer into the pool. A problem occasionally encountered was loss of fragments of the transmission line into the weld metal.

Although it is not primarily a welding technique, a method of heading aluminum rivets developed at the Ohio State University, is worthy of mention here [71]. The sonic riveter consists of a full-wave, high-Q, PZT operating at 10 kHz. A half-wave catenoidal horn is attached to the transducer by means of a threaded stud. The actual riveting tool is loosely coupled with the transducer assembly such that it receives only intermittent impacts from the horn. The rate ranges from 1000 to 3000 impacts/s. By this method, energy taken from the driving unit during impact reduces the amplitude of vibrations. Between impacts, the amplitude returns to maximum. The maximum efficiency of such a system would be obtained when the repetition rate of impacting corresponded to the exact time required for the vibrations of the horn tip to regain their maximum amplitude.

Ultrasonic bonding is used for thin-wire bonding and fabrication of electrical interconnects in flip-chip assemblies [37,38,72]. Although gold has been the preferred choice for the first-level interconnection (IC) material, there has been an increase in the use of copper in the IC packaging industry. In addition, there has been an increase in the use of aluminum in certain electronic fabrications. Thermosonic bonding has been shown to provide good bonding for these, increasingly used, alternate material combinations [73].

12.3.7.2 Welding Thermoplastic Materials

Bonding of plastic materials has become one of the most important commercial applications of high-intensity ultrasonics. Probably unawares, most people in the United States come into contact with ultrasonically welded parts every day [1]. The reasons for its wide acceptance include

1. The wide use of thermoplastic materials in industry for strapping and packaging
2. The ease with which ultrasonics is adapted to bonding and sealing thermoplastic straps and package materials
3. The fact that ultrasonic welding is fast, clean, does not require a skillful operator, and lends itself well to automation.

Ultrasonic equipment for plastics bonding is smaller and requires considerably less power than its counterpart for metals bonding. The process is designed such that losses at the interface are considerably higher than transmission losses through the plastic material and, for this reason, the heat for bonding is generated directly at the bond area. Thus, problems of heating the interface by thermal conduction or of heating the mating surfaces to above the bonding temperature prior to clamping are eliminated by the ultrasonic method. Bonding occurs when the temperature of the materials reaches the tacky level.

The terms "near field" and "far field" are used to describe the distance from the welding tip to the bonding surfaces. Near-field bonding refers to the condition in which the mating surfaces are within a short distance (6.35 mm at 20 kHz) of the welding tip, such as is the case in lap-welding of thin sheets. Absorption in soft plastics and rubber-modified and porous plastics is high. These types of plastics are more suitable for near-field welding [74]. Far-field bonding refers to the condition in which the bond line is at a distance from the welding tip (>6.35 mm at 20 kHz), as might be the case in bonding two hemispheres. Rigid, amorphous plastics are best suited to far-field welding.

Joints are designed to make the interface lossy to promote rapid bonding. In far-field welding, this is usually done by designing energy concentrators (or energy directors) into at least one of the mating surfaces and by designing the parts such that shear stresses are produced between mating surfaces. Energy concentrators may consist of a small ridge running along the center of one of the surfaces. Shear stresses are promoted by forming the parts such that mating surfaces contact at an angle, at least over part of the surfaces to be bonded. The energy is introduced in a longitudinal wave mode, because application of a shear mode (as in welding metals) would lead to surface distortion and, in such cases, transmission of energy would be poor.

In near-field welding, for instance in bonding strapping, it is preferable to move the welding tip parallel to the interface between surfaces to be bonded. The straps are laid over a rigid anvil. The welding tip is activated and brought into brief contact under pressure with the joint. Flexural bar-type welding tools are very practical for this type of application. The bar is rigidly mounted at one end in a mechanism designed to apply a pressure across the bond region on command. The transducer is coupled to the flexural bar at a position between the clamped end and the free end of the bar such that its displacement is normal to the length dimension of the bar and parallel to the mating surfaces (see Figure 12.13). To produce a weld, the tool is energized in flexural resonance in a fixed-free mode before contacting the joint. Pressure is applied as the vibrating tool contacts the joint. As pressure increases and the bond is formed, the mode of vibration in the flexural bar shifts to the fixed–fixed condition and the condition at the mating surfaces changes from sliding to stationary. The period of the entire cycle is on the order of 50–100 ms. The brief period of time wherein energy is applied is followed by a brief dwell period during which excess heat is conducted from the outer surface of the joint area and the bond solidifies to full strength. This technique provides the necessary loss at the interface by shear and, most often, leads to a very thin weld area, little distortion in the material, and strong joints. Other tool designs may be used to adapt to specific needs.

There are certain advantages to ultrasonic bonding of thermoplastics over that of adhesive bonding or thermal methods, in addition, to those mentioned previously. For example, because there is no addition of chemicals, as in the case of adhesive

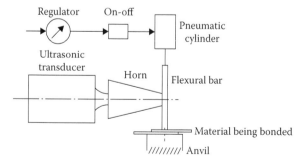

FIGURE 12.13   Flexural-bar ultrasonic bonding of thin plastics.

bonding, materials that must meet Food and Drug Administration (FDA) regulations may be joined ultrasonically. In addition, ultrasonics has an advantage over thermal methods for joining thick materials.

Ultrasonic bonding of thermoplastics is best accomplished when the relative motion of the parts is restricted to avoid chatter between them. The parts should be located in a jig in such a manner that proper alignment can be maintained and the motion of the part opposite the tool is well damped to avoid excessive vibration and loss of control. A typical industrial installation for assembling plastic products is shown in Figure 12.14.

Surfaces prepared for joining ultrasonically should be clean and free of lubricants, plasticizers, and trapped moisture.

In addition to bonding plastic parts, the same equipment has been used for inserting metal parts in plastic parts. A hole slightly smaller than the metal insert is drilled or molded into the plastic part and the metal is driven into the hole ultrasonically. During insertion (which usually requires <1 s), material encapsulates the metal piece and fills flutes, undercuts, threads, and so on.

Ultrasonic staking is another widely used application of the same equipment. Staking is a method of attaching plastics and metals (or other materials) by mechanically locking or enclosing the materials in plastic. A small protrusion is extended through a hole in the second material by a distance which is sufficient to form a suitable lock when staking is completed. The dimensions of protrusions (stud) and the mating hole form a slip fit. The end of the ultrasonic horn is contoured to the geometric design of the completed stake. The staking operation includes applying high ultrasonic amplitude under low clamping pressure to cause the plastic to flow and

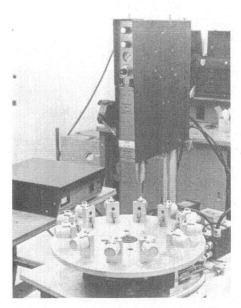

FIGURE 12.14   Assembling thermoplastic products by ultrasonic bonding. (Courtesy of Branson Sonic Power Co.)

form a locking head at temperatures that are typically less than the melting point of the plastic. This process provides tight joints because there is no material spring-back as with cold heading, and it causes only minimal degradation of crystalline materials because the temperatures attained are less than the melting points of the materials.

Kolb [75] lists typical plastics and their weldability by ultrasonic means. In addition, this reference includes recommended joint designs for best bonding. A list of materials and their welding properties can be obtained from manufacturers of ultrasonic bonding equipment. Materials that have excellent ultrasonic bonding properties in both the near field and the far field as well as for staking include polystyrene, methyl methacrylate, methyl pentene, polycarbonate, acrylonitrile–styrene–acrylate (ASA), and a copolymer of styrene and acrylonitrile (SAN). Materials with excellent near-field bonding properties and staking properties as well as good far-field properties include acrylonitrile–butadiene–styrene (ABS), acrylics, polyimide, and polyarylsulfone. Butyrates, nylon, polysulfone, acetal, phenoxy, and structural foam (poly-olefin) display excellent staking qualities, good near-field bonding qualities, and good to fair far-field characteristics. Nylon fibers can be bonded; however, in oriented fibers, ultrasonic bonding produces a disruption of the molecular orientation, thus weakening the material at the joint.

Olefins (polypropylene and polyethylene), particularly filled materials, have fair bonding properties and excellent staking properties. The unfilled materials exhibit poor bonding qualities.

Other materials that have been ultrasonically bonded successfully include cellulose acetate butyrate, cellulose propionate, cellulose acetate, ABS copolymers, polyurethane foam, polyvinyl chloride, cellulose nitrate, cellulose acetobutyrate, polyacetal resin, poly(phenylene oxide), and polyester (Mylar).

The bonding qualities of vinyls range from fair to poor.

Typical applications include splicing magnetic tape and splicing films. Ammonium nitrate fertilizers have been sealed in polyethylene-lined bags by ultrasonic means [76]. In fact, the number of applications of ultrasonic welding is growing so rapidly in the packaging and manufacturing fields that a complete listing of applications is not practical here. Applications include food packaging, packaging of pharmaceuticals, manufacturing toys, and so on.

Shoh [74] tabulates comparisons between applicability and performance features of 10, 20, and 40-kHz ultrasonic plastic welders, showing approximate power available, dimensions of components, typical peak displacements, advantages and disadvantages, and typical applications of each. Welders operating at 10 kHz are large, powerful, and useful for welding large parts requiring hermetic or continuous welds. Their major disadvantages (according to Shoh) are heavier, costlier equipment, damage to small parts through higher amplitudes and high clamping pressures, and, most important, intense audible noise, which requires noise shielding to protect personnel in the work area. The disadvantages, with the exception of noise, are not serious in that such equipment would properly be used only for welding larger parts.

Welding at 20 kHz is considered the best compromise in equipment size, cost, and performance. The fundamental frequency is inaudible to most people. Welding large parts requires the use of multiple units, except in the special case of scan welding— a progressive weld formed by passing large, flat assemblies, such as door panels, beneath an ultrasonic horn on a conveyor belt. However, even scan welding may use more than one welder. This frequency is useful for all typical plastic bonding operations: welding rigid parts, staking, metal-in-plastic insertion, spot welding, and seaming fabric and film.

Small displacement amplitudes make welding at 40 kHz safe for delicate parts and film. This frequency is increasing in popularity with operators, as it is in effect "silent" to human ears. The frequency is however, not good for far field welding, due to the higher attenuation that is typically encountered.

Oxidation reactions can occur during ultrasonic bonding of plastics [77]. Carbonyl groups from peroxides that form during local thermoxidation have been detected in polystyrene and Lucite bonds. Graft and block polymers formed in the seam during ultrasonic bonding, according to these observations.

In related areas, ultrasonics has shown potential for application as a means of reactivating adhesives. In particular, it is effective in promoting wetting of surfaces, including pores, and, in some cases, forming bonds through contaminants. Reactivation of adhesives by ultrasonics is limited to small parts and areas; however, reactivation of adhesives along seams on a continuous basis is possible using flexurally resonant, round, rolling plates to excite the seam areas [1,2].

A review of the current state of the technology and the prospects for improvements for ultrasonic welding of polymers and thermoplastic composites is provided by Nesterenko and Senchenkov [78]. The range of specialized equipment that can be used to provide continuous control for welding thermoplastic materials is reported by Khmelev et al. [79]. Numerical analysis is now being applied to improve the science base for the transducers used in cutting and welding of thermoplastic textiles and to provide better designs, optimized for new applications [80].

12.4 LIQUID ATOMIZATION AND DROPLET FORMATION

Atomization of liquids and control of droplet formation by means of ultrasonics have many practical applications. Two general types or classes of droplet formation are of interest: (1) formation of mists or fogs with some control of droplet-size distribution and (2) formation of droplets in a regular and orderly pattern by controlling the breakup of jets. Formation of mists or fogs has been applied to atomizing medical inhalants, domestic humidifiers, atomizing fuels for efficient combustion, atomizing industrial paints for use with electrostatic painting, dispersing cleansers for effective use in cleaning large tanks, and producing metallic and other types of powders. Controlling the breakup of jets to form droplets of uniform size and in an orderly pattern has found use in certain unique cases in which it is desirable to have the droplets impinge on a surface in controlled patterns. For instance, the droplets can be passed through an electric field and deflected according to a designed pattern, just as a stream of electrons is deflected when passing through the electrostatic or electromagnetic field of a cathode-ray tube. This principle is extensively applied to ink-jet printing, and micro-machined silicon has, at present, been developed to provide transducers for droplet formation and ejection [81,82].

Aerosols have been produced in several ways. Fogs can be produced in a bowl operating within the range 0.4–2 MHz when the focal region is at or near the surface of the liquid. In addition, the ultrasound produces a fountain which is indicative of the intensity. Liquids passed over an ultrasonically vibrating surface are sprayed into droplets according to the frequency and amplitude of vibration. Liquids injected into an active zone of a high-intensity stem-jet whistle are efficiently and rapidly dispersed into droplets. Controlled droplet formation from jets is accomplished by passing the liquid through a small nozzle built into a transducer vibrating in a longitudinal mode at a frequency corresponding to the frequency for maximum instability.

The production of aerosols by ultrasonic means has been attributed to at least two mechanisms: (1) production of capillary waves on the surface of the liquid and (2) cavitation. Droplets are hurled from the crests of the capillary waves and from the walls of cavitation bubbles that break open at the liquid surface. Theories of atomization based upon these mechanisms are applicable only within certain limits. The more general case is complicated, involving various dynamic factors and materials properties. To a degree, particle sizes are functions of the excitation frequency. If the conditions are correct for optimum atomization, the frequency dependence applies over a wide range of frequencies. However, on one hand, if the atomizer is overloaded with liquid, the mean particle size may be much larger than that predicted by the frequency equations that are commonly used. On the other hand, if the fluid supply is below the capacity of the atomizer, the distribution of the particles will shift to the smaller average particle sizes.

In general, cavitation is not a major cause of atomization in industrial systems. It may be a major factor in high-frequency focused nebulizers that are used for atomizing medicines in inhalation therapy. However, in horn-type systems involving atomization from a surface, cavitation can have a deleterious effect on the rate at which mists can be formed.

Various equations for predicting droplet size in an ultrasonically produced aerosol have been proposed. Lang [83] proposed the capillary-wave theory of droplet formation from the surface of an ultrasonic horn and found the number-mean particle diameter, d, of the droplets in the aerosol to be

$$d = 0.34\lambda_c \tag{12.1}$$

where λ_c is the wavelength of the capillary waves (cm) and, according to Rayleigh [84], is given by

$$\lambda_c = \left(\frac{8\pi T}{\rho f^2}\right)^{1/3} \tag{12.2}$$

where

    T is surface tension
    ρ is the density of the liquid
    f is the ultrasonic frequency (Hz)

Lang found good agreement between experimental results obtained with either oil or water and results calculated from Equation 12.2. Lang's equation applies to very low atomization rates.

Lang's work derives from Kelvin's expression for capillary wavelength as a function of forcing frequency and the assumption that the diameter of the liquid drops formed is proportional to that wavelength. Kelvin's theory assumes no coupling between the liquid-surface velocities and the forcing function driving the surface. Thus, parameters such as liquid-film thickness and forcing amplitude do not enter the expression for capillary wavelength. Lang's work is, therefore, restricted to very low flow rates and viscosities and the condition in which film thickness is fairly uniform and large as compared with the amplitude of vibration of the atomizing surface.

Popov and Goncharenko [85] proposed the following equation for predicting the size of droplets formed when using ultrasonic atomizing horns, operating at frequencies within the range 6.8–32 kHz, in spraying materials including water, transformer oil, molten salts, waxes, and other organic materials:

$$d = \left(\frac{C}{\xi} \right) \left(\frac{3Q_\eta T}{\pi D \rho^2 f^2 g \cos \alpha} \right)^{1/3} \tag{12.3}$$

where

d is the average droplet size (μm)
C is a constant varying from 0.15, for $d = 0.1$, to 0.3, for $d = 4.3$
ξ is amplitude of vibration (m)
Q is the volume flow rate of the liquid (m³/s)
η is the viscosity of the liquid (kg·s/m)
D is the nozzle diameter (m)
ρ is density (kg/m³)
g is the acceleration of gravity (m/s²) and
α is the angle between the spray cone and the vertical

Peskin and Raco [86] carried the analysis of atomization from a film covering a solid surface that is oscillating at a known frequency further than the previous work by Lang. Their analysis provides a correlation between drop size, transducer frequency, transducer amplitude, and liquid-film thickness. In addition, they considered viscosity and showed that very low viscosities cause negligible modifications of the theoretical results. Peskin and Raco derived the following equation to describe the time dependence of the amplitude of a two-dimensional perturbation on the surface of a liquid which is ultrasonically excited:

$$\frac{d^2 c}{d t^2} \left[\frac{T}{\rho} k^3 \tanh(kh) - k a_0(t) \tanh(kh) \cos \omega_0 t \right] c = 0 \tag{12.4}$$

where

$c(t)$ is the amplitude of the perturbation
T is the surface tension
ρ is the liquid density
h is the thickness of the liquid film and is related to the flow rate
a_0 is the maximum acceleration of the transducer tip

ω_o is $2\pi f$

k is π/d, the wave number of the perturbation, and is inversely proportional to the expected drop size

d is the drop diameter

Then, by substituting the quantities,

ξ = transducer amplitude

$\tau = \omega_o t$

$\alpha = (TK^3/\omega_o^2\rho) \tanh(kh)$

$\beta = ak \tanh(kh)$

Equation 12.4 reduces to

$$\frac{d^2 c}{d\tau^2} + (\alpha - \beta \cos \tau)c = 0 \tag{12.5}$$

Equation (12.5) is Matthieu's equation. Peskin and Raco investigated unstable solutions of Matthieu's equation to describe droplet formation. They followed an approach similar to that of others by assuming that the process of atomization is one in which the unstable surface perturbations become projections that eventually break off as droplets. For example, similar to Lang and earlier investigators, they assumed that the resultant droplets have a diameter which is of the same order of magnitude as the unstable capillary wave from which they were formed. This assumption leads to

$$d = \frac{\delta 2\pi}{k} \tag{12.6}$$

Under these assumptions, the most likely droplet size generated will result from the most rapidly growing unstable surface waves. From the previous assumption that $k = \pi/d$, it follows that δ is "assumed" to be 1/2. Peskin and Raco obtained a relationship based upon the first unstable solution to Equation 12.5 where droplets are most likely to be generated. The approximate equation is

$$ak \tanh kh = 2.5 - 2T$$

or

$$\frac{\xi_m \pi}{d} \tanh \frac{\pi h}{d} = 2.5 - 2\frac{T\pi^3}{\omega_o^2 d^3\rho} \tanh \frac{\pi h}{d} \tag{12.7}$$

These authors include a set of curves showing the relationship between non-dimensional groups in ultrasonic atomization. These non-dimensional groups are the ratios of (1) droplet size to transducer amplitude, (2) liquid-film thickness to transducer amplitude, and (3) inertial forces to surface tension forces. These curves, plotted as functions of constant film thickness to transducer amplitude (h/a), terminate at the lower frequency end in a curve representing the limiting value of α:

$$\alpha = \left(\frac{Tk^3}{\omega_o^2\rho}\right) \tanh(kh) \tag{12.8}$$

for the linear approximation of the maximum γ curve. The γ curves are curves plotted to represent maximum growth of instability in the first unstable region. The equation for this terminating curve is

$$\frac{d}{\pi a} = \left[\left(\frac{2T}{\rho \omega_o^2 a^3} \right) 2 \tanh \right] \left[\left(\frac{\pi a}{d} \right) \left(\frac{h}{a} \right) \right]^{1/3} \tag{12.9}$$

which, for large h, becomes

$$d = \left(\frac{4\pi^3 T}{\rho \omega_o^2} \right)^{1/3} = \left(\frac{\pi T}{\rho f^2} \right)^{1/3} = \frac{\lambda_c}{2} \tag{12.9a}$$

because tanh δ for large values of θ is 1. This equation is essentially the Kelvin capillary wave theory and represents the limit of Peskin and Raco's theory when the effects of h and a are neglected.

The theory derived by Peskin and Raco probably is as thorough as any that have been developed to date to relate droplet size, transducer frequency, transducer amplitude, and liquid-film thickness in atomization from a plane surface at the end of an ultrasonic horn. It is, however, limited in its application. Much depends upon the horn design, liquid feed-rate method, horn and fluid compatibility or surface wetting characteristics, and fluid feed rate; in addition, it would be difficult to include all factors in a simple mathematical relationship. In any feed system, as the feed rate increases, atomization will pass through three stages:

1. Low flow rate, at which atomization is attributable entirely to ultrasonic forces. Within this stage, the particles will range in size about a mean diameter which might be approximated by the method of Peskin and Raco, with the average size increasing with flow rate.
2. Intermediate flow rate, at which atomization is attributable to both ultrasonic forces and fluid dynamic forces. The particle sizes caused by the fluid dynamics are much larger than those produced by ultrasonic forces.
3. High flow rate, at which atomization is primarily a fluid dynamic phenomenon.

Wettability between the atomizing surface and the fluid is important to the distribution of the fluid over the surface and to the coupling of energy into the fluid. Good wetting is essential to efficient atomization.

An equation for determining the optimum frequency for controlled droplet formation from liquid jets is

$$f = \frac{u_j}{4,508 d_j} \tag{12.10}$$

where u_j is the velocity of the jet and d_j is the diameter of the jet. When a liquid jet is disturbed at moderate intensities and at the frequency given by Equation 12.10, the jet will break up into droplets of uniform size. Occasionally, minor satellite droplets form on the axis of the jet. Uniform droplets are also formed by discrete frequencies within a range such that $3.5d_j < \lambda < 7d_j$. At the optimum frequency, $d = 1.89d_j$ and $\lambda = 2.38d$. Here, λ is the distance the jet moves during one complete cycle of the impressed ultrasound, that is,

$$\lambda = \frac{u_j}{f}$$

Several examples of applications of ultrasonic atomization have been reported. Modified Hartmann whistles have been used to atomize fuels in turbines and combustion engines. Brockington [87] proposed ultrasonic atomization of diesel fuels to improve the efficiency of partial-aspiration engines and eliminate the combustion-delay period by achieving good air-fuel mixing without wetting the manifold walls. He claimed that the ultrasonic system enables the engine to run more economically at high loads, reduces smoke, and ensures quieter operation.

Mebes [88] proposed the use of an ultrasonic Hartmann whistle in a reaction chamber located in the exhaust stream to complete the combustion of unburned constituents. In addition, he suggested the use of catalysts in the reaction chamber, but claimed good efficiencies for the ultrasonically induced reactions even without the catalysts. His recommended frequency range of 60–120 kHz is probably an order of magnitude too high for a Hartmann whistle of practical size.

Charpenet [89] found a definite improvement in efficiency of burning both liquids and solids in an ultrasonic burner consisting of a steel combustion chamber and an external annular siren. Frequencies used were between 5 and 22 kHz. When the acoustic field was turned on, the flame temperature immediately rose by approximately 80°C and the exhaust carbon dioxide (CO_2) increased by approximately 0.8%. The fuels used were fuel oil and pulverized coal.

In commercial liquid-fuel burners with low consumption rates, high-pressure nozzles are used to atomize the fuel. The small orifices are easily plugged with dirt, which either makes them inoperative or lowers their efficiency. In addition, the compressor is usually noisy. Small, fairly inexpensive ultrasonic atomizers are an improvement over such compressor/nozzle combinations in several ways. For example, relatively large-diameter orifices can be used, which are self-cleaning under the influence of ultrasound. In addition, the efficiency of the burner is higher than that of equivalent conventional nozzle-type burners. Small atomizers for use in portable thermoelectric generators using a single transistor oscillator have been developed [90,91]. These atomizers consume approximately 5 W of power under liquid load and approximately 2 W without the load. The unit operates from a 12-V direct current supply. Feedback from the transducer to the oscillator controls the driving frequency and ensures that it is maintained at the optimum frequency of the transducer. Fuel is fed by gravity. The thermal output is approximately 21,000 kJ/h. Similar burners could be used in residential fuel-oil burners. Figure 12.15 shows the burner in operation. These burners have been described by Hazard [90] and Hunter [91].

In a similar application in which the burner is recommended for use in oil-firing installations, Pohlmann and Lierke [92] used the pumping action of the ultrasonic

FIGURE 12.15   Small ultrasonic burner.

transducer to provide oil to the atomizing surface. The oil reservoir is located beneath the transducer and one end of the transducer extends beneath the liquid surface. Throughput is between 0 and 4 L/h, but higher rates are possible by increasing the size of the unit.

Molten metals with melting points up to 700°C have been atomized at frequencies up to 0.8 MHz to produce metal powders [93]. Lead, tin, zinc, bismuth, and cadmium were successfully atomized by using a stepped half-wave-velocity transformer fixed to a magnetostrictive transducer. Aluminum was atomized by using a noncorrosive transformer of composition Ti318A with an aluminum oxide coating. The results appear to confirm the capillary-wave theory of ultrasonic atomization.

Kirsten and Bertilsson [94] claimed that using an ultrasonic atomizer in flame photometry and flame absorption spectrophotometry gives a 10-fold increase in sensitivity compared with pneumatic installations. Solutions of sodium, calcium, and magnesium were used in this application.

Research has continued to investigate the development of improved systems using high-frequency ultrasonics to atomize water. In this context, work has also continued to understand the fundamental theory for atomization, particularly as new applications are emerging. One driver for these developments is nanoparticles, formed from various fluids, which are critical to a wide range of emerging applications. In this context two approaches to atomization are being considered in nanotechnology. One approach uses droplet formation that results from cavitation phenomena [95]. In addition, this phenomena are encountered in the irrigation water applied to the outer surface of phacoemulsification tools and with some other surgical tools [96], and this is discussed in Chapter 14. A second approach uses interface destabilization and atomization driven by a surface acoustic wave (SAW) propagation along a substrate [97,98]. This process involves the propagation and interaction of a SAW (Rayleigh wave) with a drop of fluid in a way that causes leakage of energy into the fluid. The leaky energy causes streaming in and deformation of the fluid and free surface atomization at the free surface of the drop. The process for SAW-atomization is shown in schematic form in Figure 12.16.

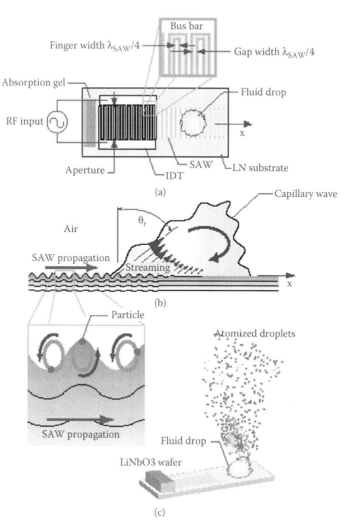

FIGURE 12.16   **(See color insert.)** Schematic representation of process for use of a surface acoustic wave for droplet atomization. (Reprinted with permission from Qi, A., L. Y. Yeo, and J. R. Friend, *Phys Fluids*, 20(7):074103, © 2008, American Institute of Physics.)

12.5   AGGLOMERATION AND FLOCCULATION

Over the years, much has been published on ultrasonically induced transport, agglomeration, and flocculation of both particles and fluids, and, for much of the time, it has remained a laboratory curiosity. There was early interest in ultrasonically assisted agglomeration or coagulation of suspended particles in air, for application to remove smoke and exhaust particles, and for processing a wide range of suspensions. However, in recent years, electrostatic precipitators proved to be more efficient for applications to clean exhaust streams, and commercialization for this application did not develop.

Gallego-Juárez [99] provided a review of new transducer designs that use a stepped-plate transducer technology which, when combined with power amplifiers, are more effective at delivering high-power ultrasonic waves into air. This type of transducer has resulted in researchers re-visiting and re-evaluating the use of ultrasound for several industrial applications, which include fine-particle removal from emulsions, agglomeration, and defoaming. This technology is presently being considered for a number of applications, particularly to meet the needs of the food industry [100].

Further, ultrasound is being used to develop a range of manipulation and processing tools. A driver has developed in the requirements within the nanotechnology community and for various biotechnology applications (which are considered further in Chapter 14), where small particles suspended in a fluid require manipulation, separation, and processing.

12.5.1 Agglomeration

The range of applications that require fine-particle agglomeration is increasing. Riera et al. [101] report the use of a standing wave field that produced agglomeration and precipitation of powders at 20 kHz. Mason [102] discusses the influence of humidity on the ultrasonic agglomeration and precipitation of submicron-sized particles in diesel exhaust. In addition, ultrasonically assisted agglomeration phenomena are, at present, being used in chemistry. For example, such a phenomenon is being used for the rapid synthesis of mesoporous catalysis. Enomoto et al. [103] demonstrated agglomeration of silica spheres from 0.3-μm spheres to 2-μm dense particles through 90 min of sonication at 20 kHz.

12.5.2 Standing Wave Separators

One approach to achieving concentration, separation, and potentially agglomeration for nanotechnology and process-industry applications is through the application of an ultrasonic standing wave field. As with other emerging applications of ultrasonic separators that can be used to agglomerate particles, they can be used to hold or to manipulate them in a fluid flow. Such fields have been used on the macro-scale in process systems (such as that reported by Gallego-Juárez [99]).

The fundamental factors for ultrasonic separation of particles using standing wave fields are well summarized in three papers by Groschl and coworkers [104–106]. These papers discuss the fundamental physics, design of separators, and some specific applications. The frequencies used go up into the low MHz range. Work by Townsend et al. [107] discusses the use of computational fluid dynamics (CFD) software and a MATLAB® model to investigate particle paths through an ultrasonic standing wave. Studies by Mandralis and Feke [108] discuss issues pertaining to the fractionation of suspensions using synchronized ultrasonic fields and flows. At present, applications routinely include separation of cells, micro-, and nanoparticles (e.g., Coakley et al. [109], Benes et al. [110], and Harris et al. [111]).

When considering separation-cell design, inadequate analysis of both the fluid dynamics and ultrasonics can easily induce turbulent flow and ineffective separation [114]. For a cell with ultrasonic illumination from one side, as shown in schematic

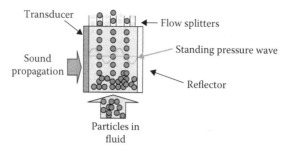

FIGURE 12.17 **(See color insert.)** Schematic representation of an ultrasonic flow-through standing wave separator. (Courtesy of Pacific Northwest National Laboratory.)

form in Figure 12.17, to achieve a standing wave at the operating frequency, one critical condition is that the transducer and the reflector should be parallel. In addition, for a given flow rate, the cell must be physically long enough to ensure a dwell time that permits the desired degree of separation to be achieved.

In all standing wave systems, there are various parameters and forces that need to be balanced. In addition, to be effective and if laminar flow is to be achieved in a cross-flow system, there are limits that need to be set for both the applied upper and lower ultrasonic power levels. The separator geometry, density contrast between fluid and suspended particles, and the interplay between thermal conduction and acoustic forces/streaming are all important and require consideration. For effective separation, it is necessary to balance the various time constants, the separation time, and the (flow) residence time, for the cell (and transducer used). In analysis, and in design balance, it is necessary to consider forces on the fluid and particles: gravitational, convective, acoustic, shear, particle repulsion and/or attraction, and diffusion.

In addition to flow cells, various particle-manipulation and agglomeration units have been devised. For example, Kozuka et al [113] have reported a three-dimensional (3D) acoustic manipulator that used four crossed fields. Using the same concepts, a diverse range of novel studies have been performed. Bauerecker and Neidhart [114] report levitation of ice particles as they form from an aerosol in a stationary ultrasonic field. Saito and Imanishi [115] report the arrangement of particles to form a two-dimensional (2D) array, which is held in place as the polymeric matrix is solidified into a composite.

12.5.3 FLOCCULATION

Flocculation precedes coalescence and droplet interactions in an emulsion. This is a phenomenon that is of interest in some chemical processes and is of particular importance in ensuring stability in some food products, including those requiring a longer shelf life [100]. The processes of creaming, in dense material combinations, are complex. In addition, it is found that ultrasound not only aids in the processing (at high powers) but also is of interest for use as a sensor (at low powers) for many of these optically opaque materials [116].

Dickinson and coworkers [117] discuss the creaming and flocculation of oil-in-water emulsions containing sodium caseinate. The interaction phenomena are shown to be complex, with a low-shear-rate apparent viscosity and a substantial shear-thinning behavior. This approach has now attracted particular interest in the food industry for more than a decade.

12.6   DRYING AND DEWATERING

12.6.1   ACOUSTICAL DRYING

Acoustic energy enhances drying. Although it is an expensive method, it does have a practical use in drying heat-sensitive materials, particularly those that degrade in a moist environment, and chemicals that have a long drying cycle. Production of pharmaceutical materials is one area that has been proposed wherein acoustic drying could be a benefit.

Although the effect appears to be related primarily to external mechanisms involving gas turbulence above the liquid interface and low-temperature boiling off of liquid vapors during the low-pressure phase of each cycle, there is some argument for internal effects as well, which involve enhancement of moisture migration to the surface of the material.

To date, air-jet types of whistles have been the source of energy in most experiments with acoustic drying. Boucher [118] enumerated the optimum conditions for acoustic drying as (1) intensities greater than 145 dB, (2) layers less than 2.5–5 cm thick to be dried, and (3) frequencies within the range 6–10 kHz. Obviously, the operator must be protected against the sound energy at these intensities and frequencies. The safest protection is to acoustically isolate the sound from the surrounding environment by locating the sound source in acoustically treated chambers.

Earlier studies of acoustic drying included a diverse range of materials, including coal [119]. An assessment of the state of the art in the mid-1980s was provided by Muralidhara and coworkers [120]. However, since then, there have been significant advances in the science and technology of high-power transducers, particularly for those that provide operation into air [121].

The application of high-power ultrasound to dehydrate porous materials, in particular, food materials, has proved to be very effective and is seeing an increase in the range of applications.

Such applications have included processing wood and an increasing range of foods that include: orange crystals, grated cheese, gelatin beads, rice grains [100], onions [122], carrot cubes, and lemon-peel cylinders [123]. One of the more novel applications has been the enhanced drying of garments in a drum-type washing machine [124].

12.6.2   ELECTROACOUSTIC DEWATERING

Although the exotic application of sonic energy to drying is interesting, it is not practical for fulfilling the general large-scale needs of industry. However, methods currently under development that combine ultrasonic or sonic energy with other forms of energy are showing much promise for dewatering sludges and other large-volume products on an industrial scale. One of the most interesting processes combines electrical and acoustical phenomena synergistically to remove water from a product without producing a phase change in the water [125,126]. In other words, the water is removed by forces produced electrically and acoustically without incurring the expense of boiling the water from the product.

In most cases, removing water from a product is a two-stage process: dewatering and drying. Dewatering is illustrated by wringing water from clothes after they have been laundered. Drying is the familiar process that follows the dewatering (wringing) operation. A significant amount of the energy for drying is consumed in supplying the heat of vaporization, which is approximately 2,321 kJ per kilogram of water evaporated under standard atmospheric conditions. Any process that can remove a large percentage of the water from a product will reduce the thermal load on the dryer accordingly. If the dewatering process costs less than the drying process, the overall result might be a financial savings in total costs.

Dewatering can be considered as a "postfiltration" process. Its effectiveness depends upon the content of the types of water associated with solid particles. These types of water are identified as

- Bulk or free water
- Micropore water
- Colloidally bound water
- Chemisorbed water

Bulk or free water is water present on or intermingled with the product but not bound chemically or mechanically to it. Conventional solid–liquid separation equipment, such as filters and centrifuges, can separate the bulk water to a limited extent.

Micropore water is water located in micropores and capillaries of the product. Filter presses remove some of the water present in the micropores.

Colloidally bound water is held by strong surface forces to the particles, particularly those of colloidal size, to which it is attached.

Chemisorbed water is bound chemically onto the surface layer of an absorbent.

Colloidally bound and chemisorbed water cannot be released by filters, centrifuges, filter presses, or similar mechanical devices.

Systems requiring dewatering may be placed into one of two categories that are related to the nature of the water to be removed and the state of the material to be dried. These major categories are (1) low-moisture-content systems in which the space between solids, such as particles and fibers, to be dried contains considerable air and (2) high-moisture-content systems or continuous-phase systems in which all potentially void space is completely filled with moisture and the systems have the characteristics of a liquid or a solid. In the first category, the water may be present in three forms: free, entrapped, or bound. The free liquid appears in a thin film on the surface of the material to be dried. The material may have the appearance of a dry product. In a second category, the liquid may also appear in the free, entrapped, or bound state, but the amount of free water is appreciably greater than that of the first category in that it does fill the available volume of the product.

It is important to understand these two conditions because the effective dewatering mechanisms differ for both conditions. In addition, to design an optimum system by combining forms of energy, it is important to know mechanisms associated with each form of energy to be applied to the system.

Researchers at Battelle's Columbus Laboratories have developed some very promising techniques for dewatering and materials separation by combining acoustical and electrical phenomena, both with and without vacuum [122,127,128]. The acoustical mechanisms may be categorized into groups of (1) mechanisms that play a major direct role in the separation process and (2) mechanisms that assist the electrical forces (electrokinetic phenomena) in their role in the separation process.

Acoustical mechanisms most closely associated with dewatering that are associated with ultrasonics in the high-moisture-content systems, and areas where the acoustical energy is expected to have a major role in dewatering include cavitation, radiation forces, and, in some combinations of media, possible localized, electrical charge generation appearing on the surfaces of particles through mechanical movement of one of the phases. Cavitation is not necessary to generate localized charges. Cavitation does produce free chemical radicals, which may increase the rate at which electrophoresis is able to separate the solid materials from the water. In general, *any* large-volume process requiring cavitation throughout the volume is not a practical application for ultrasonics for several reasons but, primarily, because cavitation is energy intensive and interferes with wave propagation, making it difficult to treat a large volume uniformly. Radiation pressure at intensity levels less than that required for cavitation can help move particles unidirectionally away from an acoustic source.

The same mechanisms are active in low-moisture-content systems but not necessarily to the same intensities. In addition, the mechanisms described in the previous section may also play a part, that is, low-temperature boiling off or accelerated evaporation during the rarefaction phase of each cycle and strong turbulence in the gas phase of the atmosphere surrounding the particles.

Among acoustical phenomena that assist the electrokinetic forces are localized cleaning of electrodes to maintain good electrical contact with the media to be separated, acceleration of moisture transfer and removal through filters or screens, and ultrasonically assisted compaction of solids as moisture is removed to maintain more sustained electrical continuity through a cake and increase the amount of fluid removal.

Two electrokinetic principles are utilized in dewatering: electrophoresis and electro-osmosis. Electrophoresis refers to the movement of solid particles through a relatively stationary liquid under the influence of an electric field. Electro-osmosis refers to the movement of a fluid through a relatively stationary permeable membrane or porous cake under the influence of an electric field. The direction in which the particle moves depends upon the electrical charge on the particle. The rate at which it moves is a function of the strength of the electrical field. Thus, in a broad sense, electrophoresis is used in a high-moisture-content system to move particles in any desired direction, particularly away from filter media to prevent or delay clogging the filter until the bulk of the free water has been removed through the filter. Electro-osmosis is used to move water through a porous cake.

By combining ultrasonics and electrokinetic principles, either with or without vacuum, it is possible to obtain synergistic effects. As with any other high-intensity application of ultrasonic energy, best results are obtained by considering several

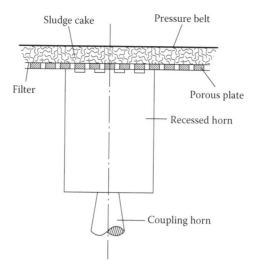

FIGURE 12.18   Belt-press dewaterer driven with ultrasonic device.

factors related to the effects on the products, the mechanisms that offer the greatest benefit to the process, and the nature of the product as it is being treated. Reviewing these factors with the various principles of electrokinetics and ultrasonics may reveal when and how to apply ultrasound to the process. Important factors include not only the mechanisms that promote dewatering but also other mechanisms, such as flocculation (particularly of fibrous materials) and degassing, and whether these might have beneficial or deleterious effects on the product.

Figure 12.18 schematically shows a plan using ultrasonic energy with a belt-press dewaterer. Vacuum may be applied on the side of the ultrasonic horn and an electric field can be applied across the sludge cake to increase the total amount of water removed from the cake by a significant amount. Probably not all of the acoustic mechanisms have been recognized, but it is possible to identify the following:

- Greater compaction of the cake through the influence of ultrasonic stresses across the cake. This maintains electrical continuity across the cake for a greater period of time and increases the amount of water that can be removed by electro-osmosis.
- Ultrasonic removal of materials that tend to coat the electrode surfaces, thus promoting better electrical contact with the product.
- Ultrasonic removal of liquid droplets from the bottom of the filter by inertial forces.
- Ultrasonic assistance in transferring moisture through the filter.

Thus, the local activity of the ultrasound, its effect on the cake, which does not require excessively high intensities, and the electric field, the vacuum, and the applied pressure in operating on the bulk of the cake cooperate to provide a very effective dewatering system.

Interest continues to develop in both acoustic and electroacoustic methods of dewatering and drying [127]. Recent advances in developing high-power ultrasound technology are making such applications more attractive. It is being considered for use in food dehydration [121] and remains of interest in the treatment of biosludge [128,129].

As with many areas of application, the availability of reliable, as well as lower power, and, for some applications, battery-powered ultrasound units has significantly increased the diversity in the range of application. The following section discusses some examples of agricultural applications and pest control.

12.7 AGRICULTURAL APPLICATIONS

12.7.1 TOMATO POLLINATION

Tomato plants grown commercially in greenhouses are pollinated by hand. Workers pass daily along the rows of blossoming plants, contacting each cluster of blossoms with a low-frequency vibrator. Contact is made daily to ensure that the blossoms are pollinated at the proper time, that is, at maturity.

Tomatoes may be pollinated sonically or ultrasonically without contact [130]. When a mass of fine dust particles is subjected to ultrasonic energy, the particles tend to disperse into a cloud. In pollinating tomatoes, the ultrasound produces a cloud of pollen. The cloud is uniformly dispersed within the blossom, resulting in uniform pollination. The sound energy does not rupture the sacs but merely scatters the available pollen within the blossoms. In fact, intensities that are high enough to rupture the sacs prematurely probably would damage the plant.

Sound intensities greater than 145 dB are necessary for good pollination. Because moisture affects the adhesive character of the pollen particles, humidity affects the ability to raise a dust cloud of pollen with a corresponding effect on the pollinating ability of ultrasound. This has been proved by experiments conducted by DeTar et al. [131], who showed that very high humidity increases the intensity required to cause pollen to fall and increases the number of blooms that fail to respond.

Research on tomato pollination was conducted at Battelle using modified Hartmann-type whistles operating at 10 and 15 kHz. The work of DeTar and coworkers was conducted by initially using a 20-W tweeter speaker with a cone mounted to concentrate the energy at the bloom. An industrial 40-W driver mounted with a 2.54×17.8-cm tube to carry the sound to the bloom proved to be more effective.

These approaches are still receiving some attention. They are presently being considered in a number of plant cross-breeding and gene-transfer studies as well as to ensure more effective airborne pollination.

12.7.2 GERMINATION OF SEEDS

Germination of seeds may be enhanced ultrasonically. The mechanisms involved are heat, transfer of moisture into the seed, and, in some cases, rupture of the seed coat. These mechanisms are variously active depending upon the seed. There is an optimum range of intensities for effective germination which also depends upon the delicacy of the seed. Very high intensities will damage many seeds, whereas low intensities will have no effect.

The application of ultrasound to seed germination might be considered relatively inexpensive. A 4-L cleaning tank could process several kiograms of seed per hour.

Wild rice, which grows in the lakes and marshes of northern Canada, is an annual crop and its seeds lie dormant during the winter. Freshly harvested seeds will not germinate until after they have been stored in nearly freezing water for up to 6 months. These seeds, however, can be artificially stimulated to germinate. Halstead and Vicario [132] ultrasonically treated wild rice which had been stored for 90 days underwater at 1–3°C. At the time of treatment, the seeds were dormant. The seeds were treated at 70 kHz for 10 min in water at 50–70°C, conditions that these experiments found to be optimum. Under these conditions, 74% germination occurred. Longer exposure times resulted in fewer germinations. The method of germination did not affect the growth. The effectiveness was attributed to cavitation, which changes the permeability of the seed coat and, perhaps, alters the response of the embryo to its environment. There was marked temperature dependence.

Further, high-power ultrasound is now being used to develop mutations in seeds. Encheva et al. [133] report a study where ultrasound was used to induce a mutant sunflower line through mutagenesis. Humara et al. [134] report that, in their treatment of *pinus pinea*, they used ultrasound to deliver a *uidA* gene that had been grown in yeast extract peptone (YEP) medium at pH 7. Although the viable plants that resulted from this study were minimal, this technology is being investigated further and applied in various forms.

12.8   PEST CONTROL

The use of sonic techniques of controlling pests, including birds, rats, and insects, has often been proposed. Possible methods include (1) use of lethal intensities, (2) use of intensities or frequencies that prove to be annoying to the pests, (3) duplicating the sounds of distress calls, and (4) duplicating the sounds of natural enemies.

Lethal intensities are impractical for several reasons. One is the need to entice the pest into a localized area where it can be exposed effectively. Another is the possible danger of annoyance to other individuals within the vicinity of the sound generator, particularly at audible frequencies.

The possibility of using annoying sounds to rid an area of pests, with the exception of birds, has not received much consideration. One possible source of annoyance in insects might be resonance of antennae or other projections on the body that would produce discomfort.

Sometimes, duplicating distress signals has proved to be effective for a short period of time. However, the animals often seem to develop an ability to distinguish between artificial or recorded calls and real distress signals, and the method loses its effectiveness. Busnel et al. [135] found that distress signals from a species in one locality may bring no response from the same species in another locality.

A report by the South Carolina Agricultural Research Service [136] claims that the bollworm (*Heliothis zea*) responds to simulated bat calls in the same manner as it does to real bat cries. When insects, such as moths, hear the cries of bats, they immediately take evasive action.

Frings and Frings [137,138] studied insect and animal sounds extensively from the viewpoint of pest control. Their conclusions are that their results allow them only to make suggestions for possible acoustical methods of pest control.

One method of repelling birds from industrial premises is sold commercially. This method utilizes emission of ultrasonic waves at 20 kHz at an intensity reported to be extremely annoying to birds such as pigeons, starlings, and sparrows. The birds vacate the protected premises to get away from the "intolerable" sounds.

With the reduction in cost of ultrasound technology, an increasing range of ultrasonic devices have appeared on the consumer market. Units are sold to remove termites, rats, and a range of other insects and rodents. Some animals, such as dogs, bats, and rodents, can hear well into the ultrasonic range and some insects, such as grasshoppers and locusts, can detect frequencies ranging from 50 to 100 kHz. Moths and lacewings can detect ultrasound waves with frequencies as high as 240 kHz that are produced by insect-hunting bats. However, the effectiveness of ultrasound for pest control is still being debated.

Extensive studies at the Kansas State University have shown that sound devices do have a repellent effect and cause a reduction in mating and reproduction of various insects. However, the results reported were mixed and ultrasonic sound had little or no effect on some pests. Various ultrasonic devices were highly effective on crickets, whereas the same devices had little or no repellent effect on cockroaches [139]. In addition, the results were mixed, with some devices being reported as effective whereas others had no effect, and this appeared to depend on the test subject. Further, the study concluded that there was no effect on ants or spiders in any of the tests. In some cases, creatures initially responded to electronic pest-control devices by moving about a bit more than usual, but did not appear overly eager to escape from the sound waves. This includes devices that emit uniform frequency as well as those that change the frequencies of ultrasound waves used. Rodents appeared to adjust to the ultrasound (or any new sound) and eventually ignore it. However, researchers reported that they were able to use the increased cockroach activity to good effect by increasing the rate at which they caught the roaches in sticky traps. At best, ultrasonic waves were reported to have only a partial or temporary effect on rodents. Several studies have actually rejected ultrasonic sound as providing a practical means of rodent control. This area of ultrasound application, does, however, remain an area of ongoing investigation.

12.9 CONTROL OF FOAMS

High-intensity ultrasonic energy in air can rapidly dissipate foams produced during various manufacturing processes [99]. Bubbles are well matched acoustically to the surrounding atmosphere. The stresses developed in a high-intensity field rupture the bubbles and release the internal gas. Only a small part of the bubble must be stressed to rupture. In a standing wave field, the high-stress regions appear periodically and most of the bubbles will experience stresses of rupturing levels at some position on their surface.

Air-jet whistles have been used for defoaming purposes. They are usually located in a cylindrical chamber above the area where the reaction occurs. They may be located in branches of a pipe and directed upstream. The intensities of the sound must exceed 145 dB. The frequency most often used is within the range 6–15 kHz. Acoustic treatment or isolation is essential to prevent exposure to damaging levels of intensity. Such treatment is not particularly difficult.

Recent advances in stepped-transducers to generate higher powers in air are being considered for defoaming, particularly in the food industry [99–101]. The wider topic of foaming, its consequences, prevention, and destruction, with particular application to bioprocesses, is discussed in detail by Vardar-Sukan [140].

12.10   COATING MATERIALS AND PARTICLES

Ultrasonics has been used effectively to enhance uniform coating of materials and to produce improved impregnation of porous or fibrous materials with resins, paints, enamels, and similar materials. Individual particles can be coated without agglomeration in air by means of high-intensity whistles. A stream of thin-resin-carrying particles to be coated is directed into the active zone but outside the main airstream of the whistle. The two constituents may be introduced separately and the particles will be coated individually. They must then be maintained in suspension until the solvent evaporates to avoid sticking to one another or to the walls of the container on contact.

Coating of fine particles can be done in liquid suspensions. The materials may be dispersed ultrasonically by placing a high-intensity probe in a beaker containing the materials to be coated and the coating resin. Sometimes, particles are coated effectively in a three-phase system in which the materials to be coated and the coating material are mixed in a liquid suspension at a temperature greater than the melting point of the coating material. The application of ultrasonics continues as the temperature drops below the melting point. The particles are then generally, uniformly, and individually coated.

Coating of glass fibers with resins or rubber compounds can be improved under the influence of ultrasound [141]. The irradiation removes the air from between the fibers and permits complete penetration of the resin between the fibers. To demonstrate the impregnating effects produced ultrasonically, bundles of glass fibers were passed through a solution, in parts per volume, of resorcinol 2.03, 37% formaldehyde (HCHO) 2.63, 10% sodium hydroxide (NaOH) 2.00, a 41% butadiene-styrene-vinyl-pyridine terpolymer latex 9.77, a 39% butadiene-styrene copolymer rubber 10.25, a 60% natural rubber latex 13.35, and water 59.97. After treatment in the solution at 25 kHz, the bundle was dried and showed complete impregnation and coating. Without ultrasonics, the bundle was not completely coated.

The emergence of nanotechnology has led to an interest in the engineering of particle surfaces for use in both catalysis and in drug delivery [142]. Ultrasound has become a critical technology that is used for both emulsification and generation of nano-sized dispersion, emulsions, and coatings [143,144].

12.11   PREPARATION OF CARBON SPHERES

Wettig and Doerre [145] claimed to form round, highly sorbent carbon particles by ultrasonically irradiating, for 20 min, 30 mg of carbon suspended in 14 ml benzene to which 0.8 mg pyrene in 1 ml benzene was added. Following the ultrasonic treatment, the suspension was mixed for 50 min. The process was said to produce carbon spheres that were 100 nm in diameter.

A study by Pol and coworkers [146] has used sonochemical deposition of air-stable iron, in the form of ~10 nm nanoparticles, on to carbon spheres. This approach is seeing an increasingly diverse range of applications as nanotechnology processes are developed, and it is being used on different material substrates of various shapes [147].

12.12   GLASSWARE TESTING

A glassware tester has been developed that is capable of detecting flaws in glass that are otherwise hard to detect, such as fire checks, abrasions, and cracks (see Figure 12.19). The glassware is moved along a conveyor belt between an air nozzle and a velocity transformer that is ultrasonically resonated. The air blast impacts the ware against the velocity transformer at a predetermined and controlled rate. If the glass contains a defect, it ruptures, with the extent of the rupture depending upon the impact force of the glass against the transformer. Inspection rates are at least 180 U/min.

FIGURE 12.19   Ultrasonic glassware tester. (Courtesy of Blackstone Corp.)

12.13   DISPERSIONS AND DE-AGGLOMERATION

When particulate matter in a liquid suspension or two immiscible liquids are subjected to ultrasonic stresses at intensities greater than the cavitation level, the particles are accelerated omnidirectionally through the liquid. This phenomenon has been applied in producing uniform dispersions of various kinds including many industrial applications.

A range of applications of de-agglomeration of nanopowders in different suspensions are now frequently carried out using ultrasound. Aggregates are broken apart by the action of local, high-velocity liquid jets (up to 100 m/s) and pressure gradients up to 20 GPa/cm [148].

Ding and Pacek [149] have reported an investigation of de-aggregation of goethite nanopowders. They reported, among other observations, that the solid concentrations were capable of being increased by up to 20 weight-percentage. Particle size can be significantly reduced using ultrasound. Li et al. [150] reported a rapid process to prepare tin nanoparticles using ultrasound. Markovic et al. [151] have reported particle-size reduction from micro to nano-sized particles in a barium titanate powder.

12.13.1   DYES AND PIGMENTS

Liquid whistles may be used to disperse pigments in enamel preparations. Best results are obtained with low-viscosity fluids. Kotlyarskii et al. [152] showed good results with chrome, helio blue, zinc white, Hansa yellow, and azure pigments. Lifshits [153] conducted comparative studies of the influence of quartz transducers, magnetostrictive transducers, liquid whistles, and mechanical mixers on the production of aqueous dispersions of dyes. The dispersions were evaluated on the basis of settling rates. In every case, ultrasonic dispersion was superior to mechanical dispersion. Lifshits recommended the liquid whistle for producing suspension dyes and for homogenizing dyes that form solutions. He claimed that ultrasonically dispersed dyes give high-quality color with low dye consumption.

The use of ultrasound for textile dyeing was reported in many papers in the 1950s and 1960s. It is a technology that is, again, receiving increased attention. Vajnhandl and Le Marechal [154] review the fundamentals and its applications, including for both dyeing and treatment of wastes from dyeing processes.

12.13.2   PREPARATION OF SPECIMENS FOR STUDY UNDER ELECTRON MICROSCOPES

Ultrasonic energy is ideally suited for preparing the evenly dispersed deposits necessary for study under the electron microscope. Aleinikov [155] used the method to prepare silicate specimens 200–400Å thick. Very fine glass powders in a 5% ethanol suspension were treated ultrasonically for 15–30 min, after which they were ready for mounting on the copper screen.

Belyaeva and Tuchkova [156] claimed that dispersion of particles is the most difficult operation in preparing certain mineralogical materials for study under the electron microscope. They stated that ultrasonic dispersion of particles appears to be more effective than any other method.

Jacke and Kolb [157] recommended the use of ultrasonics not only for dispersing the specimen in the solvent but also for applying the specimen to the grid of the microscope. The latter is accomplished by placing a drop of the dispersed mixture on the tip of a stepped mechanical amplifier and placing the microscope grid near the tip as the ultrasonic energy is turned on. The ultrasonic energy produces a fog of the dispersion, which settles on the grid to provide the necessary deposit.

The growing interest in nanotechnology and sonochemical catalysis has resulted in the increased use of electron microscopy as a tool for nanoparticle characterization. The materials of interest, including magnetic fluids, involve particles that have been prepared using ultrasound [158,159].

12.13.3 PREPARATION OF SOIL SAMPLES FOR ANALYSIS

Oxidizing agents, acids, and peptizing agents are used conventionally to promote dispersions of soil samples for analysis. Some of the soil chemicals, such as ammonium, may be lost during production of dispersions by the chemical method. Edwards and Bremner [160] have shown that ultrasonics produces more stable suspensions and produces them faster than does the conventional method. In addition, the ultrasonic method causes no significant change in pH and electrical conductivity and no dissolution of organic or inorganic material. The time required to ultrasonically disperse various types of soils varies with the soil. The major problem is to decide how long to treat the specimen. Treatment past the duration required for complete dispersion may reduce the size of primary particles, which leads to analytical errors. For example, sand-sized minerals may be reduced to clay-sized particles by excessive treatment.

Transducers equipped with stepped mechanical amplifiers are recommended for this type of soil analysis, although any system capable of producing the necessary high intensity should be satisfactory.

In addition to preparation of soils for analysis, it is found that ultrasound can assist in the extraction of materials, such as heavy metals, from both soils and sediments [161].

12.13.4 DISPERSION OF CLAY SUSPENSIONS

The liquid whistle has been used to disperse dry-clay-body scrap during the manufacture of ceramic products [162]. The whistle does the work of conventional mixing equipment at production rates, that is, 190 L/min. For this application, tungsten carbide parts are used, where this is practical. The housing is lined with rubber to resist abrasion.

Ultrasonic treatment of clay suspensions in water at cavitational intensities has an immediate effect of dispersing the clay particles uniformly throughout the water, thus increasing the rheological quantities [163–173]. These quantities reach an optimum value within a few minutes, after which the particles coagulate. The coagulated particles are so stable that further ultrasonic treatment does not destroy them. Kruglitskii et al. [170] obtained mechanical, X-ray, and electron microscopic data on clays before and after treatment. They concluded that ultrasonic energy destroys the original structure of the aqueous clay dispersion, promotes an even distribution of water coating, disperses the clay mineral particles, and causes the formation of new mineral aggregates with higher strength.

The influence of ultrasonic energy on the properties of the suspension can be varied by adding reagents prior to ultrasonic treatment. The addition of colloids, such as carboxymethylcellulose and NaCl or calcium chloride ($CaCl_2$), to the clay prior to treatment enhances the thixotropic properties and the stability of the dispersions [165,166,169].

Peptizing electrolytes and stabilizing chemicals produce various effects, depending upon the amount of chemicals added [171–173]. Large amounts of stabilizers apparently protect or shield the particles from the effects of the ultrasonic energy. Moderate amounts with ultrasonics produce beneficial effects on the mechanical properties.

Variations from low to high concentrations of peptizing electrolytes, such as NaOH, produce corresponding effects of peptization, spontaneous dispersion, and coagulation. Therefore, the addition of peptizing electrolytes may result in an improvement in the properties of the dispersion or may produce an adverse effect, depending upon the quantity used.

12.13.5 DISPERSION OF CHLORINATED PESTICIDES AND OTHER SOLUTIONS

Riggs and Biggar [174] produced ultrasonically aqueous suspensions of p,p'-dichlorodiphenylytrichloroethane (DDT), aldrin, dieldrin, and endrin that greatly exceed the theoretical solubilities of these materials in water. No emulsifiers were required and the suspensions remained stable for several days after 10–15 min of ultrasonic irradiation at cavitation levels.

Producing concentrations of solutions that exceed the theoretical solubilities is not unusual with ultrasonics. Uniform suspensions in large volumes of liquid may be produced more effectively mechanically by first ultrasonically preparing a concentrated dispersion of the material and diluting to the desired concentration. For example, in the production of a nondrying paint used to mark defects located by nondestructive testing, zinc oxide was dispersed through mineral oil. Mechanical mixing produced poor dispersions when the zinc oxide was added to the bulk liquid in the proper amount. A concentrated paste of the oxide in mineral oil produced ultrasonically was easily and uniformly dispersed through larger volumes of oil by mechanical stirring.

A 5-min treatment of calcium fluoride (CaF_2) in water at intensities greater than the cavitation level produces a dispersion of the fluoride which is four times the maximum concentration possible by other known means. The phenomenon has a possible application in water fluoridation. Calcium fluoride is a safe chemical to handle, in contrast to other chemicals used for water fluoridation. Its chief drawback is that it requires several days for a desired quantity to dissolve unaided in still water; therefore, storage is a problem in supplying the normal daily consumption of water in any city using CaF_2. Using ultrasonics, water is easily fluorinated with CaF_2 to 28 times the recommended concentration. Therefore, it may be possible to ultrasonically fluorinate water in a side branch through which 1 of every 26 L flows. The treated water could be mixed with the main stream of water, where it is dispersed, and stored for use.

12.13.6   Emulsification of Flotation Agents

Emulsions for flotation purposes have been prepared ultrasonically [175–178]. It is claimed that these preparations are superior to mechanically mixed materials and that they result in savings of 25%–50% in the amounts of reagents used. Ultrasonic treatment is now extensively used in mineral processing [179]. Applications have included studies of coal floatability, recovery of natural resources [180], and a range of waste-treatment processes, including those for materials recovery.

12.13.7   Dispersion of Sodium in Hydrocarbons

Fine-particle dispersions of sodium in hydrocarbons can be prepared ultrasonically [181]. By choosing a saturated hydrocarbon with a boiling point well above the melting point of the sodium, a homogeneous dispersion can be produced with the sodium in a liquid state. High-boiling light petroleum (boiling point [b.p.] 100–120°C) and yellow petroleum jelly are suitable dispersing media.

These dispersions can be produced with a stepped-horn or other suitable transmission line attached to either a piezoelectric or magnetostrictive driver. The material is treated in a cylindrical Pyrex cell. The transmission line is sealed into the end of the cell by a neoprene ring.

12.13.8   Dispersion of Heterogeneous Phases in Molten Metals

The use of ultrasound to disperse various materials in molten metals has been considered by several investigators. It has not been used in actual production, partly because materials used in the ultrasonic apparatus would have difficulty surviving the environments.

Seemann and Staats [182] dispersed various insoluble materials in molten aluminum at 740°C and 800°C using a sintered alumina rod to transmit ultrasonic energy into the melt. Materials, including iron and titanium, were dispersed rapidly.

Both titanium and titanium carbide caused pronounced grain refinement in the aluminum. The addition of 1% aluminum oxide doubled the high-temperature strength.

REFERENCES

1. Shoh, A. 1975. *IEEE Trans Sonics Ultrason* 22(2):60–71.
2. Shoh, A. 1988. Industrial applications of ultrasound. In *Ultrasound*, ed. K. S. Suslick, 97–122. New York: VCH Press.
3. Leighton, T. G. 1994. *The Acoustic Bubble*. San Diego, CA: Academic Press.
4. Ultrasonic Manufacturers Association, Inc. 1964. *Recommended Standard Cavitation Activity Measuring Procedure*. New York: New Rochelle.

5. Ensminger, D. 1966. Recommended Standard Cavitation Activity Measuring Procedures. Paper presented at the 5th Annual Technical Meeting and Exhibit of the American Association for Contamination Control. Houston, Texas, March 29–April 1.

6. Jacke, S. E. 1965. Cavitation Activity Measurements. Paper delivered at the *IEEE Ultrasonic Symposium*, abstract in program supplement, Boston, MA, Dec. 1–4. New York: IEEE, Sonics and Ultrasonics Group.

7. Ferris, J. A. 1962. Methods for evaluating sonic cleaning systems. In *Proceedings of Symposium on cleaning and materials processing for electronics and space applications.* Am Soc Testing and Materials (ASTM) STP # 342. Paper # STP-444485S, 8pp. West Conshohocken, PA: ASTM.

8. Hodnett, M., and B. Zeqiri. 1997. *Ultrason Sonochem* 4(4):273–88.

9. Zeqiri, B., P. N. Gelat, M. Hodnett, and N. D. Lee. 2003. *IEEE Trans UFFC* 50(10):1342–50.

10. Zeqiri, B., N. D. Lee, M. Hodnett, and P. N. Gelat. 2003. *IEEE Trans UFFC* 50(10):1351–62.

11. Hodnett, M., M. J. Choi, and B. Zeqiri. 2007. *Ultrason Sonochem* 14(1):29–40.

12. Hodnett, M., and B. Zeqiri. 2008. *IEEE Trans UFFC* 55(8):1809–22.

13. Newton, W. W., S. W. Ziegler, I.-S. Rhee, K. Bhansali, S. Whalen, D. Piscorik, and J. Menke. 2000. *Armed Services Test Protocol for Alternate Cleaner Performance Validation.* Report of the U.S. Army Aberdeen Test Center (ATC-8256).

14. Farrer, J. O. 1948. *British Pat* 602:801.

15. Balamuth, L. 1952. U.S. Methods and means for removing material from a solid body. Pat. 2, 580,716 (January 1, 1952).

16. Sherrit, S., S. A. Askins, M. Gradziol, B. P. Dolgin, X. Bao, Z. Chang, and Y. Bar-Cohen. 2002. Paper presented at the SPIE Smart Structures Conference. Vol. 4701, Paper 34.

17. Bao, X., Y. Bar-Cohen, Z. Chang, B. P. Dolgin, S. Sherrit, D. S. Pal, S. Du, and T. Peterson. 2003. *IEEE Trans UFFC* 50(9):1147–60.

18. Langenecker, B. 1969. Paper B-2 delivered at the IEEE Ultrasonics Symposium. St. Louis, Missouri, September, 24–6.

19. Balamuth, L. 1965. SAE Paper No. 650762 presented at the National Aeronautic and Space Engineering and Manufacturing Meeting. Los Angeles, October 4–8.

20. Langenecker, B. 1966. *IEEE Trans Son Ultrason* 13(1):1–8.

21. Lehfeldt, E. 1969. *VDI Z* 111(6):359–63.

22. Pozen, N. L., V. N. Semirog-Orlik, and I. A. Troyan. 1969. *Fiz Khim Mekh Mater Akad Nauk Ukr SSR* 5(1):112–3.

23. Izumi, O., K. Oyama, and Y. Suzuki. 1966. *Trans Jpn Inst Met* 7(3):162–7.

24. Kralik, G., and B. Weiss. 1967. *Z Metallkd* 58(7):471–5.

25. Langenecker, B. 1966. Method for strengthening metals. U.S. Pat. 3, 276, 918 (October 4, 1966).

26. Pines, B. Y., and I. F. Omel'yanenko. 1969. *Fiz Met Metalloved* 28(1):110–4.

27. Chachin, V. N., and A. L. Skripnichenko. 1968. *Izv. Akad. Nauk. Beloruss SSR (Fiz.-Tekh)* 3:43–6.

28. Schmid, E. 1968. *Proc Int Conf Strength Met Alloys* 1967(9):798–804.

29. Winsper, C. E., and D. H. Sansome. *J Inst Met* 97(9):274–80.

30. Pohlman, R., and E. Lehfeldt. 1966. *Ultrasonics* 4(4):178–85.

31. Kralik, G. 1965. *Acta Phys Austriaca* 20(1–4):370–5.

32. Severdenko, V. P., and V. V. Clubovich. 1963. *Dokl Akad Nauk SSSR* 7(2):95.

33. Lorant, M. 1966. *Tooling* 20(12):51.

34. Maropis, N., and J. C. Clement. 1966. ASM Technical Report No. C6–26.3. American Society for Metals, Metals Park, Ohio.

35. Anon. 1966. *Steel* 159(20):38.

36. Anon. 1968. *Machinery* 74(9):88–9.

37. Jones, J. B. 1968. *Met Prog* 93(5):103–7.
38. Dohmen, H. G. 1966. *Aluminum* 42(9):559–66.
39. Balamuth, L. 1966. *Ultrasonics* 4(3):125–30.
40. Cunningham, J. W., and R. J. Lanyi. 1965. Study of the Feasibility of Applying Ultrasonic Energy to the Rolling Process for Sheet Metals. Final Report on Contract No. w64 029J, submitted on March 15.
41. Langenecker, B., and O. Vodep. 1975. Metal plasticity in macrosonic fields. *Proc Ultrasonics Int*, ed. Z. Novak, London, March 24–26, 202–5. Guildford, UK: PIC.
42. Tan, Q., W. Zhang, B. Schaible, L. J. Bond, T.-H. Ju, and Y.-C. Lee. 1999. *IEEE Trans Adv Packag* 22(3):468–75.
43. Tan, Q., W. Zhang, B. Schaible, L. J. Bond, T.-H. Ju, and Y.-C. Lee. 1998. *IEEE Trans Compon Packag Manuf Technol B* 21(1):53–8.
44. Roark, R. J., and W. C. Young. 1975. *Formulas for Stress and Strain*. 5th ed. New York: McGraw-Hill.
45. Puskar, A. 1982. *The Use of High-Intensity Ultrasonics*. Amsterdam: Elsevier.
46. Owen, W. J., and W. J. Padgett. 2000. *IEEE Trans Reliab* 49(2):224–9.
47. Bathias, C. 1999. *Fatigue Fract Eng Mater Struct* 22(7):559–65.
48. Wu, T. Y., G. Jago, J. Bechet, C. Bathuas, and G. Guichard. 1996. *Eng Fract Mech* 54(6):891–6.
49. Yeo, S. H., B. B. K. A. Ngoi, and L. Y. Chua. 1997. *Int J Adv Manuf Technol* 13(5):333–41.
50. Lehfeldt, E., and R. Pohlman. 1968. *Planseeber Pulvermetall* 16(4):263–76.
51. Pokryshev, V. R., and V. I. Marchenko. 1969. *Poroshk Metall Akad Nauk Ukr SSR* (2):30–3.
52. Ya. Pines, B., I. F. Omel'yanenko, and A. F. Sirenko. 1967. *Poroshk Metall* (8):106–11.
53. Hoffman, R. 1967. Methods of casting different materials layers. U.S. Pat. 3, 338, 294 (August 29, 1967).
54. Gross, L., R. Hoffman, and R. Jefson. 1969. *Ultrasonics* 7(4):245–8.
55. Kostin, L. G., L. T. Buchek, V. M. Shkil', N. N. Sumaneev, M. E. Grenader, and V. A. Starkov. 1975. *Poroshk Metall* 14(9):26–9.
56. Khasanov, O. L., J. S. Lee, Yu. P. Pokholkov, V. M. Sokolov, E. S. Dvilis, M. S. Kim, and B. G. An. 1999. The use of the ultrasonic compaction method for the PZT piezoelectric ceramic fabrication. In *Proc, KORUS '99 3rd Russian–Korean Int. Symp. On Science and Technology*, Organized by Novosibirsk State Tech. Univ, Univ of Ulsan and Tomsk Polytec. Univ., 2:557–60. (Available from IEEE Xplore.)
57. Jones, J. B., and J. G. Thomas. 1954. Ultrasonic Soldering of Aluminium. Report # DP-94. E.I. Du Pont de Nemours and Co.
58. Inaba, M., K. Yamakawa, and N. Iwase. 1990. *IEEE Trans Compon Hybrids Manuf Technol* 13(1):119–23.
59. Faridi, H. R., J. H. Devletian, and H. U. E. P. Le. 2000. *Weld J* 79(9):41–5.
60. Oelschlagel, D., H. Abe, K. Yamaji, and Y. Yonezawa. 1977. *Weld J* 56(4):21–7.
61. Hulst, A. P. 1972. *Ultrasonics* 10(6):256–61.
62. Weare, N. E., J. N. Antonevich, and R. E. Monroe. 1960. *Weld J Res Suppl.*, (August 1960 issue), 39:331s–41s.
63. Ginzburg, S. K., A. M. Mitskevich, and Yu. G. Nosov. 1967. *Svar Proizvod* 5:45–7.
64. Anon. 1964. Aeroprojects, Inc., Research Report No. 65–13, December 1964.
65. Jones, J. B., and C. F. DePrisco. 1967. Welds. U.S. Pat. 3, 319, 984 (May 16, 1967).
66. Bruk, M. V. 1967. *Svar Proizvod* 10:10–1.
67. Shin, S., and H. T. Gencsoy. 1968. *Weld J* 47(9):398s–403s.
68. Varga, I. 1966. *Femip Kut Intez Kozl* 8:173–86.
69. Pfaelzer, P. F., and J. Frisch. 1967. Ultrasonic Welding of Metals in Vacuum. Final Report on Contract No. W-7405-eng-48, University of California, Berkeley, College of Engineering, December 1967.

70. Russo, V. L. 1958. *Sudostroyeniye* 4:37–41.
71. Libby, C. C. 1968. *IEEE Trans Son Ultrason* SU-15(1):57.
72. Schwizer, J., W. H. Song, M. Mayer, O. Brand, and H. Baltes. 2003. Packaging test chip for flip-chip and wire bonding process characterization. In *Proceedings of the 12th Int. Conf. Solid State Sensors*, Actuators and Microsystems, Boston, June 8–12, 2003, IEEE, Paper PE10.P, Transducers (1):440–3. New York: IEEE.
73. Srikanth, N., J. Premkumar, M. Sivakumar, Y. M. Wong, and C. J. Vath. 2007. Effect of wire purity on copper wire bonding. In *Proc. 9th Electronics Packaging Technology Conference,* Singapore, Dec. 10–12, *IEEE*, 755–9. New York: IEEE.
74. Shoh, A. 1976. *Ultrasonics* 14(5):209–17.
75. Kolb, D. J. 1967. *Mach Des* 39(7):180–5.
76. Anon. 1964. *Mod Packag* 37(11):18–9.
77. Akutin, M. S., V. A. Sorokin, and V. S. Osipchik. 1967. *Plast Massy* 12:16–8.
78. Nesterenko, N. P., and I. K. Senchenkov. 2003. *Weld Int* 17(3):232–8.
79. Khmelev, V. N., A. N. Silvin, R. V. Barsukov, S. N. Tsyganok II. Savin, A. V. Shalunov, S. V. Levin, and A. D. Abramov. 2006. Development of the new principle of batching energy at ultrasonic welding and creation of the equipment for connection of thermoplastic materials. In *Proc. 7th Annual International Workshop and Tutorials on Electronic Devices and Materials,* Erlagol, Italy *1–5th July, IEEE*, 280–8. New York: IEEE.
80. Batista de Silva, J., N. N. Franceschetti, and J. C. Adamowski. 2006. Numerical analysis off a high power piezorelectric transducer used in the cutting and welding of thermoplastic textiles. *ABCM Symp Ser Mech* 2:124–49.
81. Perçin, G., and B. T. Khuri-Yakun. 2002. *Rev Sci Instrum* 73(5):2193–6.
82. Meacham, J. M., M. J. Varady, F. L. Degertekin, and A. G. Fedorov. 2005. *Phys Fluids* 17(10) Paper # 100605.
83. Lang, R. J. 1962. *J Acoust Soc Am* 34(1):6–8.
84. Lord Rayleigh. 1945. *The Theory of Sound*. Vol. II, 343ff. New York: Dover.
85. Popov, V. F., and G. K. Goncharenko. 1965. *Izv Vyssh Uchebn Zaved Khim Khim Tekhnol* 8(2):331–7.
86. Peskin, R. L., and R. J. Raco. 1963. *J Acoust Soc Am* 35(9):1378–81.
87. Brockington, P. A. C. 1966. *Commer Motor* 30(12):33.
88. Mebes, B. 1970. Swiss Pat. 484, 359 (February 27, 1970).
89. Charpenet, M. L. 1964. *Riv Combust* 18(12):522–32.
90. Hazard, H. R. 1966. *Proc. Annu. Power Sources Conf* 20:152–5.
91. Hunter, H. H. 1969. *Ultrasonics* 7(1):63–4.
92. Pohlmann, R., and E. G. Lierke. 1966. *VDI Z* 108(34):1669–74.
93. Lierke, E. G., and G. Griesshammer. 1967. *Ultrasonics* 5:224–8.
94. Kirsten, W. J., and G. O. B. Bertilsson. 1966. *Anal Chem* 38(4):648.
95. Barreras, F., H. Amaveda, and A. Lozano. 2002. *Exp Fluids* 33(3):405–13.
96. Bond, L. J., M. Flake, B. Tucker, K. Judd, and M. Boukhyn. 2003. Physics of Phacoemulsification. In *Proceedings, 5th World Congress on Ultrasound*, WCU 2003 (Paris, France, September, 2003), French Acoustical Society, 169–72.
97. Qi, A., L. Y. Yeo, and J. R. Friend. 2008. *Phys Fluids* 20(7):074103.
98. Friend, J. R., L. Y. Yeo, D. R. Arlfin, and A. Mechler. 2008. *Nanotechnology* 19(14):145301.
99. Gallego-Jurrez, J. G. 1994. New technologies in high-power ultrasonic industrial applications. In *Proc. 1994 Ultrasonics Symposium* (IEEE), eds. M. Levy, S. C. Schneider, and B. R. McAvoy, Cannes, France, Nov. 1–4, 1343–52. New York: IEEE.
100. Povey, M. J. W., and T. J. Mason. 1998. *Ultrasound in Food Processing*. London: Blackie Academic.
101. Riera, E., J. A. Gallego-Juárez, and T. J. Mason. 2006. *Ultrasonics-Sonochemistry* 13(2):107–16.

102. Mason, T. J. 2007. *Prog Biophys Mol Biol* 93(1–3):166–75.
103. Enomoto, N., S. Maruyama, and Z. Nakagawa. 1997. *J Mater Res* 12(5):1410–5.
104. Groschl, M. 1998. *Acoustica* 84(3):423–47.
105. Groschl, M. 1998. *Acoustica* 84(4):632–42.
106. Groschl, M., W. Burger, B. Handl, O. Doblhoff-Dier, T. Gaida, and C. Schmatz. 1998. *Acoustica* 84(5):815–22.
107. Townsend, R. J., M. Hill, N. R. Harris, and N. M. White. 2004. *Ultrasonics* 42(1–9):319–24.
108. Mandralis, Z. I., and D. L. Feke. 1993. *AIChE J* 39(2):197–206.
109. Coakley, W. T., J. J. Hawkes, M. A. Sobanski, C. M. Cousins, and J. Spengler. 2000. *Ultrasonics* 38(1–8):638–41.
110. Benes, E., M. Groschl, H. Nowotny, T. Keijzer, H. Bohm, S. Radel, L. Gheradini, J. J. Hawkes, R. Konig, and Ch. Delouvroy. 2001. Ultrasonic separation of suspended particles. *IEEE Ultrason Symp*, eds. D. E. Yuhas, and S. C. Schneider, Atlanta, GA, Oct. 7–10, 649–59.
111. Harris, N. R., M. Hill, S. Beeby, Y. Shen, N. M. White, J. J. Hawkes, and W. T. Coakley. 2003. *Sensors Actuators B* 95(1–3):425–34.
112. Stenkamp, S. V., L. J. Bond, W. E. TeGrotenhuis, J. W. Grate, and M. D. Flake. 2002. Standing wave separator. In *Proc. IEEE-NANO-2002*, Washington, DC, Aug. 26–28, 453–5. New York: IEEE.
113. Kozuka, T., T. Tuziuti, H. Mitome, F. Arai, and T. Fukuda. 2000. Three-dimensional acoustic manipulation using four ultrasonic transducers. In *Proc. 2000 Int. Symp. Micromechatronics and hunan science*, Nagoya, Japan, Oct. 22–25, 201–6. New York: IEEE.
114. Bauerecker, S., and B. Neidhart. 1998. *J Chem Phys* 109(10):3709–12.
115. Saito, M., and Y. Imanishi. 2000. *J Mater Sci* 35(10):2373–7.
116. McClements, D. J. 1994. *Colloids Surf A Physicochem Eng Asp* 90(1):25–35.
117. Dickinson, E., M. Golding, M. J. W. Povey. 1997. *J Colloid Interface Sci* 185(2):515–29.
118. Boucher, R. M. B. 1959. *Ultrason News* 3(2):8–9, 14–6.
119. Fairbanks, H. V., and R. E. Cline. 1967. *IEEE Trans Sonics Ultrasonics* SU-14(4):175–7.
120. Muralidhara, H. S., D. Ensminger, and A. Putnam. 1985. *Drying Technol* 3(4):529–66.
121. Gallego-Juarez, J. G., G. Rodriguez-Corral, C. J. Galvez Monaleda, and T. S. Yang. 1999. *Drying Technol* 17(3):597–608.
122. Da-Mota, V. M., and E. Palau. 1999. *Drying Technol* 17(4):855–67.
123. Garcia-Perez, J. V., J. A. Carcel, S. De la Fuente-Blanco, and E. Riera-Franco de Sarabia. 2006. *Ultrasonics* 44:e539–43.
124. Leonov, G. V., V. N. Khmelev, I. I. Savin, R. V. Barsukov, S. N. Tsyganok, A. N. Zaborovsky, and M. V. Khmelev. 2005. Acoustic dying of garments in drum-type washing machines. In *Proc. 6th Int. Siberian workshop and tutorial*, EDM'2005, Earlagol, Russia, July 1–5, Earlagol, 106–11. New York: IEEE. Papers available from IEEEXplore.
125. Muralidhara, H. S., N. Senapati, D. Ensminger, and S. P. Chauhan. *Filtr Sep* 351–3 (November/December, 1986).
126. Muralidhara, H. S., N. Senapati, and R. B. Beard. 1986. In *Advances in Solid-Liquid Separation*, ed. H. S. Muralidhara, 335–74. Columbus, Richland, Washington: Battelle Press.
127. Ensminger, D. 1988. *Drying Technol* 6(3):473–99.
128. Yin, X., P. Han, X. LU, and Y. Wang. 2004. *Ultrason Sonochem* 11(6):337–48.
129. Dewil, R., J. Baeyens, and R. Goutvrind. 2006. *Environ Prog* 25(2):121–8.
130. Battelle Memorial Institute News Release, December 5, 1963. Columbus, OH: BMI.
131. DeTar, W. R., C. G. Haugh, and J. F. Hamilton. 1967. Acoustically Forced Vibration of Greenhouse Tomato Blossoms to Induce Pollination. Paper presented at the 1967 Winter Meeting, American Society of Agricultural Engineers, Detroit, December 12–15, 1967.
132. Halstead, E. H., and B. T. Vicario. 1969. *Can J Bot* 47(10):1638–40.

133. Encheva, J., P. Shindrova, and E. Penchev. 2008. *HELIA* 31(48):61–72.
134. Humara, J. M., M. Lopez, and R. J. Oedas. 1999. *Plant Cell Rep* 19(1):51–8.
135. Busnel, R. G., J. Giban, Ph. Gramet, H. Frings, M. Frings, and J. Jumber. 1956. Paper HA6 presented at the Second ICA (International Congress on Acoustics) Congress, Boston.
136. Anon., "Electronic Bat Sounds Drive Insects," news release from South Carolina Agricultural Research Service, Clemson, South Carolina.
137. Frings, H., and M. Frings. 1962. *Sound* 1(6):13–20.
138. Frings, H., and M. Frings. 1963. *Sound* 2(1):39–45.
139. Huang, F., and B. Subramanyam. 2006. *Insect Sci* 13(1):61–6.
140. Vardar-Sukan, F. 1998. *Biotechnol Adv* 16(5/6):913–48.
141. Iwami, I., T. Matsunaga, and K. Yoneyama. 1970. *West German Pat* 1, 934, 349 (February 19, 1970).
142. Caruso, F. 2001. *Adv Mater* 13(1):11–22.
143. Behrend, O., and H. Schubert. 2001. *Ultrason Sonochem* 8:217–76.
144. Hielscher, T. 2006. Ultrasonic production of nano-size dispersions and emulsions: In *Proc European Nano Systems (ENS'05)*, Paris, France, Dec. 14–16, 138–43. Paper at http://hdl.handle.net/2042/6282.
145. Wettig, K., and W. Doerre. 1969. *East German Pat* 68:307.
146. Pol, V. G., M. Motiei, A. Gedanken, J. Calderon-Moreno, and Y. Mastai. 2003. *Chem Mater* 15(6):1378–84.
147. Pol, V. G., H. Grisaru, and A. Gedanken. 2005. *Langmuri* 21(8):3635–40.
148. Aoki, M., T. A. Ring, and S. J. Haggerty. 1987. *Adv Ceram Mater* 2(3A):209–12.
149. Ding, P., and A. W. Pacek. 2008. *Powder Technol* 187(1):1–10.
150. Li, Z., X. Tao, Y. Cheng, Z. Wu, Z. Zhang, and H. Dang. 2007. *Ultrason Sonochem* 14(1):89–92.
151. Markovic, S., M. Mitric, G. Starcevic, and D. Uskokovic. 2008. *Ultrason Sonochem* 15(1):16–20.
152. Kotlyarskii, L. B., A. V. Zelenaya, Z. M. Kishinevskaya, and A. G. Zelenyi. 1963. *Lakokras Mater Ikh Primen* 5:51–7.
153. Lifshits, A. G. 1961. *Nauchno-Issled Tr Tsentr Nauchno-Issled Inst Khlopchatobum Prom* 25:43–7.
154. Vajnhandl, S., and A. M. Le Marechal. 2005. *Dyes Pigm* 65(2):89–101.
155. Aleinikov, F. K. 1966. *Zavod Lab* 32(2):215.
156. Belyaeva, I. D., and G. A. Tuchkova. 1965. *Eksp Metod Issled Rudn Mineralov Akad Nauk SSSR Inst Mineral Geokhim Kristallokhim Redkikh Elementov* 99–113.
157. Jacke, S. E., and D. J. Kolb. 1961. *Ultrason News* 14–17 (Winter 1961).
158. Shafi, K. V. P. M., S. Wizel, T. Prozorov, and A. Gedanken. 1998. *Thin Solid Films* 318(1–2):38–41.
159. Suslick, K. S., and G. J. Price. 1999. *Annu Rev Mater Sci* 29:295–326.
160. Edwards, A. P., and J. M. Bremner. 1967. *J Soil Sci* 18:47–63.
161. Collasiol, A., D. Pozebon, and S. M. Maia. 2004. *Anal Chim Acta* 518(1–2):157–64.
162. Walker, H. N., and W. W. McCarthy. 1967. *Am Ceram Soc Bull* 46(2):188–90.
163. Kruglitskii, N. N., and V. V. Simurov. 1964. *Ukr Khim Zh* 30(8):823–30.
164. Kruglitskii, N. N., V. V. Simurov, F. D. Ovcharenko, S. P. Nichiporenko, and A. B. Ostrovskaya. 1964. *Ukr Khim Zh* 30(12):1283–9.
165. Kruglitskii, N. N., V. V. Simurov, F. D. Ovcharenko, and S. P. Nichiporenko. 1965. *Ukr Khim Zh* 31(7):669–73.
166. Kruglitskii, N. N., V. V. Simurov, F. D. Ovcharenko, and S. P. Nichiporenko. 1965. *Ukr Khim Zh* 31(7):674–8.
167. Kruglitskii, N. N., V. V. Simurov, F. D. Ovcharenko, and S. P. Nichiporenko. 1964. *Dokl Akad Nauk SSSR* 159(6):1367–70.

168. Kruglitskii, N. N., V. V. Simurov, V. V. Minchenko, F. D. Ovcharenko, and S. P. Nichiporenko. 1965. *Dokl Akad Nauk SSSR* 162(4):861–4.
169. Kruglitskii, N. N., V. V. Simurov, F. D. Ovcharenko, and S. P. Nichiporenko. 1965. *Dopov Akad Nauk Ukr RSR* 10:1336–9.
170. Kruglitskii, N. N., V. V. Simurov, F. D. Ovcharenko, and S. P. Nichiporenko. 1966. *Fiz Khim Mekh Dispernykh Struktur Akad Nauk SSSR Sb Statei* 151–7.
171. Kruglitskii, N. N., V. V. Simurov, F. D. Ovcharenko, S. P. Nichiporenko, S. V. Barshchevskaya, and E. V. Dzhus. 1966. *Ukr Khim Zh* 32(9):971–8.
172. Ovcharenko, F. D., N. N. Kruglitskii, V. V. Simurov, S. P. Nichiporenko, and S. V. Barshchevskaya. 1966. *Dopov Akad Nauk Ukr RSR* 3:367–9.
173. Kruglitskii, N. N., and V. V. Simurov. 1967. *Khim Prom Ukr* 1:1–4.
174. Riggs, R. L., and J. W. Biggar. 1965. *Proc Soil Sci Soc Am* 29(5):629.
175. Sokolov, M. A., V. G. Varmalov, and V. M. Ermolaev. 1966. *Gorn Zh* 12:51–3.
176. Emel'yanov, D. S., N. I. Kamenev, and I. V. Trushlevich. 1963. *Peredovoi Opyt Stroit Eksp Shakht* 2:16–25.
177. Fedotov, A. M. 1966. USSR Pat. 1,453,098 (September 16, 1966).
178. Ghiani, M. 1964. *Ind Mineraria* 15:509–15.
179. Ozkan, S. G., and H. Z. Kuyumcu. 2007. *Ultrason Sonochem* 14(5):639–45.
180. Zhou, Z. A., Z. Xu, J. A. Finch, J. H. Masliyah, and R. S. Chow. 2009. *Miner Eng* 22(5):419–33.
181. Pratt, M. W. T., R. Helsby, and H. M. Stanier. 1967. *Chem Soc (London) Spec Publ* 22:284–9.
182. Seemann, H. J., and H. Staats. 1968. *Z Metallkd* 59(5):347–56.

13 Applications of Ultrasonics Based on Chemical Effects—Sonochemistry

13.1 INTRODUCTION

The applications discussed in this chapter are those in which the chemical effects are of primary interest. The chemical effects are never divorced from the mechanical activity and often both factors are important.

The influence of ultrasonic energy on chemical activity may involve any or all of the following: production of heat, promotion of mixing, promotion of intimate contact between materials, dispersion of contaminated layers of chemicals, and production of free chemical radicals. Production of free chemical radicals results from the high temperatures and stresses associated with cavitation.

This chapter is brief. Its brevity does not reflect the extent of the influence of chemical activity in applications of ultrasonics. However, the influence of chemical activity in many applications, such as those discussed in Chapter 12, may be secondary.

13.2 SONOCHEMISTRY

The history of cavitation and knowledge of its destructive power dates to the 1870s when damage to ship propellers became problematic [1]. Its use for sonochemistry reaches back to the early part of the last century. A review by Weissler [2] covers this early period, citing over 150 papers from 1926 to 1950. Reviews by Henglein [3] and Suslick [4] cover the interim period to the late 1980s. Sonochemistry and its use has increased significantly over the past 20 years.

This chapter, Chapter 11, and a multitude of references review many fascinating chemical effects that can be promoted or produced by intense ultrasonic irradiation. Accelerated chemical effects were observed early in the history of ultrasonics [5–8]. Quoting from Richards and Loomis [5], dimethyl sulfate was hydrolyzed in basic solution in the presence of an indicator, in this case "bromthymol blue." The solution contained only enough base to neutralize a quarter of the sulfuric acid liberated by hydrolysis, and consequently, when neutralization had taken place, a sharp change in hydrogen-ion concentration (roughly from pH 8 to pH 3) occurred in a very few seconds, and the indicator was almost instantaneously changed in color. This reaction is especially satisfactory for the work in hand because it is easily adjustable to a wide range of reaction times, because it is not affected by oxygen in the air, and because the gradual inevitable heating due to sound waves caused the end point only to become more sharp. The most satisfactory concentrations of constituents were obtained by adding 2 cm^3 of dimethyl sulfate to 100 cm^3 of 0.01 N sodium hydroxide strongly colored with neutral "bromthymol blue." Under these conditions, the end point of the reaction was reached in 335 seconds at 23.5°C, 155 seconds at 30.2°C, and 82 seconds at 35.3°C, the end point of a single determination being clear after 3 seconds at the lower temperature and 1 second at the higher.

In order to determine the effect of raying, the reaction was begun in a flask and as quickly as possible transferred to two scrupulously cleaned test tubes, one of these being subjected to radiation at zero time. The other, used as a control, was simultaneously dipped into a bath of water at about 50°C and held there, vigorously stirred, for a measured length of time just sufficient to bring its temperature above that of the rayed sample after 15 seconds of radiation. After radiation had been discontinued, the next 15 seconds were occupied in ascertaining the temperature of both portions of the solution with compared thermometers. Raying was then resumed for 15 seconds with one sample while the control was again warmed to follow the new rise in temperature. The process was continued until the blue color of the indicator became suddenly yellow in both, when the total time for each was recorded. By practice, it became possible to follow with the control the temperature of the rayed sample within a few tenths of a degree, keeping the former always slightly higher in temperature to make temperature errors adverse to the effect desired.

In the years following the work of Richards, Loomis, and Wood, intense interest in chemical effects caused by ultrasonic energy has occurred sporadically. Chemical compositions can be prepared under the action of ultrasound that heretofore were impractical. The primary mechanisms of reaction appear to have been identified. Again, it is interesting to review the conclusions of Richards and Loomis [5] after observing the effects of applying ultrasonic energy to the reduction of potassium iodate by sulfurous acid—the "familiar iodine clock." Their evaluation of the factors leading to their results is as follows:

Several aspects of this reaction (referring to a thermal test), in contradistinction to that preceding it, strongly suggest that a factor much more powerful than the mere deviation of its temperature coefficient from linearity is involved. In the first place, the temperature coefficient of the reaction is abnormally low, a change from 25°C to 35°C decreasing the time by only about 25%. Furthermore, the lack of variation of the accelerating effect with intensity of radiation is exceedingly puzzling if the higher temperature of the condensed region alone is the activating agent. Finally, the excessively large effect upon the reaction velocity in comparison with that of the hydrolysis of dimethyl sulfate, and its greater reproducibility, at once suggest a difference in action in the two cases. As a tentative explanation for these phenomena, it does not seem unreasonable to suppose that the frequency of the sound waves themselves had a disintegrating effect upon some compound found in the complex chain of reactions, which led to the ultimate liberation of iodine. In order to test this hypothesis, Reaction 12 was carried out at a higher frequency, no quartz crystal of lower frequency being readily available. The results show that if the frequency of vibration is a determining factor, it is a "band" rather than a "line" effect, but this does not in the least invalidate the hypothesis. A systematic potentiometric investigation into the stability of the various compounds involved in this reaction and of other unstable molecules under sound radiation of various frequencies will at once be undertaken. In this field, as well as in several others which we can only suggest in so cursory a survey of the chemical phenomena produced by high-frequency sound waves, we hope to obtain further information in the near future.

These early investigators mention cavitation and note that the "excessively large effect" attributable to ultrasonic irradiation cannot be due to heat effects. Frequency, according to the previous quotation, was not a significant factor. By the early 1950s, when there appeared a resurgence of interest in the effects of high-intensity applications of ultrasonic energy, cavitation had become recognized as the primary cause of ultrasonic effects in chemical solutions [9,2]. Weissler [2], in discussing results of an extensive survey, stated, "At the present time [1953] it is not known which aspect of cavitation is of fundamental importance in producing chemical reactions." He then mentioned the various effects associated with cavitation, emphasized by earlier investigators. Some investigators emphasized the local temperature rise of several hundred degrees. Others considered the local increase in pressure (of possibly thousands of atmospheres). Weissler attributed the estimates of the magnitudes of increases in pressure and temperature at cavitation sites to Lord Rayleigh [10]. He also mentioned production of free chemical radicals, electrical discharges and luminescence, and bubble resonances. The nature of dissolved gases has a profound influence on the course of cavitation. Weissler recalled that without dissolved gases and suspended particles, negative pressures on the order of 100 atmospheres are required to produce cavitation in water. He stated that considerable evidence exists in support of the generalization that the frequency of the sound wave has little or no effect on the chemical reaction rate. This also agrees with the earlier observations of Richards and Loomis.

Production of free chemical radicals at and near cavitation bubble walls, as mentioned in Chapter 12, is a result of cavitation, not a separate phenomenon. This phenomenon has been recognized by researchers in ultrasonics from Richards and Loomis to the present time. For example, it is the basis of the chlorine release method of evaluating cavitation in ultrasonic cleaning tanks. It also explains many of the chemical reactions, particularly those which result in new chemical formulations.

As the use of ultrasound to effect changes in chemical systems has increased greatly over the past two decades, the number of papers describing theory and practical applications has burgeoned. In 1994, this field was conferred its own journal, *Ultrasonics Sonochemistry* and an increasing number of papers on sonochemistry topics were also published in a range of journals.

Sonochemistry in a liquid generally arises when acoustic power is sufficiently intense to disrupt the intermolecular bonds, causing cavitation. In many cases, originating at seed nuclei, cavitation bubbles grow and collapse nonlinearly during the rarefaction and compression phases of the acoustic waves. The rapid collapse of a bubble subjects the vapor contents and liquid/gas interface to temperatures and pressures on the order of 5000 K and 100 MPa (1000 bar), respectively. These intense conditions, in individual bubbles, last for nanoseconds in an otherwise relatively cool liquid. It is during this period and in localized interactions occur which form sonochemistry, either from direct exposure to the extreme conditions or from reactions with chemical by-products produced within the bubbles. To obtain significant yield, in most cases multibubble cavitation is employed.

Bubble collapse near heterogeneities (e.g., particles or some classes of catalysis) can also involve mechanical effects. A jet of fluid can shooting across the bubble's interior striking a solid with tremendous, and potentially destructive, force. For suspended particulates, the collapse can agglomerate smaller particles or break polymer bonds. Often, the violent collapse of cavitation generates picosecond flashes of light by a much debated mechanism. This light, known a sonoluminescence, can contribute to interactions, and it has also been used as a probe to monitor the interior conditions of cavitation phenomena.

Sonochemistry remained relatively unknown and little used, until the 1980s, due in part to limitations with available equipment. As the situation regarding equipment changed, the field of sonochemistry began to grow. The first international conference on sonochemistry was held at Warwick University, England, in 1986. Shortly thereafter, several books became available. One of the most active early sonochemists, Mason, wrote or edited several books [11–14] in the late 1980s and early 1990s. Mason and Cordemans [15] provide a review of sonochemistry application in chemical and processing industries through 1995.

The fundamental physical-chemistry concepts behind ultrasonic cavitation and sonochemistry have become increasingly well understood and documented. Reviews have been provided by Leighton [16], Lauterborn and Ohl [17], and Crum [18]. A book by Brennen [19] covers these topics more completely. Barber et al. [20] review the physics of a single bubble collapse and discuss many aspects of sonoluminescence. Multibubble cavitation phenomena and sonochemistry theory have been discussed by both Suslick [21] and Margulis [22], who have presented somewhat different theories and interpretations of phenomena. To consolidate the then state of knowledge, a series of papers were compiled into a single volume in 1999 and these review both acoustic cavitation and sonoluminescence [23]. A further good review and assessment of sonochemistry, including listing of classes of reactions, as well as a discussion of the types of equipment employed, are given by Thompson and Doraiswamy [24].

Several key areas are discussed in the recent reviews and books, including medical applications, pharmaceuticals, biotechnology, chemical synthesis, polymer technology, minerals and powders, and metallurgy. An increasing number of papers have now looked at aspects of equipment used in sonochemistry, including both calibration and power/efficiency [25] and quantification of chemical yield and its relationship with ultrasonic system parameters. Aspects of these topics are discussed further in Chapter 11.

The mechanism of ultrasonic degassing is discussed by Eskin [26]. While this phenomenon has been studied since the 1930s, this paper discusses degassing of liquids and light alloy melts, including removal of hydrogen under industrial conditions.

Ultrasound can be used to activate transition metal complexes, so that they well catalyze organic reactions [27]. In some cases, the stoichiometry and the catalytic chemistry initiated by ultrasound differ from those of either thermally or photolyti-cally induced reactions of the same systems. Ultrasound can actually cause catalysts to undergo a structural change [27,28]. For example, sonication causes a change in an iron carbonyl catalyst in the immediate vicinity of bursting cavitation bubbles and these new and highly active chemical species diffuse into the bulk solution to catalyze olefin isomerization reactions.

In catalytic reactions, Boudjouk [28] points out the fact that ultrasound cleans the surfaces of heterogeneous metal catalysts, removing impurities such as metal oxides, and bringing the reactant materials into more intimate contact with the catalyst sur-faces. The result is increased reaction rates. Boudjouk also claims that the ultrasonic cleaning action inhibits side reactions by washing the products away from the cata-lyst before they can undergo further reaction.

Mason [29] reviews the effects of ultrasonic energy on chemical reactions, including many which do not occur without ultrasonic stimulation. He claims that ultrasonic cleaning action on the surface of metallic catalysts is helpful but it is not sufficient to explain the extent of the sonochemically enhanced reactivity. Interfacial contact area is an important parameter in the ultrasonic acceleration of chemical activity whether between reacting constituents or between the reacting materials and catalysts. Powder-type constituents are dispersed ultrasonically in the media and thus increase available contact surface. This contact surface is further increased by particle size reduction by fracture in an ultrasonic field.

Emulsification by ultrasound also enhances chemical reaction between the com-ponents of the emulsion by a dramatic increase in interfacial contact area between the liquids involved. Efficient mixing of heterogeneous reaction mixtures certainly is an important contribution of ultrasonic irradiation.

Suslick and Hammerton [30] have demonstrated that sonochemical reactions occur in both the vapor phase and the liquid immediately surrounding the cavitation event.

One of the most recent active areas of sonochemistry has been the decomposi-tion of organic contaminants by ultrasound. This has application in the treatment of wastewater. Hirai et al. [31] decomposed CFCs and HFCs in water. Chlorinated hydrocarbons and cyclohydrocarbons in water have been degraded by Kruus et al. [32], Visscher and Langenhove [33], and Shirgaonkar and Pandit [34]. Francony and Petrier [35,36] and Hoffmann et al. [37] degraded chlorinated hydrocarbons, pesti-cides, phenols, esters, and explosives in water. Pure carbon disulfide was degraded by Entezari et al. [38]. Johnston received a patent [39] for decomposing halogenated organic compounds with ultrasound.

The acceleration of certain chemical reactions by the application of an ultrasonic field can occur by direct agitation and/or new reaction intermediates. This effect has been reported by Tuulmets et al. [40] for Grignard reagent formation, Polackova et al. [41] for the Cannizzaro reaction, and Enomoto et al. [42] for iron powder processing.

Photochemical reactions were reported as being accelerated by Gaplovsky et al. [43], and Stephanis et al. [44] accelerated formaldehyde reactions with ultra-sound. During the sonication of diesel fuels, Price and McCollom [45] found alkanes shorter than C20 "crack" to shorter alkanes and alkenes, while aromatic and nitro-gen containing compounds polymerize.

The high velocity and temperature collisions of suspended particles in a cavitation field, followed by rapid cooling, can accelerate reactions and create novel products. Luche [46] discusses steric modification of cyclo additions under the influence of ultrasound. The use of ultrasound for electroless plating of ceramic materials is discussed by Zhao et al. [47]. Polymer synthesis has been enhanced and controlled by Price [48], Portenlanger and Heusinger [49], and Katoh et al. [50]. Stoffer and Sitton [51] were issued a patent for a polymerization process using ultrasound on ethylenically unsaturated monomers. Amorphous formation of iron, cobalt, molybdenum, and tungsten has been studied by Suslick et al. [52–54]. Boeing was issued two patents [55,56] for using this method to create magnetic recording media in a continuous process. The rejuvenation of a metal surface by this process can also activate and preserve catalytic activity.

Enhancement of electrolysis during sonication has been known for years. Madigan et al. [57] discuss conditions under which damage or particulate deposition occurs on electrodes. Lowering over-voltages in electrolytic processes reduces cost and by-products.

Tatsumoto et al. [58] demonstrated a lowering of the oxidation current during voltammetry. Improved electrolytic recovery of metals was reported by Walker [59].

Ultrasound has also found a number of uses in the food industry. Mason et al. [60] report on ultrasonic enzyme activation, extraction, emulsification, and several other processes. Bhatkhande and Samant [61] describe the saponification of vegetable oils assisted by ultrasound and its benefits over traditional methods.

The presence of trace contaminants in water is of great importance for the semiconductor industry. By studying sonoluminescent light from cavitating systems, qualitative information about a system can be determined. Kuhns et al. [62] measured part per thousand concentrations of alcohols in high-purity water and identified them multivariately. Ashokkumar et al. [63] quantitatively measured the effects of trace surfactants and alcohols on light emission. An ultrasonic method for detecting submicrometer particles in ultraclean liquids is reported by Madanshetty [64].

Sonoluminescence may also prove useful for identifying and quantitating species at higher concentrations. Emission from alkali and alkaline earth metals is readily observed, as shown by Flint and Suslick [65]. They also showed that emission lines from the ultrasonic destruction of larger species can also be measured [66].

Sonoluminescence is extremely sensitive to a complex set of external parameters, which has hindered its more widespread use as an analytical method. However, advancements in this area continue and are discussed in Chapter 11.

In conclusion, cavitation and intimate mixing of reactive and catalytic components have been shown to be primary mechanisms of enhanced chemical reactions under the influence of high-intensity ultrasound. Absorption leading to the general rise in temperature of the bulk material does not account for chemical synthesis by ultrasound. Neither is the accelerated activity a function of frequency.

However, some chemically related applications of ultrasonic energy are due to heat of absorption rather than cavitation when such processes as curing organic materials are considered. These materials can be cured more rapidly, uniformly, and thoroughly by ultrasonics than by ordinary thermal processes. The specimen material is placed in a mold beneath a large-area ultrasonic horn in direct contact with the material to be treated. The temperature of the mold may or may not be elevated by other means before and while the horn is activated under pressure, but the primary source of heat for curing is due to absorption of the ultrasound within the specimen material. If the material is subjected to ultrasound at a wavelength that is large compared to the thickness of the specimen material, heating throughout the specimen can be controlled more uniformly by heat of absorption than it could by thermal conductivity from outside the volume.

Throughout the history of ultrasonics, many interesting chemical reactions have been stimulated or accelerated under the influence of high-intensity ultrasonic energy, particularly under the influence of cavitation. The field of sonochemistry sees continued growth in the research laboratory and is now seeing increasing application in high value industrial processes. One issue which remains when scale-up and industrial applications are considered is that of the total energy involved, and there are also issues of the size of the interaction volume and bubble-induced screening.

13.3   INDUSTRIAL PROCESSES

In addition to what is now considered to be sonochemistry, there are a number of industrial processes, which employ chemical effects of ultrasound. Some examples of these processes are discussed below.

13.3.1   ACCELERATED ETCHING

Ultrasound applied during chemical or electrolytic etching often increases the rate of material removal by a factor. Dimensional control depends upon several factors, including intensity and directionality of the ultrasound waves, the flow characteristics of the etchant, and the selectivity of the etchant for one component material over another in a multicomponent system.

Both mechanical and chemical mechanisms appear to be operative. At low intensities, below cavitation levels, ultrasonic activity helps maintain close contact between the work surface and the etchant.

The mechanism is primarily a stirring action with the direction of the force maintained normal to the surface to be etched. The etchant is forced through any gas barrier formed during the process and contaminated material is dispersed; thus, fresh etchant is supplied to the metal surface.

Above the cavitation level, etching may proceed at a much accelerated rate. Flow of etchant becomes heterogeneous and the etching patterns may show the irregular patterns of the cavitation bubbles.

Ultrasonically assisted etching in each of the above categories, i.e., below and above cavitation levels, has been investigated. An example of the lower-intensity application is photoetching, a process in which a photographic image is developed in a coating of resist material laid over the surface of the metal to be etched. When the photograph is developed, the resist is removed from the areas where etching is to occur. For photoetching, intensities must be kept low to prevent removal of the photoresist by cavitation. The etched depression is rounded, and if etching proceeds to large depths, the resist is undercut. This type of etching has been used only for etching small surface areas, although it could be applied on a continuous basis. Ultrasound is introduced through plastic membranes attached to the transducers.

Etching core material from between layers of cladding during laboratory fabrication of nuclear fuel elements is an example of ultrasonic etching at intensities above cavitation levels. During one laboratory study, elements were fabricated in a large sheet and cut to size, exposing the core to the surrounding medium. The exposed core material was then removed to a predetermined depth, and cladding material was welded into the recess. Aqua regia was used as etchant and preferentially attacked the core material. A thin sheet attached to an active ultrasonic machine tool was guided into the slot to be etched and oscillated slowly back and forth along the length of the slot. The ultrasonic motion was normal to the surface and normal to the direction of the back and forth motion. This technique produced a straight bottom in the slot. Complete removal was obtained to the desired depth. A problem in adapting this method to large-scale fabrication of fuel elements exists in locating tool materials that resist attack by aqua regia. One material that might resist such attacks is tantalum.

Electrochemical machining of carbon steel, which ordinarily may proceed at rates to 5 mm/min at current densities of 300 A/cm^2, may be increased to 10 mm/min with the application of ultrasound [67].

Ultrasound is superior to conventional magnetic stirring for isolating precipitates and inclusions from steels for analytical purposes [68]. About 5 g of common steel samples, 1 mm thick, can be dissolved ultrasonically at 60°C within 2 hours in a 14% I_2–MeOH solution. The isolation ratios of oxide inclusions from 18 to 8 stainless steel and Fe–O system steels are higher than those obtained by magnetic stirring. Ultrasonics has also been used successfully to isolate carbides at room temperature from steels of the Fe–Mo–C, Fe–V–C, and Fe–Ti–C systems in acid.

Not all chemical etching processes are accelerated by ultrasound. Lavrent'eva and Pakhomov [69] found that ultrasound produced a severalfold increase in the rates of dissolution of gallium in HNO_3, H_2SO_4, and HCl, of indium in HNO_3 and HCl, and of arsenic in HNO_3 at room temperature and at 16–20 kHz, but it produced no significant increase in dissolution rate for tin or cadmium in H_2SO_4.

In a patent, Weinberg [70] described a system for etching metals rapidly and uniformly, using either chemical or electrolytic etching under the influence of ultrasound. A representative application and materials claimed are chemical milling of aluminum, titanium, beryllium, boron, nickel, and tungsten. The method consists of immersing the material to be etched in a rigid brittle material, such as glass, suspended in an ultrasonically activated tank of tap water. An example was given of increasing the rate of etching of 6061 Al-alloy masked by an acid resist in aqueous solutions of HNO_3, HF, and HCl from 0.0127 mm/min without ultrasound, to 0.0381 mm/min with ultrasound, accomplished without lateral undercutting of the resist edges. The etched surfaces were claimed to be smoother and flatter than those etched without ultrasound. It was proposed that ultrasound dispersed the surface sludge and hydrogen bubbles, which retard the attack during static etching. Electrolytic etching in the same solutions increased the rate of metal removal at 535 A/m^2 current density from 0.00381 mm/min without ultrasound to 0.1143 mm/min with ultrasound. The gain was attributed to ultrasonic vibrations breaking the insulating barrier that forms at the metal/liquid interface during electrolytic etching. Etching a beryllium sheet with ultrasound in an etchant of 450 ml H_3PO_4, 26.5 ml H_2SO_4, and 53 g $CrCO_3$ proceeded at 0.02286 mm/min, producing a surface finish rated at 0.20–0.41 µm roughness; without ultrasound, the rate was 5.08 µm/min and the roughness was 63–80 µm. Tungsten sheet could be etched electrolytically in a 10% NaOH solution to a depth of 889 µm in 10 minutes with ultrasound, but without ultrasound no metal was removed.

13.3.2   TREATING BEVERAGES, JUICES, AND ESSENTIAL OILS

Ultrasonics applied to newly fermented alcoholic beverages produces effects similar to those resulting from a long period of aging [71]. Singleton [72] ultrasonically treated red and white table wine, red and white port wine, and sherry base wine at intensities above the cavitation level in the presence, separately, of nitrogen, hydrogen, carbon dioxide, air, and oxygen. Some compositional changes were affected, accompanied by a readily detected odor and flavor difference. However, these changes did not necessarily improve the quality ratings of the wines. In fact, the conclusion of this author was that overall quality scores were usually decreased by application of ultrasound. This is in contrast to the conclusion of Shakhsuvaryan et al. [73] that ultrasound may be used for improvement of wine quality and appearance. They found that ultrasonically treating white wine affected with "AcOH disease" caused a reduction of nitrogen compounds and tannin contents and stopped further formation of volatile acids. These effects were accompanied by an increase in clarity.

Removal of sulfur dioxide from pear juices and wines by means of intense ultrasound has been claimed [74].

Berishvili [75] found that a higher percentage of juice could be extracted from grapes treated ultrasonically at 19–21 kHz for 5–45 minutes. The amount of the increase of juice yield ranged from 1% to 6.2%, depending upon the variety and maturity of the grapes.

Berk and Mizrahi [76] conducted laboratory experiments on the use of ultrasound in producing low-viscosity concentrated orange juices. The irradiation was combined with evaporation techniques. The treatment produced unpleasant, garlic-like odors that had to be eliminated by flash evaporation; however, the ascorbic acid content was not affected. Ultrasound did improve the concentration process, but the process is limited by the size and cost of equipment that would be required for industrial processing.

Gilyus [77] found that ultrasound could reduce the processing time for removing sugar from sugar beets by a factor of 2. The effectiveness of ultrasound decreased with increase in temperature.

Carpenter [78] patented a process utilizing ultrasound to extract fatty substances from vegetable oil-yielding cells. The ultrasound helps rupture the cells, increases diffusion through the walls of unruptured cells, and emulsifies the oils in the suspending liquid medium. A similar patent was issued to Romagnan [79] for extracting oils from vanilla beans.

Ultrasound has also been used to change the characteristics of essential oils and perfumes. The effects produced depend upon the intensity of the sound, the duration, and the atmosphere in which the treatment occurs. As an example of the effects produced, the "weedy" odor of fresh geraniol is removed by a 1-minute treatment at 1 MHz at cavitating intensities. The weedy odor is similar to that of newly mowed hay or dry grass.

In recent years, there has been a growing interest in the potential uses for ultrasound in both food characterization, including in process monitoring measurement and control (considered in Chapter 10) and in its use at higher powers for processing, including generation of emulsions, extraction, disruption of cells, enzyme inhibition, and modification of crystallization processes (e.g., McClements [80], Povey and Mason [81], Knorr et al. [82], and Mason et al. [83]). Piyasena et al. [84] provide a review of inactivation of microbes using ultrasound, and they conclude that ultrasound combined with heat and pressure is more effective than ultrasound alone at killing microorganisms.

In food products there is always the issue of the detrimental effect of any treatment on food quality: including both the texture and taste. For example, it has been found by Chemat et al. [85] that ultrasound applied to a range of edible oils (olive, sunflower, soybean, etc.) caused significant changes in their chemical composition and flavor.

13.3.3 Treatment of Sewage

There are two categories of sewage treatment: (1) treatment of industrial wastes involving discharged chemicals and other debris from industrial processes and (2) treatment of community, or residential, wastes.

The use of ultrasonics in at least three different ways has been proposed for sewage treatment. One method involves enhancing or promoting chemical reactions such as oxidation of aromatic compounds and depolymerization of organic compounds in industrial wastes [86]. A second method involves the use of ultrasonics to clean or strip masses of material from the upstream side of a filter and to concentrate the material in an area from which it can easily be removed [87]. The third method that has been proposed for treating sewage involves the effect of increasing the propagation rate of bacterial growth under the influence of ultrasound. All of these methods have shown feasibility on a small scale but are limited commercially where large volumes are involved. Supply of sufficient energy to promote the activity associated with each method does not appear to be economically feasible at the present time.

Electroacoustic dewatering systems discussed in Chapter 12 have been developed for concentrating sewage sludge for disposal in incinerators. Using the electroacoustic belt-press filter, sludges are dewatered to a level suitable for direct injection into the incinerator.

Ultrasound has been demonstrated to accelerate anaerobic digestion of sewage and assist in the disintegration of sewage sludge [88]. Yin et al. report that the use of ultrasound has positive effects on dewater and digestion behavior of activated sewage sludge, which resulted in enhanced filtration process [89]. This technology has now moved into full-scale deployment. It has been reported [90] that it has been installed in eight full-scale plants in Europe (most in Germany) and that a further eight plants are reported as, under construction. For the operating plants, it has been found that there are significant performance improvements, including significant increases in biogas production and reduction of the volume of sludge cake that remains at the end of the process.

13.3.4 EXTRACTION PROCESSES

Extraction processes actually involve several effects, including emulsification, cleaning, certain chemical effects, and diffusion. Ultrasonic energy is used to disrupt cells and to remove certain essential oils and materials that are otherwise difficult to remove from porous media, producing generally higher yields.

The beneficial effects must be weighed against the economics of the applications. Ultrasonics usually is limited to exotic, small batch extractions. The possible chemical reactions also must be considered; that is, whether the chemical to be removed will be adversely affected by the ultrasound.

Adamski and Socha [91] showed that ultrasound had about the same effect as heat in removing capsaicin from the fruit of *Capsicum annum* and did not cause any decomposition of capsaicin.

Aimukhamedova and Korneva [92] used ultrasound at 22, 300, and 750 kHz to disintegrate cells of poppies to extract morphine. Similar amounts were extracted at all three frequencies. Ultrasound did cause decomposition of the morphine with time when the process was carried out in water, so the optimum extraction time was short (10 minutes or less). However, the yield increased when an alcohol–$CHCl_3$–NH_3 mixture, aqueous HCl, $Ba(OH)_2$, or sulfamic acid solutions were used as extractants instead of pure water. When aqueous $Ba(OH)_2$ at a concentration of 50 g/liter was used, more alkaloids were removed than were removed at lower concentrations.

Extraction of tanning materials can be improved by ultrasound. The amount of tanning materials extracted ultrasonically at 800 kHz in 45 minutes is equivalent to that extracted in 8 hours by usual extraction procedures. The effectiveness is attributable both to mechanical action of the ultrasound and to heat due to absorption.

Chapman and Bradbury [93] made use of the capability of ultrasound for disrupting wool fibers in HCO_2H to separate orthocortex and paracortex from wool for analysis.

Chen and Chon [94] found that increased extraction efficiency results from applying ultrasonics at intensities above the cavitation threshold during continuous liquid–liquid extraction processes. They attributed the increased effectiveness to increased turbulence, which may also increase the interfacial area of contact between the phases, thus enhancing the mass transfer. The increase in extraction efficiency is nearly linear with increased intensity of ultrasound.

Chen and Fairbanks [95] have shown that ultrasonic energy at 20 kHz can increase the rate of flow of oil through sandstone. Additional work has demonstrated that the liquid flow in the porous media changed from normal viscous flow to a plug-type flow [96]. Ultrasound has now been used to enhance oil removal from soils [97]. Work is in progress to develop this concept and several patents have been sought for the use of ultrasound to enhance oil recovery.

Yields of alkaloids increase under the influence of ultrasound. Periods of maceration are 30–60 min. Also, yields obtained during continuous solvent extraction are greater than those obtained by conventional means [98]. A French patent has been issued [99], which claims that higher yields of sugar juice are obtained from sugar beets when the beets are subjected to an initial compression and sudden decompression followed by an ultrasonic disruption of cells. Ultrasonics at low frequencies can increase the yield of bitter substances from hops in aqueous suspension [100,101]. Increased yields of cottonseed oil [102] and of adrenaline from ground adrenals [103] have been claimed to result from the application of ultrasound. Vaisman et al. [104] claimed a 10% increase in the amount of extracted alkaloids and glycosides from belladonna leaves and lily of the valley when ultrasonics at 80 kHz and 2.5 W/cm^2 was applied for 1 hour, as compared with the extracts prepared by the percolation method.

Whiervaara [105] studied the effects of ultrasound on the extraction of dehydrogenases from yeast cells. Using a titanium probe operating at 18–20 kHz to generate an intensity of 35 W/cm^2, he found that maximum enzyme activities were liberated after 5 minutes of treatment, although the amount of protein liberated increased linearly during the first 15 minutes. Enzyme liberation increased with increased power and decreased with increased salt concentration.

Woodford and Morrison [106] found that ultrasonic irradiation increased the extraction rate of rhein glycosides, free anthraquinones, and total water-soluble solids from whole senna pericarps. The ratio of rhein glycosides to total water-soluble solids was higher in extracts prepared using ultrasound than in preparations obtained by simple maceration. The ratio was slightly lower at higher intensities than at lower intensities.

Ultrasonic processes for extracting crude drugs should be avoided where molecules that are subject to destruction by intense ultrasonics are involved.

Ultrasonics has been found to enhance flotation of mercury and antimony [107]. Ultrasound inhibits flotation of galena pulps in aqueous suspensions of the ore with small additions of ethyl xanthate "as a collector" [108]. The ultrasound apparently affects the hydrophobic properties of the galena surface.

Vinatoru [109] provides an overview of ultrasonically assisted extraction of bioactive compounds from herbs, including a discussion of some new and emerging approaches. Ultrasound-assisted extraction (UAE) is seeing use with herbs, oils, proteins, and bioactive materials from both plant and animal sources [110]. Balachandran et al. [111] have used ultrasound to enhance supercritical extraction, and employed it to obtain pungent compounds from a typical herb (ginger). Both the extraction rate and yield were found to be increased through the use of ultrasound.

13.3.5 DEMULSIFICATION OF CRUDE PETROLEUM

The idea of demulsifying oil and water is in opposition to the emulsifying effects usually expected of ultrasonic energy. However, the same principle that applies in preparing dispersions can be used to promote demulsification. This is done very effectively by ultrasonically dispersing a demulsifying agent through the emulsion. Skripnik et al. [112] showed that ultrasonic dispersion of the demulsifier in removing salt and water from crude oil gave considerable improvement over conventional methods.

Over the years, there have been a series of studies to investigate the effects of ultrasound on the properties of crude oil. Ye et al. [113] report on a crude oil pretreatment, using 10 kHz standing waves, which results in significant improvements in the dewatering and desalting rates.

13.4   MISCELLANEOUS CHEMICAL EFFECTS AND APPLICATIONS

The application of ultrasonics during the cracking of solar oil distillates has been shown to increase the yield of gasoline, with the yield, increasing with intensity and temperature [114]. At the same time, u the paraffin content decreased with intensity, but the amount of naphthenic hydrocarbons did not change. Carbene and carboid contents increased with intensity and temperature. The yield of cracking gas also increased, with considerable decrease in the amount of hydrogen in the gas at higher intensities and temperatures. The effects of raising either the temperature or the ultrasonic intensity were similar.

Ultrasound has been found to increase the dissolution of hemicellulose during the beating of softwood sulfide and sulfate pulp [115]. Studies were performed at 25 and 400 kHz, with the higher frequency being more effective.

The chlorination of unbleached beechwood kraft pulp in the presence of intense ultrasound decreased chlorination rate and lowered consumption of chlorine—but gave higher brightness and more complete delignification [116].

Application of ultrasound to the desalination of seawater has been considered on several occasions. In purification by zone freezing, impurities show a preference for the liquid phase. As the freezing progresses, a contaminated zone builds up near the liquid/solid interface. A percentage of the contaminate is trapped in the ice, requiring one or more additional cycles before the desired level of purification is reached. Ultrasound applied at the surface disperses the debris and reduces the number of passes required. Mechanical stirring produces the same effect and is less expensive.

Ultrasound has been applied to desalination of seawater in conjunction with distillation [117]. The ultrasound produces a mist, which increases the surface area in the still and increases the rate of distillation.

Another method of applying ultrasound to the desalination of seawater involves the effects of ultrasound on electrochemical processes [118]. This approach has not been used commercially.

A Czechoslovakian patent [119] claims that slaking lime is improved by activation in an ultrasonic field. The activation is accelerated by effectively increasing the access of water to lime grains. By this process, mortars are produced by simultaneously activating the lime and adding the sand.

The dissolution of steel in hydrochloric acid [120] and in phosphoric acid [121] is speeded up under the influence of ultrasound. The effect was attributed to ultrasonic generation of heat. Heating was, therefore, recommended over ultrasonics for this purpose. This probably is a good recommendation; however, it seems logical to attribute accelerated dissolution, at least in part, to other ultrasonic mechanisms such as microstreaming, cavitation, and removal of forming gas bubbles, which increases contact time between fresh solvent and the steel.

Puskar [122] reports that high-intensity ultrasound produces intensification of nearly all chemical-heat treatment processes in metals. For example, ultrasonics increases the rate of steel nitriding, depth of nitriding, and microhardness in the nitrided layer when the operation is performed at temperatures of 540–560°C.

The mechanisms contributing to the nitriding effects are reportedly dispersion of salts and delivery of fresh salt to the sample surfaces, removal of reaction residues from the surface of the metal, and increase in nitrogen diffusion into the metal.

Curing rates of epoxy materials used for potting components and as adhesives are accelerated by ultrasonic energy. The ultrasonic energy thoroughly mixes the reacting components and degasses the mixture before it sets. It is important to experiment with the types of mixtures to determine the appropriate intensity levels and application times. Excessive intensities can trigger exothermic reactions that result in internal charring.

Hodges and St. Clair [123] have developed an ultrasonic technique for mixing solid curing agents into epoxy resins. Ultrasound allows mixing of standard curing agents such as 4,4′-diaminodiphenyl sulfone (4,4′-DDS) and its 3,3′-isomer (3,3′-DDS) without prior melting. Also, curing agents with very high melting temperatures such as 4,4′-diaminobenzophenone (4,4′-DABP) can be mixed by the process without curing prematurely.

The influence of ultrasound on microbial activity in biochemistry has been studied. Srivastava and De [124] found that applying ultrasound to culture media (*Saccharomyces cerevisiae*) for time periods up to 180 seconds stimulated production of EtOH. The maximum effect occurred after 120 seconds. Adding NaCl with ultrasonic treatment caused a decrease in the production of EtOH, whereas adding KH_2PO_4 with ultrasound stimulated production of EtOH, but not as much as ultrasound alone. Adding NH_4Cl (0.5%) with ultrasound produced an increase in the production of EtOH, with the maximum increase of 51% occurring in yeast that had been treated for 240 seconds.

In similar studies of microbial production of citric acid by *Aspergillus niger*, the same authors [125] obtained similar results. They found that ultrasound alone improved the yield of citric acid by *A. niger*. In the presence of NaCl, the yield of citric acid decreased under ultrasonic treatment. However, adding 0.5% KH_2PO_4 or NH_4Cl with ultrasound caused improvements in the yields by up to 42.2% and 52.4%, respectively.

In neither of these studies did the authors mention intensities or the possible role of cavitation, which could be an important factor in either process.

13.5   ELECTROLYSIS AND ELECTROPLATING

Electrolysis is the production of chemical changes by the passage of an electric current through an electrolyte. Electroplating is plating by electrodeposition; that is, by depositing the plating on a surface by means of electrolysis.

Several studies have been conducted on the effects of ultrasonic energy during electrolysis and electroplating operations. Brown [126] reported that at intensities exceeding 1 W/cm^2, metal particles appear in suspension, probably as a result of cavitation erosion of the plating. Ultrasonics within the range 20–40 kHz permits the use of higher current densities and results in purer, harder, and stronger coatings. Brown recommended its use in plating with nickel and chromium.

Bystrov [127] confirmed Brown's claims of the use of higher current densities and the production of harder coating plus smaller grain sizes. He claimed that ultrasound at 0.02–7 W/cm^2 and 27 kHz had no effect on electrodeposition of nickel at current densities of 0.5–1 A/dm^2, while ultrasound at 0.6–1 W/cm^2 sharply increased electrodeposition by factors of 3–6 times. Increasing the intensity to 10 W/cm^2 had practically no effect. At low intensities, the effect is similar to that of chemical mixing. Similar results have been observed in plating zinc.

Kenahan and Schlain [128] reported that at all current densities, ultrasonics produces finer-grained deposits of electrolytic manganese than those obtained in normal cell operation, but it results in less adherent deposits at high current densities. The ultrasound did not affect cell voltage or cathode current efficiency.

According to Bondarenko [129], movement of gas bubbles and cavitation in an ultrasonic field can affect the acceleration of the process of electrocrystallization, but they are not determining factors in acceleration of the process. He concluded that the chief factor accelerating electrocrystallization is the steady-state microeddies that arise in the electrolyte at the surface of the cathode in the ultrasonic field.

Ultrasound reduces the cathodic and anodic polarization during electrodeposition, permitting an increase in the rate of metal deposition. With copper deposited on a helical steel wire from pyrophosphate baths, Domnikov [130] found that the deposition rate was 4–4.5 times that obtained by intensive mechanical agitation.

Electrodeposition of chromium shows similar beneficial effects of ultrasonic irradiation [131].

Simpson [132] produced striped electroplating, which is the deposition of metals in closely spaced parallel lines, by applying ultrasound at 2 MHz during electrodeposition of copper. The lines were caused by standing waves in the solution. The lines disappeared as the thickness of the metal deposit increased.

Sonoelectrochemistry, in which an electrochemical cell is irradiated with sound waves, has now become a most promising technique. Many electrochemical systems are influenced by ultrasound, and the methodology offers considerable practical benefit in a wide range of applications.

The introduction of an immersion horn into an electrochemical unit is found to have some unique actions [133], particularly with regard to effects on mass transport. Veronica et al. [134] provided an assessment of the state-of-the-art, and in the previous 2 years (2005–2006), there were over 100 papers published on the topic. The survey of this literature shows that Europe has the leading role in this effort. The contributions to the field from European laboratories amounted to over 200 papers by 2004, including several reviews and this number has increased since then.

13.6 PREPARATION OF NANOMATERIALS

Nanoparticles are very important to manufacturing of magnetic recording media and to manufacturing of permanent magnets. They are also important to the production of resinous products such as nylons, polypropylenes, and similar products, where a small addition of these materials increases the strength and impact resistance considerably. They also make these products more flame retardant and lower their permeability (resistance to thorough-passage of gases and vapors). In addition, reducing the size of solid catalysts to nanoparticle size greatly enhances their efficiencies.

There are the reasons for the amount of research in the use of ultrasonic energy producing particles of such small sizes. To produce such particles requires the material from which the materials are to be produced to be dissolved chemically and released chemically. Suslick used an ultrasonic procedure to synthesize amorphous nanoparticles of iron. He subjected an alkane solution of iron pentacarbonyl to ultrasonic irradiation using a solid titanium horn vibrating at 20 kHz [135]. Particles that precipitated from such a treatment agglomerated readily. To overcome the agglomeration, Miller et al. patented [136] a process for forming amorphous metals, using ultrasonic irradiation of metal carbonyl solution in *n*-heptane or *n*-decane and by extracting the nanoparticles from the hydrocarbon solvent by adding solvents of high vapor pressure such as ethoxyethyl alcohol. This was followed by an in situ coating process based on the addition of a polymer such as polyvinyl pyrrolidone, acrylios, or methane or related polymer precursors. The resulting thinly coated nanoparticles do not agglomerate and do maintain their miniature size.

The past decade has seen major growth in all aspects of nanotechnology, and this has included a wide range of particle synthesis, processing and manipulation studies, which have been included in discussions in other parts of this text. Nanotechnology and ultrasonics remain an emerging field with researchers investigating potential applications for ultrasound in various forms. Recent applications include work by Gedanken who has used sonochemistry to dope both polymers and ceramics [137] and that by Okitsu et al. [138] who employed sonochemistry for the synthesis of gold particles. Others are investigating the use of ultrasound for sorting, vibrating nanowires and an increasingly diverse range of biomedical applications, including gene therapy and drug delivery, which are discussed in Chapter 14.

REFERENCES

1. Young, F. R. 1989. *Cavitation*, 2. New York: McGraw-Hill.
2. Weissler, A. 1953. *J Acoust Soc Am* 25(4):651–7.
3. Henglein, A. 1987. *Ultrasonics* 25(1):6–16.
4. Suslick, K. S. 1990. *Science* 247(4949):1439–45.
5. Richards, W. T., and A. L. Loomis. 1927. *J Am Chem Soc* 49:3086–100.
6. Wood, R. W., and A. L. Loomis. 1927. *Phys Rev* 29(2):373.
7. Wood, R. W., and A. L. Loomis. 1927. *Philos Mag* 4(7):417–36.
8. Wood, R. W., and A. L. Loomis. 1931. *Proc Natl Acad Sci USA* 17:611.
9. Crawford, A. E. 1956. *Ultrasonic Engineering*. New York and Butterworth, London: Academic Press.
10. Rayleigh, L. 1917. *Philos Mag* 34:94.
11. Mason, T. J. and J. P. Lorimer. 1988. *Sonochemistry: Theory, Applications and Uses of Ultrasound in Chemistry*. Chichester, UK: Ellis Horwood.
12. Mason, T. J. 1990. *Chemistry with Ultrasound*. New York: Elsevier Applied Science.
13. Mason, T. J. 1990. *Sonochemistry: The uses of ultrasound in chemistry*. Cambridge, UK: Royal Society of Chemistry.

14. Mason, T. J. 1991. *Practical Sonochemistry: Ser's Guide to Applications in Chemistry and Chemical Engineering*. New York: Ellis Horwood.
15. Mason, T. J., and E. D. Cordemans. 1996. *Chem Eng Res Des* 74(A5):511–6.
16. Leighton, T. G. 1995. *Ultrason Sonochem* 2(2):S123–36.
17. Lauterborn, W., and C. D. Ohl. 1997. *Ultrason Sonochem* 4(2):65–75.
18. Crum, L. A. 1995. *Ultrason Sonochem* 2(2):S153–2.
19. Brennen, C. E. 1995. *Cavitation and Bubble Dynamics*. Oxford, UK: Oxford University Press.
20. Barber, B. P., R. A. Hiller, R. Loefstedt, S. J. Putterman, and K. R. Weninger. 1997. *Phys Rep* 281(2):65–143.
21. Suslick, K. S. 1988. *Ultrasound: Its Chemical, Physical and Biological Effects*. New York: VCH Publishers.
22. Margulis, M. A. 2006. *Russ J Phys Chem* 80(10):1698–702.
23. Blake, J. R., ed. 1999. *Phil Trans R Soc Lond* 357(1751):199–369.
24. Thompson, L. H., and L. K. Doraiswamy. 1999. *Ind Eng Chem Res* 38(4):1215–49.
25. Gogate, P. R., R. K. Tayal, and A. B. Pandit. 2006. *Curr Sci* 91(1):35–46.
26. Eskin, G. I. 1995. *Ultrason Sonochem* 2(2):S137–41.
27. Anon. 1983. *Chem Eng News*, (April 11), 34–5.
28. Anon. 1984. *Chem Weekly*, (January 18), 29.
29. Mason, T. J. 1986. *Ultrasonics* 24(5):245–53.
30. Suslick, K. S., and D. A. Hammerton. 1986. *IEEE Trans Ultrason Ferroelectrics Freq Control UFFC* 33(2):143–7.
31. Hirai, K., Y. Nagata, and Y. Maeda. 1996. *Ultrason Sonochem* 3(3):S205–7.
32. Kruus, P., R. C. Burk, M. H. Entezari, and R. Otson. 1997. *Ultrason Sonochem* 4(3):229–33.
33. Visscher, A. D., and H. V. Langenhove. 1998. *Ultrason Sonochem* 5(3):87–92.
34. Shirgaonkar, I. Z., and A. B. Pandit. 1998. *Ultrason Sonochem* 5(2):53–61.
35. Francony, A., and C. Petrier. 1996. *Ultrason Sonochem* 3(2):S77–82.
36. Petrier, C., and A. Francony. 1997. *Ultrason Sonochem* 4(4):295–300.
37. Hoffman, M. R., I. Hua, and R. Hochemer. 1996. *Ultrason Sonochem* 3(3):S163–72.
38. Entezari, M. H., P. Kruus, and R. Otson. 1997. *Ultrason Sonochem* 4(1):49–54.
39. Johnston, A. J. 1990. U.S. Pat. 5130031.
40. Tuulmets, A., K. Kaubi, and K. Heinoja. 1995. *Ultrason Sonochem* 2(2):S75–8.
41. Polackova, V., V. Tomova, P. Elecko, and S. Toma. 1996. *Ultrason Sonochem* 3(1):15–7.
42. Enomoto, N., J. Akagi, and Z. Nakagawa. 1996. *Ultrason Sonochem* 3(2):S97–103.
43. Gaplovsky, A., J. Donovalova, S. Toma, and R. Kubinec. 1997. *Ultrason Sonochem* 4(2):109–15.
44. Stephanis, C. G., J. G. Hatiris, and D. E. Mourmouras. 1998. *Ultrason Sonochem* 5(1):33–5.
45. Price, G. J., and M. McCollum. 1995. *Ultrason Sonochem* 2(2):S67–70.
46. Luche, J. L. 1996. *Ultrason Sonochem* 3(3):S215–21.
47. Zhao, Y., C. Bao, R. Feng, and Z. Chen. 1995. *Ultrason Sonochem* 2(2):S99–103.
48. Price, G. 1996. *Ultrason Sonochem* 3(3):S229–38.
49. Portenlanger, G., and H. Heusinger. 1997. *Ultrason Sonochem* 4(2):127–30.
50. Katoh, R., H. Yokoi, S. Usuba, Y. Kakudate, and S. Fujiwara. *Ultrason Sonochem* 5(2):69–72.
51. Stoffer, J. O., and O. C. Sitton. 1995. U.S. Pat. 5466722.
52. Long, G. J., Q. A. Hautot, D. Pankhurst, F. Vandormael, F. Grandjean, J. P. Gaspard, V. Briois, H. Taeghwan, and K. S. Suslick. 1998. *Phys Rev B* 57(17):10716–22.
53. Suslick, K. S., T. Hyeon, M. Fang, J. T. Ries, and A. Cichowlas. 1996. *Mater Sci Forum* 225–227(2):903–11.
54. Bellissent, R., G. Galli, T. Hyeon, P. Migliardo, G. Parette, and K. S. Suslick. 1996. *J Non-Cryst Solids* 205–207(2):656–9.

55. Ollie, L. K., P. K. Ackerman, R. J. Miller, and D. C. Rawlings. 1998. U.S. Pat. 5766306.
56. Ollie, L. K., D. C. Rawlings, and R. J. Miller. 1998. U.S. Pat. 5766764.
57. Madigan, N. A., C. R. S. Hagan, H. Zhang, and L. A. Coury Jr. 1996. *Ultrason Sonochem* 3(3):S239–47.
58. Tatsumoto, N., S. Fujii, and N. Kawano. 1997. *Ultrason Sonochem* 4(1):9–16.
59. Walker, R. 1997. *Ultrason Sonochem* 4(1):39–43.
60. Mason, T. J., L. Paniwnyk, and J. P. Lorimer. 1996. *Ultrason Sonochem* 3(3):S253–60.
61. Bhatkhande, B. S., and S. D. Samant. 1998. *Ultrason Sonochem* 5(1):7–12.
62. Kuhns, D., A. Brodsky, and L. Burgess. 1998. *Phys Rev E* 57(2):1702–4.
63. Ashokkumar, M., R. Hall, P. Mulvaney, and F. Grieser. 1997. *J Phys Chem B* 101(50):10845–50.
64. Madanshetty, S. I. 1997. *IEEE Trans Semicond Manuf* 10(1):11–6.
65. Flint, E. B., and K. S. Suslick. 1991. *Phys Chem* 95(3):1484–88.
66. Flint, E. B., and K. S. Suslick. 1989. *J Am Chem Soc* 111(18):6987–92.
67. Ivanov, N. I., V. P. Rassakazov, and F. V. Sedykin. 1969. *Fiz Khim Obrab Mater* 1:38–42.
68. Kammori, O., I. Taguchi, K. Takimoto, and A. Ono. 1969. *Nippon Kin-zoku Gakkaishi* 3(4):493–7.
69. Lavrent'eva, V. G., and D. A. Pakhomov. 1967. *Prom Khim Reaktivov Osobo Chist Veshchestv* 10:5–8.
70. Weinberg, H. P. 1968. Method of etching refractory metal based materials uniformly along a surface. U.S. Pat. 3411999 (November 19).
71. Bachmann, J. A., and R. Wilkins. 1937. Method of treatment for fermented and distilled beverages and the like. U.S. Pat. 2086891 (July 13).
72. Singleton, V. L. 1963. *Wines Vines* 34:31–4.
73. Shakhsuvaryan, A. V., M. G. Gulyamov, K. I. Rasulova, and V. Lapina. 1963. *Tr Tashk Politekh Inst* 22:335–40.
74. Drboglav, E. S., V. S. Maiorov, and L. M. Lipovich. 1963. *Tr Tsentij Nauchno-Issled Inst Pivovarennoi BezalkogoVnoi Vinnoi Prom* 1(11):59–60.
75. Berishvili, L. I. 1963. *Nov Fiz Metody Obrab Pishch Prod Kiev Sb* 248–51.
76. Berk, Z., and S. Mizrahi. 1965. *Conserve Deriv Agrum (Palermo)* 14(2):55–9.
77. Gilyus, I. P. 1968. *Akust UVtrazvuk Tekh* 3:40–6.
78. Carpenter, J. 1955. Installation for the extraction and treatment of fatty vegetable materials. U.S. Pat. 2717768 (June 24).
79. Romagnan, L. 1952. Process for extracting aromatic oils from vanilla beans. U.S. Pat. 2601635 (June 24).
80. McClements, D. J. 1995. *Trends Food Sci Tech* 6(9):293–9.
81. Povey, M., and T. J. Mason, eds. 1998. *Ultrasound in Food Processing*. London: Blackie Academic and Professional.
82. Knorr, D., M. Zenker, V. Heinz, and D. -U. Lee. 2004. *Trends Food Sci Tech* 15(5):261–6.
83. Mason, T. J., E. Riera, A. Vercet, and P. Lopez-Buesa. 2005. Applications of ultrasound. In *Emerging Technologies for Food Processing*, 323–51. New York: Elsevier.
84. Piyasena, P., E. Mohareb, and R. C. McKellar. 2003. *Int J Food Microbiol* 87(3):207–16.
85. Chemat, F., I. Grondin, A. Shum-Cheong-Sing, and J. Smadja. 2004. *Ultrason Sonochem* 11(1):13–5.
86. Lure, Yu. Yu., P. F. Kandzas, and A. A. Mokina. 1963. Nauchn. Soobshch. Vses. Nauchno-Issled. Inst. Vodosnabzh. Kanaliz. Gidrotekh. Sooruzh. Inzh. Gidrogeol., Ochistka Prom. Stochn Vod., Moscow, 7–9.
87. Davidson, R., W. G. Palmer, and T. H. Forrest. 1970. Ultrasonic-clarification of liquids. U.S. Pat. 3489679 (January 13).
88. Neis, U., K. Nickel, and A. Tiehm. 2001. Ultrasonic disintegration of sewage sludge for enhanced anaerobic biodegradation. In *Advanced in sonochemistry, Vol 6: Ultrasound in Environmental Protection*, ed. T. J. Mason and A. Tiehm, 59–90. Amsterdam: Elsevier.

89. Yin, X., X. Lu, P. Han, and Y. Wang. 2006. *Ultrasonics* 44(Suppl 1):e397–99.
90. Bartolomew, R. 2002. *Ultrasonic Disintegration of Sewage Sludge: An Innovative Wastewater Treatment Technology*. Harrisburg, PA: Pennsylvania Dept. Environmental Protection, Bureau of Water Supply and Wastewater Management, Division of Municipal Financial Assistance, Innovative Technology.
91. Adamski, R., and A. Socha. 1967. *Farm Pol* 23(5–6):435–6.
92. Aimukhamedova, G. B., and G. M. Korneva. 1962. *Izv Akad Nauk Kirq SSR Ser Estestv Tekh Nauk* 4(6):17–24.
93. Chapman, G. V., and J. H. Bradbury. 1968. *Arch Biochem Biophys* 127(1–3):157–63.
94. Chen, E. C., and W. V. Chon. 1967. *Chem Engr Prog Symp Ser* 63(77):44–8.
95. Chen, W. I., and H. V. Fairbanks. 1968. *IEEE Trans Son Ultrason* SU-15(1):Abstract B8, 58.
96. Fairbanks, H. V. 1976. Ultrasonic stimulation of liquid flow. In *Proc. 1976 Ultrasonics Symposium*, eds. J. deKlerk, and B. McAvoy, Annapolis, Maryland, Sept. 29–Oct. 1, 117–8. New York: IEEE.
97. Kim, Y. U., and M. C. Wang. 2003. *Ultrasonics* 41(7):539–42.
98. DeMaggio, A. E., and J. A. Lott. 1964. *J Pharm Sci* 53(8):945–9.
99. Roeder, J. O. 1964. French Pat. 1368693.
100. Emel'yanova, Z. I., and V. A. Leoninck. 1968. *Fermentn Spirt Prom* 34(7):16–8.
101. Hoggan, J. 1968. *Ultrasonics* 6(4):217–9.
102. Iskanderova, D. S., N. U. Rizaev, R. M. Mirzakarimov, A. Sultanov, and M. I. Niyazov. 1967. *Dokl Akad Nauk Uzb SSR* 24(8):30–1.
103. Kazarnovskii, L. S., V. N. Solon'ko, and L. A. Shinyanskii. 1964. *Med Prom SSSR* 18(8):29–30.
104. Vaisman, G. A., M. I. Gurevich, E. S. Skvirskaya, and V. Ya. Gorodinskaya. 1963. *Farm Zh (Kiev)* 18(4):61–5.
105. Whiervaara, K. 1967. *Acta Chem Scand* 21(1):295–7.
106. Woodford, R., and J. C. Morrison. 1969. *J Pharm Pharmacol* 21(9):595–601.
107. Glembotskii, V. A., P. M. Solozhenkin, and L. L. Ogneva. 1966. *Izv. Akad. Nauk. Tadzh. SSR Otd. Fiz. Tekh. Khim. Nauk.* 1:58–62.
108. Kolchemanova, A. E. 1964. Intensifikatsiya Flotation. Protsessa, Akad. Nauk SSSR Gos. Kom. po Toplivn. Prom, pri Gosplane SSSR Inst. Gorn. Dela, 39–43.
109. Vinitzescu, M. 2001. *Ultrason Sonochem* 8(3):303–13.
110. Vilkhu, K., R. Mawson, L. Simons, and D. Bates. 2008. *Innovat Food Sci Emerg Tech* 9(2):161–9.
111. Balachandran, S., S. E. Kentish, R. Mawson, and M. Ashokkumar. 2006. *Ultrason Sonochem* 13(6):471–9.
112. Skripnik, E. I., V. I. Dolganov, A. Z. Simileiskii, and V. G. Dyrin. 1963. *Neft Khoz* 41(7):51–6.
113. Ye, G., X. Lu, P. Han, F. Peng, Y. Wang, and X. Shen. 2008. *Chem Eng Process* 47(12):2346–2350.
114. Balakishiev, G. A., I. G. Ismailov, M. I. Korneev, and E. B. Mezhebovskii. 1965. *Izv. Vyssh. Uchebn. Zaved. Neft. Gaz.* 8(12):71–4.
115. Iwasaki, T., J. Nakano, M. Usuda, and N. Migita. 1967. *Kami-Pa Gikyoshi* 21(10):557–63.
116. Boskova, L., R. Borisek, and G. Katuscakova. 1968. *Sb. Vysk. Prac. Odboru. Celul. Pap.* 13:63–9.
117. Brown, K. D. 1967. Distillation apparatus with ultrasonic frequency agitation. U.S. Pat. 3317405 (May 2).
118. Yaeger, E. 1963. *NAS-NRC Publication No. 942*, 229–54. Washington, DC: National Academy of Sciences-National Research Council.
119. Dlabaja, M. 1963. Czech. Pat. 108767 (October 15).
120. Petrov, Khr. 1961. *God Mash Elektrotekh Inst* 10(1):163–8.
121. Petrov, Khr., and Iv. Petkov. 1961. *God Mash Elektrotekh Inst* 10(1):153–62.

122. Puskar, A. 1982. *The Use of High-Intensity Ultrasonics*. Amsterdam: Elsevier.
123. Hodges, W. T., and T. L. St. Clair. 1983. *SAMPE Quarterly* (ISSN 0036-0821) 14:46–50.
124. Srivastava, A. S., and S. K. De. 1979. *Zentralbl Bakteriol Parasit-enkd Infektionskr Hyq Abt 2; Mikrobiol Landwirtsch Technol Umweltschutzes* 134(2):149–53.
125. Srivastava, A. S., and S. K. De. 1980. *Zentralbl Bakteriol Parasit-enkd Infektionskr Hyq Abt 2, Naturwiss: Mikrobiol Landwirtsch Technol Umweltschutzes* 135(5):408–12.
126. Brown, R. 1961. *Machinery (London)* 99:415–6.
127. Bystrov, Yu. M. 1961. Prom. *Primenenie UVtrazvuka Kuibyshevsk. Aviats. Inst. Kuibyshev Sb.* 196–202.
128. Kenahan, C. B., and D. Schlain. 1962. *U.S. Bur. Mines Rep Invest No. 6073*.
129. Bondarenko, A. V. 1962. Tr. Novocherk. *Politekh Inst* 133:59–78.
130. Domnikov, L. 1969. *Met Finish* 67(4):59–61.
131. Ginberg, A. M., V. N. Nud'ga, and Yu. N. Petrov. 1967. *Elektro-khimiya* 3(6):789–90.
132. Simpson, R. J. 1965. *Electroplat Met Finish* 18(9), 305–6.
133. Compton, R. G., J. C. Eklund, F. Marken, T. O. Rebbitt, R. P. Akkermans, and D. N. Waller. 1997. *Electrochim Acta* 42(19):2919–27.
134. Veronica, S., B. Pedro, D. Esclapez-Vvicente, J. Iniesta, J. Gonzalez-Garcia, and D. Walton. 2007. Electrochemistry with ultrasound: State of the research in the field. In *Proc. 19th Int. Congress Acoustics*, Madrid, Spain, 2–7 Sept, 5. Spanish Acoustical Society (Sociedad Espanola de Acoustica). Proceedings on web: http://www.sea-acustica.es/WEB_ICA_07/fchrs/contents.htm.
135. Suslick, K. 1994. The chemistry of ultrasound. In *Year Book Of Science and the Future*, 138–55. Chicago: Encyclopedia Britannica Inc.
136. Miller, R., L. K. Olli, and D. C. Rawling. 1996. Isolating nanophase amorphous magnetic metals. U.S. Pat. 5520717 (May 28).
137. Gadanjen, A. 2007. *Ultrason Sonochem* 14(4):418–30.
138. Okitsu, K., M. Ashokkumar, and F. Grieser. 2005. *J Phys Chem B* 109(44):20673–5.

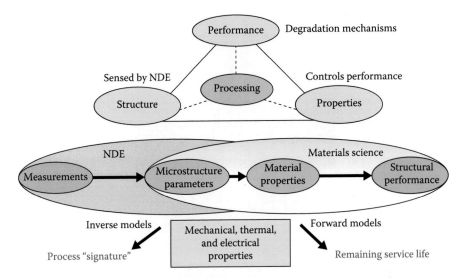

FIGURE 1.10 The intimate relationship between NDE and materials science. NDE = nondestructive evaluation. (Courtesy of Pacific Northwest National Laboratory.)

FIGURE 5.1 Example of the range of commercially available piezoelectric element transducers. (Courtesy of Piezotech LLC.)

FIGURE 5.4 Examples of compression wave transducers. (a) Sectioned 2.25 MHz, 2 cm diameter, focused immersion transducer. (b) 5 MHz, 12 mm diameter, 10 cm focused immersion transducer. (c) 2.25 MHz, 12 mm diameter contact transducer. (d) Pinducer—a small transducer-active element less than 2 mm in diameter. (e) A transducer on a wedge for shear wave generation, using mode conversion, with 1 cent coin shown for scale.

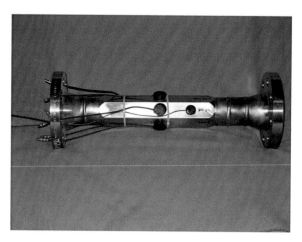

FIGURE 6.18 Field deployment configuration of density meter. (From Pappas, R. A., L. J. Bond, M. S. Greenwood, and C. J. Hostick, On-line physical property process measurements for nuclear fuel recycling, In *Proc GLOBAL 2007, Advanced Nuclear Fuel Cycles and Systems*, Boise, ID, 1808–16, © 2007 by the American Nuclear Society, La Grange Park, IL. With permission.)

(a)   (b)

FIGURE 7.4 Example of measured intensity cross section through a transducer beam. (a) Section parallel to beam axis. (b) Transverse to beam, measured near peak on-axis intensity (arrow in [a])—false color bands indicate intensity zones.

FIGURE 8.1 Ultrasonic case depth and hardness measurement system. (Courtesy of Pacific Northwest National Laboratory.)

FIGURE 8.3 Some examples of commercial electromagnetic acoustic transducers on pipes, on plates, and with rollers. (Courtesy of Sonic Sensors.)

FIGURE 8.15 Example of a commercial phased array control and display unit.

FIGURE 8.17 Phased array with mechanical scanning being set onto a composite pipe, with image display on computer monitor. (Courtesy of Pacific Northwest National Laboratory.)

FIGURE 8.33 Example of an acoustic microscope image showing delamination (red) in an IC. (Courtesy of Sonoscan Inc.)

FIGURE 8.35 Ultrasonic Intrinsic Tag, shown in use on a fuel rod assembly. (Courtesy of Pacific Northwest National Laboratory.)

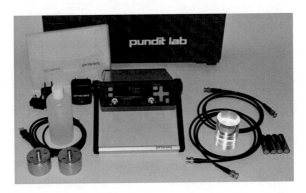

FIGURE 9.3 Commercial ultrasonic pulse-velocity meter. (Courtesy of Proceq SA.)

FIGURE 9.7 Example of automated ultrasonic inspection of a tube using a phased array system with shoe and water coupling. (Courtesy of Olympus.)

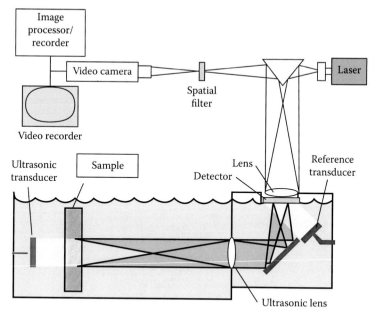

FIGURE 9.8 Schematic showing elements in RTUIS. (Courtesy of Pacific Northwest National Laboratory.)

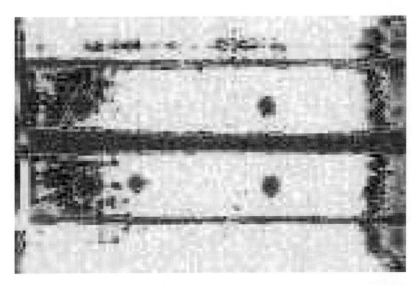

FIGURE 9.9 Image of a RAFALE CFRP self-stiffened box with simulated delaminations. (Courtesy of NDT Source.)

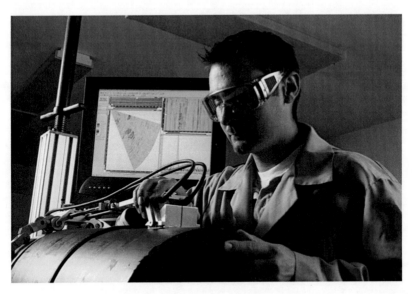

FIGURE 9.11 Phased array in use on a thick-walled polymer pipe, inspecting for lack of joint fusion, showing B-scan on computer screen behind operator. (Courtesy of Pacific Northwest National Laboratory.)

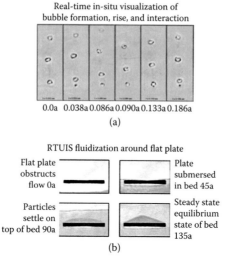

Real-time in-situ visualization of
bubble formation, rise, and interaction

0.0a   0.038a 0.086a 0.090a 0.133a 0.186a

(a)

RTUIS fluidization around flat plate

Flat plate
obstructs
flow 0a

Plate
submersed
in bed 45a

Particles
settle on
top of bed 90a

Steady state
equilibrium
state of bed
135a

(b)

FIGURE 10.8 Examples of acoustic holography visualization. (a) Real-time in situ visualization of bubble formation, rise, and interaction. (b) Fluidization of particles around a flat plate. (Courtesy of Pacific Northwest National Laboratory.)

FIGURE 10.11 The Product Acoustic Signature System (PASS) in use. (Courtesy of Pacific Northwest National Laboratory.)

FIGURE 10.13 Field deployment configuration of density meter. (Courtesy of Pacific Northwest National Laboratory.)

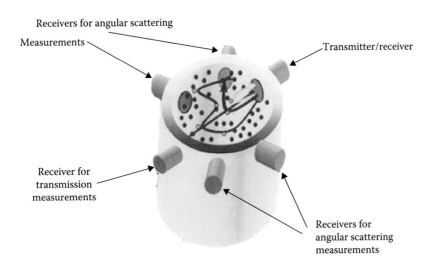

FIGURE 10.16 Schematic of experimental configuration. (From Pappas, R. A., L. J. Bond, M. S. Greenwood, and C. J. Hostick, On-line physical property process measurements for nuclear fuel recycling, In *Proc GLOBAL 2007, Advanced Nuclear Fuel Cycles and Systems*, Boise, ID, 1808–16, © 2007 by the American Nuclear Society, La Grange Park, IL. With permission.)

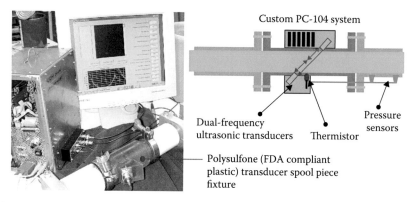

Custom PC-104 system

Dual-frequency ultrasonic transducers

Thermistor

Pressure sensors

Polysulfone (FDA compliant plastic) transducer spool piece fixture

FIGURE 10.20 Ultrasonic Doppler–based device for flow characterization. (From Pappas, R. A., L. J. Bond, M. S. Greenwood, and C. J. Hostick, On-line physical property process measurements for nuclear fuel recycling, In *Proc GLOBAL 2007, Advanced Nuclear Fuel Cycles and Systems*, Boise, ID, 1808–16, © 2007 by the American Nuclear Society, La Grange Park, IL. With permission.)

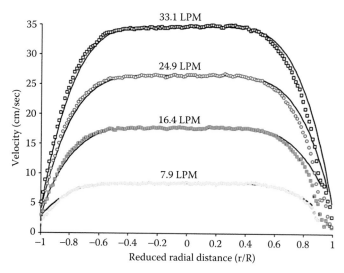

FIGURE 10.21 Velocity profiles for Carbopol solutions. (From Pappas, R. A., L. J. Bond, M. S. Greenwood, and C. J. Hostick, On-line physical property process measurements for nuclear fuel recycling, In *Proc GLOBAL 2007, Advanced Nuclear Fuel Cycles and Systems*, Boise, ID, 1808–16, © 2007 by the American Nuclear Society, La Grange Park, IL. With permission.)

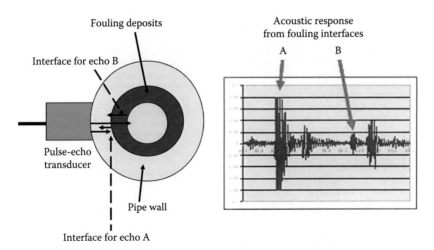

FIGURE 10.22 Schematic showing detection of impedance boundaries to provide a means to track pipe inside pipes. (From Pappas, R. A., L. J. Bond, M. S. Greenwood, and C. J. Hostick, On-line physical property process measurements for nuclear fuel recycling, In *Proc GLOBAL 2007, Advanced Nuclear Fuel Cycles and Systems*, Boise, ID, 1808–16, © 2007 by the American Nuclear Society, La Grange Park, IL. With permission.)

FIGURE 11.5 Photograph showing sonoluminescence for single-bubble cavitation. (Courtesy of L. R. Greenwood and Pacific Northwest National Laboratory.)

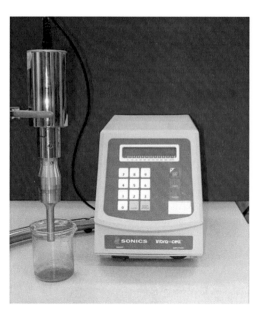

FIGURE 11.8 Commercial high-power ultrasound unit with horn of type commonly used in laboratory sonochemistry.

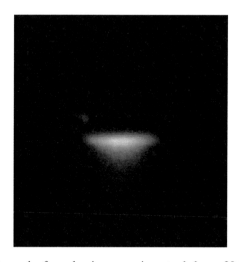

FIGURE 11.9 Photograph of sonoluminescence in water below a 20-kHz ultrasonic horn. (Courtesy of Pacific Northwest National Laboratory.)

FIGURE 11.10 Piezoelectric element used as a ring transducer. (Courtesy of Pacific Northwest National Laboratory.)

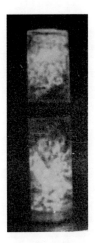

FIGURE 11.11 Example of volumetric sonoluminescence produced with a ring transducer applied to a 5-cm-diameter glass cylinder. (Courtesy of L. R. Greenwood and Pacific Northwest National Laboratory.)

FIGURE 11.12 In-line sonochemistry processing unit with a series of cylindrical ring transducers applied to a stainless steel pipe. (Courtesy of Pacific Northwest National Laboratory.)

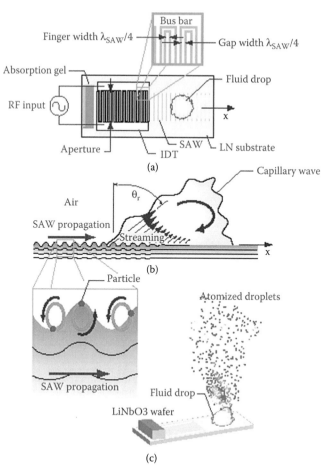

FIGURE 12.16 Schematic representation of process for use of a surface acoustic wave for droplet atomization. (Reprinted with permission from Qi, A., L. Y. Yeo, and J. R. Friend, *Phys Fluids*, 20(7):074103, © 2008, American Institute of Physics.)

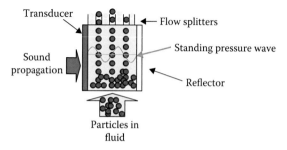

FIGURE 12.17 Schematic representation of an ultrasonic flow-through standing wave separator. (Courtesy of Pacific Northwest National Laboratory.)

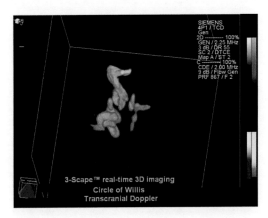

FIGURE 14.6 Three-dimensional image of a breast mass. (Courtesy of Siemens.)

FIGURE 14.9 Commercial breast-imaging unit, using acoustic holography, with example of image shown on computer monitor. (Courtesy of ADT.)

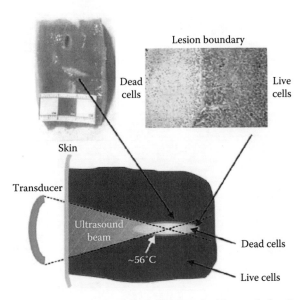

FIGURE 14.13 Focused ultrasound surgery. (Reprinted with permission from ter Haar, G., *Phys Today*, 54(12):29–34, © 2001, American Institute of Physics.)

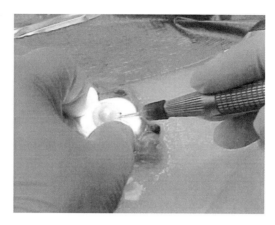

FIGURE 14.16 Laboratory phacoemulsification procedure on a porcine eye. (Courtesy of Pacific Northwest National Laboratory.)

FIGURE 14.17 Atomization and source of "hiss" from the horn-tip shoulder. (Courtesy of Pacific Northwest National Laboratory.)

Isocontour rendering at threshold T2

FIGURE 14.18 An example of a three-dimensional image of a natural biofilm. (Reprinted from Good, M. S., et al., *IEEE Trans UFFC*, 53(9):1637–48, © 2006 IEEE. With permission.)

14 Medical Applications of Ultrasonic Energy

14.1 INTRODUCTION

The brevity of this chapter is in no way indicative of the importance of ultrasonics in the field of medicine. Many tens of thousands of published articles and hundreds of books [1–5] on the subject are available. From the earliest days of the availability of modern ultrasonic technology, in the 1920s, researchers started to investigate both high- and low-power interactions with a diverse array of biological materials. Research into medical applications of ultrasound is usually said to have started with the work of Wood and Loomis [1], who made a comprehensive study of both the physical and biological effects of ultrasound on biological media. An extensive literature soon started to develop, particularly after 1954, or so, when the first compound B-scan imaging system was reported. The majority of the medical ultrasound literature, since about 1970, is written by and for medical practitioners: It focuses on the medical aspects of use, the procedures, and the clinical effectiveness in particular classes of diagnostic, therapeutic, and surgical applications. There is extensive attention focused on safety issues, particularly with regard to diagnostic imaging. However, there is far less attention directed toward investigation of the underlying physical principles of interactions and the science, engineering, and technology of the ultrasound systems. Although the literature is less well known in the medical community, there are however some early texts that do consider the biological effects and basic principles of interactions [2–7]. A good overview of the early developments of ultrasound in medicine, and its status in the mid-1970s, is provided by Erikson et al. [5]. This article is comprehensive and cites 376 references. It includes a table which provides a chronology for the major developments in the science, technology, and applications relating to ultrasound in medicine. From this literature, it is readily seen that by 1974 much of the basic science and a great many of the research areas and applications, which are important today, had been initiated and were showing early promise. In many cases, there have been advances in high-speed analog to digital conversion, modern computers (hardware and software) and displays and improved piezoelectric materials, and transducers that have enabled the developments in systems and applications since that time. Ultrasonics has found usage in all aspects of the medical field, including diagnostic, therapeutic, and surgical applications. In addition, ultrasound is now used in drug delivery and gene therapy, as well as in a range of research applications that include both cell sorting (low power) and cell disruption, lysis, at higher power. The use of ultrasound in both low- and high-intensity implementations has become an indispensable tool in many aspects of and specialties with modern medicine and biomedical research. Medical applications of ultrasonics include both low- and high-intensity applications.

Ultrasonic wave propagation in body tissues is, in large measure, controlled by working in combination with acoustic impedance contrast at boundaries and the various weaker scattering mechanisms that occur at the different size scales found within tissues velocity and attenuation factors. These are the basic factors that determine the effectiveness of both diagnostic and therapeutic applications of ultrasound.

For the purposes of this chapter, medical applications of ultrasound are considered in three groupings:

1. Diagnosis—in which clinical evaluations are made of organs or conditions
2. Therapy—in which ultrasound is intended to have a (beneficial) biological effect
3. Surgery—in which tissue disruption, destruction, or ablation occurs

There are also additional groupings of applications that are outside those used in direct clinical implementation: some of these are also identified and discussed.

In each class of application of ultrasound in medicine, particular frequency ranges, intensities, waveforms, and duration of exposure are used. The nature of the fundamental ultrasound–tissue interaction also depends on these parameters.

Diagnostic medical applications are based on the same principles that are applied in inspecting materials and in various other types of sensing and control. The merits of ultrasonic diagnosis are its safety, convenience, and due to differences in underlying contrast mechanisms, its ability to detect phenomena to which X-rays and other means of diagnosis are insensitive. For example, because of the difference in underlying contrast mechanisms, it's these factors that have led to the near-ubiquitous use of ultrasound, in developed countries, to monitor fetal development and position in pregnant women.

The principles and mechanisms involved in the industrial high-intensity applications of ultrasonics also have medical importance. The production of heat in the body has been recognized for its therapeutic value. The apparently selectively greater absorption of ultrasound in cancerous tissue has led to very interesting and positive results in its use in hyperthermic treatment of cancer. At intermediate intensities, it is now commonly used in physical therapy and sports medicine, with many millions of treatments being performed annually. Mechanical effects may include embolism destruction or increased diffusion through cell walls and better dispersion of injections. Chemical effects may be beneficial in relation to certain natural chemical reactions that occur in the body or to the activity of various injected materials. Ultrasonic therapy also includes atomization of liquids, including water, used for medicinal purposes.

Focused high-intensity ultrasonic energy can produce permanent changes deep within tissue without adversely affecting intervening intact structures. Effects produced can be selective. Body tissues are affected on either a macroscale or microscale in any number of ways, depending on both the characteristics of the tissue and the intensity of the ultrasound. Production of lesions in the brains of experimental animals for anatomic, physiological, and psychological studies is a routine procedure with appropriate instrumentation [8]. Normal structures other than the brain, including the pituitary gland and the liver, have been ultrasonically irradiated under relatively precisely controlled conditions of exposure. Carcinogenic tissues have also been irradiated. Intense ultrasound (high power) is used in several forms for surgery which can be considered as divided into two general classes: (1) bulk tissue interactions and (2) "mechanical" or surface (tool–tissue) interactions. There are further groupings that can be formed which follow from the type of ultrasound system used and the characteristics of the applied energy, defined in terms of both duration and intensity.

For internal or bulk tissue applications, focused beams of modest intensity are used in hyperthermia, and shock waves are used for kidney and gall-stone disruption (lithotripsy). In the "mechanical" interactions (e.g., tissue fragmentation with aspiration), ultrasonic tools can also be used for surface and near-surface surgery and for dental procedures. For example, an ultrasonic tool is now commonly used for the removal of cataracts (phacoemulsification). Similar instruments have been used for both soft tissue removal (e.g., tumor-debulking) and cosmetic procedures, such as ultrasound-assisted lipoplasty (UAL; liposuction).

Advances in imaging capabilities (low power) uses have been facilitated through advances in computer displays combined with new array technology and system software. These system advances have been accompanied by the development of transducers that can be deployed through all orifices and cavities in the body, both air and fluid filled, and the transducers can then be used operating either externally or internally, to image almost every part of the human body. External applied ultrasonic arrays are commonly used to give B-scan images of the developing fetus, and in numerous other diagnostic procedures. At the other end of the size scale, in miniature implementations, catheters containing arrays are inserted into blood vessels to measure wall properties and plaque buildup. Therapeutic applications have also diversified during the past 20–30 years as systems have become more affordable and available. Since 1990, there has also been a noticeable resurgence of interest in high-power ultrasonic tools, of several types, for use in surgery [9].

In this chapter, the emphasis is on illustrating the ranges of ultrasonic applications and interactions which can be utilized, the ranges of energies that are used, and the types of ultrasound systems used in medical practice and biomedical research. The discussion of the effectiveness, or otherwise, of particular medical procedures and comparisons with alternate modalities are, in general, outside the scope of this text. For discussion of the clinical aspects of ultrasound, the reader should turn to the literature written by and for the medical practitioner, rather than that which is derived from literature provided by the medical physics community.

14.2 POWER MEASUREMENTS AND DOSAGES

The intensity of the ultrasonic energy used in medical applications can be measured calorimetrically, by absorption probes, or by radiation pressure gauges (or balances) calibrated in watts. Miniature hydrophones with poly(vinylidine fluoride)-sensing elements are proving to be very useful in measuring very intense ultrasonic dosages—in the order of 10–1540 W/cm^2 [10]; power measurements are essential in laboratory work. An electronic microbalance is a popular instrument for measuring ultrasonic radiation pressure. A disk of sound-absorbing rubber attached to one arm of the balance is placed in the acoustic field. The microbalance senses the force on the disk. To evaluate the power output of a system, the transducer is placed against the bottom of a tank containing degassed water. The ultrasound is coupled into the tank through a thin plastic membrane. It travels through the water and impinges at normal incidence on the sound-absorbing disk. The acoustic power and intensity are calculated from the radiation force indicated by the electronic microbalance by the relationships and to ensure safety in clinical use.

$$P = 0.0148(A - B) \quad \text{and} \quad I = \frac{P}{S} \tag{14.1}$$

where P is total ultrasonic power (mW), A is the weight of the arm and disk measured after turning off the power switch of the device (μg), B is the weight measured with the power on (μg), I is average intensity (mW/cm²), and S is the area of the transducer (cm²).

Laser interferometry offers another method of determining ultrasonic intensities. The parameter ranges for biologic response to ultrasound, as a function of time are shown in Figure 14.1. The data in this figure is based on that reported in numerous references in the literature [11] and reviews of some of which is provided in several articles and reports, including those by Nyborg [12,13]. In diagnosis, low-intensity waves, typically less than 0.05 W/cm² and in the low-megahertz frequency range (2.5–7.5 MHz) are used. Clinical instruments available on the market are designed to produce intensities at or below the accepted limit. For general therapeutic applications, this limit is approximately 3 W/cm²; however, much higher localized dosages are used in certain types of ultrasonic treatments.

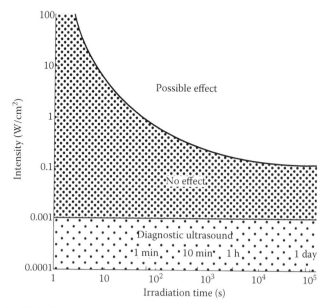

FIGURE 14.1   Biologic response to ultrasound, as a function of exposure time.

There are two aspects of dosage: intensity and time. Certain phenomena occur at intensities above certain threshold intensities that do not occur at intensities below these thresholds. Thus, if a certain beneficial effect is obtained at one intensity, doubling the intensity will not necessarily double the rate at which the beneficial effect occurs. In fact, it could prove to be harmful. The dosages should be determined for each type of application and must be carefully controlled.

Exposure criteria and the basic thermal mechanisms of interaction encountered for diagnostic ultrasound are now well documented [12]. That said, the safety aspects of ultrasound in diagnosis are periodically revisited by groups such as the American Institute of Ultrasound in Medicine (AIUM), and these reviews now also include issues relating to the use of ultrasound contrast agents [19].

An extensive discussion of the biological effects of ultrasound, particularly with regard to the development of safety guidelines was also provided by Nyborg [20]. The specific topics of tissue heating, acoustic cavitation, gas-body effects, and radiation pressure in terms of hazards, risks, and the safety of diagnostic ultrasound were reviewed most recently by Duck [21]. These various topics are discussed in more detail in the articles which he cites in his article. Although certain harmful effects in cells have been reported for laboratory settings, abnormalities in embryos and offspring of animals and humans have not been unequivocally demonstrated in any of the large number of studies that have so far appeared in the medical literature relating to the use of diagnostic ultrasound in the clinical setting.

In physical therapy, frequencies from 25 kHz, and more usually 500 kHz to low megahertz, are used at intensities between about 0.5 and 3 W/cm^2. These applications involve bulk tissue interactions and exposure must be kept below levels where tissue damage occurs. The calibration and measurement issues for therapeutic ultrasound are reviewed by Shaw and Hodnett [22]. The general characteristics of ultrasound in therapeutic applications, in terms of frequencies, peak pressures, and powers, are summarized and presented as Table 14.1. The measurement of the characteristics of

TABLE 14.1

General Characteristics of Ultrasound Used for Different Therapeutic Applications

Therapy Applications	Frequency (MHz)	Power (Spatial Peak Temporal Average) and Focal Pressures
Physiotherapy	1–3	1 W/cm^2: < 0.5 MPa
Lithotripsy	0.5	Very low: > 20 MPa
Soft tissue lithotripsy	0.25	Very low: 5–30 MPa
High-intensity focused ultrasound	0.5–5	1–10 kW/cm^2: 10 MPa
Hemostasis	1–10	0.1–5 kW/cm^2: 5 MPa
Bone growth stimulation	1.5	20 mW/cm^2: 50 kPa
Drug delivery	Up to ~2	Various: 0.2–8 MPa

Source: After Waddell, J. H., and J. W. Waddell, *Res Dev* 30:32, 1970.

the equipment used in such applications is complex and it is concluded by Shaw and Hodnett that there remains a need for further work in this field. This is, in part, because of the complex nature of the ultrasound field used, including the time-dependent and three-dimensional (3D) volumetric characteristics of the pressure field and the variations of wave-field/tissue-interactions. These interactions are also a function of the nature and structure of the tissue on which sonication is performed. The responses are further found to be significantly different when the application is performed in vivo or in vitro, on intact living tissue, on a living body with functioning blood supply or on isolated cells, including systems in which they are in a fluid suspension.

At the power levels used for therapy, and also surgery, a range of different and potentially nonlinear, ultrasound–tissue interaction phenomena can be encountered. There is in all cases an underlying thermal interaction, which exhibits volumetric variability. In addition, there are nonthermal mechanisms: these are usually considered in two classes: (1) mechanical and (2) cavitational, which all combine to cause various physical and chemical changes in tissues.

The mechanical mechanisms include radiation pressure, radiation force, radiation torque, and acoustic streaming. Cavitation can occur in both stable and inertial (or transient) forms, where the stable form involves bubbles that exist for many cycles of the driving wave form and the inertial form, and where the life cycle for an individual bubble occurs within one cycle of the driving applied wave-field. These phenomena also have spatial distributions, with the most intense forces being near the driving transducer (for a horn) or in the focal zone (for an individual piezoelectric element or an array).

In surgical applications intensities of 10 W/cm^2 and in some cases up to kW/cm^2 are used. The range of frequencies used cover the range from about 10 kHz and up to about 5 MHz, and the waveforms applied vary from very short intense pulses, used in lithotripters, to continuous sine waves, used for applications of high-intensity focused ultrasound (HIFU).

The community has developed a wide range of standards for medical ultrasound equipment. A review of medical International Electrotechnical Commission (IEC) standards that relate to protection from hazardous applications, in particular IEC Standard 60601, is provided by Duck [23]. There are also a number of other IEC standards that address hydrophones for use in ultrasonic equipment calibration, as well as the calibration of a range of classes of medical diagnostic equipment, physiotherapy systems, dental descalers, surgical and Doppler systems, among others. There are, in addition, numerous national standards that apply to medical ultrasound equipment.

The development of effective methods for full calibration of surgical units, at least in terms of full characterization of the acoustic fields delivered into tissue, remains ongoing [22]. An article by Schultheiss and Doerffel [24] reviews the standards for lithotripsy that have been developed by the IEC and the U.S. Food and Drug Administration (FDA). In addition to the existing regulations and norms for the manufacturers, special standards are discussed that were developed to address a treatment method developed in the early 1980s using extracorporeal shock waves. Initially, the FDA regulated the premarket approval process for lithotripters as a Class III device but reclassified lithotripters in 2000 to be a Class II device. The corresponding guidance document shows that there is substantial equivalence between new devices and those predated devices, and this is described in detail. The FDA guidance document is reported as being very useful in helping device manufacturers to (1) develop technical performance testing for a shock wave lithotripter within the parameters of an FDA submission and (2) conduct clinical performance testing via at least one clinical confirmation study with a small number of subjects.

14.3 BASIC MECHANISMS AND PRINCIPLES

14.3.1 MECHANISMS

Methods of medical diagnosis using ultrasonics may be based on either of the following two principles: (1) reflections from interfaces or, including scattering resulting from acoustic impedance contrast and (2) Doppler effects produced by motions within the body. Both of these are low-intensity methods. Attenuation is an important factor in medical diagnosis for it can reveal much information concerning the properties of the tissues through which the waves propagate. In the reflection method, the same principles are used that are used in nondestructive testing. Intermediate intensities are used in ultrasonic therapy. Although not all mechanisms associated with the results obtained through ultrasonic therapy are known, there are certain possibilities which must be considered in order that this form of energy may be used most beneficially and harmful effects avoided.

The generation of heat is associated with absorption, and many physicians think of ultrasound as a means of producing heat well within the body for therapeutic purposes. Undoubtedly, there are some beneficial thermal effects derived from absorption of ultrasound, but the generation of heat cannot explain all phenomena observed under the influence of ultrasound in biological tissues. Absorption within tissues varies with the type and condition of the tissue.

A review of the medical applications of ultrasonics by Wells [25] provides some more insight into thermal and mechanical effects of ultrasound for bulk tissue interactions, but little real insight into the details of biological mechanisms. It includes the statement: "The biological mechanisms to which the surgical techniques owe their success are the subject of some dispute" [25]. The biological effects of ultrasound, from the perspective of ensuring safety in diagnostic applications, have been reviewed by Nyborg [12] and several others. The use of higher power ultrasound to alter tissue is an area of growing interest. The mechanisms of higher power ultrasound interaction with tissue remain an area of research [26]. A range of possible mechanisms have been listed by authors but "the mechanism of high intensity ultrasound tissue ablation is poorly understood" [27].

The medical applications of ultrasonic have a long history. It has been known for many years that high-intensity ultrasonic waves can cause biological effects in tissue [1]. The use of ultrasound as a medical therapeutic agent has been under investigation since at least 1934 [28] and the first application in neurosurgery would appear to have been in 1954 [29].

There are a variety of early studies that report the use of low-frequency units 800–4,000 Hz to produce changes in biological systems, including heating for physical therapy [30].

A dental cutting instrument was first reported by Nielson, and the first reported study on the biological effects of ultrasonic dental equipment on oral tissue was by Hansen and Nielsen [31]. These early dental units operated at frequencies between 25 and 29 kHz [30].

At this early date two distinct types of ultrasonic unit are seen to be developing. There are the high-frequency (MHz), low-amplitude units for what was called "medical therapy" (diagnosis-imaging, etc.) and units sold for dental prophylaxis. In the dental units, the total power output did not exceed 0.1 W [32,33].

A unit operating at 29 kHz, produced by Cavitron, was reported as giving the lowest annoyance level factor, when compared with other instruments used in dental treatment [34]. This unit was modified to provide a portable unit designed for scaling procedures [35], and it could be adapted for other procedures such as curettage [36]. The modified unit operated at 25 kHz with an amplitude of 0.0015 cm (15 μm) and had tips of a variety of shapes and sizes. It is reported that the tip–tissue interaction involved is "mechanical" in nature [30].

In 1962, an International Symposium was held by the Royal College of Surgeons, London which resulted in a book *Ultrasound as a Diagnostic and Surgical Tool* [29]. This text provides a comprehensive summary of work at this time, including some studies on biological effects.

From the dental perspective, there was early work performed to investigate the clinical and biological effects of ultrasound [30]. There was also study that developed to investigate the mechanisms for ultrasound tissue interaction, mainly to ensure that ultrasound in diagnosis was safe and such studies are still being undertaken. This included consideration of the nonthermal effects of sound [37], physical and chemical aspects may simply not apply to the clinical situation. In the debate regarding the possibility of cavitation in vivo, it has been noted that in tissue the viscosity is higher than free liquids, there are cell boundaries, there are numerous organic surfactants present, and there is an absence of plentiful cavitation nuclei which are found in many in vitro studies. At the power levels used in therapy, it is not clear if cavitation can occur. From the observations that have been reported, the exact mechanisms involved in ultrasound for physical therapy remain the subject of debate and further study is required [26].

A recent review that discusses physical mechanisms for the therapeutic effects of ultrasound and cites 129 references is provided by Bailey et al. [38]. The article provides a brief overview of much of the pioneering work and then considers HIFU, ultrasound-enhanced drug delivery, and lithotripsy. For each application area, the physical mechanisms, and available models, as well as technology and challenges are summarized. The primary mechanism in HIFU is reported as being the conversion of acoustic energy into heat, which is often enhanced by various nonlinear interactions. Other mechanical effects resulting from ultrasound appear to stimulate an immune response, and the bubble dynamics play an important role in lithotripsy and ultrasound-enhanced drug delivery. This article also summarizes the development in understanding which has occurred in the past few years, particularly regarding the role of the combination of the nonlinear and mechanical interactions.

Ultrasound causes increased vascular and fluid circulation. Neurogenic effects result in changes in skeletal muscle tone. Associated with ultrasonic irradiation therapy is general relaxation and sedation. Diffusion of chemicals and ions through cell membranes increases, leading to improved organic exchanges as in metabolism and osmosis or to better and more effective utilization of injections. A 3.6-fold increase in the penetration of iodine from a hydrophilic ointment through the skin into the bodies of dogs, rabbits, and humans using therapeutic doses of ultrasound at 800 kHz for 10 minutes has been reported [39]. The early observations resulted in the investigation of ultrasound in transdermal drug delivery [40], in the controlled release of drugs deep inside the body and potentially gene therapy [41], and in the strategies that increase permeability for compound transport through the blood–brain barrier (BBB) [42]. Ultrasound applied to a dialyzer designed for extracting specific fractions of urine increased the extraction rates by several orders of magnitude [43]. Killing efficiency of chemical agents in an ultrasonic sterilizer increased owing to increased transfer rates under the influence of ultrasound.

Rupture of cells and cell structures can occur because of the high stresses generated by extremely high intensities, but such intensities are not used in routine therapeutic practice. These phenomena are, however, being investigated as a tool for lyzing cells and spores, particularly to help make DNA available for analysis [44]. Also, at these high intensities, biochemical and functional changes occur which can be related to the action of ultrasound on intracellular molecular complexes. Molecular complexes in tissues and cell structures can be broken loose, resulting in liberation or increased activity of cell enzymes. In common laboratory practice, results of this type are usually associated with cavitation; however, cavitation probably does not occur within the body at therapeutic intensities. Researchers at the University of Illinois [45] have shown that cavitation is not an important factor in the mechanism of production of paralysis of the hind legs of frogs. The same results were obtained even at pressures that were sufficiently high to suppress cavitation.

Ultrasonic energy inhibits the biological activity of fibrinogen and produces an anticoagulation factor out of its molecule [46]. If a high-intensity ultrasonic probe is brought into contact with a clot in a blood vessel, the clot will rapidly dissolve. Figure 14.2 shows an early commercial instrument for dispersing blood clots and ureteral stones. The wire is passed through a catheter to the site requiring dispersion.

As seen in the literature, in addition to devices for bulk interactions, there are also a range of ultrasonic devices used in "surface" procedures for both surgical

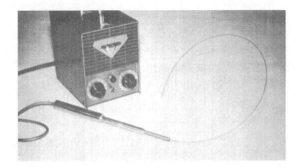

FIGURE 14.2   Instrument for the dispersion of blood clots and ureteral stones. (Courtesy of Blackstone Corporation.)

and dental applications. These use a horn or catheter to deliver a vibrating tip, or to energize a blade. In a 1991 article, Ernst et al. [27] stated for the surgical applications that use such devices, "the mechanism of high intensity ultrasound tissue ablation is poorly understood, but is most likely related to acoustic cavitation. Other potential mechanisms, most probably active in concert with cavitation, include direct the jackhammer effect, shock wave induced stress, acoustic micro-streaming shearing stress and resonance-related stress." A study by Bond and Cimino [47,48] investigated these interactions further and provide some useful insights into the fundamental mechanisms in this general class of devices and interactions termed "ultrasonic surgery using tissue fragmentation, with aspiration." That study concluded that, for the devices studied, the jackhammer effect is dominant for tissue disruption. Similar devices are also used in dentistry, in phacoemulsification (cataract removal) [49], and for a variety of soft tissue, bone cutting, lipoplasty (liposuction) and some cardiovascular procedures. These various "surface" procedures use a wide range of different designs of horns and horn tips, but in almost all cases, these units operate between 20 and 60 kHz.

The issues relating to clinical implementation of UAL, including fundamental mechanisms and safety effects are presented and discussed in detail in two volumes published in 1999 [50,51]. One aspect of the subject where published data remains limited is for investigations into the histology for tissue with ultrasonic surgery using tissue fragmentation, and insights it can give into mechanisms. An early article on the histology is that by Williams and Hodgson [52], and a more recent article is that by Blackie and Gordon [53]. Some work has also been performed to investigate the appearance (histology) of tissue fragments resulting from a procedure with an ultrasonic surgical aspirator [54]. It is found that disruption occurs almost exclusively at the tissue surface, and for tissue that remains intact, within one or two cell depths below that removed, it is viable and there is no evidence of intense shock wave interaction, as could be expected from a cavitation mechanism. One inherent challenge is that the procedures disrupt the tissues with which the tool is interacting and the interaction is a transient process. The investigation of the phenomena involved in these classes of interactions and the effects on tissue has continued to be the subject of much debate, particularly as the range of designs of horns and diversity of applications has been increased.

A recent article by O'Daly et al. [55] has provided a comprehensive critical review of the current state of knowledge with regard to this family of devices and the tissue-removal mechanisms. This article concludes that despite widespread clinical application and common device characteristics, there remains an incomplete understanding of the mechanisms of tissue failure, removal, and any related damage. This topic is revisited in some more detail in Section 14.6 of this chapter.

14.3.2 EFFECT ON HUMAN BLOOD

Extremely high intensities can cause various changes in blood. However, at non-cavitating intensity levels, corresponding to reasonable therapeutic intensities in the human body, there appears to be no permanent or significant danger to blood. Howkins and Weinstock [56] have shown that focused ultrasound below cavitation intensities can hemolyze blood if the focus is at the surface of the liquid and exposed to air. However, no damage is caused at the same intensity if the focus occurs beneath the surface. Safety aspects of ultrasonic irradiation of biological tissues in vivo are discussed in Section 14.4.

14.3.3 Effect on Tissue Regeneration

The use of ultrasonic therapy in the epiphyseal area of growing bone has been viewed with caution because high-intensity ultrasound was believed to cause damage to bone cortex and bone marrow. Vaughen and Bender [57] studied the effects of clinical doses (1 W/cm^2) of ultrasound on the epiphyses of growing rabbits. The left-knee area of the animals was treated under water for 5 minutes daily from the age of 3 months, until there was evidence (revealed radiologically) of epiphyseal closure at 6 and 8 months. The untreated hind leg was used as the control. The findings were that there was no difference in bone length, microscopic appearance, or rate or manner of epiphyseal closure between the treated and the control leg.

Regeneration of amputated tails of tritons (marine snails) has been shown to be stimulated by exposure to ultrasound [58]. The blastemal RNA in treated tritons reached levels in 12 days following amputation that corresponded to levels reached in 21 days in the controls.

Some significant work has been accomplished by Dyson et al. [59] in the anatomy department of Guy's Hospital Medical School, London. They showed accelerated tissue regeneration in adult rabbits under the influence of ultrasound. A small, 1-cm^2 hole was excised through each ear of the rabbit. One of these was used as a control and the other was treated ultrasonically.

The transducer used consisted of a lead–zirconate–titanate disk, 1 cm in diameter, operated at its third harmonic to obtain a frequency of 3 MHz. The area of the ear to be insonated was clamped lightly between the transducer holder and an absorption chamber. At the optimum intensity, around 0.5 W/cm^2, the holes closed in about 34 days in male rabbits and in about 46 days in female rabbits, while the corresponding controls closed in 50 and 70 days, respectively. Higher intensities (8 W/cm^2) caused damage.

These investigators were able to show that the effects were not because of heat. The ultrasound probably accelerates migration of cells to the site of injury and probably also increases the rate of protein synthesis.

Ultrasonically induced regenerated tissue contains a smaller proportion of collagen than normal tissue and, therefore, may have lower tensile strength.

14.4 DIAGNOSIS

Ultrasound as a diagnostic tool is much safer than X-ray, and because of the differences in contrast mechanisms, it is also sensitive to biological structures and abnormalities that are not always characterized when using X-rays. For example, radiographic detection of aneurysm (ballooning of a blood vessel in a weakened portion) is possible if an injection of materials that cause a contrast in X-ray absorption is made into the bloodstream before taking the picture. The picture must be taken within a short period

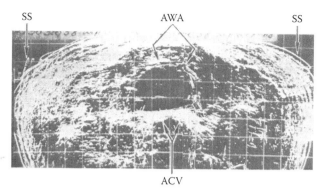

FIGURE 14.3 A B-scan transverse section of an aneurysm (ballooning out) in the wall of the aorta (large artery) in the abdomen. SS, skin surface; AWA, anterior wall of abdominal aortic aneurysm; ACV, anterior contour of vertebral bodies. This pattern was made 3 cm above the level of the upper border (crest) of the ilium (hip bone). (Courtesy of A. K. Freimanis, MD, The Ohio State University College of Medicine.)

of time following the injection. The aneurysm is indicated by a characteristic change in blood-flow pattern. Ultrasonics, however, makes possible B-scan photographs of aneurysms without injecting anything into the bloodstream and without the time limitation of the radiographic method. The actual size of the aneurysm can be measured by this means. Figure 14.3 is a B-scan indication of an aneurysm of the aorta obtained by scanning the patient in a lateral direction. In recent years, there have been great advances in the technology, including that needed for phased arrays, which has been facilitated by developments in both the hardware and software available with modern computers.

The use of ultrasound for diagnosis forms, by far, the largest area of both research and medical use [60,61]. There are a range of textbooks and several major journals devoted to diagnostic ultrasound, but in the majority of cases, the focus is on the clinical aspects of the tools, or related safety issues, rather than the fundamental physics on which it is based.

The physics of ultrasound is considered by the medical physics community, rather than the clinical community. This group has developed its own literature, but it is much more limited. For the power levels used in diagnosis, the propagation, scattering, and absorption can, in most cases, be adequately described by small signal propagation theory. Models that describe tissue interactions for many situations are now available [62,63].

There is an extensive literature, which demonstrates the intrinsic safety of diagnostic systems. For these systems, there have now been almost innumerable studies seeking to test or demonstrate the safety of commercial ultrasonic diagnostic units [64]. Although the safety of such units, through limitations on the power used has now been almost universally accepted [12,65,66], there is growing concern being expressed regarding the safety of diagnostic ultrasound because of the identification of more subtle and localized tissue interactions.

For diagnostic applications of ultrasound, any tissue damage is an unwanted and very undesirable side effect. There is particular attention focused on the possible damaging effects of ultrasound on the fetus. Typical intensities are 100 mW/cm^2, but there can be very-high acoustic pressure amplitudes (up to 4 MPa). The high-pressure amplitudes are products because many systems use short pulses (a few cycles) at relatively high frequencies (2.5–10 MHz). With the short pulses used, nonlinear, finite amplitude, propagation and tissue iterations are possible. Under these conditions for such pulses in aqueous fluids, violent transient cavitation should be expected [67,68]. The conditions in vitro are very different from those in vivo and results from one regime cannot be assumed to apply to the other case.

The possibility of acoustic cavitation and the safety of diagnostic ultrasound have recently been reviewed by Cartensen [69]. There has also been further work into gauging the likelihood of cavitation from short-pulse, low-duty cycle diagnostic ultrasound [70]. According to Crum [26], to date, no cavitation has been observed in tissue when using clinical diagnosis instruments at the intensities currently used in diagnosis, either in vitro or in vivo.

Even at low powers, the mechanisms for ultrasound–tissue interaction appear not to be completely understood. There remain two views in the research community regarding the possibility of the occurrence of cavitation in living systems, and neither has conclusive proof regarding the occurrence or absence of cavitation in vivo [26].

For medical diagnosis units, there remains current work to measure and calibrate the power output of commercial units and to establish thresholds above which tissue damage could or may occur. For such units where they are imaging bulk tissue, any nonlinear phenomena encountered can be expected to be in the focal plane or far field.

14.4.1 PRINCIPLES

The principle of reflections of acoustic energy from a boundary between two media exhibiting different acoustic impedances is the basis of one type of diagnostic procedure. Equipment used and procedures are similar to those used for nondestructive testing by the pulse-echo technique.

A second diagnostic technique is based on the Doppler effect. When an ultrasonic wave is reflected from a moving target, the reflected wave is shifted in frequency from that of the incident wave by an amount that depends on the velocity and direction of the motion. This phenomenon is finding wide usage in the medical field for measuring blood-flow rate and the movement of certain components of organs such as the heart or heart valves.

14.4.2 EQUIPMENT

There are now reported to be at least 128 companies, including both global and niche suppliers, which are providing equipment to the meet the needs of the global medical ultrasound equipment market. Growth in this market is continuing, and it is expected to exceed $6 billion by 2012 [71]. A number of major market research companies now provide regular reports on the market, with analysis by both global region and segments within the various specialized applications areas. The most current market surveys and data are best found through a search on the web.

In reflection-type units, often referred to as pulse-echo, equipment, A-scan, B-scan, and now 3D representations of body structures are presented on a computer screen or on a permanent recording.

The A-scan presents echo amplitude and distance. Its primary use is in echo-encephalography, where it has been especially useful in detecting midline shifts because of concussion or tumors. It has also been used in obstetrics, gynecology, and ophthalmology in conjunction with B-scan techniques.

With B-scan equipment, radar and sonar techniques of data processing are used to synthesize the received signals into a pattern on a screen that corresponds to a cross section of the region scanned lying in a plane parallel to the direction of beam propaga-tion. The position of the probe is synchronized with the sweep of one of the axes of the image, and the echo amplitude appears as a spot of varying intensity or false color at a position on the screen corresponding to the position of the interface causing the echo.

The M-mode (or TM-mode) is a diagnostic ultrasound presentation of temporal changes in echoes in which the depth of echo-producing interfaces is displayed along one axis and time (T) is displayed along the second axis, recording motion (M) of the interfaces toward and away from the transducer.

The transmitted pulse must be as short as possible to resolve small structural details. To do this, the transducer is highly damped just as transducers for obtaining high resolution in nondestructive testing are highly damped.

As is true in nondestructive testing applications, coupling between the transducer and the body is important. For direct-contact procedures, coupling is through a film of oil or grease which must be free of air bubbles. For immersion methods, coupling is through a bath of liquid, usually water. The acoustical match between soft tissue and water is good and little energy is lost in irradiating soft tissue. The match is poor between water and bone, and also the attenuation in bone is high. For this reason, considerable energy is lost in echoencephalography, in which contact is made almost immediately with the skull bone, and this may prevent the observation of certain abnormalities that would be observable if the bone had good transmission properties.

Instruments based on the Doppler effect are less expensive and more easily adapted by the practicing physician than are the echo-type units. These features make this type of instrument easier to market. The output from these units can be reproduced through headphones, or recorded on a computer. Some units are focused so that an indication of a blood-flow profile may be obtained using a pulse-Doppler principle. Applications range from detecting fetal heartbeat to indicating blood pressure.

Modern computer systems, which capture data and display images of structures for phased-array units, are providing real-time images of movement in organs such as the heart, and these data are now commonly combined with dimensional, volu-metric, and flow parameters.

14.4.3 ULTRASONIC CONTRAST AGENTS

To help improve diagnostic capabilities, a range of contrast agents, commonly in the form of small gas-filled beads, 2–8 μm in diameter, and galactose microcrystals to which microbubbles adhere, have been developed and are now used. FDA has approved two types of microbubbles for clinical use. These use gas-filled shells: in one, it involves an albumin shell and octafluoropropane gas core, and in the other, a lipid/galactose shell and an air core [72]. Following injection into the bloodstream, the material is transported to the body region of interest through the bloodstream and then imaged using a phased-array ultrasound unit. The contrast agent materials then disperse and dissolve, with the residuals being excreted through the kidneys.

An article by Jong [73] discusses the early history and then recent improvements in ultrasound contrast agents. These materials are reported to have resulted from chance observations by Charles Joiner in the late 1960s. Since this time, several pharmaceutical companies have worked to develop gas-filled microbubbles. Calliada et al. [74], in an article from 1998, provide a review of the basic principles of ultrasound contrast agents, which enhance backscatter intensity from the fluids in which they are introduced. The backscatter is enhanced by resonant responses that depend on material properties and bubble diameter. The article includes a discussion of the performance of then available commercial products. This area has increased in importance, and these agents are now used for investigation of cardiac structures, detection of intracardiac shunts, visualization of valvular regurgitation, analysis of congenital heart defects, and investigation of cardiac output, as well as novel applications that have combined the use of ultrasound contrast agents with both gene therapy and drug delivery [41]. The physiochemistry of microbubbles, the response under insonification together with performance in imaging and as therapy-delivery agents are important areas for ongoing research.

14.4.4 Diagnosis by Reflection Methods

In the early 1940s, attention in the medical ultrasound community turned to consider lower powers for diagnosis. Developments in RADAR during World War II were the direct precursors of two dimensional (2D) sonar and medical imaging systems. The development of compound B-scan imaging was pioneered by Douglass Howry at the University of Colorado, Denver, and an immersion tank ultrasound system was produced in 1951. This apparatus was then developed further [75] and the results achieved were featured in *Life* magazine in 1954 [76]. Although there were others also working in the field, this activity can be said to have become the basis for modern diagnostic ultrasound. A subsequent major advance, which made its use more patient friendly by eliminating the need for immersion in water, was the development of contact scanners for diagnostic applications [77].

Numerous books and articles report the effectiveness and usefulness of these technologies, from the perspective of the practitioner. Selected examples are now used to illustrate the range and variety of ultrasonic imaging in diagnosis.

14.4.4.1 Abdomen and Uterus

The history of ultrasound in gynecology from 1950 to 1980 is reported in a comprehensive review article by Levi [78]. The use of this technology has become a routine part of pregnancy management, as well as a useful tool in the diagnosis and tracking of a diverse range of other gynecological conditions. The transducers used can be applied in several ways including externally, transvaginally, and rectally. A short history of the development of ultrasound in obstetrics and gynecology, which identifies and discusses many of the key researchers in the field is provided in a website developed by Dr. Joseph Woo [79].

The ultrasonic B-scan technique for examining the abdomen has been used for many years. It is useful for diagnosing pelvic tumors, hydatidiform moles, cysts, and fibroids. It can be used to diagnose pregnancy at 6 weeks (counted from the first day of the last menstrual period) and after, and permits the obstetrician to follow the development of the fetus throughout pregnancy, including size and maturity and the location of the placenta. Also, the presence of twins or multiple pregnancy is revealed. Use of ultrasound eliminates the need for X-rays, and hence the exposure of the mother and child to X-ray irradiation. Fetal death can be confirmed much earlier by ultrasound than it can by radiography [80].

Available to the diagnostician are two methods of coupling energy from the transducer to the patient: immersion or direct coupling.

A-scan is used with B-scan, particularly to obtain specific measurements, such as biparietal diameter of the fetal head.

Equipment for automatically scanning the abdomen is available. In early work, the frequency commonly used was 2.5 MHz. To give better resolution, this has now generally increased, and commercial units most commonly operate in the 3.5–7.5 MHz range. The use of these higher frequencies, combined with the tissue properties that exhibit higher attenuation at higher frequencies, has resulted in a tendency to increase the power used and has also increased the need to consider the potential for adverse biological effects.

A hydatidiform mole produces a characteristic ultrasonic pattern that is easily recognized [80]. A radiating sunburst type of pattern of multiple echoes fills the area corresponding to the uterine cavity and no evidence of fetal structures can be seen. By taking somagrams (ultrasonic pictures) over a period of several days, it is possible to differentiate between the hydatidiform mole and a missed abortion, which sometimes is confused with the mole by its somagram.

The surface of both cystic and solid tumors can be outlined readily. A homogeneous mass such as a cystic or a solid tumor presents no internal reflections because of the absence of internal interfaces. However, the attenuation in the liquid of a cyst is lower than that in solid tumors, and this permits a larger echo from the posterior wall of the cystic tumor than is possible from a solid tumor for a given instrument sensitivity. In addition, the cystic-tumor indication shows a well-outlined, thin, uniform, and smooth surface, whereas echoes from the solid tumor are more irregular and often more dense and thickened.

Free fluid in the abdominal cavity produces no echo but can be identified by its effect on abdominal structures such as a floating bowel.

Papillations seen within a cyst and echoes from dense adhesions and metastatic nodules projecting from a tumor to surrounding organs, such as the intestines, or to the cavity wall are indications suggestive of a malignant condition.

Single cysts or polycystic kidney disease can be identified by somagram.

Early pregnancy diagnosis is performed while the patient has a full urinary bladder. The fluid in the bladder provides a useful transmission path of low attenuation and also forces the bowel out of the way for more direct access to the uterus. An embryo first appears as a small circle usually situated in the upper part of the uterus

[81]. The fetus goes through a stage of rapid growth from the sixth to tenth week, and this growth is easily followed by ultrasonic means. After this period, the gestation-sac indication begins to fragment, and this is followed by a period of about 4 weeks during which the echoes are rather random and difficult to interpret. However, at about the fourteenth week, the fetal head begins to appear, and from then on, it can be found and measured.

Following an abortion, ultrasonic diagnosis can reveal retained products of conception such as placenta, membranes, or dead fetus requiring surgical attention.

A pregnancy complicated by hydramnios is indicated by an entirely echo-free zone between the fetus and the wall of the uterus. This is because of the fact that amniotic fluid has low attenuation and contains no reflecting interfaces [82]. Acranial monsters are usually found in connection with hydramnios. In these cases, the normal ultrasound contour of the fetal skull is partly or completely missing, depending on how much of the cranial vault is missing. In such cases, the uterus is thoroughly scanned, both longitudinally and transversely, in various planes in order to verify the condition of the skull.

Prenatal hydrocephalus can be determined by oversized skull dimensions and low-echo density corresponding to thin skull bones.

Ultrasonic imaging now uses a diverse range of physical and electronic scanning technologies and data are reconstructed in many forms that utilize advances in software for data visualization. The imaging capabilities are further enhanced through improved transducers and the move toward higher operating frequencies, which provides higher spatial resolution.

Ultrasonic imaging is now a routine part of prenatal care. An example of medical B-scan is of a 25-week-old fetus, imaged using a 3.3-MHz system, is shown in Figure 14.4. In addition to B-scan images, reflection "3D" images are now available, and an example of such an image obtained with a 2.4-MHz-phased array is shown in Figure 14.5.

Imaging technologies are now used to visualize positions for biopsy needles. The differences in contrast mechanisms for the three common technologies, ultrasound, computed tomography, and magnetic resonance imaging all provide different data, and they are all now used in noninvasive and minimally invasive diagnostics; for example, in oncology, imaging is reducing the need for exploratory surgery and is enabling more effective sampling through accurate insertion of biopsy needles that are used to extract a sample of cells from organs for laboratory testing and some forms of minimally invasive surgery.

The 2D image is now being complemented by a move toward 3D structure visualization and a review of 3D imaging was provided by Fenster and Downey [83]. Real-time 3D imaging still challenges the hardware and computational capabilities of systems used in routine medical diagnosis, and this topic is discussed further in Section 14.4.4.5.

14.4.4.2 Neurology

Ultrasonic echoencephalography provides a rapid means of detecting lateral shifts in the midline septum caused by tumors or concussion. Every accident ambulance in Japan carries echoencephalographic equipment in order to identify patients with

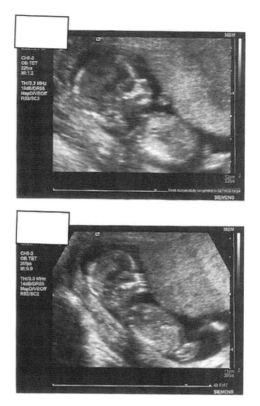

FIGURE 14.4 Examples of B-scan images 3.3 MHz of a fetus at 25 weeks.

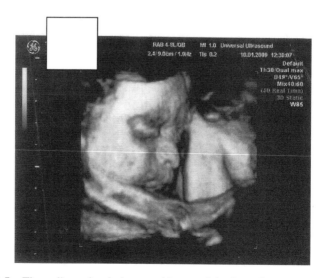

FIGURE 14.5 Three-dimensional ultrasound image of the face of a 30-week fetus.

possible subdural hemorrhage so that they may be taken directly to special neu-rological units for treatment [84]. Posttraumatic intracranial hemorrhage and skull and brain trauma are quickly diagnosed, and the lesions are located rapidly without discomfort to the patient [85]. Ultrasonic pulses are transmitted through the tem-ples. On an A-scan presentation, echo indications from the midline and the opposite temple areas of the skull are presented on the screen. A shift in the midline is easily discernible.

B-scan reproductions are being used in an attempt to improve the information capability of the ultrasonic method. High acoustic absorption in the skull, interfer-ence of the hair with the coupling contour of the skull, complexity of the brain, and complicated shape of the ventricular system are problems that interfere with obtaining high resolution. High frequencies required for good results are attenu-ated in the skull, and small echoes from interfaces between slight differences in acoustic impedance are lost by the same means. B-scan can be used effectively with small children and babies because their skulls are softer and have attenuation coef-ficients lower than those of adults [86]. Makow and Real [87] developed a 360° compound scanning device for examining the brain using an immersion method. In using this device, they observed certain phenomena which led them to believe ven-tricular pulsations could be used to identify the origin of some echoes. This was not an entirely new discovery, but it is an observation that now appears to be clinically significant. According to de Vlieger [86], echo pulsations provide significant clinical information:

- Echo-pulsation characteristics depend on the regions of the brain from which the echoes are obtained.
- Respiration affects the shape of the echo-pulsation curve. Usually, expira-tion lengthens the rise time of the curve; however, an inverse effect may appear in certain cases, for example, with patients having a predominantly thoracic type of respiration.
- Pulsations in the pressure of the cerebrospinal fluid appear to influence echo pulsations.
- Arteriovenous aneurysms cause abnormal dicrotic pulsation curves.
- Some heart diseases cause abnormal echo-pulsation curves corresponding to abnormal cerebral pulsations.

During an epileptic seizure, the ventricle echo oscillates delicately and rapidly in the prodromal stadium seizure, but the oscillation stops at the conclusion of the seizure and the echo begins a rhythmic pulsation.

In addition to the disorders mentioned earlier, echoencephalography is claimed to be useful for diagnosing hydrocephalus, brain abscess, cerebrovascular accident, hypertensive encephalography, and cerebral atrophy [88].

A comprehensive introduction and review of the early work in the field of ultra-sonic encephalography is provided by White [89]. This includes chapters that discuss the acoustic properties of the skull, and the challenges that are faced when seeking to image the brain, through the skull. A relative recent approach to through-skull imag-ing has been the investigation of time-reversal imaging [90]; however, significant challenges remain in correcting data if skull-induced distortions of internal structures are to be avoided in the resulting images.

14.4.4.3 Abdominal—Liver

Air in the lungs and gases in the intestines in addition to some structural characteristics of the liver present difficulties in ultrasonic diagnosis. Nevertheless, certain investigators have reported encouraging results in the use of ultrasonics for diagnosing liver disorders. The method of tomography used by Kimoto et al. [91] is promising for liver visualization. Also Kelly et al. [92] have shown some promising results with cat, hog, and beef liver by using a 2.5-cm-diameter focusing, lead–zirconate–titanate transducer operating at 1 MHz. Reflections from the biological tissue are reflected and focused on a 2-m-diameter lithium sulfate transducer, thus increasing the sensitivity of the system. Small, intact animals (cats and young pigs) are mounted in a chair-type device and immersed to the top of the chest in degassed distilled water. With the technique described and with sufficient amplifier sensitivity, the liver structure of both cats and young pigs can be detected with approximately equal clarity. Ultrasound methods are capable of detecting lesions in the liver that are not detected by X-ray. Holmes [93] obtained ultrasonic indications from abscesses of the liver, cirrhosis of the liver, and carcinoma of the liver.

Improvements in ultrasonic diagnostic procedures in liver have developed. Ophir and Maklad [94] have estimated attenuation coefficients in livers of living patients using a modified real-time sector scanner at 3 MHz and a narrow-band estimation technique. Scattering resulting from liver structure is one problem, but a greater variation in attenuation of signal between patients appears to be attributable to the variability in body walls, for example, layer thicknesses and fat.

O'Donnell and Reilly [95] have combined ultrasonic backscatter measurements with conventional B-scan imaging to study liver conditions. Both conventional B-scan images and B'-scan images of the right lobe of the liver were obtained on healthy volunteers, as well as on subjects with documented abnormalities of the liver. The ultrasonic backscatter coefficient represents a sensitive index of density and compressibility variations within tissue. O'Donnell and Reilly's studies showed that the backscatter coefficient from cirrhotic liver is higher than normal. The backscatter coefficient of adenocarcinoma metastasis of the liver is significantly lower than normal.

In addition to imaging the liver, abdominal ultrasound is used to investigate the condition of a number of other organs including the kidneys, gallbladder, pancreas, and spleen. These examinations are used to help diagnose a variety of conditions which are as follows:

- Abdominal pains
- Abnormal liver function
- Enlarged abdominal organ
- Stones in the gallbladder or kidney
- An aneurysm in the aorta

Doppler ultrasound images, which are discussed in Section 14.4.5, can be used to complement B-scan and other images and provide data to help the physician to see and evaluate the following conditions:

- Blockages to blood flow (such as clots)
- Narrowing of vessels (which may be caused by plaque)
- Tumors and congenital malformation

14.4.4.4   Ophthalmology

The use of ultrasonics in ophthalmology enables accurate diagnosis of conditions existing in the soft tissues of the orbit in the light-opaque portions of the eye. A typical frequency is 15 MHz, with focused transducers. The method has the capability of outlining tumors and detached retinas, measuring the length of the axis of the eye, or detecting foreign bodies close to the posterior eye wall. An instrument that combines a diagnostic transducer for locating foreign bodies with a surgical instrument for removing the object permits rapid removal of foreign bodies from the eye by directing the surgical tool to the body with least damage to the eye.

One method of scanning is the immersion method. A dam or open-ended tube containing a liquid bath is located above the eye; the ultrasonic instrument is immersed in this bath and performs the necessary search with A-scan or B-scan display of the information. A second method is to anesthetize the eye and place the transducer in direct contact with the cornea using distilled water, methylcellulose, or contact-lens wetting agents for coupling fluids [96].

Greguss [97] indicated certain errors that can result from the fact that the fluids in the eye are not homogeneous. He claimed that errors of up to 4 mm in the measured positions of interfaces can result if the heterodispersive nature of these fluids and the components of the eye are not considered. These errors can be calculated and fairly accurate measurements are possible with the present state of the art.

Greguss [98] also claimed a holographic technique for use in ophthalmology. The ultrasonic wave propagates down a liquid-filled tube into the eye. A sonosensitive plate is placed between the transducer and the eye. The initial wave passing through the plate is the reference beam, while the wave reflected from the eye is the information-bearing wave.

Studies related to detecting carotid occlusions, which often lead to stroke, have applied piezoelectric transducers in light contact with the cornea to measure ocular pressure pulses [99]. Correlations of ocular pulse amplitude and waveform shape were made initially with simulated carotid stenosis in rabbits. The technique is being adapted for use with human patients suspected or known to have carotid stenosis.

The anatomical structure of the eye can now be accurately reproduced using B-scan imaging. A review of imaging systems of the human eye, which compares ultrasound and other imaging modalities is provided by Acharya et al. [100]. The ultrasonic B-scan and the A-scan have both become a vital part of the ophthalmologist's diagnostic tool set, and they are now used to detect and characterize a number of conditions. These include investigation of ocular trauma, various melanomas, hemorrhages, and layer detachments [101,102].

There is a growing subspecialty of ultrasonic biomicroscopy, the clinical applications of higher ultrasonic frequencies, which includes the applications to the eye. A 2000 review by Frost et al. [103] discusses the technology and provides an overview of the clinical applications at higher frequency (60+ MHz) which includes ocular and 3D images.

14.4.4.5   Three-Dimensional Ultrasound

It has been found that 2D ultrasound imaging of a "slice" through 3D anatomy limits the ability to visualize diseases and interpret some of the conditions that ultrasound is used to investigate. There have been several recent advances which have exploited developments in both ultrasonic array and computer-based imaging capabilities. These developments have facilitated 3D reconstructions. Such advances have been intimately linked to the addition of various forms of advanced data processing; however, real-time 3D imaging still tests the limits for available computational and transducer capabilities. An example of a 3D fetal image of a 30-week-old fetus is shown in Figure 14.5.

A review of data processing and related computing was provided by York and Kim [104]. The most mature imaging is the "B-scan," and to this, color-flow modes have been added. In addition, the data are processed with a range of temporal, spatial, and frequency filters. Three-dimensional imaging has been investigated for more than 20 years, and 4D imaging, involving 3D spatial and temporal development, is being investigated. The early work is discussed in an article by Fenster and Downey [83]. Applications to clinical use have only been possible more recently, and the current 3D ultrasound systems do not yet provide real-time images for analysis [104]. Three-dimensional ultrasound is more complex and challenging than the 2D or B-scan, both in terms of the transducers, the electronics, and data processing. To give 3D data, the approaches that have been used include mechanical scanning of a linear or 2D array and free-hand scanning to build up images of surfaces. In the future, electronic beam steering applied to 2D arrays may become feasible for clinical use, but this requires further advances, not least in the design and fabrication of array transducers and in the supporting electronics and software for data processing and image reconstruction [83]. Following data collection, there are a range of reconstruction techniques that can be used, and a variety of computational tools are becoming more available [83]. Modern computer systems now have a wide range of visualization tools that can be used to give data display. It is, however, the sheer magnitude of the computational requirements which has, and remains, a fundamental limitation in terms of the capability for providing real-time 3D and the 4D displays.

Three-dimensional ultrasound imaging has the potential to be more accurate and reliable in an array of clinical applications. Research has been performed to investigate vascular imaging, and applications to cardiology, obstetrics, gynecology, and urology. Festner and Downey [83] highlight articles which consider studies of the eye, fetus, kidney, gall bladder, and breast. An example of a 3D ultrasound image of a breast mass is shown in Figure 14.6.

One novel study has been the combination of 3D ultrasound in a telemedicine/telepresence system, which has been developed and used for battle-field applications by Littlefield et al. [105].

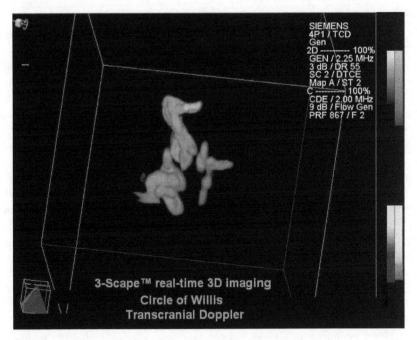

SIEMENS
4P1 / TCD
Gen
2D ----------- 100%
GEN / 2.25 MHz
3 dB / DR 55
SC 2 / DTCE
Map A / ST 2
C ----------- 100%
CDE / 2.00 MHz
9 dB / Flow Gen
PRF 867 / F 2

3-Scape™ real-time 3D imaging
Circle of Willis
Transcranial Doppler

FIGURE 14.6 **(See color insert.)** Three-dimensional image of a breast mass. (Courtesy of Siemens.)

14.4.4.6 Cardiac

Ultrasonic diagnostic procedures have generated considerable interest and application in the study of cardiac functions. Major causes of death following heart transplants have been acute rejection and infection. A-scan techniques have been used at the Stanford University Medical Center [106] in an effort to detect early signs of rejection. When rejection is indicated, dosages of immunosuppressive agents are lowered and active therapy of the rejection is begun. The method consists of measuring the dimensions of the heart as it fills with fluid. As the heart fills at the onset of rejection, the muscle walls swell and stiffen. Measurements are made using a 19.0-mm-diameter, 2.25-MHz transducer at a pulse repetition rate of 1000/s. Echo indications from the anterior wall and the posterior-wall surfaces provide the measurement information, with the distance from the anterior wall to the posterior wall indicating the overall heart size.

At the same center, echographic methods (methods using reflections of ultrasonic pulses from interfaces) have been used during cardiac catheterization to estimate the severity of valvular regurgitation in heart patients [107]. The estimations are based on calculations of end-diastolic and end-systolic volumes and stroke volumes based on ultrasonic measurements. The method has been used on several patients with the conclusion that echographic determinations of stroke volume are atraumatic, safe, and acceptable in patients who do not have valvular regurgitation and that ultrasound echocardiography does provide a reliable means of estimating the severity of valvular regurgitation in patients who have it.

Echocardiography can be used at a patient's bedside to diagnose pericardial effusion (escape of fluid through a rupture in the pericardium [108]). In such cases, the echo received from the posterior wall of the heart is split when the transducer is located on the anterior chest surface.

Echocardiography is also useful in evaluating the degree of stenosis (narrowing of the opening) in mitral-valve disease [109]. The transducer is aimed at the anterior mitral leaflet, and the echo signal is recorded on a strip chart so that an upward movement of the recorder pen corresponds to the movement toward the ultrasonic transducer and a downstroke represents movement away from the transducer. Thus, the slope of the tracing is an indication of the velocity of motion. The degree of stenosis affects the rate at which the blood flows through the opening. The speed of the diastolic downstroke during the period of ventricular filling is affected, slowing down with increased stenosis. This slowing down is readily discernible by the corresponding slope of the ultrasound cardiogram strip-chart recording. Rigidity or calcification of the valve is indicated by a decrease in the total amplitude of the movement of the anterior mitral leaflet between the closed position, during ventricular systole and the position of maximum opening in early diastole.

Intracardiac scanning has been used to obtain plan-position-indication (C-scan) displays of the interior of the heart [110,111]. This method is used to detect and measure atrial septal defects. A miniature probe is inserted into the right atrium through the external jugular vein or the femoral vein. The probe is located in the desired position with the aid of a fluoroscope. The probe may consist of a single directional element with scanning being accomplished by moving and rotating the probe. Or the probe may consist of multiple elements, with scanning being accomplished by sector techniques in which the positions of points on the recorded image are correlated with the beam direction. One such catheter-borne unit consists of four transducers spaced 90° apart in a plane normal to the axis of the catheter [112]. The transducers are pulsed sequentially at the rate of 1 kHz. The data are collected during a period of about 8 seconds, and the cardiac cycle is divided into 24 equal increments which depict the contour of the left ventricle at various stages of the cycle. The data are sorted and stored by computer and displayed after the data acquisition is completed. A-scan is used with the C-scan to provide the 3D information. The same technique is used in the inferior vena cava of the liver, thus avoiding some of the problems attributed to air in lungs and gas in intestines that are encountered when scanning the liver externally.

In order to obtain stationary patterns without the use of a computer during ultrasonic tomography of the heart, the picture must be completed within one-eighth of the cardiac cycle when the tomographs are for the systolic phase (when the movement of every part of the heart is maximum) and within one-quarter of the cardiac cycle when the tomographs are for the diastolic phase (when the movement of the heart is slowest).

Focusing transducers have been used to obtain high-resolution pictures of the heart. Movie films of the active heart using this method have been produced at the rate of seven frames per second [113], which is fast enough for examination of the movements of the internal structures of the heart.

Developments in computer methods of data processing have greatly benefited echocardiography. For example, noninvasive echocardiography is an important means of characterizing left ventricular structure and function. The ability of early techniques to identify boundaries of the heart wall was limited to the extent that data were imprecise. Delp et al. [114] developed algorithms to enable the diagnostician to detect the heart wall boundaries and to estimate the cardiac index (i.e., left ventricular wall thickness and boundary) by treating the ultrasonic images as a problem in time-varying image analysis. The beating heart is treated as an object that is moving through a scene. By tracking the endocardial boundary as it moves through the scene, the wall thickness and chamber volume can be estimated as a function of time. Abnormal bulges and deviations in ventricular shape can be determined by the technique.

A tomographic method of observing interior structures of the heart in three dimensions based on a stereoscopic display of 2D images has been developed [115]. The ultrasonic apparatus operates in synchronism with the cardiac cycle to obtain phase-selected tomograms. The system is applicable to measuring various parameters, particularly the heart volume. It can also be applied to other internal organs such as right ventricle, atrium, and kidney cyst that have irregular shapes.

14.4.4.7   Tomography and Holography

Tomographic transmission images were obtained in the early days of the development of medical ultrasound. An example of a tomographic cross section of a human neck taken in 1957 is shown in Figure 14.7. Since that time, most attention within the medical ultrasound diagnostic community has focused on reflection imaging, the B-scan, in all its multitude of forms. There has, however, remained a smaller community of researchers who have investigated both acoustic tomography and acoustic holography for medical applications.

The fundamental aspects of ultrasonic encephalography are discussed by White [89] and brain tumor detection is reported by Massa [116]. In looking for imaging

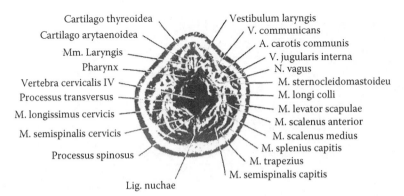

FIGURE 14.7   Ultrasonic tomographic cross section of the human neck. (Courtesy of J. Posakony.)

options, Hilderbrand and Brandon discuss the potential for medical applications of acoustic holography [117]. Such measurements give volumetric imaging of a large field-of-view in real time. It is a multiplanar imaging modality that achieves high penetration at low-insonification levels, and it has been successfully used in tests for both screening and instrument and organ visualization when guiding needles were used for performing biopsies.

One particular application that has received significant attention is through transmission ultrasound applied to breast imaging [118]. Metrics for assessment of systems performance in such applications have been provided by Garlick et al. [119] and recent experience in breast imaging is discussed by Andre et al. [120]. A schematic for a holographic imaging for breast applications is shown in Figure 14.8 and a modern commercial breast-imaging unit operating at about 3 MHz, together with an example of images on a computer monitor, is shown in Figure 14.9.

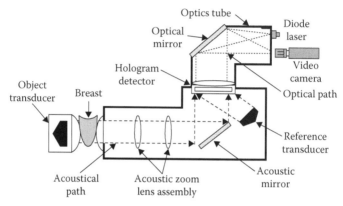

FIGURE 14.8 Schematic showing major components of a through transmission acoustic holographic breast imaging system. (Courtesy of ADI.)

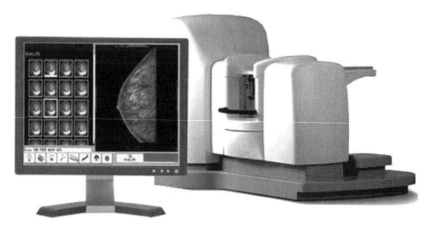

FIGURE 14.9 **(See color insert.)** Commercial breast-imaging unit, using acoustic holography, with example of image shown on computer monitor. (Courtesy of ADT.)

14.4.4.8   Miscellaneous

The potential of ultrasonic diagnosis by echo methods extends to all parts of the body. Because air and other gas have much lower acoustic impedance than either liquid or solid, air cavities produce distinct echoes. Thus, stationary air embolisms can be identified by distinct echoes. Their identification is possible because they are located in areas usually filled with fluid. Posterior-wall reflections disappear in the presence of embolisms. Moving embolisms have been located by Doppler methods [121], as discussed in Section 14.4.5. The particular interest in circulating bubbles has been in relation to decompression sickness caused by gases coming out of solution from the blood when the pressure on the body decreases too rapidly. Pulmonary embolisms can be detected and identified by echo methods. Gallstones have an acoustic impedance equivalent to that of bone and thus provide good ultrasonic echoes.

By using an aluminum transmission line, Lees and Barber [122] were able to obtain discernible echoes from the enamel–dentin junction of teeth as well as from the dentin–pulp interface. Kossoff and Sharpe [123] investigated the use of ultrasonics to detect degenerative pulpitis, a condition characterized by a degeneration of pulp tissue in which the pulp is replaced by air. It cannot be detected by radiography. Kossoff and Sharpe used excised teeth for their studies and their results appear promising.

Certain rheumatic manifestations, especially arthritis conditions, are easily diagnosed ultrasonically. However, these can also be observed radiographically and with better resolution.

The ultrasonic B-scan technique has been proved useful in diagnosing disorders of the thyroid [124,125]. Typical disorders located ultrasonically are (1) neoplastic lesions including cystic nodules, solid adenoma, and carcinoma and (2) nonneoplastic lesions such as Graves disease, subacute thyroiditis, and chronic thyroiditis. A heavily spotted pattern is usually referred to as a "malignant pattern" and should be considered a warning sign whether it is found in the thyroid area or in any other soft-tissue area.

The medical literature that reports uses of ultrasound continues to grow, and when considering an assessment of the state of the art for applications to a particular appendage or organ, the most current reviews are best identified using literature tools such as the ISI Web of Science or particular biomedical indices. Review articles are appearing almost weekly on some aspect of medical ultrasonics: examples include an assessment of ultrasound applied to the small joints of the hands and feet [126], enhanced sonography of the kidney [127], and molecular ultrasound imaging using microbubble contrast agents [128].

14.4.5   DIAGNOSIS BY DOPPLER METHODS

The Doppler principle refers to a change in frequency between the initial transmitted wave and a received wave due to relative motion between a transmitter, a target, and a receiver. Fluid moving at a continuous rate does not cause a Doppler shift in frequency, but it does affect the magnitude of the shift caused by relative motion between the three components: transmitter, target, and receiver. Also, variations in fluid motion which cause the beam of ultrasound to be deflected give the appearance of motion causing a Doppler shift in frequency.

The application of the Doppler principle to medical diagnosis is widely accepted. One particularly exciting application is for detecting fetal blood flow and heartbeat. A mother can be reassured that her unborn baby is still alive by listening to its heartbeat through a set of headphones used with such equipment.

Blood-flow rate can be measured through intact blood vessels using the Doppler principle. The carrier frequency ranges from 2 to 20 MHz, at which frequency the wavelength is short enough that significant echoes from blood particles (erythrocytes) are obtained to produce a Doppler shift. The Doppler frequency signal is calibrated in terms of velocity or detected by earphones, as mentioned earlier, and the velocity is rated from low to high according to the pitch of the Doppler signal. The method is suited for FM/FM telemetering.

Baker [129] described a pulsed ultrasonic Doppler flowmeter capable of obtaining a flow-velocity profile across a blood vessel. Ultrasonic pulses of 5 MHz and 1 μs in duration are projected into the bloodstream under study. The Doppler shift of the backscattered signals is sensed in a phase detector. The flow profile is obtained and mapped by sequential sampling at selected intervals over one period of pulsatile flow, using a comb-type gate.

The value of the Doppler principle is not limited to blood-flow measurements or indications. It is useful in detecting and monitoring motions of internal-organ components, such as heart valves and heart muscle, where the sounds generated by these motions cannot be heard. A single unit can be synchronized to record on a multichannel recorder the sequence of heart-muscle and heart-valve movements. The Doppler signal can be separated from other sound from the different parts of an organ by filtering because these signals are pitched much lower than the carrier frequency of the ultrasonic Doppler system. Thus, it has been possible to use Doppler units for measuring mitral-valve and ventricle-wall velocities [130], providing a means of detecting mitral stenosis. Opening and closing of aortic valves also can be measured [131]. The processed Doppler signals are recorded on magnetic tape simultaneously with the frequency-modulated electrocardiogram or with measurements of intracardiac pressures.

Doppler-type diagnostic units have found use in determining the severity of atherosclerosis, locating congenital heart defects and continuously monitoring fetal heartbeat during birth.

Wells [132] discusses the principles and uses of Doppler methods of diagnosis and includes a good general presentation of physical principles of ultrasonic diagnosis.

Carrigan et al. [52] report the use of 7.7-MHz pulsed Doppler measurements of blood flow in premature infants at Grady Memorial and Egleston Hospitals in Atlanta, Georgia, for evidence of patent ductus arteriosus (PDA). The ductus arteriosus is a fetal vessel which connects the left pulmonary artery with the descending aorta. Normally, it changes into a fibrous cord during the first two months after birth. A PDA is an open duct. Failure of the ductus arteriosus to close postnatally causes a definite cardiovascular handicap that can be surgically corrected. Studies have shown that approximately 29% of premature infants under 2.5 kg and 37 weeks gestational age develop PDA. The most accurate indication of PDA is obtained by measuring blood flow through the carotid artery. First indications detectable by ultrasound are zero diastolic velocities (i.e., the phase during which the heart cavities fill with blood), which later change to reverse flow velocities (i.e., flow in a direction opposite that expected in a normal circulatory system).

Deficiencies in standardization and training in the interpretation of Doppler signals corresponding to arterial flow velocity are hindrances to effective use of this valuable diagnostic technique. Rittgers et al. [134] have been able to apply modern automated quantification of flow disturbances to provide more accurate interpretation of spectral waveforms and to eliminate variations between observers. These investigators used 8-MHz Doppler recordings to investigate stenosis (narrowing or constriction) in internal carotid arteries.

Rittgers et al. [135] also have studied the use of ultrasonic Doppler flow measurements to detect renal artery stenosis in dogs. There is good reason to believe that similar measurements could be used to identify the large percentage of human victims of renovascular hypertension who could be cured by surgical or angioplasty therapy.

The importance of Doppler blood-flow measurements in the noninvasive diagnosis of peripheral vascular stenosis and occlusion has been enhanced by recent developments in instrumentation and data processing. Typical developments include (1) real-time spectrum analysis of noninvasive Doppler flow measurement signals [136], (2) automatic adjustment of the threshold for the zero-crossing FM demodulator, which simplifies setting the threshold when signal-to-noise ratios are poor [137], and (3) real-time 2D flow imaging using an autocorrelation technique which indicates the direction of blood flow and its variance by difference in color and its hue [138]. The real-time spectrum analysis increases the accuracy and specificity of noninvasive Doppler examinations [136]. In the case of poor signal-to-noise ratio conditions (for example, very small vessels or deep vessels), setting the threshold for zero-crossing FM demodulation to low results in an output that looks noisy, having an elevated baseline. Setting it too high causes a flat output between systolic peaks [137].

The earlier comment that "ultrasonics is much safer than X-rays for use in medical examination" deserves further explanation. This subject has been studied extensively throughout the world. The results of these studies have led to some differences in recommended exposure conditions, but the general consensus is that threshold limitations are advisable. According to the AIUM [14], the lowest temporal average, spatially averaged over the face of the transducer (SPTA), intensity level at which significant biological effects have been observed in mammalian tissues in vivo is 100 W/cm^2. Although the National Institutes of Health Consensus Conference [15] recommended, in 1984, ultrasound diagnosis for specific indications but not for *routine* screening of pregnancies, the European Federation of Ultrasound in Medicine [16] finds no harm in routine screening of pregnancies.

Pinamonti et al. [17] studied the effects of diagnostic-type ultrasonic irradiation of human leukocytes in vitro. Their results indicate that pulsed ultrasound at frequencies and intensities comparable with those used in diagnostic medicine produce partial breaks in cellular DNA in vitro. These breaks appear to be attributable to direct interaction between the ultrasound and nuclear structures. However, the authors deduced that the damage observed in the nuclear material is reversible as the cells were still able to multiply in the presence of a suitable stimulus. The time required for sonicated cells to repair the damage observed is not clear. Sonicated cells were able to respond to their own phytohemagglutinin reproductive stimulus, but with a 24-hour delay. This demonstrated that the pulsed ultrasound intensities were not lethal for leukocytes, but they did apparently create a metabolic imbalance. DNA extracted from sonicated cells had greater electrophoretic mobility compared with controls. Explanations proposed by the authors were a greater number of electric charges on the molecules or a decrease in molecular weight.

One may question the applicability of these in vitro results to in vivo exposures and the fact that the cells were placed in an environment during the treatment which permitted freer activity between the cells and surrounding fluids and less attenuation of ultrasound. The actual acoustic data are not clear, but intensities and exposures are such that one could anticipate the damage reported. However, these results do lend support to arguments for standards for the clinical use of ultrasonic diagnostic procedures.

Extensive studies have led to the issuance of Japanese industrial standards (JIS) for diagnostic ultrasound devices [18]. Intensity, as used in the standards, is the ratio of total (temporal average) acoustic power of the device divided by its effective radiating area. The limits these standards place on various diagnostic procedures are as follows:

- For ultrasonic Doppler fetal diagnostic equipment, the intensity shall be 10 mW/cm^2 or less (claiming a safety factor of approximately 200).
- For manual scanning B-mode ultrasonic diagnostic equipment, the intensity shall be 10 mW/cm^2 or less for each probe (for which the safety factor is estimated to be approximately 12).
- For electronic linear scanning B-mode ultrasonic diagnostic equipment, the intensity shall be 10 mW/cm^2 or less determined in a single aperture. This recommendation is only in draft form. The temporal peak intensity of the pulsed ultrasound is to be recommended when the draft is revised.
- For A-mode ultrasonic diagnostic equipment, the intensity shall be 100 mW/cm^2 or less. This standard is limited to diagnosis of the adult head and is not for pregnancy, where B-mode is the main technique used. The higher permissible intensity level is because of the attenuation through adult skull bone.
- For M-mode ultrasonic diagnostic equipment, the intensity shall be 40 mW/cm^2 or less. The M-mode is used in the clinical diagnosis of the heart. The higher permissible intensity is because of the need to obtain clear M-mode records from patients who suffer cardiac diseases. For a combination of M-mode with B-mode, the intensity is limited by the B-mode standard; that is, for the M-mode device combined with B-mode, the intensity shall be 10 mW/cm^2 or less.

The ultrasonic power in each of these cases is measured by the electronic microbalance method. Developments in scanning technologies have included combining phased-array ultrasonic 2D imaging with pulsed Doppler, and this is discussed by Halberg and Thiele [139]. False color images that display organs are now combined with images of blood movement.

14.5   THERAPY

It is commonly forgotten that the first applications of ultrasound in medicine were therapeutic. It was Loomis and his coworkers who first suggested that ultrasound could cause changes in biological tissues [1]. Since that time, there have been numerous studies that have operated across a range of powers to act to give both therapeutic and surgical effects [5,25].

Over the years, various terms and parameter regimes have been used to define domains of medical use for ultrasound in the community have evolved. Therapeutic applications of ultrasound were traditionally divided into two categories: (1) those involving inhalation therapy in which ultrasound is used to nebulizer medicine and (2) those involving direct irradiation of body tissues. The first of these is used in hospital care for treating patients with respiratory ailments such as cystic fibrosis. High-frequency nebulizers provide droplets that, when dispersed in air and inhaled, are small enough to be drawn into the alveoli and absorbed there. The second common application in therapy was for physical therapy: for the enhancement of healing in both soft tissues and bone.

Within the spectrum of physical therapies, at powers which are high enough to cause biologic effects, there are phenomena that are reported to cause both soft tissue and bone-regeneration. Paliwal and Mitragotri [140] have provided a review of the therapeutic opportunities based on the biological responses of ultrasound. These effects now see extensive use in sports medicine for physical therapy, enhancement of drug delivery, transdermal drug delivery (phonophoresis), controlled drug release, enhanced transmission through the blood–brain barrier, and in various forms of cancer therapies. Additional therapy applications have now been added which include stimulation of the immune response, thermal activation of drugs in cancer therapy and some emerging gene-based treatments.

As the power used increases, there is a continuum of interaction which transitions from "therapy," where there are beneficial effects on tissue, to "surgery," in which tissue is destroyed. In discussing these two topics there is, in some ways, an artificial distinction, as in many cases the same equipment can be used. The transition to surgery from therapy occurs simply through either increased duration of treatment or increase in the power used.

The various destructive interactions can be considered in several groupings. These are most commonly identified as (1) HIFU, which uses a heating mechanism; (2) lithotripsy, which uses focused shock waves to break up hard matter, such as kidney stones; and (3) a number of other high power–low frequency interactions, all using a horn or tip in contact with tissue to cause disruption. These include phaco-emulsification (cataract removal), soft tissue removal (with aspiration), and various cutting and dissection procedures using ultrasonic (harmonically driven) scalpels and dental instruments. These various topics are considered in the sections in this chapter.

The mechanisms of interaction for ultrasound with tissues are known to be increasingly complex as power is increased, and subtle effects, at the cellular level, do appear to be occurring in living tissues that are still not completely understood. Some of the issues and literature relating to mechanisms was discussed earlier in this chapter. That said, a large number of maladies have been treated ultrasonically and, in some cases, with a degree of apparent success not obtained by any other treatment.

As with much of medical ultrasound, the therapy literature has become extensive, with most written by and for the clinical community. The medical physics and engineering aspects of these activities have been considered in its own distinct literature, and it is this which is mostly considered as the basis for the current section.

14.5.1 EQUIPMENT

The therapeutic equipment market is smaller, and systems more diverse, than those used in ultrasonic imaging. These systems introduce considerably more energy into the body than do diagnostic instruments and higher power; in some cases, they merge into what is used in some classes of surgical application. The equipment can be found listed under suppliers who seek to meet the equipment needs of minimally invasive treatments and general "radiation" systems, which can cover both nuclear and acoustic/thermal technologies. Except in rare cases of specialized local treatment applications, commercially available therapeutic equipment is limited in maximum power output to approximately 3 W/cm^2. Coupling is usually through a heavy film of oil or grease, but immersion methods are sometimes used. Good coverage of couplant over the area to be treated is essential. Air bubbles inhibit energy transfer and overlapping an area not covered with couplant can produce a painful burning sensation.

Ultrasonic nebulization of pharmaceuticals occurs without producing destructive temperature levels. Units have been made for atomizing very small quantities of material for inhalation therapy. The particles of liquid enter the body at room temperature, which is desirable both from the standpoint of comfort to the patient and from that of nondecomposition of the pharmaceutical material being inhaled.

Drug nebulizers are designed to efficiently target drug delivery to the respiratory tract. Dennis and Hendrick [141] provide a useful review of the design characteristics for such units, including methods for accurate measurement of aerosol dose and particle (drop) size. An example of a commercial nebulizer for hospital use is shown in Figure 14.10.

A version of the same technology is also now available in the form of ultrasonic humidifier, which can also be used to disperse therapeutic agents for those with breathing problems. An example of a home humidifier is shown in Figure 14.11.

The second class of ultrasound therapy equipment is used in physical therapy, with which there is external application of energy, particularly into soft tissue. Such units come in many designs and tend to increasingly have built-in computer-based controls. An example of such a commercial unit is shown as Figure 14.12. This unit operates at frequencies between 1 and 3.3 MHz and can be set to continuous or pulsed modes, with various duty cycles. The maximum intensity delivered is 2.2 W/cm^2.

14.5.2 PHYSICAL THERAPY

Ultrasound has been used in physical therapy for more than 50 years. In early work, the powers and exposure times used were such that lesions could occur. Therapeutic

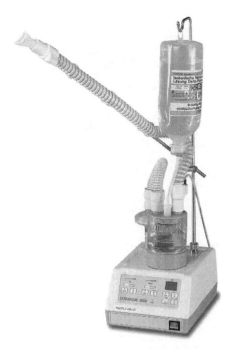

FIGURE 14.10 Example of commercial ultrasonic nebulizer, for hospital use. (Courtesy of Nouvag.)

FIGURE 14.11 Domestic ultrasonic humidifier in operation.

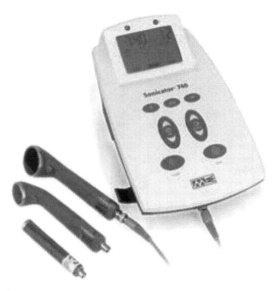

FIGURE 14.12   Example of commercial ultrasound physical therapy unit used to give externally applied treatments. (Courtesy of Mettler Electronics Corp.)

applications are now only considered to be those where advantageous tissue effects occur. It is used in the treatment of joint and soft tissue ailments such as bursitis, muscle spasms, traumatic soft tissue injuries, some types of collagen diseases, and in hyperthermia treatment where circulation is assisted [12].

Ultrasound is used as a therapeutic agent that can deliver heat and also cause nonthermal cell stimulating effects [142,143]. The acoustic intensity used varies from zero to a few watts per square centimeter: the pressure amplitude range is up to 0.3 MPa and the driving frequency is typically in the range from 0.75 to 3 MHz. In most cases, either a continuous wave (CW) or long tone-burst is used (a few milliseconds in length). The regulatory performance of ultrasonic therapy and equipment were reviewed in 1980 [144] and many of the standards cited remain in force.

In physical therapy applications, it is clear that it is the heating/absorption effects that are used. It is still debated as to whether nonthermal interactions and including cavitation occur in vivo, and if they are significant. Some changes have been reported in the membrane transport properties of frog skin after exposure to ultrasound at the power used in physical therapy [145]. There are also reports of additional free radicals being detected that may have been produced by cavitation [146]. However, much of this work has been performed in vitro where the conditions are very different from those, in vivo, so the finding may simply not apply to the clinical situation. In the debate, regarding the possibility of cavitation in vivo, it has been noted that in tissue, the viscosity is higher than free liquids, there are cell boundaries, there are numerous organic surfactants present, and there is an absence of plentiful cavitation nuclei that are found in many in vitro studies. At the power levels used in therapy, it is not clear if cavitation can occur. From the observations that have been reported, the exact mechanisms involved in ultrasound for physical therapy remain the subject of debate and further study is required [26].

In spite of the fact that millions of physical therapy treatments using ultrasound are now being performed annually, and there is a history of more than 60 years of clinical use, the effectiveness of ultrasound in treating pain, musculoskeletal injuries, and soft tissue lesions remains questionable. Physical therapists have tended to overlook the tenuous nature of the scientific basis for this form of therapeutic ultrasound provided in the scientific literature [147].

Robertson and Baker [148] have provided a review of the studies that have investigated the effectiveness of therapeutic ultrasound and concluded that, as reported in the literature, there is little evidence to show that the application of active therapeutic ultrasound is more effective than the placebo ultrasound for treating people with pain or a range of musculoskeletal injuries or for promoting soft tissue healing. Few of the studies reviewed were, however, deemed to have been performed with methods adequate to examine the wide range of patient problems and the dosages used varied considerably, often for no discernable reason. In a companion article, Baker et al. [147] have reviewed the biophysical effects in therapeutic ultrasound and conclude that there is currently insufficient biophysical evidence to provide a scientific foundation for clinical use of therapeutic ultrasound for treating pain and soft tissue injury.

In spite of the lack of adequate understanding, there are demonstrated cases, which include the personal experience of the author, with regard to its effectiveness, or at least apparent effectiveness, and the debate regarding efficacy of this treatment continues. This topic remains an area for investigation, better quantification of the treatments used in clinical settings, and investigation of the fundamental bioeffects which are resulting from the applied ultrasound to injured soft tissues. The combination of the use of ultrasound in physical therapy with its use for the delivery of therapeutic compounds is discussed further in Section 14.5.4, which considers sonicated drug delivery.

Diagnostic and therapeutic applications of ultrasound to bone are also both of importance to the biomedical ultrasound community. Two special issues of the IEEE Transactions Ultrasonics, Feroelectrics, and Frequency Control (UFFC) provide articles that give a good review of the state of knowledge [149,150]. Low-intensity ultrasound operating at various frequencies is used in the treatment of fractures and other osseous defects. Pounder and Harrison [151] have provided a review of both the clinical evidence and associated biological mechanisms of action. They report that low-intensity pulses ultrasound as having been demonstrated to accelerate in vivo all stages of the fracture repair process. An article by Kaufman et al. [152] provides an overview of the simulation of ultrasound in bone, including discussion of several examples of models of wave–bone interaction. As with the soft-tissue ultrasound interactions, those for bond remain topics for investigation and better quantification of the treatments used in clinical settings and investigation of the fundamental bioeffects that are resulting from the applied ultrasound to injured bond are needed.

14.5.2.1   Rheumatic and Related Disorders

Although ultrasound has been used for the treatment of rheumatic diseases, few articles describing techniques and results are available. Certain types apparently are benefited by such treatment, whereas in others, such as rheumatoid arthritis, the treatment produces only a transitory analgesic effect. Diseases reported to have been favorably influenced are osteoarthritis, chondrosis, and spondylosis [153]. Satisfactory results have been achieved with rheumatoid spondylitis, but other inflammatory forms of rheumatic disease show little or no response.

Kaclova et al. [154] conducted some laboratory experiments that may have clinical significance with reference to treating certain forms of arthritis. These investigators exposed the right tibias of male Wistar rats to ultrasound at therapeutic intensities for 5 minutes. Each rat was given an intraperitoneal injection of $^{45}CaCl_2$ 48 hours before it was killed. They found that the percent of ^{45}Ca retained in tibias and scapulas was lower in the treated animals killed 4–120 days following exposure to ultrasound than in nonexposed animals used for controls. The amount of ^{45}Ca retained in the incisors was the same in the treated animals and controls. However, when the incisors were irradiated directly, the intake of ^{45}Ca was less in the exposed animals than it was in the controls, as revealed by examinations 8 and 28 days following exposure. These investigators claimed that changes in intake of ^{45}Ca were not restricted to the exposed organ but occurred in all calcified tissues.

Ultrasonic therapy is effective in treating muscle spasm. When the nerve segment with which the affected limb is associated is ultrasonically irradiated, the result is relaxation of tension, significant loosening, and improved circulation in the limb. The pain is relieved, which leads to a decrease in hypersensitivity, and the improved circulation leads to regenerative processes in the limb and joint.

14.5.3   SONICATED DRUG DELIVERY

The use of ultrasound to enhance diffusion and transport of materials through polymer membranes was discussed in Chapter 11. These phenomena are also being applied by the medical community to enhance transport through biological membranes and tissues, which include the skin, the BBB and bulk organ and muscle tissue.

14.5.3.1   Phonophoresis

Phonophoresis is defined as the movement of drugs through living, intact skin, and soft tissue under the influence of an ultrasonic wave-field. There is an extensive literature that dates back to the 1950s that discusses this phenomenon which is also found reported under the names sonophoresis and ultrasophoresis.

In 1984, Skauen and Zentner [155] provided an overview of the literature, which up to that time came mostly from Eastern Europe. They concluded that phonophoresis was useful in enhancing drug therapies, including when used with ultrasound in physical therapy to deliver materials such as hydrocortisone ointment to treat tissues in inflamed joints.

Mitragotri et al. [156] reported on a mechanistic study of ultrasound transdermal drug delivery. The study used 1–3 MHz ultrasound in an intensity range up to 2 W/cm^2. There were reported to be indications that cavitation occurred, but that other phenomena were also occurring, and transport was enhanced. With the development of a growing number of peptide and protein drugs, interest in determining the effectiveness of transdermal delivery increased, and Amsden and Goosen [157] investigated the mass-transport mechanisms involved. The underlying mechanisms are, however, still not completely understood [158], but several phenomena, including cavitation, thermal effects, inductive convective transport and mechanical effects, such as those because of stress resulting from pressure variation appear to be involved. These phenomena are similar to those reported in Chapter 11, in which ultrasound enhancement of transport through polymer membranes was discussed.

It has now been demonstrated that transdermal drug delivery does provide a method to control delivery of a range of compounds including insulin, interferon, and erythopoietin through human skin. Smith (2008) [159] has provided a general review of the transdermal drug delivery literature, including a discussion of the acoustic parameters for systems using this phenomenon. Whatever the mechanisms involved, there does not appear to be damage occurring to inert polymer membranes or tissues at the powers commonly used, and the enhanced permeability of tissues, seen under the effect of ultrasound, is reported posttreatment to return to pretreatment levels.

14.5.3.2   Diffusion of Subcutaneous Injections

The increased permeability of membranes and cell walls under the influence of ultrasound has been used to good advantage. Physicians report better results from cortisone injections when they are followed by ultrasonic therapy about the area of the injection. This is also true with many other types of injections.

The exact mechanisms that occur in such treatments remain the subject of debate.

14.5.3.3   Blood–Brain Barrier

The BBB is a major hurdle encountered when seeking to deliver drugs to treat many serious neurological disorders. From early in the history of the development of ultrasound in the 1950s, there were investigations of its application to the BBB. Bernard et al. [160] considered the effects on the nervous system of a cat with ultrasound applied through a hole in the skull. Research in this field has continued, and Vykhodtseva et al. [42] provide a comprehensive review of the history of ultrasound-BBB research and recent advances in ultrasound-induced BBB disruption and the resulting changes in transport phenomena.

14.5.3.4   Ultrasonic Gene and Drug Delivery and Activation

Sonodynamic therapy is the term used for ultrasound-dependent enhancement of cyotoxic activities of certain sonosensitive compounds that interact in vitro and in tumor-bearing animals. Ultrasound, usually at 20 kHz or higher frequencies, is delivered as focused energy on malignancy sites buried deep in tissues to activate preloaded sonosensitive compounds.

This emerging field was reviewed by Rosenthal et al. [161]. It is seen in the literature that ultrasound has an increasing role in the delivery of therapeutic agents, including genetic materials, protein, and chemotherapeutic agents [162]. Cavitation, as well as heat and several other mechanical mechanisms all appear to contribute to the interactions, disrupting structures, and in making membranes more permeable to drugs. Such processes are, at a fundamental level, most probably closely related to mechanisms that occur with phonophoresis and the enhancement of membrane permeability. Several studies have sought to understand the role of cavitation in ultrasonically activated drug delivery [163], as well as the related gene therapies and these remain topics of ongoing research.

It has long been shown that therapeutic ultrasound can be used at higher powers to perform surgery and ablate tumors, and a variety of cancers are currently treated using this approach. The literature for ability of the nondestructive interactions of ultrasound with tissue to increase in drug and gene effectiveness using ultrasound is less mature, but is now growing significantly.

One approach to the delivery of both genes and therapeutic agents is through the modification and use of ultrasound microbubble contrast agents [41,164]. Frenkel [165] provides a discussion of ultrasound-mediated delivery of drugs and genes to solid tumors, and Tachibana et al. [166] also reviews the subject. They conclude that research on the bioeffects of ultrasound with the presence of various drugs in patients has only just begun. Understanding of the micro- and nano-scale interactions is still required. However, the early work is showing promise.

14.5.4 Miscellaneous Medical Therapy Applications of Ultrasound

14.5.4.1 Ophthalmic Therapy

Greguss and Bertenyi [167] applied ultrasound at 850 kHz at intensities less than 1.5 W/cm^2, whether pulsed or CW, to patients with myopia. Pulsed ultrasound was preferred to minimize heating. They found that, in some cases, the degree of myopia can be decreased by ultrasonic irradiation. However, the effect is slight, and the treatment had to be repeated every 6 months in order to achieve a constant and stable result. The probability of positive results diminishes with age, becoming zero above the age of 20 years. These authors concluded that ultrasonic treatment of myopia is indicated only for children if they exhibit increased myopy during puberty. The treatment, they claim, could slow down or even stop myopia. They attribute ultrasound's effectiveness in treating myopia to its effect on eye muscles.

14.5.4.2 Effects on Paced Hearts

Mortimer et al. [168], working with New Zealand rabbits, have shown positive inotropic effects of ultrasound on hypoxic hearts. Therapeutic ultrasound produced a marked potentiation of myocardial contractility.

14.6 SURGERY

As with the therapy application of ultrasound, surgery using this form of energy has a long history [61, 63], and it has seen a resurgence of interest in the past 10–20 years [169]. Ultrasonic surgery is any process in which the destructive nature of ultrasound, at high intensities, and its ability to rupture tissues and/or promote erosion of hard tissues, in a form similar to ultrasonic machining of brittle materials, is used.

The various destructive interactions between ultrasound and tissue can be considered in several groupings:

- HIFU which uses a heating mechanism
- Lithotripsy which uses focused shock waves to break up hard matter, such as kidney stones
- High-power, low-frequency, tissue fragmentation, using a horn or tip in contact with tissue to cause disruption
  - Phacoemulsification (cataract removal)
  - Soft tissue removal (with aspiration) for tumor-debulking and neurosurgery
  - Various cutting and dissection procedures using ultrasonic (harmonically driven) scalpels
  - Dental instruments, of several types

The powers used are typically 10 W/cm^2 or greater at the site of application and can be as high as a kW/cm^2. The waveforms used can be in the form of the short pulse, as with shock wave in lithotripsy, or the CW used in high-power focused heating, in localized hyperthermia.

14.6.1 EQUIPMENT

In discussing the two topics of ultrasonic therapy and surgery and the equipment used, there is, in some ways, an artificial distinction, as much of the same equipment can be used for both purposes. The transition to use for surgery simply occurs through either increase in duration of treatment or increase in the power used.

The ultrasonic equipment used for surgery divides into three distinct families of types of units that are each having specific applications: (1) HIFU, (2) lithotripsy, and (3) tissue fragmentation, with aspiration, using a horn or tip in contact with tissue to cause disruption. In addition, there are also a number of designs of ultrasonic scalpels and used in various cutting and dissection procedures and ultrasonically driven dental instruments.

The equipment used in surgery and related application range from small hand-held units are used in dentistry that are, in many ways, similar to those used in phacoemulsification (cataract surgery) to larger treatment units, which include a capability to hold and position a patient with, in most cases, computer control and integrated imaging capabilities, as used in lithotripsy and HIFU.

Across the various families of units, there are a growing number of manufacturers providing systems, and the numbers of surgical ultrasound units in use is reported to be increasing. However, the market remains significantly smaller than that for diagnostic ultrasound units.

14.6.2 HIGH-INTENSITY FOCUSED ULTRASOUND—HYPERTHERMIA

The use of focused high-power ultrasound to destroy tissues deep inside the body is now in routine use in treating a range of cancers. The process uses the temperature

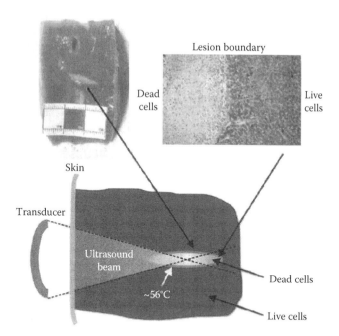

FIGURE 14.13   (See color insert.) Focused ultrasound surgery. (Reprinted with permission from ter Haar, G., *Phys Today*, 54(12):29–34, © 2001, American Institute of Physics.)

rise in the focal zone, typically to about 56°C, to kill cells. This process is illustrated with Figure 14.13, which shows a schematic for a HIFU beam, together with postexposure-sectioned tissue. The figure illustrates the sharp boundary, which is found to occur between dead and living cells [170], and which is directly related to the local maximum temperature.

This technology has been developed over a number of years, with significant levels of research and development effort.

14.6.2.1   Cancer

Some past research on the use of ultrasound in the treatment of cancer has been reported, and some of the reported results have been encouraging. Only extremely high intensities were used in the earliest work, an approach that is open to question. The high intensities may produce metastases. The problems of completely locating, identifying, and thoroughly irradiating the diseased tissue are always present.

Typical high-intensity studies of the use of ultrasound in treating cancers were conducted by Soviet scientists [171,172]. They reported treating 400 rabbits no earlier than 10 days following inoculation with Brown-Pearce tumor, by which time the tumor is very large. They claimed a minimum traumatic effect while attaining a selective effect in the tumor. There is some edema following irradiation, sometimes accompanied by slight skin maceration. Their results indicate that the controlled application of such high-level sound (on the order of 500 W/cm²) constitutes a means of destroying at least some types of malignant neoplasm in humans. The irradiation caused the "liquidation (resolution)" of the entire malignant tumor, according to the reports. Their histological investigations reportedly showed a picture of a gradual growth of dystrophic processes in the tumor tissue of the irradiated nodule and of nonirradiated metastases and a rapid reduction in the ability of tumor cells to reproduce. The treated animal not only was healed but also became immune, causing the authors to conclude that they were on the track of a serum for cancer. They claimed that they definitely found antibodies to Brown-Pearce tumors in irradiated specimens. The exposure time and frequency of application were not reported. Transplantation of the same type of tumor to the experimental animal could not be accomplished after irradiation. This implies that a type of immunity to cancer has been produced.

According to the same Soviet reports, nine human patients with melanoblastoma were treated. An irradiated tumor, with pronounced metastases in the first stage, in one patient was irradiated ultrasonically and consequently liquidated. Discussion covering the remaining eight patients was somewhat apologetic, explaining the difficulties of irradiating this type of tumor.

The objective of these earlier investigators appears to have been to cure the cancer by physically destroying the tumor cells with stress and any other phenomena associated with an ultrasonic field. Because some positive results were obtained during these early investigations, further systematic research aimed at identifying the mechanisms responsible for producing these good results is justified. At the same time, adverse and toxic effects are to be recognized, and they must be suppressed or eliminated in order to have an effective cancer therapy. Hyperthermia (artificially induced temperature rise, or fever, in body tissues) has been shown to cause tumor remission. Because acoustic absorption causes an increase in temperature in body tissues, it is logical to assume this to be a major factor in the beneficial effects previously observed.

P. P. Lele, professor of experimental medicine and director of the Harvard-MIT Hyperthermia Center, briefly reviews the history of hyperthermia in the treatment of cancer and states, "Clinically, hyperthermia has emphatically demonstrated antitumor effects in patients in whom all conventional modes of therapy had failed" [173]. In reviewing the historical development and methods of hyperthermia, Lele says, "Accounts of early medical practices indicate that hyperthermia at temperatures above 41°C has been sporadically applied in the treatment of cancer since as early as 2,000 B.C." Radiation therapy and chemotherapy are ineffective in more than half of patients with cancer. This record and the "recurring observation that cancer patients who developed high fevers often experienced spontaneous remissions" also were incentives for investigating controlled hyperthermia as a means of treating cancer. Other methods, such as electromagnetic radiation (EM) at frequencies of 8–2450 MHz (8–100 MHz for deeper tumors; the higher frequencies for shallow tumors) have been used and are widely available for local or regional hyperthermia and as laboratory systems. Ultrasound has proved to be superior to EM for localized heating of deep tumors.

Basic principles of acoustic wave propagation are used to design treatments to optimize ultrasonic absorption within tumors and to minimize adverse effects to surrounding tissues. Localization of thermal effects is accomplished by focusing. Generally, it is not possible to focus energy in a volume having dimensions smaller than a wavelength. The ratio of the wavelength of EM at typical frequencies used for hyperthermia to the wavelength of typical ultrasound is in the order of 500:1. According to Lele [173], "If one attempts to use 500 mm wavelength EM … for heating a tumor 6 cm in diameter, situated 4 cm below the skin in the upper abdomen, the entire abdomen including the liver, spleen, and the major blood vessels are subjected to radiation, producing a regional hyperthermia, rather than local hyperthermia (LHT) of the tumor alone. Since more than half of the cardiac output passes through this region, the heat carried away in the bloodstream raises the system temperature. This has been one of the major limitations in the use of annular electromagnetic arrays for heating tumors in the chest, abdomen and pelvis. On the other hand, ultrasound with a wavelength of only 1 mm can be focused within the volume of the tumor and selective heating of the tumor is possible, requiring much less energy to be radiated into the body than for EM heating."

Lele also states, "Although hyperthermia, when given alone in a new or previously treated tumor is found to have antitumor effects, its efficacy is enhanced remarkably if it is delivered in combination with other modalities of cancer therapy."

Several investigators have evaluated ultrasonic hyperthermia. The intensive work at Harvard-MIT Hyperthermia Center under the direction of Lele is especially significant. The work at this center emphasizes obtaining maximum acoustic absorption within the tumor without producing toxic effects in surrounding tissue. To accomplish this, it is first necessary to recognize the nature of a tumor and certain physical relationships between it and surrounding tissue. Again from Lele [173], Tumors of a clinically relevant size consist of three zones with different degrees of vascularity. The central zone or the core is ischemic (deficient of blood) and necrotic (dead), the intermediate zone has the perfusion "normal" to the tumor, and the margin has higher blood perfusion because the advancing edge of the tumor induces development of new blood vessels…. It is this region (the well-perfused tumor margin, infiltrating the tumor bed) which must be heated to higher, therapeutically adequate temperatures rather than the core.

Therapeutic temperatures of 42.5°C–43.0°C sustained for 20–30 minutes in this perfused tumor margin is essential to effective treatment of the tumor [173–176]. To heat this perfused margin to a therapeutic temperature for the required length of time without overheating surrounding normal tissue, the hyperthermia induction equipment must have the following capabilities [174–176]:

- Focusing or beam convergence for noninvasiveness.
- Use of appropriate frequency and aperture to localize the heating of the target volume and prevent excessive heating of deeper tissues.
- Scanning to distribute the energy through the tumor, with the scan speed sufficiently high to maintain the temperature elevation regardless of heat dissipation. The scan trajectory must be flexible to heat tumors of different shapes and adjustable to obtain the desired temperature distribution during treatment.

- Intensity modulation to compensate for different path lengths in tissues overlying the treatment field, to prevent excessive temperature rise in tissues distal to the focus, and to protect critical heat-sensitive tissues in the insonation field.
- Temperature measurement capability and feedback control of insonation intensity to monitor and maintain the temperature at specific sites at the desired level.

In order to meet these requirements, personnel at the Harvard-MIT Hyperthermia Center developed a system they call "the scanned, intensity modulated focused ultrasound (SIMFU) system." This system is capable of a large variety of scan patterns controlled either through keyboard entries or through a joystick programmer. The patterns most commonly used include concentric ellipses and circles and inverse spirals, that is, spirals with progressively closer spacing at increasing distances from their center. The scan pattern cycle time ranges between 1.0 and 8.5 seconds. The system is capable of scan patterns of up to 23 × 23 cm; however, to date, the maximum scan pattern used is a 15-cm ellipse. The SIMFU system uses transducers of three different diameters up to 16 cm operating at 10 frequencies between 500 kHz and 6 MHz and a number of interchangeable lenses from "ultrawide angle" to "extreme telephoto" focal length in order to provide the versatility to meet various clinical needs. The numbers of frequencies and lenses are more than are needed to fill the clinical needs [173]. The energy is coupled into the body either (1) through a fluid-filled bag within which the transducer and lens system is located or (2) through an open water bath, which offers greater flexibility and is, therefore, preferred [174].

Because the effectiveness of hyperthermia in treating cancer is highly dependent on operation within an optimum temperature range, monitoring temperature in the tumor is essential for proper therapy. Temperature monitoring in the SIMFU system is provided by using miniature thermocouples or thermistors inserted by means of a hypodermic needle. Although viscous heating results from relative motion between the metallic components of the heat sensor and surrounding tissues, this heating "is a transient problem with a short relaxation time and is not of much practical significance under steady state conditions" [173]. Metallic sheaths (such as hypodermic needles) are, therefore, preferred to sheathing composed of biologically inert and tissue-compatible plastic material, which may have ultrasonic absorption coefficients significantly higher than those of the metallic components or of the tissues.

One very important factor contributing to the successful use of the SIMFU system in treating deep tumors (depths from 3 to 12 cm) at the Harvard-MIT Hyperthermia Center is their individualized therapy planning before treatment [173–176]. This procedure is used to optimize the insonation conditions to maximize the time-averaged ultrasonic intensity in the treatment volume to minimize the intensity in the intervening and surrounding tissues and to preclude insonation of any heat-sensitive, critical normal tissues. It also helps determine the location and placement of thermometric probes. Optimum insonation parameters are determined for each tumor by a Therapy Planning Program (TPP) unique to the SIMFU system.

Therapy planning with the TPP is a closed-loop, iterative process which includes the following steps [175]:

1. Visualization: This is the first step, which includes developing an accurate model of the patient's internal anatomy. A 3D model is built from 2D computed tomographic scans. A polygonal mesh model of surface, organs, tumor, reflective interfaces, and sensitive tissues is constructed. Values of ultrasonic attenuation and impedance are assigned to each set of tissues in the model.
2. Parameter optimization: The functions of this step include (1) choosing frequency appropriate to the tumor depth and size, aperture and focal length of lens, scanning pattern, and scanner setup parameters; (2) checking the ultrasonic field, setup geometry, scanner motion, and ultrasound coupling; and (3) modifying parameters until all are satisfied. Distal and proximal tissues that must not be heated are identified, and these factors control the TPP's choice of beam aperture and angle of incidence. The focal length of the lens is determined by the depth of the tumor and the transducer clearance above the portal. The ultrasound intensity within the focused beam is calculated considering the effects of reflection and refraction of the ultrasound rays at tissue interfaces and also the effect of the difference between the focal length within the attenuating tissue and the geometric focal length, the difference being dependent on frequency and coefficient of attenuation.
3. Thermal modeling: The temperature field and hyperthermia dose are predicted from iterative calculation of time-averaged specific absorption. The ultrasound beams are modeled as superposed sources, producing heat at rates depending on the local intensity. The ultrasound intensity at the transducers is modulated throughout the scan cycle to accommodate differences in the ultrasonic and thermal properties of the various tissues and to avoid insonating critical tissues or reflecting interfaces. A mean specific equivalent dose throughout the tumor volume, as well as for the surrounding tissues, is calculated using the temperature–duration relationship for morphological injury to solid mammalian tissues in vivo as well as comparable data for cells in suspension cultures in vitro.
4. Treatment: The hyperthermia treatment is delivered based on the plan developed in the previous steps. Tissue temperatures are measured at locations determined from the predicted isotherms developed in step 3. The actual measurements, or extrapolated isotherms if appropriate, are compared with the predictions of the TPP and, if the predictions and the measurements do not agree, then the tissue perfusion and heat conduction parameters are adjusted appropriately and used in planning the patient's next treatment.

Controlling the intensity by focusing in order to localize the hyperthermia treatment within the planned volume without excessive exposure to other tissues includes careful consideration of attenuation properties (including convergence and divergence effects and absorption within tissues) and wavelength. For a focused transducer, the maximum depth to which selective heating can be achieved is a function both of the frequency and of the aperture. Ideally, when rays from a focused source do not bend, gain in intensity with progressive convergence toward the focus follows the inverse square law. However, if attenuation is constant, energy lost by attenuation follows an exponential function. Thus, for a beam of a given angle of convergence and for a given attenuation in the tissue, there is a certain depth beyond which there is no effective gain in intensity. Therefore, the choice of the proper frequency, which is crucial to localization of hyperthermia, and of the aperture is an exercise in optimizing these conflicting requirements for the size and depth of each specific tumor and the ultrasonic attenuation/absorption properties of the tumor and overlying tissues.

The results of Lele's work were very encouraging. Sustained heating of tumors for 20–30 minutes at temperatures of 42.5°C–43.0°C is essential for effective therapy. Heating to subtherapeutic temperatures leads to the development of thermotolerance and eventual recurrence [176]. Reports of treatments using the SIMFU system to produce recommended therapeutic temperatures in tumors consistently refer to "extremely low incidence of toxicity" or "the absence of local or systemic toxicity" [174]. After reporting clinical results with the SIMFU system [177], Lele concluded, "The SIMFU system offers the flexibility and versatility for rapidly heating both superficial and deep tumors in a noninvasive, safe and adequate manner. To have any significant impact in cancer therapy, it is essential that hyperthermia be used in combination with the other primary modalities at an earlier stage of the disease, rather than as the last modality after everything else has failed."

14.6.2.2 Neurosonic Surgery

Very high-intensity ultrasound produces, under properly controlled dosage conditions, selective changes in the central nervous system. Pioneering research at the Bioacoustics Laboratory of the University of Illinois led to applications in humans, such as the alleviation of tremors because of Parkinson's disease. The researchers conducted thorough investigations into mechanisms and dosages and developed electronic generators, transducers, and devices for proper orientation of the patient's head with respect to the location of the transducers. With the University of Illinois apparatus, the skull cap was removed and the skin was attached to a special metallic pan, which was then filled with degassed physiological saline. The transducers were then precisely located in the saline bath so that the energy would be focused in the desired part of the brain.

There are other, definitely less expensive, methods of treating Parkinson's disease, which include the use of probes that either pierce the region to be treated and generate heat at the point to be treated or freeze the tissue in the region. These methods damage intervening tissue and also present the risk of rupturing blood vessels. The advantage of the ultrasonic method is in the focusing so that tissue between the focal region and the transducer suffers little damage and no instrument needs to penetrate the brain. The main disadvantages of the method are the requirements for removal of the skull cap and specialized equipment to perform the operation.

Fry [179] considers the selective action of ultrasound and its possible application in fundamental neurological research, in neurosurgery, and in gaining an understanding of basic cellular mechanisms to be among the benefits to be derived from the use of focused ultrasound in forming lesions in the central nervous system.

Fry points out, as does Lele [180], that white matter (fiber tracts) is more easily disrupted (i.e., lower intensities are required) than gray matter. Fry also states that the most desirable condition of application is to use an irradiation procedure which combines the choice of dosage conditions for selective action and focusing, rather than focusing irrespective of selectivity.

Lele [180] claims that focused ultrasound is used successfully in several laboratories of basic neural sciences without the continual assistance of physicists and electronics technicians. Applications of focused ultrasound bringing satisfactory results at the Massachusetts General Hospital include transdural irradiation of superficial cortical epileptogenic foci in cases of posttraumatic epilepsy, commissural myelotomy of the spinal cord for intractable pain and for anterior horn neuronolysis in a case of multiple sclerosis, and irradiation of painful subcutaneous neuromata [180].

Focusing of ultrasound for neurosurgery may be accomplished either by shaping the piezoelectric transducer in the form of a bowl or by using focusing lenses with either a single transducer element or an array of elements. Focusing lenses of Lucite have proved to be satisfactory for this purpose in spite of high attenuation within the lens material.

The size of the smallest lesion that can be made is inversely related to frequency because the size of the focal region depends on the wavelength of the ultrasound. Lele claims that at a frequency of 2.7 MHz, the smallest lesion that could be made with certainty was 1.2 mm in length and 0.4 mm in diameter. Lesions that could be made consistently with single pulses were 11 mm long, while approximately 17 mm was the maximum length of lesions that could be produced consistently using multiple pulses.

In related research, Fry et al. [181] were able to visualize ultrasonically essentially all soft-tissue fluid-filled space interfaces and, in some instances, interfaces between gray and white matter by using an acoustically transparent window in the skull for transferring the ultrasound. By using this diagnostic technique, lesions produced ultrasonically or by any other means can be visualized immediately after placement.

14.6.2.3　Applications of Hyperthermia

This technology and its application have matured significantly. Hyperthermia is in routine clinical use to manage a range of cancers and benign diseases. Ultrasound can be used either alone, with increase in temperature to 50°C or more or in combination with other therapies, including chemotherapy, immunotherapy, and gene therapy [181]. The frequency ranges used in such treatments are typically in the range 500 kHz to 3 MHz. The dose is typically in the range where power levels can be 20 kW•cm^2, applied for 300 μs, to 200 W•cm^2 applied for 10 seconds. With this technology, there is the requirement to provide good focusing and beam control and to be able to deliver power at higher frequencies, which enable better focusing. The higher frequencies have the added challenge that as frequency increases, so does the attenuation in surrounding tissue and that between the source and the site of treatment. Recent advances in this technology have included the development of the transducers and equipment to operate at both higher and multifrequencies. Chopra et al. [182] demonstrated transducers operating at both 4.7 and 9.7 MHz for use in thermal therapy.

In hyperthermia, the interaction is based predominantly on a thermal mechanism, and models of ultrasonically induced heating have been developed. These were reviewed by Lizzi and Ostromogilsky [183] and Meaney et al. [184]. Several studies have also now shown, using histology, that experimental results and the models are in good agreement and the required lesion can be well predicted. Additional confirmation of the performance of hyperthermia has been provided through the use of MRI technology, which has been used to investigate and quantify temperature effects during HIFU therapy [185].

14.6.3 SHOCK WAVE LITHOTRIPSY

Early experiments proved that it is possible to pulverize ureteral stones by ultrasonic means by placing these stones in a beaker of water and irradiating them with high-intensity ultrasound. The disintegration was caused by cavitation-induced stresses and relative motion and impact between particles, and between particles and the beaker walls. Similar mechanisms are now made to operate within a body to pulverize and discharge stones by a technique, which focuses a high-intensity burst of ultrasound through the body onto the stones. Another approach involves the insertion of probes into the ureteral tract through a catheter. The stones are disintegrated by impact and shear action by the probe tip. Extreme care is necessary in either operation to avoid damaging the surrounding tissues.

In shock wave lithotripsy, a repetitive pulse excitation is used; this is focused using an elliptical mirror set in a water bath as shown in Figure 14.13. The ultrasonic is concentrated near the object which is to be destroyed. As the ultrasonic field comes to focus, transient cavitation occurs. This results in shock waves which can breakup solid particles, such as kidney stones as shown in Figure 14.14.

For "clinical" or bulk interactions, with lithotripsy reparative pulses with a large positive pressure peak, as large as 100 MPa, the negative peak seldom exceeds 15 MPa (because of cavitation). The pulse repetition frequency is normally synchronized with the heart beat, and this keeps the peak acoustic intensity quite small.

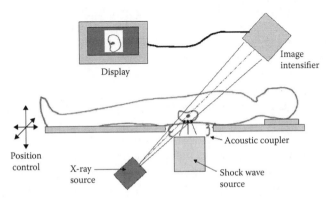

FIGURE 14.14 A schematic for a typical shock wave lithotripsy system used for treating kidney stones.

The cavitation and subsequent shock wave process in the solid are now reasonably well understood. Analytical models are available, such as those reported by Coleman et al. [186] for cavitation generated by an extracorporeal shock wave lithotripter.

The characterization in vivo of the acoustic cavitation during lithotripsy was investigated and reported by Cunningham et al. [187]. Bailey et al. [38] provide a review of the current understanding of the science and the technology for lithotripsy as part of a wider review of physical mechanisms of the full range of therapeutic ultrasound methodologies.

There remains significant ongoing research into both the technologies, including the transducers, and also into other applications for these systems. One novel application has been the investigation in vivo into transfection of melanoma cells. Bao et al. [188] report the demonstration of this approach, and it seems to show that a combined gene and shock wave therapy might be advantageous.

14.6.4 TISSUE DISSECTION AND ABLATION

In addition to HIFU and lithotripsy, there is also a third family of ultrasonic surgical tools that use the direct interactions between an ultrasonic horn and tissue. This class of ultrasonic surgical interactions exploits the direct effects of mechanical vibrations, most commonly in the frequency range 20–60 kHz [63]. They have been used as dental scalers, for root canal debridement, emulsification of cataracts (phacoemulsification), decalcification of heart tissue, arterial heart ablation, thrombolysis, in ultrasonic scalpels, and in a wide range of soft tissue procedures.

Much of the literature that reports ultrasonic "surface surgery" concentrates on the clinical aspects of the subject including evaluation of the outcomes of procedures, and this is usually classified under the particular procedure or the organ on which it is performed. As with so much else in medical ultrasound, the literature is written by and for clinical practitioners. In terms of discussing the devices, in most cases, there is a standard paragraph on "Device and Mechanism" that includes a statement to the effect that: "the ultrasonic device induces tissue fragmentation." There may also be a comment that these devices exhibit tissue selectivity: soft tissue is removed and that with high collagen content is left almost undamaged. A final sentence may also relate tissue-removal rates (gram per second) to water content. Despite widespread clinical application, there is still much ongoing debate surrounding mechanisms of interaction, and some of what is reported in the literature is simply incorrect.

On examination of the literature, it is seen that these devices have now evolved into a series of families of units that address requirements of particular applications. They are of varying size, operating frequency and power, which is now, in most cases, delivered with a piezoelectric stack attached to the horn. A schematic for an ultrasonic unit of this type is shown in Figure 14.15. The units come in a variety of sizes; for example, for phacoemulsification, the device uses a tip that is typically 1 mm in diameter, whereas lipoplasty uses much larger and longer cannula, typically 3–5 mm in diameter and up to about 30 cm in length; all with an array of tip designs. For some applications, the units include integrated irrigation and suction (aspiration)

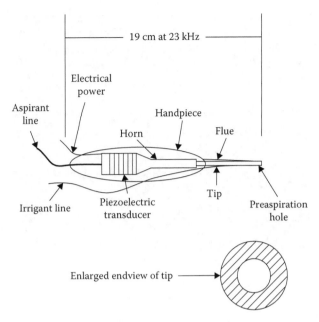

FIGURE 14.15 Ultrasonic unit used for tissue fragmentation, with both irrigation and aspiration. (Reprinted from Bond, L. J., and W. W. Cimino, Physics of ultrasonic surgery using tissue fragmentation, In *Ultrasonics*, 34:579–85, 1996. With permission.)

capabilities. Some applications for this type of instrument are discussed, and the topic of the physics of tissue dissection and ablation for this class of instruments is then discussed in Section 14.6.4.6.

14.6.4.1 Phacoemulsification

Charles D. Kelman introduced what was then called "Kelman phacoemulsification," in 1967, and this became the impetus for today's outpatient cataract surgery. The procedure uses a small ultrasonic tip whose vibrations break up the mass of the cataractous lens within its capsule, and the debris is removed using suction though the same small needle. Over one million operations of this type have been performed in this country alone. In 1975, Kelman also began designing lens implants for use in conjunction with cataract surgery; since then, several companies have entered this market.

Units used for phacoemulsification tend to operate at higher frequencies, and clearly the horns are small. A hand piece being used in a laboratory procedure on a porcine eye is shown in Figure 14.16. A study by Bond et al. [49] showed that the tip will fragment both a hard-tissue phantom and the lens in porcine eyes, and the measured ultrasonic spectra for the ultrasonic signals in the eye show no indications of significant cavitation noise. However, a cavitation "hiss" was audible during the procedure. This was investigated and the cavitation hiss was observed to be resulting from atomization of irrigation fluid from the horn-tip shoulder. The resulting spray of atomized drops is shown in Figure 14.17. This finding was

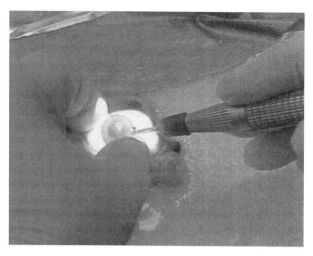

FIGURE 14.16 **(See color insert.)** Laboratory phacoemulsification procedure on a porcine eye. (Courtesy of Pacific Northwest National Laboratory.)

FIGURE 14.17 **(See color insert.)** Atomization and source of "hiss" from the horn-tip shoulder. (Courtesy of Pacific Northwest National Laboratory.)

further confirmed when the hand piece was operated (briefly) without the irrigation, and there was no cavitation audible hiss, and yet the tissue in the porcine eye was disrupted.

This finding was in agreement with the findings of earlier work by the authors on larger scale units used for soft-tissue applications that cavitation is not a major mechanism at tip–tissue interactions, at least for the classes of instruments investi-gated. For phacoemulsification, it was therefore concluded that the primary mode of tissue disruption using a straight bevel-ended tip was because of a jackhammer effect.

14.6.4.2 Ultrasonic Surgery Using Tissue Fragmentation

Neurosurgeons, and others, have adopted and developed the Kelman phacoemul-sification machine for use in a range of surgical procedures. The early instrument of this type was produced by Cavitron and called the Cavitron Ultrasonic Surgical Aspirator or CUSA. For this general instrument, there are attachments (horns) designed for specific classes of procedures that include neurosurgery, general sur-gery, gynecological surgery, gastrointestinal and affiliated organ surgery, urologi-cal surgery, plastic and reconstructive surgery, laparoscopic surgery, orthopedic surgery, thoracic surgery, thoracoscopic surgery, and wound debridement. Several manufacturers now provide equipment of this general type, and it is seeing growing clinical use.

14.6.4.3 Ultrasonic-Assisted Lipoplasty

The UAL procedure was introduced in Europe in the late 1970s, and it began to attract attention in the United States in the early 1980s. Since this time, several hun-dred thousand procedures have been performed. These procedures use much larger and longer cannular to access the body than is used in phacoemulsification, and significant irrigation fluids are inserted into the tissue to be removed [189]. Although the equipment is in many ways similar to both phacoemulsification and CUSA-type surgery, the conditions and interaction zone around the cannular are somewhat dif-ferent, particularly given the presence of the irrigation fluid and the power levels used. Cavitational erosion has been seen on cannula used in this procedure, and some have raised safety questions around this observation. A conference on UAL safety and effects was held in 1998 [190]. This was organized to address questions raised in an article by Moris Topaz [191] regarding possible negative effects because of cavitation and resulting byproducts of sonochemical and sono luminescence phenomena.

The horn–tissue interactions involved UAL are more intense than those used for phacoemulsification. There is also a significant high-power ultrasonic field that can, and has been reported to, cause local heating if the horn is not kept in motion during the procedure. The acoustic fields generated by the cannulas used for UAL are clearly sources of cavitation and sonoluminescence. The localized stress fields and heating caused by cavitation are strong enough to lyze cells, but the UAL mechanism remains the subject of debate. By analogy, with the phenomena seen with tissue liquefaction in biological sample preparation under a horn, various effects can occur. There is also the potential for the jackham-mer effect to disrupt tissue, together with fluid-pumping motion resulting from cannular vibrations. The phenomena and effects that occur have been debated extensively and are discussed in detail in two issues of *Clinics in Plastic Surgery* [50,51].

14.6.4.4   Ultrasonic Scalpels

Ultrasonically driven scalpels have been available for more than 20 years and typically operate at 55 kHz. These devices are in many ways similar to those used in phacoemulsification, except that the horn is formed into a surgical knife or other cutting edge. These scalpels are now used in a range of procedures on both soft tissue and bone. They provide enhanced coagulation, and in many cases, they are used when the alternates would be electrocautery or use of a laser.

The ultrasonic scalpel divides tissue using high-frequency ultrasonic energy, which is stated to disrupt protein hydrogen bonds within the tissue. Blood vessels are coated and sealed by denatured proteins in the vessel lumen. The relatively low temperature (80°C) at which this takes place results in minimal thermal injury (<1.5 mm) [192]. With these devices, there is seen to be "surgical smoke" or particulates that are generated. The health issues, for the medical staff, have been discussed extensively [193] and measures need to be taken to minimize exposure and to prevent potential adverse effects.

14.6.4.5   Intravascular Surgery (Thrombolysis)

Ultrasonic arterial plaque ablation again became of interest in the late 1980s [194, 195]. In this work, devices predominantly use solid catheters, with no suction. The significance of suction in the interaction is receiving minimal consideration; suction is considered only as a means for debris removal.

In the work by Siegel et al. [194], an ultrasonic drive unit operated through a flexible catheter with a solid was used. It operates at 20 kHz giving tip motion of 25 to 30 μm and an acoustic output of 20–50 W. A tone-burst mode of operation was used with a 50% duty cycle, and a 20 ms on-time. The "g" force at the tip is estimated to be between 0 and 82,000 g. Saline solution is used as irrigation. The mode of operation is attributed to both mechanical and cavitational effects. Some histology is provided which shows minimal tissue damage. In Siegel's paper, there is most interest expressed in determination of the range of the debris size in the ablated material. An ultrasonic B-scan unit was used to monitor the interaction zone and cavitation bubbles are clearly seen to have been generated near the unit tip.

In a second article, the recanalization of atherosclerotic total occlusions was reported [196]. The device used in the earlier study was used with a solid wire probe, 20 kHz operating frequency, tip amplitude of 25–75 μm, power of 25–50 W, and the g force at the tip from 0 to 82,000 g. The unit was operated in both a continuous and a 50% duty cycle mode, with a 20-ms period. In all cases, irrigation was provided and after several minutes of exposure, there was heating. The unit tip reached 50°C after 5 minutes continuous operation in the laboratory. In a blood vessel, the maximum temperature recorded was 39°C. Some consideration is given to the mode of tissue interaction and biological effects. Four possible mechanisms for interaction are "identified" as follows: (1) mechanical vibration, (2) cavitation, (3) thermal, and (4) formation of intracellular microcurrents. However, no real data to estimate the magnitude or significance for each of the classes of interaction are given.

A third article by the same team [197] used a modified version of the unit used previously. It operates at 20 kHz, power is 8–25 W, and in this case, it has a ball-tipped solid wire. Irrigation with saline is provided for cooling, and there is no suction. The article provides some histology, which shows minimal damages to surrounding tissue. The ball tip is reported to provide an improved plaque-removal device.

There are other studies that follow similar approaches, and a 20-kHz flexible solid horn was used by Rosenschein et al. [198]. The unit operated in a tone-burst mode with a 30% duty cycle to keep heating less than 10°C. The power is 10–30 W with the contact area irrigated. This gives a longitudinal mode amplitude of 150 ± 25 μm. The plaque-removal rates were measured and debris size range was determined. The mechanism for ultrasound-tip/tissue interaction is reported to be "mechanical fragmentation." It also states that "cavitation has been implicated as the mechanism of ultrasound tissue disruption." These authors conclude that both of these mechanisms probably combine to disrupt tissue. It was observed that tissues heavily embedded with collagen and elastin matrix are especially resistant to ultrasound. Tissues lacking normal collagen and elastin fiber support such as atherosclerotic plaques, fat, and calcified deposits especially susceptible to ultrasonic disruption.

A second article by the same group Rosenschein et al. [199] considered the use of an improved prototype ultrasonic angioplasty device; and its use in human patients. A conically shaped field of cavitation bubbles was emitted from the wire tip and clearly seen in diagnostic ultrasonic images of the zone near the tip. A cavitation threshold at the unit 8 W level was found, and the unit was operated at 12 W in the clinical setting. On the basis that the microbubbles were seen, the mode of action was stated to be cavitation. Some histology and debris analysis were also reported.

Clot disruption has been considered by Hong et al. [195]. This article provides a useful review of earlier studies. A 7F catheter (2.3 mm) diameter solid titanium wire probe is used. It operates at a power of 50 W at a fixed 20-kHz frequency. Irrigation is provided, and 31-, 56-, and 106-cm long probes were used. The study demonstrates clot disruption and concludes that mechanical and cavitational mechanisms are used.

For ablate of arterial plaques, there have been several other studies. In the work by Ernst et al. [27], the main interest is to minimize debris size. A unit operated at 20 kHz was used with a 2.2-mm-diameter titanium alloy tip. A calorimeter is used to estimate acoustic power output. A normalization of the power setting for 0–5 W divided by tip area (tip area 0.038 cm^2) gives up to 150 W/cm^2 with a mean amplitude up to 300 μm. Histology is included to show that thrombosis disruption is provided. This group report hearing cavitation noise, but it is stated that the mechanism of ultrasound-tissue ablation is poorly understood, and they conclude that a range of mechanisms may be significant. The various mechanisms listed in this article are considered further in Section 14.4.

The further study by Philippe et al. [200] again used a 20-kHz device with a 130-cm long flexible titanium probe, at a maximum power of 250 W. A 0.8-mm wire was used with a 2-mm-diameter ball tip. The main focus of this article is to determine the range in the resulting debris size and not the mode of the tip–tissue interaction.

In addition to the direct surgical studies of the use of ultrasonic surface surgery, there has been one study that has compared a motor-driven probe operating at (250 Hz) with the CUSA operating at 23 kHz. In the article by Chan et al. [201], it is reported that similar tissue-removal rates are obtained with the two devices and that the rate with ox-liver of 7 g/min was obtained. In addition, it was stated that with the motor-driven aspirator, the same tissue-removal rate was also obtained at 1-, 0.5-, and 0.3-mm stroke.

14.6.4.6 Physics of Tissue Dissection and Ablation

A review of the early studies to investigate the biological effects associated with the dental instruments is provided by Green and Sanderson [30]. References are given to work where ultrasonic waves with powers from 0.5 to 1.25 W/cm^2 were applied to a variety of tissues, including bone. Histology was performed to investigate tissue damage. However, reports that describe the physical acoustic phenomena encountered are very limited.

In the early dental work, the loading pressure was identified as being more important than duration in determining the nature of the effects produced [202]. To avoid adverse heating and tissue damage effects, a water coolant was used. The units used appear to have been derived from industrial ultrasonic machining devices that used a grit paste to achieve cleaning and material removal through a grinding action. The cleaning interaction was assumed to be caused by effects of friction and absorbed heat and not by any unidentified type of energy produced by the ultrasound [30]. The first reported use of low-frequency ultrasound for the surgical removal of human tissue is in an article by Kelman [203]. This work considered the emulsification of cataracts, and these studies are reported in a series of three articles [203–205]. The study used a Cavitron phacoemulsifier with a special tip that had an oscillating needle, combined with controlled irrigation and suction. The articles focus on the surgical procedure, and little information is provided regarding the nature of the ultrasound–tissue interaction. It was reported that size, stroke and frequency combinations, and the parameters used to characterize industrial power ultrasonic devices were not considered to be optimized [203]. The details of optimization of the parameters were stated as to be reported in a later publication, but this seems never to have been published.

The unit used by Kelman [204] has a tip diameter is approximately 1 mm O/D. In the first article, Kelman [203] states that the suction did not exceed 25 mm Hg. In the article published in 1969, a more detailed description of the hand piece is provided, and it is stated that an operating frequency of 40 kHz was used with tool amplitude 50–127 µm. Two tip types are mentioned, with these being for (1) emulsifying and (2) cutting. In these articles, almost nothing is said about the mechanism of tissue interaction or the differences between the two types of tip.

In an article by Shpuntoff and Wuchinich [206], reference is made to a discussion of Soviet work reported at a conference by Goliamina [207], who described "ultrasonic surgery," with applications in orthopedics, neurology, and ophthalmology. There are some further references to Russian language work in the literature.

Work using similar technology for ultrasonic angioplasty was pioneered in the late 1960s with the development of the first catheter-based systems for intravascular ultrasound disruption of atherosclerotic plaques and thrombi [208,209]. There was some interest in this work in the mid-1970s. There was also some initial work by Brown and Davies [210] to decalcify calcified cardiac valves. In this study, a "Dental

Sonic" scaling unit was reported as being used. It operates at 25 kHz and provides power of 2.5 W. Irrigation and a range of dental tips were used. Suction was used, and it is stated that ultrasound induces cavitation effect in the water which helps to break up the calculus.

The introduction of the CUSA in neurosurgery was reported by Flamm et al. [211]. The unit operated at 23 kHz, with tip amplitude of 100 μm. Irrigation was provided for cooling and suction was used for debris removal. This unit was a larger version of phacoemulsifier used by Kelman [203]. In the surgical work, both "aspirator" and "dissector" tips are reported. It would appear that a dissector tip did not have suction.

In the article by Flamm et al. [211], a series of cat brain and spinal cord studies are reported that were designed to assess the spread of energy from the CUSA. The work reports the clinical effects that were observed and also histology is included. However, no evaluation of the acoustic energy delivered or the mode of interaction is reported.

Following from the work by Flamm et al. [211], there are then a number of articles that report the clinical application of the CUSA and similar units. This literature reports work from the viewpoint of the surgeon clinical procedures and their effectiveness, and these articles appear to be focused on applications in neurosurgery and liver resections.

There are numerous articles that present other clinical reports on the use of a CUSA and similar units, but there is almost no discussion of the detail of the tissue ablation-fragmentation mechanism involved. A review by Epstein [212] considers the specific applications in tumor surgery for which the CUSA Ultrasonic Surgical System has been used, defined in clinical terminology. A paragraph is included in most of these articles which says that "ultrasonic fragmentation" or simply "fragmentation" is the mechanism for tissue disruption. In some cases, supporting histological studies looking close to the damage site are reported that show a very limited collateral damage zone. In most cases, there is a tissue-disruption zone of 1–2 mm diameter which is the diameter of the tip, but the possibility of more remote effects is not investigated.

A study by Cimino and Bond that investigated the physics of the interaction and the performance of a CUSA unit [47,48] for soft-tissue fragmentation concluded that the jackhammer effect was the dominant mechanism occurring in tissue disruption. A subsequent study by Bond et al. [49], which focused on phacoemulsification, again concluded that it was the jackhammer effect that was dominant.

Packer et al. [213] have reviewed the phacoemulsification literature and state that despite its unparalleled success in the field of surgery, the precise mechanism of ultrasonic phacoemulsification cataract extraction remains controversial. They still conclude that phacoemulsification most likely operates by a combination of mechanisms, including direct action of the vibrating tip against tissue (jackhammer) and indirect cavitational effects. However, they do not provide experimental data to support the conclusion regarding cavitation.

As discussed in Section 14.3 of this chapter, there is an ongoing debate with regard to mechanisms that occur in ultrasonic surgery, particularly that with regard to the physics in phacoemulsification and the related family of procedures. O'Daly et al. [55] have provided a comprehensive critical review of the current state of knowledge with regard to this family of devices and the tissue-removal mechanisms. That article concludes that, despite widespread clinical application and common device characteristics, there remains an incomplete understanding of the mechanisms of tissue failure, removal, and any related damage. The issue of the presence, or absence, of cavitation with some ultrasonic surgery tools remains controversial, but the ability of such devices to cut holes in dry wooden tongue depressors and plastic tissue surrogates, or varying hardness, clearly confirms a mechanical mechanism is involved.

14.6.5 ULTRASOUND IN DENTISTRY

The use of ultrasound to prepare cavities in teeth before filling received considerable publicity during the 1950s and in the early 1960s. Ultrasonic dental units were small (about the size of a large fountain pen) magnetostrictive transducers operating within the range of 20–30 kHz in a longitudinal mode. The drilling method was patterned after that with the industrial ultrasonic machine tools which utilized abrasive slurry for cutting. The slurry used with ultrasonic dentistry consisted of a fine-grit aluminum oxide in water flooding the tooth. The slurry was recirculated by means of a small pump.

Tools were blunt and shaped according to the pattern of the cavity to be prepared. Only light pressure was applied, and there was no danger of damaging soft tissue. Heavy pressures could cause excessive heating and pain. Not only are these high pressures undesirable physiologically, but also they have an adverse effect on the operation of the tool and the drilling rate. Proper light pressure should permit good drilling and have no adverse effect on the tissues.

The ultrasonic method of dental-cavity preparation could not compete with more modern developments of conventional dental equipment. The ultrasonic equipment developed has not gone to waste, however. It is finding wide acceptance as a clinical tool for dental hygiene. Removing scale from teeth, both supragingival and subgingival, requires considerable effort on the part of the dentist using conventional scrapers. The scaling implement gouges the gums and causes bleeding, and the overall operation is uncomfortable to the patient. The ultrasonic dental drill has been adapted for use in removing dental calculus. The tips used to irradiate the teeth are interchangeable and water-irrigated. The water couples energy to the teeth and at the same time cools the tooth and tool. The cavitation and light chiseling action combine to remove the calculus. The method also provides a good means of treating the gums. Patients sometimes complain of sore gums following ultrasonic treatment. The exact cause of this pain is not certain; however, if light pressure only is applied and movement of the tip is maintained during operation, such pain should be minimal or nonexistent.

Some Czechoslovakian scientists [214] have investigated the possibility of making use of the increased permeability of tissues under the influence of ultrasonic energy for the application of local anesthetics in dentistry. A mixture of procaine, adrenaline, synderman, adepslanae, and oleum olivae was applied to the skin or gingiva of 50 patients before tooth extraction and ultrasound applied to the area to promote diffusion. In most cases, the anesthesia was ineffective, and in difficult cases, it had to be supplemented by a local injection.

Aluminum wires reportedly have been successfully bonded to excised teeth embedded in a plastic block [215]. These bonds were produced using a magnetostrictive transducer operating at 60 kHz which applied an ultrasonic shear stress to the interface between the dentin and the wire. Pressure was applied across the wire and tooth in a direction normal to the surface of the tooth as in conventional ultrasonic welding. The energy was applied for a brief period of time (usually less than 2 seconds). A good bond was claimed for the process, with the wire breaking before the bond. Whether the break occurred at the point where wire deformation started as a result of the application of ultrasound and, hence, at a point of stress concentration was not reported. The method has not evolved to clinical practicality.

The mode of interaction for dental tools used for debridement of root canals has recently been considered in an extensive study of the Cavi-endo file. Both acoustic streaming and cavitation have been identified as being present near the oscillating tip of the solid tool, [216–218].

The status of the use of ultrasound in endodontics was reviewed by Stock [219]. Units operating in the range 25–40 kHz are used. Cavitation, heating, microstreaming are all reported to have been clearly identified for ultrasonic drills, which are Group II devices, where there is not integral suction. With dental files the conditions are considered to be unlikely to induce transient cavitation, and it is acoustic streaming which is considered to be the active erosion mechanism. For scaler tips, there is reported to be evidence of cavitation.

When the modes of interaction for dental tools and that for the CUSA are compared, the acoustic field and the mode of tip–issue interaction are simply not the same. The dental units are more similar to the dissectors or ultrasonic scalpels.

Ultrasonic tools have now become well established in dentistry [220], and this technology has many applications. At the heart of the technology are the biophysical interactions, which were reviewed by Laird and Walmsley [221]. The most common use is in the ultrasonic dental scaler, which is a valuable tool in the prevention of periodontal disease [222]; however, this equipment has a number of potential hazards associated with it. Issues can include heating of the tooth, vibrational interactions that cause cell disruption, platelet damage induced by cavitation, auditory damage to patient and clinical staff, and the release of aerosols with the ability to contain dangerous bacteria. Trenter and Walmsley [223] provide an analysis of these issues and suggest guidelines to aid in reducing the hazards. Arabaci et al. [224] review the mechanisms, safety, efficacy, role, and deleterious side effects from both sonic and ultrasonic scalers in mechanical periodontal therapy. Endosonics utilizes an ultrasonically driven file to clean and shape the root canal before obturation [224]. A review of the literature that discusses passive ultrasonic irrigation of the root canal is provided by Sluis et al. [225].

Walmsley, in a 1997 article [226], provided a review of the literature for the ultrasonic toothbrush. He discusses the potential contributions to the mechanisms of interaction of cavitation and microstreaming, and the relationship between the vibratory motion and the ability of such motion to remove plaque and improve gingival health. Developments in the technology, including both the rotary head and higher frequency of operation, are discussed. These devices have now become a common consumer product.

Some other more novel applications of ultrasound in dentistry is to use it to set cements [227] and in an imaging modality to use acoustic microscopy as a diagnostic tool [228].

14.6.6 Selected Other Ultrasonic Surgical Procedures

14.6.6.1 Laryngeal Papillomatosis

Laryngeal papillomatosis is a rare juvenile disease involving wart-like growths on the larynx. The disease may strike in infancy before the age of 1 year and persist until adolescence. Rarely does it disappear on its own before adolescence. Frequent surgery is required to keep the child alive. Until ultrasonic treatments were adopted, there was no known cure.

The ultrasonic treatment of laryngeal papillomatosis was an outgrowth of a search by Birck and Manhart [229] for a more effective means of treating a patient who had repeatedly been treated surgically but in whom the disease showed signs of spreading into the tracheobronchial tree. Reports of successful therapy of plantar warts (other warts are also treated successfully by means of ultrasound) were the incentive for using ultrasound on the patient. Treatment with a commercially available therapeutic unit (see Figure 14.6) in the throat area daily for 1 week brought favorable results, as it did in several succeeding cases. The objectionable aspect of this treatment was the amount of anesthesia given to the patient. For this reason, Birck went to a high-intensity probe-type transducer, with the transducer located in the tip of the probe in order to bring the source into direct contact with the diseased tissue. The application is now administered immediately following surgical removal of the diseased tissue. The relatively large number of patients who appear to have been cured of the disease following this procedure attests to the efficacy of the technique. Others also are treating this disease with ultrasound with apparent success [230–232]. The use of frequencies above 1 MHz helps localize the intensity by absorption, and surrounding tissues suffer no adverse effects.

14.6.6.2 Meniere's Disease

Meniere's disease is a disease of the inner ear producing disabling attacks of giddiness and vomiting. These symptoms have been attributed to an imbalance in the two fluid-filled cavities of the semicircular canals. In a large percentage of cases, the disease can be controlled by diet and medication. Formerly, medication was used to bring relief to the patient with an acute case, but patients suffered repeated attacks of vertigo, and the frequency and severity of these attacks were unpredictable. Medical treatment would often lose its effectiveness, with surgery being the only alternative. However, surgery involved the destruction of both the auditory and the vestibular portions of the inner ear, which produced permanent deafness. For this reason, there was a reluctance to operate on patients with good hearing in the affected ear, and patients in whom both ears were affected were offered no chance of surgical cure.

It is claimed that Meniere's disease offered the first opportunity for surgical application of ultrasound [232]. The treatment involved irradiation of the semicircular balance canals. The equipment used in the initial applications was designed and manufactured in Italy by the Federicci Company of Milan [177]. The equipment used in this early work operated at approximately 1 MHz and was capable of a maximum acoustical output of 15 W/cm^2. It was equipped with a dial supposedly indicating the irradiation intensity.

Ultrasonic treatment in the early history involved giving the patient light sedation and local anesthesia in order to obtain patient cooperation in identifying the treatment stages, particularly the end point [177]. Access to the horizontal semicircular canal was obtained by means of a postauricular simple mastoidectomy. The canal was skeletonized, its surface flattened, and its wall thinned until the blue line was visible. Fresnel glasses were then placed over the patient's eyes so that an assistant could monitor the patient for induced nystagmus (rapid, involuntary oscillation of the eyeballs). The tip of the applicator was placed on the posterior half of the flattened horizontal semicircular canal, and the beam (adjusted by dial for 6–9 W/cm^2) was directed into the labyrinth in such a direction as to avoid irradiating the facial nerve. (Irradiation of this nerve could result in either temporary or permanent facial paralysis of the affected side.) An irritative nystagmus directed toward the operated ear would develop within a few minutes of irradiation. After about 15–30 minutes of irradiation, the nystagmus would subside with no recurrence. Irradiation was continued until a paralytic nystagmus directed toward the opposite ear was noted, at which time the intensity, according to the dial on the intensity meter, was raised to 9–12 W/cm^2. The end point was indicated if no further irritative nystagmus was induced and if the nystagmus remained. A fair percentage of success was reported for this procedure, and failures were attributed to insufficient radiation of the canal during surgery, which was perhaps occasionally resulting from insufficient thinning of the horizontal canal but was more probably the result of improper tuning.

Dial readings in electronic circuitry cannot be relied on for indicating radiation intensities. Newell [233] reported that other investigators measured the output of the Federicci equipment calorimetrically and found it could vary from 6 to 10 W with as many as six maxima and minima occurring over a frequency change of 10%.

Improvements in the technique of treating Meniere's disease include the use of 3 MHz instead of the 1 MHz used with the original Federicci apparatus. The higher frequency is more directional and therefore makes irradiation of the facial nerve easier to avoid. (In the few cases in which paralysis of this nerve has been caused by ultrasonic irradiation, the effect usually is temporary.)

The Commonwealth Acoustic Laboratories in Australia developed an ultrasonic unit for the treatment of Meniere's disease that has received general approval by those who have had the opportunity to evaluate it [234]. The tip of the applicator contains a flat 3.5-MHz transducer, 1.5 mm in diameter, mounted at an angle of 20° to the axis of the 1-mm-diameter stainless steel tube to which it is attached. It is capable of output intensities variable up to 25 W/cm^2 (70 mW). With the slender construction of the applicator, the transducer can be inserted into the middle ear, and the energy is directed in a cylindrical beam directly into the vestibule through the round-window membrane [235]. The beam is then directed through the oval window to the saccule, the utricle, and the ampulla of the lateral and the superior canals. The cochlear duct is at an angle and receives energy only after multiple reflections and after attenuation of the ultrasound has occurred. A large area of the membranous labyrinth is exposed, and because ultrasound increases the permeability of membranes, the result is prevention of the accumulation of hydrops. The niche beside the sound window is filled with saline for coupling energy from the transducer into the window. During treatment, the ultrasound is turned off every 5 minutes, and one or two drops of saline are added to the niche. During this period, at 25 W/cm^2, the temperature rise is 6°C in the facial nerve. The increase in the niche in 2 minutes is 15°C.

With the Commonwealth Acoustic Laboratories (CAL) unit, selective treatment of the membranous labyrinth is possible with minimal side effects. The treatment requires only reflection of the tympanic membrane—a surgical procedure that inflicts little trauma. The patient is discharged in 2 or 3 days.

With this treatment, irritative horizontal nystagmus develops within the first 3 minutes of treatment time. As the treatment continues, vertical nystagmus develops and either alternates or combines with the horizontal, producing a rotary nystagmus. The intensity of the motion decreases with continued exposure, becoming latent after 5–10 minutes of irradiation. Following this period, both types of nystagmus reverse their direction, with the horizontal one becoming paralytic before the vertical one reverses. Irradiation continues for 3 minutes following reversal of the vertical nystagmus.

A device for ultrasonically treating Meniere's disease in much the same manner as the CAL unit has been developed in Sweden [236]. This unit uses a concave-focused transducer which uses a hollow water-filled cone as a transmission line. The liquid in the cone is continuously recirculated to maintain cooling. The tip of the cone is at an angle to the axis of the cone for convenient irradiation of the labyrinth. Only a small drop of coupling liquid is used between the tip and the labyrinth, which permits operation with a completely dry cavity.

In its present state, ultrasonic treatment may totally replace destructive surgical treatment of Meniere's disease.

14.6.6.3 Stapedectomy

Otosclerosis is a common cause of deafness. A spongy bone growing in the inner ear gradually obstructs the vestibular or cochlear window or both and causes progressively increasing deafness. In recent years, stapes surgery has become fairly common and has been successful in restoring at least partial hearing to patients. Curettes, hooks, and microdrills are generally used for this type of surgery. These instruments produce trauma to the inner ear, with drills causing the greatest damage because of heat, lesions from dust particles thrown into the vestibule, and microfractures in the petrous bone. At one time, the ultrasonic dental drill was suggested as a possible aid to stapes surgery. Early attempts were disappointing in that they resulted in the destruction of the organ of Corti and other neurosensorial elements in the inner ear. This should not have been entirely unexpected from an instrument that produces the amplitude of vibration that is characteristic of the dental drill, considering the relative sensitivity of these components of hearing to damage by loud audible sounds. Much of the failure in using these instruments has been attributed to the low frequency (which was only slightly above the hearing range), high amplitude, and heat resulting from high absorption in the bone. A persistent, continuous application would certainly be destructive to the hearing organs. However, a procedure is possible that avoids the severe acoustic trauma of these early experiments.

Arslan, one of the pioneers in the use of ultrasound for treating Meniere's disease, has had considerable success in stapes surgery using a specially constructed probe operating at 40 kHz [237]. The amplitude of vibration is lower than it is at lower frequencies for a given intensity. Tests with laboratory animals showed that no injury was caused when the probe was used on the edges of the window of the periosteum, but injury to the inner ear was caused by applying the probe to the promontory or apical turns. Arslan has used the probe clinically to disintegrate the footplate down to the vestibule endosteum, the most difficult and dangerous step in stapes surgery, and to make an intercrural section of the plate. Arslan applies the tool for intervals of 4 seconds at a time and permits a 5-second cooling period between applications. As in dentistry, the heat generated increases rapidly with increased pressure. For such a delicate operation, one would expect to apply very light pressure, and the effectiveness would be expected to be better than that of curettes, hooks, and drills. Arslan's results show no injury to cochlear receptors, and none of his patients showed greater hearing loss than those operated on with conventional stapedectomy techniques.

14.6.6.4 Selective Hypophysectomy

Hypophysectomy refers to the surgical removal of the pituitary body. It is prescribed for patients with advanced cancer of the breast with bone metastases, diabetic retinitis, malignant exophthalmus (abnormal protrusion of the eyeball), Cushing's syndrome, and acromegaly (a chronic disease characterized by permanent enlargement of the head, thorax, hands, and feet).

Arslan [238] reported favorable results using ultrasound for selective hypophysectomy with patients having cancer of the breast and prostate, Cushing's syndrome, acromegaly, and retinitis diabetica. The operation consists of a surgical procedure to expose the dural sheath of the pituitary to firm contact by the ultrasonic probe through the sphenoid sinuses, and irradiation of the pituitary at 3 MHz and an ultrasonic intensity of 2.5 W/cm^2 for 20 minutes or more. Following irradiation, the anterior septum is replaced in its original position (i.e., its position before displacement in preparing access to the pituitary), and the columellar incision necessary for access is sutured. The operative cavity is loosely packed. The irradiation may be repeated, in a completely bloodless manner, by introducing the probe through one nasal fossa, in a month or more following the initial operation.

14.7 TISSUE CHARACTERIZATION

Ultrasonic tissue characterization involves the determination of propagation characteristics (attenuation and velocity of sound) of ultrasonic energy in various body tissues. These propagation characteristics are of particular importance for the interpretation of diagnostic data, but they also are factors in therapeutic applications of ultrasound. The objective of tissue characterization studies is to find properties that enable the diagnostician to discriminate between healthy and diseased tissue and to determine the degree or extent of the disease.

Many measurements of the velocity of sound in body tissues have been made, and the results have been published in technical literature [239,240]. Ophir [241] lists these and other important references and discusses problems associated with conventional pulse-echo or transmission methods of measuring velocity of sound in vivo. Similar problems are associated with in vivo measurements of attenuation. Ophir discusses a beam-tracking technique studied in an attempt to provide a more accurate in vivo measurement procedure, but the method itself has weaknesses that are likely to introduce errors in measurement data.

The reported speed of sound in soft tissues averages approximately 1540 m/s, varying within 5% of this value depending on the type of tissue involved. Speed of sound in muscle may be as high as 1640 m/s, whereas in fat, it may be as low as 1440 m/s. Most velocity of sound data have been obtained from in vitro tissue, using conventional pulse-type measurements referred to in Section 7.5 of Chapter 7.

Attenuation of sound is a very important acoustical property of body tissue in the diagnostic process. It is highly frequency-dependent. It varies with type of tissue and the condition of the tissue. For example, normal liver, pancreas, and spleen can be differentiated on the basis of their frequency-dependent attenuations [242]. Attenuation in cirrhotic liver is nearly twice that of normal liver.

Various methods of measuring attenuation in vivo have been proposed and tried by different investigators [242–245]. Kuc [243] proposes a procedure for estimating the K (or kurtosis) value from reflected radio-frequency (rf) ultrasonic signals as an additional parameter for tissue characterization. The K parameter measures the "peakiness" of the reflected signal, illustrated by the observation that signals reflected from a normal liver tend to contain more distinct peaks than signals reflected from livers that are heavily infiltrated with fat. The clinical potential of the method lies in the fact that the K value is an indirect measure of the distribution of reflecting structures within the liver. This distribution changes as the diseased condition of the liver changes.

Salomonsson and Bjorkman [244] describe and illustrate a method of separating the *attenuation* and the *texture* of tissue using a pulse-echo technique. The method recursively estimates the *attenuation*, and this estimate controls the parameters of an inverse time-varying network in which the output depends on the *texture*. The method not only provides amplitude compensation of the reflected signal but also restores the frequency content of the signal. It is hoped that the method will improve characterization of tissue, particularly at large depths.

Thomas et al. [245] studied real-time integrated backscatter of ultrasound for real-time characterization of myocardium in vivo. These investigators developed an analog system used in conjunction with a 2D M-mode echocardiographic image to measure the cyclic variation in integrated backscatter and the time-averaged integrated backscatter in anesthetized open-chested dogs. The system measures the energy contained in a portion of the received rf signal from a selected myocardial segment using a cadmium sulfide transducer. The authors claim that "the system developed offers promise for real-time-diagnostic ultrasonic tissue characterization of heart muscle in patients with cardiomyopathic disorders or manifestations thereof."

Ultrasonic backscatter from canine myocardium varies significantly throughout the cardiac cycle [246]. The magnitude of the effect depends on the part of the heart being interrogated. Although the mechanisms causing such variations have not been identified with certainty, the nature of the variations certainly implies various possibilities. For example, the areas of the heart corresponding to the greatest contractile activity show the largest variations in amplitude of backscatter signals. Stresses and the variation in directions of stresses in the heart tissues during a cycle can be expected to affect the shape, size, and relative orientation of cells and muscle fibers. These are elements that affect attenuation, and how attenuation is affected is frequency-sensitive, and they would also be factors in backscattering of ultrasonic pulses. The authors also point out variations in muscle elasticity which would affect the velocity of sound in the tissue and ultimately the reflected energy.

Shore et al. [247] use a pulse-transmission method for measuring attenuation in bovine skeletal muscle, claiming results that are lower than most previous measurements of skeletal muscle but comparable with recent measurements of canine heart muscle. These measurements, of course, were on in vitro tissues. Measurements were made at $-20°C$, $0°C$, $20°C$, and $40°C$. The ratio of attenuation over frequency varied slowly with frequency, and at 4 MHz and $20°C$ mean values were 1.3 dB/cm-MHz along the fibers and 0.55 dB/cm-MHz across fibers.

The literature now contains many hundreds of articles that report data for ultrasonic tissue properties: some measured in clinical situations and others characterizing samples in the laboratory. A comprehensive review of the ultrasonic methods and parameters for tissue characterization measured and reported in the literature is given in an article by Jones and Leeman [248]. There is also now a growing body of work which is using acoustic microscopy, but such studies are performed at significantly higher frequencies than those used in clinical studies.

Two comprehensive compilations of ultrasonic properties of mammalian tissues were provided in 1978 and 1980 in articles by Gross, Johnston, and Dunn [249–250]. Probably the most comprehensive text on physical properties of tissue, which includes ultrasonically derived data, is in the book by Duck [251].

The ultrasonic data that are available for normal, aged, and diseased tissues, both in vitro and in vivo, is now simply vast. For some parameters, there are well-established bands for values; however, there is a natural variation that occurs within an individual, with a population and also between normal and diseased or abnormal states. Significant challenges remain when it comes to the identification of some classes of pathological changes: for example, for many single, local, ultrasonic property measurements, a particular tissue conclusion cannot be drawn. Both normal and pathological conditions can have properties within the total population that overlap. To conclusively identify changes, there is a need for a reference state against which a comparison can be made for that particular individual.

The use of imaging can be of more value and diagnostic ultrasound is an exceptionally good tool for identification of many structural anomalies and both hard and fluid-filled masses. For some changes and condition that occur within some organs because of disease, the signatures which result are quite simply within the bounds for normal for the population, and detection of these changes within an individual can, and in many cases is, hard or in some cases even impossible. Novel ultrasonic tools for tissue imaging and characterization remain areas of ongoing research.

14.8   HIGH-FREQUENCY IMAGING/ACOUSTIC MICROSCOPY

From early in the development of acoustic microscopy, which is discussed in Chapter 10, there was interest in the potential for applications in medical research. Acoustic microscopy provides images based on tissue contrast mechanisms that are different from those which are seen optically, with MRI or induced through the use of stains or other compounds. Here, we will just consider some aspects of high-frequency imaging for medical and biomedical applications.

The acoustic microscopy is now, from an engineering and physics perspective, well established [252]. The contrast mechanisms are understood, and there are several commercial instruments on the market which operate, at least to modest frequencies ~500 MHz. However, the community who truly understand the science and technology of acoustic microscopy is far smaller than that for other microscopy modalities, such as optical or scanning electron microscopy. The largest challenge faced by the potential user is, most probably, defining the requirements and then obtaining a transducer that is well matched to meet the needs of the desired imaging task.

It was interest in understanding biological systems that was a major driver in the development of the highest resolution early acoustic microscopes. Microscopes that operate using water as a couplant have been demonstrated in the 1–4 GHz frequency range, where resolution of less than 0.2 μm have been obtained. To give higher resolution, instruments were developed using liquid helium as the couplant. An instrument which operated at 8 GHz gave a resolution of less than 200 Å. Some stunning images were obtained, such as that of Myxobacteria in liquid helium that enabled interesting internal structures to be seen. An overview of many of the key developments from this period was provided by Quate in a 1985 article [253].

Ultrahigh frequency acoustic microscopy remains a field of interest for a small research community. With the commercial availability of instruments, interest has moved from that being focused on novel instrument development (based mostly in the electrical engineering and the applied physics community) to becoming more applications orientated working with routing products such as those encountered in the semiconductor industry.

The medical community has increasingly looked to other ultrasonic capabilities for the potential insights that can be given in biomedical systems through moving above the current diagnostic ultrasonic frequency ranges, the 3–10 MHz range, and in some cases up to about 20 MHz, used in current phased-array systems. Some of the earlier applications of high-frequency imaging, beyond 30 MHz, were reviewed by Lockwood et al. [254]. The work reported included studies of the eye at 60 MHz, imaging of the skin to 100 MHz, intravascular and intra-articular imaging, and some early work at between 40 and 60 MHz investigating mouse embryonic development. Other studies used the acoustic microscope as a tool to determine properties, such as those in fresh tissue at 100 MHz, which were measured by Scherba et al. [255].

Properties of live human smooth-muscle cells were investigated at 100, 450, and 600 MHz by Kinoshita et al. [256] and properties of egg yolk and albumen were measured across the range from 20 to 400 MHz by Akashi et al. [257].

Among several ultrasound-biomedical research areas that have been developing are imaging of various cancers and biological colonies of cell. Such studies include work by Sherar et al. [258] to investigate the internal structure of living tumor spheroids at 100 MHz and by Gertner [259] at 30–70 MHz who studied changes in multicellular spheroids, during heating. Saijo et al. [260] have imaged populations of endothelial cell at 100 and 210 MHz. Kundu et al. [261] performed measurements in the 25–400 MHz range to determine cell properties for single and multilayered cell systems. A living, growing, biofilm was periodically imaged at 70 MHz by Good et al. [262] and an image displayed as a 3D intensity contoured image is shown in Figure 14.18.

Acoustic microscopy is starting to see applications in developmental biology: for example, for the analysis of early mouse embryonic brain development [263]. The emerging field of biomicroscopy, with most applications in the 40–60 MHz frequency range, was discussed in 2000 by Foster et al. [101]. They provide an assessment of state of the art in ophthalmology, intravascular ultrasound, dermatology, and in both mouse embryonic and tumor-development studies. The potential for application of acoustic microscopy in developmental biology was further discussed in a later article by Turnbull and Foster [264]. There is a growing community that is using ultrasonic imaging for fetal mouse imaging. Spurney et al. [265] discuss the use of both a 15-MHz-phased array and a 20–55 MHz scanned acoustic microscope for this application.

A further overview of developments in both high-frequency biomedical and industrial application was given in an article by Lewin [266]. He considers many examples of imaging at frequencies more than 20 MHz, including for dermatology, odontology, cardiologic, intraluminal imaging, and bone imaging. He also discusses the growing interest in applications to small animal imaging, and in preclinical applications using medical (diagnostic) contrast agents.

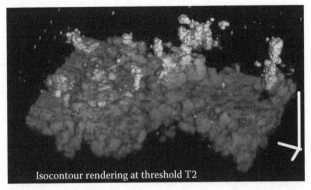

Isocontour rendering at threshold T2

FIGURE 14.18   (See color insert.) An example of a three-dimensional image of a natural biofilm. (Reprinted from Good, M. S. et al., *IEEE Trans UFFC*, 53(9):1637–48, © 2006 IEEE. With permission.)

14.9 ANCILLARY APPLICATION OF BIOMEDICAL AND RESEARCH APPLICATIONS

Hospital and clinical use of ultrasonic cleaning tanks for cleaning surgical instruments is generally accepted. Other ancillary uses include preparation of emulsions, cell disruptions, and analytical purposes.

Andrianova [267] reported using ultrasound for the preparation of 10% emulsions of corn oil in sterilized 5% glucose solution for intravenous injection.

A large number of high-intensity ultrasonic probes are on the market for use in cell disruption for the preparation of cell extracts. These have varied uses, particularly in the study of certain blood diseases. The probes consist of a mechanical amplifier (tapered solid horn usually made of an alloy of titanium) driven by a high-powered transducer and capable of irradiating the suspension of cells at very-high intensities.

Ultrasound is useful for other analytical purposes. For instance, exposure of seminal stains to ultrasonics facilitates the separation of spermatozoa from the material. Growchowska [268] used ultrasound to improve the extraction of poisons from human tissue for analytical purposes. Poston [269] used ultrasound to treat sputum for cytologic diagnosis of cancer of the lung.

One of the manufacturers of high-intensity ultrasonic probes for cell disruption, Ramco Equipment Corporation, lists many biological uses for their instrument. These are typical of similar probes from other manufacturers on the market and include the following:

Extracting enzyme from coagulase globulin, *Aerobacter aerogenes*, and so on.
Disrupting materials such as *Colletotrichum capsici* spores, *Corynebacterium, Blastomyces dermatitidis,* red and white blood cells.
Brevibacterium, Caryophanon latum, chloroplasts, actinomycin D, heart muscle, herpesvirus, kidney, *Lactobacillus,* lung tissue, poliovirus, trypanosomes, and uterus muscle breaking DNA chains.
Manufacturers can supply information on many other applications of these probes and users may develop many of their own.

14.9.1 CELL AND SPORE DISRUPTION

There is a long and extensive literature that reports tissue, cell and spore disruption by ultrasound. The disruption of microorganisms by ultrasound was reported by Davis in 1959 [270]. He used 26-kHz ultrasound at powers up to 500 W. There were several studies that considered the effects of ultrasound on blood and cells in suspension, which included work by Hughes and Nyborg [271]. Thacker reported work using 20 kHz and 1 MHz to killing cells in suspension. Hawley et al. [272] reported DNA degradation using intense, noncavitating, ultrasound at 1 MHz, and an intensity of 30 W/cm^2. Several researches investigated the mechanisms involved in cell and spore disruption, including in noncavitating systems. Rooney [273] proposed shear as a mechanism, for at least some of the induced biological effects.

Over the years, there has been some ongoing interest in an application of ultrasound for lysis, since the work that was performed in the 1970s. However, in recent years, significant work has developed in this area based on three themes: gene transaction; enhanced spore (and cell) lysis for DNA extraction; and interest in the use of ultrasound as a biocide.

Miller et al. [274] demonstrated the feasibility of the use of ultrasound to give enhancement of gene transfection in murine melanoma tumors. This study used a lithotripter system to deliver a shock wave pulse. In a study by Brayman et al. [275], cell surface damage was reported, and cell surface receptors were found to be removed when exposed to 1 MHz at powers up to 8 W/cm^2.

Tachibana et al. [276] sonicated a cell suspension with continuous 255-kHz ultrasound and SEM pictures showed pores in irradiated HL-60 cells, and these appeared to offer the opportunity for enhanced gene transfection. Chandler et al. [277] developed a flow through unit, operating at 1 MHz for spore lysis. This technology offers the possibility for enhancing sensitivity for real-time field analysis of DNA in spores, and other biopathogens [278]. Feril et al. [279] reported cell lysis enhancement through the addition of contrast agents, when operating at 1 MHz and with powers up to 4 W/cm^2.

In addition to lysis for enhanced DNA extraction and testing, ultrasound offers the possibility of a "green" biocide to inactivating bacteria. This has been investigated for applications in the food industry [280]. Also the potential for this approach applied to water-borne bacteria was demonstrated by Joyce and Mason [281].

REFERENCES

1. Wood, R. W., and A. L. Loomis. 1927. *Phil Mag* 4:417–36.
2. Brown, B., and D. Gordon. 1967. *Ultrasonic Techniques in Biology and Medicine.* Springfield, Illinois: Charles C. Thomas.
3. Gordon, D. 1964. *Ultrasound as a Diagnostic and Surgical Tool.* Baltimore: Williams & Wilkins.
4. Van Went, J. M. 1954. *Ultrasonic and Ultrashort Waves in Medicine.* Amsterdam: Elsevier.
5. Erikson, K. R., F. J. Fry, and J. P. Jones. 1974. *IEEE Trans Sonics Ultrason* SU-21(3):144–70.
6. Nyborg, W. L., and M. C. Ziskin, eds. 1985. *Biological Effects of Ultrasound.* New York: Churchill Livingstone.
7. Wells, P. N. T. 1969. *Physical Principles of Ultrasonic Diagnosis.* London: Academic Press.
8. Fry, W. J., and F. J. Fry. 1965. In *Paper C-1 presented at the 1965 IEEE Ultrasonics Symposium*, Boston, December 1–4, 1965. Abstract supplement, 5. New York: IEEE, Sonics and Ultrasonics Group.
9. ter Haar, G. 2000. *Ultrasound in Med and Biol* 26(Supplement 1):S51–4.
10. Dohnalek, J., I. Hrazdira, J. Cecava, F. Novak, and J. Svoboda. 1965. Cesk. *Dermatol* 40(3):173–6.
11. Bushong, S. C., and B. R. Archer. 1991. *Bilogical effects of ultrasound*, 155–62. St. Louis: Mosby Year Book.
12. Nyborg, W. L. (Chairman). 1983. *Biological effects of ultrasound: mechanisms and clinical implications*, National Council on Radiation Protection and Measurement, Report 74.

13. Nyborg, W. L. 1985. Mechanisms. In *Biological Effects of Ultrasound*, ed. W. L. Nyborg and M. C. Ziskin, 23–33. New York: Churchilll Livingstone.

14. 1984. *Safety Considerations for Diagnostic Ultrasound*. Prepared by the AIMU. Bethesda, Maryland: American Institute of Ultrasound in Medicine) Bio-effects Committee.

15. National Institute of Health. 1984. *Diagnostic Ultrasound Imaging in Pregnancy*. No. 894–667. Bethesda, Maryland: Department of Health and Human Services, Public Health Service, National Institute of Health.

16. Bang, J. 1985. In *Proceedings 4th Meeting WFUMB (World Federation of Ultrasound in Medicine and Biology) and 1st World Congress Sonographers*, eds. R. W. Gill and M. J. Dadd, 481. Sydney, Australia. New York: Pergamon.

17. Pinamonti, S., A. Caruso, V. Mazzeo, E. Zebini, and A. Rossi. 1986. *IEEE Trans Ultrason Ferroelectrics, and Frequency Control UFFC-33*(2):179–85.

18. Maeda, K., and M. Ide. 1986. *IEEE Trans Ultrason Ferroelectrics Frequency Control UFFC-33*(2):241–4.

19. Fawlkes, J. B. 2008. (and 8 others) Bioeffects Committee AIUM. *J Ultrasound Med* 27(4):503–15.

20. Nyborg, W. L. 2001. *Ultrasound in Med & Biol* 27(3):301–33.

21. Duck, F. A. 2008. *Med Eng Phys* 30(10):1338–48.

22. Shaw, A., and M. Hodnett. 2008. *Ultrasonics* 48(4):234–52.

23. Duck, F. A. 2007. *Prog Biophys mol biol* 93(1–3):176–91.

24. Schltheiss, R., and M. Doerffel. 2008. Standards for lithotripter performance. In *Renal Stone Disease 2: 2nd International Urolithiasis Research Symposium. AIP Conference Proceedings*, eds. J. C. Williams, A. P. Evans, J. E. Lingeman and J. A. Mcateer, Indianapolis, Indiana, April 17-18, 2008, Vol. 1049, 226–37. Melville, NY: AIP.

25. Wells, P. N. T. 1970. *Rep Prog Phys* 33(1):45–99.

26. Crum, L. A. 1989. Acoustic cavitation and medical ultrasound. In *Proceedings of Ultrasonics International*, Madrid, Spain. July 3–7, 1989, 852–8. Guildford, Surry, UK: Butterworth.

27. Ernst, A., E. A. Schenk, S. M. Gracwski, T. J. Woodlock, F. G. Murant, H. Alliger, and R. S. Meltzer. 1991. *Am J Card* 68(2):242–6.

28. Nakahara, W., and R. Kobayashi. 1934. *Jpn J Exp Med* 12:137.

29. Gordon, D. 1964. *Ultrasound as a Diagnostic and Surgical Tool*. Baltimore: The Williams & Williams Co.

30. Green, G. H., and A. Sanderson. 1965. *J Preidontestry* 36(3):232–6.

31. Hanson, I. S., and A. G. Nielson. 1956. *J Am Dental Ass* (J.A.D.A.) 52:131.

32. Ewen, S. J. 1960. *J Peridont* 31:101.

33. Ewen, S. J. 1962. New York *J Den* 32:278.

34. Hartley, J. L., D. C. Hudson, and F. A. Brogan. 1959. *J Am Dental Ass* (J.A.D.A.) 59:72.

35. Wilson, J. R. 1958. *J Pros Den* 8:161.

36. Ewen, S. J. 1959. New York State *Sen J* 25:189.

37. Nyborg, W. L. 1968. *J Acoust Soc Am* 44(5):1302–9.

38. Bailey, M. R., V. A. Khokhlovs, O. A. Sapozhnikov, S. G. Kargl, and L. A. Crum. 2003. *Acoustical Phys* 49(3):369–88.

39. Shombert, D. G., and G. R. Harris. 1986. *IEEE Trans Ultrason Ferroelectrics Frequency Control* UFFC-33(3):287–94.

40. Joshi, A., and J. Raje. 2002. *J Control Release* 83(1):13–22.

41. Ferrara, K., R. Polland, and M. Borden. 2007. *Annu Rev Biomed Eng* 9:415–47.

42. Vykhodtseva, N., N. McDannold, and K. Hynynen. 2008. *Ultrasonics* 48(4):279–96.

43. Lasner, P., I. Embreckt, Dr. Miyake, and H. H. Zinsser. 1966. In *Paper J-6 presented at the 1966 IEEE Ultrasonics Symposium*, Abstract in Program Supplement, Cleveland, Ohio, October 12–14, 1966. New York: IEEE, Sonics and Ultrasonics Group.

44. Chandler, D. P., J. Brown, C. J. Bruckner-Lea, L. Olson, G. J. Posakony, J. R. Stults, N. B. Valentine, and L. J. Bond. 2001. *Anal Chem* 73(15):3784–89.
45. Fry, W. J., D. Tucker, F. J. Fry, and V. J. Wulff. 1951. *J Acoust SocAm* 23(3):364–8.
46. Bilkova, B. 1969. *Scr Med (Brno)* 42(1):11–16.
47. Cimino, W. W., and L. J. Bond. 1996. *Ultrasound Med Biol* 22(1):89–100.
48. Bond, L. J., and W. W. Cimino. 1996. *Ultrasound Med Biol* 22(1):101–17.
49. Bond, L. J., M. Flake, B. Tucker, K. Judd, and M. Boukhyn. 2003. Physics of Phacoemulsification. In *Proceedings, 5th World Congress on Ultrasound, WCU 2003* (Paris, France, September, 2003), French Acoustical Society, 169–172.
50. DiSpaltro, F. L., ed. 1999. Ultrasound-Assisted lipoplasty. *Part 1: Clin Plast Surg* 26(2): 187–339.
51. DiSpaltro, F. L., ed. Ultrasound-Assisted Lipoplasty. *Part II: Clin Plast Surg* 26(2): 341–527.
52. Williams, J. W., and W. J. B. Hodgson. 1979. *Mt Sinai J Med* 46(2):105–6.
53. Blackie, R. A. S., and A. Gordon. 1984. *J Clin Path* 37(10):1101–4.
54. Malhotra, V., R. Malik, R. Gondal, P. C. Beohar, and B. Parkash. 1986. *Acta Neurochir* 81(3–4):132–4.
55. O'Daly, B. J., E. Morris, G. P. Gavin, J. M. O'Byrne, and G. B. McGuinness. 2008. *J Mater Process Technol* 200(1–3):38–58.
56. Howkins, S. D., and A. Weinstock. 1970. *Ultrasonics* 8(3):174–6.
57. Vaughen, J. L., and L. F. Bender. 1957. The effects of ultrasound on growing bone. In *Paper K9 presented at the 54th Meeting of the Acoustical Society of America*. Ann Arbor, Michigan: University of Michigan, October 24–26, 1957. Abstract published in *J Acoust Soc Am*, Vol. 29, Part 11, page 1262.
58. Preda, V., C. Craciun, M. Cristea, M. Gocan, and M. Georgescu. 1965. Stud. Cercet. *Embriol Citol Ser Citol* 1(2):155–60.
59. Dyson, M., J. B. Pond, J. Joseph, and R. Warwich. 1969. In *Paper presented at the IEEE Ultrasonics Symposium*. St. Louis, Missouri, September 24–26, 1969. Final Program (Abstracts book). Sonics and ultrasonics Group, Institute of Electronics and Electrical Engineers, New York.
60. DiStatsio, J. I. ed. 1980. Ultrasonics as a medical diagnostic tool. In *Bioeffects of ultrasonics*. Noyer Data Corp (New Jersey) 1189–249.
61. Wells, P. T. N. 1992. *Ultrasonics* 30(1):3–7.
62. Duck, F. A. 1990. *Physical Properties of Tissue: A Comprehensive Reference Book*. London: Academic Press.
63. Wells, P. T. N. 1984. *IEE Proc Pt A* 131(4):225–32.
64. Ziskin, M. C., and D. B. Petitti. 1988. *Ultrasound Med & Biol* 14(2):91–97.
65. Miller, D. L. 1991. *J Clin Ultrasound* 19(9):531–40.
66. IEC (1992) Publication 61157: Requirements for the declaration of the acoustic output of medical diagnostic equipment: TC-87 (Technical Committee) International Electrotechnical Commission. Paris, France.
67. Flynn, H. G. 1982. *J Acoust Soc Am* 72:1926.
68. Apfel, R. E. 1982. *Br J Cancer* 45(suppl. V):140–6.
69. Cartensen, E. L. 1987. *Ultrasound Med Biol* 13(10):597–606.
70. Apfel, R. E., and C. K. Holland. 1991. *Ultrasound Med Biol* 17(2):179–85.
71. Harris, S. 2009. In *Ultrasound market presents opportunities for growth despite recession: Medical News Today*, 30 Jan 2009: (http://www.medicalnewstoday.coom/articles/137292.php)
72. Lindner, J. R. 2004. *Nat Rev Drug Discov* 3:527–32.
73. de Jong, N. 1996. *IEEE Eng Med Biol* 15(6):72–82.
74. Calliada, F., R. Campani, O. Bottinelli, A. Bozzini, and M. G. Sommaruga. 1998. *Eur J Radiol* 27(2):S157–60.

75. Holmes, J. H., D. H. Howry, G. J. Posakony, and C. R. Cushman. 1954. *Trans Am Clin Climatol Assoc* 66:208–25.

76. *Life Magazine*, September (1954). Sound-wave portrait in the flesh, 37(12):71–72.

77. Holmes, J. H., W. Wright, E. P. Meyer, G. J. Posakony, and D. H. Howry. 1965. *Am J Med Electron* 4(4):147.

78. Levi, S. 1997. *Ultrasound in Med Biol* 23(4):418–522.

79. Woo, J. 2003. A short history of the development of ultrasound in obstetrics and gynecology: Parts 1–3. http://www.ob-ultrasound.net/history1.html (accessed 2008–2011).

80. Thompson, H. E. 1968. *J Acoust Soc Am* 44(5):1365–72.

81. Donald, I., and U. Abdulla. 1967. *Ultrasonics* 5(1):8–12.

82. Sunden, B. 1967. *Ultrasonics* 5(1):67–71.

83. Fenster, A., and D. B. Downey. 1996. *IEEE Eng Med Biol* 15(6):41–51.

84. Smyth, C. N. 1967. *Ultrasonics* 5(1):59.

85. Lahoda, F. 1969. *Wehrmed Monatsschr* 13:8–11.

86. de Vlieger, M. 1967. *Ultrasonics* 5(2):91–7.

87. Makow, D. M., and R. R. Real. 1965. *Ultrasonics* 3(2):75–80.

88. Schlagenhauff, R. E., J. A. Mazurowskii, and M. S. Megahed. 1968. *IEEE Trans Son Ultrason* SU-15(3):173–8.

89. White, D. N. 1970. *Ultrasonic Encephalography*. Kingston Ontario, Canada: Medical Ultrasonic Laboratory, Queen's University.

90. Aubry, J.-F., D. Cassereau, M. Tanter, T. Pellegrini, and M. Fink. 2002. Skull surface detection algorithm to optimize time reversal focusing through a human skull. *Proc. 2002 IEEE Ultrasonic Symp (IEEE)*, eds. D. E. Yuhas and S.C. Schneider, Munich, Germany, Oct. 8–11, 1451–54. New York: IEEE.

91. Kimoto, S., R. Omoto, M. Tsunemoto, T. Muroi, K. Atsumi, and R. Uchida. 1964. *Ultrasonics* 2(2):82–6.

92. Kelly, E., F. Fry, and D. Okuyama. 1968. Ultrasonic differentiation of normal liver structure as a function of age and species. In *Proc. 6th International Congress on Acoustics, Paper M-l-1*, ed. Y. Kohasi, M-1–4. Tokyo, August 21–28, 1968. Paris: International Council Scientific Unions.

93. Holmes, J. H. 1964. *Digest* 6:27–34.

94. Ophir, J., and N. F. Maklad. 1985. *IEEE Trans Son Ultrason* SU-32(3):465–72.

95. O'Donnell, M., and H. F. Reilly Jr. 1985. *IEEE Trans Son Ultrason* SU-32(3):450–7.

96. Leary, G. A. 1967. *Ultrasonics* 5(2):84–7.

97. Greguss, P. 1964. *Ultrasonics* 2(3):134–6.

98. Greguss, P. 1968. Ultrasonic holography in ophthalmology. *Opt Technol* 1(1):40–1.

99. LaCourse, J. R., and D. A. Sekel. 1986. *IEEE Trans Biomed Eng* BME-33(4):381–5.

100. Acharya, U. R., W. L. Yun, E. Y. K. Ng, W. W. Yu, and J. S. Suri. 2008. *J Med Syst* 32(4):301–15.

101. Foster, F. S., C. J. Pavlin, K. A. Harasiewicz, D. A. Christopher, and D. H. Turnball. 2000. *Ultrasound Med Biol* 26(1):10–27.

102. Byrne, S. F., and R. L. Green. 2002. *Ultrasound of the Eye and Orbit*. 2nd ed. St. Louis, MO: Mosby Year Book.

103. Coleman, D. J., R. H. Silverman, F. L. Lizzi, and M. J. Rondeau. 2006. *Ultrasonography of the Eye and Orbit*. 2nd ed. Philadelphia, PA: Lippincott Williams & Wilkins.

104. York, G., and Y. Kim. 1999. *Annu Rev Biomed Eng* 1:559–88.

105. Littlefield, R. J., C. R. Macedonia, and J. D. Coleman. 1998. MUSTPAC 3-D ultrasound telemedicine/telepresence system. *Proc IEEE Ultrason Symp* 2:1669–75.

106. Anon. 1969. *Ultrasonics* 7(4):217–18.

107. Popp, R. L., and D. C. Harrison. 1970. *Circulation* 41:493–502.

108. Soulen, R. L., M. S. Lapayowker, and J. L. Gimenez. 1966. *Radiology* 86(6):1047–51.

109. Edler, I. 1967. *Ultrasonics* 5(2):72–9.

110. Kimoto, S., R. Omoto, M. Tsunemoto, T. Muroi, K. Atsumi, and R. Uchida. 1964. *Ultrasonics* 2(2):82–6.
111. Omoto, R. 1967. *Ultrasonics* 5(2):80–3.
112. Eggleton, R. C., C. Townsend, J. Herrick, G. Templeton, and J. H. Mitchell. 1970. *IEEE Trans Son Ultrason* 17(3):143–53.
113. Asberg, A. 1967. *Ultrasonics* 5(2):113–7.
114. Delp, E. J., A. J. Buda, D. N. Smith, J. M. Jenkins, F. Splittberger, C. R. Meyer, and B. Pitt. 1982. In *Proceedings of the 35th Annual Conference on Engineering in Medicine and Biology*, ed. M. ter-Pogossian, Vol. 24, 206, September 22–24, 1982. Bethesda, Maryland: Alliance for Engineering in Medicine and Biology.
115. Nakatani, H., S. Tamura, K. Tanaka, A. Kitabatake, and M. Inoue. 1979. *IEEE Trans Biomed Eng* BME-26(2):65–7.
116. Massa, F. 1962. *Proc IRE* 50(5):1382–92.
117. Hilderbrand, B. P., and B. B. Brenden. 1972. *An Introduction to Acoustical Holography*. New York: Plenum.
118. Lehman, C. D., M. P. André, B. A. Fecht, J. M. Johansen, R. L. Shelby. 2000. *Acad Radiol* 7(2):1–8.
119. Garlick, T. F., B. A. Fecht, J. O. Shelby, and V. I. Neeley. 2002. Metrics program for assessing diffractive energy imaging system performance. In *Medical Imaging 2002, Physics of Medical Imaging*, ed. L. E. Antonuk and M. J. Yaffe, Vol. 2682, 267–76. *SPIE Proceedings*. Bellingham, WA: International Society for Optics and Photonics.
120. André, M. P., C. D. Lehman, V. Neely, B. A. Fecht, J. O. Shelby, and T. Nguyen. 2001. Recent experience with acoustical holography for breast imaging. In *Acoustical Imaging*, ed. M. Halliwell and P. N. T. Wells, Vol. 25, 309–16. Berlin: Springer.
121. Evans, A., and D. N. Walder. 1970. *Ultrasonics* 8(4):216–7.
122. Lees, S., and F. E. Barber. *Science*, 161:477–88.
123. Kossoff, G., and C. J. Sharpe. 1966. *Ultrasonics* 4(2):77–83.
124. Fujimoto, Y., A. Oka, R. Omoto, and M. Hirose. 1967. *Ultrasonics* 5(3):177–88.
125. Damascelli, B., N. Cascinelli, T. Livraghi, and U. Veronesi. 1968. *Ultrasonics* 6(4):242–3.
126. McNally, E. G. 2008. *Skeletal Radiol* 37(2):99–113.
127. Setola, S. V., O. Catalano, F. Sandomenico, and A. Siani. 2007. *Abdom Imaging* 32(1): 21–8.
128. Dayton, P. A., and J. J. Rychak. 2007. *Front Biosci* 12:5124–42.
129. Baker, D. W. 1969. Pulse ultrasonic doppler flowmeter biological and engineering applications: Abstract paper J-6, Final Program. In *Paper J6 presented at the 1969 IEEE Ultrasonics Symposium*, St. Louis, Missouri, September 24–26, 1969. New York: IEEE Sonics and Ultrasonics Group.
130. Yoshitoshi, Y., K. Machi, H. Sekiguchi, Y. Mishina, S. Ohta, Y. Hanoaka, Y. Kohashi, S. Shimizu, and H. Kuno. 1966. *Ultrasonics* 4(1):27–8.
131. Kostis, J. B. 1969. *Circulation* 40:197–207.
132. Wells, P. N. T. 1969. *Physical Principles of Ultrasonic Diagnosis*. London: Academic Press.
133. Carrigan, T. A., W. D. Wilcox, and D. P. Giddens. In *Proceedings of the 35th Annual Conference on Engineering in Medicine and Biology*, Vol. 24, 204, September 22–24, 1982. Bethesda, Maryland: Alliance for Engineering in Medicine and Biology.
134. Rittgers, S. E., B. M. Thornhill, and R. W. Barnes. In *Proceedings of the 35th Annual Conference on Engineering in Medicine and Biology*, Vol. 24, 198, September 22–24, 1982. Bethesda, Maryland: Alliance for Engineering in Medicine and Biology.
135. Rittgers, S. E., J. S. Pfeifer, and R. W. Barnes. In *Proceedings of the 35th Annual Conference on Engineering in Medicine and Biology*, Vol. 24, 199, September 22–24, 1982. Bethesda, Maryland: Alliance for Engineering in Medicine and Biology.

136. Ingle, F. W., J. Brennan, and N. Rogers. In *Proceedings of the 35th Annual Conference on Engineering in Medicine and Biology*, Vol. 24, 200, September 22–24, 1982. Bethesda, Maryland: Alliance for Engineering in Medicine and Biology.

137. Ingle, F. W. 1982. In *Proceedings of the 35th Annual Conference on Engineering in Medicine and Biology*, Vol. 24, 201, September 22–24, 1982. Bethesda, Maryland: Alliance for Engineering in Medicine and Biology.

138. Kasai, A., K. Namekawa, A. Koyano, and R. Omoto. 1985. *IEEE Trans Son Ultrason* SU-32(3):458–64.

139. Halberg, L. I., and K. Thiele. 1986. *Hewlett-Packard J* 37(6):35–40.

140. Paliwal, S., and S. Mitragotri. 2008. *Ultrasonics* 48(4):271–8.

141. Dennis, J. H., and D. J. Hendrick. 1992. *J Med Eng Technol* 16(2):63–8.

142. Lehmann, J. F., C. G. Warren, and A. W. Guy. 1978. Therapy with continuous wave ultrasound. In *Ultrasound: its applications in medicine and biology*, Part II, Chapter 10, ed. F. J. Fry, New York: Elsevier.

143. Lehmann, J. F., ed. 1982. *Therapeutic Heat and Cold*. 3rd ed. Baltimore: Williams and Wilkins.

144. Stewart, H. F., J. L. Abzug, and G. R. Harris. 1980. *Phys Ther* 60(4):424–8.

145. Dinno, M. A., L. A. Crum, and J. Wu. 1989. *Ultrasound Med Biol* 15(5):461–70.

146. Idom, C. B. 1989. The effect of free radical scavengers on ultrasonically induced biophysical changes in frog skin. MS Thesis, Department of Physics and Astronomy, University of Mississippi.

147. Baker, K. G., V. J. Robertson, and F. A. Duck. 2001. *Phys Ther* 81(7):1351–8.

148. Robertson, J., and K. G. Baker. 2001. *Phys Ther* 81(7):1339–50.

149. Laugier, P., K. A. Wear, and K. R. Waters, eds. 2008. *IEEE Trans UFFC* 55(6): 1177–327.

150. Laugier, P., K. A. Wear, and K. R. Waters, eds. 2008. *IEEE Trans UFFC* 55(7):1415–554.

151. Pounder, M., and A. J. Harrison. 2008. *Ultrasonics* 48(4):330–8.

152. Kaufman, J. J., G. M. Luo, and R. S. Siffert. 2008. *IEEE Trans UFFC* 55(6):1205–18.

153. Tschannen, F. 1957. *Ultrason News* 1(4):25.

154. Kaclova, J., J. Kolar, A. Babicky, and J. Kacl. 1965. *Cesk Stomatol* 65(6):437–42.

155. Skauen, M., and G. M. Zentner. 1984. *Int J Pharmaceutics* 20(3):235–45.

156. Mitragotri, S., D. A. Edwards, D. Blankschtein, and R. Langer. 1995. *J Pharm Sci* 84(6):697–705.

157. Amsden, B. G., and M. F. A. Goosen. 1995. *AIChE J* 41(8):1972–996.

158. Mitragotri, S., D. Blankschtein, and R. Langer. 1995. *Science* 269(5225):850–3.

159. Smith, N. B. 2008. *Expert Opin Drug Deliv* 5(10):1107–20.

160. Barnard, J. W., W. J. Fry, F. J. Fry, and R. F. Krumins. 1955. *J Comp Neurol* 103(3):459–84.

161. Rosenthal, I., J. Z. Sostaric, and P. Riesz. 2004. *Ultrason Sonochem* 11(6):349–63. Ref 42.

162. Pitt, W. G., G. A. Husseini, and B. J. Staples. 2004. *Expert Opin Drug Deliv* 1(1):37–56.

163. Husseini, G. A., M. A. Diaz de la Rosa, E. S. Richardson, D. A. Christensen, and W. G. Pitt. 2005. *J Control Release* 107(2):253–61.

164. Pichon, C., K. Kaddur, P. Midoux, F. Tranquart, and A. Bouakaz. 2008. *J Exp Nanosci* 3(1):17–40.

165. Frenkel, F. 2008. *Adv Drug Deliv Rev* 60(10):1193–208.

166. Tachibana, K., L. B. Feril, and Y. Ikeda-Dantsuji. 2008. *Ultrasonics* 48(4):253–9.

167. Greguss, P., and A. Bertenyi. 1976. *Ultrasonics* 14(2):81–2.

168. Mortimer, A. J., G. V. Forester, O. Z. Roy, and W. J. Keon. 1982. In *Proceedings of the 35th Annual Conference on Engineering in Medicine and Biology*, Vol. 24, 205, September 22–24, 1982. Bethesda, Maryland: Alliance for Engineering in Medicine and Biology.

169. ter Haar, G. 2008. *Ultrasonics* 48(4):233.

170. ter Haar, G. 2001. *Phys Today* 54(12):29–34.

171. Burov, A. K. 1956. *Dokl Akad Nauk SSSR* 106:239.

172. Burov, A. K., and G. D. Andreewskaya. 1956. *Dokl Akad Nauk SSSR* 106:445.
173. Lele, P. P. 1988. Ultrasound hyperthermia. In *Encyclopedia of Medical Devices and Instrumentation*, ed. J. G. Webster, Vol. 3, 1599–1612. New York: Wiley.
174. Lele, P. P., J. Goddard, and M. Blanter. 1987. Clinical results with intensity modulated, scanned, focused ultrasound (SIMFU) system. *Syllabus: A Categorical Course in Radiation Therapy: Hyperthermia*, ed. R. A. Steeves and B. R. Palival. Oak Brook, Illinois: Radiological Society of North America.
175. Lele, P. P., and J. Goddard. 1987. Optimizing insonation parameters in therapy planning for deep heating by SIMFU. In *IEEE/Ninth Annual Conference of the Engineering in Medicine and Biology Society*. Boston, November 13–17, 1987. New York: IEEE. (Cat. No. 87CH2513–0).
176. Lele, P. P. 1986. In *IEEE/Eighth Annual Conference of the Engineering in Medicine and Biology Society*, 1435–40. New York: IEEE. (Cat. No. 86CH2368–9).
177. Wollfson, R. J. 1963. *Eye Ear Nose Throat Digest* 73–4, 76–8, 80–1.
178. Fry, W. J. 1956. *J Acoust Soc Am* 28(4):719–21.
179. Lele, P. P. 1967. *Ultrasonics* 5(2):105–12.
180. Fry, F. J., R. F. Heimburger, L. V. Gibbons, and R. C. Eggleton. 1970. *IEEE Trans Son Ultrason* 17(3):165–9.
181. Diederich, C. J., and K. Hynynen. 1999. *Ultrasound Med Biol* 25(6):871–87.
182. Chopra, R., C. Luginbuhl, F. S. Foster, and M. J. Bronskill. 2003. *IEEE Trans UFFC* 50(7):881–9.
183. Lizzi, F. L., and M. Ostromogilsky. 1987. *Ultrasound Med Biol* 13(10):607–18.
184. Meaney, P. M., R. L. Clarke, G. R. ter Haar, and I. H. Rivens. 1998. *Ultrasound Med Biol* 24(9):1489–99.
185. Bohris, C., W. G. Schreiber, J. Jenne, I. Simiantonakis, R. Rastert, H.-J. Zabel, P. Hauber, R. Bader, and G. Brix. 1999. *Magn Reson Imaging* 17(4):603–10.
186. Coleman, A. J., J. E. Daunders, L. A. Crum, and M. Dyson. 1987. *Ultrasound Med Biol* 13(2):69–76.
187. Cunningham, K. B., A. J. Coleman, T. G. Leighton, and P. R. White. 2001. *Acoust Bull* 26(5):10–6.
188. Bao, S., B. D. Thrall, R. A. Gies, and D. L. Miller. 1998. *Cancer Res* 58(2):219–21.
189. Thornton, L. K., and F. Nahai. 1999. *Clin Plast Surg* 26(2):299–304.
190. Young, V. L., and M. W. Schorr. 1999. *Clin Plast Surg* 26(3):481–524.
191. Topaz, M. 1998. *Plast Reconstr Surg* 102(1):280.
192. Armstrong, D. N., W. L. Ambroze, M. E. Schertzer, and G. R. Orangio. *Dis Colon Rectum* 44(4):558–564.
193. Barrett, W. L., and S. M. Garber. 2004. *Business Briefing: Global Surgery*, ed. E. Boulton, (October 2004), 7 pages, Touch Briefings, www.touchbriefings.org.
194. Siegel, R. J., M. C. Fishbein, J. Forrester, K. Moore, E. DeCastro, L. Daykhovsky, and T. A. Donmichael. 1988. *Circulation* 78(6):14443–8.
195. Hong, A. S., J.-S. Chae, S. B. Dubin, S. Lee, M. C. Fishbein, and R. J. Siegel. 1990. *Am Heart J* 120(2):418–22.
196. Siegel, R. J., T. A. Don. Michael, M. C. Fishbein, J. Brookstein, L. Adler, T. Reinsvolt, E. DeCastro, and J. S. Forrester. 1990. *J Am Coll Cardiology* 15:345–51.
197. Ariani, M., M. C. Fishbein, J. S. Chae, M. Sadeghi, A. D. Michael, S. B. Dubin, and R. J. Seigel. 1991. *Circulation* 84(4):1680–8.
198. Rosenschein, U., J. J. Bernstein, E. Disegni, E. Kaplinsky, J. Brenheim, and L. A. Rosenszajn. 1990. *J Am Coll Cardiology* 15(3):711–7.
199. Rosenschein, U., L. A. Rosenszajn, L. Kraus, C. C. Marboe, J. F. Watkins, E. A. Rose, D. David, P. J. Cannon, and J. S. Weinstein. 1991. *Circulation* 83(6):1976–86.
200. Philippe, F., G. Drobinski, C. Bucherer, A. Ankri, C. Lancombe, D. Kremer, D. Brisset, and G. Montanescot. 1993. *Catherization and Cardiovascular Diagnosis* 28(2):173–8.

201. Chan, K. K., D. J. Watmough, D. T. Hope, and K. Mori. 1986. *Ultrasound Med Biol* 12(6): 279–83.
202. Allen, E. F., and R. H. Rhoads. 1963. *J Periodont* 34(4):352–56.
203. Kellman, C. D. 1967. *Am J Opthalmol* 64:23–35.
204. Kellman, C. D. 1969. *Am J Opthalmol* 67:464–77.
205. Kellman, C. D. 1973. *Am J Opthalmol* 75:764–8.
206. Shpuntoff, H., and D. Wuchinich. 1984. *Ultrasound Med Biol* 10(6):697–700.
207. Goliamina, I. P. 1974. Ultrasonic surgery. In *Proc 8th Int Congress Acoustics* (Invited Lecture) London, July, 63–9. Guilford, UK: IPC.
208. Kruis, A. 1968. U.S. Patent 3,565,062, June 3.
209. Delaney, L. J. 1965. U.S. Patent 3,352,303, July 28.
210. Brown, A. H., and P. G. H. Davis. 1972. *British Med J* 3:274–7.
211. Flamm, E. S., J. Ransohoff, D. Wuchinich, and D. Baldwin. 1978. *Neurosurgery* 2(3): 240–5.
212. Epstein, F. 1984. *Clin Neurosurg* 31:487–505.
213. Packer, M., W. J. Fishkind, H. Fine, B. S. Seibel, and R. S. Hoffman. 2005. *J Cataract Refract Surg* 31(2):424–31.
214. Novak, F., J. Cecava, J. Dohnalek, and I. Hrazdira. 1966. *Cesk Stomatol* 66(4):271–4.
215. Anon. 1967. *Ultrasonics* 5:201.
216. Ahmad, M., T. R. P. Ford, and L. A. Crum. 1987. *J Endod* 13(3):93–101.
217. Ahmad, M., T. R. P. Ford, and L. A. Crum. 1987. *J Endod* 13(10):490–9.
218. Ahmad, M., T. R. P. Ford, L. A. Crum, and A. J. Walton. 1987. *J Endod* 14(10):486–93.
219. Stock, C. J. 1991. *Int Dent J* 41(3):175–82.
220. Walmsley, A. D. 1988. *Ultrasound Med Biol* 14(1):7–14.
221. Laird, W. R., and A. D. Walmsley. 1991. *J Dent* 19(1):14–7.
222. Walmsley, A. D., W. R. Laird, and P. J. Lumley. 1992. *J Dent* 20(1):11–7.
223. Trenter, S. C., and A. D. Walmsley. *J Clin Periodontol* 30(2):95–101.
224. Arabaci, T., Y. Cicek, and C. F. Canakci. 2007. *Int J Dent Hyg* 5(1):2–12.
225. van der Sluis, L. W. M., M. Versluis, M. K. Wu, and P. R. Wesselink. 2007. *Int Endod J* 40(6):415–26.
226. Walmsley, A. D. 1997. *Br Dent J* 182(6):209–18.
227. Tanner, D. A., N. Rushe, and M. R. Towler. 2006. *J MaterSci Mater Med* 17(4):313–8.
228. Wang, R. R., E. Meyers, and L. Katz. 1998. *J Biomed Mat Res Part A* 42(4):508–16.
229. Birck, H. G., and H. E. Manhart. 1963. *Arch Otolaryngol* 77:603–8.
230. Jenkins, I. C. 1967. *J Laryng Otol* 81:385–90.
231. Preibisch-Effenberger, R. 1966. *Arch Klin Exp Ohren Nasen Kehlkopfheilkd* 186:146–52.
232. Schlappi, P. 1967. *Pract Oto-Rhino-Laryngol* 29:365–70.
233. Newell, J. A. 1963. *Phys Med Biol* 8(3):241–64.
234. Kossoff, G. 1964. *IEEE Trans Son Ultrason* SU-11(2):95.
235. Anon. 1968. *Ultrasonics* 6(1):3.
236. Johnson, S. J. 1967. *Ultrasonics* 5(3):173–6.
237. Anon. 1967. *Med World News*, January 27, 1967, 33.
238. Arslan, M. 1967. *Ultrasonics* 5(2):98–101.
239. Venrooij, G. E. P. M. 1971. *Ultrasonics* 9(4):240–2.
240. Kossoff, G., E. K. Fry, and J. Jellins. 1973. *J Acoust Soc Am* 57:1730–6.
241. Ophir, J. 1986. *IEEE Trans Ultrason Ferroelectr Freq Control* UFFC-33(4):359–68.
242. Ferrari, L., J. P. Jones, and V. M. Gonzalez. 1986. *Ultrasonics* 24(2):66–72.
243. Kuc, R. 1986. *IEEE Trans Ultrason Ferroelectr Freq Control* UFFC-33(3):273–79.
244. Salomonsson, G., and L. Bjorkman. 1986. *IEEE Trans Ultrason Ferroelectr Freq Control* UFFC-33(3):280–6.
245. Thomas III, L. J., S. A. Wickline, J. E. Perez, B. E. Sobel, and J. G. Miller. 1986. *IEEE Trans Ultrason Ferroelectr Freq Control* UFFC-33(1):27–32.

246. Wear, K. A., T. A. Shoup, and R. L. Popp. 1986. *IEEE Trans Ultrason Ferroelectr Freq Control* UFFC- 33(4):347–53.

247. Shore, D., M. O. Woods, and C. A. Miles. 1986. *Ultrasonics* 24(2):81–7.

248. Jones, J. P., and S. Leeman. 1984. *Acta Electronica* 26(1–2):3–31.

249. Goss, S. A., R. L. Johnston, and F. Dunn. 1978. *J Acoust Soc Am* 64(2):423–57.

250. Goss, S. A., R. L. Johnston, and F. Dunn. 1980. *J Acoust Soc Am* 68(1):93–108.

251. Duck, F. A. 1990. *Physical Properties of Tissue*. London; Academic Press.

252. Quate, C. F., A. Atalar, and H. K. Wickramasinghe. 1979. *Proc IEEE* 67(8):1092–114.

253. Quate, C. F. 1985. *IEEE Trans Sonics Ultrason* SU-32(2):132–5.

254. Lockwood, G. R., D. H. Turnbull, D. A. Christopher, and F. S. Foster. 1996. *IEEE Eng Med Biol* 15(6):60–71.

255. Scherba, G., P. A. Hoagland, and W. D. O'Brien. 1994. *IEEE Trans* UFFC 41(4): 451–7.

256. Kinoshita, A., S. Senda, K. Mizushige, H. Masugata, S. Sakamoto, H. Kiyomoto, and H. Matsuo. 1998. *Ultrasound Med Biol* 24(9):1397–405.

257. Akashi, N., J. Kushibiki, and F. Dunn. 1997. *J Acoust Soc Am* 102(6):3774–8.

258. Sherar, M. D., M. B. Noss, and F. S. Foster. 1987. *Nature* 330(6147):493–5.

259. Gertner, M. R., B. C. Wilson, and M. D. Sherar. 1998. *Ultrasound Med Biol* 24(3):461–8.

260. Saijo, Y., H. Sasaki, M. Sato, S. Nitta, and M. Tanaka. 2000. *Ultrasonics* 38(1–8):396–9.

261. Kundu, T., J.-P. Lee, C. Blase, and J. Breiter-Hahn. 2006. *J Acoust Soc Am* 120(3): 1646–54.

262. Good, M. S., C. F. Wend, L. J. Bond, J. S. McLean, P. D. Panetta, S. Ahmed, S. L. Crawford, and D. D. Daly. 2006. *IEEE Trans* UFFC 53(9):1637–48.

263. Turnbull, D. H., T. S. Bloomfield, H. S. Baldwin, and F. S. Foster. 1995. *Proc Natl Acad Sci* 92(6):2239–43.

264. Turnbull, D. H., and F. S. Foster. 2002. *Trends Biotechnol* 20(8):S29–33.

265. Spurney, C. F., C. W. Lo, and L. Leatherbury. 2006. *Echardiography – A: J Cardiovasc Ultrasound Allied Tech* 23(10):891–9.

266. Lewin, P. A. 2007. High frequency biomedical and industrial ultrasound applications. In *Proc Int Congress Ultrasonics*, 1–8, ed. E. Benes, Vienna, April 6–13, 2007, Paper # 1796, Session K02.

267. Andrianova, I. G. 1967. *Khim-Farm Zh* 1(7):55–6.

268. Growchowska, Z. 1967. *Mikrochim Acta* 1(1):1–6.

269. Poston, F. 1965. *Am J Med Technol* 31:279–84.

270. Davis, R. 1959. *Polynucleotides VII* 33:481–93.

271. Hughes, D. E., and W. L. Nyborg. 1962. *Science* 138(3537):108–14.

272. Hawley, S. A., R. M. Macleod, and F. Dunn. 1963. *J Acoust Soc Am* 35(8):1285–7.

273. Rooney, J. A. 1972. *J Acoust Soc Am* 52(6):1718–24.

274. Miller, D. L., S. Bao, R. A. Giles, and B. D. Thrall. 1999. *Ultrasound Med Biol* 25(9): 1425–30.

275. Brayman, A. A., M. L. Coppage, S. Vaidya, and M. W. Miller. 1999. *Ultrasound Med Biol* 25(6):999–1008.

276. Tachibana, K., T. Uchida, K. Ogawa, N. Yamashita, and K. Tamura. 1999. *The Lancet* Vol. 353(April 24):1409.

277. Chandler, D. P., J. Brown, C. J. Bruckner-Lea, G. J. Posakony, J. R. Stults, N. B. Valentine, and L. J. Bond. 2001. *Anal Chem* 73(15):3784–9.

278. Warner, C. L., C. J. Bruckner-Lea, J. W. Grate, T. Straub, G. J. Posakony, N. Valentine, R. Ozanich et al. 2009. *JALA* 14(5):277–84.

279. Feril, L. B., T. Kondo, Q. L. Zhao, R. Ogawa, K. Tachibana, N. Kudo, S. Fujimoto, and S. Nakamura. 2003. *Ultrasound Med Biol* 29(2):331–7.

280. Povey, M., and T. J. Mason, ed. 1998. *Ultrasound in Food Processing*. London: Blackie Academic and Professional.

281. Joyce, E. M., and T. J. Mason. 2008. *Chimica Oggi-Chemistry Today* 26(6):22–6+.

Glossary

This glossary includes terms that are commonly used in ultrasonics and are used throughout this book.

Absorption (of sound): The dissipation or conversion of sound energy into other forms of energy, such as heat, either within the medium or attendant on a reflection.

Acoustical attenuation constant: The real part of the acoustical propagation constant.

Acoustical propagation constant: The natural logarithm of the complex ratio of steady-state particle velocity, volume velocities, or pressures either at two points separated by unit distance in a uniform system (assumed to be of infinite length) or at two successive corresponding points in a system of recurrent structures (assumed to be of infinite length). The ratio is determined by dividing the value at the point near to the transmitting end by the corresponding value at the more remote point.

Acoustic impedance: The complex quotient of a sound pressure (force per unit area) on a surface lying in the wavefront divided by the flux (volume velocity or linear velocity multiplied by the area) through the surface. When concentrated rather than distributed impedances are considered, the impedance of a portion of the medium is defined by the complex quotient of the pressure difference effective in driving that portion by the flux (volume velocity).

Acoustics: The science of sound including its production, transmission, and effects.

Active acoustic device: Device that generates sound for the purpose of detecting an object.

Attenuation: Loss of intensity in an ultrasonic wave with distance or time.

Beats: Periodic reinforcement of a sound produced by the interference of sound waves of slightly different frequencies.

Couplant: Material through which an ultrasonic wave may pass from a source to another material with minimum loss of energy.

Cylindrical wave: A wave in which the wavefronts are coaxial cylinders.

Decibel: A term used in designating relative intensity. The number of decibels corresponding to a given intensity, I, is determined by $10 \log_{10}(I/I_0)$, where I_0 is a reference intensity.

Doppler effect: The phenomenon evidenced by the change in observed frequency of a wave in a transmission system caused by a time rate of change in the effective length of the path of travel between the source and the point of observation.

Doppler shift: The magnitude of change in the observed frequency of a wave due to the Doppler effect.

Echolocation: A process for locating distant or invisible objects by means of reflection of sound waves to the emitter by the object.

Elastic waves: A wave sustained by the mass and elasticity of a medium.

Electromyography: The process of measuring electrical responses of nerves to various stimuli.

Flexural wave: A wave characterized by a bending, or flexing, type of motion.

Free progressive wave: A wave in a medium free from boundary effects.

Frequency: For a periodic quantity, for which time is the independent variable, the number of periods occurring in a unit of time.

Hertz (Hz): Cycle per second.

Intensity: The average rate of flow of energy through a unit area normal to the direction of wave propagation (kilohertz [kHz]).

Kilohertz (kHz): 1000 Hz.

Lamb wave: A type of wave associated with plates, which propagates in a direction parallel to the surface of a plate at velocities that depend on the elastic properties of the plate, plate thickness, and frequency of the wave. There are two types of Lamb waves: (1) symmetrical and (2) asymmetrical. In a symmetrical wave, the particle motion along the center of the plate is longitudinal, thus producing a series of bulges in the surface of the plate at regular intervals. An asymmetrical wave is a flexural wave.

Longitudinal wave: A wave in which the direction of displacement at each point of the medium is normal to the wavefront.

Love wave: A surface wave in which the direction of displacement at each point of the surface is parallel to the wavefront and to the surface of the medium.

Magnetostriction: A change in dimensions in ferromagnetic materials, such as iron, nickel, cobalt, and alloys of these metals, which occurs when these materials are subjected to a change in magnetic state.

Megahertz (MHz): 1 million Hz.

Mode: The vibration pattern of waves.

Oscillator: A device for producing alternating current or motion. In ultrasonics, the radiofrequency or audiofrequency generator is usually referred to by this term. Occasionally, this term is used to refer to an oscillating mechanical device such as a tuning fork, a pendulum, or an active transducer.

Passive acoustic device: A listening device for detecting sounds generated by a source under investigation.

Piezoelectric effect: Appearance of electric charges on the parallel faces of a slab cut from a crystal having an axis of nonsymmetry when these faces are lying normal to an axis of nonsymmetry and when the slab is subjected to a mechanical stress.

Piezoelectricity: Electric charges that appear on the parallel surfaces of a piezoelectric crystal when the crystal is subjected to a mechanical stress.

Plane wave: A wave in which the wavefronts are parallel planes normal to the direction of propagation everywhere.

Plate wave: A Lamb wave.

Radar: The art of detecting by means of radio echoes the presence of objects, determining their direction and range, recognizing their character, and employing the data thus obtained in the performance of various functions such as military, naval, or meteorological observations.

Rayleigh wave: Type of surface wave in which particle motion is elliptical about axes that are parallel to the wavefront.

Reflection coefficient: The ratio of acoustic intensity of a reflected wave to that of the incident wave at a boundary between two media of different acoustic impedances.

Resonance: Condition of vibration at which the absolute value of the driving-point impedance is a minimum. Velocity resonance exists between a body, or system, and an applied sinusoidal force if any small change in the frequency of the applied force causes a decrease in velocity at the driving point. Except for simple systems, such as pendulums, it is usually necessary to specify the type of resonance or resonant frequency.

Response: Ratio of electric output to the acoustic input of a transducer (microphone or hydrophone).

Sensitivity: Response of a transducer for a sine wave at a prescribed frequency.

Shear wave: Transverse wave; a wave in which particle motion is normal to the direction of propagation.

Sonar: Acoustical analog of radar; the art of detecting by acoustical means the presence of objects, determining their direction and range, recognizing their character, and employing the data thus obtained in the performance of various functions, such as submarine detection, navigation, and distance and depth measurements. In underwater applications, sonar refers to the human use of sound for a variety of purposes in the use and exploration of the seas. Both active and passive sonars are used.

Sonics: The general subject of sound.

Spherical wave: A wave in which the wavefronts are concentric spheres.

Stress wave: Any wave in a medium that is dependent on the elastic nature of the medium.

Surface waves: Waves that propagate along the surface of a medium and penetrate the medium only to shallow depths.

Torsional wave: A type of shear wave in which, within any given plane parallel to the wavefront, particle motion is normal to the direction of propagation and the plane itself oscillates through an angle about an axis that is parallel to the direction of propagation.

Transducer: An instrument that converts one energy form into a second energy form. The types of transducers most commonly used in ultrasonics convert electrical energy to ultrasonic energy and vice versa.

Transmission coefficient: The ratio of intensity of the transmitted wave to that of the incident wave at a boundary between two media of different acoustic impedances.

Ultrasonics: The general subject of sound in the frequency range above the average audible range of human beings.

Ultrasound: Sound that is pitched too high for a human being to hear.

Wave: A disturbance propagated within a medium in such a manner that at any point in the medium the displacement is a function of the position of the point.

Wavefront: For a progressive wave in space, a continuous surface that is a locus of points having the same phase at a given instant; for a progressive surface wave, a continuous line that is a locus of points having the same phase at a given instant.

Appendix A

For convenience, the acoustic properties for selected materials are presented in tabular form (see Tables A.1 through A.4). In addition, a selected bibliography to sources of additional data including some hard to find properties and some particularly useful web resources, are given as Appendix B.

TABLE A.1
Acoustic Properties of Solids

| Material | Velocities (10⁵ cm/s) | | | Density (ρ) (g/cm³) | Characteristic Impedence (ρc_B) (10⁶ g/cm²·s) | Melting Point | | | Poisson's Ratio (σ) | References |
| | Longitudinal | | Shear (c_S) | | | °K | °C | °F | | |
	Bar (c_0)	Bulk (c_B)								
Metals										
Aluminum pure	5.06			2.7		933.602	660.452	1222.8		A14
Aluminum 2S0	5.10	6.35	3.10	2.71	1.73				0.33	A1–A3
Aluminum 17ST	5.08	6.25	3.10	2.80	1.75				0.33	A1, A2
Aluminum 6061-T6	5.05	6.27	3.08	2.70	1.70				0.33	–
Aluminum 7075-T6	5.07	6.35	3.10	2.79	1.78				0.33	–
Antimony	3.40	–	–	6.70	–				–	A5
Arsenic	3.64			5.73	5.08	876	603	1117.4		A14
Barium	1.86			3.63	0.682	1002	729	1344.2		A14
Beryllium	12.75	12.80	8.71	1.82	2.33				0.024–0.030	A1, A2
Beryllium	12.87	12.89	8.88	1.87	2.41					A4
Bismuth	1.79	2.18	1.10	9.80	2.14					A5
Brass 70-30	3.40	4.37	2.10	8.50	3.70				0.35	A4
Brass 70-30	3.48	4.70	2.11	8.60	4.04				0.331	A7
Brass, Naval	3.49	4.43	2.12	8.10	3.61					A1, A2, A5
Bronze, phosphor 5%	3.43	3.53	2.23	8.86	3.12				0.35	A1, A2
Cadmium	2.40	2.78	1.50	8.60	2.40					A5
Cadmium	2.53			8.66	2.18	594.258	321.108	609.994	0.35	A5, A14
Calcium	3.66			1.55		1113	840	1544		A14
Carbon	1.48			2.22		4100	3,826	6918.8		A14
Chromium	5.88			7.20		2133	1860	3380		A14

Material										
Cobalt	4.83			8.86	4.60	1768	1495	2723		A14
Columbium	3.48			8.58	4.18					
Constantan	4.30	5.24	2.64	8.80	4.25				0.35	A5
Copper	3.71	4.66	2.26	8.90	4.25					A1, A2
Copper, annealed	3.81	4.76	2.33	8.93	4.10					A7
Copper	3.65	4.80	2.33	8.90	4.47					A4
Copper	3.60	4.60	–	8.20	4.18					A3
Copper, rolled	3.75	5.01	2.27	8.93						A7
Copper	3.71	4.70	2.26	8.90						A5
Duralumin 17S	5.15	6.32	3.13	2.79	1.76					A7
Gallium	1.09			5.98						
German silver	3.58	4.76	2.16	8.40	4.00					A5
Germanium	1.94			5.31	6.26	1211.5	938.3	1720.94		A14
Gold	2.03	3.24	1.20	19.30	6.26	1337	1064.43	1947.974		A2, A4, A5, A14
Gold	2.03	3.24	1.20	19.30	4.00					A2, A4, A5, A14
German silver	3.58	4.76	2.16	8.40	4.77					A5
Hastelloy X	–	5.79	2.74	8.23	5.22					A6
Hastelloy C	–	5.84	2.90	8.94						A5
Hafnium	3.43			13.09	4.71	2504	2231	4047.8		A14
Iron	5.18	5.76	2.05	7.90		1.808	1535	2785	0.28	A14
Lanthanum	2.37			6.19		1191	918	1684.4		A14
Iridium	4.79	–	–	22.40	4.68					A5
Iron	5.18	5.96	2.05	7.90	4.70					A4
Iron, electrolytic	5.12	5.95	3.24	7.90	4.56					A7
Iron	5.17	5.85	3.23	7.80	4.69					A5
Iron, armco	5.20	5.96	3.24	7.85	2.50–4.00					A4
Iron, cast	3.0–4.7	3.50–5.60	2.2–3.2	7.20	2.96				0.28	A4
Iron, cast	–	4.50	2.40	7.70						A2

(Continued)

TABLE A.1
Acoustic Properties of Solids (*Continued*)

Material	Velocities (10⁵ cm/s)			Density (ρ) (g/cm³)	Characteristic Impedence (ρc_B) (10⁶ g/cm²·s)	Melting Point			Poisson's Ratio (σ)	References
	Longitudinal		Shear (c_S)			°K	°C	°F		
	Bar (c_0)	Bulk (c_B)								
Metals										
Iron, cast	3.70	4.35	–	7.70	3.35					A3
Iron, wrought	5.00	5.90	–	7.80	4.60				0.28	A3
Lead, pure	1.20	2.16	0.70	11.40	2.46				0.44	A1, A2, A5
Lead, annealed	1.19	2.16	0.70	11.40	2.46					A7
Lead, pure	1.25	2.40	0.79	11.30	2.72				0.40–0.45	A4
Lead, pure	1.20	2.20	–	11.30	2.50					A3
Lead, rolled	1.21	1.96	0.69	11.40	2.23					A7
Lead, antimony 6%	1.37	2.16	0.81	10.90	2.36					A1
Lithium	4.72			0.02						A14
Magnesium	4.90	5.74	3.08	1.70	0.99				0.35	A4
Magnesium (Am 35)	5.00	5.79	3.10	1.74	1.01				0.35	A1, A2
Magnesium, drawn Annealed	4.94	5.77	3.05	1.74	1.00					A5
Manganese	4.62			7.42		1519	1246	2274.8		A14
Molybdenum (used in Ultrasonic treatment of glass)	5.45	6.29	3.35	10.09	6.35	2696	2623	4753.4	0.31	A1, A2, A14
Nickel (Curie pt. = 358°C [676.4°F])	4.79	5.63	2.96	8.80	4.95	1728	1455	2651		A1, A2, A5, A14

Material										Ref.
Inconel, wrought	5.08	5.82	3.02	8.25	6.45					A1
Manganin	3.83	4.66	2.35	8.40	3.90					A5
Molybdenum	5.45	6.29	3.35	10.09	3.35				0.32	A1, A2
Molybdenum	5.40	6.25	3.35	10.10	6.31					A7
Molybdenum	5.82	6.25	3.35	10.20	6.37					A4
Monel metal	4.40	5.35	2.72	8.90	4.76				0.315	A7
Nickel	4.90	5.60	–	8.80	4.90					A3
Nickel	4.90	6.04	3.00	8.90	5.38				0.309	A7
Nickel	4.79	5.63	2.96	8.80	4.95				0.31	A1, A2, A5
Inconel X	–	5.94	3.12	8.30	4.93					A6
Inconel, wrought	4.52	6.02	2.72	8.83	5.31					A1
Monel, wrought	4.52	6.02	2.72	8.83	5.31					A1
Silver-nickel (18%)	3.83	4.62	2.32	8.75	4.03					A1
Osmium	4.95			22.5		3306	3033	5491.4		A14
Palladium	3.12			12.01		1828	1555	2831		A14
Platinum	2.80	3.96	1.67	21.40	8.46				0.39	A4, A5
Pottassium	2.00			0.86		336.34	63.19	145.742		A14
Rhenium	3.80			21.18		3459	3186	5766.8		A14
Rhodium	5.47			12.44		2236	1963	3565.4		A14
Radon						202	−71	−95.4		A14
Rubidium						2607	2334	4233.2		A14
Selenium	3.47			4.82		494	221	429.8		A14
Silicon	6.89			2.33		1687	1414	2577.2		A14
Steel, stainless 410	5.03	5.76	2.99	7.67	4.42	3293	3020	5468	0.29	A1, A2, A14
Tantalum	3.35	4.10	2.90	16.60	5.48				0.35	A5, A8, A15
Terfenol-D (Curie temperature = 357°C)	(1.395–2.444)			(9.21–9.25)						A13
Titanium	5.08	6.07	3.125	4.50	2.73	1943	1670	30.8	0.32	A7, A14
Titanium	5.08	6.07	3.125	4.50	2.73					A7

(Continued)

TABLE A.1
Acoustical Properties of Solids (Continued)

| | Velocities (10⁵ cm/s) | | | | Characteristic | Melting Point | | | Poisson's | |
| | Longitudinal | | Shear (c_s) | Density (ρ) | Impedence (ρc_B) | °K | °C | °F | Ratio (σ) | References |
Material	Bar (c_0)	Bulk (c_B)		(g/cm³)	(10⁶ g/cm²·s)					
Metals										
Titanium	–	5.99	3.12	4.50	2.70				0.323	A4
Titanium 6A1 4V	5.08	6.23	–	–	–				0.323	
Tungsten (Ti 150A)	5.08	6.10	3.12	4.54	2.77				0.323	A1, A2
Tungsten	4.60	5.18	2.87	19.25	9.98				0.28	A1, A2
Tungsten, annealed	4.62	5.22	2.89	19.30	10.07					A7
Tungsten	4.31	5.46	2.62	19.10	10.42					A5
Tungsten, drawn	4.32	5.41	2.64	19.30	10.44					A7
Tungsten	–	5.17	2.88	19.30	10.00					A4
Uranium	3.28	3.37	2.02	18.70	6.30				0.21	A4
Uranium	–	3.37	1.98		6.30					A2, A4, A5
Zinc	3.81	4.17	2.41	7.10	2.96					A2, A4, A5
Zinc, Rolled	3.85	4.21	2.44	7.10	2.99					A7
Zirconium	3.40	4.65	2.30	6.40	2.98					A4
Zirconium	3.45	4.65	2.22	–	3.01					A8
Nonmetals										
Alumina dense	–	9.78	–	–	–					
Concrete	3.10	–	–	2.60	–					A3
Cork	0.50	–	–	0.24	–					A3, A5
Fused silica	5.76	5.968	3.764	2.20	1.31					A7
Glass, crown	5.30	5.66	3.42	2.50	1.41					A4, A5

Material							
Glass, borate crown, light	4.54	5.10	2.84	2.24	1.14		A7
Flint	4.00	4.26	2.56	3.60	1.54		A5
Flint, heavy silicate	3.72	3.98	2.38	3.88	1.54		A1
Plate	–	5.77	3.43	2.51	1.45		A1
Pyrex	5.17	5.64	3.28	2.32	1.31		A7
Pyrex	–	5.57	3.44	2.23	1.24		A1
Quartz	5.37	5.57	3.51	2.60	1.45		A5
Quartz	–	5.84	3.68	–	–		A8
Soft	5.00	5.40	–	2.40	1.30		A8
Granite[a]	3.95	–	–	2.75	–		A4
Ice	–	3.98	1.99	0.90	0.36		A4
Ice	3.20	–	–	0.92	–		A3
Ice	3.28	3.98	1.99	1.00	0.40		A1, A5
Lucite	–	2.68	1.27	1.18	0.316		A7
Lucite	1.84	2.68	1.10	1.18	0.316		
Nylon	–	1.8–2.2	–	1.1–1.2	0.2–0.27		A4
Nylon 6-6	1.80	2.62	1.07	1.11	0.29		A7
Paraffin, hard	–	2.20	–	0.83	0.18		A4
Plastic, acrylic resin	–	2.67	1.12	1.18	0.32	0.34	A1
Plexiglas	1.80	2.68	1.32	1.20	0.32		A1
Plexiglas	–	2.67	1.12	1.18	0.32		A5
Polyethylene	0.92	1.95	0.54	0.90	0.176		A7
Polystyrene	–	2.67	–	1.06	0.28		A4
Polystyrene	2.24	2.35	1.12	1.06	0.23		A5
Porcelain	4.88	5.34	3.12	2.41	1.29		A5
Quartz, natural	–	5.73	–	2.65	1.52		A1
Quartz, fused	5.37	5.57	3.52	2.60	1.45		A4
Quartz, fused	–	5.93	3.75	2.20	1.30		A1

(Continued)

TABLE A.1
Acoustical Properties of Solids (*Continued*)

Material	Velocities (10^5 cm/s) Longitudinal Bar (c_0)	Bulk (c_B)	Shear (c_s)	Density (ρ) (g/cm³)	Characteristic Impedance (ρc_B) (10^6 g/cm²·s)	Melting Point °K	°C	°F	Poisson's Ratio (σ)	References
Nonmetals										
Quartz, X-cut	5.45	5.75	–	2.65	1.52					A3
Rubber, butyl	–	1.83	–	1.07	0.196					A7
Rubber, gum	–	1.55	–	0.95	0.15					A7
Rubber, hard	1.45	–	–	1.10	–					A3
Rubber, India	–	1.48	–	0.90	0.14					A5
Rubber, neoprene	–	1.60	–	1.33	0.21					A3
Rubber, soft	0-07	–	–	0.95	–					A3
Teflon	–	1.35	–	2.20	0.30					A4
Titanium carbide	–	8.27	5.16	–	4.26					A8
Tungsten carbide	–	6.66	3.98	10.0–15.0	6.65–9.85					A4
Tungsten carbide	6.22	6.655	3.98	13.80	9.18					A7
Wood, oak	4.10	–	–	0.72	–					A3
Wood, oak	4.10	–	–	0.80	–					A4
Wood, pine	3.65	–	–	0.45	–					A3

[a] For an extensive listing of velocities of sound in rocks at various pressures, see [A9].
(For an extensive listing of acoustic properties of biological tissues, see [A10].)

TABLE A.2
Acoustic Properties of Common Liquids and Molten Materials

Materials	Temperature (°C)	Density (ρ) (g/cm^3)	Velocity (c) (10^5 cm/s)	Characteristic Impedance (ρc) (10^5 g/cm$^2 \cdot$ s)	References
Acetone	25	0.79	1.174	0.93	A7
Alcohol (ethyl)	20	0.79	1.17	0.92	A3
Benzene	25	0.87	1.295	1.13	A7
Carbon disulfide	25	1.26	1.149	1.45	A7
Carbon tetrachloride	25	1.595	0.926	1.48	A7
Castor oil	25	0.969	1.477	1.43	A7
Chloroform	25	1.49	0.987	1.47	A7
Ethanol	25	0.79	1.207	0.95	A7
Ethanol amide	25	1.018	1.724	1.76	A7
Ethyl ether	25	0.713	0.985	0.70	A7
Ethylene glycol	25	1.113	1.658	1.85	A7
Glycerine	20	1.26	1.92	2.40	A3, A4
Glycerine	25	1.26	1.904	2.40	A7
Kerosene	25	0.81	1.324	1.07	A7
Mercury	20	13.546	1.451	19.72	A3, A5
Mercury	25	13.5	1.45	19.58	A7
Mercury	50	13.6	1.440	–	A5
Methanol	25	0.791	1.103	0.87	A7
Methyl iodide		3.23	0.98	3.17	A7
Nitrobenzene	25	1.20	1.463	1.76	A7
Oil, transformer	20	0.92	1.38	1.27	A1
Oil, castor	20	0.95	1.54	1.45	A3
Oil, SAE 20	–	0.87	1.74	1.50	A4

(Continued)

TABLE A.2
Acoustic Properties of Common Liquids and Molten Materials (*Continued*)

Materials	Temperature (°C)	Density (ρ) (g/cm³)	Velocity (c) (10⁵ cm/s)	Characteristic Impedance (ρc) (10⁵ g/cm²·s)	References
Turpentine	20	0.87	1.33	1.15	A3
Turpentine	25	0.88	1.255	1.10	
Water, fresh[a]	4	–	1.422	–	A11
Water, fresh	10	–	1.448	–	A11
Water, fresh	15	–	1.466	–	A11
Water, fresh	20	0.998	1.483	1.48	A3, A11
Water, fresh	50	–	1.543	–	A11
Water, fresh	75	–	1.555	–	A11
Water, fresh	90	–	1.551	–	A11
Water, sea	25	1.025	1.531	1.57	A7
Xylene hexafluoride	25	1.37	0.879	1.20	A7
Molten Materials (Low Melting Point)					
Bismuth	271.0	–	1.635	–	A5
Cadmium	321.0	–	2.222	–	A5
Cesium	28.5	–	0.967	–	A5
Gallium	29.5	–	2.740	–	A5
Indium	156.0	–	2.215	–	A5
Lead	327.0	–	1.79	–	A5
Potassium	64.0	–	1.820	–	A5
Rubidium	39.0	–	1.260	–	A5
Sodium	98.0	–	1.350	–	A5
Sulfur	115.0	–	1.350	–	A5
Thallium	302.0	–	1.625	–	A5

Material					Source
Tin	232.0	—	2.270	—	A5
Zinc	419.0	—	2.790	—	A5
85% sodium carbonate + 15% sodium sulfide	896	—	2.274	—	Battelle
85% sodium carbonate + 15% sodium sulfide	1015	—	2.180	—	Battelle
70% sodium carbonate + 30% sodium sulfide	860	—	2.250	—	Battelle
70% sodium carbonate + 30% sodium sulfide	1000	—	2.148	—	Battelle
Sodium carbonate	903	—	2.300	—	Battelle
Sodium carbonate	1001	—	2.220	—	Battelle
Sodium chloride	864	—	1.726	—	Battelle
Sodium chloride	998	—	1.593	—	Battelle
Sodium sulfide	960	—	2.037	—	Battelle
Sodium sulfide	1020	—	2.000	—	Battelle
Sodium sulfide	1060	—	1.963	—	Battelle
50% sodium carbonate + 25% sodium chloride + 25% sodium sulfide	860	—	2.070	—	Battelle
50% sodium carbonate + 25% sodium chloride + 25% sodium sulfide	947	—	1.988	—	Battelle
50% sodium carbonate + 25% sodium chloride + 25% sodium sulfide	1002	—	1.941	—	Battelle
70% sodium carbonate + 25% sodium sulfide + 5% sodium aluminate	840	—	2.280	—	Battelle
70% sodium carbonate + 25% sodium sulfide + 5% sodium aluminate	1000	—	2.160	—	Battelle
70% sodium carbonate + 25% sodium sulfide + 5% sodium hydroxide	840	—	2.278	—	Battelle

(Continued)

TABLE A.2
Acoustic Properties of Common Liquids and Molten Materials (*Continued*)

Materials	Temperature (°C)	Density (ρ) (g/cm³)	Velocity (c) (10⁵ cm/s)	Characteristic Impedance (ρc) (10⁵ g/cm²·s)	References
Molten Materials (*Low Melting Point*)					
70% sodium carbonate + 25% sodium sulfide + 5% sodium aluminate	840	—	2.280	—	Battelle
70% sodium carbonate + 25% sodium sulfide + 5% sodium aluminate	1000	—	2.160	—	Battelle
70% sodium carbonate + 25% sodium sulfide + 5% sodium hydroxide	840	—	2.278	—	Battelle
70% sodium carbonate + 25% sodium sulfide + 5% sodium hydroxide	998	—	2.161	—	Battelle
Liquid Gases					
Argon	−186.0	1.404	0.837	1.175	25
Helium	−269.1	0.125	0.1798	0.0225	25
Helium	−271.5	0.146	0.2314	0.0338	25
Hydrogen	−252.7	0.355	1.127	0.400	25
Nitrogen	−197.0	0.815	0.869	0.708	25
Oxygen	−183.6	1.143	0.911	1.041	25

a For an extensive list of the velocities of sound in fresh water over a wide temperature range in both metric and English units, see [A11].

TABLE A.3
Acoustic Properties of Organic Liquids[a]

Liquid	Chemical Formula	Temperature (°C)	Density (ρ) (g/cm³)	Velocity of Sound (c) (10⁵ cm/s)	Characteristic Impedance (ρc) (10⁵ g/cm²·s)
Acetic anhydride	$(CH_3CO)_2O$	24	1.087	1.384	1.51
Acetone	$CH_3 \times CO \times CH_3$	20	0.792	1.192	0.94
Acetonylacetone	$CH_3CO(CH_2)_2COCH_3$	20	0.970	1.416	1.38
Acetophenone	$C_6H_5 \times CO \times CH_3$	20	1.026	1.496	1.54
Acetylacetone	$CH_3COCH_2COCH_3$	20	0.976	1.383	1.35
Acetyl chloride	C_2H_3OCL	20	1.105	0.060	1.17
Acetylene dichloride	$CHCL_1 \times CHCL_1$	25	1.265	1.025	1.30
Acetylene tetrabromide	$CHBr_2 \times CHBr_2$	20	2.963	1.041	3.09
Acetylene tetrachloride	$CHCl_2-CHCl_2$	28	1.578	1.155	1.82
Acrolein	$CH_2:CHCHO$	20	0.841	1.207	1.02
Amyl acetate	$CH_3COO(CH_2)_4CH$	26	0.879	1.168	1.03
tert-Amyl alcohol (dimethylethylcarbinol)	$CH_3CH_2C(CH_3)OHCH_3$	28	0.809	1.204	0.97
n-Amyl bromide	$CH_3(CH_2)_3CH_2Br$	20	1.246	0.981	1.22
Amyl ether	$[CH_3(CH_2)_3CH_2]_2O$	26	0.774	1.153	0.89
Amyl formate	$HCOO(CH_2)_4CH_3$	26	0.893	1.201	1.07
Aniline	$C_6H_5NH_2$	20	1.022	0.656	1.69
Biacetyl	$CH_3COCOCH_3$	25	0.990	1.236	1.22
Benzaldehyde	C_6H_5CHO	20	1.050	1.479	1.55
Benzene	C_6H_6	20	0.879	1.326	1.17
Benzylacetone	$C_6H_5CH_2CH_2COCH_3$	20	0.989	1.514	1.50
Benzyl alcohol	$C_6H_5CH_2OH$	20	1.050	1.540	1.62
Benzyl chloride	C_7H_7Cl	20	1.1026	1.420	1.57

(Continued)

TABLE A.3
Acoustic Properties of Organic Liquids[a] (*Continued*)

Liquid	Chemical Formula	Temperature (°C)	Density (ρ) (g/cm^3)	Velocity of Sound (c) (10^5 cm/s)	Characteristic Impedance (ρc) (10^5 g/cm^2·s)
Bromal	$CHBr_2CHO$	20	2.30	0.966	2.22
Bromoform	$CHBr_3$	20	2.890	0.928	2.68
1-Bromonaphthalene	$C_{10}H_7Br$	20	1.487	1.372	2.04
2-Butanone (ethyl methyl ketone)	$CH_3COC_2H_5$	20	0.805	1.207	0.97
n-Butyl alcohol	$CH_3(CH_2)_2CH_2OH$	20	0.810	1.268	1.03
sec-Butyl alcohol	$CH_3CH_2CHOHCH_3$	20	0.808	1.222	0.99
tert-Butyl alcohol	$(CH_3)_3COH$	20	0.789	1.155	0.91
n-Butyl bromide	$CH_3(CH_2)_2CH_2Br$	20	1.299	0.990	1.29
n-Butyl chloride	C_4H_9Cl	20	0.884	1.133	1.00
Butyl formate (butyl methanoate)	$HCOOC_4H_9$	24	0.911	1.199	1.09
n-Butyl iodide	$CH_3(CH_2)_2CH_2I$	20	1.617	0.977	1.58
Butyric acid	C_3H_7COOH	20	0.959	1.203	1.15
Caprylic acid	$CH_3(CH_2)_6COOH$	20	0.910	1.331	1.21
Carbon disulfide	CS_2	20	1.263	1.158	1.46
Carbon tetrachloride	CCl_4	20	1.595	0.938	1.50
Carvacrol	$CH_3(C_3H_7)C_6H_3OH$	20	0.976	1.475	1.44
Chlorobenzene	C_6H_5Cl	20	1.107	1.291	1.43
Chloroform	$CHCl_3$	20	1.498	1.005	1.51
1-Chlorohexane	$CH_3(CH_2)_4CH_2Cl$	20	0.872	1.221	1.06
1-Chloronaphthalene	$C_{10}H_7Cl$	20	1.194	1.481	1.77
o-Chlorotoluene	$ClC_6H_4CH_3$	20	1.0817	1.344	1.45
m-Chlorotoluene	$ClC_6H_4CH_3$	20	1.072	1.326	1.42
p-Chlorotoluene	$ClC_6H_4CH_3$	20	1.0697	1.316	1.41

Cinnamaldehyde	C_9H_8O	25	1.112	1.554	1.73
Citral b	$C_{10}H_{16}O$	20	0.888	1.442	1.28
o-Cresol	$CH_3C_6H_4OH$	25	1.046	1.506	1.58
Crotonaldehyde	C_4H_6O	20	0.858	1.344	1.15
Cumene	$C_6H_5CH(CH_3)_2$	20	0.862	1.442	1.28
Cyclohexane	C_6H_{12}	20	0.779	1.284	1.00
Cyclohexanol	$C_6H_{12}O$	20	0.945	1.493	1.41
Cyclohexanone	$CO(CH_2)_4CH_2$	20	0.948	1.449	1.37
Cyclohexene	$CH_2CH_2CH_2CH_2CH:CH$	20	0.811	1.305	1.06
Cyclohexylamine	$C_6H_{11}NH_2$	20	0.819	1.435	1.18
Cyclohexyl chloride	$C_6H_{11}Cl$	20	1.016	1.319	1.34
1,3-Cyclopentadiene	$CH:CHCH:CHCH_2$	20	0.805	1.421	1.14
Cyclopentanone	$COCH_2CH_2CH_2CH_2$	24	0.948	1.474	1.40
p-Cymene (p-isopropyltoluene)	$C_6H_4CH_3CH(CH_3)_2$	28	0.857	1.308	1.12
1-Decene	$C_{10}H_{20}$	20	0.741	1.250	0.93
n-Decyl alcohol	$C_{10}H_{21}OH$	20	0.829	1.402	1.16
2-Dibromoethylene	$CHBr \times CHBr$	20	2.271	0.957	2.17
m-Dichlorobenzene	$C_6H_4Cl_2$	28	1.285	1.232	1.58
o-Dichlorobenzene	$C_6H_4Cl_2$	20	1.305	1.295	1.69
1,1-Dichloroethane	$CH_3CH_2Cl_2$	20	1.174	1.034	1.21
Diethyl aldehyde	$CH_3CH(OC_2H_5)_2$	24	0.825	1.378	1.14
N,N-Diethylaniline	$C_6H_5N(C_2H_5)_2$	20	0.935	1.482	1.39
Diethylene glycol	$O(CH_2CH_2OH)_2$	25	1.132	1.586	1.80
Diethyl ester malonic acid	$CH_2(COOC_2H_5)_2$	22	1.055	1.386	1.46
2,3-Dimethylbutanoic acid	$C_5H_{11}COOH$	20	0.928	1.280	1.19
Di-n-propyl ether	$(CH_3CH_2CH_2)_2O$	20	0.736	1.112	0.82
Diphenyl ether	$C_6H_5OC_6H_5$	24	1.0728	1.469	1.57
Diphenylmethane	$C_6H_5-CH_2C_6H_5$	28	1.001	1.501	1.50
n-Dodecyl alcohol	$CH_3(CH_2)_{11}OH$	30	0.831	1.388	1.15

(Continued)

TABLE A.3
Acoustic Properties of Organic Liquids[a] (*Continued*)

Liquid	Chemical Formula	Temperature (°C)	Density (ρ) (g/cm³)	Velocity of Sound (c) (10^5 cm/s)	Characteristic Impedance (ρc) (10^5 g/cm²·s)
Ethyl acetate	$CH_3COOC_3H_5$	20	0.901	1.176	1.06
Ethyl acetoacetate	$CH_3 \times CO \times CH_2 \times COOC_2H_5$	25.5	1.025	1.417	1.45
Ethyl acetylmalonate	$CO \times (CH_2COOC_2H_5)_2$	22.5	1.085	1.348	1.46
Ethyl adipate	$CH_2CH_2COOC_2H_5$ $\mid$ $CH_2CH_2COOC_2H_5$	22	1.009	1.376	1.39
Ethyl alcohol	C_2H_5OH	20	0.789	1.180	0.93
Ethyl bromide	C_2H_5Br	28	1.428	0.892	1.27
Ethyl butyrate	$C_3H_7COOC_2H_5$	23.5	0.879	1.1171	0.98
Ethyl caprylate	$CH_3(CH_2)_6COOC_2H_5$	28	0.867	1.263	1.10
Ethyl carbonate	$(C_2H_5)_2CO_3$	28	0.975	1.173	1.14
Ethyl chloroethanoate	$CH_2ClCOOC_2H_5$	25.5	1.159	1.234	1.43
Ethylene bromide	$C_2H_4Br_2$	20	2.1701	1.009	2.19
Ethylene chloride	$CH_2Cl \times CH_2Cl$	23	1.257	1.240	1.56
Ethylene glycol	CH_2OHCH_2OH	20	1.115	1.616	1.80
Ethyl ester propionic acid	$CH_3CH_2COOC_2H_5$	23.5	0.8846	1.185	1.05
Ethyl ether	$C_4H_{10}O$	20	0.714	1.008	0.72
Ethyl formate	$H \times COOC_2H_5$	24	0.924	1.724	1.59
Ethyl iodide	CH_3CH_2I	20	1.940	0.869	1.69
Ethyl-o-tolyl ether	$CH_3C_6H_4OC_2H_5$	25	0.959	1.315	1.26
Ethyl oxalate	$(COOC_2H_5)_2$	22	1.084	1.392	1.51
Ethyl phenyl ketone (propiophenone)	$C_2H_5COC_6H_5$	20	1.012	1.498	1.51
Ethyl p-phthalate	$C_6H_4(COOC_2H_5)_2$	23	1.123	1.471	1.65
Ethyl succinate	$(CH_2-COOC_2H_5)_2$	22	1.040	1.378	1.43

Formamide	HCONH$_2$	20	1.134	1.550	1.76
Formic acid	HCOOH	20	1.226	1.287	1.58
Furfuryl alcohol	C$_4$H$_3$O × CH$_2$OH	25	1.1296	1.450	1.64
Geranyi acetate	C$_{12}$H$_{20}$O$_2$	28	0.917	1.328	1.22
Gylcerine	CH$_2$OHCHOHCH$_2$OH	20	1.261	1.923	2.42
n-Heptane	C$_7$H$_{16}$	20	0.684	1.162	0.78
Heptanoic acid	CH$_3$(CH$_2$)$_5$COOH	20	0.913	1.312	1.20
1-Heptene	CH$_2$CH(CH$_2$)$_4$CH$_3$	20	0.697	1.128	0.79
n-Heptyl alcohol	CH$_3$(CH$_2$)$_5$CH$_2$OH	20	0.823	1.341	1.10
Hexane	C$_6$H$_{14}$	20	0.660	1.083	0.71
n-Hexyl alcohol	CH$_3$(CH$_2$)$_4$CH$_2$OH	20	0.820	1.322	1.08
Hexyl methyl ketone	CH$_3$COC$_6$H$_{13}$	24	0.818	1.324	1.08
Indane (hydrindene)	C$_6$H$_4$CH$_2$CH$_2$CH$_2$	20	0.965	1.403	1.35
Indene	C$_6$H$_4$CH$_2$CH:CH	20	1.006	1.475	1.48
Iodobenzene	C$_6$H$_5$I	20	1.832	1.113	2.04
1-Iodohexane	CH$_3$(CH$_2$)$_4$CH$_2$I	20	1.441	1.081	1.56
α-Ionon	C$_{13}$H$_{20}$O	20	0.930	1.432	1.33
d-Linalool	C$_{10}$H$_{17}$OH	20	0.863	1.341	1.16
Mesityl oxide	(CH$_3$)$_2$C:CHCOCH$_3$	20	0.854	1.310	1.12
3-Methoxy toluene	CH$_3$C$_6$H$_4$OCH$_3$	26	0.976	1.385	1.35
Methyl alcohol	CH$_3$OH	20	0.796	1.123	0.89
Methyl acetate	CH$_3$COOCH$_3$	25	0.928	1.154	1.07
N-Methylaniline	C$_6$H$_5$NHCH	20	0.986	1.586	1.56
Methyl chloroethanoate	CH$_2$ClCOOCH$_3$	26	1.227	1.331	1.64
Methyl cyanide	CH$_3$CN	20	0.783	1.304	1.02
Mehtylcyclohexane	CH$_3$C$_6$H$_{11}$	20	0.7864	1.247	0.98
2-Methylcyclohexanol	CH$_3$C$_6$H$_{10}$OH	25.5	0.937	1.421	1.33
3-Methyl-1-cyclohexanol	CH$_3$C$_6$H$_{10}$OH	25.5	0.914	1.406	1.28
4-Methylcyclohexanol	CH$_3$C$_6$H$_{10}$OH	25.5	0.913	1.387	1.27

(Continued)

TABLE A.3
Acoustic Properties of Organic Liquids[a] (*Continued*)

Liquid	Chemical Formula	Temperature (°C)	Density (ρ) (g/cm^3)	Velocity of Sound (c) (10^5 cm/s)	Characteristic Impedance (ρc) (10^5 g/cm$^2 \cdot$ s)
2-Methylcyclohexanone	$C_7H_{12}O$	25.5	0.924	1.353	1.25
4-Methyicyclohexanone	$C_7H_{12}O$	25.5	0.913	1.348	1.23
Methylene bromide	CH_2Br_2	24	2.4953	0.971	2.42
Methylene chloride	CH_2Cl_2	20	1.336	1.092	1.45
Methylene iodide	CH_2I_2	24	3.325	0.977	3.25
Methyl ester salicylic acid	$HOC_6H_4COOCH_3$	28	1.184	1.408	1.67
Methyl iodide	CH_3I	20	2.279	0.834	1.90
Methyl phenol ether (anisole)	$C_6H_5OCH_3$	26	0.9988	1.353	1.35
Methyl propionate	$CH_3CH_2COOCH_3$	24.5	0.915	1.215	1.11
Methyl propyl ketone (diethyl ketone)	$C_2H_5COOC_2H_5$	24	0.816	1.314	1.06
Monoethyl ether	$C_2H_5OCH_2CH_2OCH_2CH_2OH$	25	0.990	1.458	1.44
Morpholine	$OCH_2CH_2NHCH_2CH_2$	25	1.000	1.442	1.44
Nicotine	$C_{10}H_{14}N_2$	20	1.009	1.491	1.50
Nitrobenzene	$C_6H_5NO_2$	20	1.199	1.473	1.77
2-Nitroethanol	$NO_2C_2H_4OH$	20	1.270	1.578	2.00
Nitromethane	CH_3NO_2	20	1.130	1.346	1.52
o-Nitrotoluene	$NO_2C_6H_4CH_3$	20	1.163	1.432	1.67
m-Nitrotoluene	$NO_2C_6H_4CH_3$	20	1.164	1.489	1.73
Nonane	$CH_3(CH_2)_7CH_3$	20	0.718	1.248	0.89
1-Nonene	$CH_3(CH_2)_6CH:CH_2$	20	0.729	1.218	0.89
n-Nonyl alcohol	$CH_3(CH_2)_7CH_2OH$	20	0.828	1.391	1.15
n-Octane	$CH_3(CH_2)_6CH_3$	20	0.703	1.197	0.84

Name	Formula	Temp			
1-Octanol	CH$_3$(CH$_2$)$_6$CH$_2$OH	20	0.825	1.358	1.12
n-Octyl bromide	CH$_3$(CH$_2$)$_6$CH$_2$Br	20	1.116	1.182	1.32
n-Octyl chloride	CH$_3$(CH$_2$)$_6$CH$_2$Cl	20	0.875	1.280	1.12
Oleic acid	C$_8$H$_{17}$CH:CH(CH$_2$)$_7$COOH	45	0.873	1.333	1.16
Paraldehyde	C$_6$H$_{12}$O$_3$	20	0.994	1.204	1.20
Pentachloroethane	CHCl$_2$CCl$_3$	20	1.709	1.113	1.90
Pentane	C$_5$H$_{12}$	20	0.626	1.008	0.63
1-Pentanol	C$_5$H$_{11}$OH	20	0.815	1.294	1.06
Perchloroethylene	CC$_{12}$:CC$_{12}$	20	1.623	1.066	1.73
Phenetole	C$_2$H$_5$OC$_6$H$_5$	26	0.967	1.153	1.12
Phenylethane	C$_6$H$_5$ × C$_2$H$_5$	20	0.867	1.338	1.16
Phenylhydrazine	C$_6$H$_5$NHNH$_2$	20	1.098	1.738	1.91
Phenyl mustard oil	C$_6$H$_5$NCS	27	1.131	1.412	1.60
α-Picoline	CH$_3$C$_5$H$_4$N	28	0.951	1.453	1.38
β-Picoline	CH$_3$C$_5$H$_4$N	28	0.961	1.419	1.36
dl-Pinene	C$_{10}$H$_{16}$	24	0.858	1.247	1.07
Piperidine	(CH$_2$)$_5$NH	20	0.862	1.400	1.20
Propionic acid	CH$_3$CH$_2$COOH	20	0.992	1.176	1.17
n-Propionitrile	CH$_3$CH$_2$CN	20	0.783	1.271	1.00
n-Propyl acetate	CH$_3$COOC$_3$H$_7$	26	0.887	1.182	1.05
n-Propyl alcohol	CH$_3$CH$_2$CH$_2$OH	20	0.804	1.223	0.98
n-Propyl chloride	CH$_3$CH$_2$CH$_2$Cl	20	0.890	1.091	0.97
n-Propyl iodide	C$_3$H$_7$I	20	1.747	0.929	1.62
Pseudocumene	(CH$_3$)$_3$C$_6$H$_3$	20	0.876	1.368	1.20
Pyridine	C$_6$H$_5$N	20	0.982	1.445	1.42
Quinaldine	CH$_3$C$_9$H$_6$N	20	1.1013	1.575	1.73
Quinaline	C$_6$H$_4$N:CHCH:CH	20	1.095	1.600	1.75
Resorcinol dimethyl ether	C$_6$H$_4$(OCH$_3$)$_2$	26	1.080	1.460	1.58
Salicylaldehyde	HOC$_6$H$_4$CHO	27	1.166	1.474	1.72

(Continued)

TABLE A.3
Acoustic Properties of Organic Liquids[a] (Continued)

Liquid	Chemical Formula	Temperature (°C)	Density (ρ) (g/cm^3)	Velocity of Sound (c) (10^5 cm/s)	Characteristic Impedance (ρc) (10^5 g/cm$^2 \cdot$ s)
1,1,2,2-Tetrabromoethane	$CHBr_2CHBr_2$	20	2.963	1.041	3.08
1,1,2,2-Tetrachloroethane	$C_2H_2Cl_4$	20	1.6	1.171	1.87
Tetrachloroethylene	CCl_2:CCl_2	28	1.623	1.027	1.67
Tetraethylethylene glycol	$(C_2H_5)_2COHCOHC(C_2H_5)_2$	25	1.123	1.586	1.78
1,2,3,4-Tetrahydronaphthalene (tetralin)	$C_{10}H_{12}$	20	0.971	1.492	1.45
Tetranitromethane	$C(NO_2)_4$	20	1.650	1.039	1.71
Thiolacetic acid	CH_3COSH	20	1.074	1.168	1.25
Thiophene	SCH:$CHCH$:CH ˙	20	1.065	1.300	1.38
Toluene	$C_6H_5CH_3$	20	0.867	1.328	1.15
o-Toluidine	$CH_3C_6H_4NH_2$	20	1.004	1.634	1.64
m-Toluidine	$CH_3C_6H_4NH_2$	20	0.989	1.620	1.60
1,2,4-Trichlorobenzene	$C_6H_3Cl_3$	20	1.456	1.301	1.89
Trichloroethylene	$CHCl$:CCl_2	20	1.477	1.049	1.55
Tridecylene	$C_{13}H_{26}$	20	0.798	1.313	1.05
Triethylene glycol	$(CH_2OCH_2CH_2OH)_2$	25	1.125	1.608	1.81
Trimethylene bromide	$BrCH_2CH_2CH_2Br$	23.5	1.979	1.144	2.26
Triolein	$(C_{18}H_{33}O_2)_3C_3H_5$	20	0.915	1.482	1.36
o-Xylene	$C_6H_4(CH_3)_2$	20	0.880	1.360	1.20
m-Xylene	$C_6H_4(CH_3)_2$	20	0.864	1.340	1.46
p-Xylene	$C_6H_4(CH_3)_2$	20	0.861	1.330	1.15

[a] In general, the velocities used in this table are from [A5]. Densities are from Handbook of Chemistry and Physics [A12], and ρc is corrected to these values of ρ.

TABLE A.4
Acoustic Properties of Gases

Gas	Temperature (°C)	Density (ρ) (10^3 g/cm^3)	Velocity (c) (10^4 cm/s)	Characteristic Impedance (ρc) (g/cm$^2 \cdot$ s)	Reference
Air[a]	0	1.293	3.316	42.8	23
Air	20	1.21	3.43	41.5	23
Argon	20	1.72	3.21	55.2	25
Carbon dioxide (low frequency)	0	1.98	2.58	51.0	23
Carbon dioxide (high frequency)	0	1.98	2.68	53.0	23
Carbon dioxide	16.6	1.935	2.77	53.6	25
Carbon monoxide	18.7	1.215	3.49	42.4	25
Helium	17.5	0.174	9.97	17.35	25
Hydrogen	0	0.09	12.65	11.5	23
Hydrogen	19.9	0.0868	13.29	11.54	25
Neon	19.0	0.872	4.50	39.25	25
Nitric oxide	16.3	1.311	3.34	43.8	25
Nitrogen	19.9	1.207	3.49	42.1	25
Oxygen	0	1.43	3.14	45.0	23
Oxygen	19.6	1.38	3.27	45.1	25
Steam	100	0.6	4.75	28.5	23

[a] For an extensive list of the velocity of sound in gases and vapors and of the absorption and velocity of sound in still air as functions of frequency and humidity, see CRC Handbook of Chemistry and Physics, E43–E48 [A7].

REFERENCES

A1 R. C. McMaster, ed., Nondestructive Testing Handbook, Vol. II, Ronald Press, New York, 1959.

A2 General Dynamics Convair Division, Nondestructive Testing—Ultrasonic Testing, Classroom Training Handbook, San Diego, California, 1967.

A3 L. E. Kinsler and A. R. Frey, Fundamentals of Acoustics, Wiley, New York, 1950.

A4 J. R. Frederick, Ultrasonic Engineering, Wiley, New York, London, Sydney, 1965.

A5 L. Bergmann, Der Ultraschall, Hirzel Verlag, Stuttgart, 1954.

A6 W. D. Jolly, "Materials Characterization Study—Ultrasonic Longitudinal and Shear Velocities of Candidate Isotope Encapsulation Materials," USAEC Report BNWL-547, Pacific Northwest Laboratories, Richland, Washington; available from Clearinghouse for Federal Scientific and Technical Information, National Bureau of Standards, U.S. Department of Commerce, Springfield, Virginia, 1967.

A7 R. C. Weast, editor-in-chief, Handbook of Chemistry and Physics, 66th ed., pp. E43–E48, CRC Press, Boca Raton, Florida, 1985.

A8 Anon., Mater. Eng., 99, 30–37 (May 1984).

A9 F. Birch, J. Geophys. Res., 65(4), 1083–1102 (1960).

A10 D. E. Goldman and T. F. Hueter, J. Acoust. Soc. Am., 28(1), 35–37 (1956).

A11 M. Greenspan and C. E. Tschiegg, J. Acoust. Soc. Am., 31(1), 75–76 (1959).

A12 C. D. Hodgman, editor-in-chief, Handbook of Chemistry and Physics, 37th ed., Chemical Rubber Publishing Co., Cleveland, Ohio, 1955.

A13 G. Engdahl, Handlook of Giant Magnetostriction Materials, Academic Press, 525 B Street, Suite 1900, San Diego, CA 92101-4495 USA and Academic Press, 24–28 Oval Road, London, NW1 7DX UK (2000).

A14 M. Bauccio, ed., ASM Metals Reference Book, ASM International, Materials-Park, OH 44073–0002, 1993, p.139.

Appendix B

B.1 SELECTED BIBLIOGRAPHY FOR ACOUSTIC PROPERTIES OF MATERIALS

B.1.1 SELECTED SOLIDS, LIQUIDS, AND GASES

Gray, D. E., ed. 1972. *AIP Handbook*. 3rd ed. New York: McGraw-Hill.
  Beranek, L. L. Acoustic Properties of Gases, 3-68–3-85, Section 3d.
  Greenspan, M. Acoustic Properties of Liquids, 3-86–3-98, Section 3e.
  Mason, W. P. Acoustic Properties of solids, 3-98–3-117, Section 3f.
Kino, G. S. 1987. *Acoustic Waves*. Englewood Cliffs, NJ: Prentice Hall. Appendix B: acoustic parameters of common materials, 548–562.
Lynnworth, L. C. 1989. Numerical and graphical databases—acoustical properties of selected media. In *Ultrasonic Measurements for Process Control*, 224–42. Boston: Academic Press.

B.1.2 APPROXIMATE PROPERTIES—ISOTROPIC MATERIALS

Selfridge, A. R. 1985. Approximate material properties in isotropic materials. *IEEE Trans Sonics Ultrason* SU-32(3):381–94.

B.1.3 ABSORPTION IN FLUIDS

Bahtia, A. B. 1967. *Ultrasonic Absorption*. Oxford: Clarendon Press.
Markham, J. J., R. T. Beyer, and R. B. Lindsay. 1951. Absorption of sound in fluids. *Rev mod phys* 23(4):353–411.
Matheson, A. J. 1971. *Molecular Acoustics*. London: Wiley-Interscience.
Povey, M. J. W. 1997. *Ultrasonic Techniques for Fluids Characterization*. San Diego: Academic Press.

B.1.4 GASES AT ATMOSPHERIC AND ELEVATED PRESSURES

Method for the calculation of the absorption of sound by the atmosphere. ANSI S1.26-1978. Acoustical Society of America.
Wickramasinghe, H. K., and C. R. Petts. 1980. Acoustic microscopy in high pressure gases. *Proc 1980 IEEE Ultrason symp*, ed. B. R. McAvoy, Boston, MA, Nov. 5–7, 1:668–72. New York: IEEE, Sonics and Ultrasonics Group.

B.1.5 MAMMALIAN TISSUES

Gross, S. A., R. L. Johnston, and F. Dunn. 1978. Comprehensive compilation of empirical ultrasonic properties of mammalian tissues. *J Acoust Soc Am* 64(2):423–75.
Gross, S. A., R. L. Johnston, and F. Dunn. 1980. Compilation of empirical ultrasonic properties of mammalian tissues II. *J Acoust Soc Am* 68(1):93–108.

B.2   SELECTED WEB-BASED DATABASES FOR ACOUSTIC
PROPERTIES OF MATERIALS

With the advent of the World Wide Web, new resources have become available, which provide acoustic/ultrasonic properties of materials. In addition to those resources listed below (e.g., for piezoelectric material), many vendors also provide extensive data on their products.

Five Web resources found to be particularly useful by the authors are identified.

1. ONDA Corporation, Sunnyvale, CA

   A widely used Web resource for specialty engineering (originally published by Selfridge [1985]) is now part of ONDA Corporation. They offer our list of material properties, complete with all of the updates. Materials are listed in order of hardness within data tables given for the following: solids, longitudinal piezoelectric materials, shear piezoelectric materials, plastics, rubbers, liquids, and gases.

   In addition, they currently post extensive data on materials, gathered by Laust Pedersen, which cover hundreds of materials, and the data are given as a standalone spreadsheet that can be downloaded.

   The URL for ONDA Corporation is: http://www.ondacorp.com/tecref_acoustictable.html.

2. NDT Resource Center: Center for NDE, Iowa State University, IA

   The NDT Resource Center provides a Website that was designed to be a comprehensive source of information and materials for NDT and NDE technical education. The Website was created by NDT professionals and educators from around the world. It includes extensive list of material properties. The acoustic data cover the following: metals (solids, liquids, and powdered), ceramics, rubber, wood, liquids, vapors, gases (and gases as liquids), piezoelectric materials, and body tissues.

   The URL for the NDT–Education Resource Center material database is: http://www.ndt-ed.org/GeneralResources/MaterialProperties/UT/ut_matlprop_index.htm.

3. National Institute of Standards and Technology (NIST) (for general material properties and some acoustic data)

   NIST has several material databases, including those for thermophysical properties of fluids and standard materials. A particularly valuable resource is the NIST Chemistry WebBook (NIST Standard Reference Database Number 69).

   This database provides chemical and physical property data on more than 40,000 compounds.

   In addition, these data give thermophysical properties of fluid systems.

   Accurate thermophysical properties are available for several fluids, including speed of sound over a wide range of temperatures and pressures. These data include the following: density, Cp, enthalpy, internal energy, viscosity, Joule–Thomson coefficient, specific volume, Cv, entropy, speed of sound, thermal conductivity, and surface tension (saturation curve only).

   The URL for NIST is: http://webbook.nist.gov/chemistry/fluid/.

4. IEEE Electronic Libraries

The IEEE has an extensive library with more than two million documents. It provides access to all IEEE/IET journals, conference proceedings, and active standards.

This material is accessed via a robust, user-friendly interface and powerful search tools powered by the IEEE Xplore digital library.

The URL for IEEE Xplore is: http://ieeexplore.ieee.org/Xplore/guesthome.jsp.

In addition, the IEEE Ultrasonics, Ferroelectrics and Frequency Control Society (UFFC) has its own digital library that provides access to the proceedings of the IEEE Ultrasonic Symposia and the IEEE transactions for UFFC.

The URL for the IEEE UFFC digital archive is: http://www.ieee-uffc.org/main/publications.asp.

5. Acoustical Society of America

The ASA has a digital library that provides access to the journal of the society (JASA) and also standards.

The URL for the ASA digital library is: http://asadl.org/.

Author Index

Subject Index